Introduction to Chemical Processes

Principles, Analysis, Synthesis

Regina M. Murphy

University of Wisconsin, Madison

Higher Education

Boston   Burr Ridge, IL   Dubuque, IA   Madison, WI   New York
San Francisco   St. Louis   Bangkok   Bogotá   Caracas   Kuala Lumpur
Lisbon   London   Madrid   Mexico City   Milan   Montreal   New Delhi
Santiago   Seoul   Singapore   Sydney   Taipei   Toronto

INTRODUCTION TO CHEMICAL PROCESSES: PRINCIPLES, ANALYSIS, SYNTHESIS

2 3 4 5 6 7 8 9 0 DOC/DOC 0 9 8 7 6

ISBN-13:   978-0-07-284960-8
ISBN-10:   0-07-284960-6

Publisher: *Suzanne Jeans*
Senior Sponsoring Editor: *Bill Stenquist*
Developmental Editor: *Amanda J. Green*
Executive Marketing Manager: *Michael Weitz*
Senior Project Manager: *Mary E. Powers*
Lead Production Supervisor: *Sandy Ludovissy*
Associate Media Producer: *Christina Nelson*
Senior Coordinator of Freelance Design: *Michelle D. Whitaker*
Interior Designer: *Rokusek Design*
(USE) Cover Image: Section of Molecular Model, © *Roz Woodward/Getty Images*
Lead Photo Research Coordinator: *Carrie K. Burger*
Photo Research: *Pam Carley/Sound Reach*
Supplement Producer: *Tracy L. Konrardy*
Compositor: *International Typesetting & Composition*
Typeface: *10.5/12 Times Roman*
Printer: *R. R. Donnelley Crawfordsville, IN*

Library of Congress Cataloging-in-Publication Data

Murphy, Regina M.
    Introduction to chemical processes : principles, analysis, synthesis / Regina M. Murphy.—1st ed.
      p.   cm.   (McGraw-Hill chemical engineering series)
    Includes bibliographical references and index.
    ISBN 978-0-07-284960-8   ISBN 0-07-284960-6 (acid-free paper)
    1. Chemical processes—Textbooks. I. Title. II. Series.

TP155.7.M87   2007   2005054372
660'.28 dc22   CIP

www.mhhe.com

McGraw-Hill Chemical Engineering Series

The Founding of a Discipline: The McGraw-Hill Companies, Inc. Series in Chemical Engineering

Over 80 years ago, 15 prominent chemical engineers met in New York to plan a continuing literature for their rapidly growing profession. From industry came such pioneer practitioners as Leo H. Baekeland, Arthur D. Little, Charles L. Reese, John V. N. Dorr, M. C. Whitaker, and R. S. McBride. From the universities came such eminent educators as William H. Walker, Alfred H. White, D. D. Jackson, J. H. James, Warren K. Lewis, and Harry A. Curtis. H. C. Parmlee, then editor of *Chemical and Metallurgical Engineering*, served as chairman and was joined subsequently by S. D. Kirkpatrick as consulting editor.

After several meetings, this committee submitted its report to the McGraw-Hill Book Company in September 1925. In the report were detailed specifications for a correlated series of more than a dozen texts and reference books which became the McGraw-Hill Series in Chemical Engineering—and in turn became the cornerstone of the chemical engineering curricula.

From this beginning, a series of texts has evolved, surpassing the scope and longevity envisioned by the founding Editorial Board. The McGraw-Hill Series in Chemical Engineering stands as a unique historical record of the development of chemical engineering education and practice. In the series one finds milestones of the subject's evolution: industrial chemistry, stoichiometry, unit operations and processes, thermodynamics, kinetics, and transfer operations.

Textbooks such as McCabe et al., *Unit Operations of Chemical Engineering,* Smith et al., *Introduction to Chemical Engineering Thermodynamics,* and Peters et al., *Plant Design and Economics for Chemical Engineers* have taught generations of students principles that are key to success in chemical engineering. Juan de Pablo, Jay Schieber, and Regina Murphy, McGraw-Hill's next band of classic authors, will lead students worldwide toward the latest developments in chemical engineering.

Chemical engineering is a dynamic profession and its literature continues to grow. McGraw-Hill, with its in-house editors and consulting editors Eduardo Glandt [Dean, University of Pennsylvania], Michael Klein [Dean, Rutgers University], and Thomas Edgar [Professor, University of Texas at Austin] remains committed to a publishing policy that will serve the needs of the global chemical engineering profession throughout the years to come.

Dedication

To my wonderful family

Contents

CHAPTER 3   Mathematical Analysis of Material Balance Equations and Process Flow Sheets   169

CHAPTER 5   **Selection of Separation Technologies**
and Synthesis of Separation
Flow Sheets   **365**

Preface

Introduction to Chemical Processes: Principles, Analysis, Synthesis is intended for use in an introductory, one-semester course for students in chemical engineering and related disciplines. The text assumes that the students have had one semester of college-level general chemistry and one or two semesters of college-level calculus. Although student understanding of the material would be deeper with greater background in linear algebra or organic chemistry, the text is organized so that this background is not required for successful completion.

Course Trends

Introductory chemical engineering courses traditionally focus on chemical process calculations. Material and energy balances are taught, a few concepts in thermodynamics are introduced and miscellaneous information on units, dimensions, and curve fitting are included. By the end of the semester most students, given a well-defined problem, can set up and solve material and energy balance equations, but they do not have a good understanding of how these calculations are related to actually designing chemical processes to make products.

Several years ago the chemical engineering faculty at UW—Madison decided to redesign our introductory course. Our goals were twofold: (1) to give the students a better flavor of how chemical processes convert raw materials to useful products and (2) to provide the students with an appreciation for the ways in which chemical engineers make decisions and balance constraints to come up with new processes and products. At the end of the semester, we wanted students to be able, with a minimum amount of information, to synthesize a chemical process flowsheet that would approximate real industrial processes. This includes selection of appropriate separation technology, determination of reasonable operating conditions, optimization of key process variables, integration of energy needs, and calculation of material and energy flows. This becomes possible at the introductory level through use of limiting cases, idealizations, approximations, and heuristics.

The modern approach equips students with the tools necessary for thinking about the creative strategies of chemical process synthesis and greatly enhances students' understanding of the connection between the *chemistry* and the *process*. It provides the students a framework for much of the rest of the curriculum: Students are more motivated to struggle through the rigor and abstraction of engineering science courses in thermodynamics, transport, and kinetics, because the connection between fundamental concepts and practical engineering problem solving has been made. Senior process design courses revisit the same terrain but at a more sophisticated level. Students learn that the principles of chemical processes, and the strategies of process synthesis and analysis, can be advantageously applied to an enormous diversity of problems, from intracellular trafficking of a drug to accumulation of pollutants in the ecosystem. The ready availability of easy-to-use computational tools means that students in an introductory course can tackle challenging and complex problems.

Organization

Many times, students decide to major in chemical engineering because they like chemistry and math, and are interested in practical applications. In designing this text, we have tried to keep this motivation in mind. We start right off the bat, in Chapter 1, providing a link to freshman chemistry courses. We show how simple stoichiometric concepts are used to make informed choices about raw materials and reaction pathways. Students should understand that engineering is not simply about doing calculations, but about using calculations wisely to make good choices. The idea of combining calculations, data and heuristics to make choices is a central theme throughout the text.

Chapter 2 introduces the simple but powerful idea of process flow sheeting as the chemical engineer's means to communicate ideas about raw materials, reaction chemistry, processing steps, and products. Here students learn the 10 Easy Steps for process flow calculations, and are introduced, in a very conceptual manner, to system variables, system and stream specifications, and material balances. Many example problems, drawn from a wide diversity of applications, are worked out in detail.

In Chapter 3 we revisit material balance equations, reaction stoichiometry, and process flow sheeting, but with a more rigorous and mathematical approach. Throughout, the text retains this spiral organization, in which we first reinforce concepts introduced in earlier chapters, and then expand and deepen student understanding of these concepts. In this chapter, material balance equations are derived from conservation-of-mass principles, using a format students will see in more advanced classes, and we do not shy away from transient processes. In optional sections, we demonstrate the power of linear algebra to find independent systems of chemical reactions, to determine the existence and uniqueness of solutions to systems of linear equations, and to develop flexible linear models of chemical processes.

Chapters 4 and 5 delve in greater depth into reactors and separators. Heuristics for synthesizing reactor and separation-train flow sheets are discussed. Quantitative measures of reactor and separator performance are introduced, and students learn how performance specifications influence process flow calculations and process design. Within this context, chemical reaction equilibrium and phase equilibrium are discussed in some detail. Students learn that equilibrium places constraints on the performance of reactors and separators, but also learn how to select process operating conditions and how to design around these constraints. Additionally, students gain considerable experience with using physical property data, graphs, and model equations.

Finally, Chapter 6 covers energy balance equations and process energy calculations. A 12 Easy Step strategy for attacking these problems is developed; the strategy is illustrated in many example problems. Students learn techniques for conserving energy resources wisely and safely.

Features of the Text

The text is written to encourage students to:

- **Link to chemistry.** The text provides a clear link to freshman chemistry courses. Students will remain more interested in the processes and get a better flavor of what chemical processes do if they understand how chemistry relates to processing.
- **Synthesize chemical processes.** The text treats process calculations as a means to an end: the design of safe, reliable, environmentally sound, and economical chemical processes. The author's approach gives students a good understanding of how these calculations inform choices that must be made in designing chemical processes to make desired products.
- **Develop solid problem-solving strategies.** Developing good problem-solving strategies is an important outcome of this introductory course. Readers will find a systematic approach to deriving equations and accounting for specifications. A novel feature of this text is the use of heuristics, introducing beginning students to the notion that practicing engineers rely not just on calculations but also on collected experiences.
- **Invent and analyze.** The text integrates the best of the "process synthesis" philosophy with modern approaches, problems, and techniques. Students learn that principles of process synthesis are gainfully applied to problems in biotechnology, medicine, materials science, and environmental protection.
- **Let pedagogy lead.** The text is heavily laden with pedagogy, tools to guide the reader and enhance the subject matter. A few of the pedagogical elements in this text include *Helpful Hints* !, *Chemistory, Did You Know* ?, and *Case Study* sections. For a complete overview of the pedagogical elements see the Guided Tour section.
- **Explore software.** This text is not directly tied to one software program, allowing students to use software as a common tool to solve problems. An appendix illustrates the use of spreadsheets to find roots of equations, scientific calculators to solve matrix equations, and graphing software to fit model equations to data.

Additional Resources

Available to adopters of the text:

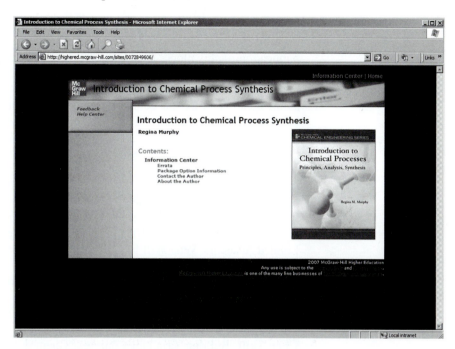

Text Web site. Helpful resources are available for instructors and students at http://www.mhhe.com/murphy.

Information Center

   –Errata

   –Package Option Information

   –Contact the Author

   –About the Author

Instructors

   –Solutions Manual

   –Images in PowerPoint

   –Transition Guide (Felder to Murphy)

Students

   –Physical Property Data Tables

   –Helpful Web links

The following resources are available with the text as *package options* (under separate ISBNs).

- **Engineering Equation Solver (EES).** The basic function provided by EES is the numerical solution of a set of algebraic equations. EES can also be used to solve differential and integral equations, do optimization, provide uncertainty analyses and linear and nonlinear regression, and generate publication-quality plots. There are two major differences between EES and other equation-solving programs. First, EES allows equations to be entered in any order with unknown variables placed anywhere in the equations; EES automatically reorders the equations for efficient solution. Second, EES provides many built-in mathematical and thermophysical property functions useful for engineering calculations. For example, the steam tables are implemented such that any thermodynamic property can be obtained from a built-in function call in terms of any two other properties. Similar data are provided for many other substances including R134a, R22, ammonia, carbon dioxide, hydrogen and air. See the following web sites for more information: www.fchart.com and www.mhhe.com/engcs/mech/ees/.

  Order ISBN 9780073428024 or 0073428027 to receive the text with the Limited Academic Version of EES on CD-ROM. Departments at educational institutions that adopt the text book with the Limited Academic Version CD will also have access to the full academic version of EES. Contact your sales representative for a password to download it to your school computer lab or a similar location. Students in the class may then access and copy the full academic version.

- **ChemSkill Builder** (available online). ChemSkill Builder is a highly regarded tutorial and electronic homework program that provides more than 1500 algorithmically generated questions for students for every topic in the general chemistry course, with tutorial feedback. The questions are presented in a randomized fashion with a constant mix of variables so that no two students will receive the same questions. The application provides feedback for students when incorrect answers are entered, and the answers can be submitted online to an instructor for grading. Order ISBN 9780073424095 or 0073424099 to receive the Murphy text with access to ChemSkill Builder. http://www.chemskillbuilder.com.

- **Perry's Chemical Engineering Handbook** (available on CD-ROM). This world-famous handbook provides unrivaled, state-of-the-art coverage of every aspect of chemical engineering—from the fundamentals to details on computer applications and control. Offering convenience, portability, and outstanding search capacity, this CD-ROM version is fully printable and includes a Table of Contents and Index hyperlinked to the appropriate section for speed of access. This product will run on both a Mac and IBM platform and offers unparalleled authority and comprehensiveness. Order ISBN 9780073430423 or 0073430420 to receive the Murphy text with a Perry's CD-ROM.

About the Author

Regina Murphy received her S.B. in Chemical Engineering in 1978 from MIT, then took a job at Chevron's Richmond Refinery to learn about real engineering. During her 5 years at Chevron she wore several hats, all of them hard. She returned to MIT in 1983 where she spent many happy hours playing basketball and softball and hoisting a pint at the Muddy Charles, and obtained her Ph.D. under the guidance of Clark Colton and Martin Yarmush. In 1989 she joined the faculty in the Department of Chemical Engineering at the University of Wisconsin—Madison. Her current research interests are in protein misfolding and aggregation. She has taught several courses throughout the undergraduate curriculum, with a particular interest in chemical processing, and is currently the departmental Associate Chair for Undergraduate Affairs. She lives in an old Victorian house in Madison with her husband Mark Etzel, also a professor at UW, and their twin sons.

Acknowledgements

Eray S. Aydil
University of California, Santa Barbara

Chelsey D. Baertsch
Purdue University

Paul Blowers
University of Arizona

Paul C. Chan
University of Missouri—Columbia

Wayne R. Curtis
The Pennsylvania State University

Janet De Grazia
University of Colorado at Boulder

Jeffrey Derby
The University of Minnesota

Gregory L. Griffin
Louisiana State University

Sarah W. Harcum
Clemson University

Joseph Kisutcza
New Jersey Institute of Technology

Dana E. Knox
New Jersey Institute of Technology

Douglas Lloyd
The University of Texas at Austin

Teng Ma
Florida State University

Michael E. Mackay
Michigan State University

Susan Montgomery
University of Michigan

James F. Rathman
Ohio State University

James T. Richardson
University of Houston

Richard L. Rowley
Brigham Young University

Michael S. Strano
University of Illinois at Urbana-Champaign

Eric Thorgerson
Northeastern University

Timothy M. Wick
Georgia Institute of Technology

Lale Yurttas
Texas A&M University

The author would like to acknowledge the many people who have contributed in various ways to this project. In particular:

Dale Rudd, who co-authored *Process Synthesis* and provided wise counsel about dealing with publishers.

Michael Mohr, a warm and witty instructor who provided my first introduction to chemical engineering.

Thatcher Root, who cheerfully taught for many semesters from earlier versions of this text, and provided not only many end-of-chapter problems but also practical insights and moral support.

Holly Ferguson, for her unflagging efforts to check all the example and end-of-chapter problems, for finding and fixing many errors, and for her friendship over many years.

Harvey Spangler, a friendly and down-to-earth person who crafts remarkable wooden pens, for his sponsorship of the named chair that provided funds to make completion of this project feasible.

Many students at UW, who suffered through early, error-ridden versions of this text, and despite this were enthusiastic participants in this experiment.

Many teachers at Lapham and Marquette Elementary Schools, who gave me a fresh perspective on learning and teaching, who taught me that learning is risky and that the best teachers provide an environment where students are unafraid to take big leaps into the unknown.

Amanda Green, Suzanne Jeans, Mary E. Powers and all the folks at McGraw-Hill for supporting this project through its ups and downs and pushing me to actually finish the book.

Kevin and Nick, who grew from toddlers to teenagers during the course of this project, who inspired at least one cartoon figure, one quick quiz answer, and one example problem (guess which ones!), who typed some of the tables in App. B (for a fee), and who are still willing to shoot hoops with their mom.

Mark, for his contributions of problems and ideas, but mostly for his unwavering support and love over many, many years.

Guided Tour

Tools that Reinforce Concepts

In this Chapter

Words to Learn

An *In this Chapter* section provides a brief introduction of the subject matter and a bulleted list of questions that are addressed in each chapter. A list of *Words to Learn* is also outlined at the beginning of each chapter. These elements help the reader to focus on the fundamental points as they read each chapter.

Quick Quizzes

The *Quick Quizzes* are sprinkled within the chapters and are intended to test student understanding of the topics covered in each chapter. Answers to the quizzes are provided at the end of each chapter.

Helpful Hints

Did you Know?

Helpful Hint
The ideal gas law does not apply to liquids and solids!

Did You Know?
Chemical engineers tend to prefer continuous processes. But it's not clear that consumers do. For example, the label on a bag of gourmet potato chips brags "made in small batches." "Batch" manufacturing is used to imply more lovingly made, higher-quality products; such products command a premium price at the grocery store.

Helpful Hints ! and *Did You Know* ? sections can be found in the margins sprinkled throughout the text. *Helpful Hints* are designed to help students with difficult points.

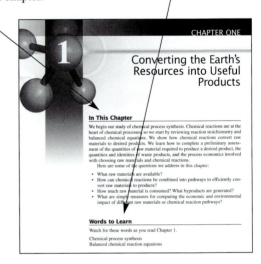

CHAPTER ONE

Converting the Earth's Resources into Useful Products

In This Chapter

We begin our study of chemical process synthesis. Chemical reactions are at the heart of chemical processes, so we start by reviewing reaction stoichiometry and balanced chemical equations. We show how chemical reactions convert raw materials to desired products. We learn how to complete a preliminary assessment of the quantities of raw material required to produce a desired product, the quantities and identities of waste products, and the process economics involved with choosing raw materials and chemical reactions.
Here are some of the questions we address in this chapter:

- What raw materials are available?
- How can chemical reactions be combined into pathways to efficiently convert raw materials to products?
- How much raw material is consumed? What byproducts are generated?
- What are simple measures for comparing the economic and environmental impact of different raw materials or chemical reaction pathways?

Words to Learn

Watch for these words as you read Chapter 1.

Chemical process synthesis
Balanced chemical reaction equations

Example 3.10 | **Integral Equation with Unsteady Flow and Chemical Reaction: Controlled Drug Release**

Patients with certain diseases are treated by injection of proteins or drugs into their bloodstream. Upon injection there is a sudden increase in the protein or drug concentration in the blood to very high levels, but then the concentration rapidly falls. A steadier blood concentration is often desirable, to reduce toxic side effects and increase therapeutic efficacy. Controlled-release technology reduces the variability in drug concentration in the blood. With controlled release, the protein or drug is encapsulated in a polymer and is released slowly into the bloodstream. This maintains the concentration of drug or protein in the bloodstream at a lower, more constant level.

Examples

Over 100 worked examples indicate the conceptual idea the problem is designed to illustrate as well as the specific application chosen. Classical and modern topics are used in the example problems.

Case Studies

Case Studies are provided at the end of each chapter. These in-depth examples illustrate the application of key concepts from that chapter to modern problems. Case studies integrate analysis and synthesis, and boost student confidence in their ability to tackle complex problems and issues.

End-of-Chapter Summaries

The **Summary** sections appear at the end of each chapter and provide an overview of the key definitions and equations from that chapter.

ChemiStories

ChemiStories describe historical events in the lives of the people who contributed to the chemical industry and its products. The stories bring to life the chemical products we take for granted, illustrate the humanity of the heroes of chemical technology, demonstrate that social and political forces drive scientific and engineering progress, and caution readers that technological breakthroughs sometimes have unwanted adverse effects.

Homework Problems

Homework Problems are broken into four categories:

-**Warm-Ups:** Short-answer questions that cover basic definitions and straightforward calculations. Minimal proficiency.

-**Drills and Skills:** Drills and Skills problems cover the fundamental skills and concepts learned in that chapter. Average proficiency.

-**Scrimmage:** Scrimmage problems require application of more than one skill or concept and may involve material from multiple (previous) chapters. Creativity is needed and some problems require library research.

-**Game Day:** Game Day problems are best suited for use in the classroom or as term projects and can be used to promote teamwork and improve communication skills.

List of Nomenclature
(Typical Units)

a_i	activity of compound i (dimensionless)
C_p	heat capacity at constant pressure, (J/gmol °C or J/g K)
C_v	heat capacity at constant volume, (J/gmol °C or J/g K)
E_k	kinetic energy (kJ)
E_p	potential energy (kJ)
f_{Ci}	fractional conversion of reactant i (dimensionless)
f_{Rij}	fractional recovery of component i in stream j (dimensionless)
f_{Sj}	fractional split to stream j (dimensionless)
g	acceleration due to gravity (m/s²)
$\Delta \widehat{G}_f^\circ$	standard molar Gibbs energy of formation, usually 298 K and 1 atm (kJ/gmol)
$\Delta \widehat{G}_r$	molar Gibbs energy of reaction (kJ/gmol)
$\Delta \widehat{G}_r^\circ$	standard molar Gibbs energy of reaction, usually at 298 K and 1 atm (kJ/gmol)
h	height above a reference plane (m)
H_i	Henry's law constant (atm)
H	enthalpy (kJ)
$\dot{H}$	enthalpy flow (kJ/s)
$\widehat{H}$	molar or specific enthalpy (kJ/gmol or kJ/g)
$\Delta \widehat{H}_c^\circ$	standard enthalpy of combustion (kJ/gmol)
$\Delta \widehat{H}_f^\circ$	standard molar enthalpy of formation, usually at 298 K and 1 atm (kJ/gmol)
$\Delta \widehat{H}_m$	molar or specific enthalpy of melting (phase change from solid to liquid) (kJ/gmol or kJ/g)
$\Delta \widehat{H}_{\mathrm{mix}}$	molar or specific enthalpy of mixing (kJ/gmol or kJ/g)

$\Delta \hat{H}_r$	molar enthalpy of reaction (kJ/gmol)
$\Delta \hat{H}_r^{\circ}$	standard molar enthalpy, usually 298 K and 1 atm (kJ/gmol)
$\Delta \hat{H}_{\text{soln}}$	molar or specific enthalpy of solution (kJ/gmol or kJ/g)
$\Delta \hat{H}_v$	molar or specific enthalpy of vaporization (kJ/gmol or kJ/g)
K_a	chemical reaction equilibrium constant
M	molar mass (g/gmol)
m_{sys}	mass in system (g)
$\dot{m}$	mass flow rate (g/s)
n_{sys}	moles in system (gmol)
$\dot{n}$	molar flow rate (gmol/s)
P	pressure (atm, N/m², bar)
p_i	partial pressure of compound i (atm, N/m², bar)
P_i^{sat}	saturation pressure of compound i (atm, N/m², bar)
Q	heat (kJ)
$\dot{Q}$	rate of heat transfer (kJ/s)
R	ideal gas constant (J/gmol-K, liters-bar/gmol K)
$\dot{R}_{ik}$	mass rate of reaction of compound i in reaction k (g/s)
$\dot{r}_{ik}$	molar rate of reaction of compound i in reaction k (gmol/s)
$s_{A \rightarrow P}$	fractional selectivity for conversion of reactant A to product P (dimensionless)
t	time (s)
T	temperature (°C, K)
T_b	normal boiling point temperature (°C, K)
T_m	normal melting point temperature (°C, K)
U	internal energy (kJ)
$\hat{U}$	molar or specific internal energy (kJ/gmol or kJ/g)
v	velocity (m/s)
V	volume (m³)
$\hat{V}$	molar or specific volume (m³/gmol, m³/kg)
w_i	weight fraction of i (dimensionless)

W	work (kJ)
W_s	shaft work (kJ)
$\dot{W}$	rate of work transfer (kJ/s, kW, hp)
$\dot{W}_s$	rate of shaft work transfer (kJ/s, kW, hp)
x_i	mole fraction of i, typically in the liquid or solid phase (dimensionless)
y_i	mole fraction of i in the vapor phase (dimensionless)
$y_{A \rightarrow P}$	fractional yield for conversion of reactant A to product P (dimensionless)
z_i	mole fraction of i in the feed (dimensionless)

Subscripts

f	final
h	element
i	compound or stream component
j	stream
k	reaction
0	initial

Greek Letters

α_{AB}	separation factor for components A and B (dimensionless)
ε_{hi}	number of atoms of element h in compound i
ν_{ik}	stoichiometric coefficient of compound i in reaction k
ρ	density (kg/m^3 or gmol/m^3)
ξ	extent of reaction (gmol)
$\dot{\xi}$	extent of reaction (gmol/s)
χ_k	multiplying factor for reaction k
η	efficiency (dimensionless)

List of Important Equations

Material Balance Equations

Differential form:

Total mass:

$$\frac{dm_{sys}}{dt} = \sum_{\text{all } j \text{ in}} \dot{m}_j - \sum_{\text{all } j \text{ out}} \dot{m}_j$$

Mass of i:

$$\frac{dm_{i,\,sys}}{dt} = \sum_{\text{all } j \text{ in}} \dot{m}_{ij} - \sum_{\text{all } j \text{ out}} \dot{m}_{ij} + \sum_{\substack{\text{all } k \\ \text{reactions}}} \nu_{ik} M_i \dot{\xi}_k$$

Total moles:

$$\frac{dn_{sys}}{dt} = \sum_{\text{all } j \text{ in}} \dot{n}_j - \sum_{\text{all } j \text{ out}} \dot{n}_j + \sum_{\substack{\text{all } k \\ \text{reactions}}} \sum_{\substack{\text{all } i \\ \text{compounds}}} \nu_{ik} \dot{\xi}_k$$

Moles of i:

$$\frac{dn_{i,\,sys}}{dt} = \sum_{\text{all } j \text{ in}} \dot{n}_{ij} - \sum_{\text{all } j \text{ out}} \dot{n}_{ij} + \sum_{\substack{\text{all } k \\ \text{reactions}}} \nu_{ik} \dot{\xi}_k$$

Integral form

Total mass:

$$m_{sys,f} - m_{sys,0} = \sum_{\text{all } j \text{ in}} \left(\int_{t_0}^{t_f} \dot{m}_j \, dt \right) - \sum_{\text{all } j \text{ out}} \left(\int_{t_0}^{t_f} \dot{m}_j \, dt \right)$$

Mass of i:

$$m_{i,sys,f} - m_{i,sys,0} = \sum_{\text{all } j \text{ in}} \left(\int_{t_0}^{t_f} \dot{m}_{ij} \, dt \right) - \sum_{\text{all } j \text{ out}} \left(\int_{t_0}^{t_f} \dot{m}_{ij} \, dt \right) + \sum_{\substack{\text{all } k \\ \text{reactions}}} \left(\int_{t_0}^{t_f} \nu_{ik} M_i \dot{\xi}_k \, dt \right)$$

Total moles:

$$n_{sys,f} - n_{sys,0} = \sum_{\text{all } j \text{ in}} \left(\int_{t_0}^{t_f} \dot{n}_j \, dt \right) - \sum_{\text{all } j \text{ out}} \left(\int_{t_0}^{t_f} \dot{n}_j \, dt \right) + \sum_{\substack{\text{all } k \\ \text{reactions}}} \sum_{\substack{\text{all } i \\ \text{compounds}}} \left(\int_{t_0}^{t_f} \nu_{ik} \dot{\xi}_k \, dt \right)$$

Moles of i:

$$n_{i,\text{sys},f} - n_{i,\text{sys},0} = \sum_{\substack{\text{all } j \text{ in}}} \left(\int_{t_0}^{t_f} \dot{n}_{ij}\, dt \right) - \sum_{\substack{\text{all } j \text{ out}}} \left(\int_{t_0}^{t_f} \dot{n}_{ij}\, dt \right) + \sum_{\substack{\text{all } k \\ \text{reactions}}} \left(\int_{t_0}^{t_f} \nu_{ik}\, \dot{\xi}_k\, dt \right)$$

System Performance Specifications

Splitter

Fractional split:

$$f_{Sj} = \frac{\text{moles of } i \text{ leaving in stream } j}{\text{moles of } i \text{ fed to splitter}} = \frac{\dot{n}_{ij}}{\dot{n}_{i,\text{in}}}$$

Reactor

Fractional conversion:

$$f_{Ci} = \frac{\text{moles of } i \text{ consumed by reaction}}{\text{moles of } i \text{ fed to reactor}} = \frac{-\sum\limits_{\text{all } k \text{ reactions}} \nu_{ik}\, \dot{\xi}_k}{\dot{n}_{i,\text{in}}}$$

Selectivity:

$$s_{A \to P} = \frac{\text{moles of reactant A converted to product P}}{\text{moles of reactant A consumed}} = \frac{\nu_{A1}}{\nu_{P1}} \frac{\sum\limits_{\text{all } k \text{ reactions}} \nu_{Pk}\, \dot{\xi}_k}{\sum\limits_{\text{all } k \text{ reactions}} \nu_{Ak}\, \dot{\xi}_k}$$

Yield:

$$y_{A \to P} = \frac{\text{moles of reactant A converted to desired product P}}{\text{moles of reactant A fed}}$$

$$= -\frac{\nu_{A1}}{\nu_{P1}} \frac{\sum\limits_{\text{all } k \text{ reactions}} \nu_{Pk}\, \dot{\xi}_k}{\dot{n}_{A,\text{in}}}$$

Separator

Fractional recovery:

$$f_{Rij} = \frac{\text{moles of } i \text{ leaving in stream } j}{\text{moles of } i \text{ fed to separator}} = \frac{\dot{n}_{ij}}{\dot{n}_{i,\text{in}}}$$

Separation factor:

$$\alpha_{AB} = \frac{x_{A1}}{x_{A2}} \frac{x_{B2}}{x_{B1}}$$

Chemical Reaction Equilibrium

$$K_a = \prod_{\text{all } i} a_i^{\nu_i}$$

where, to a first approximation,

$$a_i = \frac{y_i P}{1 \text{ atm}} \quad \text{for a gas}$$

$$a_i = x_i \quad \text{for a liquid}$$

$$a_i = 1 \quad \text{for a solid}$$

$$\ln K_{a,T} = -\frac{1}{R}\left[\frac{\Delta \widehat{G}_r^\circ - \Delta \widehat{H}_r^\circ}{298} + \frac{\Delta \widehat{H}_r^\circ}{T}\right]$$

where $\Delta \widehat{G}_r^\circ = \sum \nu_i \Delta \widehat{G}_{i,f}^\circ$ and $\Delta \widehat{H}_r^\circ = \sum \nu_i \Delta \widehat{H}_{i,f}^\circ$.

Phase Equilibrium

Raoult's law:

$$y_i = \frac{P_i^{\text{sat}}}{P} x_i$$

Henry's law:

$$y_i = \frac{H_i}{P} x_i$$

Energy Balance Equations

Differential form:

$$\frac{d(E_{k,\text{sys}} + E_{p,\text{sys}} + U_{\text{sys}})}{dt}$$

$$= \sum_{\text{all } j \text{ in}} \dot{m}_j \left(\widehat{E}_{k,j} + \widehat{E}_{p,j} + \widehat{H}_j\right) - \sum_{\text{all } j \text{ out}} \dot{m}_j \left(\widehat{E}_{k,j} + \widehat{E}_{p,j} + \widehat{H}_j\right) + \sum_j \dot{Q}_j + \sum_j \dot{W}_{s,j}$$

Integral form:

$$(E_{k,\text{sys}} + E_{p,\text{sys}} + U_{\text{sys}})_f - (E_{k,\text{sys}} + E_{p,\text{sys}} + U_{\text{sys}})_0$$

$$= [E_{k,\text{in}} - E_{k,\text{out}}] + [E_{p,\text{in}} - E_{p,\text{out}}] + [H_{\text{in}} - H_{\text{out}}] + Q + W_s$$

Converting the Earth's Resources into Useful Products

In This Chapter

We begin our study of chemical process synthesis. Chemical reactions are at the heart of chemical processes, so we start by reviewing reaction stoichiometry and balanced chemical equations. We show how chemical reactions convert raw materials to desired products. We learn how to complete a preliminary assessment of the quantities of raw material required to produce a desired product, the quantities and identities of waste products, and the process economics involved with choosing raw materials and chemical reactions.

Here are some of the questions we address in this chapter:

- What raw materials are available?
- How can chemical reactions be combined into pathways to efficiently convert raw materials to products?
- How much raw material is consumed? What byproducts are generated?
- What are simple measures for comparing the economic and environmental impact of different raw materials or chemical reaction pathways?

Words to Learn

Watch for these words as you read Chapter 1.

Chemical process synthesis
Balanced chemical reaction equations
Stoichiometric coefficients
Generation-consumption analysis
Atom economy
Basis
Scale factor
Process economy

1.1   Introduction

Why do humans synthesize, design, build, and operate chemical processes?

To make a product that has a specific desired function. Many children bring lunch to school every day. Wouldn't it be great to have a lightweight, safe, easy-open packaging material for carrying juice or milk? Aseptically packaged drink boxes fulfill these product requirements and have replaced heavy, bulky thermoses in the nation's lunch bags. But, although throwaway products are convenient, they carry with them waste disposal concerns.

To convert waste materials into useful products. It takes about 10 pounds of milk to make 1 pound of cheese. The other 9 pounds end up as whey. Whey used to be simply a waste product, dumped in nearby waterways or sprayed on farmers' fields. Now processes have been developed that recover the useful components of whey. For example, the protein lactoferrin is purified from whey and used in infant formula to improve iron uptake. Similarly, whey sugars may serve as a feedstock for production of biodegradable polymers.

To improve the performance of a natural material. Vincristine is a vinca alkaloid present in minute quantities in the periwinkle plant. Concentrated and purified, vincristine has proved to be a powerful drug for treating leukemia and lymphomas. Its success has led to synthesis in the laboratory of structurally related molecules, any of which might serve as effective medicines to treat cancers.

To convert material into energy. Huge quantities of energy are used every day to heat or cool our homes, power our motor vehicles, and cook our food.

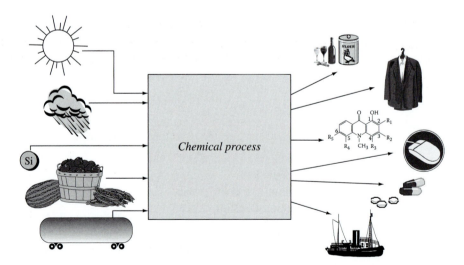

Figure 1.1 Chemical processes convert raw materials into desired products. In synthesizing chemical processes, we choose appropriate raw materials, then select chemical reactions and physical operations to change the properties of the raw materials to those of the desired products. We aim to design a chemical process that is safe to operate, that uses raw materials efficiently and economically, that reliably produces the desired products, and that has minimal environmental impact.

Manufacturing industries also require energy for their processes to work. Much of this energy is derived from combustion of fossil fuels—natural gas, oil, or coal. In this process, the raw material (natural gas, for example) reacts with oxygen to form carbon dioxide and water. In this instance, it is the energy released by the reaction, not the reaction products, that is useful.

Chemical processes convert raw materials into needed products by changing the chemical and/or physical properties of the materials (Fig. 1.1). An enormous breadth of industries—paper, foods, plastics, fibers, glass, electronic materials, fuels, pharmaceuticals, to name a few—depend on chemical processes. The art and science of **chemical process synthesis** is in choosing appropriate raw materials and chemical reaction pathways, and in developing an *efficient, economical, reliable,* and *safe* chemical process. Articulation of product requirements must be made before process development can begin, thus product engineering and process engineering are inextricably linked. The quality and availability of raw materials, economic forecasts, product safety and reliability, marketing concerns, patents, and proprietary technology all influence process design.

1.2   Raw Materials

Ultimately, we derive all of our raw materials from the earth (Fig. 1.2). The fundamental raw materials are:

Air. Plentiful, readily available, and cheap, air serves as the source of oxygen and nitrogen in many chemical processes. Oxygen is used widely for oxidation reactions, the most important of which is the burning of fuels to generate heat and electricity. Discovery of a method to "fix" atmospheric nitrogen by converting it to liquid ammonia spawned the whole agricultural fertilizer industry, with enormous repercussions for production of sufficient food to feed the growing world population.

Water. Water is used as a reactant in many chemical processes and serves an important role as a solvent. This is especially true for the biotechnology industries—old (e.g., beer making), middle-aged (antibiotic production by fermentation), or new (insulin production from genetically engineered cells). Water may eventually serve as a source of hydrogen, a clean-burning fuel, although this is too expensive with current technology.

Minerals. Minerals are solid inorganic elements or compounds. One important mineral is salt, which, besides its use as a preservative and a flavoring, serves as the raw material for the enormous chlor-alkali industry. Minerals are the feedstocks for the inorganic chemicals industries, which produce silicon chips for computers and aluminum for bicycles.

Fossil fuels. Natural gas, crude oil, and coal are all hydrocarbon materials produced by the decay of once-living things. Besides providing us with heat, light, and electricity, fossil fuels serve as the raw material for the synthesis of carbon-based products like polymers for plastic soft drink bottles and contact lenses, fibers for clothing and furnishings, medicines, and pesticides.

Figure 1.2 Cartoon from Calvin & Hobbes "There's Treasure Everywhere."
Calvin and Hobbes © 1995 Watterson. Dist. By UNIVERSAL PRESS SYNDICATE. Reprinted with permission. All rights reserved.

Agricultural and forest products. Living plants, are carbon-based, but they also contain a significant quantity of fixed oxygen and (sometimes) nitrogen. Our food, of course, is produced from these raw materials. Other products derived from agricultural raw materials include paper, natural fibers such as wool or cotton, natural rubber, and medicines. There is an increasing interest in using agricultural materials (also called biomass) as raw materials for production of carbon-containing chemicals, thus reducing our reliance on non-renewable fossil fuels.

Many different kinds of companies are chemical processors. For example (Fig. 1.3):

- An oil company extracts crude oil from underground reservoirs.
- A petroleum refining company processes the oil to recover benzene.

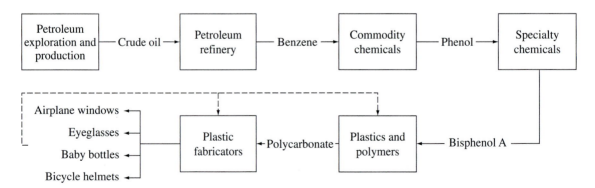

Figure 1.3 Many companies and processes are needed to convert a raw material such as crude oil to products such as bicycle helmets. Companies and municipalities are trying to close the loop, by recovering consumer products at the end of their useful life and reprocessing the materials into new products.

- A commodity chemicals company reacts the benzene to phenol.
- A fine chemicals company converts phenol to bisphenol A.
- A plastics company polymerizes bisphenol A to polycarbonate.
- Fabricators use polycarbonate to make airplane windows, bullet-proof glass, eyeglasses, baby bottles, compact discs, and football helmets.
- Consumers purchase eyeglasses, baby bottles, and compact discs, use them, and then discard them to the landfill or recycling bin.

A humble chemical may serve as the raw material for many products. For example, common table salt, NaCl, is the basis for the huge chlor-alkali business, one of the three largest inorganic chemical industries (the other two are ammonia and sulfuric acid). The chlor-alkali industry includes three basic chemicals: soda ash (sodium carbonate, Na_2CO_3), caustic soda (sodium hydroxide, NaOH) and chlorine (Cl_2) (Fig. 1.4).

Carbon-based products—from acetaminophen to zoledronic acid—are primarily synthesized from nonrenewable fossil fuels. Interest is rising in developing novel chemical processes and products based on renewable raw materials: corn, grass clippings, even municipal wastes. For example, DuPont and partners developed new processes in which corn-derived glucose is fermented, using engineered bacteria, to make 1,3-propanediol. The 1,3-propanediol is purified and then reacted to form a polymer called 3GT, which is spun into a fabric sold as Sorona® (Fig. 1.5). Many challenges await solution in the move to develop new chemical processes for converting renewable raw materials to useful products.

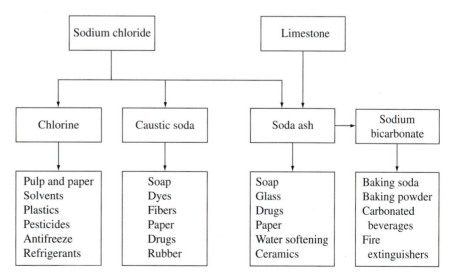

Figure 1.4 Important commodity chemicals as well as common household products are made from sodium chloride and limestone, as part of the chlor-alkali industry. Adapted from *Chemical Process Industries,* 4th ed. by R. N. Shreve and J. A. Brink, 1977.

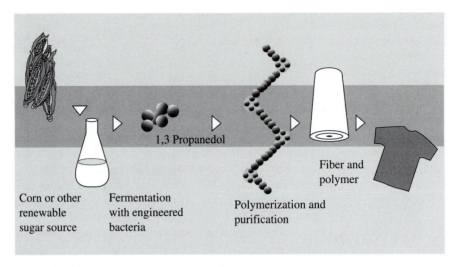

Figure 1.5 New processes to make chemical products from renewable resources are being developed, like this DuPont process to synthesize fiber from corn.

1.3   Balanced Chemical Reaction Equations

At the heart of most chemical processes lies one or more chemical reactions. If A and B are compounds that undergo a chemical reaction to form products C and D, we write:

$$A + B \rightarrow C + D$$

In this notation, we are simply indicating that C and D are products of a chemical reaction in which A and B are the reactants. As an example, in making electronics-grade silicon, silicon tetrachloride ($SiCl_4$, also called tetrachlorosilane) reacts with hydrogen to make pure silicon and hydrogen chloride:

$$SiCl_4 + H_2 \rightarrow Si + HCl$$

This reaction as written is not balanced. If we are interested in showing not only the identity but also the quantity of compounds taking part in a chemical reaction, we write a **balanced chemical reaction equation**. In a balanced chemical equation, the *relative* number of moles of each reactant and product are indicated by coefficients a, b, c, and d:

$$aA + bB \rightarrow cC + dD$$

A chemical equation is balanced if the number of atoms of each element on the left-hand side of the equation equals the number of atoms of that element on the right-hand side. For example, the reaction of tetrachlorosilane with hydrogen to make silicon and hydrogen chloride is balanced if we write

$$SiCl_4 + 2H_2 \rightarrow Si + 4HCl$$

Because the coefficients are relative rather than absolute quantities, one of the coefficients can be chosen arbitrarily; we can equally write

$$2SiCl_4 + 4H_2 \rightarrow 2Si + 8HCl$$

or even

$$\tfrac{1}{2}SiCl_4 + H_2 \rightarrow \tfrac{1}{2}Si + 2HCl$$

because the coefficients do not have to be integers.

Now let's define **stoichiometric coefficients** ν_i for each chemical compound i, and specify that ν_i *is negative for compounds that are reactants and positive for compounds that are products*. For example, in the previous reaction equation,

$$\nu_{SiCl_4} = -\tfrac{1}{2}$$

and

$$\nu_{HCl} = +2$$

We define

$$\varepsilon_{hi} \equiv \text{ number of atoms of the element } h \text{ in compound } i.$$

Then, a chemical equation is stoichiometrically balanced with respect to the hth element if and only if

$$\sum_i \varepsilon_{hi}\nu_i = 0 \tag{1.1}$$

Helpful Hint
Find stoichiometric coefficients by first balancing the element that appears in the fewest compounds.

where we indicate that the summation is taken over all I compounds. In our example, the element Cl appears in two compounds, $SiCl_4$ ($\varepsilon_{Cl, SiCl_4} = 4$) and HCl ($\varepsilon_{Cl, HCl} = 1$), and Eq. (1.1) for $h = $ Cl is simply

$$\varepsilon_{Cl,SiCl_4}\nu_{SiCl_4} + \varepsilon_{Cl,HCl}\nu_{HCl} = 4(-\tfrac{1}{2}) + 1(+2) = 0$$

To balance a chemical equation where the stoichiometric coefficients are unknown, we write Eq. (1.1) for each element. We end up with a system of H equations (one for each element) in I unknowns (the stoichiometric coefficients—one for each compound).

For example, suppose the reaction of interest is oxidation of methane (CH_4) to CO_2 and water. Written in an *unbalanced* form the reaction is:

$$CH_4 + O_2 \rightarrow CO_2 + H_2O$$

Quick Quiz 1.1

Why did we set $\nu_{CH_4} = -1$ and not $\nu_{CH_4} = 1$?

Instead of setting $\nu_{CH_4} = -1$, suppose we had chosen to set $\nu_{O_2} = -1$. What would be the balanced chemical reaction equation?

There are three elements, carbon (C), hydrogen (H) and oxygen (O), and four compounds, so there are three equations involving four unknown stoichiometric coefficients:

$$C: \quad 1\nu_{CH_4} + 0\nu_{O_2} + 1\nu_{CO_2} + 0\nu_{H_2O} = 0$$

$$H: \quad 4\nu_{CH_4} + 0\nu_{O_2} + 0\nu_{CO_2} + 2\nu_{H_2O} = 0$$

$$O: \quad 0\nu_{CH_4} + 2\nu_{O_2} + 2\nu_{CO_2} + 1\nu_{H_2O} = 0$$

Since there are four unknowns but only three equations, there are many possible solutions. To proceed, we arbitrarily set one of the stoichiometric coefficients. For example, we can pick ν_{CH_4} as the basis, and set $\nu_{CH_4} = -1$. There are now only three unknowns, and we solve to find $\nu_{O_2} = -2$, $\nu_{CO_2} = 1$, $\nu_{H_2O} = 2$. The balanced chemical reaction equation is

$$(1)CH_4 + 2O_2 \rightarrow (1)CO_2 + 2H_2O$$

Example 1.1   **Balanced Chemical Reaction Equation: Nitric Acid Synthesis**

Nitric acid (HNO_3) is an important industrial acid used, among other things, in the manufacture of nylon. In one of the reactions for making nitric acid, ammonia (NH_3) and oxygen (O_2) react to form NO and H_2O. Write the balanced chemical equation.

Solution
We'll write the unbalanced chemical equation as

$$NH_3 + O_2 \rightarrow NO + H_2O$$

There are three elements and four compounds, so there are three equations in four unknowns:

$$N: \quad 1\nu_{NH_3} + 1\nu_{NO} = 0$$

$$H: \quad 3\nu_{NH_3} + 2\nu_{H_2O} = 0$$

$$O: \quad 2\nu_{O_2} + 1\nu_{NO} + 1\nu_{H_2O} = 0$$

We'll set $\nu_{NH_3} = -1$. Starting with the N balance, we solve to get $\nu_{NO} = 1$. From the H balance, $\nu_{H_2O} = \frac{3}{2}$. Finally, from the O balance, $\nu_{O_2} = -\frac{5}{4}$. The balanced equation is:

$$NH_3 + \tfrac{5}{4}O_2 \rightarrow NO + \tfrac{3}{2}H_2O$$

Example 1.2   **Balanced Chemical Reaction Equations: Adipic Acid Synthesis**

Adipic acid is an intermediate used in the manufacture of nylon. (We'll discuss this process in greater detail later in this chapter, and again in Chap. 3.) Several chemical reaction steps are involved in synthesis of adipic acid:

Reaction 1.   Cyclohexane (C_6H_{12}) reacts with oxygen (O_2) to produce cyclohexanol ($C_6H_{12}O$).

Reaction 2.   Cyclohexane (C_6H_{12}) reacts with oxygen (O_2) to produce cyclo-hexanone ($C_6H_{10}O$).

Water (H_2O) is a byproduct of one of these reactions.

Reaction 3.   Cyclohexanol reacts with nitric acid to produce adipic acid, ($C_6H_{10}O_4$).

Reaction 4.   Cyclohexanone reacts with nitric acid to produce adipic acid ($C_6H_{10}O_4$).

NO and H_2O are byproducts of both Reactions 3 and 4.

Write the four balanced chemical equations corresponding to these four reactions.

Solution

Reaction 1.   From the problem statement, water may be a byproduct of this reaction. Let's assume it is, and see what happens. The unbalanced chemical equation is

$$C_6H_{12} + O_2 \rightarrow C_6H_{12}O + H_2O$$

There are three elements and four compounds, so we can set one stoichiometric coefficient arbitrarily. C appears in two compounds, O and H in three each, so we start with a balance on the element that appears in the fewest number of compounds:

$$\text{C:}\quad 6\nu_{C_6H_{12}} + 6\nu_{C_6H_{12}O} = 0$$

> **Helpful Hint**
> If one of the element balance equations has only two stoichiometric coefficients, choose one of those stoichiometric coefficients to set to a fixed value.

If we choose to set one of these two stoichiometric coefficients, we can solve for the other immediately. Let's set $\nu_{C_6H_{12}O} = +1$. Then

$$\nu_{C_6H_{12}} = -1$$

We then move on to the other two elements:

$$\text{H:}\quad 12\nu_{C_6H_{12}} + 12\nu_{C_6H_{12}O} + 2\nu_{H_2O} =$$
$$12(-1) + 12(1) + 2\nu_{H_2O} = 0$$
$$\text{O:}\quad 2\nu_{O_2} + \nu_{C_6H_{12}O} + \nu_{H_2O} = 2\nu_{O_2} + 1 + \nu_{H_2O} = 0$$

These are easily solved to yield

$$\nu_{H_2O} = 0$$
$$\nu_{O_2} = -\tfrac{1}{2}$$

The balanced chemical equation is

$$C_6H_{12} + \tfrac{1}{2}O_2 \rightarrow C_6H_{12}O$$

Finding the stoichiometric coefficients led us to the conclusion that water is not a byproduct of Reaction 1 after all.

Reaction 2. The unbalanced reaction of cyclohexane with oxygen to produce cyclohexanone, with water as a possible byproduct, is

$$C_6H_{12} + O_2 \rightarrow C_6H_{10}O + H_2O$$

Proceeding in a manner similar to that used for Reaction 1, we write three equations:

$$C: \quad 6\nu_{C_6H_{12}} + 6\nu_{C_6H_{10}O} = 0$$

$$H: \quad 12\nu_{C_6H_{12}} + 10\nu_{C_6H_{10}O} + 2\nu_{H_2O} = 0$$

$$O: \quad 2\nu_{O_2} + \nu_{C_6H_{10}O} + \nu_{H_2O} = 0$$

We arbitrarily set $\nu_{C_6H_{12}} = -1$ and solve the equations in order to find the other three stoichiometric coefficients. The balanced chemical equation is

$$C_6H_{12} + O_2 \rightarrow C_6H_{10}O + H_2O$$

Reaction 3. In the third chemical reaction, cyclohexanol ($C_6H_{12}O$) and nitric acid (HNO_3) react to make adipic acid ($C_6H_{10}O_4$), with nitric oxide (NO) and water (H_2O) as byproducts. The unbalanced reaction is

$$C_6H_{12}O + HNO_3 \rightarrow C_6H_{10}O_4 + NO + H_2O$$

There are four elements and five compounds:

$$C: \quad 6\nu_{C_6H_{12}O} + 6\nu_{C_6H_{10}O_4} = 0$$

$$H: \quad 12\nu_{C_6H_{12}O} + \nu_{HNO_3} + 10\nu_{C_6H_{10}O_4} + 2\nu_{H_2O} = 0$$

$$O: \quad \nu_{C_6H_{12}O} + 3\nu_{HNO_3} + 4\nu_{C_6H_{10}O_4} + \nu_{NO} + \nu_{H_2O} = 0$$

$$N: \quad \nu_{HNO_3} + \nu_{NO} = 0$$

Starting with either the C or the N balance is OK. Let's set $\nu_{C_6H_{12}O} = -1$. We immediately solve the C balance to find $\nu_{C_6H_{12}O_4} = 1$. Substituting these values into the remaining three equations yields

$$H: \quad -12 + \nu_{HNO_3} + 10 + 2\nu_{H_2O} = 0$$

$$O: \quad -1 + 3\nu_{HNO_3} + 4 + \nu_{NO} + \nu_{H_2O} = 0$$

$$N: \quad \nu_{HNO_3} + \nu_{NO} = 0$$

We can't immediately solve any of the remaining equations. To solve "by hand," we

1. Subtract the N balance from the O balance to eliminate ν_{NO}:

$$3 + 2\nu_{HNO_3} + \nu_{H_2O} = 0$$

2. Subtract the H balance from 2× this equation to eliminate ν_{H_2O}:

$$8 + 3\nu_{HNO_3} = 0$$

3. Solve for ν_{HNO_3} and work backwards to find the other stoichiometric coefficients:

$$\nu_{HNO_3} = -\tfrac{8}{3}$$

$$\nu_{NO} = \tfrac{8}{3}$$

$$\nu_{H_2O} = +\tfrac{8}{3}$$

The balanced chemical equation is:

$$C_6H_{12}O + \tfrac{8}{3}HNO_3 \rightarrow C_6H_{10}O_4 + \tfrac{8}{3}NO + \tfrac{7}{3}H_2O$$

Reaction 4.   The balanced chemical equation is (details are left for you)

$$C_6H_{10}O + 2HNO_3 \rightarrow C_6H_{10}O_4 + 2NO + H_2O$$

Quick Quiz 1.2

In the balanced chemical equation for Reaction 3 of Example 1.2, noninteger coefficients appear. Rewrite the equation, using only integer coefficients.

To learn how to balance chemical equations using matrices, see Appendix A.1.

1.4   Generation-Consumption Analysis

Choosing the raw materials and writing balanced chemical reaction equations are early steps in chemical process synthesis. Often we combine multiple chemical reactions in a reaction pathway, in order to make the most product out of the least (and least expensive) raw material, with the fewest byproducts. **Generation-consumption analysis** is a systematic method for synthesizing reaction pathways involving multiple chemical reactions with these goals in mind. This analysis allows us to calculate the moles of raw materials consumed in generating a given quantity of product, and the moles of byproducts generated per mole of product.

Suppose we have K reactions involving I compounds. We define

$$\nu_{ik} \equiv \text{stoichiometric coefficient of compound } i \text{ in reaction } k$$

To complete a generation-consumption analysis, we

1. Write balanced chemical equations for all K reactions.
2. List all I compounds (reactants and products) in a column.
3. For each reaction k, write ν_{ik} associated with each compound in a column. There will be K columns, one for each reaction.

4. Sum up ν_{ik} in each row, corresponding to each compound i, to get the net stoichiometric coefficient:

$$\nu_{i,\text{net}} = \sum_{\substack{\text{All } k \\ \text{reactions}}} \nu_{ik}$$

where the summation is taken over all K reactions. This sum is the net generation or consumption of that compound. Compounds that have negative sums are raw materials, indicating net consumption. Compounds that have positive sums are products, indicating net generation. Compounds that have zero sums are intermediates, indicating no net generation or consumption.

5. If desired, adjust the net generation or consumption of compounds by finding multiplying factors χ_k for reactions involving those compounds. For example, if we wish to have zero net generation or consumption of compound i, we find χ_k such that

$$\nu_{i,\text{net}} = \sum_{\substack{\text{All } k \\ \text{reactions}}} \chi_k \nu_{ik} = 0 \tag{1.2}$$

Why *can* we do step 5? Because stoichiometric coefficients give relative quantities, or ratios, of reactants and products, not absolute quantities. If we multiply a balanced chemical equation by a common factor, the equation is still balanced.

Why *should* we do step 5? There are many reasons; for example, we may want to avoid net consumption of an expensive compound or net generation of a toxic or hazardous byproduct.

Note: it is not always possible to find χ_k that will satisfy Eq. (1.2). In that case, we may need to search for additional or different chemical reactions to achieve our goals.

Generation-consumption analysis is illustrated in the next two examples.

Example 1.3   **Generation-Consumption Analysis: The LeBlanc Process**

In the LeBlanc process, salt (NaCl) and sulfuric acid (H_2SO_4) are heated in batches to produce sodium sulfate (Na_2SO_4) and hydrochloric acid (HCl). The balanced chemical equation is

$$2NaCl + H_2SO_4 \rightarrow Na_2SO_4 + 2HCl \tag{R1}$$

The sodium sulfate is then cooked in furnaces with coal (C) and limestone ($CaCO_3$) to produce a "black ash." Sodium carbonate (Na_2CO_3) is washed out of the black ash with water, leaving behind calcium sulfide (CaS). These reactions are

$$Na_2SO_4 + 2C \rightarrow Na_2S + 2CO_2 \tag{R2}$$

$$Na_2S + CaCO_3 \rightarrow Na_2CO_3 + CaS \tag{R3}$$

How many moles of what reactants are consumed per mole of sodium carbonate generated? How many moles of what byproducts are generated per mole of product generated?

Solution

There are three reactions involving 10 compounds, so $I = 10$ and $K = 3$. The reactions are already balanced. We list the compounds in the first column, and write the stoichiometric coefficients for each reaction in the next three columns:

Compound	R1 ν_{i1}	R2 ν_{i2}	R3 ν_{i3}
NaCl	−2		
H_2SO_4	−1		
Na_2SO_4	+1	−1	
HCl	+2		
C		−2	
Na_2S		+1	−1
CO_2		+2	
$CaCO_3$			−1
Na_2CO_3			+1
CaS			+1

Then we calculate the net generation or consumption by adding up the coefficients across each row (Table 1.1). Negative numbers in the Net column indicate net consumption of that compound; positive numbers indicate net generation.

The Net column of Table 1.1 tells us that, for every mole of desired product (Na_2CO_3) generated, we consume 2 moles of NaCl, 1 mole of H_2SO_4, 2 moles of C, and 1 mole of $CaCO_3$. We also generate several waste products: 2 moles of HCl, 2 moles of CO_2, and 1 mole of CaS. We can write this information compactly as a net chemical reaction:

$$2NaCl + H_2SO_4 + 2C + CaCO_3 \rightarrow 2HCl + 2CO_2 + Na_2CO_3 + CaS$$

The LeBlanc process produces the desired product, sodium carbonate, from inexpensive and readily available raw materials. However, 5 moles of waste byproducts are generated for each mole of desired product. Is it possible to synthesize a better process by using different raw materials, or combining different reactions? In Example 1.4, we use generation-consumption analysis to showcase a process developed by Ernest Solvay that addresses these concerns. The ChemiStory at the end of this chapter describes more of the history of LeBlanc and Solvay, and explains what sodium carbonate manufacturing has to do with the French Revolution.

Table 1.1	**Generation-Consumption Analysis of the LeBlanc Process**			
Compound	**R1**	**R2**	**R3**	**Net**
	ν_{i1}	ν_{i2}	ν_{i3}	$\nu_{i,net} = \sum \nu_{ik}$
NaCl	−2			−2
H_2SO_4	−1			−1
Na_2SO_4	+1	−1		0
HCl	+2			+2
C		−2		−2
Na_2S		+1	−1	0
CO_2		+2		+2
$CaCO_3$			−1	−1
Na_2CO_3			+1	+1
CaS			+1	+1

Example 1.4   **Generation-Consumption Analysis: The Solvay Process**

Limestone ($CaCO_3$) decomposes to lime (CaO), and lime reacts with water to form "milk of lime," $Ca(OH)_2$

$$CaCO_3 \rightarrow CaO + CO_2 \tag{R1}$$

$$CaO + H_2O \rightarrow Ca(OH)_2 \tag{R2}$$

Milk of lime reacts with ammonium chloride to make ammonia and calcium chloride:

$$Ca(OH)_2 + 2NH_4Cl \rightarrow 2NH_3 + CaCl_2 + 2H_2O \tag{R3}$$

Ammonia dissolved in water makes ammonium hydroxide, which reacts with CO_2 to make ammonium carbonate and then ammonium bicarbonate:

$$NH_3 + H_2O \rightarrow NH_4OH \tag{R4}$$

$$2NH_4OH + CO_2 \rightarrow (NH_4)_2CO_3 + H_2O \tag{R5}$$

$$(NH_4)_2CO_3 + CO_2 + H_2O \rightarrow 2NH_4HCO_3 \tag{R6}$$

Ammonium bicarbonate reacts with sodium chloride to produce sodium bicarbonate and generate more ammonium chloride:

$$NH_4HCO_3 + NaCl \rightarrow NH_4Cl + NaHCO_3 \tag{R7}$$

Finally, sodium bicarbonate ($NaHCO_3$, common baking soda) decomposes to the desired product, sodium carbonate, releasing carbon dioxide and water as byproducts:

$$2NaHCO_3 \rightarrow Na_2CO_3 + CO_2 + H_2O \qquad \text{(R8)}$$

Can we use these reactions to come up with a process for making sodium carbonate from limestone and salt that uses no other raw materials and generates less waste than the LeBlanc process?

Solution

These 8 reactions involve 14 different compounds. Let's use the generation-consumption analysis in Table 1.2a to evaluate this set of chemical reactions:

Table 1.2a	Generation-Consumption Analysis of the Solvay Process (first try)							
Compound	**R1** ν_{i1}	**R2** ν_{i2}	**R3** ν_{i3}	**R4** ν_{i4}	**R5** ν_{i5}	**R6** ν_{i6}	**R7** ν_{i7}	**R8** ν_{i8}	**Net** $\nu_{i,\,\text{net}} = \sum \nu_{ik}$
$CaCO_3$	-1								-1
CaO	$+1$	-1							0
CO_2	$+1$				-1	-1		$+1$	0
H_2O		-1	$+2$	-1	$+1$	-1		$+1$	$+1$
$Ca(OH)_2$		$+1$	-1						0
NH_4Cl			-2				$+1$		-1
NH_3			$+2$	-1					$+1$
NH_4OH				$+1$	-2				-1
$(NH_4)_2CO_3$					$+1$	-1			0
NH_4HCO_3						$+2$	-1		$+1$
$NaCl$							-1		-1
$NaHCO_3$							$+1$	-2	-1
$CaCl_2$			$+1$						$+1$
Na_2CO_3								$+1$	$+1$

We are using 1 mole $CaCO_3$ and 1 mole $NaCl$ to make 1 mole Na_2CO_3 (the desired product), but we are also consuming or generating a lot of other chemicals. Could we synthesize a reaction pathway with no net consumption of any raw materials other than $CaCO_3$ and $NaCl$, and no net generation or consumption of any of the ammonia-containing compounds? In other words, can we find multiplying factors χ_k such that the entry in the Net column equals zero for NH_4Cl, NH_3, NH_4OH, $(NH_4)_2CO_3$, NH_4HCO_3, and $NaHCO_3$? Applying Eq. (1.2) to these six compounds gives:

$$NH_4Cl: \quad -2\chi_3 + \chi_7 = 0$$
$$NH_3: \quad 2\chi_3 - \chi_4 = 0$$
$$NH_4OH: \quad \chi_4 - 2\chi_5 = 0$$
$$(NH_4)_2CO_3: \quad \chi_5 - \chi_6 = 0$$
$$NH_4HCO_3: \quad 2\chi_6 - \chi_7 = 0$$
$$NaHCO_3: \quad \chi_7 - 2\chi_8 = 0$$

One solution that satisfies all these constraints is

$$\chi_3 = \chi_5 = \chi_6 = \chi_8 = 1$$
$$\chi_4 = \chi_7 = 2$$

Let's see what happens if we multiply the stoichiometric coefficients for reactions (R4) and (R7) by 2 (Table 1.2b):

Table 1.2b	Generation-Consumption Analysis of the Solvay Process (second try)*								
Compound	**R1**	**R2**	**R3**	**R4**	**R5**	**R6**	**R7**	**R8**	**Net**
	ν_{i1}	ν_{i2}	ν_{i3}	$\chi_4\nu_{i4}$	ν_{i5}	ν_{i6}	$\chi_7\nu_{i7}$	ν_{i8}	$\nu_{i,net} = \sum_k \chi_k\nu_{ik}$
$CaCO_3$	-1								-1
CaO	$+1$	-1							0
CO_2	$+1$				-1	-1		$+1$	0
H_2O		-1	$+2$	-2	$+1$	-1		$+1$	**0**
$Ca(OH)_2$		$+1$	-1						0
NH_4Cl			-2				$+2$		**0**
NH_3			$+2$	-2					**0**
NH_4OH				$+2$	-2				**0**
$(NH_4)_2CO_3$					$+1$	-1			0
NH_4HCO_3						$+2$	-2		**0**
$NaCl$							-2		-2
$NaHCO_3$							$+2$	-2	**0**
$CaCl_2$		$+1$							$+1$
Na_2CO_3								$+1$	$+1$

*Changes are shown in bold.

Perfect! All of the ammonia-containing compounds are now strictly intermediates, with no net generation or consumption. Furthermore, there is no net consumption of $NaHCO_3$. Remarkably, the net effect of this pathway of 8 chemical reactions involving 14 compounds is simply (from the last column of Table 1.2b):

$$CaCO_3 + 2NaCl \rightarrow Na_2CO_3 + CaCl_2$$

You may have been able to determine the multiplying factors for Example 1.4 by simple inspection of Table 1.2a. In the next example, it is harder to find the multiplying factors without applying Eq. (1.2). To find the correct values of χ_k, we write Eq. (1.2) for all intermediate compounds in terms of the known stoichiometric coefficients ν_{ik} and the unknown multiplying factors χ_k. Then we try to solve the system of equations. If we have more unknowns than equations, we arbitrarily set one of the multiplying factors equal to a fixed value (say, $\chi_1 = 1$) and then proceed.

Example 1.5

Generation-Consumption Analysis: Ammonia Synthesis

Ammonia, one of the largest-tonnage chemicals produced today, is a simple chemical found under the kitchen sink in many homes. Ammonia synthesis proceeds by reacting steam (H_2O) with methane (CH_4) to make carbon monoxide and hydrogen. Then CO and water react to make CO_2 and more H_2. Finally nitrogen (N_2) and hydrogen combine to produce ammonia, NH_3.

How can we combine these reactions so there is no net generation or consumption of CO or H_2?

Solution

We start with balanced chemical equations:

$$CH_4 + H_2O \rightarrow CO + 3H_2 \tag{R1}$$
$$CO + H_2O \rightarrow CO_2 + H_2 \tag{R2}$$
$$N_2 + 3H_2 \rightarrow 2NH_3 \tag{R3}$$

Let's look at the generation-consumption table, using these balanced chemical equations as written.

Compound	R1	R2	R3	Net
	ν_{i1}	ν_{i2}	ν_{i3}	$\nu_{i,\,net}$
CH_4	−1			−1
H_2O	−1	−1		−2
CO	+1	−1		0
H_2	+3	+1	−3	+1
CO_2		+1		+1
N_2			−1	−1
NH_3			+2	+2

This solution does satisfy the constraint that net CO = 0, but doesn't satisfy the requirement that net H_2 = 0. Let's write Eq. (1.2) for these two intermediates:

$$\text{CO:} \qquad \chi_1 - \chi_2 = 0$$

$$\text{H}_2\text{:} \qquad 3\chi_1 + \chi_2 - 3\chi_3 = 0$$

All we need to do is find a set of values for (χ_1, χ_2, χ_3) that satisfies these two equations. Since there are two equations but three variables, there is more than one valid solution. Because there is more than one valid solution, we can pick any number greater than zero for the value of one of the multiplying factors, and then solve for the other two. Let's pick

$$\chi_1 = 1$$

Then,

$$\chi_2 = 1, \; \chi_3 = \tfrac{4}{3}$$

By multiplying all entries in the (R1), (R2), and (R3) columns by these values for χ_1, χ_2, and χ_3, respectively, we get the result we want:

Quick Quiz 1.3

In Example 1.5, what's the net reaction if you set $\chi_2 = 3$ and solve for χ_1 and χ_3?

Would it be possible to combine the set of reactions in Example 1.5 such that there is no net generation or consumption of CO_2?

Compound	$\chi_1 \nu_{i1}$	$\chi_2 \nu_{i2}$	$\chi_3 \nu_{i3}$	$\nu_{i,\,net}$
CH_4	-1			-1
H_2O	-1	-1		-2
CO	$+1$	-1		0
H_2	$+3$	$+1$	-4	0
CO_2		$+1$		$+1$
N_2			$-4/3$	$-4/3$
NH_3			$+8/3$	$+8/3$

The net reaction for ammonia synthesis is read from the last column:

$$CH_4 + 2H_2O + \tfrac{4}{3}N_2 \rightarrow CO_2 + \tfrac{8}{3}NH_3$$

1.5   A First Look at Material Balances and Process Economics

In this section, we'll examine how to use the results from a generation-consumption analysis to calculate the mass of raw materials needed to produce a specified mass of product, the mass of byproducts produced per mass of desired product, and the cost of raw materials per mass of desired product. These are simple but essential calculations in the early stages of chemical process synthesis, as we evaluate alternative choices of raw materials and chemical reaction pathways.

1.5.1 Mass, Moles, and Molar Mass

Let's briefly review a few definitions.

Atomic mass is expressed in terms of atomic mass units (amu). One amu is equal to one-twelfth the mass of a carbon ^{12}C atom, or $1.66053873 \times 10^{-27}$ kg. The atomic mass reported in periodic tables is the compositional average mass of that element, averaged over the distribution of isotopes in nature. The atomic mass of carbon = 12.011 amu (taking into account ^{12}C, ^{13}C, and ^{14}C isotopes), while the relative atomic mass (dimensionless) of ^{12}C = 12. Refer to Appendix B for a listing of atomic mass and number of the elements.

Molecular mass is the sum of the atomic masses of all the atoms in a molecule. **Molar mass** is the mass in grams of one mole ($6.02214199 \times 10^{23}$) of atoms or molecules. The molar mass is numerically equivalent to molecular mass but has units of [grams/gram-mole], abbreviated as [g/gmol]. (Molecular weight is often used interchangeably with molecular mass and with molar mass, although this term is considered out of date.)

Illustration: Glucose ($C_6H_{12}O_6$) contains 6 carbons (atomic mass of 12.011 amu), 12 hydrogens (atomic mass of 1.0079 amu) and 6 oxygens (atomic mass of 15.9994 amu). The molecular mass of glucose is $6(12.011) + 12(1.0079) + 6(15.9994) = 180.157$ amu. The molar mass of glucose is 180.157 g/gmol. For calculations that do not require a high level of accuracy, it is common practice to approximate the molecular mass as $6(12) + 12(1) + 6(16) = 180$ amu and the molar mass as 180 g/gmol.

To convert from *moles* to *mass,* multiply the total moles by the molar mass. To convert from *mass* to *moles,* divide the total mass by the molar mass.

Illustration: $104.2 \text{ gmol glucose} \left(\dfrac{180.157 \text{ g glucose}}{\text{gmol glucose}} \right) = 18{,}770 \text{ g glucose}$

$104.2 \text{ g glucose} \left(\dfrac{\text{gmol glucose}}{180.157 \text{ g glucose}} \right) = 0.5784 \text{ gmol glucose}$

You will encounter many different systems of units throughout your career. Although SI units (kilograms, meters, seconds, K) are used in most scientific venues, many industries still use the British system (pounds, feet, seconds, °F).

To convert from one unit of mass to another, use the following conversion factors.

1 lb = 453.59 g = 16 oz = 0.45359 kg = 5×10^{-4} short ton
1 kg = 1000 g = 35.274 oz = 2.2046 lb = 10^{-3} metric ton
1 short ton = 907,180 g = 2000 lb = 907.18 kg = 0.90718 metric ton
1 metric ton = 10^6 g = 2204.6 lb = 1000 kg = 1.1023 ton

How many gmol of ethanol (C_2H_5OH) are contained in 104 grams of ethanol?

How many grams of ethanol are contained in 104 gmol of ethanol?

How many kg of ethanol are in 104 lb?

How many kgmol of ethanol are in 104 lb?

Illustration: $5.075 \text{ tons}\left(\dfrac{2000 \text{ lb}}{\text{ton}}\right) = 10{,}150 \text{ lb}$

$$5.075 \text{ ton}\left(\dfrac{907.18 \text{ kg}}{\text{ton}}\right)\left(\dfrac{35.274 \text{ oz}}{\text{kg}}\right) = 162{,}400 \text{ oz}$$

Since many of our calculations will give mass in units of lb, kg, tons—units other than grams—it is useful to define molar mass in different units. The molar mass may be written as [lb/lbmol], [kg/kgmol], [ton/tonmol], or any other convenient units. The numerical value of the molar mass of a compound in any of these units is identical. The conversion factors given for mass units can be used to convert from one molar unit to another.

1 lbmol = 453.59 gmol = 16 oz.mol = 0.45359 kgmol = 5×10^{-4} tonmol
1 kgmol = 1000 gmol = 35.274 oz.mol = 2.2046 lbmol = 10^{-3} metric tonmol
1 tonmol = 907,180 gmol = 2000 lbmol = 907.18 kgmol
    = 0.90718 metric tonmol
1 metric tonmole = 10^6 gmol = 2204.6 lbmol = 1000 kgmol = 1.1023 tonmol

Illustration: The molar mass of glucose ($C_6H_{12}O_6$) is 180.157 g/gmol; it is also 180.157 lb/lbmole, 180.157 kg/kgmol, or 180.157 ton/tonmol.
The mass of 104.2 kgmol of glucose is

$$104.2 \text{ kgmol glucose}\left(\dfrac{180.157 \text{ kg glucose}}{\text{kgmol glucose}}\right) = 18{,}770 \text{ kg glucose}$$

1.5.2 Atom Economy

In the early 1990s, Roger Sheldon and Barry Trost proposed that raw materials and chemical reaction pathways be evaluated using the concept of "atom economy" (also called "atom utilization"). **Atom economy** gives a rapid and simple measure of the efficiency of a reaction pathway in converting reactants to products:

$$\text{Fractional atom economy} = \dfrac{\text{mass of desired product}}{\text{total mass of reactants}}$$

We are accustomed to writing balanced chemical equations and to analyzing chemical reaction pathways in terms of moles rather than mass. Thus, a convenient mathematical expression for atom economy is:

⚠Helpful Hint
The sum in the denominator includes only the compounds with net consumption.

$$\text{Fractional atom economy} = \frac{\nu_P M_P}{- \sum_{\text{All reactants}} \nu_i M_i} \qquad (1.3)$$

where ν_P is the stoichiometric coefficient, and M_P is the molar mass, for the desired product P, while ν_i is the stoichiometric coefficient, and M_i is the molar mass, of reactant i. In the case where multiple reactions are combined, the stoichiometric coefficients in Eq. (1.3) are the net coefficients.

Calculating the fractional atom economy for a reaction pathway is straightforward once a generation-consumption analysis has been completed. Notice that the atom economy tells you the best you could ever do, given the chosen reaction pathway. A real process will never achieve quite as good utilization of raw materials as the calculated atom economy. (We will learn why in Chap. 4.)

Example 1.6

Atom Economy: LeBlanc versus Solvay

Compare the atom economy of the LeBlanc process to that of the Solvay process for making sodium carbonate.

Solution

We completed a generation-consumption analysis of the LeBlanc process in Example 1.3. The net reaction is:

$$2NaCl + H_2SO_4 + 2C + CaCO_3 \rightarrow Na_2CO_3 + 2HCl + 2CO_2 + CaS$$

For convenience, we list the stoichiometric coefficient ν_i and the molar mass M_i of all reactants and the desired product, Na_2CO_3, in table form:

Compound	ν_i	M_i	$\nu_i M_i$
NaCl	-2	58.5	-117
H_2SO_4	-1	98	-98
C	-2	12	-24
$CaCO_3$	-1	100	-100
Na_2CO_3	$+1$	106	$+106$

We then calculate the fractional atom economy, using Eq. (1.3):

$$\frac{\nu_p M_p}{- \sum_{\text{Reactants}} \nu_i M_i} = \frac{106}{- [(-117) + (-98) + (-24) + (-100)]} = 0.31$$

The net reaction of the Solvay process (Example 1.4) is

$$2NaCl + CaCO_3 \rightarrow Na_2CO_3 + CaCl_2$$

From the stoichiometric coefficients and the molar masses, we calculate

$$\frac{\nu_p M_p}{- \sum\limits_{\text{All reactants}} \nu_i M_i} = \frac{106}{-[(-117) + (-100)]} = 0.49$$

The Solvay process makes much better use of its raw materials.

All else being equal, reaction pathways with high fractional atom economy are preferable; these should have fewer waste products and, by making good use of the raw materials, should be more cost-efficient.

Example 1.7   Atom Economy: Improved Synthesis of 4-ADPA

4-ADPA (4-aminodiphenylamine, $C_6H_5NHC_6H_4NH_2$) is used to make compounds that reduce degradation of rubber tires. The traditional process required four reactions: chlorination of benzene to chlorobenzene, reaction with nitric acid to make PNCB (*p*-nitrochlorobenzene), reaction of PNCB with formaniline to make 4-NDPA, and hydrogenation of 4-NDPA to 4-ADPA. The balanced chemical equations are:

$$C_6H_6 + Cl_2 \rightarrow C_6H_5Cl + HCl \tag{R1}$$

$$C_6H_5Cl + HNO_3 \rightarrow C_6H_4ClNO_2 + H_2O \tag{R2}$$

$$C_6H_4ClNO_2 + C_6H_5NHCHO + 0.5K_2CO_3 \rightarrow \tag{R3}$$
$$C_6H_5NHC_6H_4NO_2 + KCl + CO + 0.5CO_2 + 0.5H_2O$$

$$C_6H_5NHC_6H_4NO_2 + 3H_2 \rightarrow C_6H_5NHC_6H_4NH_2 + 2H_2O \tag{R4}$$

In the early 1990s, a new process was developed and commercialized. The new process requires only two reaction steps, starting with nitrobenzene and aniline:

$$C_6H_5NO_2 + C_6H_5NH_2 \rightarrow C_6H_5NHC_6H_4NO + H_2O \tag{R1}$$

$$C_6H_5NHC_6H_4NO + 2H_2 \rightarrow C_6H_5NHC_6H_4NH_2 + H_2O \tag{R2}$$

What is the difference in atom economy between the traditional and the new process?

Solution

First let's complete a generation-consumption analysis of the traditional scheme:

Compound	ν_{i1}	ν_{i2}	ν_{i3}	ν_{i4}	$\nu_{i,\,net}$
C_6H_6	-1				-1
Cl_2	-1				-1
C_6H_5Cl	$+1$	-1			0
HCl	$+1$				$+1$
HNO_3		-1			-1
$C_6H_4ClNO_2$		$+1$	-1		0
H_2O		$+1$	$+0.5$	$+2$	$+3.5$
C_6H_5NHCHO			-1		-1
K_2CO_3			-0.5		-0.5
$C_6H_5NHC_6H_4NO_2$			$+1$	-1	0
KCl			$+1$		$+1$
CO			$+1$		$+1$
CO_2			$+0.5$		$+0.5$
H_2				-3	-3
$C_6H_5NHC_6H_4NH_2$				$+1$	$+1$

Now let's calculate the atom economy, using the stoichiometric coefficients from the "net" (last) column. We need to consider only the reactants (negative stoichiometric coefficient) and the desired product (4-ADPA) in the atom economy calculation.

Compound	ν_i	M_i	$\nu_i M_i$
C_6H_6	-1	78	-78
Cl_2	-1	71	-71
HNO_3	-1	63	-63
C_6H_5NHCHO	-1	121	-121
K_2CO_3	-0.5	138	-69
H_2	-3	2	-6
$C_6H_5NHC_6H_4NH_2$	$+1$	184	$+184$

$$\frac{\nu_P M_P}{-\sum\limits_{\text{All reactants}} \nu_i M_i} = \frac{184}{-[(-78) + (-71) + (-63) + (-121) + (-69) + (-6)]} = 0.45$$

Now let's complete a generation-consumption analysis of the new scheme:

Compound	ν_{i1}	ν_{i2}	$\nu_{i,\,net}$
$C_6H_5NO_2$	-1		-1
$C_6H_5NH_2$	-1		-1
$C_6H_5NHC_6H_4NO$	$+1$	-1	0
H_2O	$+1$		$+1$
H_2		-2	-2
$C_6H_5NHC_6H_4NH_2$		$+1$	$+1$

and let's calculate the atom economy of the new scheme:

Compound	ν_i	M_i	$\nu_i M_i$
$C_6H_5NO_2$	-1	123	-123
$C_6H_5NH_2$	-1	93	-93
H_2	-2	2	-4
$C_6H_5NHC_6H_4NH_2$	$+1$	184	$+184$

$$\frac{\nu_P M_P}{-\sum\limits_{\text{All reactants}} \nu_i M_i} = \frac{184}{-[(-123) + (-93) + (-4)]} = 0.84$$

Quick Quiz 1.5

Refer to Example 1.5. What is the fractional atom economy for ammonia synthesis?

Converting from the traditional to the new process increases the atom economy from 0.45 to 0.84. This is a remarkable achievement, which was recognized by a Presidential Green Chemistry Challenge Award.

1.5.3 Process Economy

In Examples 1.3 and 1.4 we completed generation-consumption analyses for two different chemical reaction pathways for making sodium carbonate: the LeBlanc and Solvay processes. In Example 1.6 we compared the atom economy of the two processes. Now we want to compare the **process economy** of these two alternatives; we ask, what is the monetary difference between the value of the products and the costs of the raw materials?

Suppose we want to calculate net annual value of producing 1000 tons sodium carbonate per day via the LeBlanc and Solvay processes. We can't directly use

Tables 1.1 and 1.2b to do this, because those tables give only the relative molar quantities of raw materials consumed and products generated. Rather, we need to do a few simple steps:

1. Convert moles to mass.
2. Scale up or scale down.
3. Convert mass to money.

Let's discuss each one of these steps in turn.

1. *Convert moles to mass.* To convert moles to mass, we simply multiply the stoichiometric coefficient ν_i by its molar mass M_i. This calculation gives a *relative* mass quantity, since the stoichiometric coefficients give a relative rather than absolute molar quantity. For example,

$$\nu_{Na_2CO_3} M_{Na_2CO_3} = (+1)\left(106 \frac{g\ Na_2CO_3}{gmol\ Na_2CO_3}\right) = 106\ g\ Na_2CO_3$$

$$\nu_{NaCl} M_{NaCl} = (-2)\left(58.5 \frac{g\ NaCl}{gmol\ NaCl}\right) = -117\ g\ NaCl$$

2. *Scale up or scale down.* The desired production rate provides a **basis** for all subsequent calculations. *A basis is a quantity or flow rate that indicates the size of the process.* Either a raw material or a product can serve as the basis compound. In our example, the basis is 1000 tons Na_2CO_3 produced/day. We *scale up* from a relative mass quantity to the basis quantity by using a **scale factor**, which is simply the ratio of the basis to the relative mass quantity. For our example,

$$\text{Scale factor} = \frac{\text{basis}}{\text{relative mass quantity}} = \frac{1000\ tons/day}{106\ g}$$

Now, the quantity of *any* raw material or product is calculated by simply multiplying the relative quantity of the raw material or product by the scale factor. For example, NaCl is consumed at a rate of $(-117) \times (1000/106) = -1104$ tons per day.

3. *Convert mass to money.* In the last step, we calculate the raw material costs and the product value by multiplying the quantity of each compound by its unit price. Suppose, for example, that the current price for NaCl is $95/ton. How much will we pay every day for NaCl if our process requires 1104 tons NaCl per day?

$$\left(\frac{-1104\ tons\ NaCl}{day}\right)\left(\frac{\$95}{ton\ NaCl}\right) = -\$105,000/day$$

The negative sign indicates a cost to the process. To evaluate the overall process economy, we simply sum up the cost of all raw materials and the value of all products. For a process to be profitable, this sum must be greater than zero.

In Example 1.8, we analyze the process economy of the Solvay process. Current prices for many commodity chemicals are available in the biweekly publication *Chemical Market Reporter*.

Example 1.8	**Process Economy: The Solvay Process**

The Solvay process (Example 1.4) consumes limestone ($CaCO_3$) and salt (NaCl) to produce soda ash (Na_2CO_3), with calcium chloride ($CaCl_2$) as a byproduct. If we wish to produce 1000 tons soda ash/day, what are the required feed rates of limestone and salt?

Suppose current prices for bulk quantities are $87/ton for $CaCO_3$, $95/ton for NaCl, $105/ton for Na_2CO_3, and $250/ton for $CaCl_2$. Does the Solvay process make economic sense if the byproduct $CaCl_2$ cannot be sold? How does the economic picture change if there is a market for the byproduct?

Solution

From Table 1.2b of Example 1.4, we see that there is net consumption or generation of 4 compounds: NaCl, $CaCO_3$, Na_2CO_3 and $CaCl_2$. These compounds, along with their stoichiometric coefficients from their net reaction, are listed in the first two columns of Table 1.3.

To convert *moles to mass,* we multiply the stoichiometric coefficients ν_i (column 2) by the molar mass M_i (column 3) to get the relative mass (column 4). Then, we *scale up*. The relative mass of Na_2CO_3 is 106, and the desired production rate is 1000 tons/day of Na_2CO_3, so the scale factor $SF = 1000/106$. We multiply the numbers in column 4 by the scale factor to get the tons/day consumed or generated for all compounds. Finally, we convert *mass to money.* The cost of raw materials and the selling price of products, per ton, are listed in column 6. By multiplying column 5 (tons/day) by column 6 ($/ton), we get column 7 ($/day).

The mass of raw materials consumed should equal the mass of products made. We check this by summing up all the numbers in the tons/day column (pay attention to the sign of each number). It should sum to zero. It is a good habit to get into checking your numbers to be sure your solution is reasonable.

Considering *just* raw materials costs, if we were unable to sell the calcium chloride, there would be a net *loss* of $82,000/day! This does not include energy costs, labor costs, or capital equipment costs—all of which contribute substantially to the overall process economics. If there is a market for calcium chloride, then we make a profit (again, ignoring other processing costs) of $180,000/day of Na_2CO_3. In fact, at these prices, we might consider the Solvay process as a way to make $CaCl_2$, with soda ash as a byproduct!

!
Helpful Hint
The sum of the mass quantities of all materials consumed should equal the mass quantities of all materials generated.

Quick Quiz 1.6

In Example 1.8 we evaluated a process for making 1000 tons/day sodium carbonate, which had a value of $105/ton. Use Table 1.4 to categorize this process as "commodity" or "specialty".

1.5.4 Process Capacities and Product Values

Chemical process facilities vary enormously in scale; some are small enough to fit in your hand while others occupy several city blocks. Chemical products vary enormously in value; some are bought with the spare change in your pocket while others are more precious than gold. Table 1.4 gives some useful order-of-magnitude numbers regarding the scale of chemical processes and the value of chemical products. How well do the examples we've worked through in this chapter fit these order-of-magnitude estimates?

Table 1.3   Raw Material Requirements and Process Economics for 1000 tons/day Soda Ash Production

Compound	ν_i	M_i	$\nu_i M_i$	$\nu_i M_i \times SF$ tons/day,	$/ton	$/day, tons/day $\times$ $/ton
NaCl	−2	58.5	−117	−1104	95	−105,000
CaCO$_3$	−1	100	−100	−943	87	−82,000
Na$_2$CO$_3$	+1	106	+106	1000	105	+105,000
CaCl$_2$	+1	111	+111	1047	250	+262,000
Sum (w/o CaCl$_2$)						−82,000
Sum (w/ CaCl$_2$)			0	0		+180,000

*A negative number indicates a material consumed or a cost. A positive number indicates a material produced or income.

Table 1.4   Typical Plant Capacities, Product Values, and Waste Generation for Chemical Processes

Chemical category	Typical plant capacity, lb/year	Typical product value, $/lb	Typical waste generation, lb waste/lb product
Petroleum	1 billion–100 billion	0.1	0.1
Bulk (commodity)	10 million–1 billion	0.1–2	<1–5
Fine (specialty)	100 thousand–10 million	2–10	2–50
Pharmaceuticals	1 thousand–100 thousand	10–infinity	10–100

CASE STUDY   Six-Carbon Chemistry

In this case study we illustrate how the concepts introduced in Chap. 1 are used to make decisions about raw materials, products, and reaction pathways, by looking in some depth at specific processes of importance in the organic chemicals business. These processes are linked by their connection to 6-carbon compounds. We'll look at two questions:

1. Benzene is a 6-carbon compound purified from petroleum. Suppose we have available 15,000 kg/day benzene. What are some useful 6-carbon products we might make from benzene?

2. Could we replace benzene with a raw material from a renewable resource to make the same 6-carbon products?

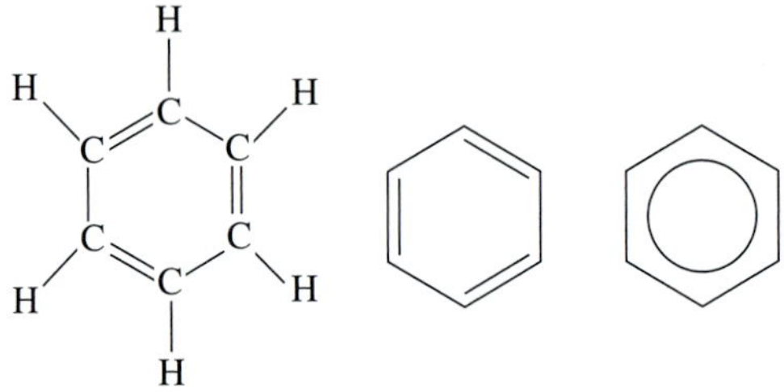

Figure 1.6 Three different representations of the structure of benzene, C_6H_6, one of the most important raw materials in the synthetic organic chemicals industry.

Simple organic compounds like benzene serve as raw materials in the production of the plastics, detergents, pharmaceuticals, and fibers that are ubiquitous in modern societies. Think, for example, of nylon. Nylon was first sold commercially in 1940, in the early days of World War II. The fiber rapidly became an indispensable element in the war effort, as it was used for parachutes, tents, ropes, airplane tire cords, and other military essentials. Perhaps nylon's greatest commercial success was in women's hosiery, as nylon stockings replaced the silk stockings formerly supplied by the Japanese.

There are several kinds of nylon, of which one of the most important is called nylon 6,6. Nylon 6,6 is a polymer—a very large macromolecule containing many small repeating units linked by covalent bonds. Nylon 6,6 contains two repeating units, both of which are 6-carbon compounds: hexamethylene diamine and adipic acid. (That is where the 6,6 comes from.) We will focus our attention on the manufacture of adipic acid from benzene. The structures of adipic acid and of important intermediates are shown; notice that the 6-carbon structure is conserved.

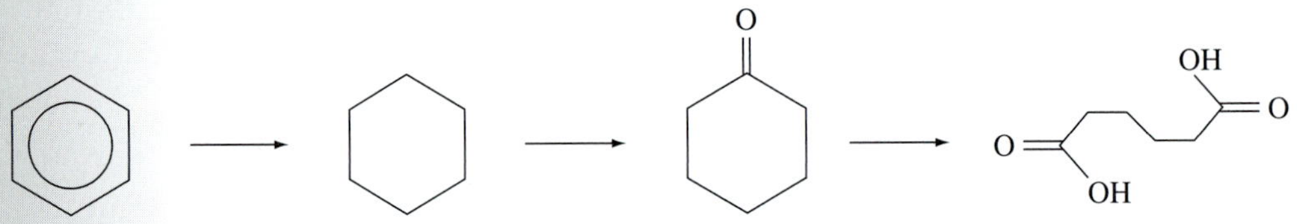

Benzene, C_6H_6 Cyclohexane, C_6H_{12} Cyclohexanone, $C_6H_{10}O$ Adipic acid, $C_6H_{10}O_4$

Figure 1.7 Benzene is converted to adipic acid through a series of chemical reactions involving intermediates cyclohexane and cyclohexanone.

Reaction 1. Benzene is hydrogenated to cyclohexane:

$$C_6H_6 + 3H_2 \rightarrow C_6H_{12} \tag{R1}$$

Reaction 2. Cyclohexane is partially oxidized with oxygen. Although in reality there are multiple products of this reaction, for simplicity we assume here that only one reaction occurs, producing cyclo-hexanone ($C_6H_{10}O$) and water:

$$C_6H_{12} + O_2 \rightarrow C_6H_{10}O + H_2O \tag{R2}$$

Reaction 3.   Cyclohexanone is oxidized with nitric acid to make adipic acid:

$$C_6H_{10}O + 2HNO_3 \rightarrow C_6H_{10}O_4 + 2NO + H_2O \tag{R3}$$

The generation-consumption analysis is shown in Table 1.5. There is zero net generation/consumption of the intermediates cyclohexane and cyclohexanone, so no further adjustments are needed.

The net reaction is:

$$C_6H_6 + 3H_2 + O_2 + 2HNO_3 \rightarrow C_6H_{10}O_4 + 2NO + 2H_2O$$

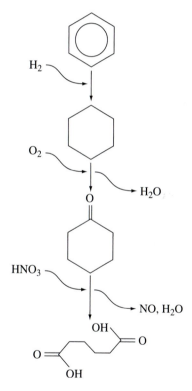

Figure 1.8 Reaction pathway from benzene to adipic acid, showing other raw materials and byproducts.

We consume one mole of benzene, 3 moles of hydrogen, 1 mole of oxygen and 2 moles of nitric acid to produce one mole of adipic acid. There are two waste products: nitric oxide, which is released to the atmosphere, and water, which goes down the drain (via a water treatment system, of course!). Release of

Table 1.5	Generation-Consumption Analysis of Benzene-to-Adipic Acid Process			
Compound	ν_{i1}	ν_{i2}	ν_{i3}	$\nu_{i,\,net}$
C_6H_6	-1			-1
H_2	-3			-3
C_6H_{12}	$+1$	-1		0
O_2		-1		-1
$C_6H_{10}O$		$+1$	-1	0
HNO_3			-2	-2
$C_6H_{10}O_4$			$+1$	$+1$
NO			$+2$	$+2$
H_2O		$+1$	$+1$	$+2$

nitrogen oxide compounds is an environmental concern, but so far no commercial process has been developed that avoids the nitric acid oxidation that leads to generation of nitrogen oxides.

Adipic acid, of which about 85 percent is used to make nylon 6,6, is one possible value-added product to make from benzene. Are there other options?

One idea is catechol, an important feedstock for fine-chemical production. Catechol is used to make pharmaceuticals like L-Dopa (used to treat Parkinson's disease) and flavorings like vanillin. Catechol is one of three *isomers* of dihydroxybenzene $C_6H_6O_2$; the other two, hydroquinone (*p*-dihydroxybenzene) and resorcinol (*m*-hydroxybenzene) are also industrially important chemicals. (*Isomers* have identical molecular formulas, but the atoms are arranged in different geometries.) From the structure of catechol, it is easy to see why benzene makes sense as a raw material.

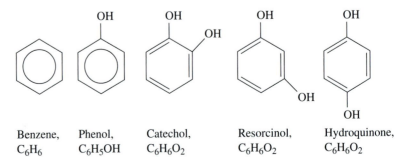

Benzene,   Phenol,   Catechol,   Resorcinol,   Hydroquinone,
C_6H_6 C_6H_5OH $C_6H_6O_2$ $C_6H_6O_2$ $C_6H_6O_2$

Figure 1.9 Dihydroxybenzenes and their precursors, benzene + phenol.

Let's look at the reaction pathway from benzene to catechol.

Reaction 1:   Benzene and propylene (C_3H_6) combine to make isopropylbenzene (C_9H_{12}, also called cumene):

$$C_6H_6 + C_3H_6 \rightarrow C_9H_{12} \tag{R1}$$

Reaction 2:   Cumene reacts with oxygen to give the unstable intermediate cumene hydroperoxide ($C_9H_{12}O_2$):

$$C_9H_{12} + O_2 \rightarrow C_9H_{12}O_2 \tag{R2}$$

Reaction 3:   Cumene hydroperoxide breaks down into phenol (C_6H_6O) and the byproduct acetone (C_3H_6O):

$$C_9H_{12}O_2 \rightarrow C_6H_6O + C_3H_6O \tag{R3}$$

Reaction 4:   Phenol reacts with hydrogen peroxide (HOOH), a strong oxidizing agent, to produce catechol:

$$C_6H_6O + H_2O_2 \rightarrow o\text{-}C_6H_6O_2 + H_2O \tag{R4}$$

(Hydroquinone is also produced in significant quantities; for simplicity this reaction will be neglected.)

The generation-consumption analysis is shown in Table 1.6.

Table 1.6	**Generation-Consumption Analysis for Benzene-to-Catechol Process**				
Compound	ν_{i1}	ν_{i2}	ν_{i3}	ν_{i4}	$\nu_{i,\,net}$
C_6H_6	-1				-1
C_3H_6	-1				-1
C_9H_{12}	$+1$	-1			0
O_2		-1			-1
$C_9H_{12}O_2$		$+1$	-1		0
C_6H_6O			$+1$	-1	0
C_3H_6O			$+1$		$+1$
H_2O_2				-1	-1
$o\text{-}C_6H_6O_2$				$+1$	$+1$
H_2O				$+1$	$+1$

The net result is:

$$C_6H_6 + C_3H_6 + O_2 + H_2O_2 \rightarrow o\text{-}C_6H_6O_2 + H_2O + C_3H_6O$$

Overall, we've consumed 1 mole of benzene, 1 mole of propylene, and two different oxygen sources, O_2 and H_2O_2, to make 1 mole of catechol. Unlike the adipic acid case, we've produced a byproduct that is valuable: acetone is a useful solvent and feedstock for synthesis of other organic chemicals.

We've identified two useful products we might make from benzene. How do the two processes compare on atom economy? Considering *only* the cost of the raw material and the value of the products, what is the best course of action? Assume benzene is valued at $0.41/kg. (The price of benzene changes dramatically with changes in crude oil prices.)

Option 1: Sell the benzene. Selling 15,000 kg/day benzene at this price generates

$$\frac{15{,}000 \text{ kg benzene}}{\text{day}} \times \frac{\$0.41}{\text{kg benzene}} = \frac{\$\,6150}{\text{day}}$$

Option 2: Make adipic acid. The generation-consumption analysis for this option was shown in Table 1.5. We obtain pricing information from the *Chemical Market Reporter* or other sources and complete the analysis in tabular form, as shown below.

Table 1.7		Atom Economy and Process Economy for Benzene-to-Adipic Acid Process				
Compound	ν_{i1}	M_i	$\nu_{i1}M_i$	**kg/day (SF = 192.3)**	**$/kg**	**$/day**
C_6H_6	-1	78	-78	$-15{,}000$	0.41	$-6{,}150$
H_2	-3	2	-6	$-1{,}154$	0.2	-230
O_2	-1	32	-32	$-6{,}154$	~0	0
HNO_3	-2	63	-126	$-24{,}230$	0.40	$-9{,}700$
$C_6H_{10}O_4$	$+1$	146	$+146$	$+28{,}076$	1.54	$+43{,}200$
NO	$+2$	30	$+60$	$+11{,}538$	~0	0
H_2O	$+2$	18	$+36$	$+6{,}923$	~0	0
Sum			0	0		$+27{,}100$

*The scale factor SF is (15,000 kg benzene/day)/(78 g benzene) = 192.3.

The basis for the calculation is 15,000 kg/day benzene consumed.

The fractional atom economy is

$$\frac{\nu_P M_P}{- \sum\limits_{\text{Reactants}} \nu_i M_i} = \frac{146}{(78 + 6 + 32 + 126)} = 0.60$$

The process economics are attractive: we could make a tidy $27,000/day, a considerable increase over the value of the benzene itself. Of course, we've neglected the cost of building and operating the facility, and we've assumed that the price of adipic acid will remain stable despite the increase in worldwide plant capacity that would occur if such a plant were built. This very preliminary assessment simply tells us that it is worth considering this process in greater detail.

Option 3:   Make catechol. Now, consider the possibility of producing catechol from benzene (Table 1.8).

If we consider both acetone and catechol as useful products, the atom economy is very high at 0.90. The net profit is a whopping $89,300/day! (Of course, we've neglected the cost of building and operating the facility.)

Table 1.8   Atom Economy and Process Economy of Benzene-to-Catechol Process

Compound	ν_i	M_i	ν_i	kg/day (SF = 192.3)	$/kg	$/day
C_6H_6	-1	78	-78	$-15,000$	0.41	$-6,150$
C_3H_6	-1	42	-42	$-8,077$	0.26	$-2,100$
O_2	-1	32	-32	$-6,154$	~0	0
H_2O_2	-1	34	-34	$-6,538$	1.49	$-9,740$
C_3H_6O	$+1$	58	$+58$	$+11,154$	0.86	$+9,600$
o-$C_6H_6O_2$	$+1$	110	$+110$	$+21,153$	4.62	$+97,700$
H_2O	$+1$	18	$+18$	$+3,462$	~0	0
Sum				0		$+89,300$

Let's take a step back and consider the raw material, benzene, which is a widely used reactant. Why benzene? Benzene is derived from crude oil, and is plentiful and relatively cheap; decades of research and development in the petroleum industry have made it that way. We know how to recover crude oil from the ground, how to purify benzene from crude oil, and how to use all the other components of crude oil for numerous functions.

So what's the problem? First, petroleum is a nonrenewable resource. Second, benzene is carcinogenic. Third, it is volatile, so some of it ends up in the air and contributes to smog. Additionally with the benzene-to-adipic acid process, nitrogen oxides are produced, which may contribute to ozone depletion and the greenhouse effect.

Is there another raw material that might substitute for benzene? What other 6-carbon compounds are readily available, perhaps from renewable resources? Glucose ($C_6H_{12}O_6$) is one such compound. It's nontoxic, and is produced from renewable resources like corn. Compare the structure of glucose to those of adipic acid and catechol: glucose is chemically more similar to these two products than is benzene.

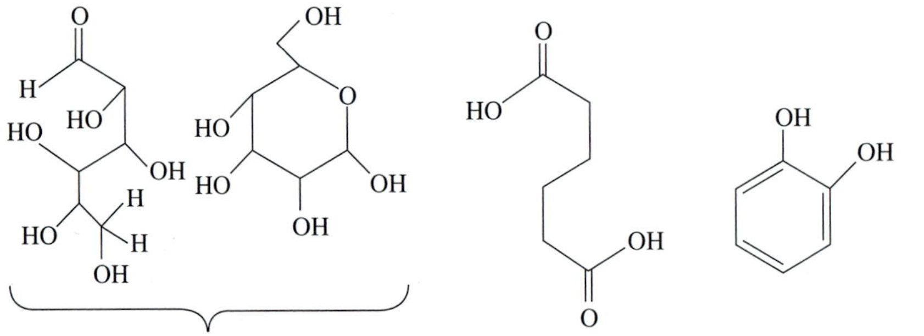

Glucose, $C_6H_{12}O_6$ Adipic acid, $C_6H_{10}O_4$ Catechol, $C_6H_6O_2$

Figure 1.10 Linear and cyclic structures of glucose, compared to adipic acid and catechol. Not all hydrogens are shown.

Is glucose a suitable substitute for benzene as a raw material in adipic acid and catechol production? The first challenge is to identify reaction pathways that convert glucose to the desired products. Unfortunately, glucose does not have the same chemical reactivity as benzene. It cannot withstand the high pressures and temperatures, frequently used with benzene chemistry, without degrading. On the other hand, glucose is a very useful feedstock for microorganisms like yeast and bacteria (not to mention humans!). Bacteria and yeast consume glucose for energy, maintenance, growth, and reproduction. With modern genetic engineering methods, microorganisms can often be tricked into converting some of the glucose into products that are useful for humans.

The bacteria *E. coli* has been genetically engineered in a research laboratory to convert glucose to muconic acid ($C_6H_6O_4$). *E. coli* needs $\frac{7}{3}$ mole of glucose and $\frac{17}{2}$ moles oxygen to produce 1 mole of muconic acid; carbon dioxide and water are the byproducts:

$$\frac{7}{3} C_6H_{12}O_6 + \frac{17}{2} O_2 \rightarrow C_6H_6O_4 + 8CO_2 + 11H_2O \tag{R1}$$

Muconic acid can then be hydrogenated to adipic acid in a more conventional chemical reactor:

$$C_6H_6O_4 + 2H_2 \rightarrow C_6H_{10}O_4 \tag{R2}$$

The generation-consumption analysis for conversion of glucose to adipic acid is shown in Table 1.9.

The same researchers genetically engineered *E. coli* to convert glucose directly to catechol. Bacterial conversion of glucose to catechol requires $2\frac{1}{3}$ moles of glucose plus oxygen to produce 1 mole of catechol, with carbon dioxide and water as byproducts:

$$\frac{7}{3} C_6H_{12}O_6 + \frac{15}{2} O_2 \rightarrow C_6H_6O_2 + 8CO_2 + 11H_2O \tag{R1}$$

How do the atom and process economies compare for glucose versus benzene as a raw material? The comparison must be based on the same rate of production of desired products: 21,150 kg catechol/day, or 28,100 kg adipic acid/day. (Rates were rounded off to reflect level of accuracy of these calculations.) The price of glucose fluctuates somewhat with crop prices, purity, and location. Let's use a price of $0.60/kg glucose. We'll assume that oxygen is free, and that carbon dioxide and water have no value.

Table 1.10 shows that glucose is clearly not a good choice as a raw material for adipic acid production. The fractional atom economy is only 0.21, because

Table 1.9	Generation-Consumption Analysis of Glucose-Adipic Acid Process		
Compound	ν_{i1}	ν_{i2}	$\nu_{i,\,net}$
$C_6H_{12}O_6$	$-7/3$		$-7/3$
O_2	-8.5		-8.5
$C_6H_6O_4$	$+1$	-1	0
CO_2	$+8$		$+8$
H_2O	$+11$		$+11$
H_2		-2	-2
$C_6H_{10}O_4$		$+1$	$+1$

Table 1.10   Atom Economy and Process Economy of Glucose-to-Adipic Acid Process

Compound	ν_i	M_i	$\nu_i M_i$	kg/day (SF = 192.5)	$/kg	$/day
$C_6H_{12}O_6$	$-7/3$	180	-420	$-80,850$	0.60	$-48,500$
O_2	-8.5	32	-272	$-52,360$	~0	0
CO_2	$+8$	44	$+352$	$+67,760$	~0	0
H_2O	$+11$	18	$+198$	$+38,120$	~0	0
H_2	-2	2	-4	-770	0.2	-150
$C_6H_{10}O_4$	$+1$	146	$+146$	$+28,100$	1.54	$+43,300$
Sum				0		$-5,400$

*The scale factor SF = 28,100/198 = 192.5.

so much of the carbon is consumed to make CO_2 (to produce the energy for bacterial survival and growth). The process loses money.

Table 1.11 shows that the glucose-to-catechol process is poor in atom economy (0.17), but is profitable because of the high value of the catechol product. Still, glucose is not competitive with benzene if *just* raw material costs are considered. Other considerations (such as environmental impact, reliability of raw material source, patent protection, energy costs, cost of equipment, safety, technical feasibility, and projected changes in raw material costs) may swing a decision toward the more expensive raw material. In catechol manufacture, for example, a significant amount of the isomer hydroquinone is made as a by-product when benzene is used as the raw material, but not when glucose is used.

Table 1.11   Atom Economy and Process Economy of Glucose-to-Catechol Process

Compound	ν_i	M_i g/g-mol	$\nu_i M_i$	kg/day (SF = 192.3)	$/kg	$/day
$C_6H_{12}O_6$	$-7/3$	180	-420	$-80,770$	0.60	$-48,500$
O_2	-7.5	32	-240	$-46,150$	~0	0
CO_2	$+8$	44	$+352$	$+67,690$	~0	0
H_2O	$+11$	18	$+198$	$+38,080$	~0	0
o-$C_6H_6O_2$	$+1$	110	$+110$	$+21,150$	4.62	$+97,700$
sum				0		$+49,200$

Table 1.12. Waste Generation from Four Processes

Raw material	Product	kg/day product	kg/day byproduct	Identity of byproduct	kg/day wastes	Identity of wastes	kg waste per kg product
Benzene	Adipic acid	28,100	0		18,500	NO, H_2O	0.66
Benzene	Catechol	21,150	11,150	acetone	3,460	H_2O	0.16
Glucose	Adipic acid	28,100	0		105,880	CO_2, H_2O	3.77
Glucose	Catechol	21,150	0		105,770	CO_2, H_2O	5.00

If it is expensive to separate the hydroquinone from the catechol, the glucose process becomes more economically competitive.

In comparing different processes, besides considering the costs of raw materials and the value of the products, we need to consider waste production. Production of wastes means that some of our valuable raw materials, for which we've paid good money, have been converted into things we didn't want. At best, "waste" products are valuable byproducts. At worst, if the waste products are toxic, costly disposal is required. Let's compare waste generation for four processes: benzene to adipic acid, benzene to catechol, glucose to adipic acid, and glucose to catechol (Table 1.12).

Remember that these calculations are *minimum* waste generation; we have not accounted for any inefficiencies in the process. The processes using benzene produce less waste than those using glucose. A lot of the carbon in glucose ends up as CO_2 rather than as product (as we already saw in the atom economy calculations). Why? One reason is this: in fermentation, glucose conversion to CO_2 produces energy for bacterial survival and growth. For a fairer comparison, we should see if energy needs for the benzene processes are met by burning fuels and thus producing CO_2. If so, then the waste calculations must consider energy requirements as well as raw material requirements.

Summary

- Chemical processes convert raw materials into useful products. In the initial stages of **chemical process synthesis,** we choose raw materials to make a specific product, or products to make from a specific raw material. We choose a chemical reaction pathway for converting the chosen raw materials into desired products. These choices all have profound consequences on the technical and economic feasibility of the process.

- Balanced chemical equations are needed to begin process calculations. Chemical equations are balanced if

$$\sum_{\substack{\text{All } i \\ \text{compounds}}} \varepsilon_{hi}\nu_i = 0$$

for all elements, where ε_{hi} is the number of atoms of element h in compound i, and v_i is the **stoichiometric coefficient** for compound i; v_i is negative for compounds that are reactants and positive for compounds that are products.

- A **generation-consumption analysis** is a systematic way to analyze chemical reaction pathways involving I compounds and K reactions. To complete a generation-consumption analysis:

  (1) Write balanced chemical equations for all K reactions.
  (2) List all I compounds (reactants and products) in a column.
  (3) For each reaction k, write the stoichiometric coefficient v_{ik} associated with each compound i in a column. There will be K columns, one for each reaction.
  (4) Calculate

  $$v_{i,\,net} = \sum_{\substack{\text{All } k \\ \text{reactions}}} v_{ik}$$

  This sum is the net generation or consumption of compound i. Compounds that have negative sums are raw materials, indicating net consumption. Compounds that have positive sums are products, indicating net generation. Compounds that have "zero" sums are intermediates, indicating no net generation or consumption.

  (5) If desired, adjust the net generation or consumption of compounds by finding multiplying factors χ_k for reactions involving that compound. For example, if we wish to have zero net generation or consumption of compound i, we find all χ_k such that

  $$\sum_{\substack{\text{All } k \\ \text{reactions}}} \chi_k v_{ik} = 0$$

- A **basis** is a quantity or flow rate that indicates the size of a process. A scale factor provides an easy way to scale a process up or down to the desired basis. A **scale factor** is a ratio of the basis quantity to a relative quantity.

- **Atom economy** is a simple indicator of the efficiency of utilization of raw materials in a given reaction pathway

  $$\text{Fractional atom economy} = \frac{v_P M_P}{-\sum_{\text{All reactants}} v_i M_i}$$

- A simple measure of the **process economy** is made by starting from the generation-consumption analysis, scaling up or down to a desired production rate, and then computing the difference between the product values and the raw material costs.

ChemiStory: Changing Salt into Soap

Soap is made by combining fats or oils from animals or plants with an alkaline material. Today caustic soda (sodium hydroxide, NaOH) is the alkali used for making soap, but in the past sodium carbonate (Na_2CO_3), potassium hydroxide (KOH), and potassium carbonate (K_2CO_3) were common choices.

In the 1700s in Europe, soap was a luxury reserved for the wealthy. But technology to make cheaper cotton clothing was rapidly developing. Cotton needed to be cleaned before it could be dyed and sold, so demand for soap for textile manufacture increased dramatically. (Use of soap for personal hygiene was still uncommon.) At the same time, glass and paper factories were expanding; both products required sodium carbonate for their manufacture. Growth of the textile, glass, and soap industries led to a huge demand for new sources of alkali.

© Vol. 18/Corbis

King Louis XVI of France issued a proclamation offering an award—the equivalent of half a million dollars—to the person who invented a process to turn common table salt (NaCl) into "washing soda," better known now as sodium carbonate or soda ash. Why was sodium carbonate so valuable to the French King? At that time, alkali for French factories was imported from two sources: Spanish and Irish peasants who harvested and burned seaweed and recovered the ashes, and New England settlers who burned brush to clear land and to make "potash" (a mix of mainly KOH, NaOH, K_2CO_3, and Na_2CO_3). A few problems arose. First, there wasn't enough seaweed to meet the increasing demand. Second, France supported the American War of Independence, which led Britain to block potash exports from New England to France. French industry was threatened because of the loss of access to raw materials.

Luckily for the King, chemistry was quite fashionable at this time. The Duke of Orleans, the wealthiest man in France and an outspoken critic of the French absolute monarchy, established a large chemistry research lab on his palace grounds. One of the talented but poor young men who benefited from the Duke's patronage was Nicolas LeBlanc. LeBlanc knew that common salt (NaCl) was very stable, but it could be converted to more reactive sodium sulfate by treating it with sulfuric acid. It took him 5 more years to stumble upon the idea of reacting sodium sulfate with limestone to make sodium carbonate. In 1789, LeBlanc planned to collect his prize. Unfortunately for him, in 1789 the French Revolution happened. LeBlanc never received his award from King Louis. The king was even more unfortunate—he lost his head.

Despite these setbacks, with financial help from the Duke, LeBlanc built his first factory in 1791. The next few years were not easy, however. In 1793, the Duke of Orleans was arrested and executed. Then, LeBlanc's factory was

(*continued*)

seized by the government, he lost his job, and his daughter died. At the age of 63, depressed and broke, LeBlanc killed himself with a gunshot to the head.

The story doesn't end here, however. The LeBlanc process survived and thrived in the next decades; demand for washing soda exploded, and his process was the only reliable and economical way to make it on a large scale. However, the LeBlanc process was an environmental disaster. Acid gas spewed out over the landscape, wasting forests, destroying farmland, and poisoning workers. HCl gas combined with the waste sulfur solids to make hydrogen sulfide, a deadly and smelly gas. In 1863, disgusted Britons passed the Alkali Act, one of the earliest pieces of legislation dealing with chemical pollution. Tall chimneys were installed to disperse the gases over a broader area. Facilities were built next to the soda ash plant to recover and reuse the waste products.

In the mid-1860s, Ernest Solvay figured out a way to exploit a different series of chemical reactions. Solvay used the same raw materials, limestone and salt, to make the same product, soda ash. However, the reaction pathway

Courtesy of Jacques Solvay

was different, no acid was consumed or generated, and a useful by-product, $CaCl_2$, was made. Furthermore, whereas the LeBlanc process operated in batch mode, the Solvay process was continuous. One might argue that Ernest Solvay was to large-scale chemical manufacturing what Henry Ford was to large-scale automaking. The Solvay process completely displaced the LeBlanc process by about 1915. Ernest Solvay made millions, and gave it all away to charity.

Today in the United States, it is cheaper to mine soda ash, mainly in Wyoming, than to make it from salt and limestone. In countries where there are no natural sources of soda ash, the Solvay process is still used. In the 1890s, the availability of cheap hydroelectric power led to the development of an electrolytic process to convert NaCl to produce NaOH—the alkali used today to make soap.

Quick Quiz Answers

1.1 Because methane is a reactant, not a product. $0.5\ CH_4 + O_2 \rightarrow 0.5CO_2 + H_2O$.

1.2 $3C_6H_{12}O + 8HNO_3 \rightarrow 3C_6H_{10}O_4 + 8NO + 7H_2O$.

1.3 $3CH_4 + 6H_2O + 4N_2 \rightarrow 3CO_2 + 8NH_3$. No, because CO_2 is in only one reaction.

1.4 2.26 gmol; 4790 g; 47.2 kg; 1.024 kgmol.

1.5 0.51.

1.6   At 730 million lb/yr and \$0.05/lb, clearly commodity.
1.7   0.73 lb/person/year, about 0.3 percent, commodity at \$0.70/lb.

References and Recommended Readings

1. *Chemical Market Reporter* is a weekly periodical that contains current pricing information on commodity chemicals as well as stories about the chemical business. Published by Schnell Publishing Co., New York. Available online through subscription and at many university libraries.

2. *Chemical and Engineering News* is a highly readable weekly news magazine published by the American Chemical Society. Research news, business developments, and policy issues of interest to chemists and chemical engineers are covered. Regular features include salary surveys, data on performance of specific companies in various industry sectors, and information on chemical production. Available online at http://pubs.acs.org/, at most university libraries, and to all members of the American Chemical Society.

3. *Chemical Engineering Progress* is a news magazine published by the American Institute of Chemical Engineers. Technology and business developments, practical solutions to engineering problems, and career advice of interest to chemical engineers are covered. Available at many university libraries and to all members of the American Institute of Chemical Engineers.

4. *The Kirk-Othmer Encyclopedia of Chemical Technology* is a multi-volume treasure trove of information about chemical products and processes.

5. *Real World Cases in Green Chemistry,* by Michael C. Cann and Mark E. Connelly, documents several specific examples in which consideration of atom economy and other "green chemistry" concepts have sparked changes in industrial practice, leading to processes with less waste and minimal adverse environmental impact. The book was published in 2000 by the American Chemical Society.

6. The genetic engineering of *E. coli* to produce muconic acid and catechol, described in the case study, was published in *J. Am. Chem. Soc.* (1994), 116:399–400, and *J. Am. Chem. Soc.* (1995), 117:2395–2400.

7. *Prometheans in the Lab: Chemistry and the Making of the Modern World,* by Sharon Bertsch McGrayne, is a balanced and engaging study of the people and places behind important chemical advances. The book served as a major reference for all the ChemiStories. Published in 2001 by McGraw-Hill.

Chapter 1 Problems

Warm-Ups

P1.1   Silicon carbide (SiC), also called carborundum, is one of the hardest known materials in the world. It is used as an industrial abrasive for coating grinding wheels. It is made by reacting SiO_2 (essentially sand) with C (coke); CO is the only byproduct. Write the stoichiometrically balanced chemical equation.

P1.2 Nitroglycerin $[C_3H_5(NO_3)_3]$ is an explosive that decomposes to CO_2, H_2O, N_2, and O_2. Write the balanced chemical equation for explosive decomposition of nitroglycerin.

P1.3 Platinum is used in catalytic converters in automobiles, in fuel cells, and in industrial reactors for increasing the octane of gasoline. Platinum metal is produced by the decomposition of $(NH_4)_2PtCl_6$ to Pt, NH_4Cl, N_2, and HCl. Write the balanced chemical equation.

P1.4 One cause of upset stomachs is excess stomach acid (HCl). Over-the-counter antacids contain various compounds such as sodium bicarbonate ($NaHCO_3$), calcium carbonate ($CaCO_3$), or magnesium carbonate ($MgCO_3$) that react with acid to neutralize it. Write stoichiometrically balanced chemical equations for the reaction of these antacid compounds with HCl to produce a salt, CO_2, and water. Which of these three compounds has the most acid neutralizing ability per gram?

P1.5 Urea $[(NH_2)_2CO]$ is used extensively as the nitrogen source in inorganic fertilizers. What is the mass in pounds of 10 gmol urea? What is the mass in grams of 10 lbmol urea?

P1.6 Urea $[(NH_2)_2CO]$ is a solid at room temperature, so it can be mixed into lawn fertilizer and packaged in bags. When applied to the ground, urea spontaneously decomposes to ammonia (NH_3), which is the active ingredient, and the byproduct CO_2. Write down the stoichiometrically balanced chemical equation. What is the fractional atom economy of the reaction in which urea decomposes to ammonia?

P1.7 Complete oxidation of hexane (C_6H_{14}) or of glucose ($C_6H_{12}O_6$) with oxygen (O_2) produces carbon dioxide (CO_2) and water (H_2O). Hexane is purified from crude oil, while glucose is made from renewable sources such as cornstarch. Write down the stoichiometrically balanced equation for complete oxidation of hexane and of glucose. How many grams of CO_2 and H_2O are generated per gram of hexane consumed by reaction? Per gram of glucose consumed by reaction?

P1.8 If 1 gmol of N_2 and 3 gmol of H_2 are consumed to make 2 gmol of NH_3, how many pounds of N_2 and H_2 are required to make 1 billion lb of ammonia? (The worldwide annual production of ammonia is close to 40 billion lb.)

P1.9 If chlorine (Cl_2) costs \$0.016/gmol and ammonia (NH_3) costs \$0.0045/gmol, what do these chemicals cost per ton?

P1.10 300 million lb of 4-ADPA are produced per year. Compare the waste production (lb/year) for using the conventional versus the newer reaction pathway, as described in Example 1.7.

P1.11 About 45 million tons of sulfuric acid (H_2SO_4) are manufactured every year, topping the list of all chemicals in quantity produced. Calculate
   (a) The ton-moles of sulfuric acid produced per year.

(b) The grams of sulfuric acid produced per year.

(c) The pounds of sulfuric acid per person in the world, assuming the world population is 6 billion.

What is the dollar value of the worldwide market if sulfuric acid sells for about $75/ton?

P1.12. In the case study, we evaluated the economics of using either benzene or glucose as a raw material for making adipic acid or catechol. Prices of raw materials can change, sometimes dramatically, as supply and demand fluctuate. Suppose the price of all raw materials stays the same except glucose. How much would the price of glucose ($/kg) have to drop for the glucose-to-adipic acid process to be economically competitive with the benzene-to-adipic acid process? What about for the catechol process?

P1.13 Why might waste generation (lb waste/lb product) be much higher for pharmaceutical products than for commodity chemicals?

P1.14 A gallon of milk sells for $2.89. Would milk be classified as a commodity, a specialty, or a pharmaceutical chemical on the basis of price? (Assume that a gallon of milk weighs about 8 lb.) A 12-oz. bottle of water in a vending machine sells for $1.75. Should that bottle of water be classified as a commodity, a specialty, or a pharmaceutical chemical on the basis of price?

Drills and Skills

P1.15 Excess nitrates in well water are responsible for "blue baby" syndrome (also called methemoglobinemia). Some bacteria are capable of carrying out a reaction that removes nitrates from water; the reaction requires methanol and can be written as

$$HNO_3 + CH_3OH \rightarrow C_3H_7NO_2 + CO_2 + H_2O$$

The equation as written is not balanced. Find the correct stoichiometric coefficients. Calculate the grams of methanol required to reduce the nitrate content (as HNO_3) of 10 L of well water from 54 mg/L to 10 mg/L.

P1.16 Chlorofluorocarbons (CFCs) have been widely used as refrigerants. They are particularly attractive because they are nontoxic and stable. But their stability means that they persist for a long time in the atmosphere. CFCs have been linked to depletion of the upper ozone layer. Some researchers have proposed destroying CFCs by reaction with sodium oxalate ($Na_2C_2O_4$) to produce three solids: sodium fluoride (NaF), sodium chloride (NaCl), and coke (C), along with CO_2 gas. Write the stoichiometrically balanced chemical equation for destruction of one CFC [Freon-12, CF_2Cl_2] with sodium oxalate. Calculate the grams of sodium oxalate required, and the total grams of solid products produced, per gram of Freon-12 destroyed.

P1.17 Carbon nanotubes have been proposed as a novel way to store hydrogen safely and compactly in a solid form at room temperature and pressure. Several researchers have reported synthesizing nanotubes containing 1% to 8% by weight of hydrogen. One group claimed that certain graphite nanofibers can store hydrogen at levels exceeding 50 wt%, but others were skeptical. What is the wt% hydrogen in common liquids and solids such as ethanol (C_2H_5OH), methane (CH_4), water (H_2O) and glucose ($C_6H_{12}O_6$)? Use the results of your calculations to evaluate the importance and validity of the claims regarding the hydrogen storage potential of nanotubes.

P1.18 The following chemical equations are not balanced. In each set, the first equation uses traditional chemical reagents and the second uses newer catalysts. Balance each equation. Compare the fractional atom economy for the (a) conventional and (b) catalytic reaction schemes in each set.

Hydrogenation:

$$C_6H_5COCH_3 + NaBH_4 + H_2O \rightarrow C_6H_5CH(OH)CH_3 + NaB(OH)_4$$

$$C_6H_5COCH_3 + H_2 \rightarrow C_6H_5CH(OH)CH_3$$

Oxidation

$$C_6H_5CH(OH)CH_3 + CrO_3 + H_2SO_4 \rightarrow C_6H_5COCH_3 + Cr_2(SO_4)_3 + H_2O$$

$$C_6H_5CH(OH)CH_3 + H_2O_2 \rightarrow C_6H_5COCH_3 + H2O$$

Carbon-carbon bond formation

$$C_6H_5CH(OH)CH_3 + Mg + CO_2 + HCl \rightarrow C_6H_5CH(CH_3)COOH + MgCl_2 + H_2O$$

$$C_6H_5CH(OH)CH_3 + CO \rightarrow C_6H_5CH(CH_3)COOH$$

P1.19 Polycarbonates are strong, lightweight, impact-resistant, and transparent polymers used to make products like football helmets, eyeglass lenses, airplane windows, and compact discs. You've been told that the synthesis of polycarbonate involves the following reactions:

$$C_6H_5OH \text{ (phenol)} + CH_3COCH_3 \rightarrow C_{15}H_{16}O_2 \text{ (bisphenol A)}$$

$$CH_4 + H_2O \rightarrow CO + H_2$$

$$Cl_2 + CO \rightarrow COCl_2 \text{ (phosgene)}$$

$$C_{15}H_{16}O_2 + COCl_2 + NaOH \rightarrow [-OC_6H_4C(CH_3)_2C_6H_4OCO-]_n + NaCl$$

where n indicates that the chemical unit repeats itself n times in the polymer chain. Assume that $n = 50$.

Bisphenol A

Not all the equations as written are balanced, and there may be minor byproducts such as CO_2 or H_2O generated or consumed as well. Write the balanced chemical equations. Complete a generation-consumption analysis to determine the moles of raw materials required per mole of polycarbonate produced. CO and $COCl_2$ are toxic, so the reaction pathway should have no net generation or consumption of these compounds.

P1.20 We take in nitrogen when we eat protein. After digestion, some of that nitrogen ends up as ammonium ions. Ammonium ions (NH_4^+) are extremely toxic, so our bodies work hard to get rid of them. Here is a simplified version of the detoxification reactions that happen inside our cells:

$HCO_3^- + NH_4^+ + C_5H_{12}O_2N_2$ (ornithine) $\rightarrow C_6H_{13}O_3N_3$ (citrulline)

Citrulline $+ C_4H_7O_4N$ (aspartic acid) $\rightarrow C_{10}H_{18}O_6N_4$ (arginosuccinate)

Arginosuccinate $\rightarrow C_4H_4O_4$ (fumarate) $+ C_6H_{14}O_2N_4$ (arginine)

Arginine $\rightarrow CH_4ON_2$ (urea) $+ C_5H_{12}O_2N_2$ (ornithine)

The reactions have not been balanced. As usual in biological systems, water is ubiquitous and can be a reactant or a product in any reaction, but is not explicitly shown. Balance each chemical reaction. Use a generation-consumption analysis to figure out the overall reaction to convert ammonium ions to less toxic materials. What is the product(s)? Is there net consumption or generation of water?

P1.21 Magnetic nanoparticles are under investigation for medical imaging and drug delivery applications. In one proposed synthesis scheme, 1.52 mmol $Fe(CO)_5$ is mixed with 1.28 g oleic acid ($C_{17}H_{33}COOH$) in an octyl ether solvent to form an iron oleate complex (detected by a change in color from orange to black). Then, 0.34 g trimethylamine oxide [$(CH_3)_3NO$] is added to the mixture. Ethanol precipitation yields monodisperse and highly crystalline nanoparticles of γ-Fe_2O_3. Calculate the atom economy of this scheme.

P1.22 In Example 1.8, raw material requirements for production of 1000 tons/day soda ash by the Solvay process were calculated. Determine the raw material requirements for the LeBlanc process at the same soda ash production rate. Evaluate the profit or loss of the LeBlanc process, assuming the following prices: \$95/ton NaCl, \$80/ton H_2SO_4, \$87/ton $CaCO_3$, \$105/ton Na_2CO_3. All other raw materials and byproducts are assumed to be essentially free.

P1.23 Vinegar is a solution of acetic acid (CH_3COOH) in water. Vinegar is made by fermentation of wine or juice, or by chemical synthesis. The key reactions are oxidation (with O_2) of ethanol (C_2H_5OH) to acetaldehyde (CH_3CHO) and then further oxidation of acetaldehyde to acetic acid. (These reactions happen in the stomach, too.) What are the likely byproducts of these reactions? Write balanced chemical equations and complete a generation-consumption analysis on the conversion of ethanol to acetic acid. Calculate the kilograms of ethanol and oxygen consumed, and the kilograms of byproducts, per kilogram of acetic acid. If ethanol sells for \$0.29/kg, what is the minimum selling price of acetic acid?

P1.24 High-purity silicon for semiconductor manufacture is made from cheap ingredients—sand (SiO_2) and coke (C). The unbalanced chemical reaction is:

$$SiO_2 + C \rightarrow Si + CO$$

The silicon produced by this reaction is pure enough for making silicone-based polymers and for mixing with metals like aluminum, but it is not pure enough for electronics applications. For those applications, the crude silicon is reacted with chlorine (Cl_2) to make tetrachlorosilane, $SiCl_4$. $SiCl_4$ is then reduced with hydrogen (H_2) to make Si and HCl. Write the balanced chemical equations for making high-grade silicon crystals from SiO_2 and C. Complete a generation-consumption analysis to calculate the quantity (grams) of SiO_2 and other reactants required to make 100 grams of high-purity silicon. Calculate the quantities of byproducts. What is the fractional atom economy?

P1.25 Citral is extracted from lemongrass oil and is blended into many consumer products, such as perfumes, soaps, ice creams, and soft drinks, imparting a pleasant lemon-lime fragrance. A company has announced a new route for chemical synthesis of citral from butene (C_4H_8) and formaldehyde (CH_2O) that involves three reactions. The unbalanced reactions are given below; water, which may participate in any of these reactions as a reactant or product, is not shown.

$$C_4H_8 + CH_2O \rightarrow C_5H_{10}O$$
$$C_5H_{10}O + O_2 \rightarrow C_5H_8O$$
$$C_5H_{10}O + C_5H_8O \rightarrow C_{10}H_{16}O \text{ (citral)}$$

Balance the chemical equations, adding water as a reactant or product as needed. Complete a generation-consumption analysis of the reaction pathway. Calculate the kilograms of raw materials required per kilogram of citral produced, and the kg of waste materials generated per kilogram of citral.

P1.26 Plants need phosphate. Although phosphate rock $[(CaF)Ca_4(PO_4)_3]$ is used extensively by organic farmers, it does not release phosphate rapidly enough for some agricultural and landscaping applications. Rapid release of phosphate is possible with monocalcium phosphate $(CaH_4(PO_4)_2H_2O)$, which is produced by reacting phosphate rock with sulfuric acid and water. Other byproducts are calcium sulfate $(CaSO_4)$ and hydrogen fluoride (HF). Superphosphate fertilizer is the mix of calcium sulfate and monocalcium phosphate that is produced by this reaction; the HF is not recovered. Write the balanced chemical equation. Per ton of monocalcium phosphate generated, calculate the tons of phosphate rock and sulfuric acid that are consumed, and calculate the tons of byproducts generated. If phosphate rock sells for \$128/ton, sulfuric acid for \$80/ton, and superphosphate fertilizer for \$320/ton, how much (if any) profit can you make?

P1.27 An elemental analysis of yeast shows that the yeast contains 50 wt% C, 6.94 wt% H, 9.72 wt% N, and 33.33 wt% O. If the yeast is modeled as a single chemical compound of molecular formula $CH_xO_yN_z$, find the values of x, y, and z. In bread dough, yeast is added to a mix of water and flour. The yeast feed on sugars and protein in the flour and produce CO_2, which causes the dough to rise. In addition, the yeast grow and reproduce, which consumes glucose $(C_6H_{12}O_6)$ and ammonia from protein (NH_3). In one experiment, 3.9 g of CO_2 was produced per gram of yeast produced. With this information, write a balanced chemical equation with glucose, oxygen, and ammonia as reactants, and yeast, CO_2, and water as products. What fraction of the glucose was used to make CO_2, and what fraction was consumed in making more yeast?

P1.28 Copper occurs in nature in many forms; chalcocite (Cu_2S) is one form found in copper ores located in the western United States. To recover metallic Cu (for pennies, wire, and many other uses) chalcocite is leached in an acid solution of ferric sulfate to form cupric sulfate. Several reactions take place:

$$Cu_2S + Fe_2(SO_4)_3 \rightarrow CuS + CuSO_4 + 2\ FeSO_4$$

$$Cu_2S + 2\ Fe_2(SO_4)_3 \rightarrow 2\ CuSO_4 + 4\ FeSO_4 + S$$

$$CuS + Fe_2(SO_4)_3 \rightarrow CuSO_4 + 2\ FeSO_4 + S$$

Copper sulfate is then reduced with iron (Fe) to produce metallic copper (Cu) and iron sulfate $(FeSO_4)$

$$CuSO_4 + Fe \rightarrow Cu + FeSO_4$$

Combine reactions to get the most metallic copper per ton of chalcocite. Per ton of metallic copper generated, calculate the tons of raw materials consumed and byproducts generated.

P1.29 Lactose, or milk sugar, is a byproduct of cheese making. Lactose $(C_{12}H_{22}O_{11})$ is a disaccharide—a dimer of glucose and galactose. Galactose is an *isomer* of glucose—it has the same molecular formula but a different structure. Lactose is hydrolyzed to glucose and galactose with an enzyme called lactase.

$$C_{12}H_{22}O_{11} + H_2O \xrightarrow{\text{Lactase}} C_6H_{12}O_6(\text{glucose}) + C_6H_{12}O_6(\text{galactose})$$

A series of three enzymes will catalyze the conversion of galactose to glucose:

$$C_6H_{12}O_6(\text{galactose}) \xrightarrow{\text{Enzyme}} C_6H_{12}O_6(\text{glucose})$$

(Infants have a lot of lactase, but the amount decreases in adults. Some adults lack this enzyme and are lactose intolerant—they can't consume dairy products. Lactase is available in most supermarkets as a dietary supplement for people who love ice cream and cheese but are lactose intolerant.) At \$0.22/lb, lactose is somewhat cheaper than glucose (\$0.60/kg). The market for galactose is small. Evaluate the economic incentive for (a) designing a process to use lactase to convert lactose to glucose, with galactose as a byproduct, and (b) designing a process to use lactase plus the three isomerizing enzymes to convert lactose fully to glucose.

P1.30 Sulfuric and nitric acid are two important commodity chemicals.

Sulfuric acid (H_2SO_4) is made by burning sulfur S with O_2 to make SO_2, then further oxidizing SO_2 to SO_3 over a catalyst, using air as the source of O_2, then finally cooling and absorbing the gases into water to produce H_2SO_4.

Nitric acid (HNO_3) is made by oxidation of ammonia (NH_3). Three reactions occur. In reaction 1, ammonia reacts with oxygen (O_2) to form nitric oxide (NO) and water (H_2O). In reaction 2, NO is reacted further with O_2 to nitrogen dioxide (NO_2). In reaction 3, NO_2 is bubbled through water (H_2O) to produce HNO_3 and NO. (*P.S.:* A Nobel Prize was awarded for figuring out this process.)

For each acid, write out three stoichiometrically balanced chemical equations, one for each of the reactions described. Use generation-consumption analysis to synthesize a reaction pathway for sulfuric acid with no net generation or consumption of SO_2 or SO_3, and a reaction pathway to make nitric acid from ammonia and oxygen with generation or consumption of NO or NO_2. Water is allowed as a byproduct.

Sulfuric acid is one of the cheapest chemicals around, selling for about \$0.04/lb. Nitric acid sells for about 3 times that amount. What do you think accounts for the difference in price?

P1.31 Soaps are the sodium or potassium salts of various fatty acids (such as stearic acid), derived from natural products such as animal fat. In a typical process, glycerol stearate $[(C_{17}H_{35}COO)_3C_3H_5]$ is contacted with hot water to produce stearic acid ($C_{17}H_{35}COOH$) and glycerine ($C_3H_5(OH)_3$):

$$(C_{17}H_{35}COO)_3C_3H_5 + 3H_2O \rightarrow 3C_{17}H_{35}COOH + C_3H_5(OH)_3$$

The stearic acid and glycerine are separated. The stearic acid is then neutralized with caustic soda (NaOH) to produce sodium stearate soap ($C_{17}H_{35}COONa$):

$$C_{17}H_{35}COOH + NaOH \rightarrow C_{17}H_{35}COONa + H_2O$$

If glycerol stearate costs $1.00/lb, caustic soda costs $0.50/lb, and glycerine can be sold for $1.10/lb, what is the lower bound on the sales price of a pound of sodium stearate soap? Compare to what you might pay at the supermarket or drugstore for a bar of soap.

Scrimmage

P1.32 Consider the following everyday products: paper, laundry detergent, plastic soda bottle, wine glass, ceramic flowerpot, bleach. Look up information on the manufacture of these products in the *Kirk-Othmer Encyclopedia of Chemical Technology* or other sources. What raw materials are used in the manufacture of each product? Categorize the raw materials as to source: air, water, agricultural products, fossil fuels, minerals.

P1.33 Hexamethylenediamine (HMD, $H_2N(CH_2)_6NH_2$), is one of the two reactants used to make nylon-6,6. Connie Chemist has proposed two alternative reaction pathways for making HMD. Eddie Engineer has to decide which one to use for a process making 116,000 lb HMD/day.

Reaction pathway 1: React butadiene (C_4H_6) with hydrogen cyanide (HCN) to make adiponitrile ($NC(CH_2)_4CN$). Then react adiponitrile with hydrogen (H_2) to produce HMD.
Reaction pathway 2: React acrylonitrile (CH_2CHCN) with hydrogen to make adiponitrile. Then react adiponitrile with hydrogen to make HMD.

Which pathway should Eddie Engineer recommend?

Compound	Formula	Cost, $/lb
Butadiene	C_4H_6	0.21
Hydrogen cyanide	HCN	0.93
Hydrogen	H_2	0.09
Acrylonitrile	CH_2CHCN	0.65
Adiponitrile	$NC(CH_2)_4CN$	??
Hexamethylenediamine	$H_2N(CH_2)_6NH_2$	??

P1.34 Estimate the size of the market (tons/year) in the United States for: gasoline, high fructose corn syrup, and aspirin. For each of the following categories—organic chemicals, inorganic chemicals, minerals, plastics, and synthetic fibers—guess which chemical has the highest U.S. production rates. Don't look anything up at first—simply make an estimate based on your own daily consumption and that of your family and friends. Assume that the U.S. population is 300 million. Then, find information on the market size, and compare to your estimates and to the range of numbers given in Table 1.4. Good sources of information include government websites (USDA, DOE), the *Kirk-Othmer Encyclopedia of Chemical Technology,* and *Chemical and Engineering News.*

P1.35 Acrylonitrile (C_3H_3N) is the basic building block of synthetic rubbers and orlon fiber. Consider three different proposed reactions:

Pathway 1: $C_2H_2 + HCN \rightarrow C_3H_3N$

Pathway 2: $C_3H_6 + NH_3 + 1.5\ O_2 \rightarrow C_3H_3N + 3\ H_2O$

Pathway 3: $C_2H_4O + HCN \rightarrow C_3H_3N + H_2O$

Look up current prices for raw materials and products in the *Chemical Market Reporter.* Which pathway would you choose? Consider safety, atom economy, and environmental consequences as well as process economy.

P1.36 Agricultural raw materials include: wood, cornstarch, vegetable oils, natural rubber, cellulose, and lignin. Look up information on each of these raw materials and list products made from it. Good sources include the U.S. Department of Energy website on renewable resources and the *Kirk-Othmer Encylopedia of Chemical Technology.*

P1.37 Amino acids are the constituents of proteins and an important class of chemicals. All amino acids have the chemical formula $H_2NCH(R)$-COOH, where R is a different chemical group for each different amino acid. Some amino acids are added to inexpensive feeds like soybean meal to make nutritionally balanced animal feedstocks. Others are added to foods to enhance flavor, or serve as starting materials for the synthesis of compounds such as the artificial sweetener aspartame or the Parkinson's medicine L-DOPA. Some biodegradable polymers synthesized from amino acid constituents have been developed.

Amino acids are made commercially either by chemical synthesis or by fermentation. Here we'll evaluate two alternative chemical synthesis routes to alanine

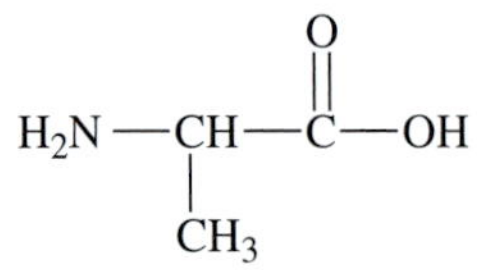

$$H_2N - CH - \overset{\overset{\displaystyle O}{\|}}{C} - OH$$
$$\underset{\displaystyle CH_3}{|}$$

Figure 1.11   Alanine, with R = CH_3

Strecker synthesis:

$$RCHO + NH_3 \rightarrow H_2NCH(R)OH$$
$$H_2NCH(R)OH + HCN \rightarrow H_2NCH(R)CN$$
$$H_2NCH(R)CN \rightarrow H_2NCH(R)COOH + NH_3$$

Bucherer synthesis:

$$RCHO + NaCN + (NH_4)_2CO_3 \rightarrow RC_3H_3O_2N_2 + NaOH + NH_3$$
$$RC_3H_3O_2N_2 \rightarrow H_2NCH(R)COOH + CO_2 + NH_3$$

Water may be a reactant or product in any of the equations but is not explicitly written.

Evaluate each synthesis pathway. First, balance the chemical equations for the synthesis of the amino acid alanine ($R = CH_3$) starting from acetaldehyde (CH_3CHO). Complete a generation-consumption analysis, and calculate the fractional atom utilization of each pathway. Look up the prices of the starting materials in the *Chemical Market Reporter,* and calculate the potential profit or loss for a plant that produces 200 metric tons/year alanine, if the selling price is $4.75/kg. Unless byproducts can be used within the process, they have no sales value.

P1.38 Microorganisms contain a complex mix of proteins, carbohydrates, and fats that are sometimes lumped together as a pseudochemical compound. For example, bacterial biomass has an empirical formula of $CH_{1.666}N_{0.20}O_{0.27}$. Under aerobic conditions, bacteria take in glucose ($C_6H_{12}O_6$), oxygen (O_2) and ammonia (NH_3) and make more bacteria ($CH_{1.666}N_{0.20}O_{0.27}$), CO_2, lactic acid (($CH_3CH(OH)COOH$) and water.

Write three balanced chemical reaction equations for glucose reacting to form (1) CO_2, (2) lactic acid, and (3) bacteria. In each reaction, O_2 and/or NH_3 can be additional reactants, and CO_2 and/or H_2O can be additional products.

We mixed 1.0 g bacteria, 18.0 g glucose and some ammonia in a large vessel of water and bubbled air through the solution. After some time we found that all the glucose had been consumed and the vessel contained 2.10 g bacteria, 3.6 g lactic acid (along with water). How many grams of CO_2 were generated? What fraction of the glucose was consumed to make bacteria? lactic acid? Assume that there are no other reactions for glucose consumption other than the 3 you wrote.

P1.39 Ethylene oxide is an important intermediate in the production of a number of compounds including ethylene glycol, used as antifreeze. Ethylene glycol is quite toxic, however, and there is interest in replacing it with propylene glycol, which is made from propylene oxide (C_3H_6O). The problem is that ethylene oxide is cheaper and easier to make. Ethylene oxide (C_2H_4O) is made by direct oxidation with O_2 of ethylene (C_2H_4). Propylene oxide is made commercially by using an old process called the chlorohydrin process, but the research department of your company has come up with a new process in which propylene is oxidized with hydrogen peroxide, H_2O_2, which is generated in situ. The reactions involve compounds called RAHQ and RAQ:

$$RAQ + O_2 \rightarrow RAHQ + H_2O_2$$

$$RAHQ + H_2 \rightarrow RAQ$$

$$C_3H_6 + H_2O_2 \rightarrow C_3H_6O + H_2O$$

Hydrogen is made by a reaction between methane (CH_4) and water (H_2O), producing CO as a byproduct.

Compare the economics of producing 1000 kg of ethylene oxide by the conventional process versus 1000 kg of propylene oxide with the new process, assuming that both sell for $0.60/lb.

Compound	Price, $/kg
Oxygen	0.033
Methane	0.110
Ethylene	0.57
Propylene	0.42

Look up information on conventional methods of producing propylene oxide in *Kirk-Othmer Encyclopedia of Chemical Technology* or other sources. Do you think the new process using H_2O_2 could be competitive?

P1.40 In the case study, we described the chemical reactions for making adipic acid (a precursor for nylon-6,6 manufacture), using benzene as the raw material, and oxygen and nitric acid as oxidizing agents.

It's been reported that cyclohexene (C_6H_{10}) is oxidized with hydrogen peroxide (H_2O_2) to adipic acid over a sodium tungstate catalyst in an aqueous solution. Water is a byproduct. Write down the stoichiometrically balanced reaction. Use a generation-consumption analysis to evaluate

(a) raw material costs,
(b) environmental impact,
(c) safety of this new reaction pathway, compared to the benzene process, at a production rate of 28,100 kg/day of adipic acid.

Assume that cyclohexene sells for $0.20/kg, and 35 wt% H_2O_2 aqueous solution sells for $0.25/lb.

A British research group reported that they discovered a remarkable new catalyst that facilitates a new chemical route to adipic acid from *n*-hexane (C_6H_{14}) and oxygen (O_2). Write down the stoichiometrically-balanced reaction. Use a generation-consumption analysis to evaluate

(a) raw material costs.
(b) environmental impact.
(c) safety of this new reaction pathway, compared to the benzene process, at a production rate of 28,100 kg adipic acid/day.

Assume *n*-hexane costs $0.33/kg and oxygen is free. One of the problems with this process is that only one-third of the *n*-hexane reacts to form adipic acid. The rest of the *n*-hexane forms other byproducts by other chemical reactions. How does this affect the economics?

Adipic acid sells for $1.54/kg, Your company is interested in developing new routes to adipic acid. Write a memo to your supervisor explaining why your company should/should not invest in developing a new route to adipic acid, starting from cyclohexene or *n*-hexane.

P1.41 Polyvinyl chloride (PVC) is produced by the catalytic polymerization of vinyl chloride and is used extensively to make products like plastic pipe and film. Your assignment is to design a process for making vinyl chloride (C_2H_3Cl). A brief survey of the synthetic chemistry literature unearths the following reactions involving vinyl chloride or similar molecules:

$$C_2H_2 + HCl \rightarrow C_2H_3Cl$$

$$C_2H_4 + Cl_2 \rightarrow C_2H_4Cl_2$$

$$C_2H_4Cl_2 \rightarrow C_2H_3Cl + HCl$$

$$2\ HCl + 1/2\ O_2 + C_2H_4 \rightarrow C_2H_4Cl_2 + H_2O$$

$$C_2H_4Cl_2 + NaOH \rightarrow C_2H_3Cl + H_2O + NaCl$$

Come up with several different reaction pathways for the production of vinyl chloride by mixing and matching these five reactions. Using the prices given below, analyze which of your pathways look most promising. Consider atom economy and process economy. Since you work for a PVC manufacturer, management is not interested in setting up a business to sell HCl or $C_2H_4Cl_2$.

Compound	Price, $/lb
Ethylene (C_2H_4)	0.27
Dichloroethane ($C_2H_4Cl_2$)	0.17
Acetylene (C_2H_2)	1.22
Chlorine (Cl_2)	0.10
Hydrogen chloride (HCl)	0.72
Sodium hydroxide (NaOH)	1.13
Vinyl chloride (C_2H_3Cl)	0.22

P1.42 Round-up® is a popular biodegradable nonselective herbicide. A key intermediate in the synthesis of Round-up is DSIDA (disodium iminodiacetate, $C_4H_5NO_4Na_2$). DSIDA is made by mixing ammonia (NH_3), formaldehyde (CH_2O), and hydrogen cyanide (HCN) under acidic conditions to make IDAN (iminodiacetonitrile). (This is called a Strecker synthesis, and similar reactions are used to synthesize amino acids, the building blocks of proteins.) Then an aqueous solution of sodium hydroxide (NaOH) is mixed with IDAN to make DSIDA, with NH_3 as a byproduct.

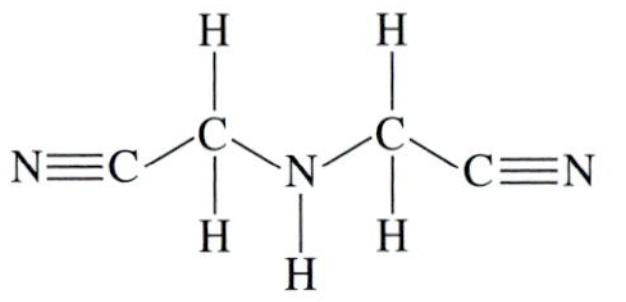

Iminodiacetonitrile (IDAN), $C_4H_5N_3$ 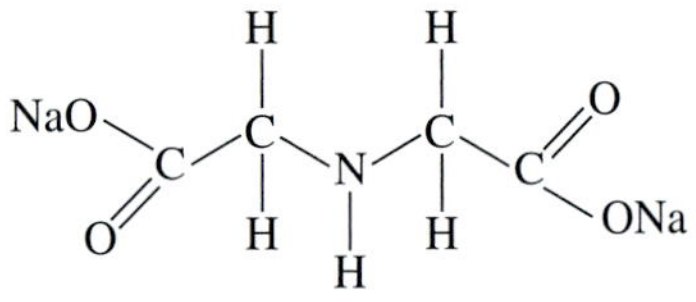   Disodium iminodiacetate (DSIDA), $C_4H_5O_4NNa_2$

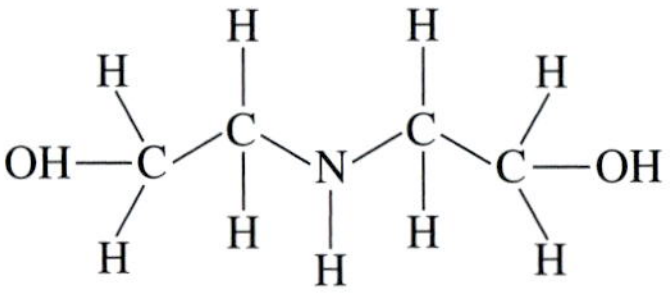

Diethanolamine, $C_4H_{11}O_2N$

There are several problems with this reaction pathway. HCN is highly toxic. The reaction is exothermic (gives off heat), and without tight control of the reactor a runaway reaction situation can develop. Because of some side reactions (not considered here), the process generates 1 kg cyanide-contaminated waste per 7 kg DSIDA produced.

Your job is to investigate an alternative, safer approach, that uses DEA, diethanolamine. Over a copper catalyst in a NaOH solution, DEA reacts to form DSIDA. The structure of DEA is shown above.

First, figure out the two stoichiometrically balanced reactions (with byproducts) for production of DSIDA from ammonia, formaldehyde, hydrogen cyanide, and sodium hydroxide. Develop a generation-consumption table for the overall process. Determine the cost per day, in raw materials, for production of 1770 lb/day DSIDA.

Next, figure out the reaction stoichiometry for synthesis of DSIDA from DEA. What is the byproduct? Compare the cost of producing DSIDA for this process to the cost for the conventional HCN process. Is the safer process economically competitive?

DEA is synthesized by oxidation of ethylene (C_2H_4) with oxygen to ethylene oxide (C_2H_4O), then reaction of C_2H_4O with NH_3. Assume that air is used as the source of O_2 and it is free. Write the stoichiometrically balanced reactions, complete the production-consumption analysis and develop the input-output table. From a raw materials cost point of view, would it be a good idea to make the DEA inhouse instead of purchasing it?

Assume the following values for raw materials used in this process:

Compound	Price
Formaldehyde (37 wt% in water)	$0.12/lb solution
Ammonia	$145/ton
Hydrogen cyanide	$0.70/lb
Sodium hydroxide	$0.32/lb
Diethanolamine	$0.58/lb
Ethylene	$0.38/lb
Hydrogen	~0

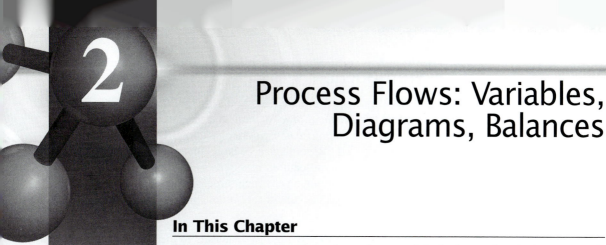

Process Flows: Variables, Diagrams, Balances

In This Chapter

We take the next step along the path toward designing a chemical process. We describe the key process variables used to characterize a chemical process. We illustrate how to translate a generation-consumption analysis into an input-output flow diagram. We introduce two important visual representations of chemical processes, the block flow diagram and the process flow diagram. We briefly describe the major processing units in a block flow diagram, and we begin our study of material balance calculations.

Here are some of the questions we address in this chapter:

- What are the important process variables?
- What are the four major processing units common to most chemical processes?
- What are three different ways to diagram a process flowsheet? Which diagram is used for which purpose?
- How do I begin to synthesize a process flow sheet?
- How are process flow rates and compositions calculated?
- What kind of information is useful for completing process flow calculations?
- How do I know I have just enough information to calculate process flows?

Words to Learn

Watch for these words as you read Chapter 2.

Process flow sheet	Material balance equations
Process variables	Basis
Process streams	System
Process units: Mixers, Reactors, Separators, Splitters	Components
	Stream composition specification
Input-output diagrams	System performance specification
Block flow diagrams	Degree of freedom
Process flow diagrams	
Batch or continuous-flow	
Steady-state or transient	

2.1   Introduction

In order to manufacture products, we need to design, build, and operate a chemical process plant. *The chemical process plant is a physical facility in which the raw materials undergo chemical and physical changes in order to make the desired products* (Fig. 2.1). Chemical process plants come in all shapes and sizes, but share many common features.

The flow of materials through a chemical process plant is shown visually on **process flow sheets.** In general, a chemical process plant contains some or all of the following (Fig. 2.2):

- *Feed preparation facilities.* Bring raw materials to the correct composition or physical state.

Figure 2.1 Chemical process plants come in many shapes and sizes. Left: clean diesel plant at the Danube Refinery in Hungary, made to produce diesel fuel with ultralow sulfur content to meet new European standards. *Courtesy of Emerson Process Management.* Center: part of a pharmaceutical plant. *Courtesy of Emerson Process Management.* Right: chicken eggs, besides providing an important food source, are under study as a possible factory for producing high-value proteins. © *David Frazier/Photo Researchers, Inc.*

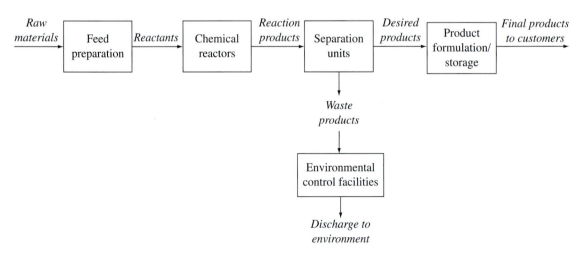

Figure 2.2 Chemical process plants share many features. The flow of raw materials to desired products is illustrated on process flow sheets.

- *Reactors.* Provide conditions to allow desired chemical reactions to take place under control.
- *Separators.* Separate desired products from raw materials, byproducts, and wastes.
- *Environmental control facilities.* Handle wastes for safe reuse or disposal.
- *Product formulation facilities.* Mix, formulate, package, and store the final product.
- *Material transfer equipment.* Move solids, liquids, and gases.
- *Energy transfer equipment.* Supply refrigeration, heat and work.

In this chapter, we identify the key process variables needed to describe chemical processes, and we describe three kinds of flow sheets used to illustrate chemical processes. We learn simple heuristics for choosing processing units and connecting them together in a viable flow sheet. We learn how to complete simple process flow calculations, using material balance equations and process specifications.

2.2   Process Variables

Process flow sheets incorporate quantitative information about **process variables.** The process variables that we are interested in now are: moles, mass, composition, concentration, pressure, temperature, volume, density, and flow rate. In Chap. 6 we will review energy-related process variables. Before we proceed to discuss the process variables of interest, we will first briefly review what you have already learned about dimensions and units.

2.2.1   A Brief Review of Dimensions and Units

Quick Quiz 2.1

What is the derived dimension of volume and of density in terms of the base dimensions?

A *dimension* is a fundamental quantity, a property of a physical entity. There are only a few base dimensions; the ones we concern ourselves with in this book are: mass M, length L, time t, thermodynamic temperature T, and amount of substance N. Quantities such as area, volume, density, and pressure are all derived from these base units. For example, area has the derived dimension of L^2 and pressure has the derived dimension of M/Lt^2.

A *unit* is a specific magnitude of a dimension, either base or derived. You have probably used the Système Internationale (SI) system (kilograms, meters, seconds) in your science classes, but you will encounter many different units in the chemical process industry. For example:

Base dimension	Units
Mass M	Kilogram, gram, ounce, ton, long ton, short ton, . . .
Length L	Meter, foot, centimeter, mile, light-year, . . .
Time t	Day, hour, second, year, semester, . . .
Temperature T	Degree Celsius, degree Fahrenheit, Kelvin, degree Rankine, . . .
Amount of substance N	Gram-mole, kilogram-mole, ton-mole, . . .

There are all kinds of crazy units. One of my personal favorites is the EFOB—equivalent fuel oil barrel—which is not a measure of volume but of energy! There's not much you can do about unit profusion—except to work hard to avoid unit confusion.

Helpful hints to avoid unit confusion: A quantity is meaningless without units. If an equation is dimensionally consistent, all the terms in the equation are of the same dimension. Check that the dimensions of quantities on both sides of any equals, plus, or minus sign are the same.

> ! **Helpful Hint**
> Whenever you write a number, write down the associated units.

Illustration: When you go to the store and buy groceries, the cashier doesn't say "that'll be 52 and 47", she says "that'll be 52 dollars and 47 cents".)

> ! **Helpful Hint**
> Whenever you do a calculation, check first for dimensional consistency.

Illustration: Every Monday you buy two gallons of milk and five boxes of cookies. Since you have to carry your purchases home, you'd like to know how many pounds of food you've bought. Can we just add together gallons of milk plus boxes of cookies to get pounds of food?

$$(2 \text{ gallons of milk}) + (5 \text{ boxes of cookies}) = ? \text{ pounds of food}$$

Gallons has dimension of volume, or $[L^3]$. Boxes has dimension of amount of substance, or $[N]$. The result is in pounds, which has dimension of mass, or $[M]$. The dimensions are not the same, so the equation is not dimensionally consistent. We want every term to have dimension of mass. To convert volume $[L^3]$ to mass $[M]$, we see that we need to multiply by $[M]/[L^3]$, or density. In other words, dimensional analysis has guided us to determine that we need to multiply the volume of milk by the density of milk (8.3 lb/gallon). To convert amount (boxes, N) to mass M, we need to know the mass per unit box. We check the cookie package for information and find out that there are 2.0 lb of cookies per box. Now we can solve:

> ! **Helpful Hint**
> Whenever you need to convert units, cross out units to be sure you have done the conversions correctly.

$$\left(2 \text{ gal milk} \times \frac{8.3 \text{ lb}}{\text{gal}} \right) + \left(5 \text{ boxes of cookies} \times \frac{2.0 \text{ lb}}{\text{box}} \right)$$

$$= (16.6 \text{ lb milk}) + (10.0 \text{ lb cookies}) = 26.6 \text{ lb food}$$

> ! **Helpful Hint**
> Choose units so that you can work with numbers that are not too large or too small to comprehend.

Illustration: If one gallon of milk costs 11 U.S. quarters, you can trade in one euro for $1.20, and you can trade in 107 yen per euro, how much will you spend, in yen, for a gallon of milk?

$$\frac{11 \text{ U.S. quarters}}{\text{gal milk}} \times \frac{\$\,0.25}{\text{U.S. quarter}} \times \frac{1 \text{ euro}}{\$\,1.20} \times \frac{107 \text{ yen}}{\text{euro}} = \frac{245 \text{ yen}}{\text{gallon milk}}$$

Illustration: 15 kg/h, not 0.004167 kg/s, 15 metric tons/h or 15,000 kg/h, not 1.5×10^{10} mg/h or 15,000,000,000 mg/h.

2.2.2 Mass, Moles, and Composition

In Sec. 1.5.1, we reviewed mass and mole units and unit conversions. Recall that **molar mass** is the mass in grams of 1 mole (6.02214×10^{23}) of atoms or molecules, and has units of [g/gmol]. For convenience, the molar mass may be written as [lb/lbmol], [kg/kgmol], [ton/tonmol], or any other similar units with dimension [M/N]. The numerical value of the molar mass of a compound in any of these units is identical. To convert from *moles* to *mass*, multiply the moles by the molar mass. To convert from *mass* to *moles*, divide the mass by the molar mass. In equation form, if n_i is the moles of compound i, m_i is the mass of compound i, and M_i is its molar mass, then

$$m_i = n_i M_i$$

We will sometimes be interested in the total moles of an element in a given number of moles or given mass of a compound. If n_{hi} is the moles of element h present in compound i, then

$$n_{hi} = \varepsilon_{hi} n_i = \varepsilon_{hi} \frac{m_i}{M_i}$$

where ε_{hi} is the moles of element h per mole of compound i.

Illustration: 12 gmol glucose ($C_6H_{12}O_6$) sits in a beaker. We'd like to know how many grams of glucose, and how many gram-moles of carbon (C) are in the beaker.

$$m_{\text{glucose}} = n_{\text{glucose}} M_{\text{glucose}} = 12 \text{ gmol glucose} \times \frac{180 \text{ g glucose}}{\text{gmol glucose}} = 2160 \text{ g glucose}$$

$$n_{\text{C in glucose}} = \varepsilon_{\text{C in glucose}} n_{\text{glucose}} = \frac{6 \text{ gmol C}}{\text{gmol glucose}} \times 12 \text{ gmol glucose} = 72 \text{ gmol C}$$

Process streams are often mixtures of compounds. If m is the total mass and n is the total moles, then

$$m = \sum_i m_i$$

$$n = \sum_i n_i$$

where the summation sign indicates that the sum is taken over all compounds in the stream.

With mixtures, it is important to characterize their **composition,** or the relative quantities of compounds in a mixture. Process stream composition is most usefully quantified in terms of mass fraction (mass percent) or mole fraction (mole percent). **Mass fraction** of compound i in a mixture is the mass of i divided by the total mass of the mixture. **Mass percent** (sometimes called weight percent) of compound i is 100% $\times$ the mass fraction of i. **Mole fraction** of compound i in a mixture is the moles of i divided by the total moles of the mixture. **Mole percent** of species i in a mixture is $100\% \times$ the mole fraction of i. We will generally use the convention of indicating mass fraction by w_i and mole fractions by x_i, y_i, or z_i.

$$\text{Mass fraction of } i = \frac{\text{mass of } i}{\text{total mass}} = w_i = \frac{m_i}{\sum_i m_i} = \frac{m_i}{m}$$

$$\text{Mass percent of } i = \frac{\text{mass of } i}{\text{total mass}} \times 100\% = w_i \times 100\%$$

$$\text{Mole fraction of } i = \frac{\text{moles of } i}{\text{total moles}} = x_i = \frac{n_i}{\sum_i n_i} = \frac{n_i}{n}$$

$$\text{Mole percent of } i = \frac{\text{moles of } i}{\text{total moles}} \times 100\% = x_i \times 100\%$$

Mass and mole fractions are dimensionless. Notice that the mass or mole fractions in a stream must sum to 1:

$$\sum_i w_i = 1$$

$$\sum_i x_i = 1$$

Converting between mass fraction and mole fraction requires knowledge of the molar mass of all the species in the mixture:

$$w_i = \frac{x_i M_i}{\sum_i x_i M_i}$$

$$x_i = \frac{w_i / M_i}{\sum_i (w_i / M_i)}$$

Illustration: 12 g of glucose ($C_6H_{12}O_6$) and 3 g sodium chloride (NaCl) are dissolved in 85 g of water. Molar masses are 180, 58.5, and 18 g/gmol, respectively.

$$w_{\text{glucose}} = \frac{12 \text{ g glucose}}{12 \text{ g glucose} + 3 \text{ g NaCl} + 85 \text{ g water}} = 0.12$$

$$\text{mass percent glucose} = 0.12 \times 100\% = 12 \text{ mass\% } (12 \text{ wt\%})$$

$$x_{\text{glucose}} = \frac{12 \text{ g glucose}/180 \text{ g/gmol}}{12 \text{ g glucose}/180 \text{ g/gmol} + 3 \text{ g NaCl}/58.5 \text{ g/gmol} + 85 \text{ g water}/18 \text{ g/gmol}} = 0.014$$

$$\text{mole percent glucose} = 0.014 \times 100\% = 1.4 \text{ mol\%}$$

Quick Quiz 2.2

A bottle contains
100 lb of 37 wt% HCl
in water. How many lb
of HCl and H_2O are in
the bottle?

How many lbmol of
HCl and H_2O are in
the bottle? (Molar
mass of HCl is
36.5 lb/lbmol.)

Illustration: 12 gmol glucose ($C_6H_{12}O_6$) and 3 gmol sodium chloride (NaCl) are dissolved in 85 gmol water.

$$x_{glucose} = \frac{12 \text{ gmol glucose}}{12 \text{ gmol glucose } + \text{ 3 gmol NaCl } + \text{ 85 gmol water}} = 0.12$$

$$\text{mole percent glucose } = 0.12 \times 100\% = 12 \text{ mol}\%$$

$$w_{glucose} = \frac{12 \text{ gmol glucose} (180 \text{ g/gmol})}{12 \text{ gmole glucose} (180 \text{ g/gmol}) + 3 \text{ g NaCl} (58.5 \text{ g/gmol}) + 85 \text{ g water} (18 \text{ g/gmol})} = 0.56$$

$$\text{mass percent glucose } = 0.56 \times 100\% = 56 \text{ mass}\% \ (56 \text{ wt}\%)$$

2.2.3 Temperature and Pressure

Temperature and pressure are critical process variables that affect process performance—for example, how far a chemical reaction will proceed to completion, or how pure the products obtained from a separation device will be.

Temperature T is a base dimension. Kelvin (K) and Rankine (°R) scales are both absolute scales (no negative temperatures). Celsius (°C) and Fahrenheit (°F) are displaced from Kelvin and Rankine scales by a constant number. To convert among various temperature scales, use the equations below.

$$T(K) = T(°C) + 273.15$$
$$T(°R) = T(°F) + 459.67$$
$$T(°R) = 1.8T(K)$$
$$T(°F) = 1.8T(°C) + 32$$
$$T(°C) = \tfrac{5}{9}[T(°F) - 32]$$

Illustration:

$$25°C = 77°F = 298.15 \text{ K} = 536.67°R$$

Pressure P has dimension of $[M/Lt^2]$. You might never know this from looking at the plethora of units used. Use the conversion factors below to convert from one system of pressure units to another.

1 bar $= 0.1$ MPa $= 100$ kPa $= 10^5$ Pa $= 10^5$ N/m^2 $= 10^6$ dyn/cm^2 $= 750.062$ mm Hg (at 0°C) $= 33.4553$ ft H_2O (at 4°C) $= 14.50377$ lb$_f$/in^2 (psi) $= 0.9869233$ atm

1 atm $= 1.01325 \times 10^5$ Pa $= 101.325$ kPa $= 1.01325$ bar $= 0.101325$ MPa $= 760$ mm Hg (at 0°C) $= 33.89854$ ft H_2O (at 4°C) $= 14.69595$ lb$_f$/in^2 (psi)

Illustration:

$$5.075 \text{ bar} \left(\frac{1 \text{ atm}}{1.01325 \text{ bar}} \right) \left(\frac{760 \text{ mm Hg}}{1 \text{ atm}} \right) = 3807 \text{ mm Hg}$$

A pressure gauge open to the atmosphere reads 0, not 1, so **gauge pressure** = (absolute pressure) − (atmospheric pressure). Gauge pressure is indicated by a "g" after the pressure unit, for example, "psig" (pounds of force per square inch gauge). When you pump up a bicycle tire, the pressure you read is the gauge pressure. If you have a bicycle pump with a pressure gauge, try reading the pressure with the pump disconnected from the tire.

Illustration:

$$5.07542 \text{ bar} - 1.01325 \text{ bar} = 4.06217 \text{ barg}$$

2.2.4 Volume, Density, and Concentration

Volume has dimension of $[L^3]$. Use the conversion factors below to convert between different units of volume.

$$1 \text{ cm}^3 = 1 \text{ mL} = 0.001 \text{ L} = 0.033814 \text{ fl. oz. (U.S.)} = 0.06102374 \text{ in}^3$$
$$= 2.6418 \times 10^{-4} \text{ gallons (U.S.)} = 3.532 \times 10^{-5} \text{ ft}^3$$

$$1 \text{ liter (L)} = 1000 \text{ cm}^3 = 1 \text{ dm}^3 = 0.001 \text{ m}^3 = 61.02374 \text{ in}^3 = 0.03531467 \text{ ft}^3$$
$$= 33.814 \text{ fl. oz. (U.S.)} = 2.11376 \text{ pints (U.S. liquid)}$$
$$= 1.056688 \text{ qt (U.S. liquid)} = 0.26417205 \text{ gallons (U.S.)}$$
$$= 0.21997 \text{ gallons (U.K.)}$$

$$1 \text{ ft}^3 = 28316.847 \text{ cm}^3 = 28.316847 \text{ L} = 0.028316847 \text{ m}^3 = 1728 \text{ in}^3$$
$$= 7.480519 \text{ gallons (U.S)} = 0.803564 \text{ bushels (U.S. dry)} = 0.037037 \text{ yd}^3$$

$$1 \text{ barrel (oil)} = 158.987 \text{ L} = 42 \text{ gallons (U.S.)} = 1.333 \text{ barrels (U.S. liquid)}$$
$$= 5.614583 \text{ ft}^3 = 0.15899 \text{ m}^3$$

Illustration:

$$5.075 \text{ ft}^3 \left(\frac{7.480519 \text{ gal}}{\text{ft}^3} \right) = 37.96 \text{ gal (U.S.)}$$

It is often convenient to work with volume per mass or volume per mole rather than total volume. The volume per mass is called the **specific volume** and has dimension of $[L^3/M]$. The volume per mole is called the **molar volume** and has dimension of $[L^3/N]$. We will use $\hat{V}$ to denote either specific or molar volume; you will know which is meant by looking at the units.

Specific density is simply the inverse of the specific volume and has dimension of $[M/L^3]$. **Molar density** is the inverse of the molar density and has dimension of $[N/L^3]$. We will use ρ to denote either specific or molar density.

To convert from mass to volume, divide the mass by the density, or multiply the mass by the specific volume.

$$V = \frac{m}{\rho} = m\hat{V}$$

The conversion between moles and volume is similar.

Illustration: Liquid water at 4°C has a specific density ρ of 62.43 lb/ft³. Its specific volume $\hat{V}$ is (1/62.43 lb/ft³) = 0.160 ft³/lb. 234 lb of water occupy a volume of (234 lb)/(62.43 lb/ ft³) = 3.75 ft³. The molar mass of water is 18 lb/lbmol. The molar density of liquid water at 4°C is 62.43 lb/ft³/ 18 lb/lbmol = 3.47 lbmol/ft³.

Specific gravity is the ratio of the specific density of a substance to the specific density of water at 4°C. (Sometimes temperatures other than 4°C are used as reference. If so, this will be indicated.) Specific gravity is dimensionless. The *CRC Handbook of Chemistry and Physics* is a good source of data for specific gravity or density of solids and liquids. The density of solids is nearly independent of temperature and pressure. The density of liquids is nearly independent of pressure but somewhat dependent on temperature.

Use the conversion factors below to convert between different units of density.

$$1 \text{ g/cm}^3 = 1000 \text{ kg/m}^3 = 1 \text{ kg/L} = 62.42796 \text{ lb/ft}^3$$

$$= 8.345404 \text{ lb/gal (U.S.)} = 0.0361279 \text{ lb/in}^3$$

Illustration: The specific gravity of liquid benzene is reported as $0.8765^{20/4}$, indicating that this is the specific gravity of benzene at 20°C relative to liquid water at 4°C. Liquid water at 4°C has a density of 1.00 g/cm³ = 62.43 lb/ft³. The density of benzene at 20°C is therefore 0.8765 g/cm³ or 54.72 lb/ft³.

The density of gases is strongly dependent on both pressure and temperature. For most gases and gas mixtures at moderate temperatures and pressures, the molar density can be calculated with reasonable accuracy from the **ideal gas law**

$$\frac{n}{V} = \frac{1}{\hat{V}} = \rho = \frac{P}{RT}$$

where T is the absolute temperature and R is the ideal gas constant.

R = 83.144 bar cm³/gmol K = 82.057 atm cm³/gmol K = 62.361 mmHg L/gmol K
= 1.314 atm ft³/lbmol K = 0.083144 bar L/gmol K = 0.082057 atm L/gmol K
= 555.0 mm Hg ft³/lbmol °R = 10.73 psi ft³/lbmol °R = 0.7302 atm ft³/lbmol °R

The ideal gas law is not a law at all. Rather, the "law" is a *model equation,* which relates one physical property of a material (in this case, molar density of a gas) to process variables (in this case, temperature and pressure). In this text we will assume that gases obey the ideal gas law, because this simple model equation is sufficiently accurate for our purposes. See App. B for other model equations for more accurately estimating the molar density of a gas from its pressure and temperature.

Specific volumes and densities of gases are often reported at **standard temperature and pressure (STP),** 0°C and 1 atm pressure.

Illustration: A gas at 100°C and 3.50 atm pressure has a specific density of

$$\rho = \frac{3.50 \text{ atm}}{0.082057 \text{ L atm/gmol K } (373.15 \text{ K})} = 0.114 \frac{\text{gmol}}{\text{L}}$$

Helpful Hint
The ideal gas law does not apply to liquids and solids!

At STP, the specific density is

$$\rho = \frac{1 \text{ atm}}{0.082057 \text{ L atm/gmol K } (273.15 \text{ K})} = 0.0446 \frac{\text{gmol}}{\text{L}}$$

Quick Quiz 2.3

Calculate the molar density of an ideal gas (gmol/L) at 0°C and 1 atm pressure.

Johnnie Genius says that pure H_2O at 77°F and 1 atm has a molar density of 0.0019 ft³/lbmol. He says he calculated this from the ideal gas law. Is Johnnie right?

The specific molar volume of an ideal gas at STP is 22.414 L/gmol or 359 ft³/lbmol.

Concentration gives the mass (or moles) of a *solute* per volume of *solution.* It has dimension of $[M]/[L^3]$ or $[N]/[L^3]$. Concentration has the same dimension as density, but the meaning is distinct: The density of a liquid solution is the mass (or moles) of *solution* per volume of *solution.*

2.2.5 Flowrates

Flow rates have dimension of $[M/t]$ (mass flow rate), $[N/t]$ (molar flow rate) or $[L^3/t]$ (volumetric flow rate). Units for mass, molar, and volumetric flow rates are interconverted in the same way as those for mass, moles, and volumes.

Illustration: The mass flow rate of a process stream is 115.0 lb oxygen/min. The molar mass of O_2 is 31.9988 g/gmol, and the gas behaves

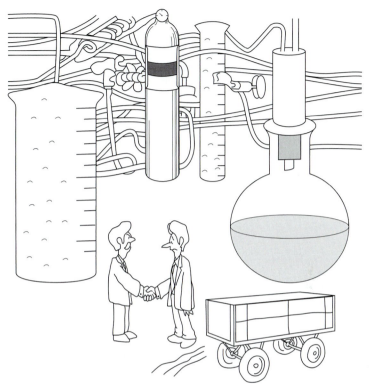

Figure 2.3 Taking a process from the lab into full-scale commercial production requires more than just purchase of bigger beakers!

the ideal gas law, so $\hat{V} = 22.414$ L/gmol at STP. The volumetric flow rate in cm^3/s at STP is:

$$\left(\frac{115.0 \text{ lb}}{\text{min}}\right)\left(\frac{453.59237 \text{ g}}{\text{lb}}\right)\left(\frac{1 \text{ gmol}}{31.9988 \text{ g}}\right)\left(\frac{22.414 \text{ L}}{\text{gmol}}\right)\left(\frac{1 \text{ min}}{60 \text{ s}}\right) = \frac{608.97 \text{ cm}^3}{\text{s}}$$

In process flow rates, time may be given in units of seconds, minutes, hours, days, or years. Commodity chemical process plants operate 24 hours per day, 7 days per week. Typically they operate year-round except for a short shutdown period of 1–2 weeks for cleaning, maintenance and upgrades. Thus, a typical operating year is approximately 350 to 360 days. Hours and days of operation of specialty-chemical process plants are more variable.

2.3   Chemical Process Flow Sheets

Process flow sheets are compact and precise diagrams that present a large amount of technical information about chemical processes. They are the language that chemical process engineers use, and you should become fluent at translating from words to flow sheet and back again.

Table 2.1 Major Types of Chemical Process Flow Sheets	
Diagram	**Information**
Input-output flow diagram	Raw materials Reaction stoichiometry Products
Block flow diagram	Everything above, plus Material balances Major process units Process unit performance specifications
Process flow diagram (PFD)	Everything above, plus Energy balances Process conditions (T and P) Major process equipment specifications

The three types of chemical process flow sheets that we will use are listed in Table 2.1. Every flow sheet has three features in common: raw materials enter, physical and/or chemical changes take place within the process, and products leave. The diagrams differ in their level of detail and complexity (and in the cost and difficulty of generating them!)

In the following sections, we first describe the key features of each type of flow sheet, then illustrate each, using ammonia synthesis as an example. Generation-consumption analysis of the reaction pathway from methane and air to ammonia was the subject of Example 1.5. Worldwide annual production is on the order of 100 million tons, or roughly 30 lb per person per year. For more on the long and checkered history of this simple and ubiquitous chemical, read the ChemiStory at the end of this chapter.

2.3.1 Input-Output Flow Diagrams

An **input-output diagram** (Fig. 2.4) is the simplest process flow sheet. It has the following features:

- A single block represents all of the physical and chemical operations in the process.
- Lines with arrows represent material moving into and out of the process.
- Raw materials enter on left.

Figure 2.4 A generic input-output diagram simply shows the raw materials consumed and the products (and byproducts) generated.

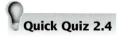

In Figure 2.5, does the total mass of raw materials into the process equal the total mass of products out? Should it?

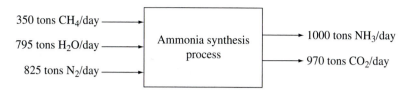

Figure 2.5   This input-output diagram for ammonia synthesis was developed by scaling up the generation-consumption analysis of Example 1.5.

- Products leave on right.
- Raw material and product flow rates (or quantities) may be shown.

Figure 2.5 shows an input-output diagram for ammonia synthesis, using a basis of production of 1000 metric tons ammonia/day. Quantities of raw materials and products were calculated by using the methods described in Chap. 1.

An input-output diagram is the simplest process flow sheet we can imagine. Despite its simplicity, the input-output diagram might still trigger a few thoughts. For example, in developing the input-output flow diagram for ammonia synthesis, we might consider such questions as: What is the source of nitrogen? Should we use air? If so, what should we do with the oxygen? Asking and answering these questions constitute vital steps in the journey from a balanced chemical equation to a functioning chemical process.

2.3.2   Block Flow Diagrams

Block flow diagrams represent the next step up in complexity and detail. We will use block flow diagrams extensively in this text. These diagrams have the following features:

- Chemical processes are represented as a group of connected blocks, or **process units.**
- Each process unit represents a specific *process function,* in which materials undergo chemical and/or physical changes.
- Lines with arrows connect the blocks and represent **process streams:** inputs (material moving into each process unit) and outputs (material moving out of each unit).
- Raw materials enter on left.
- Products leave on right.
- Quantities or flow rates of inputs and outputs may be indicated, either directly on the diagram or in an accompanying table.

There are only four kinds of process units that are included in a block flow diagram: **Mixers, Reactors, Splitters,** and **Separators** (Fig. 2.6). (This is truly remarkable, considering the complexity and diversity of chemical processes!) Keep in mind that these process units are categorized on the basis of their *function.* With process units, we simply show what important jobs need to be done to convert raw

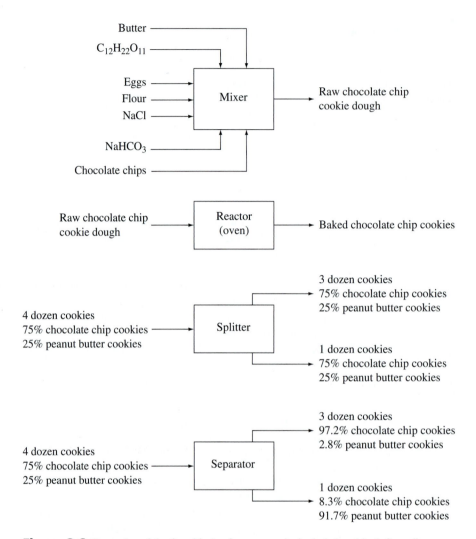

Figure 2.6 Examples of the four kinds of process units included on block flow diagrams.

materials to desired products. We ignore what happens inside the process units and say nothing about how the process units are to achieve their function. A process unit may, but does not necessarily, correspond one-to-one with a specific piece of process equipment; one process function may require multiple pieces of process equipment, or one piece of equipment may accomplish multiple functions.

Mixers.   Mixers combine two or more inputs into a single output.

Reactors.   The input streams contain reactants. One or more chemical reactions take place inside a reactor. The output streams contain reaction products as well as unconsumed reactants. In the simplest case, there is one input and one output stream.

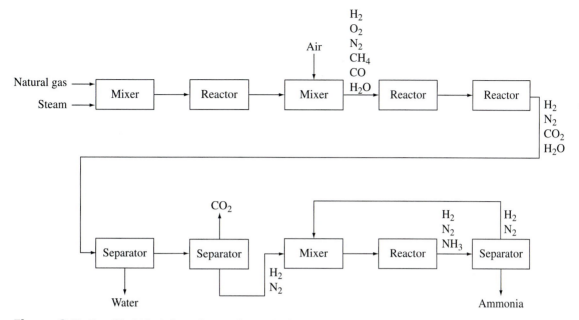

Figure 2.7 Simplified block flow diagram for synthesis of ammonia from natural gas. Natural gas and steam are mixed and then reacted to CO and H_2. Air is added to facilitate further oxidation of methane and to supply nitrogen. A water-gas shift reactor converts CO and H_2O to CO_2 and more H_2. Excess water and CO_2 are removed in separators, then the H_2/N_2 mix is sent to the synthesis reactor. Ammonia is separated from unreacted gases, which are recycled back to the reactor.

Splitters. Splitters split a single input into two or more outputs. If the input is a mixture of two or more components, then the output streams of a splitter all have the same composition as the input.

Separators. An input stream is separated into two or more outputs. The outputs have different compositions from each other and from the input. The change in composition is due to physical operations, not chemical reaction. In the simplest case, there is one input and two output streams.

These four kinds of process units are connected in myriad ways in a block flow diagram. Output streams from one process unit become input streams to other units. A simplified block flow diagram for the ammonia synthesis process is shown in Figure 2.7. Read the description carefully and compare with the flow diagram.

2.3.3 Process Flow Diagrams (PFD)

Block flow diagrams are useful sketches that show the major processing units and the materials flowing between the units, but they are far from realistic pictures of a functional chemical process. **Process flow diagrams** (PFDs) are a step up in complexity and information. Most PFDs have these features:

- Chemical processes are represented as a connected group of process equipment.
- All major pieces of process equipment are drawn representationally. Reactor type and separation methods have been chosen.
- Equipment used to move material around (e.g., pumps, compressors, conveyer belts) and to heat or cool material (e.g., heat exchangers, furnaces) is included.
- Each piece of equipment is assigned a number and given a descriptive name.
- Major utilities (steam, cooling water, etc.) are shown.
- Lines with arrows connect the process equipment and represent inputs (material moving into each block) and outputs (material moving out of each block).
- Process streams are numbered.
- Materials enter from the left and leave from the right.
- Generally, gas streams are at the top, liquid streams are in the middle, solid streams are toward the bottom.
- The flow rate or quantity, composition, temperature, pressure, and/or phase of process streams are indicated, usually in an accompanying table.

Figure 2.8 illustrates a simplified process flow diagram for an ammonia plant. Some typical process equipment representational icons are shown in Figure 2.9.

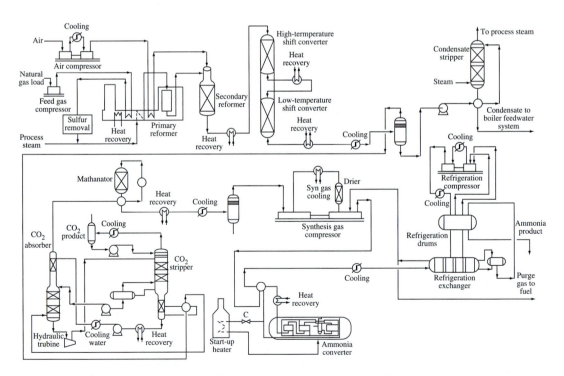

Figure 2.8 Simplified process flow diagram, for an ammonia synthesis facility. Adapted from a figure in *Kirk-Othmer Encyclopedia of Chemical Technology.*

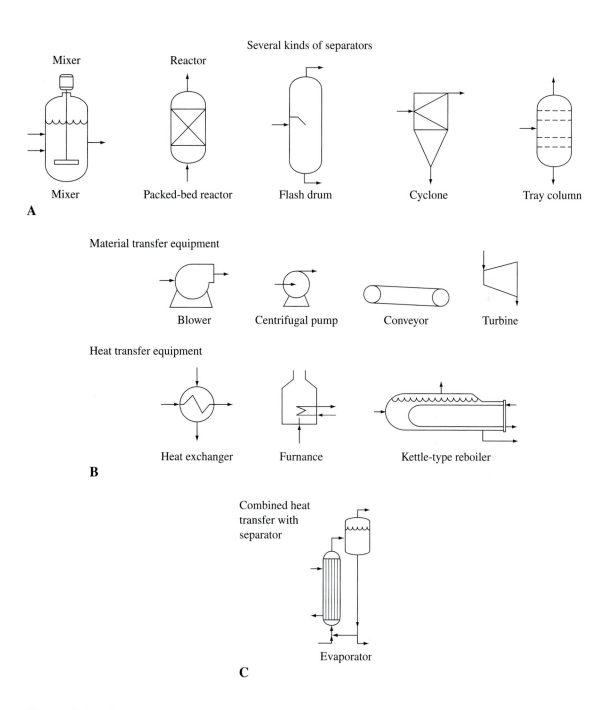

Figure 2.9A–C   A selection of process equipment icons used in process flow diagrams. Many other icons are used in addition to the ones shown. The level of detail and accuracy in representation is variable.

Chemical processes are brought from idea to reality by methodically moving from the simplest to the most complex diagrams. The engineering cost to produce each diagram increases dramatically as we move down the list in Table 2.1, because so much more information is required. Therefore, at each stage, the process is re-evaluated for economics and feasibility. For example, initial screening of process economics considers only the difference between product value and raw material costs; input-output diagrams are sufficient at this stage. A clearer picture of the required process operations as well as alternative process schemes emerges after generation of block flow diagrams. A more accurate estimate of potential profit requires information about capital costs (cost of purchasing land, equipment, buildings, etc.) and operating costs (energy and labor costs, for example). Preliminary estimates of capital and operating costs require completion of a PFD. More accurate cost estimates are calculated after generation of detailed construction drawings.

2.3.4 Modes of Process Operation

Chemical processes can function under several different modes of operation. In this section we discuss two ways to categorize the mode of operation: (1) by how streams enter and leave a process unit and (2) by their time dependence.

We categorize process operations by how input and output process streams are handled. In **batch** processes, input streams enter the process unit all at once, and at some later time output streams are removed from the process unit all at once. Input and output streams are quantified in dimensions of mass or moles. In **continuous-flow** processes, input streams continuously flow into the process unit and output streams flow continuously out of the unit. Input and output streams are quantified in dimensions of mass flow rate or molar flow rate. **Semibatch** processes are some combination of batch and continuous; for example, input streams might be added all at once with output streams removed continuously. Any of the process units—Mixers, Reactors, Separators or Splitters—can be operated as batch, continuous, or semibatch. In a process flow sheet it is possible to have mixed modes of operation; for example, a reactor might be operated in batch mode while a separator may be operated continuously. Mixed modes of operation require intermediate storage facilities.

Batch processes are common in the pharmaceutical, specialty polymer, and personal care product industries. In these industries, the annual production rate is often low (50 to 500 tons/yr or less), and the same equipment can be used over and over again for different products. Home cooking is an example of batch processing. Continuous-flow processes are common in the petroleum and industrial chemicals businesses, where annual production rates are large (1000 to 5000 tons/yr or more at a single manufacturing site). In continuous-flow processes,

? Did You Know?

Chemical engineers tend to prefer continuous processes. But it's not clear that consumers do. For example, the label on a bag of gourmet potato chips brags "made in small batches." "Batch" manufacturing is used to imply more lovingly made, higher-quality products; such products command a premium price at the grocery store.

equipment is dedicated to a single purpose and may be in operation 24 hours a day, 7 days a week.

We also categorize process operations on the basis of their dependence on time. **Steady-state** processes are time-independent: Process variables do not change with time. For example, if the flow rate to a reactor is the same today at 9 A.M. as it will be tomorrow at 10 P.M. (and at every time in between), then that flow rate is at steady state. In **transient** or **unsteady-state** processes, one or more process variables change with time. Batch and semibatch processes by their nature are unsteady state. Continuous-flow processes are almost always operated under steady-state conditions, except during start-up and shutdown, when they are unsteady state. Product quality is more consistent, and operating costs are generally lower, with steady-state operation.

The different modes of operation for a reactor are illustrated (Figure 2.10). Suppose the job of the reactor is to take compounds A and B and convert them to compound C. In a batch reactor, A and B are added to the reactor at time $t = 0$. Then, from time $t = 0$ to $t = t_f$, no material enters or leaves the reactor, the amount of A and B in the reactor decreases over time, and the amount of C increases over time. Then, at time $t = t_f$, the product is removed all at once from the reactor. In a continuous-flow reactor operated at steady state, there is a constant flow of Reactants A and B into the reactor, the reaction takes place inside the reactor at a constant rate, and there is a constant flow of product C out of the reactor. One way in which a semibatch reactor might be operated is as follows: Reactant A is added all at once at time $t = 0$, then from $t = 0$ to $t = t_f$ reactant B flows into the reactor and nothing flows out of the reactor, until at $t = t_f$ the flow of B is discontinued and the reactor contents are removed.

All three kinds of flowsheets—input-output diagram, block flow diagram, and PFD—are used whether the process is operated in batch, semibatch, or continuous mode, or whether the process is at steady or unsteady state.

Quick Quiz 2.5

Consider your digestive system as a block flow diagram. Identify the type of processing unit (mixer, splitter, reactor, separator) of (a) mouth, (b) stomach, (c) intestine. Does your digestive system operate in batch, continuous, or semibatch mode? Steady state or unsteady state?

2.4   Process Flow Calculations

In Sec. 2.2 you reviewed units and dimensions of important process variables and in Sec. 2.3 you learned about the various kinds of process flow sheets and modes of process operation. In this section, we will put together what we know about generation-consumption analysis, process flow sheeting, and process variables, so that we can complete preliminary process flow calculations. We focus on calculations for input-output diagrams and block flow diagrams. These calculations are needed to synthesize process flow sheets and to evaluate alternative processing schemes.

There are several important definitions required to complete process flow calculations. We will briefly describe each in turn.

Batch operation

Continuous steady-state operation

Semibatch operation

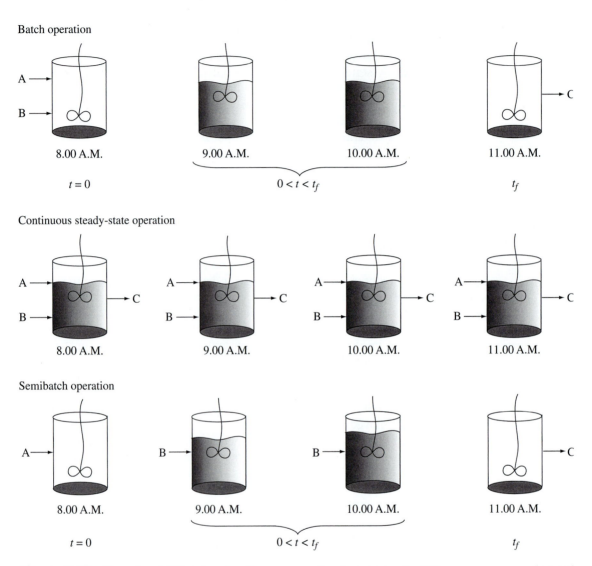

Figure 2.10 Examples of different modes of process operation for a reactor. In all three cases, compounds A and B are mixed and reacted to C, but the way in which inputs and outputs are handled is different, as is the time dependence of the process.

2.4.1 Definitions

Basis A **basis** is a flow rate or quantity that indicates the size of a process. Most often, the basis of a process flow calculation is the quantity (mass or moles) or flow rate (mass/time or moles/time) of either a raw material entering the process or a desired product leaving the process. However, this is not always the case, and the flow rate (or quantity) of any process stream on the flow diagram could serve as the basis. You were already introduced to the idea of a basis in Chap. 1. Recall that processes can be scaled up or scaled down from an old basis to a new basis by using a scaling factor, applied to all process streams.

Systems and streams   A **system** is a specified volume with well-defined boundaries. Within these boundaries we define what material is inside the system, what material is outside the system, and what material is crossing the system boundaries. On a block flow diagram, a system might correspond to a process unit. Or we might draw a boundary around several process units and group them into a single system. (See Fig. 2.11.) **Streams** are inputs to and

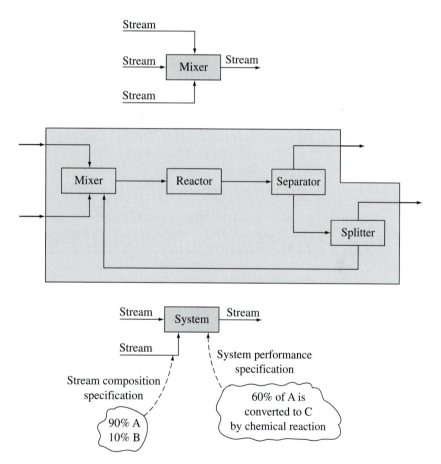

Figure 2.11   Examples of streams, systems and specifications. Top: A single process unit is defined as the system (shaded). Middle: Several process units are grouped together in a single system (shaded). Only the streams shown with heavy lines are inputs or outputs to the system. Bottom: Stream composition specifications describe a single stream, while system performance specifications describe the chemical and/or physical changes occurring inside the system.

outputs from the system. A **stream variable** describes the quantity or flow rate of a material in a stream, while a **system variable** describes the change in a quantity inside a system. Both stream and system variables have dimensions of mass, moles, mass/time, or moles/time. (In Chap. 6 we will expand this definition to include energy variables.)

Specifications   There are two kinds of specifications of importance to process flow calculations:

- stream composition specifications
- system performance specifications

Stream composition specifications provide information about the composition of a process stream. This information could be in the form of mass or mole percent, mass or mole fraction, mass or mole ratio, or concentration.

Illustration:

- Glucose is fed to a process as a *10 wt% glucose* solution in water.
- A customer requires that a hydrogen product must be at least *99.9 mol% pure.*
- Environmental regulations state that a wastewater stream must contain no more than *1 g acetic acid per 1000 kg water.*

System performance specifications describe quantitatively the extent to which chemical and/or physical changes have occurred inside the system. This is a topic we will return to in much greater detail in Chaps. 4 and 5.

Illustration:

- *65% of the nitrogen fed to a reactor is converted to ammonia*
- *Two-thirds of a fruit juice stream fed to a splitter is sent to a bottling plant.*
- *98% of the fat in milk processed in a centrifuge is recovered in the fluid skimmed off the top.*

2.4.2   Material Balance Equations

Consider the cartoon in Fig. 2.12. The sketch shows a large lake with fish. We'll call the lake our *system*. The lake contains material—water, fish, perhaps some plants—and has defined boundaries—the surface in contact with the air, the surface in contact with the earth, and the points at which streams and rivers enter and exit the lake. Material enters and leaves the system through its boundaries: Water and fish enter the lake through a mountain stream, and both water and fish

Figure 2.12

leave the lake through the river. Water enters the lake when it rains, and water leaves by evaporation when the sun shines. If a hungry bear comes along, more fish may leave the lake. Inside the lake, fish generate more baby fish, and some smaller fish are consumed by bigger fish.

Suppose we want to know whether the number of fish in the lake is increasing or decreasing. The number of fish in the lake increases because fish swim into the lake from the mountain stream and because fish reproduce. The number of fish in the lake decreases because fish swim out of the lake into the river and because some fish are eaten by other fish. The net change in the number of fish in the lake is the sum of all these factors:

> Number of fish entering the lake from the stream
> — the number of fish leaving the lake in the river
> + the number of fish born (generated) in the lake
> — the number of fish eaten (consumed) in the lake
> = change in number of fish in lake

This is just a **material balance** on the number of fish in the lake. There are a number of different kinds of material balances we could write. We could write a balance on all fish, or on each species of fish. We could account for other sources or sinks of fish, like that hungry bear. We could write a balance on water too, with terms accounting for water entering the lake from rain, or leaving the lake by evaporation. We could write balances on nitrogen, phosphates, or oxygen. In all cases we would consider the same items—material entering or leaving the lake through its boundary, and material generated or consumed inside the lake. Notice that this balance does not tell us the total number of fish in the lake, only the change in the number, and that the change in the number of fish in the lake is due to inputs and outputs (streams) as well as processes happening inside the lake (system). Notice too that we could alternatively have written a balance on the mass of fish rather than on the number of fish. The balance on mass would be different from the balance on number. For example, when a big fish eats a little fish, the number of fish changes but the total mass of fish does not.

A general form of the **material balance equation** is:

$$\text{Input} - \text{Output} + \text{Generation} - \text{Consumption} = \text{Accumulation} \qquad (2.1)$$

where

Input	=	material that *enters* the system by crossing system boundaries
Output	=	material that *leaves* the system by crossing system boundaries
Generation	=	material that is *generated* inside the system
Consumption	=	material that is *used up* inside the system
Accumulation	=	change in material inside the system

By our definitions above of stream and system variables, input and output are *stream variables,* while generation, consumption, and accumulation are *system variables.* The *dimension* of each term in the equation is either [*N*] (usually moles) or [*M*] (mass). The *units* of each term in the equation must be the same for all the variables.

The material balance equation is written for a chosen **component.** If we chose to apply the material balance equation to number of fish, for example, the variables in the equation would be input of fish, ouput of fish, generation of fish, consumption of fish, and accumulation of fish. Alternatively, if we chose to apply the material balance equation to moles of water, the variables in the equation would be input of water, output of water, generation of water, consumption of water, and accumulation of water. In process flow calculations, the material balance equation is most commonly used with one of three kinds of **components:**

Elements: Such as carbon C, oxygen O, hydrogen H, or arsenic As.

Compounds: With defined molecular formulas, such as sucrose $C_{12}H_{22}O_{11}$, oxygen O_2, water H_2O, or gallium arsenide GaAs.

Composite materials: Mixtures of compounds of defined composition, such as candy bars, air, seawater, or computer chips.

Most of the time, *compounds* are the most convenient, particularly if there is a chemical reaction of known stoichiometry. *Elements* may be more convenient when there is a chemical reaction of unknown stoichiometry, or when

Quick Quiz 2.6

You deposit $5000 in a savings account at a bank at the beginning of the year. The bank pays 7% interest on all deposits. At the end of the year you withdraw $400 to purchase a new bike.

Identify an appropriate system, and define input, output, generation, and consumption terms.

What is the accumulation?

there are a multitude of chemical reactions to consider. You should write a material balance equation in terms of composite materials *only* if there is no change in the composition or chemical makeup of the material anywhere in the process.

If there are N components in the streams and system of interest, then there are N independent material balance equations, one for each component. The material balance equation can also be applied to the total mass. In that case, the material balance equation is simply a restatement of the law of conservation of mass.

For chemical processes, generation and consumption are almost always due to chemical reaction. Elements cannot be generated or consumed (except with nuclear reactions!), only compounds. Generation and consumption terms are related by the stoichiometric coefficients of the compounds (just as we saw in Chap. 1). If the reaction is

$$aA + bB \rightarrow cC$$

then

$$\frac{\text{Moles of A consumed}}{\text{Moles of B consumed}} = \frac{\nu_A}{\nu_B} = \frac{a}{b} \quad \text{and} \quad \frac{\text{Moles of C generated}}{\text{Moles of B consumed}} = \frac{-\nu_C}{\nu_B} = \frac{c}{b}$$

These relationships between generation and consumption provide links between the material balance equations written for compound A, for compound B, and for compound C. We could write a third equation relating moles of C generated to moles of A consumed, but it would just be a combination of the two other equations, so it is not independent. In general, if there are N reactants and products, we can write $N - 1$ independent equations that relate Generation and Consumption of different compounds.

Accumulation can be positive, negative, or zero. Accumulation is nonzero when there is an imbalance between the rate at which materials enter and are produced and the rate at which materials exit and are consumed. For steady-state processes, accumulation is zero.

2.4.3. A Systematic Procedure for Process Flow Calculations

Now we wish to pull together all these strands—variables, flow sheets, specifications, material balances—to complete process flow calculations. These calculations form the cornerstone upon which the synthesis and analysis of chemical processes are built. A systematic approach is the only approach that will reliably ensure successful and accurate completion of process flow calculations. Here is a highly recommended procedure to follow. Study this procedure carefully.

Process Flow Calculations in 10 Easy Steps

Step 1.   Draw a *flow diagram.*

Step 2.   Define a *system.*

Step 3.   Choose *components* and define *stream variables* for all material streams entering or leaving the system.

Step 4.   Convert all numerical information into consistent *units* of mass or moles.

Step 5.   Define a *basis.* Write an equation describing the basis in terms of the defined stream variables.

Step 6.   Define *system variables* for generation, consumption, and accumulation. If there are chemical reactions of known stoichiometry, write equations using system variables, which relate generation of products to consumption of reactants. If the system is not at steady state, define a system variable for accumulation.

Step 7.   List all *stream composition* and *system performance specifications.* Write these specifications as equations, using the stream and system variables you defined in steps 3 and 6.

Step 8.   Write *material balance* equations for each component entering or leaving the system, using the stream and system variables you defined in steps 3 and 6.

Step 9.   *Solve* the equations you wrote in steps 5 to 8. Convert units if necessary.

Step 10.   *Check* your solutions.

It goes without saying (but we'll say it anyway) that this list includes one additional step.

Step 0.   *Understand the problem.* Solving the wrong problem correctly can be worse than solving the right problem incorrectly.

2.4.4. Helpful Hints for Process Flow Calculations

Here are some Helpful Hints that should assist you as you apply the 10 Easy Steps to solve process flow calculations. Scan the list quickly now, and refer back to it if you get stuck while working a problem.

Step 1. Draw a diagram.
- Don't skip this step!
- Draw one box for each process unit. Label each box as a Mixer, Reactor, Splitter, or Separator.

- Draw one line for each input and output stream. Draw only one line for a stream that is a mixture of compounds—don't draw a line for each compound in the mixture.

Step 2.   Define a system.
- A system can be a single process unit, a group of units, or an entire process.
- Group together several process units into a single system if you do not need to know anything about the process streams that connect the units.

Step 3.   Choose components and define stream variables.
- Choose compounds as components if there is a chemical reaction of known stoichiometry.
- Choose elements as components if there is a chemical reaction of unknown stoichiometry.
- Choose composite materials as components only if they do not undergo any changes in composition.
- If there are N components in a stream, there are N stream variables. The N stream variables can be the quantities or flow rates of each of the N components, or they can be the total quantity or flow rate of the stream plus $(N-1)$ mole or mass fractions.

Step 4.   Convert all units to moles or mass.
- Use moles if there is a reaction of known stoichiometry.
- Use either mass or moles if there is no reaction or if elements are chosen as components.

Step 5.   Define a basis.
- The input of a raw material or the output of a desired product is often a convenient basis.
- You can change a basis if it makes the problem easier to solve, then scale up or down the solution to get back to the original basis.
- If no basis is specified, define any convenient basis.

Step 6.   Define system variables.
- Elements are not consumed or generated, only compounds.
- For each reaction of known stoichiometry with N reactants and products, there are N system variables for generation + consumption, and $N-1$ equations relating generation and consumption variables through stoichiometric coefficients.
- Accumulation is zero for continuous steady-state processes.

Step 7.   List specifications.
- A hidden specification with a splitter is that all output streams have the same composition.

Step 8.   Write material balance equations.
- If there are N components in the system, there are N independent material balance equations.
- A balance equation on total mass can replace one of the component material balance equations.

Step 9.   Solve equations.
- Solve the equation with the fewest number of unknown variables first.

Step 10.   Check the solution.
- Don't skip this step!
- If you've used the component material balance equations to solve the problem, use the total mass balance equation to check the solution.

2.4.5. A Plethora of Problems

In this section is a veritable cornucopia of example problems involving process flow calculations. We *strongly* recommend that, for each example, you try to solve the problem by yourself before looking at the worked-out solution. If you have difficulty, study the solution, then cover it up and try to work it out by yourself. If you still have trouble, be sure you (1) understand the question, (2) correctly identify the different terms in the material balance equation, (3) follow the 10 Easy Steps, and (4) consult the Helpful Hints. The examples progress from simple to complicated, so be sure you are able to *solve on your own* each problem before proceeding to the next.

Warning! The single biggest mistake you can make with process flow calculations is to think that if you can follow along with the solution then you have demonstrated an understanding of the material. That is like thinking that because you have watched Michael Jordan shoot basketballs you too can be an NBA star (or because you have listened to Vladimir Horowitz you too can play Liszt, or. . . .). The second biggest mistake you can make is to not follow a systematic procedure like the 10 Easy Steps, even for problems that are easy and "intuitively obvious." Intuition is great; joined with logic it's unstoppable.

Example 2.1	**Mixers: Battery Acid Production**

Your job is to design a mixer to produce 200 kg/day of battery acid. The mixer will operate continuously and at steady state. The battery acid product must contain 18.6 wt% H_2SO_4 in water. Raw materials available include a concentrated sulfuric acid solution at 77 wt% H_2SO_4 in water, and pure water. What is the flow rate of each raw material into the mixer?

Solution

Step 1.   *Draw a diagram.* We draw a block to indicate a mixer. There are two inputs available, the concentrated sulfuric acid and the pure water, so we draw two lines with arrows entering the unit. There is one output, the battery acid, so we draw only one line leaving the unit.

Step 2.   *Define a system.* The system is the mixer.

Step 3.   *Choose components, define stream variables.* Components might be elements, compounds, or composite materials. It makes the most sense to choose two compounds, sulfuric acid (H_2SO_4) and water (H_2O), as components for the following reasons. The acid solutions are compos- ite materials, but the composition changes in the system, therefore we should not choose composite materials as components. There is no chemical reaction in the system, so either elements (H, S, O) or com- pounds (H_2SO_4 and H_2O) are reasonable choices. Since there are only two compounds but three elements, and since information on composi- tion of the acid solutions is given in terms of the compounds and not the elements, it is simplest to choose compounds as components. We'll use S to indicate sulfuric acid and W to denote water.

We'll number the process streams 1, 2, and 3. Since there are two com- ponents in stream 1, there should be two stream variables associated with stream 1. By the same reasoning, there is one stream variable for stream 2 and two for stream 3, for a total of five stream variables.

Raw material and product specifications are given as mass per- cents, and we know the total product flow rate. Given this, we'll use mass fractions and total mass flow as the stream variables. We'll use F to indicate the total mass flow rate of a stream, w_S to indicate the mass fraction of sulfuric acid, and w_W to indicate the mass fraction of water. With this notation, $w_S F$ is the mass flow rate of sulfuric acid in a stream. We'll use a subscript of 1, 2, or 3 to indicate the stream. We redraw the flow diagram to reflect our variable-naming scheme.

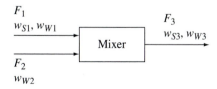

Steps 4 and 5.   *Check units and define basis.* All our information is given as kg/day or wt%. We convert wt% to mass fraction by dividing by 100. With this, the units of the stream variables are

$$F[=]\frac{kg}{day}$$

$$w_S[=]\frac{\text{kg } H_2SO_4}{\text{kg solution}}$$

$$w_W[=]\frac{\text{kg } H_2O}{\text{kg solution}}$$

Notice that the units of $w_S F$ and $w_W F$ are

$$w_S F[=]\frac{\text{kg } H_2SO_4}{\text{kg solution}} \times \frac{\text{kg solution}}{\text{day}} [=] \frac{\text{kg } H_2SO_4}{\text{day}}$$

$$w_W F[=]\frac{\text{kg } H_2O}{\text{kg solution}} \times \frac{\text{kg solution}}{\text{day}} [=] \frac{\text{kg } H_2O}{\text{day}}$$

We are given a desired production rate of 200 kg of battery acid produced per day, which serves as a convenient basis. In terms of our stream variables:

$$F_3 = 200 \frac{\text{kg}}{\text{day}}$$

Step 6. *Define system variables.* There are no chemical reactions, so no generation or consumption variables are needed. The system is at steady state, so the accumulation variable equals zero.

Step 7. *List specifications.* There are several stream composition specifications. The concentrated acid raw material is 77 wt% H_2SO_4, or the mass fraction of acid in stream 1 is 0.77. In terms of our stream variables:

$$w_{S1} = 0.77 \frac{\text{kg } H_2SO_4}{\text{kg solution}}$$

The product quality is also specified—the battery acid must contain 18.6 wt% H_2SO_4. In terms of our stream variables:

$$w_{S3} = 0.186 \frac{\text{kg } H_2SO_4}{\text{kg solution}}$$

Since sulfuric acid and water are the only two components in the system, we also know that $w_{S1} + w_{W1} = 1.0$, or $w_{W1} = 1.0 - 0.77 = 0.23$. By the same reasoning, $w_{W3} = 0.814$. Stream 2 is pure water, so $w_{W2} = 1.0$. There are no additional specifications regarding the performance of the mixer. So we are done with step 7. We incorporate the information from steps 5 to 7 onto the flow diagram.

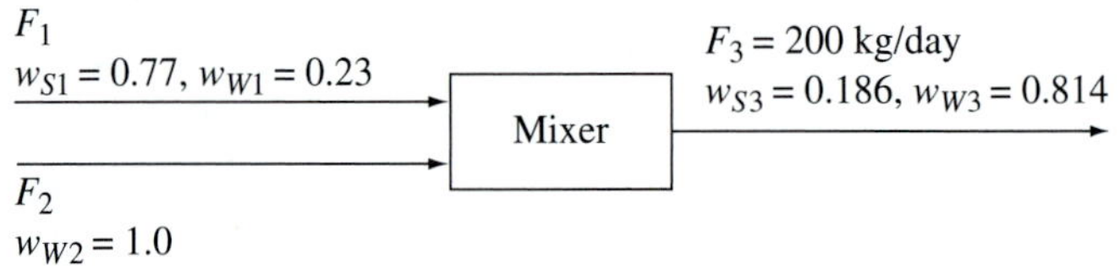

F_1
$w_{S1} = 0.77, w_{W1} = 0.23$

$F_3 = 200$ kg/day
$w_{S3} = 0.186, w_{W3} = 0.814$

Mixer

F_2
$w_{W2} = 1.0$

Helpful Hint
If there are *N* components in a system, there are *N* material balance equations.

Step 8.   *Write material balance equations.* We identified two components—sulfuric acid and water—so there are two independent material balance equations. Let's apply the material balance equation, Eq. (2.1), to each component. There are no generation, consumption, or accumulation variables (see step 6) for either component. Therefore the material balance equation is simply Input = Output.

For sulfuric acid, there is one input (stream 1), which is expressed in terms of our stream variables as $w_{S1}F_1$, and one output (stream 3), which is $w_{S3}F_3$. Therefore the material balance equation for sulfuric acid becomes simply:

$$w_{S1}F_1 = w_{S3}F_3$$

Substituting in the known numerical values we get

$$0.77F_1 = 0.186(200)$$

For water there are two inputs and one output:

$$w_{W1}F_1 + w_{W2}F_2 = w_{W3}F_3$$

Substituting in the known numerical values, we get (in units of kg H_2O/day)

$$0.23F_1 + F_2 = 0.814(200)$$

Step 9.   *Solve the system of equations.* We solve the equations to find

$$F_1 = 48.3\frac{kg}{day}$$

$$F_2 = 151.7\frac{kg}{day}$$

The updated flow diagram is shown:

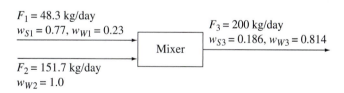

$F_1 = 48.3$ kg/day
$w_{S1} = 0.77, w_{W1} = 0.23$

$F_2 = 151.7$ kg/day
$w_{W2} = 1.0$

Mixer

$F_3 = 200$ kg/day
$w_{S3} = 0.186, w_{W3} = 0.814$

Quick Quiz 2.7

In Example 2.1, we wrote the sulfuric acid material balance equation as $w_{S1}F_1 = w_{S3}F_3$. Why didn't we just write $w_{S1} = w_{S3}$? The latter equation is dimensionally consistent . . .

Step 10.   *Check the answer.* To check the answer, use the material balance on total mass:

$$F_1 + F_2 \overset{?}{=} F_3$$

$$48.3\frac{kg}{day} + 151.7\frac{kg}{day} = 200\frac{kg}{day}$$

| **Example 2.2** | **Reactors: Ammonia Synthesis** |

A gas mixture of hydrogen and nitrogen is fed to a reactor, where they react to form ammonia (NH_3). The nitrogen flow rate into the reactor is 150 gmol /h and the hydrogen is fed at a ratio of 4 gmol H_2 per gmol N_2. The balanced chemical reaction is:

$$N_2 + 3H_2 \rightarrow 2NH_3$$

Of the nitrogen fed to the reactor, 70% is consumed by reaction. The reactor operates at steady state. What is the flow rate (gmol/h) of N_2, H_2, and NH_3 in the reactor outlet?

Solution

Steps 1 and 2. *Draw a diagram and choose a system.* The reactor is the system. Don't neglect to include the unreacted raw materials in the reactor outlet stream!

Step 3. *Choose components and define stream variables.* Since there is a reaction of known stoichiometry, we'll choose the compounds N_2, H_2, and NH_3 as the components. We'll use N to symbolize N_2, H to symbolize H_2, and A to symbolize ammonia. We'll call the streams 1 and 2. There are two components in stream 1, so there are two stream variables, $N_{(1)}$ and $H_{(1)}$. There are three components in stream 2, so there are three stream variables: $N_{(2)}$, $H_{(2)}$, and $A_{(2)}$. (We put the brackets around the stream number subscripts to avoid any confusion between N_2, which is a chemical compound containing two N atoms, and $N_{(2)}$, which is the flow rate of N_2 in stream 2.)

Steps 4 and 5. *Check units and define basis.* All numerical information is given in units of gmol or gmol/h. We'll use as the basis the nitrogen feed rate,

$$N_{(1)} = 150 \frac{\text{gmol}}{\text{h}}$$

Step 6. *Define system variables.* The chemical reaction involves three compounds—two reactants and one product—so there are three system variables related to generation and consumption: N_{cons} (rate of nitrogen consumption by reaction), H_{cons} (rate of hydrogen consumption by reaction), and A_{gen} (rate of ammonia generation by reaction). Using the known reaction stoichiometric coefficients, we write two equations that relate these system variables as:

$$\frac{H_{cons}}{N_{cons}} = \frac{3 \text{ gmol } H_2 \text{ consumed}}{1 \text{ gmol } N_2 \text{ consumed}} = 3$$

$$\frac{A_{gen}}{N_{cons}} = \frac{2 \text{ gmol } NH_3 \text{ generated}}{1 \text{ gmol } N_2 \text{ consumed}} = 2$$

The system is at steady state, so there is no accumulation system variable.

Step 7. *List specifications.* There is one stream composition specification: Hydrogen is fed at a ratio of 4 gmol H_2 per gmol N_2, or

$$H_{\langle 1 \rangle} = 4N_{\langle 1 \rangle} = 600 \frac{\text{gmol}}{\text{h}}$$

There is one system performance specification, describing the reaction taking place inside the system: 70% of the nitrogen fed to the reactor is consumed, or

$$N_{cons} = 0.7N_{\langle 1 \rangle} = 105 \frac{\text{gmol}}{\text{h}}$$

Helpful Hint
A system performance specification may describe a relationship between the system and an input or output stream.

Step 8. *Write material balances.* Since there are 3 components (N_2, H_2, and NH_3), there are three independent material balance equations, which simplify to

$$N_{\langle 1 \rangle} - N_{\langle 2 \rangle} - N_{cons} = 0$$

$$H_{\langle 1 \rangle} - H_{\langle 2 \rangle} - H_{cons} = 0$$

$$-A_{\langle 2 \rangle} + A_{gen} = 0$$

Steps 9 and 10.   *Solve and check.* We substitute in known values for variables and then proceed to solve, working first with the equations with the fewest unknown variables. The solution is:

$$N_{\langle 2 \rangle} = N_{\langle 1 \rangle} - N_{cons} = 150 - 105 = 45 \frac{\text{gmol}}{\text{h}}$$

$$H_{cons} = 3N_{cons} = 3 \times 105 = 315 \frac{\text{gmol}}{\text{h}}$$

$$H_{\langle 2 \rangle} = H_{\langle 1 \rangle} - H_{cons} = 600 - 315 = 285 \frac{\text{gmol}}{\text{h}}$$

$$A_{gen} = 2N_{cons} = 2 \times 105 = 210 \frac{\text{gmol}}{\text{h}}$$

$$A_{\langle 2 \rangle} = A_{gen} = 210 \frac{\text{gmol}}{\text{h}}$$

Helpful Hint
Checking your solution may be as simple as checking that the total mass in equals the total mass out.

To check the solution, we see whether the total mass flow in equals the total mass flow out. The mass flow rate is simply the sum of the molar flow rate of each compound times its molar mass, or

$$\text{total mass in} = 150\frac{\text{gmol } N_2}{h} \times 28\frac{\text{g } N_2}{\text{gmol } N_2} + 600\frac{\text{gmol } H_2}{h} \times 2\frac{\text{g } H_2}{\text{gmol } H_2} = 5400\frac{g}{h}$$

$$\text{total mass out} = 45\frac{\text{gmol } N_2}{h} \times 28\frac{\text{g } N_2}{\text{gmol } N_2} + 285\frac{\text{gmol } H_2}{h} \times 2\frac{\text{g } H_2}{\text{gmol } H_2}$$

$$+ 210\frac{\text{gmol } NH_3}{h} \times 17\frac{\text{g } NH_3}{\text{gmol } NH_3} = 5400\frac{g}{h}$$

$$5400 = 5400$$

The results can be nicely summarized in table form:

	Input gmol/h	Generated − consumed gmol/h	Output gmol/h
Nitrogen	150	− 105	45
Hydrogen	600	− 315	285
Ammonia	0	+ 210	210

Quick Quiz 2.8

Does the total molar flow rate into the reactor in Example 2.2 equal the total molar flow rate out? Why or why not?

Example 2.3   **Separators: Fruit Juice Concentration**

Fruit juice is a complex mixture of water, fructose (fruit sugar), pulp, citric and other acids, acetates, and other chemicals. Fresh fruit juice from the Fruity-Fresh Farm contains 88 wt% water. A fruit juice processor buys a batch of 2680 lb fresh juice from Fruity-Fresh, and makes concentrated juice by filling an evaporator with the fresh juice, evaporating 75% of the water, and then removing the concentrated juice. How much water (lb) must the evaporator remove? If the processor pays $0.09 per pound for the fresh juice, and sells the concentrated juice for $0.50 per pound, can he make a profit?

Solution

Steps 1 and 2.   *Draw a diagram and choose a system.* The system is the evaporator, which performs as a separator.

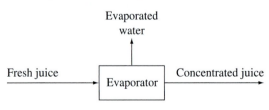

!
▄Helpful Hint
Choose a com-
posite material as
a component if
the material acts
as a single entity
throughout the
process.

Step 3:   *Choose components and define stream variables.* We do not have much information about the composition of the juice other than that it contains 88 wt% water and unspecified quantities of a lot of other things. Since the "other things" stay together—they all come out in the concentrated juice—and they do not undergo any chemical reaction, then we can lump together the fructose, acids, pulp, acetates, etc. as one composite material that we'll call "solids". (Solids include dissolved solutes as well as suspended solids.) We'll indicate water with W and solids as S. We have three streams, one input and two outputs, which we will number 1, 2, and 3. Stream variables will be denoted as, for example, W_1 = water flow in stream 1.

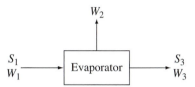

Steps 4 and 5.   *Check units and define basis.* Everything is in mass units, which is fine because there is no chemical reaction. The basis is the fresh juice into the evaporator, or

$$S_1 + W_1 = 2680 \text{ lb}$$

Steps 6 and 7.   *Define system variables and list specifications.* There are no chemical reactions, so no generation or consumption system variables are needed. All the material put into the evaporator is removed as either water vapor or concentrated juice, so there is no accumulation of material inside the evaporator. There is one stream composition specification—the fresh juice is 88 wt% water, or

$$W_1 = (0.88)2680 \text{ lb} = 2360 \text{ lb}$$

Combining this equation with the previous one we find:

$$S_1 = 320 \text{ lb}$$

We have one system performance specification: We know that the evaporator removes 75% of the water in the fresh juice:

$$W_2 = 0.75 \times W_1 = 0.75 \times 2360 \text{ lb} = 1770 \text{ lb}$$

!
▄Helpful Hint
A system per-
formance specifi-
cation may
describe a change
between input
and output
streams caused
by the system.

We'll update our flow diagram, as an easy way to keep track of our calculations.

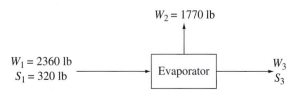

Step 8. *Write material balances.* There are two components, so there are two independent material balance equations. In both cases the material balance equation simplifies to input = output, or

$$W_1 = W_2 + W_3$$

$$S_1 = S_3$$

Steps 9 and 10.   *Solve and check.* Now we simply plug in known numerical values into the equations and solve:

$$W_1 = W_2 + W_3$$

$$2360 \text{ lb} = 1770 \text{ lb} + W_3$$

$$W_3 = 590 \text{ lb}$$

$$S_1 = S_3 = 320 \text{ lb}$$

The total quantity of concentrated juice is $W_3 + S_3 = 590 + 320 = 910$ lb.

Now we have a completed block flow diagram.

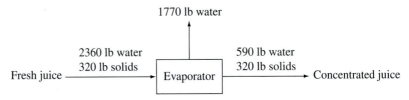

1770 lb water

2360 lb water
320 lb solids

590 lb water
320 lb solids

Fresh juice — Evaporator — Concentrated juice

Considering only raw material costs, the processor would make a profit of

$$\left(910 \text{ lb concentrated juice} \times \frac{\$0.50}{\text{lb}}\right) - \left(2680 \text{ lb fresh juice} \times \frac{\$0.09}{\text{lb}}\right)$$

$$= \$455 - \$240 = \$215$$

We can check the material balance on total mass. This is left for the reader.

Example 2.4   **Splitter: Fruit Juice Processing**

Mr. and Mrs. Fruity squeeze 275 gallons of juice per day at the Fruity-Fresh Farm. They plan to sell 82% of their juice to a processor, who will make frozen concentrated juice. The processor pays $0.75 per pound of juice solids. Some 17% of the juice will be bottled for sale as fresh juice at a local farmers' market, where it sells for $3 per 2-L bottle. Mr. and Mrs. Fruity will keep the remainder for all the little Fruitys. What are the total annual sales ($/year) for the Fruity-Fresh Farm?

Solution

Steps 1 and 2. *Draw a diagram, define system.* The fruit juice producer needs a splitter, with one input and three outputs.

Steps 3 to 5. *Choose components, define stream variables. Check units and define basis.* Since the processor purchases the juice on the basis of its solids content, we'll consider two components: water, which we'll denote as W, and solids, which we'll denote as S. There are four streams, identified as 1, 2, 3, and 4. Stream variables will be named accordingly; for example, S_3 is the flow rate of solids in stream 3.

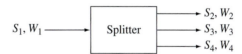

The juice flow rate is in gallons per day, which is a volumetric flow rate. We always work with mass or molar quantities. For this problem, let's choose a mass flow rate, because the composition is given in mass (wt) percent. We'll work with pounds, because the price is given as $/lb. To convert volumetric flow rate to mass flow rate, we need a density. It might be hard to find the density of juice, but we can find the density of a similar solution—12 wt% fructose in water—in the *CRC Handbook of Chemistry and Physics*. The density of a 12 wt% fructose-water solution at 20°C is 1.047 g/mL. Good enough.

The juice flow rate into the splitter, in lb/day, is therefore

$$275 \; \frac{\text{gal juice}}{\text{day}} \times \frac{3.78 \text{ L}}{\text{gal}} \times \frac{1000 \text{ mL}}{\text{L}} \times \frac{1.047 \text{ g}}{\text{mL}} \times \frac{1 \text{ lb}}{454 \text{ g}} = 2400 \; \frac{\text{lb}}{\text{day}}$$

The juice flow rate is a convenient basis; in terms of stream variables we write:

$$S_1 + W_1 = 2400 \text{ lb/day}$$

Steps 6 and 7. *Define system variables and list specifications.* There are no chemical reactions and the system is steady state, so generation, consumption, and accumulation variables are all equal to zero.

For stream composition specifications, we know that the juice is 88 wt% H_2O, or in terms of stream variables:

$$\frac{W_1}{S_1 + W_1} = 0.88 \frac{\text{lb } H_2O}{\text{lb juice}}$$

or:

$$W_1 = 0.88 \frac{\text{lb } H_2O}{\text{lb juice}} \times 2400 \frac{\text{lb juice}}{\text{day}} = 2112 \frac{\text{lb } H_2O}{\text{day}}$$

Therefore,

$$S_1 = 288 \frac{\text{lb solids}}{\text{day}}$$

❗Helpful Hint
A splitter has "hidden" stream composition specifications: All streams in and out of a splitter must have the same composition.

With a splitter, all input and output streams have the same composition. In other words,

$$\frac{W_2}{S_2 + W_2} = \frac{W_3}{S_3 + W_3} = \frac{W_4}{S_4 + W_4} = 0.88 \frac{\text{lb } H_2O}{\text{lb juice}}$$

(We wrote down three equations describing the composition of the output streams, but only two are independent. See if you can prove this to yourself.)

The splitter must meet the system performance specifications that 82% of the juice goes to the processor, and 17% is bottled for sale. In terms of stream variables,

$$S_2 + W_2 = 0.82 \times (S_1 + W_1)$$

$$S_3 + W_3 = 0.17 \times (S_1 + W_1)$$

We substitute in our known values of W_1 and S_1 to get

$$S_2 + W_2 = 0.82 \times \left(2400 \frac{\text{lb}}{\text{day}}\right) = 1968 \frac{\text{lb}}{\text{day}}$$

Combining this with the stream composition specification yields

$$W_2 = 1732 \frac{\text{lb}}{\text{day}}$$

$$S_2 = 236 \frac{\text{lb}}{\text{day}}$$

We proceed in the same manner to find

$$W_3 = 359 \frac{\text{lb}}{\text{day}}$$

$$S_3 = 49\frac{\text{lb}}{\text{day}}$$

Steps 8 and 9. *Write material balances and solve equations.* There are two components, so there are two material balance equations, which simplify to input = output, or

$$S_1 = S_2 + S_3 + S_4$$

$$W_1 = W_2 + W_3 + W_4$$

Now substitute in the known numerical values to get

$$W_4 = 21\frac{\text{lb}}{\text{day}}$$

$$S_4 = 3\frac{\text{lb}}{\text{day}}$$

The problem asked for the total income to the juice producer. To calculate the daily sales receipts of the producer, we sum the sales to the processor and farmers' market:

$$\text{Sales to processor} = S_2 \times \frac{\$}{\text{lb solids}} = \frac{236 \text{ lb solids}}{\text{day}} \times \frac{\$0.75}{\text{lb solids}} = \frac{\$177}{\text{day}}$$

At the farmers' market, the product is sold per liter of juice, not per pound of solid, but it is interesting to convert from one cost basis to the other:

$$\left(\frac{\$3.00}{2 \text{ L juice}}\right)\left(\frac{\text{L}}{1000 \text{ mL}}\right)\left(\frac{\text{mL juice}}{1.047 \text{ g juice}}\right)\left(\frac{454 \text{ g juice}}{\text{lb juice}}\right)\left(\frac{1 \text{ lb juice}}{0.12 \text{ lb solids}}\right) = \frac{\$5.42}{\text{lb solids}}$$

(The Fruitys should sell as much juice as possible at the market rather than to the processor.)

$$\text{Sales at farmers' market} = S_3 \times \frac{\$}{\text{lb solids}} = \frac{49 \text{ lb solids}}{\text{day}} \times \frac{\$5.42}{\text{lb solids}} = \frac{\$266}{\text{day}}$$

Product sales total $443/day or nearly $162,000/year, if the family is able to sell this much product every day of the year.

$$\text{Value of juice consumed by the little Fruitys} = S_4 \times \frac{\$}{\text{lb solids}} = \frac{3 \text{ lb solids}}{\text{day}}$$

$$\times \frac{\$5.42}{\text{lb solids}} = \frac{\$16.26}{\text{day}}$$

(At nearly $6000/year, the little Fruitys can drink water!)

Step 10.   *Check.* One way to check the answer is to see if the solids content of the juice consumed by the little Fruitys is 12 wt%, since we did not use that information in our calculations yet.

$$\frac{S_4}{S_4 + W_4} = \frac{3}{3 + 21} = 0.125 \text{ (close enough)}$$

Notice that in this problem we carried along more digits than significant in our intermediate calculations. This is often a good idea, to avoid round-off errors. Try recomputing all flow rates carrying just significant digits. What do you find to be the solids content of the juice in stream 4?

Example 2.5   **Elements as Components: Ibuprofen Analysis**

You have a summer job with a company that is interested in building a plant to manufacture the painkiller ibuprofen [2-(p-isobutylphenyl)propionic acid, $C_{13}H_{18}O_2$] using a new reaction scheme. Your assignment is to collect some data on one of the reactions, as a first step in designing a full-scale reactor. In one experiment, you mix 134 g isobutylbenzene (IBB, $C_{10}H_{14}$) with 134 g acetic anhydride (AAn, $C_4H_6O_3$) in a laboratory-scale batch reactor, adjust the temperature, and wait 1 hour. At the end of the hour you stop the reaction, collect all the material in the pot, and send it for chemical analysis. The report comes back that the pot contains IBB, acetic anhydride, isobutylacetophenone (IBA, $C_{12}H_{16}O$), and acetic acid (AAc, CH_3COOH). Unfortunately, someone spilled coffee on the report and all you can read is the amount of acetic acid: 68 g. Your boss is upset—he needs the data right away. Here's your opportunity to come to the rescue and impress your boss. Can you?

Solution

Steps 1 and 2. *Draw a diagram, choose system.* This is a batch reactor. Material is placed in a vessel all at once at the start, the contents change over time due to the reaction, and the contents are removed at the end. We can still draw a flow diagram, remembering that the sketch indicates what happens over time. The system is the reactor.

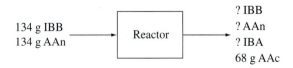

!

Helpful Hint
Choose elements as components when there are reactions of unknown stoichiometry.

Steps 3 to 6. *Choose components, define stream variables, check units, define basis.* There is a chemical reaction but we don't know the stoichiometry, so it will be easier to work with elements rather than compounds. These compounds contain three elements: C, H, and O. There are two streams, so we'll define six stream variables:

$$C_{(1)} = \text{C put into reactor} \qquad C_{(2)} = \text{C taken out of reactor}$$

$$H_{(1)} = \text{H put into reactor} \qquad H_{(2)} = \text{H taken out of reactor}$$

$$O_{(1)} = \text{O put into reactor} \qquad O_{(1)} = \text{O taken out of reactor}$$

We know the grams of IBB and AAn placed in the reactor, and the grams of AAc removed. But we don't know the moles of each element. We calculate these values from the grams of compounds and the molar mass and the chemical formula of each compound:

$$n_{hi} = n_i \varepsilon_{hi} = \frac{m_i}{M_i} \varepsilon_{hi}$$

where n_{hi} = moles of element h in compound i, n_i = moles of compound i, ε_{hi} = moles of element h per mole of compound i, m_i = mass of compound i, and M_i = molar mass of compound i.

It is easy to summarize these calculations in table form.

For stream 1:

Compound	m_i (g)	M_i (g/gmol)	ε_{Ci}	n_{Ci} (gmol)	ε_{Hi}	n_{Oi} (gmol)	ε_{Oi}	n_{Oi} (gmol)
IBB, $C_{10}H_{14}$	134	134	10	10	14	14	0	0
AAn, $C_4H_6O_3$	134	102	4	5.255	6	7.88	3	3.94

We simply sum up the correct column to find our stream variables:

$$C_{(1)} = 10 + 5.255 = 15.255 \text{ gmol}$$

$$H_{(1)} = 14 + 7.88 = 21.88 \text{ gmol}$$

$$O_{(1)} = 0 + 3.94 = 3.94 \text{ gmol}$$

For stream 2:

Compound	m_i (g)	M_i (g/gmol)	ε_{Ci}	n_{Ci} (gmol)	ε_{Hi}	n_{Oi} (gmol)	ε_{Oi}	n_{Oi} (gmol)
IBB, $C_{10}H_{14}$	$m_{IBB,(2)}$	134	10	$\dfrac{10 m_{IBB,(2)}}{134}$	14	$\dfrac{14 m_{IBB,(2)}}{134}$	0	0
AAn, $C_4H_6O_3$	$m_{AAn,(2)}$	102	4	$\dfrac{4 m_{AAn,(2)}}{102}$	6	$\dfrac{6 m_{AAn,(2)}}{102}$	3	$\dfrac{3 m_{AAn,(2)}}{102}$
IBA, $C_{12}H_{16}O$	$m_{IBA,(2)}$	176	12	$\dfrac{12 m_{IBA,(2)}}{176}$	16	$\dfrac{16 m_{IBA,(2)}}{176}$	1	$\dfrac{m_{IBA,(2)}}{176}$
AAc, $C_2H_4O_2$	68	60	2	2.267	4	4.533	2	2.267

In terms of our stream variables,

$$C_{(2)} = \frac{10m_{\text{IBB},(2)}}{134} + \frac{4m_{\text{AAn},(2)}}{102} + \frac{12m_{\text{IBA},(2)}}{176} + 2.267$$

$$H_{(2)} = \frac{14m_{\text{IBB},(2)}}{134} + \frac{6m_{\text{AAn},(2)}}{102} + \frac{16m_{\text{IBA},(2)}}{176} + 4.533$$

$$O_{(2)} = \frac{3m_{\text{AAn},(2)}}{102} + \frac{m_{\text{IBA},(2)}}{176} + 2.267$$

(All we've done is define our desired stream variables in terms of other unknowns—the grams of each of the compounds at the end of the reaction.)

Steps 6, 7, and 8. *Define system variables, list specifications, write material balances.* Even though there are chemical reactions, we have chosen elements as our components, and elements cannot be generated or consumed. Therefore generation = consumption = 0. There is nothing in the reactor before the beginning of the experiment, and all is removed at the end, so accumulation = 0. There are no additional specifications given. The material balance equations are quite simple:

$$C_{(1)} = C_{(2)}$$

$$H_{(1)} = H_{(2)}$$

$$O_{(1)} = O_{(2)}$$

Steps 9 and 10. *Solve and check.* By substituting into the material balance equations the expressions we derived for each stream variable, we end up with three equations in three unknowns.

$$15.225 = \frac{10m_{\text{IBB},(2)}}{134} + \frac{4m_{\text{AAn},(2)}}{102} + \frac{12m_{\text{IBA},(2)}}{176} + 2.267$$

$$21.88 = \frac{14m_{\text{IBB},(2)}}{134} + \frac{6m_{\text{AAn},(2)}}{102} + \frac{16m_{\text{IBA},(2)}}{176} + 4.533$$

$$3.94 = \frac{3m_{\text{AAn},(2)}}{102} + \frac{m_{\text{IBA},(2)}}{176} + 2.267$$

We solve simultaneously to find

$$m_{\text{IBB},(2)} = 1.6 \text{ g}$$

$$m_{\text{AAn},(2)} = 22.9 \text{ g}$$

$$m_{\text{IBA},(2)} = 175.5 \text{ g}$$

To check the solution, we check that the total mass in equals the total mass out:

$$134 \text{ g} + 134 \text{ g} = 1.6 \text{ g} + 22.9 \text{ g} + 175.5 \text{ g} + 68 \text{ g}$$
$$268 \text{ g} = 268 \text{ g}$$

You report your findings to your boss, who applauds your resourcefulness and gives you a raise!

So far, we've always set the accumulation term equal to zero. This happens either when the process is continuous and steady state, or batch over a fixed time interval with all materials added to the system at the beginning and all materials removed from the system by the end. Now, let's turn to two semibatch problems, where material accumulates in the system during the process.

Example 2.6	**Separation with Accumulation: Air Drying**

Air is used throughout a process plant to move control valves (special valves that regulate flow). If the air is humid, it needs to be dried before being used. To produce dry air for instrument use, filtered and compressed humid room air at 83°F and 1.1 atm pressure, containing 1.5 mol% H_2O (as vapor), is pumped through a tank at a flow rate of 100 ft³/min. The tank is filled with 60 lbs of alumina (Al_2O_3) pellets. The water vapor in the air adsorbs (sticks) onto the pellets. Dry instrument air, containing just 0.06 mol% H_2O, exits from the tank. The maximum amount of water that can adsorb to the alumina pellets is 0.22 lb H_2O per lb alumina. How long can the tank be operated before the alumina pellets need to be replaced?

Solution

Steps 1 and 2.   *Draw a diagram, choose a system.* The system is the separator—the tank containing the alumina pellets.

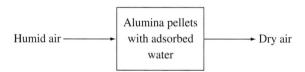

Step 3.   *Choose components, define stream variables.* Air is a composite material: it contains nitrogen, oxygen, argon, carbon dioxide, water vapor, and other gases. The alumina pellets remove only water from the air; all the other gases stay together as the stream passes through the separator. Therefore, we will choose as our components water (W) and water-free air (A). In other words, we lump together everything in the air other than H_2O as a single composite material. Stream variables

are: A_1 = water-free air into tank, W_1 = water vapor into tank, A_2 = water-free air leaving tank, and W_2 = water vapor leaving tank.

Steps 4 and 5. *Check units, define basis.* The units are not consistent—the air flow rate is volumetric, the water content of the air is given as mol%, and the adsorption capacity of the pellets is given as a mass ratio. We need to convert everything to consistent mass or mole units—let's use lbmol.

First let's convert the volumetric air flow rate to a molar flow rate. For that, we need a density. We'll assume air at these conditions behaves as an ideal gas and calculate the molar density from the ideal gas law:

$$\frac{n}{V} = \frac{P}{RT} = \frac{(1.1 \text{ atm})}{(0.7302 \text{ ft}^3 \text{ atm}/\text{lbmol } °R)(83 + 459 °R)} = 0.00278 \frac{\text{lbmol}}{\text{ft}^3}$$

The molar flow rate is simply the volumetric flow rate times the molar density:

$$100 \frac{\text{ft}^3}{\text{min}} \times 0.00278 \frac{\text{lbmol}}{\text{ft}^3} = 0.278 \frac{\text{lbmol}}{\text{min}}$$

This is the total molar flow rate of humid air fed to the separator, and will serve as our basis. In terms of system variables,

$$A_1 + W_1 = 0.278 \frac{\text{lbmol}}{\text{min}}$$

Steps 6 and 7. *Define system variables, list specifications.* There are no chemical reactions, so no system variables of generation or consumption are needed. The air does not accumulate inside the system, but the water does. Therefore we will have one system variable, W_{acc}, which describes the rate of water accumulation inside the tank.

Stream composition specifications include that the humid air is 1.5 mol% water and the dry air is 0.06 mol% water, or

$$\frac{W_1}{A_1 + W_1} = 0.015 \frac{\text{lbmol water}}{\text{lbmol}}$$

$$\frac{W_2}{A_2 + W_2} = 0.0006 \frac{\text{lbmol water}}{\text{lbmol}}$$

A system performance specification reflects physical and/or chemical changes occurring within the system. In this case, we know that the tank contains 60 lb alumina, which can adsorb, at most, 0.22 lb water/lb alumina. The total allowable accumulation of water in the tank is therefore

$$60 \text{ lb alumina} \times \frac{0.22 \text{ lb water}}{\text{lb alumina}} \times \frac{\text{lbmole water}}{18 \text{ lb water}} = 0.73 \text{ lbmol water}$$

The stream variables have dimension of [mol/time], so we need to express the system accumulation variable in the same dimension—as a *rate* of accumulation. This will be equal to the total accumulation divided by the time t over which water is allowed to accumulate in the tank, or

$$W_{acc} = \frac{0.73 \text{ lbmol water}}{t \, (\text{min})}$$

Steps 8 to 10. *Write material balances, solve equations, and check.* There are two components, water-free air and water vapor, so two material balance equations are written:

$$A_1 = A_2$$
$$W_1 - W_2 = W_{acc}$$

We work through this system of equations (the details are left to you) to find

$$W_1 = 4.17 \times 10^{-3} \text{ lbmol/min}$$
$$A_1 = 0.2738 \text{ lbmol/min}$$
$$A_2 = 0.2738 \text{ lbmol/min}$$
$$W_2 = 1.64 \times 10^{-4} \text{ lbmol/min}$$
$$W_{acc} = 4.01 \times 10^{-3} \text{ lbmol/min}$$
$$t = 180 \text{ min}$$

Check your answer by checking that the total mass of water adsorbed plus the instrument air produced equals the mass of room air fed, over the 3-h period.

After 3 hours, the separator would no longer have the capacity to adsorb more water. (Think of a bucket being filled with water—eventually its capacity is reached.) Two tanks are used in most such operations. The air flows through one of the tanks for 3 hours, then the flow is switched to the second tank. The wet alumina pellets are not thrown away, but are regenerated, probably by heating, to drive off the accumulated water. Then the regenerated pellets are reused.

| **Example 2.7** | **Reaction with Accumulation: Light from a Chip** |

Light-emitting diodes (LEDs) are used in all kinds of lighted displays, from small handheld electronic games to huge billboards. LEDs are made from semiconductor material; several thin layers of material are built up on a substrate.

By changing the chemical composition of the semiconductor material, different colors are produced.

A researcher is interested in making blue LEDs. She places a 1 cm × 1 cm chip of Al_2O_3 in a reactor. In a process called MOCVD (metalorganic chemical vapor deposition), trimethyl gallium [$(CH_3)_3Ga$] and ammonia (NH_3) are pumped continuously into the reactor at a 1:1 molar ratio, along with a carrier gas. The two reactants form gallium nitride (GaN), which deposits as an even layer on the Al_2O_3 chip, and methane (CH_4), which is swept out of the reactor continuously by the carrier gas. The balanced chemical reaction is:

$$(CH_3)_3Ga + NH_3 \rightarrow GaN + 3CH_4$$

The researcher would like to develop a method to estimate the rate of growth of the height of the GaN layer on the chip, in micrometers per hour, by measuring the methane flow rate, in μmol/h, out of the reactor. Can you help?

Solution

Steps 1 to 3. *Draw diagram, define system, choose components, define stream variables.* The system is the reactor. The sketch shows the GaN layer growing on the substrate.

Since there is a chemical reaction of known stoichiometry, we'll use the four compounds as components: $(CH_3)_3Ga$ (*T*), NH_3 (*A*), GaN (*G*) and CH_4 (*M*). There is also a carrier gas, which does not take part in the reaction. We'll call it *I*, for inert. There are two streams that enter and leave the system; we'll designate them as streams 1 and 2. The stream variables are therefore: T_1 = trimethylgallium flow into reactor, A_1 = ammonia flow into reactor, I_1 = inert gas flow into reactor, M_2 = methane flow out of reactor, and I_2 = inert gas flow out of reactor. There is no *stream* variable in GaN because GaN is not present in a stream, only in the system.

Steps 4 to 6. *Check units, define basis, define system variables.*

No basis is specified in the problem statement. No problem! We just choose *any* basis that is convenient. Since a chemical reaction is involved, we choose moles rather than mass. Since the researcher will measure the methane flow rate out of the reactor in μmol/h (10^{-6} gmol/h), it makes sense to use units of μmol/h. We want a relationship involving the methane flow rate out, so it makes sense to set M_2 as the basis. So let's choose:

$$M_2 = 100 \ \mu mol/h$$

Since there are four reactants and products, there are four system variables for generation and consumption, T_{cons}, A_{cons}, G_{gen} and M_{gen}. There are three equations relating these four system variables through their stoichiometric coefficients.

$$\frac{A_{cons}}{T_{cons}} = 1$$

$$\frac{G_{gen}}{T_{cons}} = 1$$

$$\frac{M_{gen}}{T_{cons}} = 3$$

The GaN accumulates on the chip, while other compounds do not. Therefore we have one system variable for accumulation, G_{acc}.

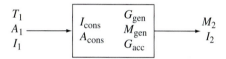

Finally, we chose μmol/h for units, but the researcher requested a relationship involving growth of the GaN layer in micrometers/h. How do we convert from one unit to the other? First, we recognize that the layer is three dimensional, with length and width defined by the size of the Al_2O_3 chip, so the growth in thickness of the layer is really a volumetric growth rate. Second, we relate a volumetric growth to a molar growth rate by finding a density. We look up the density of GaN in the *CRC Handbook of Chemistry and Physics* and find that the mass density is 6.1 g/cm^3. That, plus the molar mass of 84 g/gmol for GaN gives us the conversion factor that we need, where the brackets indicate the units:

$$\text{Growth rate}\left[\frac{\mu m}{h}\right] = \frac{\left(\dfrac{10^4\,\mu m}{cm}\right)\left(\dfrac{cm^3}{6.1g}\right)\left(\dfrac{84\,g}{gmol}\right)\left(\dfrac{gmol}{10^6\,\mu mole}\right)}{1\,cm \times 1\,cm}\,G_{acc}\left[\frac{\mu mol}{h}\right]$$

$$= 0.138 G_{acc}\left[\frac{\mu mol}{h}\right]$$

Steps 7 and 8. *List specifications, write material balances.* We have one stream composition specification: The molar ratio in the feed gas is specified as 1:1, or

$$T_1 = A_1$$

There are five components so there are five material balance equations, simplified from input − output + generation − consumption = accumulation:

$$T_1 - T_{cons} = 0$$

$$A_1 - A_{cons} = 0$$

$$G_{gen} = G_{acc}$$

$$-M_2 + M_{gen} = 0$$

$$I_1 - I_2 = 0$$

Steps 9 and 10. *Solve and check.* We solve this system of equations by starting with the methane balance, because this equation has only one unknown. From the methane balance, we find $M_{gen} = 100$ μmol/h. Then we use the stoichiometric relationships to find $T_{cons} = 33.3$ μmol/h $= G_{gen} = G_{acc}$. Finally, we use the unit conversion factor that we derived to find that, if $G_{acc} = 33.3$ μmol/h, then the growth rate = 4.60 μm/h. (We are unable to solve for the carrier gas flows I_1 and I_2 because we don't have any information about these streams. That's OK, as long as the researcher measures methane flow in the reactor output, and not just total gas flow.)

We calculated the growth rate of 4.6 μm/h given our chosen basis of 100 μmol/h methane. The researcher wants a relationship that applies for *any* measured methane flow, which we get by simply scaling our results:

$$\text{growth rate} \left[\frac{\mu m}{h} \right] = 0.046 \times M_2 \left[\frac{\mu mol}{h} \right]$$

(We could have gotten this answer by symbolic manipulation of our equations, without choosing a basis at all. Try it!)

2.5   Degree of Freedom Analysis

Except for Example 2.7, we were always able to calculate all the stream and system variables. In Example 2.7, we were unable to find the inert gas flow rate, and we needed to define a basis to calculate a growth rate. Is there a quick way to predict whether there will be a complete solution to a process flow problem, without actually solving the problem?

A problem is going to be "solvable" if there is just the right *amount* of information—not too much, not too little—and just the right *kind* of information. There are rigorous mathematical methods for ascertaining whether a problem has a unique solution. What we'll present here is a simpler method, called **degree of freedom (DOF) analysis.** DOF analysis provides a rapid means for determining if a specific process flow calculation problem is solvable, without actually solving the problem, and even without setting up the equations. If you can count, you can complete a DOF analysis.

Here is the basic procedure (along with some comments).

1. Draw a flow diagram and choose a system.
2. Choose components.
3. Label the flow diagram to show the components in each stream and any accumulation in the system. Write down any chemical reactions.
4. Count the *number of independent variables:*

	Do this by counting
a. Count the number of independent stream variables.	The number of components in each stream, then adding them all up.
b. Count the number of independent system variables.	The number of independent chemical reactions, plus the number of components that accumulate (or are depleted) in the system.

5. Count the *number of independent equations:*

	Comments
a. Count the number of specified flows.	Usually there is only one flow specified, that serves as the basis.
b. Count the number of stream composition specifications.	If there are N components in a stream, there are at most $N - 1$ specified stream compositions. For a splitter, add $(N - 1)*(N_s - 1)$, to the count, where N_s is the number of output streams.
c. Count the number of system performance specifications.	
d. Count the number of material balance equations.	If there are N components in a system, then there are N material balance equations.

6. Calculate DOF = *number of independent variables* − *number of independent equations.*

   If DOF = 0: The problem has an equal number of variables and equations, and there is probably a solution.

If DOF < 0: The problem is overspecified. There are more equations than variables. Some of the equations are either redundant or inconsistent.

If DOF > 0: Problem is underspecified. There are more variables than equations. This is an opportunity for optimization; you can add more equations that might describe desired outcomes, such as minimal cost.

One of the biggest difficulties in DOF analysis is establishing whether equations are independent. An independent equation provides unique information that cannot be surmised by combining two or more other equations or specifications. If you find a problem where DOF < 0, look carefully at all equations and specifications. Here are two common pitfalls:

If a reactant is completely (100%) consumed, then there will be no stream variable for the reactant in the reactor outlet, and therefore 100% consumption is *not* an independent specification.

If a stream is specified as containing 10% A, 30% B, and 70% C, then only two of these three composition specifications are independent, because the third can be calculated from knowing that the percentages must add up to 100%.

In the next four examples, we illustrate DOF analyses of problems similar to the ones you have already worked. Determining whether a problem is solvable is faster than actually solving the problem, once you have a bit of experience with this technique.

Example 2.8   **DOF Analysis: Ammonia Synthesis**

A gas mixture of hydrogen and nitrogen is fed to a reactor, where they react to form ammonia, NH_3. The N_2 flow rate into the reactor is 150 gmol/h and the hydrogen is fed at a ratio of 4 gmol H_2 per gmol N_2. The balanced chemical reaction is

$$N_2 + 3H_2 \rightarrow 2NH_3$$

Of the nitrogen fed to the reactor, 30% leaves in the reactor outlet stream; the rest is consumed by reaction. The reactor operates at steady state. Determine the DOF.

Solution
This problem is a repeat of Example 2.2. Here is a flow diagram, labeled to indicate the presence of hydrogen (*H*), nitrogen (*N*), and ammonia (*A*). Notice we haven't worried about numbering the streams.

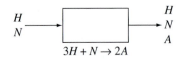

Count the number of independent variables:

	Answer	**Explanation**
Stream variables	5	2 components in input stream (N, H)
		3 components in output stream (N, H, A)
System variables	1	1 reaction,
		no accumulation (steady state)
Total	6	

Count the number of independent equations:

	Answer	**Explanation**
Specified flows	1	Feed rate of 150 gmol N_2/h
Specified stream compositions	1	N_2:H_2 ratio in feed = 1:4
Specified system performance	1	30% of N_2 fed does not react
Material balance equations	3	3 components in system (N, H, A), so 3 material balance equations
Total	6	

DOF = 6 − 6 = 0. Completely specified.

If you compared Examples 2.2 and 2.8, you may have noticed what appears to be a discrepancy. In Example 2.2 we had three system variables, N_{cons}, H_{cons}, and A_{gen}. But, in Example 2.8, we counted only one variable for the reaction. Why? Because we can write two equations that relate N_{cons}, H_{cons}, and A_{gen} through their stoichiometric coefficients. Since these three system variables are related through two equations, only one of them is independent. Therefore, we count one system variable for each chemical reaction.

Example 2.9   **DOF Analysis: Light from a Chip**

A researcher is interested in making blue LEDs. She places a 1 cm × 1 cm chip of Al_2O_3 in a reactor. In a process called MOCVD (metalorganic chemical vapor deposition), trimethyl gallium [$(CH_3)_3Ga$] and ammonia (NH_3) are pumped continuously into the reactor at a 1:1 molar ratio, along with a carrier gas. The two reactants form gallium nitride (GaN), which deposits as an even layer on the Al_2O_3 chip, and methane (CH_4), which is swept out of the reactor continuously by the carrier gas. The balanced chemical reaction is

$$(CH_3)_3Ga + NH_3 \rightarrow GaN + 3CH_4$$

The researcher would like to develop a method to estimate the rate of growth of the height of the GaN layer on the chip, in micrometers per hour, by measuring the methane flow rate, in $\mu mol/h$, out of the reactor. Is this problem completely specified?

Solution
This problem is a repeat of Example 2.7. The flow diagram is labeled to indicate the presence of trimethyl gallium (T), ammonia (A), gallium nitride (G), methane (M), and inert carrier gas (I). We've indicated that there is accumulation of G in the system, and that one reaction takes place.

Count the number of independent variables:

	Answer	Explanation
Stream variables	5	3 components in input stream (T, A, I) 2 components in output stream (M, I)
System variables	2	1 reaction, G accumulation
Total	7	

Count the number of independent equations:

	Answer	Explanation
Specified flows	0	
Specified stream compositions	0	The $T{:}A$ feed ratio of 1:1 is *not* independent! See below for explanation.
Specified system performance	0	All the T reacts, but we have already accounted for complete conversion by having no T in the outlet stream nor accumulated in the system. Similarly, since there is no A in the outlet stream or accumulated in the system, then the $T{:}A$ feed ratio *must* be 1:1

		to match the reaction stoichiometry. That ratio is *not* independent in this case!
Material balance equations	5	5 components in system (T, A, I, G, M), so 5 material balance equations
Total	6	

DOF $= 7 - 5 = 2$. The problem is underspecified. We saw this when we tried to solve Example 2.7: We defined a basis of our own choosing, which provided an additional equation, and we found we were unable to solve for the inert gas flow.

This DOF analysis is tricky! It takes some practice to identify whether stream composition and system performance specifications are independent. If you are not sure, try to imagine whether the solution would change if that specification were not given in the problem statement.

Example 2.10   ### DOF Analysis: Battery Acid Production

Your job is to design a mixer to produce 200 kg/day of battery acid. The mixer will operate at steady state. The battery acid product must contain 18.6 wt% H_2SO_4 in water. Raw materials available include a concentrated sulfuric acid solution at 77 wt% H_2SO_4 in water, a dilute acid solution that contains 4.5 wt% H_2SO_4, and pure water. Is this problem completely specified?

Solution
This problem is very similar to Example 2.1, except there is an extra input stream—the dilute acid solution.

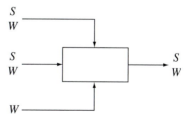

Count the number of independent variables:

	Answer	Explanation
Stream variables	7	For you to provide
System variables	0	
Total	7	

Count the number of independent equations:

	Answer	Explanation
Specified flows	1	200 kg/day product
Specified stream compositions	3	The %S in 3 streams is given. The %W in each stream is NOT independent.
Specified system performance	0	
Material balance equations	2	For you to provide
Total	6	

DOF $= 7 - 6 = 1$. The problem is underspecified.

Underspecified problems often provide opportunities. In this case, for example, you could choose the relative quantities of dilute acid and water to add, depending on the costs of the raw materials. By having an additional degree of freedom, you can optimize the solution to minimize cost.

Example 2.11   DOF Analysis: Eat Your Greens!

Salad greens are washed to remove dirt, bugs, and other debris (dbd) before being packaged for sale. A facility processes 1500 16-oz. packages of greens per day. The fresh-picked greens contain 1 lb dbd per 12 lb greens. The salad greens are mixed with 150 gal water per day and washed, then spun to separate the dirty water from the greens. The process removes 99.9% of the dbd and all of the washwater from the salad greens. The washed greens are sent for packaging. The dirty water is dumped to a river. The plant is limited by environmental constraints to dumping a maximum of 4 barrels (bbl) of water per day, at a maximum dbd content of 1.5 vol%. Is this problem completely specified?

Solution
We'll choose as components salad greens (S), water (W), and dirt, bugs and debris (D). There are two process units, a mixer (the washer) and a separator (the spinner), but we'll choose the entire processing facility as our system.

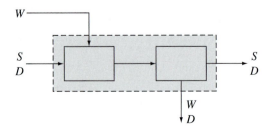

Count the number of independent variables:

	Answer	Explanation
Stream variables	7	For you to provide
System variables	0	
Total	7	

Count the number of independent equations:

	Answer	Explanation
Specified flows	3	1500 bags salad per day 150 gal washwater/day 4 bbl dirty water dumped to river per day
Specified stream compositions	2	1 lb dbd to 12 lb salad greens 1.5% dbd in dirty water dumped to river
Specified system performance	1	99.9% removal of dbd (100% removal of water is *not* independent—we have accounted for that by having no stream variable for water on the salad product stream)
Material balance equations	3	For you to provide
Total	9	

Quick Quiz 2.9

For Example 2.3, count the number of stream variables and the number of material balance equations.

DOF $= 7 - 9 = -2$. The problem is overspecified.

Or is it? There are two constraints—a *maximum* of 4 bbl/day dirty washwater, at a *maximum* of 1.5 vol% dirt—that are one-sided inequalities but not absolute equalities. Such one-sided constraints place minimum or maximum constraints on the numerical value of a variable, but do not require that the variable have that exact value. For this example problem, it is possible that we meet both of these constraints, neither of them, or one but not the other. In this situation, the best way to proceed is to solve the problem in the absence of the one-sided constraints, then check to see if the solution complies with these constraints. If not, we need to redesign the process.

2.6   Process Flow Calculations with Multiple Process Units

In all the previous examples, we worked with a single system. What if we have the task of calculating process flows for a block flow diagram with many process units? Although this may seem like a daunting task, when broken down into parts it is not. The 10 Easy Steps still apply; but there are a few new Helpful Hints.

Step 2.   *Choose system(s)*
- You can choose to treat each process unit as a separate system in turn. As you complete calculations for one system, you gain information that allows you to proceed to the next system. This procedure is necessary if you plan to calculate all process stream variables and all system variables in the block flow diagram.
- You can group two (or more) process units together and choose the group as your system. If you do this, then draw a box around the grouped units. Input and output streams are only those streams that cross the boundaries of your box—not the streams that are internal to the box. This is advantageous if you have insufficient information about the streams that are internal to your box. If you group all of the process units together, you've just converted the block flow diagram into an input-output diagram! This is often a good place to start.

Step 8.   *Write material balance equations.* For *each* system in your diagram, if there are *N* components in that system, write *N* material balance equations around that system.

Step 9.   *Solve the equations.* Set up a table to keep track of your results.

The next few examples illustrate how to apply the 10 Easy Steps to multi-unit problems.

Example 2.12   **Multiple Process Units: Toxin Accumulation**

An immunotoxin (IT) is a drug designed to kill cancer cells. An IT is constructed from two proteins: One protein (usually an antibody) specifically targets cancer cells and leaves healthy cells alone, while the second protein (the toxin) kills the cell once it is inside. An experimental IT is internalized into cancer cells at a rate of 185,000 molecules per minute. Inside the cell, all the IT first enters a compartment called an endosome, which acts like a splitter. 97% of the IT that enters the endosome is spit back out of the cell. Much of the rest is sent to a compartment called a lysosome, where it is degraded into harmless byproducts, at a degradation rate of 5500 molecules/min. Anything remaining moves to another compartment, called the cytosol, where it accumulates. It is estimated that 500 IT molecules must accumulate inside the cell before it will be killed. How long will that take?

Solution

Steps 1 and 2.   *Draw a diagram and define a system.*

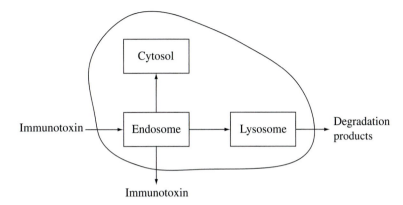

The cellular compartments are shown as boxes; the endosome is a splitter, the lysosome is a reactor, and the cytosol is a storage tank. We can choose each compartment in turn as our system. Or we can group all compartments together and consider the entire cell as our system, with the boundary as depicted. Let's try the latter approach.

Steps 3 to 5.   Choose components and define stream variables, convert units, define basis. *IT* is our component, and we'll use units of molecules/min. Given our choice of system, there are only two streams (we'll call them in and out) that have *IT* in them. (The third stream has degradation products only.) Our stream variables are:

$$IT_{in} = \text{immunotoxin entering the cell}$$

$$IT_{out} = \text{immunotoxin leaving the cell}$$

(We will not worry about the degradation products, as we do not need to know anything more about them.)
    The basis is

$$IT_{in} = 185{,}000 \, \frac{\text{molecules}}{\text{min}}$$

Steps 6 to 10.   *Define system variables, list specifications, write material balance equations, solve, check. IT* is consumed by degradation reactions inside the cell. We don't know anything about the reaction stoichiometry, but we do know the net rate of degradation, so we write one system variable as

$$IT_{cons} = 5500 \, \frac{\text{molecules}}{\text{min}}$$

We wish to know when the total accumulation inside the cell equals 500 molecules. This equals the rate of accumulation IT_{acc} multiplied by the time of accumulation t.

$$IT_{acc} \times t = 500 \, \text{molecules}$$

We have one system performance specification, because we know the splitter ratio:

$$IT_{out} = 0.97IT_{in}$$

The material balance equation over the entire cell is

$$IT_{in} - IT_{out} - IT_{cons} = IT_{acc}$$

We easily combine and solve to find

$$IT_{acc} = 50 \frac{\text{molecules}}{\text{min}}$$

$$t = 10 \, \text{minutes}$$

Example 2.13   **Multiple Process Units: Adipic Acid Manufacture from Glucose**

A preliminary block flow diagram for production of 12,000 kg/h of adipic acid from glucose, based on the technology described in Chap. 1's case study, is shown in Figure 2.13. Available raw materials are air (assumed to be 21 mol% O_2 and the remainder N_2), glucose solution (10 mg/mL in water), and hydrogen. In Reactor 1, genetically engineered *E. coli* convert glucose to muconic acid, with CO_2 as a byproduct. The reaction is

$$\tfrac{7}{3}C_6H_{12}O_6 + \tfrac{17}{2}O_2 \rightarrow C_6H_6O_4 + 8CO_2 + 11H_2O$$

The outlet from Reactor 1 is sent to the separator, where it is separated into three streams: One stream contains only gases, another contains pure muconic acid, and the third contains water and bacteria. The muconic acid is mixed with hydrogen and fed to Reactor 2, where it is hydrogenated to adipic acid:

$$C_6H_6O_4 + 2H_2 \rightarrow C_6H_{10}O_4$$

Assume 100% conversion of reactants to products in both reactors.

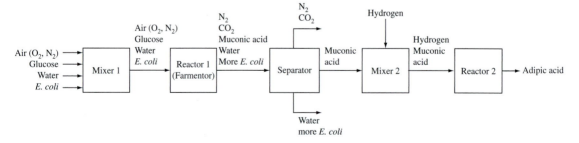

Figure 2.13 A preliminary block flow diagram for a process that converts glucose, air, and hydrogen to adipic acid, using genetically engineered bacteria. Refer to the case study of Chap. 1 for details.

Calculate component and total flow (kgmol/h) for all process streams, at steady-state operation. Summarize results in table form. (Ignore *E. coli.* contributions to process flows.)

Solution

Steps 1 to 3.   *Draw diagram, define system, choose components, and define stream variables.* This looks like a really challenging problem. If we break the problem down into smaller chunks, however, you will see that it is quite manageable.

We know that we want the process to produce 12,000 kg adipic acid/h, and we know what raw materials are available as inputs to the process. Therefore, a good place to start is with an input-output diagram, because this involves *only* raw materials and products, and no intermediate streams. We just draw a box around the entire block flow diagram:

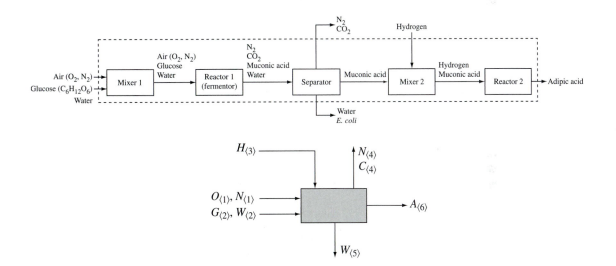

Everything inside the dashed box is considered inside the system. The only streams that enter or leave the system are those that cross the dashed lines.

We'll choose compounds as components, because there are chemical reactions of known stoichiometry. We'll use O, N, G, W, H, and A for oxygen, nitrogen, glucose, water, hydrogen, and adipic acid, respectively. We'll number the streams from 1 to 6, as shown on the redrawn diagram.

Steps 4 to 10.   *Convert units, define basis, define system variables, list specifications, write material balances, solve, and check.*

Since chemical reactions occur, we'll convert everything to moles (kgmol or kgmol/h). The adipic acid production rate will serve as our basis:

$$A_{\langle 6 \rangle} = \left(\frac{12,000 \text{ kg}}{h} \right) \left(\frac{1 \text{ kgmol}}{146 \text{ kg}} \right) = \frac{82.2 \text{ kgmol adipic acid}}{h}$$

We are given the mass concentration of glucose (mg glucose/mL solution) in the feed and need to convert this to a mole ratio (kgmol glucose/kgmol water). For this we need the density of glucose solution, which we find in the *CRC Handbook of Chemistry and Physics* to be 1.0038 g/mL at 20°C.

$$\frac{10 \text{ mg glucose}}{1 \text{ mL solution}} \times \frac{1 \text{ mL solution}}{1.0038 \text{ g solution}} \times \frac{1 \text{ g glucose}}{1000 \text{ mg glucose}}$$

$$= \frac{0.00996 \text{ g glucose}}{1 \text{ g solution}}$$

1 g solution therefore contains 0.00996 g glucose and 0.99004 g water. The molar masses of water and glucose are 18 and 180 kg/kgmol, respectively. Converting to mole fractions gives us one of the stream composition specifications:

$$\frac{G_{\langle 2 \rangle}}{W_{\langle 2 \rangle}} = \frac{0.00996 \text{ kg glucose}}{0.99004 \text{ kg water}} \times \frac{18 \text{ kg/kgmol water}}{180 \text{ kg/kgmol glucose}}$$

$$= 0.001006 \frac{\text{kgmol glucose}}{\text{kgmol water}}$$

There is one additional stream composition specification:

$$\frac{O_{\langle 1 \rangle}}{N_{\langle 1 \rangle}} = \frac{21 \text{ kgmol O}_2}{79 \text{ kgmol N}_2}$$

There are no additional system performance specifications. (We have already accounted for 100% conversion of reactants to products by not having any stream variables involving reactants in outlet streams 4, 5, or 6.)

Two chemical reactions take place inside the process, in two different reactors. But, since there is no muconic acid that enters or leaves this system, we care only about the *net* generation and consumption:

$$\tfrac{7}{3} C_6 H_{12} O_6 + \tfrac{17}{2} O_2 + 2 H_2 \rightarrow C_6 H_{10} O_4 + 8 CO_2 + 11 H_2 O$$

We now write system variables for generation and consumption. Let's write everything in terms of adipic acid generation (simply because it is the only component with a stoichiometric coefficient of 1). There are six reactants and products, so we write five equations:

$$\frac{G_{\text{cons}}}{A_{\text{gen}}} = \frac{7}{3}$$

$$\frac{O_{\text{cons}}}{A_{\text{gen}}} = \frac{17}{2}$$

$$\frac{H_{\text{cons}}}{A_{\text{gen}}} = 2$$

$$\frac{C_{\text{gen}}}{A_{\text{gen}}} = 8$$

$$\frac{W_{\text{gen}}}{A_{\text{gen}}} = 11$$

The process is at steady state, so no accumulation term is necessary.

There are seven components, so there are seven material balance equations. Simplified from input − output + generation − consumption = accumulation, these seven equations are

$$O_{\langle 1 \rangle} = O_{\text{cons}}$$

$$N_{\langle 1 \rangle} = N_{\langle 4 \rangle}$$

$$G_{\langle 2 \rangle} = G_{\text{cons}}$$

$$W_{\langle 2 \rangle} + W_{\text{gen}} = W_{\langle 5 \rangle}$$

$$H_{\langle 3 \rangle} = H_{\text{cons}}$$

$$C_{\text{gen}} = C_{\langle 4 \rangle}$$

$$A_{\text{gen}} = A_{\langle 6 \rangle}$$

Starting from the material balance equation on adipic acid and the basis equation, we solve for A_{gen}. We then proceed to solve for the remaining generation/consumption variables, then move through the remaining material balance equations and specifications. The solution is given in tabular form, with all values given in kgmol/h:

Component	Input	Consumption	Generation	Output
Oxygen	$O_{\langle 1 \rangle} = 699$	$O_{cons} = 699$		
Nitrogen	$N_{\langle 1 \rangle} = 2630$			$N_{\langle 4 \rangle} = 2630$
Glucose	$G_{\langle 2 \rangle} = 192$	$G_{cons} = 192$		
Water	$W_{\langle 2 \rangle} = 190{,}800$		$W_{gen} = 904$	$W_{\langle 5 \rangle} = 191{,}700$
Hydrogen	$H_{\langle 3 \rangle} = 164$	$H_{cons} = 164$		
Carbon dioxide			$C_{gen} = 658$	$C_{\langle 4 \rangle} = 658$
Adipic acid			$A_{gen} = 82.2$	$A_{\langle 6 \rangle} = 82.2$

This is a good start. Now we need to go back and look at each individual process unit as a separate system, so we can calculate intermediate flows (e.g., flow from Reactor 1 to the Separator). To decide how to proceed, we examine the information that we have. Notice that we now know the flow rates and compositions of the two input streams for Mixer 1. One strategy for proceeding is this: Choose Mixer 1 as the system, complete process flow calculations to find Mixer 1 output; then choose Reactor 1 as the system, complete process flow calculations to find Reactor 1 output; etc. Notice that we also know the flow rate out of Reactor 2. Therefore, another workable strategy is this: Choose Reactor 2 as the system, complete process flow calculations to find Reactor 2 input; then choose Mixer 2 as the system, complete process flow calculations to find Mixer 2 input; etc. Either strategy works fine!

There is a third strategy, more suitable for computer-aided solution than manual solution. In this strategy, we solve the entire process at once. We define stream variables for every stream on the block flow diagram, define system variables for every process unit, choose a single basis, list all specifications, and then solve the entire system of equations simultaneously. This set of equations will include those we derived previously as well as additional equations describing the intermediate streams (streams 7, 8, 9, and 10) and the intermediate component muconic acid ("M"). In our equations we distinguish between generation/consumption in Reactor 1 and Reactor 2. We'll illustrate this approach in an abbreviated form.

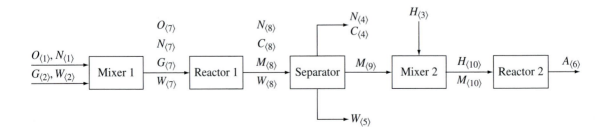

Basis:

$$A_{\langle 6 \rangle} = 82.2 \text{ kgmol}/\text{h} \tag{1}$$

Stream composition specifications:

$$\frac{O_{\langle 1 \rangle}}{N_{\langle 1 \rangle}} = \frac{21}{79} \tag{2}$$

$$\frac{G_{\langle 2 \rangle}}{W_{\langle 2 \rangle}} = 0.001006 \tag{3}$$

Reaction stoichiometry equations:
Reaction 1:

$$\frac{G_{\text{cons},1}}{M_{\text{gen},1}} = \frac{7}{3} \tag{4}$$

$$\frac{O_{\text{cons},1}}{M_{\text{gen},1}} = \frac{17}{2} \tag{5}$$

$$\frac{C_{\text{gen},1}}{M_{\text{gen},1}} = 8 \tag{6}$$

$$\frac{W_{\text{gen},1}}{M_{\text{gen},1}} = 11 \tag{7}$$

Reaction 2:

$$\frac{M_{\text{cons},2}}{A_{\text{gen},2}} = 1 \tag{8}$$

$$\frac{H_{\text{cons},2}}{A_{\text{gen},2}} = 2 \tag{9}$$

Material balance equations:
Mixer 1:

$$O_{\langle 1 \rangle} = O_{\langle 7 \rangle} \tag{10}$$

$$N_{\langle 1 \rangle} = N_{\langle 7 \rangle} \tag{11}$$

$$G_{\langle 2 \rangle} = G_{\langle 7 \rangle} \tag{12}$$

$$W_{\langle 2 \rangle} = W_{\langle 7 \rangle} \tag{13}$$

Reactor 1:

$$O_{\langle 7 \rangle} = O_{\text{cons},1} \tag{14}$$

$$N_{\langle 7 \rangle} = N_{\langle 8 \rangle} \tag{15}$$

$$G_{\langle 7 \rangle} = G_{cons,1} \tag{16}$$

$$W_{\langle 7 \rangle} + W_{gen,1} = W_{\langle 8 \rangle} \tag{17}$$

$$C_{gen,1} = C_{\langle 8 \rangle} \tag{18}$$

$$M_{gen,1} = M_{\langle 8 \rangle} \tag{19}$$

Separator:

$$N_{\langle 8 \rangle} = N_{\langle 4 \rangle} \tag{20}$$

$$W_{\langle 8 \rangle} = W_{\langle 5 \rangle} \tag{21}$$

$$C_{\langle 8 \rangle} = C_{\langle 4 \rangle} \tag{22}$$

$$M_{\langle 8 \rangle} = M_{\langle 9 \rangle} \tag{23}$$

Mixer 2:

$$M_{\langle 9 \rangle} = M_{\langle 10 \rangle} \tag{24}$$

$$H_{\langle 3 \rangle} = H_{\langle 10 \rangle} \tag{25}$$

Reactor 2:

$$M_{\langle 10 \rangle} = M_{cons,2} \tag{26}$$

$$H_{\langle 10 \rangle} = H_{cons,2} \tag{27}$$

$$A_{gen,2} = A_{\langle 6 \rangle} \tag{28}$$

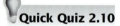

Quick Quiz 2.10

List the eight system variables, and their numerical values, for Example 2.13.

We've derived 28 equations. We have 20 stream variables, plus 8 system variables. (Can you list them?) We can employ an equation-solving program to solve the system of equations, *as long as they have been correctly set up.*

Here's the solution, in table form, showing the stream variables in kgmol/h. (The stream numbers are arranged with the flow of material through the process, not in numerical order.)

Stream no.	<1>	<2>	<7>	<8>	<4>	<5>	<9>	<3>	<10>	<6>
Oxygen, O	699			699						
Nitrogen, N	2,628			2,628	2,628	2,628				
Glucose, G			192	192						
Water, W			190,800	190,800	191,704		191,704			

Stream no.	<1>	<2>	<7>	<8>	<4>	<5>	<9>	<3>	<10>	<6>
Carbon dioxide, C				658	658					
Muconic acid, M				82.2			82.2		82.2	
Hydrogen, H								164.4	164.4	
Adipic acid, A										82.2
Total	3,327	191,000	194,300	195,100	3,286	191,700	82.2	164.4	246.6	82.2

There are several ways to check the solution. One good way is to convert all molar flow rates to mass flow rates, and check that the conservation of mass is satisfied for each process unit. This is left to the reader to do.

2.6.1 Synthesizing Block Flow Diagrams

In Example 2.13, you were handed a preliminary block flow diagram and asked to complete process flow calculations. But what if you need to first synthesize a block flow diagram, given just a reaction pathway and a basis? Here is one logical approach that will get you started:

1. Start with the known chemical reactions in the reaction pathway. For each reaction, draw a reactor process unit, with all reactants entering in one input stream and with all products and byproducts leaving in one output stream.
2. Add mixers before each reactor. The output from the mixer is the input to the reactor. The inputs to the mixer are all the reactants needed, in the form in which they are available.
3. If the raw materials are not pure and contain components that are not needed for the reaction, consider adding separators before the mixers to remove unnecessary components.
4. Add separators after the reactors. The input to the separator is the reactor output. The outputs from the separator include the desired product, any unreacted reactants, and byproducts.
5. Add splitters if the quantity of a stream is greater than that needed in the downstream process units.
6. If unreacted reactants leave the reactor and are separated from the product streams, add a mixer upstream of the reactor, to mix these recycled reactants with fresh reactor feed.

Given this strategy, there are still usually multiple ways to connect together all the process units. Synthesizing the best preliminary block flow diagram requires further analysis of costs, feasibility, and safety. A clever engineer uses **heuristics** to eliminate clearly undesirable or unworkable options, leaving fewer options that require more detailed analysis. Heuristics are simply rules-of-thumb—useful guidelines based on experience and logic. They are not laws, and are not *always* true. Here is a heuristic that many people find useful in predicting the weather:

Red sky at night, sailors delight.

Red sky in the morning, sailors take warning.

And here is a heuristic about time management:

A stitch in time saves nine.

Or perhaps you prefer

Better late than never.

Here are a couple of useful heuristics for synthesizing block flow diagrams:

- Mix raw materials together just before the reactor, and not earlier.
- Remove byproducts and waste products as soon after they are formed as possible.
- If possible, split rather than separate.
- If possible, mix together streams of similar composition.

Example 2.14 | **Synthesizing Block Flow Diagrams: Adipic Acid Process**

Sketch a block flow diagram for production of adipic acid from glucose, given the reaction pathway:

$$\tfrac{7}{3}C_6H_{12}O_6 + \tfrac{17}{2}O_2 \rightarrow C_6H_6O_4 + 8CO_2 + 11H_2O$$

$$C_6H_6O_4 + 2H_2 \rightarrow C_6H_{10}O_4$$

Air is the source of oxygen. Glucose is purchased as a dry product, but is dissolved in water on-site before being fed to the process.

Solution

Production of adipic acid from glucose involves two reaction steps, so our block flow diagram includes two reactor process units. In the first reaction, glucose and oxygen are fermented to muconic acid ($C_6H_6O_4$), in the presence of air. CO_2 is a byproduct. In the second reactor, muconic acid is hydrogenated to adipic acid.

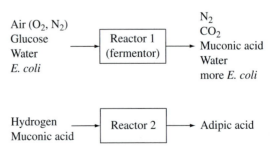

The input to Reactor 1 is a mixture of raw materials that are purchased separately. We'll add in a mixer: the output from Mixer 1 is the input to Reactor 1

The output from Reactor 1 contains materials that are not needed in the input to Reactor 2. We'll put a separator downstream of Reactor 1 to recover the muconic acid and remove everything else. We'll separate the gases out as one stream, the muconic acid as another, and the water as a third stream.

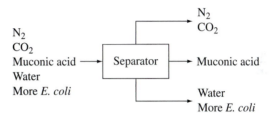

The input to Reactor 2 is a mixture of hydrogen and muconic acid. We've got the muconic acid from the separator, but we need a source of hydrogen and a mixer:

Now we can put the pieces together in a logical arrangement, matching outputs from one process unit to inputs to the next.

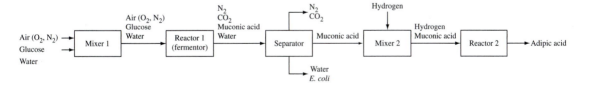

This is the logic that led to the development of the flow sheet that we worked with in Example 2.13.

The block flow diagram is simple, yet it is invaluable in facilitating the generation of alternative schemes. At this stage, we can (and should) consider questions regarding the need for each processing unit, and the appropriate pattern of connecting them together. For instance, in Example 2.14 we might ask the following kinds of questions:

- Nitrogen is not needed in Reactor 1 feed. Should we install a separator before Mixer 1 to separate oxygen from the nitrogen in the air?
- Is all the oxygen and glucose really consumed in Reactor 1? What if it's not?
- Should we remove the water before sending the muconic acid to Mixer 2, or is it better to remove it later?
- Water is a byproduct. Will the water need to be cleaned up before disposal? Water is also used as a solvent for the glucose. Could the water be recycled for use as a solvent for glucose?
- Should the *E. coli* in the Separator output be recycled to Mixer 1?
- Should we add the hydrogen to Mixer 1 and eliminate the need for Mixer 2?
- Is all the hydrogen and muconic acid consumed in Reactor 2? What if it's not?

Answering these questions will take some additional information and study. The point is this: the preliminary block flow diagram provides the essential framework for developing a list of key questions to be answered.

2.6.2 The Art of Approximating

In Example 2.14, several simplifying approximations were made in generating the preliminary block flow diagram. For example, we assumed that air contains only nitrogen and oxygen, that all the oxygen fed to Reactor 1 was completely consumed by the reaction, and that no muconic acid was lost to the water stream leaving the Separator. In reality, air contains argon and other gases besides oxygen and nitrogen, it may be difficult to ensure that all the oxygen is consumed in Reactor 1, and some of the muconic acid is likely to be discharged with the water stream. Still, at this stage of process development, making useful approximations is the wise thing to do. These approximations are not made arbitrarily, but are chosen for good reason.

Approximations that are frequently made early in chemical process synthesis fall into 3 classes:

1. *Stream composition approximations.* For example,

   The raw material is pure.
   The product is pure.
   Air contains only nitrogen and oxygen.
   Reactants are fed at stoichiometric ratio.

2. *System performance approximations:* For example,

   The reactants are completely consumed by reaction.
   No unwanted side reactions take place in the reactor.
   The separator separates all components into pure streams.

3. *Physical property approximations;* For example

Gases behave as ideal gases.
Liquids behave as ideal solutions.
Solid density is independent of temperature.

These approximations make calculations much simpler, and are usually very appropriate early in the design process. (If you look back through the worked examples in this chapter, you will find cases where each of these approximations was made.) Whenever you make an approximation, ask yourself two questions:

- *Does the approximation have a resemblance to reality?* It's reasonable to approximate the shape of a strand of uncooked spaghetti as a thin cylinder, a tortilla as a disk, and even a turkey as a sphere. But it would be unreasonable to approximate a spaghetti strand as a sphere. It's reasonable to use the ideal gas law to estimate the density of air at room temperature and pressure, but it's unreasonable to use the ideal gas law to estimate the density of gold at room temperature.
- *If I made a more realistic approximation, would I make a different decision?* In deciding whether to use benzene or glucose as a reactant, it would be reasonable to consider each as a pure raw material. But in choosing between two different manufacturers of glucose, the purity of the material must be considered carefully.

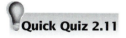

Quick Quiz 2.11

What simplifying approximations were made in Example 2.13?

With experience, you will develop a knack for knowing what approximations can be safely made, and when. You will also learn when you must reanalyze a problem, using more stringent conditions.

2.6.3 Degree of Freedom Analysis for Block Flow Diagrams with Multiple Process Units

After a complicated problem like Example 2.13, you may start to wonder: How do I know that I'll be able to complete the process flow calculations? How do I know that there's one, and only one, answer?

This is the same question we asked—and answered—in Sec. 2.5, when we introduced DOF analysis for single process units. DOF analysis is especially helpful when there are multiple process units, because of the increased complexity of the problem. We can determine not only if we have enough information, but also where information might be lacking, and we can gain insight into setting up a calculation strategy.

DOF analysis for multi-unit block flow diagrams proceeds in much the same manner as for single-unit diagrams. However, we need to take extra care in attributing specifications to particular systems. When completing a DOF analysis of a block flow diagram, keep the following two points in mind.

1. In determining DOF of an individual unit, count stream variables and stream specifications only for the streams that are directly entering or leaving that unit.
2. Count material balance equations for each unit.

Remember

> If DOF > 0, the problem is underspecified—there is not enough information.
>
> If DOF < 0 the problem is overspecified—there is inconsistent or redundant information.
>
> If DOF = 0, the problem is (probably) completely specified.

The procedure is illustrated in the following example.

Example 2.15   ## DOF Analysis: Adipic Acid Production

A preliminary block flow diagram for production of 12,000 kg/h of adipic acid from glucose, based on the technology described in Chap. 1's case study, was given in Fig. 2.13. Available raw materials are air (assumed to be 21 mol% O_2 and 79 mol% N_2), glucose solution (10 mg/mL in water), and hydrogen. Show that the block flow diagram is completely specified, neglecting *E. coli*.

Solution

There are eight components in this process, but not all components are in or out of all process units. We've redrawn and labeled the flow diagram, showing each component in each stream and listing the reactions taking place in each reactor. (Refer back to Example 2.13 for the component naming system.)

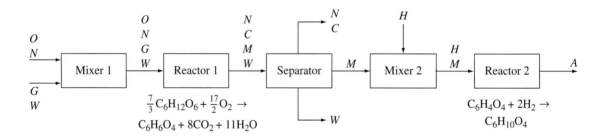

The DOF analysis for each process unit, and for the entire process, is summarized in the table. To generate the table, we analyzed each process unit as a separate system, considering *only* the inputs, outputs, and reactions for that unit. A few points are worth noting:

Mixer 1: Two stream compositions are specified: the oxygen content in air, and the glucose : water ratio in the glucose solution. The nitrogen content in the air is not an independent specification.

Reactor 1: One chemical reaction takes place in Reactor 1. Total (100%) conversion of oxygen and glucose do *not* count as system performance specifications, because we have already taken care of that information by putting no oxygen or glucose in the Reactor 1 output stream.

Reactor 2: One chemical reaction takes place in Reactor 2. The adipic acid flow rate is specified; since this is the output stream from reactor 2, it is counted for this process unit.

To complete the process DOF analysis, we counted every component in every stream, all stream composition and system performance specifications, all independent chemical reactions, and all material balance equations. There is one specified flow rate (12,000 kg/h adipic acid in Reactor 2 output) and two specified compositions (21% O_2 in air, glucose concentration in glucose solution into mixer 1). The number of independent material balance equations is calculated by simply adding up those for the individual process units. For stream variables, however, you can't simply add up across the row. This is because the output stream variable from one process unit becomes the input stream variable for another unit. If you simply added up the stream variables across the row to determine the total number of stream variables in the process, you would be double-counting.

	Mixer 1	Reactor 1	Separator	Mixer 2	Reactor 2	Process
No. of variables						
Stream variables	8	8	8	4	3	20
Chemical reactions	0	1	0	0	1	2
No. of equations						
Flows	0	0	0	0	1	1
Stream composition	2	0	0	0	0	2
System performance	0	0	0	0	0	0
Material balances	4	6	4	2	3	19
Sum	8 6	9 6	8 4	4 2	4 4	22 22
DOF	2	3	4	2	0	0

This DOF analysis indicates that the process is completely specified. Only one of the individual process units is completely specified—Reactor 2. An analysis of where the entire process is lumped together as a single system (in other words, as in input-output diagram) would also produce DOF = 0. (Try it!) The best place to start solving equations is therefore with either the input-output diagram (as we did in Example 2.13) or with Reactor 2.

CASE STUDY   Evolution of a Greener Process

Round-up® is a popular biodegradable non-selective herbicide; glyophosphate [*N*-(phosphonomethyl)glycine] is the active ingredient. When sprayed on foliage, glyophosphate is absorbed by the plant and blocks the action of a specific enzyme, which prevents the plant from making essential amino acids. Plants wither and die within a week of being sprayed.

A key intermediate in the synthesis of glyophosphate is DSIDA (disodium iminodiacetate), $C_4H_5NO_4Na_2$. In the conventional process, DSIDA is produced by reaction of formaldehyde, ammonia, and hydrogen cyanide. A new process is proposed that uses safer and more environmentally benign chemicals. In this process, diethanolamine (DEA) is synthesized from ethylene, oxygen, and ammonia in a two-step reaction pathway: ethylene (C_2H_4) is oxidized to ethylene oxide (C_2H_4O):

$$2C_2H_4 + O_2 \rightarrow 2C_2H_4O$$

and then ethylene oxide reacts with ammonia (NH_3) to DEA:

$$2C_2H_4O + NH_3 \rightarrow C_4H_{11}O_2N$$

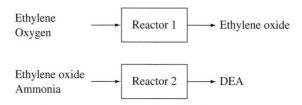

Diethanolamine, $C_4H_{11}O_2N$

Finally, DEA reacts with sodium hydroxide to make DSIDA. A very preliminary economic analysis looks favorable, so we'd like to pursue some ideas for designing this new process.

In this case study, we'll focus on the manufacture of DEA. (DSIDA manufacture was the subject of Problem 1.42.) Our goal is to synthesize a reasonable block flow diagram for production of 105,000 kg/h (1000 kgmol/h) DEA.

We'll start by considering the two reactions in the DEA synthesis pathway. The two reactions require different catalysts and are carried out in separate reactors. Therefore our block flow diagram must include two reactors.

Ethylene
Oxygen   ———→ | Reactor 1 | ———→ Ethylene oxide

Ethylene oxide
Ammonia   ———→ | Reactor 2 | ———→ DEA

Let's add in mixers and connect the mixers and reactors into a single preliminary block flow diagram.

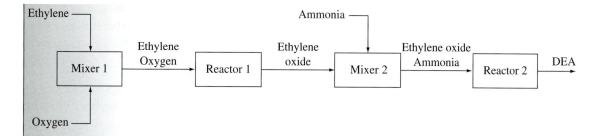

We chose to introduce NH_3 into Mixer 2, following one of our heuristics for synthesizing block flow diagrams (Sec. 2.6.1). We could feed the ammonia to Mixer 1 along with C_2H_4 and O_2, but (1) this increases the required volume (and hence cost) of Reactor 1 and (2) there could be other unwanted reactions, like oxidation of ammonia.

Notice an assumption we made implicitly—that the reaction goes to completion in both Reactor 1 and Reactor 2. After consulting with some chemist friends, we learn that it is best to carry out the first reaction such that all the oxygen but only 25% of the ethylene is converted to ethylene oxide. (This avoids unwanted side reactions, like complete oxidation of ethylene to CO_2. These issues will be discussed in more detail in Chap. 4.) This means that the outlet stream from Reactor 1 contains *both* ethylene and ethylene oxide. Furthermore, our chemist colleagues tell us that the second reaction must be carried out in the absolute absence of oxygen. Our preliminary block flow diagram is in need of some modification to account for these new concerns.

Our second heuristic suggests removing byproducts as soon as possible. The unreacted ethylene isn't a byproduct, but it isn't necessary further downstream. Therefore, according to this heuristic, we insert a Separator right after Reactor 1 to remove ethylene. If the Separator can also ensure that there is no trace of oxygen in the ethylene oxide stream, we're even better off. Now the preliminary block flow diagram evolves:

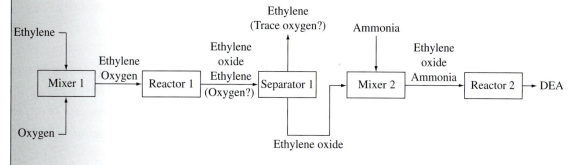

Are there alternative block flow diagrams that might be better? For example, could the process units be connected in a different arrangement? Perhaps the Separator could be placed after Reactor 2? However, this increases the volume of ethylene that must be processed in Reactor 2, thus increasing the size (and cost) of the reactor. Plus, we would still need some sort of Separator to remove any traces of oxygen from the feed to Mixer 2. We'll stick with the arrangement we have.

At this stage in the synthesis of the preliminary block flow diagram, we might think a bit deeper about raw materials. Specifically, what should we use as a source of O_2? Pure oxygen is expensive. Air is much cheaper, but it contains a lot of N_2—in fact, more N_2 than O_2. One idea is to feed air to a Separator placed upstream of Mixer 1, which would remove N_2 from air before feeding O_2 to the rest of the process. A second idea is to feed air to Mixer 1; all of the O_2 would be consumed in Reactor 1 and the N_2 would pass through. The N_2 could then either be separated along with the ethylene from the ethylene oxide in Separator 1, or it could be separated from DEA at the tail end of the process by adding a new Separator 2 after Reactor 2.

The first alternative has the extra expense of the O_2/N_2 separator, but reduces the volumetric flow through Reactor 1, hence reducing reactor costs. The second alternative requires no additional separator, but requires a large flow of N_2 through Reactor 1. The third alternative requires an additional separator, and requires a large flow of N_2 through both Reactors. This is the least attractive option. It is cheaper to separate N_2 from ethylene oxide than from O_2 (we'll learn why in Chap. 5), so the best option is the second alternative. The modified block flow diagram is:

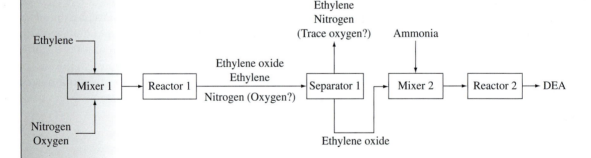

Let's complete process flow calculations. We'll choose compounds as components, number the streams, use as a basis the desired production rate of 105,000 kg DEA/h (1000 kgmol DEA/h), approximate air composition as 21 mol% O_2 and 79 mol% N_2, and specify 25% conversion of ethylene in Reactor 1. We'll leave the details to the reader and summarize the results in Fig. 2.14.

Now is a good time to review what simplifying approximations have been made in getting this far. We have:

- Assumed the air was only N_2 and O_2, and neglected argon and other gases that are present in air.
- Assumed the ethylene was pure, and neglected any contaminants that might be present.

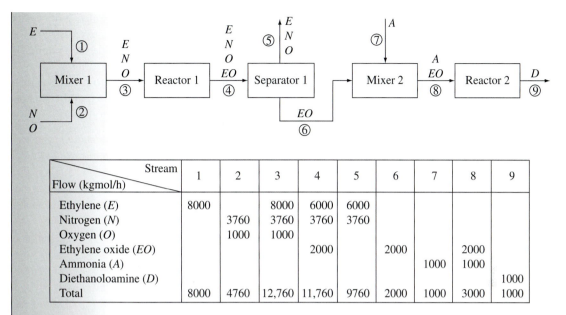

Figure 2.14 Preliminary block flow diagram for production of DEA. First alternative.

- Assumed Separator 2 perfectly separated all ethylene oxide from the other gases.
- Assumed 100% conversion of reactants to products in Reactor 2.
- Neglected any side reactions (e.g., oxidation of C_2H_4 to CO or CO_2 in Reactor 1).

Making these approximations has greatly simplified the calculations. It's a good idea to explicitly list all approximations. As the design progresses, more realistic approximations will be incorporated.

Now let's look again at Fig. 2.14. Perhaps you have noticed something odd about the proposed block flow diagram. We are feeding 8000 kgmol ethylene/h to Reactor 1, then discarding 75% of it! This is very wasteful. Can you come up with a solution? What if we recycle this stream back to Mixer 1?

The table in Figure 2.14:

Flow (kgmol/h) \ Stream	1	2	3	4	5	6	7	8	9
Ethylene (E)	8000		8000	6000	6000				
Nitrogen (N)		3760	3760	3760	3760				
Oxygen (O)		1000	1000						
Ethylene oxide (EO)				2000		2000		2000	
Ammonia (A)							1000	1000	
Diethanoloamine (D)									1000
Total	8000	4760	12,760	11,760	9760	2000	1000	3000	1000

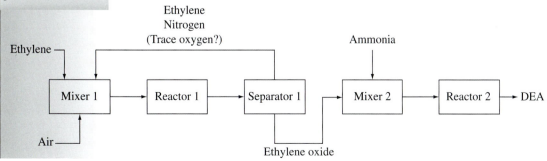

Do you see what the problem is with this block flow diagram? Notice that nitrogen enters the process with the air, but never leaves the process. Since nitrogen is not consumed by reaction, then it would accumulate inside the system. (If you don't see this, try working through the process flow calculations.)

Let's put in Separator 2, to separate unreacted ethylene from the N_2, then recycle the unreacted ethylene back to Reactor 1. We'll assume that the separator works perfectly. The block flow diagram has now evolved to this:

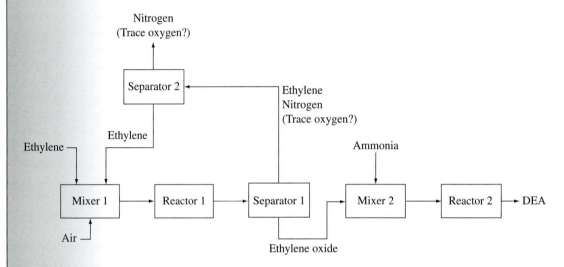

We'll want to calculate process flow calculations for our new design, but first let's check whether the problem is completely specified. Refer to Fig. 2.15 for the labeled flow diagram. The DOF analysis, summarized in Table 2.2, shows that the process is completely specified. This gives us courage to continue with process flow calculations. Further, the DOF analysis tells us how many of what kinds of equations we need (e.g., four material balance equations around Reactor 1). Finally, from the DOF analysis we see that the best way to solve the equations is to start with Reactor 2 balances, because that is the only individual process unit that is completely specified.

Now we proceed to apply the 10 Easy Steps as we calculate all the process flows. The equations, in abbreviated form, are:

Units: All in kgmol/h

Basis: $D_9 = 1000$

Reaction stoichiometry:

$$\frac{O_{cons,1}}{E_{cons,1}} = \frac{1}{2}$$

$$\frac{EO_{gen,1}}{E_{cons,1}} = 1$$

Table 2.2 Summary of DOF Analysis

	Mixer 1	Reactor 1	Separator 1	Separator 2	Mixer 2	Reactor 2	Process
No. of variables							
Stream variables	7	6	6	4	4	3	18
Chemical reactions	0	1	0	0	0	1	2
No. of equations							
Flows	0	0	0	0	0	1	1
Stream compositions	1	0	0	0	0	0	1
System performance	0	1	0	0	0	0	1
Material balances	3	4	3	2	2	3	17
Total	7 4	7 5	6 3	4 2	4 2	4 4	20 20
DOF	3	2	3	2	2	0	0

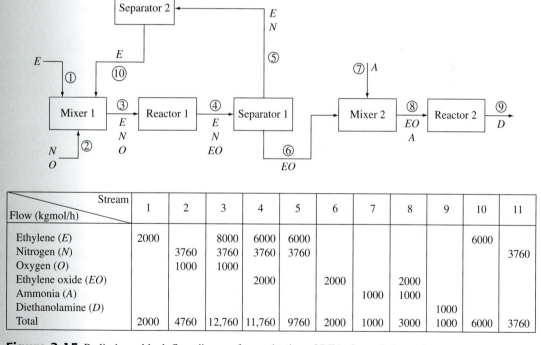

Flow (kgmol/h) \ Stream	1	2	3	4	5	6	7	8	9	10	11
Ethylene (E)	2000		8000	6000	6000					6000	
Nitrogen (N)		3760	3760	3760	3760						3760
Oxygen (O)		1000	1000								
Ethylene oxide (EO)				2000		2000		2000			
Ammonia (A)							1000	1000			
Diethanolamine (D)									1000		
Total	2000	4760	12,760	11,760	9760	2000	1000	3000	1000	6000	3760

Figure 2.15 Preliminary block flow diagram for production of DEA. Second alternative.

$$\frac{D_{\text{gen},2}}{EO_{\text{cons},2}} = \frac{1}{2}$$

$$\frac{A_{\text{cons},2}}{EO_{\text{cons},2}} = \frac{1}{2}$$

Stream (air) composition specification:

$$O_1 = \tfrac{21}{79} N_1$$

System (Reactor 1) performance specification:

$$E_{\text{cons},1} = 0.25 E_3$$

Material balances equations:

Mixer 1	Reactor 1	Sep 1	Sep 2	Mixer 2	Reactor 2
$E_1 + E_{10} = E_3$	$E_3 - E_{\text{cons},1} = E_4$	$E_4 = E_5$	$E_5 = E_{10}$		
$O_2 = O_3$	$O_3 - O_{\text{cons},1} = 0$				
$N_2 = N_3$	$N_3 = N_4$	$N_4 = N_5$	$N_5 = N_{11}$		
	$EO_{\text{gen},1} = EO_4$	$EO_4 = EO_6$		$EO_6 = EO_8$	$EO_8 - EO_{\text{cons},2} = 0$
				$A_7 = A_8$	$A_8 - A_{\text{cons},2} = 0$
					$D_{\text{gen},2} = D_9$

We've systematically come up with 24 equations describing the block flow diagram of Fig. 2.15, and we have 18 stream variables plus 6 system variables. (Count them! Why do we have 24 total variables when our DOF analysis lists only 20?) We start with the Reactor 2 balances and work our way systematically through the remaining equations (or we use equation-solving software). Results are summarized along with the flow diagram in Fig. 2.15.

Compare stream 1 in Fig. 2.15 with stream 1 in Fig. 2.14. With the new and improved process, we make much better use of our raw material! How? By recycling the unreacted ethylene from Reactor 1 back to Mixer 1, we are ensuring that *all* of the ethylene is eventually consumed by reaction, and none leaves the process.

Our block flow diagram is taking shape. At this point the diagram acts as a springboard for further questions. We might ask, for example, how easy is it to separate ethylene from nitrogen? We do a little investigating and find out that separating ethylene from nitrogen isn't cheap. Once again, we head back to the drawing board. We need to get the nitrogen out of the process, while still recycling the ethylene. Yet, we don't want to pay for separating the nitrogen from the oxygen (as discussed before) or for separating the nitrogen from the ethylene. Can we have our cake and eat it too?

Here's a compromise solution: Bleed off *part* of the nitrogen/ethylene stream, and recycle the rest. We do this by replacing Separator 2 with Splitter 1. Some ethylene leaves with the bleed stream, but some is recycled.

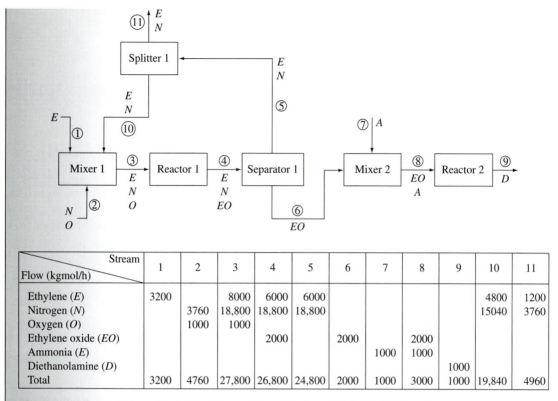

Figure 2.16 Preliminary block flow diagram for production of DEA. Third alternative.

Stream / Flow (kgmol/h)	1	2	3	4	5	6	7	8	9	10	11
Ethylene (E)	3200		8000	6000	6000					4800	1200
Nitrogen (N)		3760	18,800	18,800	18,800					15040	3760
Oxygen (O)		1000	1000								
Ethylene oxide (EO)				2000		2000		2000			
Ammonia (E)							1000	1000			
Diethanolamine (D)									1000		
Total	3200	4760	27,800	26,800	24,800	2000	1000	3000	1000	19,840	4960

To proceed with process flow calculations for this latest design, we need to first specify the fraction of the ethylene/nitrogen stream fed to the splitter that is recycled back to Mixer 1. (Through a DOF analysis, we could prove this!) Let's provide a system performance specification for Splitter 1: that 80% of the feed to Splitter 1 is recycled to Mixer 1. We could then proceed to complete process flow calculations, using a strategy similar to that illustrated earlier. We'll skip the details, and just summarize the results in Fig. 2.16.

Compare Fig. 2.16 to Fig. 2.15. By installing Splitter 1, we've saved the cost of building and operating an expensive Separator 2. (A splitter can be as simple as a three-way control valve.) But this has come at a price—we are throwing away 1200 kgmol ethylene/h, a raw material that we've paid for. And we've increased the flow through Reactor 1, which will increase its size and hence its cost. Which alternative is better? We don't know until we've completed a more detailed analysis of equipment and raw material costs. But we've made a good start at laying out the best alternatives.

Summary

- Chemical processes are represented schematically by three types of flowsheets:

  - **Input-output diagrams**
  - **Block flow diagrams**
  - **Process flow diagrams**

  The diagrams differ in the level of detail, the amount of information needed to generate them, and the cost to produce them.

- Block flow diagrams contain four basic **process units: Mixers, Reactors, Separators, and Splitters**. Each process unit represents operations in which material undergoes physical and/or chemical changes. The block flow diagram shows how **process streams** connect the process units by transferring material between units.

- Process operation is categorized according to how input and output streams are handled. In **batch** processes, input streams enter the process unit all at once, and output streams are removed from the process unit all at once at some later time. In **continuous-flow** processes, input streams continuously flow into the process unit and output streams flow continuously out of the unit. **Semibatch** processes are some combination of batch and continuous flow.

- **Steady-state** processes are time independent: Process variables do not change with time. In **transient,** or **unsteady-state,** processes, one or more process variables change with time. Batch and semibatch processes by their nature are unsteady state. Continuous-flow processes are operated under steady-state conditions, except during start-up and shutdown.

- A **basis** is a flow rate or quantity that indicates the size of a process. A **system** is a specified volume with well-defined boundaries. **Streams** are inputs to and outputs from the system. A **stream variable** describes the quantity or flow rate of a material in a stream, while a **system variable** describes the change in a quantity inside a system. **Stream composition specifications** provide information about the composition of a process stream. **System performance specifications** describe quantitatively the extent to which chemical and/or physical changes occur inside the system.

- The **material balance equation** is

  $$\text{Input} - \text{output} + \text{generation} - \text{consumption} = \text{accumulation}$$

  The material balance equation is written for a **component** and around a system. A component can be an element, a compound, or a composite material.

- The "10 Easy Steps" is a systematic approach that, if followed, will lead to successful completion of process flow calculations. The steps are:

(1) Draw a flow diagram
(2) Choose a system
(3) Choose components and define stream variables
(4) Convert units
(5) Define a basis
(6) Define system variables
(7) List stream composition and system performance specifications
(8) Write material balance equations
(9) Solve
(10) Check

- **Degree of freedom (DOF) analysis** is a systematic method for determining if a process flow calculation problem is completely specified. To complete a DOF analysis we:

(1) Draw and label a flow diagram.
(2) Count the number of variables by adding together the number of stream variables and the number of chemical reactions.
(3) Count the number of equations by adding together the number of flow, stream composition, and system performance specifications with the number of material balance equations.
(4) Calculate DOF as the number of variables minus number of equations. If DOF = 0, the problem is completely specified. If DOF > 0, the problem is underspecified. If DOF < 0, the problem is overspecified.

ChemiStory: Guano and the Guns of August

In 1804 the geographer Alexander von Humboldt introduced Europeans to a marvelous substance from the New World: Peruvian guano. For untold years, fish-eating sea birds had deposited their nitrogen-rich wastes on rocky islands off the South American coast. The dry climate preserved the guano deposits in layers 150 feet deep. Access to these bird droppings was so important that the United States passed the Guano Island Act in 1856. This law allowed U.S. citizens who discovered a rock or island covered with guano to take possession of the land and harvest the material. This was perhaps the first and only time in which sovereignty over land was determined by chemistry rather than by history or geography.

Enterprising Americans made fortunes selling Peruvian guano to Europe, as did the European importers. Still, there wasn't nearly enough of the nitrogen-rich fertilizer to satisfy the food needs of a rapidly growing population. Nitrogen in air is plentiful but cannot fertilize crops unless it is converted to liquid or solid substances, like ammonia or ammonium nitrate. Unfortunately, the triple bond in nitrogen is extremely stable. No one could figure out how to break the bond and then get N to combine with H to make liquid ammonia. No one, until Fritz Haber came along.

Chemical synthesis of nitrogen-containing fertilizers from N_2 dramatically increased crop yields. *Courtesy of U.S. Department of Agriculture.*

Fritz Haber was born on December 9, 1858, in Breslau, Germany. His mother died shortly after his birth; his father left him to be raised by an assortment of relatives. Rather aimless as an adolescent, he attended 6 universities in six years. Although he wanted to be a chemist, he found chemistry classes either too boring or too hard. He finally earned a Ph.D. from the University of Berlin in 1891, and studied chemical technology: at an alcohol distillery in Hungary, a Solvay soda factory in Austria, and a salt mine in Poland. He was particularly interested in the new field of physical chemistry, and applied to study under the great Wilhelm Ostwald, but was rejected. (Ostwald didn't seem to have much of an eye for young talent—he also rejected an application from Albert Einstein). He finally gained a position at the Karlsruhe Institute of Technology. Chemical engineering had not yet emerged as a separate discipline, but Haber thought and acted as an engineer, solving many practical problems in chemistry. He was a charming and energetic man who wrote nonsense poetry in his spare time.

In 1901, Fritz met and married Clara Immerwahr. She was the first woman to earn a doctorate (in chemistry) from the University of Breslau, and a Jew. (Fritz was born Jewish, but had converted to Christianity in 1892 because that was the only way he could get a university position.) Early in their marriage, Clara kept her hand in science by translating chemical literature and helping Fritz write his book *Thermodynamics of Technical Gas Reactions*. But children came along. Fritz was a thoughtless husband and father, often bringing home unannounced large groups of friends for dinner parties, and leaving for a 5-month trip to the United States shortly after the birth of their first son.

Haber had a difficult time gaining the recognition his technical contributions deserved. It could have been his Jewish background, or perhaps his penchant for moving headlong and recklessly (and successfully) into research areas already being studied by famous professors. (Ostwald told him "Achievements generated at greater than the customary rate raise instinctive opposition amongst one's colleagues.") Eventually, in 1906, Fritz snagged a promotion to an elite German professorship, and became increasingly interested in the nitrogen fixation problem. Germany at the turn of the century was ripe to solve the problem. Its chemists and chemical technicians were the best in the world, its chemical industry was large and diversified, its farms needed fertilizer, and continued access to natural fertilizers was uncertain. Ostwald and electrochemist Walther Nernst both worked, unsuccessfully, on the problem of nitrogen fixation. Haber had some advantages over these more established physical chemists: He had experience working in chemical plants and with mechanical equipment. Haber realized that higher pressures were needed to drive the reaction toward ammonia production. He and Robert Le Rossignol designed and built a high-pressure experimental chamber. They discovered that hydrogen and nitrogen would convert to ammonia only under then unheard of conditions: 200°C and 200 atm. At that time, 7 atm was considered high pressure! Carl Bosch, BASF's chief chemist, was intrigued. Three top executives from BASF marched into Haber's lab to see for themselves. As luck would have it, one of the seals on the high-pressure chamber broke, and the experiment was a disaster. But one of the executives stuck around long enough to see the seal fixed and was rewarded with the amazing sight of a tiny spoonful of liquid ammonia. BASF quickly signed a contract with Haber to commercialize the process.

Many problems remained to convert the lab experiment into an industrial-scale process. For example, BASF chemists tested 4000 different catalysts, finally discovering that iron was the best. The process, patented in 1908, was commercialized within 5 years. The first plant produced 30 metric tons per day. Haber became rich beyond belief. The invention of the Haber-Bosch process ushered in 20th century industrial chemical processing, introducing such concepts as metallic catalysts and high-pressure, high-temperature gas reactions.

Fritz Haber with Albert Einstein. *Courtesy of Archiv zur Geschichte der Max-Planck-Gesellschaft, Berlin-Dahlem.*

Germany was freed from its dependence on imported fertilizer. Haber became a national hero. He was appointed a director at the Kaiser Wilhelm Institute in Berlin, and socialized with the wealthy and powerful of Berlin, including scientists such as Einstein and Lise Meitner. Meanwhile, Fritz's long-suffering wife Clara moved in very different social circles—embracing the Reform Movement, wearing loose clothing, doing her own marketing, making friends with the servants, eating simple food. One visiting scientist mistook her for a cleaning woman.

In 1914, Germany invaded Belgium. The war quickly spread to engulf much of Europe. Germany's intelligentsia—including Haber and other scientists like Max Planck—saw the war as an "act of purification and a means of redemption." Haber directed his scientific talents toward making nitrogen-based munitions for the war effort. He convinced Carl Bosch and BASF to make nitric acid from his ammonia. Without the Haber-Bosch process, Germany would have run out of explosives within 6 months and the war would likely have ended quickly with German defeat.

Allied forces and Germany fought ferociously for 3 years, at a cost of millions of lives, without the frontline budging by more than a few miles. The war's stalemate led German leaders (as well as leaders in France, Britain, and the United States) to consider chemical and other unconventional weapons. The Hague Conventions, signed in 1899 and 1907, prohibited the use of unconventional weapons that would cause unnecessary suffering. But many (including the United States), did not consider chemical weapons any worse than shrapnel or explosives. The patriotic Haber agreed to develop chemical weapons for Germany. He personally supervised the burial of 6000 cylinders of liquid chlorine near the front in Belgium; The cylinders released 150 tons of chlorine and poisoned about 7000 French soldiers. This was the first systematic use of chemical weapons in warfare. By the end of the war, both sides had used chemical weapons extensively, although there was no evidence that their use provided any military advantages. Clara despised her husband's work on chemical warfare, and pleaded with him to stop, to no avail. A week after the first use of chlorine, the morning after hosting a dinner party, Clara shot herself through the heart. Her 13-year-old son Hermann found her, alive but near death. Fritz left the teenager alone the next morning, heading for the Eastern Front.

Germany's surrender in 1918 pushed Fritz into a deep depression. Haber's name reportedly appeared on a draft list of war criminals; he sent his second wife and their two children to Switzerland and he himself escaped in disguise. However, his name was not on the final list, and he escaped prosecution as a war criminal. His fortunes changed dramatically once again, when he was awarded the Nobel Prize for his development of the ammonia synthesis process. The award caused a storm of controversy, because of his work on chemical weapons. Almost all the non-German Nobel Prize winners boycotted the ceremony.

Despite prohibitions of chemical and biological weapons by the Versailles Peace Treaty and the 1925 Geneva Protocol, Haber continued to push for poison gas and chemical warfare research. He helped build poison gas plants in the Soviet Union and Spain, and remained a dedicated German nationalist. The rise of the Nazi party took him by surprise. As a leader of the Kaiser Wilhelm Institute, he was ordered in 1933 to simultaneously fire all Jews working there and to keep all important senior scientists. This was impossible, as many of the leading scientists of the day were Jewish. He was torn apart by the conflict, and did not want to continue doing poison gas work for the Nazis, nor did he want to discharge his scientists. His health deteriorated, his financial situation became shaky, his friends deserted him, and the chemical industry (except for Carl Bosch) dropped their support. Finally, he acted. In a letter that infuriated the Nazis, he claimed his right to remain in his post, but his unwillingness to use racial makeup as a deciding characteristic in employment. The strident German nationalist left Germany, never to return. He died of a heart attack, in a hotel in Switzerland, in 1934.

The Nazis, trying to discredit Haber, claimed that others had invented the ammonia synthesis process. Still, they were willing to use another offshoot of Haber's work. The pesticide Zyklon B, developed in post–WW I Germany under Haber's direction, was used to gas prisoners at concentration camps, including some of Haber's own relatives.

Quick Quiz Answers

2.1 L^3, M/L^3.

2.2 37 lb HCl, 63 lb H_2O, 1.014 lbmol HCl, 3.5 lbmol H_2O.

2.3 0.0446 gmol/L. Johnnie is no genius; H_2O is liquid at room temperature and pressure.

2.4 Yes, and yes.

2.5 Mouth: mixer. Stomach: reactor. Intestine: separator. Semibatch, unless you're a teenage boy, in which case the input is nearly continuous. Unsteady state.

2.6 System: savings account. Input: $5000. Output: $400. Generation: $350. Consumption: none. Accumulation: $4950.

2.7 Because the material balance equation applies to the quantity of material, not the composition!

2.8 No, because there is a change in the number of moles with reaction.

2.9 5 stream variables, 2 material balance equations.

2.10 $O_{cons,1} = 699$; $G_{cons,1} = 192$; $W_{gen,1} = 904$; $M_{gen,1} = 82.2$; $C_{gen,1} = 658$; $M_{cons,2} = 82.2$; $H_{cons,2} = 164.4$; $A_{gen,2} = 82.2$; all in kgmol/h.

2.11 Air is oxygen and nitrogen and nothing else; all oxygen and glucose are consumed in reactor 1; the separator works perfectly; all reactants are consumed in reactor 2.

2.12 Total mass flow in = 8000 × 28 + 3760 × 28 + 1000 × 32 + 1000 × 17 = 378,280 kg/h. Total mass flow out = 6000 × 28 + 3760 × 28 + 1000 × 105 = 378,200 kg/h. Looks good.

References and Recommended Readings

1. *CRC Handbook of Chemistry and Physics* (latest edition), is an invaluable desktop reference containing many pages of physical property data such as molar masses and formulas of organic and inorganic compounds, and densities of pure compounds as well as mixtures. It also contains extensive listings of unit conversion factors. The *CRC Handbook* is published by CRC Press, Boca Raton, Florida.

2. *Perry's Chemical Engineers' Handbook* (latest edition) is another invaluable desktop reference that is worth purchasing. Perry's has chapters covering unit conversion factors, a review of mathematics, and some physical and chemical data. The bulk of the handbook is concerned with chemical engineering principles and methods; the book includes many sketches of chemical process equipment. Perry's is published by McGraw-Hill, New York.

3. The *Kirk-Othmer Encyclopedia of Chemical Technology* is a multivolume compendium of information on chemicals and chemical processes. The coverage is truly encyclopedic, and includes data on process economics, market size, physical and chemical properties, and process technology. It is published by Wiley, New York. Two other books in the same vein are

Shreve's Chemical Process Industries (latest edition), McGraw-Hill, New York, and the *McGraw-Hill Encyclopedia of Science and Technology,* (latest edition), McGraw-Hill, New York.

4. The *Knovel Engineering and Scientific Online Database* is a comprehensive source of searchable information. *Perry's Handbook* and *CRC Handbook* are a few of the many authoritative references that are searchable from the Knovel database. Access requires a subscription, which is carried by many universities and companies.

5. Round-up®, the subject of our case study, is widely used by homeowners and farmers alike. It is generally considered safe, but this safety has been challenged by some. Observed toxicity of Round-up to some amphibians was traced to a surfactant used in formulation of the herbicide product, and not the active ingredient glyophosphate. Monsanto, the manufacturer of Round-up, also sells Round-up Ready® corn, soybeans, and other seeds; these plants are genetically modified to contain an enzyme that is resistant to glyophosphate action. Round-up Ready crops have engendered controversy over concerns about excessive use of herbicide, patent rights, ownership of crop materials, and impact on poorer agricultural economies. *Chemical and Engineering News* has carried several stories discussing the controversies surrounding Round-up.

Chapter 2 Problems

Warm-Ups

P2.1   (a) 1 g hydrogen (H_2) is mixed with 1 g benzene (C_6H_6) and 1 g cyclohexane (C_6H_{12}). The mixture is all gas. What is the mass fraction and mole fraction of hydrogen, benzene, and cyclohexane in the mixture?

   (b) 1 gmol hydrogen (H_2) is mixed with 1 gmol benzene (C_6H_6) and 1 gmol cyclohexane (C_6H_{12}). The mixture is all gas. What is the mole fraction and mass fraction of hydrogen, benzene, and cyclohexane in the mixture?

P2.2   You go into the lab and put a 100-mL volumetric flask on a balance. You tare the balance so that it reads 0 g. Then you measure out anhydrous fructose ($C_6H_{12}O_6$—the major sugar in fruit) into the flask until the balance reads 15.90 g. Then you fill the flask up to the 100-mL line with water. The balance now reads 105.97 g. Calculate the wt% fructose and mol% fructose of the solution. The molar mass of fructose is 180 g/gmol and that of water is 18 g/gmol.

P2.3   If air contains 79 mol% N_2 and 21 mol% O_2, what is the mass (in grams) of 1 gmol of air? What is the wt% N_2 and O_2 in air?

P2.4   The human body contains 63% H, 25.5% O, 9.45% C, 1.35% N, 0.31% Ca, and 0.22% P, plus several elements present in smaller quantities.

Are these mass or mole percents? About how many pounds and pound-moles of each element do you carry around?

P2.5 Epsom salt is just hydrated magnesium sulfate ($MgSO_4 \cdot 7H_2O$). If you heat 100 g of Epsom salts to a high temperature, all the water is removed. What will the weight of the anhydrous salt be?

P2.6 I bought a pack of plain white paper that contains 500 sheets of paper, each 8.5 × 11 in. On the package it says the paper is 75 g/m^2. What is the mass of each sheet, in grams? How much does the pack weigh in lb?

P2.7 Yeast for home bakers is sold in $\frac{1}{4}$ oz. packages. If one yeast cell weighs about 6×10^{-5} μg, about how many yeast cells are in a package?

P2.8 About how many grams H_2O are there in a cup of water? How many gram-moles?

P2.9 Assume air is an ideal gas. What is the density of air at 4°C and 1 atm, in units of g/cm^3? What is the approximate mass (in tons) of a cubic mile of air? (If someone says a cake is "lighter than air," is it really a compliment?) The density of water at 4°C is 1.00 g/cm^3 and that of gold is 19.31 g/cm^3. If a typical textbook were made out of air, water, or gold, about what would it weigh in kilograms?

P2.10 In Germany, auto fuel consumption is legislated to be 5.97 L per 100 km. In the United States, the government requires that the automobile fleet average 27.5 miles per gallon. If the average American drives 20 miles per day, what is the savings in gasoline usage (gallons per year per driver) that would be achieved if the United States switched to the German standard?

P2.11 In 1970, the year the Clean Air Act was passed by the U.S. Congress, the average hydrocarbon emission was about 14 grams per mile traveled. That number dropped to about 1 gram per mile by the year 2000. The total number of motor vehicle miles traveled increased in the same period from about 1100 billion miles in 1970 to about 2600 billion miles in 2000. Calculate the total hydrocarbon emissions in 1970 and in 2000, in tons. What is the percent increase or decrease over the 30-yr period?

P2.12 A chemical reactor operates at 10 atm pressure (absolute) and 250°C. Determine the pressure and temperature of this reactor in the following units: kPa and K, psig and °F, bar and °R.

P2.13 Oxygen at 100°C and 75 psia flows through a pipe at 115 lb/min. Calculate the molar flow rate (lbmol/min), and the volumetric flow rate (m^3/min) at both the actual temperature and pressure and at STP.

P2.14 Nitrogen gas costs 25 cents per 100 standard cubic feet. Liquid nitrogen costs 28 cents per liter. The specific gravity of liquid nitrogen is 0.808. Compare the costs of gas and liquid nitrogen on a dollars per kilogram basis. Why is one so much cheaper than the other?

P2.15 Titanium dioxide (TiO_2) is used extensively in paint, but a new use for this chemical is in the works: windows you don't need to wash. Glass coated with TiO_2 stays clean in two ways: (1) TiO_2 works as a catalyst in the presence of ultraviolet light to react organic dust and grime to CO_2

and water and (2) TiO_2 reduces surface tension of glass, causing rainwater to sheet on the surface and wash dirt away. To make this glass, a 50- to 60-nm-thick layer of TiO_2 is laid on the glass while it is still in its molten state. How many grams of TiO_2 would you need for a 6 × 4 ft window? The density of titanium dioxide is 3.84 g/cm^3.

P2.16 Draw a block flow diagram for making milk, starting with grass. Be sure to include all raw materials and byproducts on your diagram. Indicate mixers, splitters, reactors, and separators as needed. You do not need to include any quantitative information.

P2.17 Draw an input-output diagram for an automobile engine. Identify raw materials and products. Is this a batch, semibatch, or continuous-flow process? Is this a steady-state or transient process?

P2.18 Draw simple block flow diagrams of each of the four types of process units. List a common household item that functions as each type of unit.

P2.19 Simplify the material balance equation for each of the following situations by eliminating terms that are equal to zero.

(a) Water is pumped into a large tank. System: tank. Component: water.

(b) Water is pumped into a large tank that contains sugar crystals. The sugar dissolves, and a sugar solution is pumped out of the tank. System: tank. Component : sugar.

(c) Ethylene and air are pumped into a reactor operating at steady state, where 30% of the ethylene reacts with the oxygen to form ethylene oxide. System: reactor. Component: ethylene.

(d) Ethylene and air are pumped into a reactor operating at steady state, where 30% of the ethylene reacts with the oxygen to form ethylene oxide. System: reactor. Component: ethylene oxide.

(e) Ethylene and air are pumped into a reactor operating at steady state, where 30% of the ethylene reacts with the oxygen to form ethylene oxide. System: reactor. Component: nitrogen.

P2.20 Air (assumed to contain 79 mol% nitrogen and 21 mol% oxygen) is fed to a separator operating at steady state. Two products are made. One product contains 98 mol% oxygen; 80% of the oxygen in the air fed is recovered in this product. Draw a block flow diagram corresponding to this description, and correctly label all streams. Identify (a) stream composition specification(s), and (b) system performance specification(s). You do not need to do any calculations.

Drills and Skills

P2.21 Turn on a faucet in your bathroom or kitchen full blast. Then measure the water flow rate, using a bucket and watch. Report your measurement in the following units: gallons per minute, grams per second, pound-moles per hour, and tons per year. At this flow rate, would your faucet have a capacity similar to a typical commodity, specialty, or pharmaceutical plant? (Refer to Table 1.4.)

P2.22 Oxygen flows into a reactor at 115 lb/min. Plot the volumetric flow rate (ft³/min) at a range of temperatures from 0°C to 150°C at 1 atm pressure, and at a range of pressures from 1 psia to 100 psia at 25°C. Use the ideal gas law to model the specific volume of oxygen as a function of pressure and temperature.

P2.23 Ore from gold mines contains about 0.02 wt% gold. Seawater contains gold too, at about 5 parts per trillion (e.g., 5 g gold per 10^{12} g seawater). Calculate the (short) tons of gold ore and of seawater required to produce 1 oz. of gold. What would you estimate to be the total quantity of gold (tons) dissolved in seawater worldwide? Assume the density of seawater is 1.05 g/cm³, the earth's diameter is 25,000 miles, and the average depth of ocean water is 2.4 miles.

P2.24 Your company needs on-site storage for 60,000 lb ammonia. You decide to build a spherical vessel. What is the radius of the vessel needed if you store the ammonia (a) as gas at STP, (b) as gas, at 80°F and 5 atm, (c) as liquid at –30°F and 1 atm, or (d) as liquid, at 80°F and 11 atm? Considering costs, safety, and operability, what temperature and pressure would you pick, and why? You may calculate the gas density from the ideal gas law. The density of liquid ammonia is 42.6 lb/ft³ at –30°F and 37.5 lb/ft³ at 80°F.

P2.25 Methane (CH_4) and oxygen (O_2) are mixed and heated before being sent to a burner, as shown. What is the volumetric flow rate, the mass flow rate, and the mass fraction of methane in the stream leaving the preheater? You can assume the mixture obeys the ideal gas law.

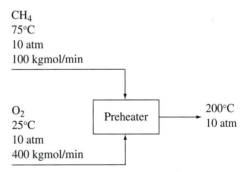

P2.26 Semiconductor devices are fabricated from single-crystal silicon rods. To make these rods, a single-crystal silicon "seed" is dipped into a container of molten Si. The silicon seed is slowly rotated and pulled up from the melt as the silicon freezes onto the seed. In one setup, a rod 100 mm in diameter is grown for 50 h to a final length of 60 cm (constant diameter). The density of solid silicon is 2.329 g/cm². What is the mass rate of accumulation (g/h) of the silicon onto the rod?

P2.27 Benzene is an important chemical intermediate. It can be manufactured from toluene ($CH_3C_6H_5$) by a process known as hydrodealkylation. In this process, toluene is mixed with hydrogen along with recycled gases from a high-pressure gas-liquid phase separator. The stream is heated to 600°C in a furnace. The hot stream is fed to a reactor, where 75% of the toluene is converted to benzene. The reactor effluent, at 654°C and 24 bar, is cooled to 38°C and sent to a separator where it is separated into vapor and liquid streams. The liquid stream, containing primarily benzene and toluene but also 0.014 mol% hydrogen and 0.062 mol% methane, is further separated into components by distillation. 99.6 mol% benzene from the distillation column is further cooled and sent to off-site storage. Toluene from the distillation column is recycled back to the reactor feed. About 80% of the gas stream from the high-pressure phase separator is recompressed; most is recycled back to be mixed with fresh toluene and hydrogen feed, but about 5% is sent directly to the reactor. The remaining gas from the high-pressure separator is sent off site as fuel gas. The process produces 8.21 tons/h benzene. (Adapted from R. Turton et al., *Analysis, Synthesis and Design of Chemical Processes,* Prentice-Hall, 1998.)

From this description, *sketch a block flow diagram* and identify:
(a) A basis
(b) Three stream composition specifications
(c) Three system performance specifications

P2.28 Vinyl chloride (C_2H_3Cl) is used to make polyvinylchloride (PVC) for piping and other products. In the direct chlorination process, ethylene (C_2H_4) and chlorine (Cl_2) are used as raw materials. Liquid chlorine as received at the plant is contaminated with a viscous liquid, so it is first sent to an evaporator, where the contaminant is separated from the chlorine. Chlorine vapor and ethylene vapor are then mixed and fed to a reactor, where they are completely reacted to dichlorethylene ($C_2H_2Cl_2$). The dichloroethylene is then fed to another reactor, where some of it reacts to HCl and C_2H_3Cl. The reaction mix is fed to a separator, where the HCl is separated from the two chlorinated ethylene products. The latter stream is fed to another separator, where unreacted dichloroethylene is removed and recycled back to join the feedstream to the second reactor. The vinyl chloride product is sent to storage. From this description, draw a block flow diagram. Identify process units such as mixers, reactors, or separators. List the compounds present in each process stream.

P2.29 In the Siemens process for making high-purity electronic-grade silicon, metallurgical-grade silicon powder, Si, is first converted to tetrachlorosilane and trichlorosilane, which are then purified. The trichlorosilane is reduced with hydrogen back to Si. The reduction reaction occurs on a solid silicon rod; thus, the silicon rod grows inside the silicon reaction chamber. Below is a simplified block flow diagram of a representative Siemens process. Write a paragraph describing the features of the process. In your description, specifically identify the purpose of each process unit, and carefully describe how material flows between units.

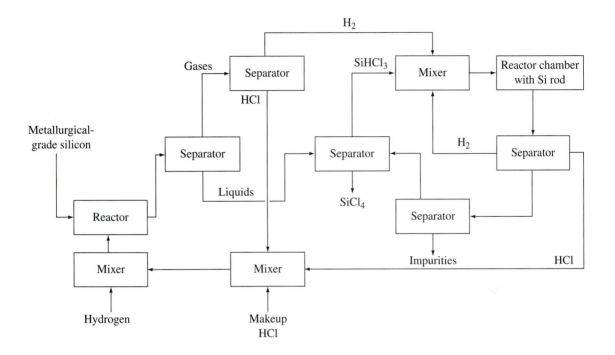

P2.30 Here is a description of a turkey-parts-to-oil plant from the June 2003 issue of *Technology Review*. "Unused turkey parts and water are dumped into a grinder and pulverized into a slurry that has the consistency of peanut butter. The slurry is then heated to 260°C and subjected to 275 kilograms of pressure. In the first of two reactor stages, the heated and pressurized mixture is depolymerized—that is, cooked for between 15 minutes and an hour to break apart the molecular structure of the organic material. The mixture is sent to a flash tank, where the pressure is released. The resulting steam is recaptured to power the system. Minerals sink to the bottom and flow into a separate tank. Organic materials move to the second reactor. Temperatures of almost 500°C further break down the organic materials. An auger moves carbon particles into a drum. The hot fluid moves into distillation tanks, where it cools and condenses. Organic materials and water separate. The water sinks to the bottom. A fuel gas is taken off the top, leaving a crude oil similar to a mix of diesel fuel and gasoline. The crude oil moves into storage tanks for later sale." From this description, draw a block flow diagram. Identify process units such as mixers, reactors, or separators. Label the streams to indicate the compounds present in each process stream and, where known, the temperature and pressure.

P2.31 Milk is pasteurized by heating it rapidly in a pasteurizer to temperatures high enough that most of the bacteria in the milk are killed. Is the pasteurizer a Mixer, Reactor, Splitter, or Separator? Write material balance equations for milk, live bacteria, and dead bacteria. Simplify each equation by crossing out terms that you assume are zero. Explain your reasoning.

P2.32 "Hard" water contains calcium, magnesium, and other mineral salts that tend to deposit on piping, coffeepots, and bathtubs. Hard water is "softened" in water softeners. In a water softener, hard water flows over ion exchange beads in a vessel. The beads carry Na^+ ions. Ca^{++} and Mg^{++} from the water preferentially stick to the beads, displacing Na^+, which then dissolves in the water, The "soft" water flows out of the softener. Is the water softener a Mixer, Reactor, Splitter or Separator? Write material balance equations for H_2O, Ca^{++}, Mg^{++}, and Na^+, using the water softener as the system. Simplify each equation by crossing out terms that are zero. Explain your reasoning.

For Problems P2.33 to P2.50, complete a DOF analysis and *explicitly* use the 10 Easy Steps to solve the problem.

P2.33 Your chemical engineering class is taught in Room 1234. At 9:00 A.M., six students are in the classroom. Between 9:00 and 10:00, 37 students enter the classroom, and one student leaves. How many students are in the classroom at 10:00?

P2.34 Sixty percent of the students (new students plus students taking the course again) that register for Chemical Engineering 101 pass. Of the students who do not pass, 70% register again the following semester. Draw a block flow diagram illustrating the process. If 100 new students register for Chemical Engineering 101 each semester, what's the total number of students taking the class?.

P2.35 To make cherry jam, cherries (18 wt% solids, 82 wt% water) are mixed with sugar at a 1:2 (lb/lb) cherry:sugar ratio. Then the mixture is fed to an evaporator, where two-thirds of the water is boiled off.
(a) What is the jam production rate if 10 lb/h cherries are fed to the process?
(b) What feed rate of cherries (lb/h) is required to produce 10 lb jam/h?
(c) Suppose we now make the same jam by a semibatch process: 10 lb cherries and 20 lb sugar are mixed in a pot all at once. Then, water is boiled off at 0.20 lb/min. How many pounds are in the pot after 30 minutes?

P2.36 You're a witch in need of a new magic potion. You've got three flasks containing the ingredients listed below. You'd like to mix these together in your cauldron, heat the cauldron over a fire to evaporate off excess water, and make 100 g of a liquid potion containing 27 wt% toe of frog, 22 wt% eye of newt, and 11 wt% wool of bat. How many grams from each flask should you add to your cauldron? How many grams of water should you evaporate off?

	Flask A, wt%	Flask B, wt%	Flask C, wt%
Toe of frog	10	0	50
Eye of newt	0	30	0
Wool of bat	40	0	10
Water	50	70	40

P2.37 Air (assumed to contain 79 mol% nitrogen and 21 mol% oxygen) is separated into two product streams. The separator operates at steady state. One product stream is 98 mol% oxygen, and it contains 80% of the oxygen in the air fed to the column. The other product stream is mostly nitrogen. Calculate the quantity of air required (tons/day) to produce 1 ton/day of the oxygen product. Calculate the mol% nitrogen in the second product stream.

P2.38 High-fructose corn syrup is a popular and inexpensive sweetener used in many carbonated beverages. When manufactured, the syrup contains a small amount of colored impurities; before it is mixed into beverages the impurities are removed by pumping the syrup over a bed of activated charcoal. The impurities adsorb to the charcoal while the rest of the syrup passes through. The charcoal adsorbs 0.4 kg impurities per kg charcoal. You are in charge of designing a charcoal adsorption facility that can process 115 lb of high-fructose corn syrup per hour. The syrup typically contains 0.5 wt% impurities. You'd like the charcoal to last for 7 days before needing replacement. How much charcoal (lb) should be used?

P2.39 Nitrogen and hydrogen are fed at a 1:3 molar ratio to an ammonia synthesis reactor operating at 1440°R and 100 atm, where the following reaction takes place:

$$N_2 + 3H_2 \rightarrow 2NH_3$$

25% of the N_2 fed is converted to ammonia, and the reactor produces 1000 lbmol/h NH_3. Calculate the volumetric gas feed rate to the reactor (ft^3/h), at the reactor temperature and pressure.

P2.40 Silicon tetrachloride ($SiCl_4$, also called tetrachlorosilane) reacts with magnesium metal (Mg) to make pure solid silicon Si and magnesium chloride ($MgCl_2$). A researcher places a mixture of 255 g $SiCl_4$ with 48 g Mg into a small empty laboratory-scale reactor. The next day, the researcher removes the entire contents of the reactor and finds only Si, $MgCl_2$, and $SiCl_4$. How many grams of each should he find?

P2.41 A mixture containing 84.2 wt% $SiCl_4$ and 15.8 wt% Zn is fed continuously at a rate of 303 g/h into a small laboratory-scale reactor. The reactor operates at steady state. Inside the reactor, Si is produced by the following reaction:

$$SiCl_4 + 2Zn \rightarrow Si + 2ZnCl_2$$

The reactor exit stream contains Si, $ZnCl_2$, and $SiCl_4$. Calculate the flow rate and composition of the exit stream. (This reaction formed the basis for the earliest commercial production of electronics-grade silicon, but has been replaced by the Siemens process.)

P2.42 Silica aerogels are some of the lightest synthetic materials known. To make them, liquid silicon is mixed at a 1:1 (wt:wt) ratio with a fast-evaporating solvent to form a gel. Then the gel is dried under pressure, which removes 95% of the solvent. The solvent is recovered for reuse. After the gel thickens, the pressure is released and air is incorporated into the mixture. During this step the remainder of the solvent evaporates into the atmosphere. This process produces a porous sponge of silicon and air. One cubic centimeter of the lightest silica aerogel weighs only 3.0 mg. The density of solid silicon is 2.33 g/cm^3. Calculate the quantity of solvent (g) and the volume of air (cm^3) used to produce 10 cm^3 of this silica aerogel. What is the mass% and vol% of air in the aerogel?

P2.43 Mixtures of hydrocarbon gases and air will burn only if the gas:air ratio is within the flammability limit. For example, propane gas-air mixtures will not ignite, even if exposed to a flame, if the propane composition of the mixture is greater than 11.4 mol%. Calculate the pounds of propane (C_3H_8) to mix with 100 L of air (at 1 atm pressure and 32°F) to just exceed the flammability limit and avoid ignition of the propane-air mixture. Air can be assumed to be 79 mol% N_2 and 21 mol% O_2. The density of air or propane gas can be calculated from the ideal gas law.

P2.44 Thermostability of vaccines is of serious concern, especially in less-industrialized countries, where access to refrigeration may not be reliable. Research indicates that replacing H_2O with deuterium oxide (D_2O) in the vaccine formulation increases the thermostability of some viruses. However, some concerns have been raised regarding the safety of injecting D_2O into infants. Deuterium naturally constitutes about 0.014% of all hydrogen atoms. Suppose a newborn weighs 3 kg, human elemental analysis is as given in Problem 2.4, and a vaccine dose is 0.1 mL containing 87 wt% D_2O. After a single injection of the vaccine, what is the deuterium loading (percent deuterium of all hydrogen atoms) in the infant? Do you think this increase is likely to constitute a significant safety issue?

P2.45 In baking bread, yeast is added to flour and water. An elemental analysis of yeast shows that the yeast contains 50 wt% C, 6.94 wt% H, 9.72 wt% N, and 33.33 wt% O. The yeast feed on glucose and protein supplied by the flour and produce CO_2, water, and more yeast. The reaction consumes glucose ($C_6H_{12}O_6$) and ammonia from protein (NH_3). Two grams of CO_2 are produced per gram of yeast produced. If you start with 120 in^3 of bread dough and put it at room temperature to rise, how many grams of CO_2 must be produced for the dough to double in size? How many grams of glucose will be consumed? You may assume that all the CO_2 produced is trapped as an ideal gas by the dough.

P2.46 Soybean meal contains about 44% protein. Through a series of processing steps a spun soy protein filament is produced that can be mixed and shaped into soy "bacon," "burgers," and similar meat substitutes. (The processing steps are very similar to those used to produce synthetic fibers for clothing and carpeting.) About 70% of the soybean protein fed to this process is recovered as edible protein product. Alternatively, the soybean meal can be fed to cows. About 7% of the soy protein fed to a cow is converted to beef burgers, steaks, and similar. Estimate the cost of soy protein and of beef protein, if the soybean meal sells for $0.40/lb.

P2.47 Manufacturers of lawn and garden fertilizers use a standardized method to indicate the quantity of active ingredients—nitrogen, phosphorus, and potassium. Gro-Right fertilizer is labeled 5-10-5, which means that it contains 5 wt% nitrogen as N, 10 wt% phosphorus as P_2O_5, and 5 wt% potassium as K_2O. The fertilizer is prepared by mixing ammonium nitrate (NH_4NO_3), calcium phosphate [$Ca(H_2PO_4)_2$], potassium chloride (KCl), and filler. In a 100-lb bag, calculate the mass (lb) of ammonium nitrate, calcium phosphate, and potassium chloride. What fraction of the bag's contents is filler?

P2.48 As part of the process of producing sugar crystals from sugar cane, raw sugar cane juice is sent to a series of three evaporators to remove water. The sugar cane juice, which is 85 wt% water, is fed to the first evaporator at 10,000 lb/h. Equal amounts of water are removed in each evaporator. The concentrated juice out of the last evaporator is 40 wt% water. Calculate:
(a) The flow rate of the concentrated juice out of the last evaporator
(b) The amount of water removed in each evaporator
(c) The concentration of water in the juice fed to the second evaporator

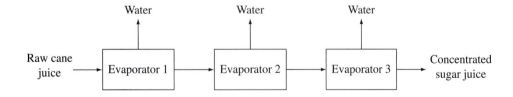

P2.49 Sorbitol is an ingredient in "sugar-free" candy. It is sweet, but does not promote tooth decay because bacteria cannot metabolize it for food, and it is considered diet food because humans don't metabolize it well either. Sorbitol ($C_6H_{14}O_6$) is made from glucose $C_6H_{12}O_6$ (which does cause tooth decay) and hydrogen. 100 kg/day of a 30 wt% glucose solution is mixed with a stoichiometric flow rate of hydrogen and sent to a reactor; 80% of the glucose is converted to sorbitol. The hydrogen is then separated from the sugar solution as a gas stream. How much hydrogen (kg/day) is fed to the process? What is the composition (wt%) and flow rate (kg/day) of the liquid stream leaving the process?

P2.50 Cheese whey contains a number of proteins that may have specific uses when purified. For example, glycomacropeptide (GMP) contains no phenylalanine, and is therefore a protein source that could be safely consumed by people with the disease phenylketonuria (PKU). GMP must be separated from other cheese whey proteins, in particular beta-lactoglobulin (BLG), before it can be consumed by people with PKU.

In a process under development, whey containing 1.2 g GMP/L and 0.8 g BLG/L is fed to a separator at a flow rate of 150 mL/min. The separator contains an ion exchange resin, to which some of the protein adsorbs. 89% of the GMP and 24% of the BLG adsorbs to the resin while the remainder of the whey passes through. After 30 minutes, the whey feed is discontinued, and a buffer containing 0.25M NaCl is pumped through the separator at 150 mL/min for 10 min. During this time all the protein on the resin is desorbed ("unstuck") and comes off with the buffer. The protein-containing buffer is collected and all the water is evaporated off, producing a dried product. What is the final purity (wt% GMP) and quantity (grams) of dry product?

Scrimmage

P2.51 Choose one of the high-volume commodity chemicals listed below. Find a description of the chemical in one of the reference books listed at the end of this chapter. Sketch out a block flow diagram showing the major process units. Write a brief (1- to 2-page) report describing the process for making the chemical and the major uses of the chemical. Your description of the process should include: the raw materials, chemical reactions, major process units, any byproducts or wastes, size of a typical facility, size of the worldwide market, and the value of the chemical. Be sure to properly reference your sources.

Ethylene	Vinyl chloride	Sodium silicate	Bis-phenol A
Lime	Methanol	Acetone	Vinyl acetate
Ammonia	Cumene	Styrene	Phosphoric acid
Sodium hydroxide	Aluminum sulfate	Ethylene glycol	Terephthalic acid
Propylene	Ethylene oxide	Glucose	Isopropanol
Urea	Acetic acid	Caprolactam	Xylene

P2.52 Describe a commercial process for making *one* (your choice) of the following industrial products: sulfuric acid, Teflon, synthetic detergent, lactic acid, porcelain, acrylonitrile, polyethylene. Trace the history of the development of the technology for producing your product. Your report should be typed double-spaced and should include a process flow sheet. An appropriate length would be 3 to 6 pages (including flow sheets). Points to consider include: Are there alternative choices of raw material? If so, what are the advantages and disadvantages of each alternative? Are there competing process technologies? If so, what are the advantages and disadvantages of each alternative? What is the product used for? What fundamental scientific advances were needed before the process/product could be developed? What people contributed significantly to the development of the product/process technology? How was the development of the process influenced by historical or political forces (e.g., wars)? Be sure to correctly reference all sources of information.

P2.53 Curious Carl is very curious: He wants to know if nitrogen really follows the ideal gas law. To test this, Curious Carl puts 1.000 gmol N_2 into a rigid container with a volume of 22.414 L. Then Curious carefully changes the temperature of the container and measures the pressure. After repeating this at several different temperatures, he obtains these data.

T (K)	100	200	300	400	500	600	700	800	900	1000
P (atm)	0.363	0.732	1.099	1.466	1.833	2.201	2.568	2.935	3.302	3.669

Then Curious puts 1.000 gmol N_2 into a rigid container with a volume of 2.24 L and repeats the experiment. Here are the data:

T (K)	100	200	300	400	500	600	700	800	900	1000
P (atm)	3.375	7.234	11.00	14.75	18.49	22.24	25.98	29.73	33.47	37.21

Curious is not sure what to do next, but he thinks plotting the data as *P* vs. *T* is a good idea. Can you help? Is nitrogen really an ideal gas?

Curious repeats the two experiments, but with 1.000 gmol CO_2 instead of N_2. Here are the data:

In the 22.414 L container:

T (K)	100	200	300	400	500	600	700	800	900	1000
P (atm)	<0.01	0.721	1.092	1.461	1.828	2.195	2.562	2.929	3.296	3.663

In the 2.24 L container:

T (K)	100	200	300	400	500	600	700	800	900	1000
P (atm)	<0.01	1.12	10.49	14.63	16.27	21.94	25.7	29.45	33.2	36.95

Plot the data as P vs. T. Is carbon dioxide an ideal gas? What's happened at 100 K?

Dubious Dora thinks the ideal gas law is all wet. She repeats the experiments, but with 1.000 gmol H_2O instead of CO_2. Here are the data:

In the 22.414 L container:

T (K)	100	200	300	400	500	600	700	800	900	1000
P (atm)	<0.01	<0.01	0.035	1.461	1.847	2.231	2.597	3.00	3.34	3.67

In the 2.24 L container:

T (K)	100	200	300	400	500	600	700	800	900	1000
P (atm)	<0.01	0.01	0.035	2.437	17.45	21.6	25.0	28.67	32.25	35.8

Plot the data as P vs T. Is Dora right to be dubious? Is H_2O an ideal gas?

P2.54 Soaps are the sodium or potassium salts of various fatty acids (such as stearic acid), derived from natural products such as animal fat. In a typical process, glycerol stearate [$(C_{17}H_{35}COO)_3C_3H_5$] is contacted with excess hot water to produce stearic acid ($C_{17}H_{35}COOH$) and glycerine ($C_3H_5(OH)_3$):

$$(C_{17}H_{35}COO)_3C_3H_5 + 3\ H_2O\ \rightarrow\ 3\ C_{17}H_{35}COOH + C_3H_5(OH)_3$$

The stearic acid is separated from the glycerine-water stream. Water in the glycerine stream is evaporated off to produce a crude glycerine product that is further processed elsewhere. The stearic acid is then sent to a separator, where trace amounts of unreacted glycerol stearate and high-molecular-weight reaction byproducts are removed. The stearic acid is then neutralized with caustic soda (NaOH) to produce sodium stearate soap ($C_{17}H_{35}COONa$):

$$C_{17}H_{35}COOH + NaOH\ \rightarrow\ C_{17}H_{35}COONa + H_2O$$

The soap-water mixture is then heated and water is removed by evaporation. The soap is then mixed with air, frozen, and cut into aerated soap bars.

Draw a block flow diagram showing the major steps in production of soap from glycerol stearate. In Prob. 1.31 of Chap. 1, you determined a lower bound on the sales price of sodium stearate soap. On the basis of the information above and your process flow sheet, can you think of reasons why the price must be greater than this lower bound?

P2.55 Sulfuric acid (H_2SO_4) is made by burning sulfur, (S), in air (79 mole% N_2 and 21 mol% O_2) to make SO_2, then further oxidizing SO_2 to SO_3 over a catalyst, using air as the source of O_2, then finally cooling and absorbing the gases into water to produce H_2SO_4.

We want to design a plant that produces 200 tons/day of concentrated H_2SO_4 (98 wt% H_2SO_4 and 2 wt% water), using water, sulfur, and air as raw materials. Use a generation-consumption analysis to develop a reaction pathway such that there is no net generation or consumption of SO_2 or SO_3. Develop a block flow diagram. Calculate the raw materials required as: tons sulfur/day, standard cubic feet of air/h, and tons water /day.

P2.56 You are interested in developing a process for making high-purity silicon from cheap ingredients—sand (SiO_2) and coke. Here are the main reactions:

$$SiO_2(s) + C(s) \rightarrow Si(s) + CO(g)$$

$$Si(s) + Cl_2(g) \rightarrow SiCl_4(g)$$

$$SiCl_4(g) + Mg(s) \rightarrow MgCl_2(s) + Si(s)$$

The reactions as written are not balanced. Figure out the appropriate stoichiometric coefficients. Then, complete a generation-consumption analysis to determine the mass (grams) of sand and other reactants required to make one gram of silicon. Calculate the quantities of byproducts. Which reactant do you think is the most expensive? Why is the combination of three reactions used instead of just reaction 1? Sketch out a block flow diagram. (*Note:* Magnesium chloride is water soluble but silicon powder is not.)

P2.57 Gold is recovered from rock using sodium cyanide (NaCN). Rock containing solid gold (Au) is mixed with an aqueous solution of NaCN (0.05 wt%, pH 11) to make a 50% solids slurry. Air is bubbled through the solution; the gold reacts with the sodium cyanide, oxygen, and water to make $NaAu(CN)_2$, which is soluble, and NaOH. The solution (containing $NaAu(CN)_2$) is separated from the solids and pumped across a zinc powder. Zinc reacts with $NaAu(CN)_2$, displacing Au and causing the gold to precipitate as a solid. The gold precipitate is then recovered as nuggets for sale or further processing. Sketch a block flow diagram for recovery of gold nuggets from one ton of rock that contains 0.019 wt% gold. Calculate the quantities of sodium cyanide and water consumed, per ounce of gold produced. What is the volume of gold nuggets produced from one ton of rock, if the density of a nugget is 19.3 g/cm^3?

P2.58 You are to prepare a broth for a fermenter to grow antibiotic-producing cells. (A fermentor is just a special reactor vessel for growing microorganisms.) The broth should be an aqueous solution containing 15 wt% glucose, 6 wt% phosphate, 6 wt% nitrate, and various trace nutrients. A custom supplier can produce the required blend for $15/kg, but your boss suspects you can produce the blend in-house for much less. Several commercial powders are available as raw ingredients, to be mixed with water as needed. Their mass compositions and costs are:

	"Fast-Feed"	**"Super-Gro"**	**"Formula N"**
Glucose	45%	0%	0%
Phosphate	2%	35%	12%
Nitrate	1%	15%	58%
Trace nutrients	3%	8%	0%
Filler	49%	42%	30%
Cost/kg	$20	$10	$15

(a) Suggest a combination of these powders that will produce the desired broth composition. Give the masses of powders and water required per kilogram of broth.

(b) What is the maximum savings per kilogram broth? (Assume water is free.) If the mixing equipment and storage tanks for preparing a 20-kg batch cost $2000, how many batches will you need to make to pay for the equipment before you begin to see real savings?

(c) What is the wt% trace nutrients in the home-made broth? If you now require 3 wt% trace nutrients, can you mix up a broth from the available powders that meets all specifications? What combination of powders would you recommend, and how does meeting this new requirement affect the economics?

P2.59 Antibiotics are typically produced by fermentation, where the fermenter is operated under what is called fed-batch culture. In fed-batch culture, the fermenter is initially filled with fermentation broth and cells. As the cells grow, they consume nutrients (glucose, phosphate, nitrate, etc.) and produce products. During cell growth, additional broth is added continuously to feed the cells.

You are planning to run a fermenter to produce antibiotics from fungi, using the fermentation broth you mixed up in Prob. 2.58a. The fermentor is filled with 6000 mL broth and some antibiotic-producing fungi. Additional broth is added continuously to the fermenter at a rate of 200 mL/h. The cells consume glucose at a rate of 35 g/h, phosphates at the rate of 13 g/h, and nitrates at the rate of 12 g/h. The fermentation is stopped when the concentration of one of these three nutrients goes to zero (because the cells can no longer grow). Which nutrient will be depleted first? How long will the fermentation run? What is the concentration (g/L) of the other two nutrients in the fermenter at the end of the run?

P2.60 An environmental group's website claims that about 5.6 billion metric tons of CO_2 are pumped into the atmosphere every year, about a quarter of which come from U.S. energy use. If there are 300 million people in the United States, how many pounds of CO_2 are produced per person per year? Does this claim seem reasonable? To check, calculate the CO_2 generated by combustion of gasoline, assuming gasoline is C_8H_{18}, has a specific gravity of 0.7, and reacts completely with O_2 to form CO_2 and H_2O, and that the average U.S. automobile gets 27.5 miles per gallon. Make any reasonable assumptions, and/or look up additional information, as needed.

P2.61 Natural gas usually contains some hydrogen sulfide (H_2S), a very toxic and smelly gas. One way to remove H_2S from sour natural gas streams is to pass the natural gas over a bed of zinc oxide, where the reaction

$$ZnO + H_2S \rightarrow ZnS + H_2O$$

takes place. Suppose you need to treat 1 million standard cubic feet of natural gas per day, and the gas contains 0.82 mol% H_2S. Sketch a block flow diagram, assuming water is not allowed in the natural gas product. You'd like to have to remove and discard the ZnS only once per day. How much ZnO is required? How large is the vessel to hold this quantity? (You will need to look up some physical property data.) How much water must be removed (kg/day)?

P2.62 (Adapted from D.F. Rudd and C.C. Watson, *Strategy of Process Engineering*, Wiley, 1968.) You are the engineer in charge of ensuring the safety of the municipal water supply in a small town. A local newspaper has charged that the municipal well water is contaminated with wastewater from a paper mill. The paper mill says that the contamination comes from an underground pipeline used to supply irrigation water to local farms. Well water, mill wastewater, groundwater, and water from the irrigation pipeline were sampled and sent for analysis to a water chemistry lab at the state university. The results reported by the laboratory were (in ppm – grams/million grams, $\pm15\%$ accuracy):

	TDS	**Ca**	**Mg**	**SO$_4$**	**Cl**	**NO$_3$**
Municipal well	245	22	15	13	8.0	15
Irrigation water	187	52	37	2	2.0	0.7
Papermill wastewater	250	18	12	8	18	18
Local ground water	236	25	16	14	3.5	12

TDS is total dissolved solids.

Write a polite but convincing memo to the newspaper editor in response to their charges.

P2.63 Automotive air bags contain a cylinder packed with a mixture of three solids: sodium azide (NaN_3), potassium nitrate (KNO_3), and silicon dioxide (SiO_2—sand). In a collision, an electrical signal is sent to the cylinder, causing rapid decomposition of NaN_3 to sodium (Na) and nitrogen (N_2). Na metal reacts with KNO_3 to produce K_2O and Na_2O, as well as more N_2. K_2O and Na_2O fuse with SiO_2 to produce an amorphous glassy solid. Write down the stoichiometrically balanced reactions for decomposition of sodium azide, for reaction of sodium metal with potassium nitrate, and for the overall reaction. If you want 6 ft^3 of nitrogen at 1 atm pressure in the bag when filled, how much NaN_3 (in lb and in ft^3) should be packed into the cylinder? How much potassium nitrate and silicon dioxide should be in the cylinder, if the final solids mixture should be 50 wt% SiO_2? (You will need to look up some physical property data.)

P2.64 Your job is to design a mixer to produce 200 kg/day of battery acid. The battery acid must contain 18.6 wt% H_2SO_4 in water. Raw materials available include a concentrated sulfuric acid solution at 77 wt% H_2SO_4 in water, a dilute acid solution at 10.8 wt% H_2SO_4, and pure water. The concentrated solution costs 10 cents/kg, the dilute acid solution costs 2 cents/kg, and the water is free. Complete a DOF analysis and determine whether this problem is completely specified. Recommend the best flow rates of each solution into the mixer.

P2.65 Titanium dioxide (TiO_2) is by far the most widely used pigment in white paint. Specifications for the white pigment powder used in paint making require that it contain at least 60% TiO_2, 5% ZnO, and 25% fillers (such as SiO_2 or $CaCO_3$).

The following powders are available from suppliers. Sketch out a block flow diagram for mixing these powders to make white pigment powder. Complete a DOF analysis of the process to determine if the problem is completely specified.

	Powder A (wt%)	Powder B (wt%)	Powder C (wt%)	Powder D (wt%)
TiO_2	90	0	0	0
ZnO	0	50	0	0
$CaCO_3$	0	50	75	0
NaCl	10	0	25	10
SiO_2	0	0	0	90

Come up with a recipe using these powders for making 1000 kg white pigment powder that meets the specifications. What is the lowest wt% NaCl product you could make? What is the highest wt% NaCl product?

P2.66 Titanium dioxide (TiO_2) is a major pigment in white paint. You are in charge of designing a washer for processing raw pigment. The washer should be designed to handle 10,000 lb/h raw pigment (40 wt% TiO_2, 20 wt% salt, and 40 wt% water). In the washer, the raw pigment is mixed with washwater. Pilot plant studies show that the optimum washwater feed rate is 6 lb washwater per lb raw pigment feed. 100% of the TiO_2 in the raw pigment is recovered in the washed pigment. The washed pigment product leaving the washer contains 50 wt% TiO_2, and the salt content in the washed pigment must be no more than 100 parts salt per 1 million parts TiO_2. The plant manager would like to dump the wastewater into a nearby river. The local water quality control regulatory agency will allow at most a discharge of 30,000 lb/h at a maximum salt content of 0.5%. Draw a flow diagram. Complete a DOF analysis and state whether this process is completely specified. If the problem is incorrectly specified, explain how you might change the design to fix the situation. Clearly explain your reasoning. (Adapted from G. V. Reklaitis, *Introduction to Material and Energy Balances*, Wiley, 1983.)

P2.67 Portland cement is made by mixing a variety of raw materials, grinding them, then heating the mixture in a kiln. Lots of changes occur in the kiln, including evaporation of water and decomposition of magnesium and calcium carbonates. Ignitable materials are burned and exit the kiln as gases. The cement leaving the kiln must contain 21 wt% SiO_2 and 65 wt% CaO. The Fe_2O_3 content must be between 1 and 5 wt%.

The compositions and costs of the raw materials available are listed below:

	Limestone (wt%)	Clay (wt%)	Mill scale (wt%)	Oyster shells (wt%)
SiO_2	1	68	0	1
CaO	54	0	0	54
Al_2O_3	0	20	0	0
Fe_2O_3	0	4	100	0
MgO	4	0	0	0
Ignitable material	41	8	0	45
Cost ($/ton)	88	35	Free	60

Your job is to develop a process for making 100 metric tons/day of Portland cement. Come up with the *best* recipe for mixing the raw materials to make a cement product that meets all the specifications. Calculate the raw materials cost per ton of cement produced ($/ton).

P2.68 Containers made of polystyrene and paper are used commonly in the food service industry to serve food and beverages, and there is a great deal of controversy over which type of container is less environmentally damaging. Although both materials can theoretically be reprocessed, at present neither type of container is recycled. Instead, the waste is either sent to a landfill or incinerated.

Two different sources of data will be used to estimate the relative environmental impact of paper versus polystryene containers. The first is a paper by Martin Hocking of the University of Victoria published in the respected journal *Science* (*Science,* 251:504–505, 1991). This table summarizes some of the results reported by Hocking.

	Paper cup	**Polystyrene foam cup**
Raw materials (per cup)		
Wood and bark (g)	33	0
Petroleum (g)	4.1	3.2
Other chemicals (g)	1.8	0.05
Finished weight (g)	10.1	1.5
Cost	2.5x	1x
Water emissions (per metric ton)		
Suspended solids (kg)	35-60	Trace
Dissolved biodegradable solids (kg)	30–50	0.07
Organochlorines (kg)	5–7	0
Metal salts (kg)	1–20	20
Air emissions (per metric ton)		
Chlorine (kg)	0.5	0
Chlorine dioxides (kg)	0.2	0
Reduced sulfides (kg)	2	0
Particulates (kg)	5–15	0.1
Pentane (kg)	0	35–50
Sulfur dioxide (kg)	10	10

	Paper cup	**Polystyrene foam cup**
Recycle potential		
Reuse as cup	Possible	Easy
Reuse as material	Low potential	High potential
Ultimate disposal		
Mass to landfill	10.1	1.5
Biodegradable?	Yes	No

The second is a report by a consulting firm, Franklin Associates Ltd., which was prepared for the Council of Waste Solutions, a group that is partially funded by plastics manufacturers. The data from their report is summarized in the next table. (Franklin Associates Ltd., "Resource and environmental profile analysis of foam polystyrene and bleached paper-board containers," final report to the Council for Solid Waste Solutions, Prairie Village, Kansas.

Environmental Impact per 10,000 Cups

	Polystyrene foam	**LDPE-coated paper**	**Wax-coated paper**
Air emissions (lb)	11.8	18.1	21.8
Water emissions (lb)	2.1	2.9	4.5
Industrial solid waste (lb)	18.6	54.3	71.0
Postconsumer solid waste (lb)	120.3	218.3	266.2

Franklin Associates reports that 10,000 16-oz polystyrene cups weigh 96.9 lb, while LDPE-coated paper and wax-coated paper cups weigh 229.1 and 287.9 lb, respectively.

On the basis of the data in the tables, estimate the mass of solid wastes and the atmospheric emissions generated per paper cup and per polystyrene cup. Compare the two different sets of data. Use your analysis as the basis for a short letter to your local coffee shops to argue that they use only one kind of cup.

There are major differences between the solid wastes reported for paper cups in the two reports. Hocking states in his paper that "paper cups are made from bleached pulp, which in turn is obtained in yields of about 50% by weight from wood chips." Similar inefficiencies are assumed in the Franklin Associates report, but these inefficiencies are not counted as solid wastes. Which approach do you think is better, and why?

P2.69 Refer to the block flow diagram below. A 95 mol% propylene P (C_3H_6)/5 mol% propane I (C_3H_8) feed is mixed with benzene B (C_6H_6) feed at a molar ratio of 1.2:1 propylene:benzene (P:B). These fresh feeds are mixed with recycled streams, then fed to a reactor. The stream entering the reactor contains 10 mol% propane. Two reactions occur simultaneously in the reactor, producing the desired product cumene C (C_9H_{12}) and an undesired byproduct diisopropylbenzene D ($C_{12}H_{18}$):

$$C_3H_6 + C_6H_6 \rightarrow C_9H_{12}$$
$$C_3H_6 + C_9H_{12} \rightarrow C_{12}H_{18}$$

Under the reaction conditions, propane I is an inert. 80% of the propylene and 72% of the benzene fed to the reactor are converted to products. The reactor effluent is cooled and sent to a separator, where a vapor stream containing propylene and propane is taken off the top and a liquid stream containing benzene, cumene, and diisopropylbenzene is taken off the bottom. A fraction of the vapor stream leaving the separator is purged and the remainder is recycled to be mixed with the incoming feed stream. The liquid stream leaving the reactor is sent to a series of two distillation columns, where benzene is recovered and recycled, and cumene and diisopropylbenzene are separated and sent to storage tanks. The cumene production rate is 25 gmol/s.

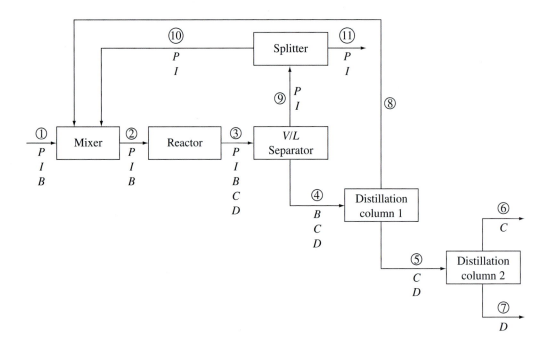

Complete a DOF analysis on the process. On the basis of the DOF analysis, immediately write down the set of variables and equations that you would use to calculate flows of all streams. You do not have to solve the equations.

P2.70 In petroleum refining, crude oil is separated into several different mixtures of hydrocarbons. One product stream is called the light alkanes, which contain mainly methane, ethane, propane, butane, and pentane. In one facility, the light alkane stream [1000 kgmol/h, 10 mol% methane (*M*), 30 mol% ethane (*E*), 15 mol% propane (*P*), 30 mol% butane (*B*), and 15 mol% isopentane (*I*)] is processed to produce five different product streams The separation is done in a series of distillation columns, as shown. Stream 4, 6, and 11 flow rates are 40, 100, and 300 kgmol/h, respectively. Complete a DOF analysis. Calculate flow rates and compositions of all process streams. Report your results in both mole and mass units, using a table format, with stream numbers as column headings and components as row headings.

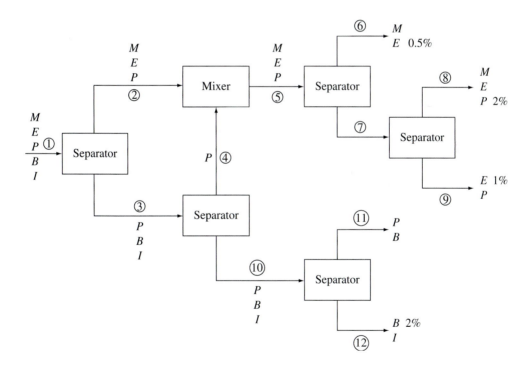

P2.71 The sugar cane industry is a big industry on Hawaii. Raw sugar cane is first cut from the fields, then chopped and shredded. The raw cane contains 15 wt% sucrose, 25 wt% solids, and water, along with some additional impurities that can be assumed to have negligible wt%.

To produce raw sugar for shipment to California, the chopped sugar cane is mixed with some water and macerated in a mill. About 93% of the sugar juice in the cane is recovered in the mill. The spent cane (called *bagasse*) contains about 20 wt% water and is burned for fuel along with the unrecovered juice. The recovered juice is sent to a clarifier, where lime is added to precipitate impurities. You can assume that all the lime precipitates with the settled-out "mud," and that essentially no juice is lost to the mud. The juice leaving the clarifier, which contains 85 wt% water, is sent to an evaporator, where the water content of the juice is reduced to 40 wt%. The thickened juice is sent to an evaporator/crystallizing pan, where more water is removed and sugar crystals start to form. The crystals and syrup leaving the pan contain 10 wt% water. The crystals are separated from the syrup in a centrifuge. The raw sugar crystals are 97.8 wt% sucrose, and the syrup (called blackstrap molasses) is 50 wt% sugar.

Assume that 10,000 lb/h of sugar cane is processed. Sketch a simplified process flow diagram. Calculate the flow rate of raw sugar crystals and molasses, the flow rate of water added to the cane fed to the mill, and the flow rate of water removed in the evaporator.

Game Day

P2.72 Mr. Big, a senior executive at ABC Industrial Alcohols Inc., has decided that new products are needed for continuing corporate growth. The market for ethyl acetate (a commercial solvent) is expanding, and your group, as the process engineering team for ABC, has been assigned the job of coming up with a process for production of 700 million lb/yr ethyl acetate (assuming 330 days/yr operation). After an initial discussion with company chemists and environmental engineers, you've come up with the following information:

Raw materials available:
  Solution containing 70 mol% ethanol and 30 mol% water.
  Air (79 mol% nitrogen, 21 mol% oxygen).

Reaction pathways available:
  Oxidation of ethanol to produce acetic acid and water

$$C_2H_5OH + O_2 \rightarrow CH_3COOH + H_2O$$

Reaction takes place at high pressure in the vapor phase over a catalyst. At least 50 mol% nitrogen is needed in the feed as a diluent. Ethyl acetate must not be present in the feed. Water can be present in the feed. Oxygen must be in a 20% excess of the stoichiometric amount to allow for complete consumption of the ethanol.

Esterification of ethanol and acetic acid to produce ethyl acetate

$$C_2H_5OH + CH_3COOH \rightarrow CH_3COOC_2H_5 + H_2O$$

Reaction takes place in liquid solution at ambient conditions. Only 60% conversion can be achieved because of equilibrium constraints. Oxygen is prohibited in the feed. Water and nitrogen are allowed contaminants.

Waste products:
   No acetic acid, ethanol, or ethyl acetate is allowed to leave the plant in a waste stream.

Come up with a process flow sheet for ethyl acetate production, using this information. You should submit your design in the form of a proposal to Mr. Big. Your report should include: (1) a 1 to 2 page executive summary describing the key features of your process, the assumptions that went into your design, any uncertainties or additional information you would need to finalize the design, and what action you recommend Mr. Big should take, (2) completed flow sheet showing all flows and compositions that you were able to specify, and (3) an appendix showing detailed supporting calculations. (Mr. Big is too busy to look at this, but documentation of your results would be important for any follow-on engineering work).

P2.73 Metallurgy companies use a lot of acid to wash metal surfaces. For example, when steel is "pickled," it is dunked in an acid bath to remove rust on its surface. This is good for the steel, but a lot of metal-contaminated acid wastes are produced. This waste can't be simply dumped into the nearest river. Small processors in particular have difficulty figuring out what to do with the acid waste. The U.S. Department of Energy came up with a skid-mounted process. (In a skid-mounted process, all the equipment is mounted on a skid, so it can be trucked from location to location.) In the process, acid waste containing sulfuric, phosphoric, hydrochloric, and/or nitric acids and dissolved metal ions is heated. The acids vaporize, and the metal ions remain in liquid. The vapor is purified to yield clean acid. Up to 90% of the spent acid is recovered for reuse; the remainder must be disposed of as chemical waste. The metal ion solution is collected and treated by adding salts that precipitate the heavy metal ions. The metal salts can be processed to reclaim the metals. Savings of $1 to $5 per gallon are claimed. (Reference: Viatec Recovery Systems, *Chemical and Engineering News,* vol. 71, p. 27, 1993.)

You work for a small metal-plating company that produces about 500 gal of acid waste every week. Right now you pay $0.25/gal to dispose of the acid waste. (Disposal costs are calculated on the basis of the concentration of acid in the waste.) You are thinking about investing in the skid-mounted unit. Your acid waste typically contains 12 M HCl and some soluble $FeCl_3$. To justify making the investment, the cost of purchasing the skid-mounted unit must be totally recovered in savings in operating costs within one year. What is the maximum price you will pay for the unit? Don't forget to consider what salts you will add to precipitate the iron; you will need to look up data on iron salts in the *CRC Handbook* or a similar source. You will need to look up current prices of acid and salts.

P2.74 The idea of humans traveling to Mars is being seriously considered. Current estimates suggest that a Mars mission should last 3 years: 6 months travel, 2 years on the planet, and 6 months return travel. With such extended voyages, it would be impossible to carry enough food and oxygen to support the astronauts, so spacecraft must be designed to use wastes to regenerate food.

The diagram shows the flow sheet for a simplified system that uses plants to make food and oxygen from the carbon dioxide and water produced by the astronauts. The astronauts are represented by the metabolism reactor, and the plants are represented by the photosynthesis reactor. The carbon dioxide and water in stream 3 are removed by a freezer. The nitrogen and oxygen from the freezer are returned to the metabolism reactor and a small amount is directed to the photosynthesis reactor. For this analysis, assume that food has the molecular formula of glucose $(C_6H_{12}O_6)$, and that the reactions of metabolism can be simplified to the overall reaction:

$$C_6H_{12}O_6 + 6O_2 \rightarrow 6CO_2 + 6H_2O \tag{1}$$

The reactions occurring in the photosynthesis chamber are:

$$6CO_2 + 6H_2O \rightarrow C_6H_{12}O_6 + 6O_2 \tag{2}$$

$$C_6H_{12}O_6 \rightarrow 3CO_2 + 3CH_4 \tag{3}$$

Reaction 3 produces the undesired waste product methane. To prevent the build up of methane, a small amount of gas from the photosynthesis chamber must be vented.

For this analysis, assume:
There are 9 crew members, each requiring 2 kg of food per day.
1% of the glucose produced by photosynthesis is consumed in the production of methane.
Air contains 21 mol% oxygen and 70 mol% nitrogen.
The number of moles of methane and carbon dioxide in stream 9 are equal.
48 g of methane is purged per day.
Methane is 2 mol% of stream 9.

What is the rate of glucose production in the photosynthesis chamber of the system? Express your answer in kmol/day. Determine the minimum mass of glucose and air that must be stored for a 1000-day mission to Mars.

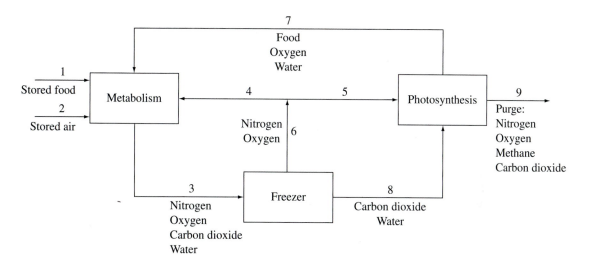

Mathematical Analysis of Material Balance Equations and Process Flow Sheets

In This Chapter

We revisit some key ideas from Chap. 2, but take a more mathematical and rigorous approach. We develop generalized expressions for the material balance equation, and use these expressions to solve problems of increasing complexity. In an optional section, we construct linear models of steady-state process flow sheets, and illustrate how linear algebra techniques are used in evaluating these flow sheets.

Some questions we address include:

- How is the material balance equation related to the law of conservation of mass?
- How do I handle transients in the material balance equation?
- What do I do when there are multiple reactions occurring simultaneously?
- What are useful specifications for system performance?
- How do I develop a well-defined system of linear independent equations to describe a process flow sheet?

Words to Learn

Watch for these words as you read Chapter 3.

Material balance equations
Extent of reaction
Linear equations
Linear model
Fractional split
Fractional conversion
Fractional recovery
Process topology

3.1   Introduction

Chemical process synthesis is evolutionary. We start with the basics, as we did in Chap. 1, asking: What product do I want to make? What raw materials are available? What reaction chemistries are feasible? Then we move on to sketches of simple block flow diagrams, as we did in Chap. 2. We identify the key process units required, and we consider how to connect these units together. We make simplifying approximations, quickly complete process flow calculations, and make preliminary assessments of alternative arrangements.

Once one (or a few) preliminary block flow diagrams have been sketched, we move further into the details. At this stage in the *tour de* process synthesis, we continue to work with block flow diagrams. We include more realistic specifications of stream composition and system performance. We consider time-varying processes. We develop methods that allow examination of how the overall performance of a process might change with changes in specific process variables or specifications. In short, we take a more rigorous, mathematical, and systematic look at process flow calculations (Fig. 3.1).

3.2   The Material Balance Equation—Again

In Chap. 2, we introduced the material balance equation as

$$\text{Input} - \text{output} + \text{generation} - \text{consumption} = \text{accumulation}$$

You learned the importance of clearly choosing a system, identifying components, and defining stream and system variables. In this section we will revisit

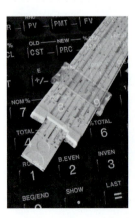

Figure 3.1 Until the advent of pocket calculators in the 1970s, engineers relied on tools like slide rules and their own numerical prowess. Modern scientific calculators and personal computers place incredible calculational power at the engineer's fingertips, making it much easier to find mathematically rigorous solutions to engineering problems.
Source: Left: © Vol. 41/PhotoDisc/Getty; Right: © Vol. 11/PhotoDisc/Getty

these ideas with three goals in mind: (1) to reiterate the importance of mastering these concepts, (2) to develop more complete and rigorous expressions for the material balance equation, and (3) to illustrate use of the material balance equation in solving more challenging process flow problems.

3.2.1 Conservation of Mass and the Material Balance Equation

Let's first recall the meaning of the word **system.** A system is three-dimensional: It has a well-defined boundary, or surface area, which encloses a volume. If material enters and leaves the system, it does so by crossing the system boundary. Within the system, physical and chemical changes may take place.

Now recall the law of conservation of mass: Mass is neither created nor destroyed. (This law required revision with Einstein's statement that $E = mc^2$. Throughout this text we ignore Einstein and obey the mass conservation law.) We'll apply the law to the system illustrated in Fig. 3.2. The system contains a mixture of compound A (of molar mass M_A) and compound B (of molar mass M_B).

We'll denote the mass of A and B inside the system as $m_{A,\text{sys}}$ and $m_{B,\text{sys}}$, respectively. If the total mass in the system is m_{sys}, then

$$m_{\text{sys}} = m_{A,\text{sys}} + m_{B,\text{sys}} \tag{3.1}$$

Stream 1 flows into the system and stream 2 flows out, with compounds A and B in both. We'll denote the total mass flow rate of streams 1 and 2 as $\dot{m}_1$ and $\dot{m}_2$, respectively, where we use the convention that the dot distinguishes a rate

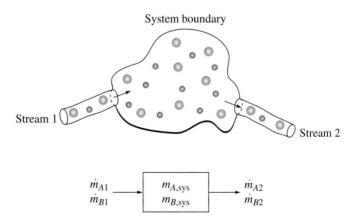

Figure 3.2 (Top) A system containing A (larger shaded spheres) and B (smaller dark spheres), with one stream carrying material into the system and another stream carrying material out of system. (Bottom) Block flow representation.

(e.g., mass/time) from a quantity (e.g., mass). Then $\dot{m}_{A1}$, $\dot{m}_{B1}$, $\dot{m}_{A2}$, and $\dot{m}_{B2}$ denote the mass flow rates of compounds A and B in streams 1 and 2, and

$$\dot{m}_1 = \dot{m}_{A1} + \dot{m}_{B1} \tag{3.2a}$$

$$\dot{m}_2 = \dot{m}_{A2} + \dot{m}_{B2} \tag{3.2b}$$

If $\dot{m}_1$ and $\dot{m}_2$ are constant, then over a time interval Δt, the total mass entering the system is $\dot{m}_1 \Delta t$ and the mass leaving the system is $\dot{m}_2 \Delta t$. If the mass entering the system over this time interval is different than the mass leaving the system, then by the law of conservation of mass, the system mass must change. We'll use Δm_{sys} to denote the *change* in mass inside the system over the time interval Δt:

$$\Delta m_{\text{sys}} = \dot{m}_1 \Delta t - \dot{m}_2 \Delta t \tag{3.3}$$

If we now divide both sides by Δt and take the limit as $\Delta t \to 0$, we get

$$\frac{dm_{\text{sys}}}{dt} = \dot{m}_1 - \dot{m}_2 \tag{3.4}$$

Eq. (3.4) is a restatement of the law of conservation of mass as applied to the system in Fig. 3.2. If we are at steady state, there is no change in the mass in the system; therefore, $dm_{\text{sys}}/dt = 0$ and $\dot{m}_2 = \dot{m}_1$ at steady state.

Illustration: Water flows into a tank at a rate of 14 g/s and out of the tank at a rate of 12 g/s. The rate of change in mass of water in the tank is $dm_{\text{sys}}/dt = \dot{m}_1 - \dot{m}_2 = 14 - 12 = 2$ g/s.

Let's substitute Eqs. (3.1) and (3.2) into Eq. (3.4):

$$\frac{d(m_{A,\text{sys}} + m_{B,\text{sys}})}{dt} = (\dot{m}_{A1} + \dot{m}_{B1}) - (\dot{m}_{A2} + \dot{m}_{B2}) \tag{3.5}$$

Would it be OK to split Eq. (3.5) into two equations, one equation including only the terms referring to compound A and the other including only the compound B terms? No! Why not? Because the law of conservation of mass applies to the *total* mass, not necessarily to the mass of compound A alone (or compound B alone). In particular, suppose compounds A and B participate in a chemical reaction inside the system, A $\to$ B, where the reaction as written is not necessarily balanced. We'll define the mass rate of reaction of compound A (in units of mass/time) as $\dot{R}_A$ and the mass rate of reaction of compound B as $\dot{R}_B$. If A is consumed and B is generated, then $\dot{R}_A$ is negative and $\dot{R}_B$ is positive. Since total mass is conserved, it must be true that

$$\dot{R}_A = -\dot{R}_B \quad \text{or} \quad \dot{R}_A + \dot{R}_B = 0 \tag{3.6}$$

Since it is OK to add zero to one side of an equation, we add Eq. (3.6) to Eq. (3.5) and get:

$$\frac{d(m_{A,\text{sys}} + m_{B,\text{sys}})}{dt} = (\dot{m}_{A1} + \dot{m}_{B1}) - (\dot{m}_{A2} + \dot{m}_{B2}) + (\dot{R}_A + \dot{R}_B) \tag{3.7}$$

If we've accounted for all the chemical reactions occurring in the system, we *can* split Eq. (3.7) into two:

$$\frac{dm_{A,sys}}{dt} = \dot{m}_{A1} - \dot{m}_{A2} + \dot{R}_A \tag{3.8a}$$

$$\frac{dm_{B,sys}}{dt} = \dot{m}_{B1} - \dot{m}_{B2} + \dot{R}_B \tag{3.8b}$$

Eq. (3.8) is the material balance equation applied to a single compound, in units of mass/time.

In earlier chapters we've generally found it convenient to consider reactions in molar, rather than mass, units. If we denote the molar reaction rates of A and B as $\dot{r}_A$ and $\dot{r}_B$, respectively, then

$$\dot{R}_A = M_A \dot{r}_A \tag{3.9a}$$

$$\dot{R}_B = M_B \dot{r}_B \tag{3.9b}$$

where M_A and M_B are the molar masses of A and B, respectively. Recall that Eq. (3.6) provided the simple relationship $\dot{R}_A + \dot{R}_B = 0$. But, if $M_A \neq M_B$, then $\dot{r}_A + \dot{r}_B \neq 0$! Is there a simple way to relate $\dot{r}_A$ and $\dot{r}_B$? Yes! Through stoichiometry. Suppose the stoichiometrically balanced chemical reaction is

$$2A \rightarrow 3B$$

then for every 2 mol A consumed, 3 mol B is generated, or

$$\frac{\dot{r}_A}{\dot{r}_B} = -\frac{2}{3}$$

or, more generally,

$$\frac{\dot{r}_A}{\dot{r}_B} = \frac{\nu_A}{\nu_B} \tag{3.10}$$

where ν_i is the stoichiometric coefficient for compound i; ν_i is negative for reactants and positive for products.

We can rewrite Eq. (3.10) as a ratio of the molar reaction rate of a compound to its stoichiometric coefficient. This ratio is called the **extent of reaction** $\dot{\xi}$:

$$\frac{\dot{r}_A}{\nu_A} = \frac{\dot{r}_B}{\nu_B} = \dot{\xi} \tag{3.11}$$

$\dot{\xi}$ is always a positive number, and is written here with a dot to remind us that it is a *rate*, with units of moles/time. $\dot{\xi}$ tells us the rate at which a reaction event occurs. It is a very useful number because it links together the rates of consumption and generation of all compounds that participate in the reaction. Note that the value of $\dot{\xi}$ depends on the way in which the stoichiometrically balanced equation is written. For $2A \rightarrow 3B$, if $\dot{\xi} = 2$ gmol/s then $\dot{r}_B = 6$ gmol/s, but for $A \rightarrow \frac{3}{2}B$ and $\dot{\xi} = 2$ gmol/s, then $\dot{r}_B = 3$ gmol/s.

Quick Quiz 3.1

For the reaction
$2C_2H_4 + O_2 \rightarrow 2C_2H_4O$,
if $\dot{r}_{ethylene} = -4$
gmol/s, what is $\dot{\xi}$?

For the reaction
$C_2H_4 + 0.5O_2 \rightarrow C_2H_4O$, if
$\dot{r}_{ethylene} = -4$ gmol/s,
what is $\dot{\xi}$?

Eq. (3.11) combined with Eq. (3.9) yields:

$$\dot{R}_A = M_A \nu_A \dot{\xi} \tag{3.12a}$$

$$\dot{R}_B = M_B \nu_B \dot{\xi} \tag{3.12b}$$

Using these expressions, we find alternative versions of Eq. (3.8):

$$\frac{dm_{A,\text{sys}}}{dt} = \dot{m}_{A1} - \dot{m}_{A2} + M_A \nu_A \dot{\xi} \tag{3.8c}$$

$$\frac{dm_{B,\text{sys}}}{dt} = \dot{m}_{B1} - \dot{m}_{B2} + M_B \nu_B \dot{\xi} \tag{3.8d}$$

Eq. (3.8) is a differential material balance equation written in units of mass/time. Now we'd like to derive a similar equation written in units of moles/time. Let's define $n_{A,\text{sys}}$ and $n_{B,\text{sys}}$ as the moles of A and B in the system, such that

$$m_{A,\text{sys}} = n_{A,\text{sys}} M_A$$

$$m_{B,\text{sys}} = n_{B,\text{sys}} M_B$$

and let's define $\dot{n}_{A1}$, $\dot{n}_{B1}$, $\dot{n}_{A2}$, and $\dot{n}_{B2}$ as the molar flow rates of the individual compounds in streams 1 and 2, such that:

$$\dot{m}_{A1} = \dot{n}_{A1} M_A$$

$$\dot{m}_{B1} = \dot{n}_{B1} M_B$$

$$\dot{m}_{A2} = \dot{n}_{A2} M_A$$

$$\dot{m}_{B2} = \dot{n}_{B2} M_B$$

Substituting these expressions into Eq. (3.8c) and Eq. (3.8d) produces

$$M_A \frac{dn_{A,\text{sys}}}{dt} = M_A \dot{n}_{A1} - M_A \dot{n}_{A2} + M_A \nu_A \dot{\xi}$$

$$M_B \frac{dn_{B,\text{sys}}}{dt} = M_B \dot{n}_{B1} - M_B \dot{n}_{B2} + M_B \nu_B \dot{\xi}$$

Now we divide through by the molar mass to get

$$\frac{dn_{A,\text{sys}}}{dt} = \dot{n}_{A1} - \dot{n}_{A2} + \nu_A \dot{\xi} \tag{3.13a}$$

$$\frac{dn_{B,\text{sys}}}{dt} = \dot{n}_{B1} - \dot{n}_{B2} + \nu_B \dot{\xi} \tag{3.13b}$$

Eq. (3.13) is the material balance equation applied to a compound, written in units of moles/time.

If we now add Eq. (3.13a) and (3.13b) together and substitute in the total moles $n_{\text{sys}} = n_{A,\text{sys}} + n_{B,\text{sys}}$, and the total molar flow rates $\dot{n}_1 = \dot{n}_{A1} + \dot{n}_{B1}$ and $\dot{n}_2 = \dot{n}_{A2} + \dot{n}_{B2}$, we get

$$\frac{dn_{\text{sys}}}{dt} = \dot{n}_1 - \dot{n}_2 + (\nu_A + \nu_B)\,\dot{\xi} \tag{3.14}$$

Compare Eq. (3.14) to Eq. (3.4). Notice that total moles are *not* conserved! (unless $\nu_A = -\nu_B$).

Example 3.1   ### Decomposition Reactions

N_2O_4 is fed to a reactor at a flow rate of 84 gmol/min, where some of it decomposes to NO_2. The reactor operates at steady state. The stream leaving the reactor flows at 126 gmol/min. What is the extent of reaction? What is the molar flow rate of each component in the reactor outlet stream?

Solution
We start as always by sketching and labeling a flow diagram.

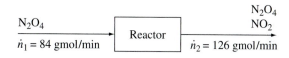

$$N_2O_4 \rightarrow 2NO_2$$

If N_2O_4 is identified as compound A and NO_2 as B, then $\nu_A = -1$ and $\nu_B = 2$.

We start with Eq. (3.14). Since the reactor operates at steady state, there is no *change* in the number of moles in the system, or $dn_{sys}/dt = 0$. Using the information provided, that $\dot{n}_1 = 84$ gmol/min, and $\dot{n}_2 = 126$ gmol/min, we substitute into Eq. (3.14) to find:

$$0 = 84 - 126 + (-1 + 2)\dot{\xi}$$

$$\dot{\xi} = 42 \text{ gmol/min}$$

To find the molar flow rate of each component in the outlet stream we return to Eq. (3.13), applied first to N_2O_4 and then to NO_2.

$$\frac{dn_{A,sys}}{dt} = \dot{n}_{A1} - \dot{n}_{A2} + \nu_A\dot{\xi}$$

$$0 = 84 - \dot{n}_{A2} - 42$$

$$\dot{n}_{A2} = 42 \text{ gmol/min}$$

$$\frac{dn_{B,sys}}{dt} = \dot{n}_{B1} - \dot{n}_{B2} + \nu_B\dot{\xi}$$

$$0 = 0 - \dot{n}_{B2} + 2(42)$$

$$\dot{n}_{B2} = 84 \text{ gmol/min}$$

!
Helpful Hint:
For a system at steady state,
$$\frac{dn_{i,sys}}{dt} = \frac{dm_{i,sys}}{dt}$$
$$= 0$$

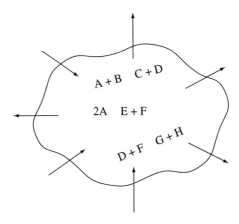

Figure 3.3 In the general case, many streams enter and leave the system, carrying many components. Multiple reactions occur within the system.

3.2.2 General Forms of the Differential Material Balance Equations

Eqs. (3.4), (3.8), (3.13), and (3.14) were derived for a specific case: two components, one input stream, one output stream, and one reaction. Now we'd like to generalize. (See Fig. 3.3.) Let's suppose we have I components entering, leaving, and/or inside the system. Recall that components can be elements (e.g., C), compounds with a defined molecular formula (e.g., fructose, $C_6H_{12}O_6$), or a composite material that behaves as a single entity (e.g., fruit juice solids). Let's suppose we have J process streams in or out of the system. Finally, let's suppose we have K chemical reactions occurring within the system boundaries. For bookkeeping, we'll use three subscripts to indicate a specific component, stream, or chemical reaction:

i = component
j = stream
k = chemical reaction

We'll consider first the material balance equations written in units of mass and mass/time, with

m_{sys} = total mass in system [mass]
$\dot{m}_j$ = mass flow rate of stream j [mass/time]
$m_{i,sys}$ = mass of component i in system [mass]
$\dot{m}_{ij}$ = mass flow rate of component i in stream j [mass/time]
$\dot{R}_{ik}$ = mass reaction rate of component i by reaction k [mass/time] = $M_i \nu_{ik} \dot{\xi}_k$

Eq. (3.4) included only one input and one output stream. To generalize, we must sum up all mass flows entering the system and all mass flows leaving the system. The differential total mass balance equation is:

$$\frac{dm_{sys}}{dt} = \sum_{\text{all } j \text{ in}} \dot{m}_j - \sum_{\text{all } j \text{ out}} \dot{m}_j \tag{3.15a}$$

The component mass balance equation must include flows of that component in all entering and leaving streams. Additionally, all chemical reactions involving component i must be included. This general form of Eq. (3.8) becomes

$$\frac{dm_{i,\text{sys}}}{dt} = \sum_{\substack{\text{all } j \text{ in}}} \dot{m}_{ij} - \sum_{\substack{\text{all } j \text{ out}}} \dot{m}_{ij} + \sum_{\substack{\text{all } k \\ \text{reactions}}} M_i \nu_{ik} \dot{\xi}_k \qquad (3.16\text{a})$$

where

$$\sum_{\substack{\text{all } k \\ \text{reactions}}} M_i \nu_{ik} \dot{\xi}_k = \sum_{\substack{\text{all } k \\ \text{reactions}}} \dot{R}_{ik}$$

is simply the *net* generation/consumption of compound i, on a mass basis.

If the process operates at steady state, then $dm_{\text{sys}}/dt = dm_{i,\text{sys}}/dt = 0$ and

$$\sum_{\substack{\text{all } j \text{ out}}} \dot{m}_j = \sum_{\substack{\text{all } j \text{ in}}} \dot{m}_j \qquad (3.15\text{b})$$

$$\sum_{\substack{\text{all } j \text{ out}}} \dot{m}_{ij} = \sum_{\substack{\text{all } j \text{ in}}} \dot{m}_{ij} + \sum_{\substack{\text{all } k \\ \text{reactions}}} M_i \nu_{ik} \dot{\xi}_k \qquad (3.16\text{b})$$

| **Example 3.2** | **Mass Balances: Sugar Dissolution** |

A bucket holds a large quantity of sugar. Water at 3 kg/min is pumped into the bucket. The sugar dissolves in the water. Sugar water, at 84 wt% sugar, leaves the bucket at 3 kg/min. The labeled flow diagram is shown. Apply Eqs. (3.15) and (3.16) to this situation.

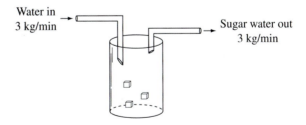

Solution
For total mass (Eq. 3.15):

$$\frac{dm_{\text{sys}}}{dt} = \dot{m}_{\text{in}} - \dot{m}_{\text{out}} = 3 \text{ kg/min} - 3 \text{ kg/min} = 0$$

There are two components: sugar S and water W. In the bucket there is a physical change as the sugar dissolves in the water, but no chemical reaction, so

$\dot{R}_S = \dot{R}_W = 0$. Applying Eq. 3.16 to each of the components in turn yields:

$$\frac{dm_{S,sys}}{dt} = \dot{m}_{S,in} - \dot{m}_{S,out} = 0 \text{ kg sugar/min} - (0.84 \times 3) \text{ kg sugar/min} = -2.52 \text{ kg sugar/min}$$

$$\frac{dm_{W,sys}}{dt} = \dot{m}_{W,in} - \dot{m}_{W,out} = 3 \text{ kg water/min} - (0.16 \times 3) \text{ kg water/min} = +2.52 \text{ kg water/min}$$

Notice that $dm_{S,sys}/dt + dm_{W,sys}/dt = dm_{sys}/dt$, as it should.

Example 3.3   Mass Balance Equation: Glucose Consumption in a Fermentor

A broth containing 20 wt% glucose is fed to a fermentor at a constant rate of 100 g/h. Yeast in the fermentor consume glucose at a rate of 12.9 g/h. The fermentor becomes contaminated with bacteria, which consume glucose at a rate of 1.4 g/h. What is the rate of change of glucose in the fermentor?

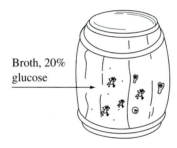

Broth, 20% glucose

Solution

> **! Helpful Hint**
> $\dot{R}_{ik}$ is negative for reactants and positive for products.

We will write the mass balance equation for glucose. There is one input stream ($\dot{m}_{g,in} = 0.2\,(100) = 20$ g/h), no output streams ($\dot{m}_{g,out} = 0$ g/h) and two reactions ($\dot{R}_{g1} = -12.9$ g/h and $\dot{R}_{g2} = -1.4$ g/h).

Eq. 3.16 simplifies to:

$$\frac{dm_{g,sys}}{dt} = 20 + [(-12.9) + (-1.4)] = 5.7 \text{ g/h}$$

Now we'll quickly develop the general form of the differential material balance equation written in units of moles and moles/time, with

> **! Helpful Hint**
> Use a differential material balance equation when you are interested in what happens at a single instance in time.

n_{sys} = total moles in system [moles]

$\dot{n}_j$ = molar flow rate of stream j [moles/time]

$n_{i,sys}$ = moles of component i in system [moles]

$\dot{n}_{ij}$ = molar flow rate of component i in stream j [moles/time]

$\dot{r}_{ik}$ = molar reaction rate of component i by reaction k [moles/time]

We convert from mass to mole units by dividing each term in Eqs. (3.15) and (3.16) by the molar mass of component i, M_i, and recognizing that $\dot{m}_i/M_i = \dot{n}_i$ and $m_i/M_i = n_i$. We define an extent of reaction $\dot{\xi}_k$ for each reaction k, such that

$$\dot{r}_{ik} = \nu_{ik}\,\dot{\xi}_k$$

Quick Quiz 3.2

How is

$$\sum_{\substack{\text{all } k \\ \text{reactions}}} \nu_{ik}\,\dot{\xi}_k$$

related to the "net" column in generation-consumption analysis of Chapter 1?

where ν_{ik} is the stoichiometric coefficient for component i in reaction k. Using this notation, we write the material balance equation on component i, in molar units, as

$$\frac{dn_{i,\text{sys}}}{dt} = \sum_{\text{all } j \text{ in}} \dot{n}_{ij} - \sum_{\text{all } j \text{ out}} \dot{n}_{ij} + \sum_{\substack{\text{all } k \\ \text{reactions}}} \nu_{ik}\,\dot{\xi}_k \tag{3.17a}$$

Finally, Eq. (3.14) is generalized to:

$$\frac{dn_{\text{sys}}}{dt} = \sum_{\text{all } j \text{ in}} \dot{n}_j - \sum_{\text{all } j \text{ out}} \dot{n}_j + \sum_{\substack{\text{all } k \\ \text{reactions}}} \left(\sum_{\substack{\text{all } i \\ \text{compounds}}} \nu_{ik} \right) \dot{\xi}_k \tag{3.18a}$$

Helpful Hint
Use the steady-state differential material balance equation to analyze continuous-flow steady-state processes.

Many times we are interested in analyzing systems at steady state, so it is useful to rewrite the material balance equations for the special case where $dn_{\text{sys}}/dt = dn_{i,\text{sys}}/dt = 0$:

$$\sum_{\text{all } j \text{ out}} \dot{n}_{ij} = \sum_{\text{all } j \text{ in}} \dot{n}_{ij} + \sum_{\substack{\text{all } k \\ \text{reactions}}} \nu_{ik}\,\dot{\xi}_k \tag{3.17b}$$

$$\sum_{\text{all } j \text{ out}} \dot{n}_j = \sum_{\text{all } j \text{ in}} \dot{n}_j + \sum_{\substack{\text{all } k \\ \text{reactions}}} \left(\sum_{\substack{\text{all } i \\ \text{compounds}}} \nu_{ik} \right) \dot{\xi}_k \tag{3.18b}$$

Example 3.4   ### Mole Balances: Manufacture of Urea

Urea, $(NH_2)_2CO$, is a widely used fertilizing agent made from ammonia. (It is also a component of urine, made during metabolism of proteins and amino acids.) Commercially, urea is manufactured from ammonia and carbon dioxide:

$$2NH_3 + CO_2 \rightarrow (NH_2)_2CO + H_2O \tag{R1}$$

Two gas streams, one at 230 gmol/min and containing 85 mol% NH_3 and 15 mol% CO_2, and the other at 100 gmol/min and containing 80 mol% CO_2 and 20 mol% H_2O, are mixed and fed to a reactor operating at steady state. Inside the reactor, ammonia is consumed at a rate of 180 gmol/min. (See sketch.) Use Eqs. (3.17) and (3.18) to find the flow rate (gmol/min) of each component out of the reactor as well as the total flow out.

Solution

We'll use subscripts A, C, U, and W to indicate components ammonia, carbon dioxide, urea, and water, respectively. Streams are indicated by number, as shown on the flow diagram. First we calculate the extent of reaction (R1), $\dot{\xi}_1$, from $\dot{r}_{A1}$, the rate of consumption of ammonia by reaction (R1):

$$\dot{r}_{A1} = -180 \text{ gmol/min} = v_{A1}\dot{\xi}_1 = (-2)\dot{\xi}_1$$

$$\dot{\xi}_1 = 90 \text{ gmol/min}$$

Now we use Eq. (3.17b) to calculate the outlet flows of the four components, noting that the reactor is operating at steady state:

$$\dot{n}_{A3} = \dot{n}_{A1} + v_{A1}\dot{\xi}_1 = (0.85)230 + (-2)90 = 15.5 \text{ gmol NH}_3\text{/min}$$

$$\dot{n}_{C3} = \dot{n}_{C1} + \dot{n}_{C2} + v_{C1}\dot{\xi}_1 = (0.15)230 + (0.8)100 + (-1)90 = 24.5 \text{ gmol CO}_2\text{/min}$$

$$\dot{n}_{U3} = v_{U1}\dot{\xi}_1 = (+1)90 = 90 \text{ gmol urea/min}$$

$$\dot{n}_{W3} = \dot{n}_{W2} + v_{W1}\dot{\xi}_1 = (0.2)100 + (+1)90 = 110 \text{ gmol H}_2\text{O/min}$$

To calculate the total molar flow in the reactor outlet stream, we can simply add up the component molar flows:

$$\dot{n}_3 = \dot{n}_{A3} + \dot{n}_{C3} + \dot{n}_{U3} + \dot{n}_{W3} = 15.5 + 24.5 + 90 + 110 = 240 \text{ gmol/min}$$

Alternatively, we can try Eq. (3.18b)—we should get the same answer:

$$\dot{n}_3 = \dot{n}_1 + \dot{n}_2 + (v_{A1} + v_{C1} + v_{U1} + v_{W1})\dot{\xi}_1$$

$$= 230 + 100 + (-2 - 1 + 1 + 1)90 = 240 \text{ gmol/min}$$

Example 3.5	**Mole Balances: Urea Manufacture from Cheaper Reactants**

Urea can be manufactured from methane (CH_4), water, and nitrogen via a pathway requiring four chemical reactions:

$$CH_4 + H_2O \rightarrow CO + 3H_2 \tag{R1}$$

$$CO + H_2O \rightarrow CO_2 + H_2 \tag{R2}$$

$$N_2 + 3H_2 \rightarrow 2NH_3 \tag{R3}$$

$$NH_3 + 0.5CO_2 \rightarrow 0.5(NH_2)_2CO + 0.5H_2O \tag{R4}$$

We'd like to design a process to make 90 gmol/min urea at steady state, from the raw materials methane, water, and nitrogen. Furthermore, there should be no reactants (CH_4, H_2O, or N_2) nor any CO, CO_2, or NH_3 leaving the process. (See sketch.) What should be the reactant feed rates? What are the byproducts?

Solution

In Chap. 1, we would have used a generation-consumption analysis to solve this problem. Here we'll use Eq. (3.17) instead. We'll use subscripts M, W, CO, CD, H, N, A, and U to indicate components methane, water, carbon monoxide, carbon dioxide, hydrogen, nitrogen, ammonia and urea, respectively. Because the process is at steady state, $dn_{i,\text{sys}}/dt = 0$ for all components. Eq. (3.17) simplifies to:

！Helpful Hint
With flowrates $\dot{n}_{ij}$, the second subscript is the stream number. With stoichiometric coefficients ν_{ik}, the second subscript is the reaction number.

$$0 = \dot{n}_{M1} + \nu_{M1}\dot{\xi}_1 = \dot{n}_{M1} - \dot{\xi}_1$$

$$0 = \dot{n}_{W1} + \nu_{W1}\dot{\xi}_1 + \nu_{W2}\dot{\xi}_2 + \nu_{W4}\dot{\xi}_4 = \dot{n}_{W1} - \dot{\xi}_1 - \dot{\xi}_2 + 0.5\dot{\xi}_4$$

$$0 = \nu_{CO1}\dot{\xi}_1 + \nu_{CO2}\dot{\xi}_2 = \dot{\xi}_1 - \dot{\xi}_2$$

$$0 = \nu_{CD2}\dot{\xi}_2 + \nu_{CD4}\dot{\xi}_4 = \dot{\xi}_2 - 0.5\dot{\xi}_4$$

$$\dot{n}_{H2} = \nu_{H1}\dot{\xi}_1 + \nu_{H2}\dot{\xi}_2 + \nu_{H3}\dot{\xi}_3 = 3\dot{\xi}_1 + \dot{\xi}_2 - 3\dot{\xi}_3$$

$$0 = \dot{n}_{N1} + \nu_{N3}\dot{\xi}_3 = \dot{n}_{N1} - \dot{\xi}_3$$

$$0 = \nu_{A3}\dot{\xi}_3 + \nu_{A4}\dot{\xi}_4 = 2\dot{\xi}_3 - \dot{\xi}_4$$

$$\dot{n}_{U2} = 90 \text{ gmol/min} = \nu_{U4}\dot{\xi}_4 = 0.5\dot{\xi}_4$$

Now, we work backwards, first to find the extents of reaction and then to find the flowrates:

$$90 \text{ gmol/min} = 0.5\dot{\xi}_4 \qquad \Rightarrow \qquad \dot{\xi}_4 = 180 \text{ gmol/min}$$

$$2\dot{\xi}_3 = \dot{\xi}_4 \qquad \Rightarrow \qquad \dot{\xi}_3 = 90 \text{ gmol/min}$$

$$\dot{\xi}_2 = 0.5\dot{\xi}_4 \qquad \Rightarrow \qquad \dot{\xi}_2 = 90 \text{ gmol/min}$$

$$\dot{\xi}_1 = \dot{\xi}_2 \qquad \Rightarrow \qquad \dot{\xi}_1 = 90 \text{ gmol/min}$$

$$\dot{n}_{N1} = \dot{\xi}_3 \qquad \Rightarrow \qquad \dot{n}_{N1} = 90 \text{ gmol/min}$$

$$\dot{n}_{W1} = \dot{\xi}_1 + \dot{\xi}_2 - 0.5\dot{\xi}_4 \qquad \Rightarrow \qquad \dot{n}_{W1} = 90 \text{ gmol/min}$$

$$\dot{n}_{M1} = \dot{\xi}_1 \qquad \Rightarrow \qquad \dot{n}_{M1} = 90 \text{ gmol/min}$$

$$\dot{n}_{H2} = 3\dot{\xi}_1 + \dot{\xi}_2 - 3\dot{\xi}_3 \qquad \Rightarrow \qquad \dot{n}_{H2} = 90 \text{ gmol/min}$$

Equations (3.15), (3.16), (3.17), and (3.18) may look complicated, but they have the same meaning as an equation that you are already very familiar with. In Table 3.1, these four differential material balance equations are compared to the material balance equation that you learned in Chap. 2.

Table 3.1	Material Balance Equations: Differential Form			
	Accumulation =	**Input**	**−Output**	**+Generation − Consumption**
Total mass	$\dfrac{dm_{sys}}{dt} =$	$\displaystyle\sum_{\text{all } j \text{ in}} \dot{m}_j$	$- \displaystyle\sum_{\text{all } j \text{ out}} \dot{m}_j$	
Mass of i	$\dfrac{dm_{i,sys}}{dt} =$	$\displaystyle\sum_{\text{all } j \text{ in}} \dot{m}_{ij}$	$- \displaystyle\sum_{\text{all } j \text{ out}} \dot{m}_{ij}$	$+ \displaystyle\sum_{\substack{\text{all } k \\ \text{reactions}}} M_i \nu_{ik}\, \dot{\xi}_k$
Total moles	$\dfrac{dn_{sys}}{dt} =$	$\displaystyle\sum_{\text{all } j \text{ in}} \dot{n}_j$	$- \displaystyle\sum_{\text{all } j \text{ out}} \dot{n}_j$	$+ \displaystyle\sum_{\substack{\text{all } k \\ \text{reactions}}} \left(\displaystyle\sum_{\substack{\text{all } i \\ \text{compounds}}} \nu_{ik} \right) \dot{\xi}_k$
Moles of i	$\dfrac{dn_{i,sys}}{dt} =$	$\displaystyle\sum_{\text{all } j \text{ in}} \dot{n}_{ij}$	$- \displaystyle\sum_{\text{all } j \text{ out}} \dot{n}_{ij}$	$+ \displaystyle\sum_{\substack{\text{all } k \\ \text{reactions}}} \nu_{ik}\, \dot{\xi}_k$

3.2.3   Degree of Freedom Analysis

In the 10 Easy Steps method of Chap. 2, you defined system variables for reactions to account for generation and consumption of each component. For example, if the reaction was

$$CH_4 + 2O_2 \rightarrow CO_2 + 2H_2O$$

you would have written four system variables: methane consumed, oxygen consumed, CO_2 generated, and water generated. But, in the DOF analysis, you counted the reaction as just one system variable. Now you see why: because the system variable is really the extent of reaction, $\dot{\xi}_k$, and there is only one $\dot{\xi}_k$ per reaction. (Since $\dot{R}_{ik} = \nu_{ik} M_i \dot{\xi}_k$, this is true for either mass or mole units.) The stoichiometric coefficients and molar masses are not system variables, because they are a function of the chemistry and do not change when the process changes.

Example 3.6   **DOF Analysis: Urea Synthesis from Cheaper Reactants**

Complete a DOF analysis of the following problem:

Urea can be manufactured from methane (CH_4), water, and nitrogen via a pathway requiring four chemical reactions:

$$CH_4 + H_2O \rightarrow CO + 3H_2 \tag{R1}$$
$$CO + H_2O \rightarrow CO_2 + H_2 \tag{R2}$$
$$N_2 + 3H_2 \rightarrow 2NH_3 \tag{R3}$$
$$NH_3 + 0.5CO_2 \rightarrow 0.5(NH_2)_2CO + 0.5H_2O \tag{R4}$$

We'd like to design a process to make 90 gmol/min urea at steady state, from the raw materials methane, water, and nitrogen. Is this process completely specified?

$$\begin{array}{c} CH_4 \\ H_2O \\ N_2 \end{array} \xrightarrow{\dot{n}_1} \boxed{\text{Process}} \xrightarrow{\dot{n}_2} \begin{array}{l} \text{90 gmol/min urea} \\ H_2,\ CH_4,\ H_2O,\ N_2,\ CO,\ CO_2,\ NH_3 \end{array}$$

Solution

This problem is almost the same as Example 3.5, except we no longer restrict the allowable components in the exit stream.

Number of Independent Variables

	Answer	Explanation
Stream variables	11	3 components in input stream 8 components in output stream
System variables	4	4 reactions no accumulation (steady-state)
Total	15	

Number of Independent Equations

	Answer	Explanation
Specified flows	1	90 gmol urea/min
Specified stream compositions	0	
Specified system performance	0	
Material balance equations	8	8 components in system
Total	9	

DOF $= 15 - 9 = 6$.

The problem is greatly underspecified. We were able to solve Example 3.5 by specifying that 6 components do not appear in the output stream.

Example 3.7   **Differential Material Balance Equation with Multiple Chemical Reactions at Steady State: Benzene into Catechol**

Catechol is used to make pharmaceuticals like L-Dopa (used to treat Parkinsons' disease) and flavorings like vanillin. In the synthetic route to catechol, benzene and propylene (C_3H_6) combine to make cumene (also called isopropylbenzene, C_9H_{12}):

$$C_6H_6 + C_3H_6 \rightarrow C_9H_{12} \tag{R1}$$

Then cumene reacts with oxygen to give the unstable product cumene hydroperoxide:

$$C_9H_{12} + O_2 \rightarrow C_9H_{12}O_2 \tag{R2}$$

Cumene hydroperoxide breaks down into phenol and a byproduct, acetone (C_3H_6O):

$$C_9H_{12}O_2 \rightarrow C_6H_6O + C_3H_6O \tag{R3}$$

Phenol reacts with hydrogen peroxide (HOOH), a strong oxidizing agent, to produce catechol (o-$C_6H_6O_2$)

$$C_6H_6O + HOOH \rightarrow o\text{-}C_6H_6O_2 + H_2O \tag{R4}$$

Your job is to design a process using this reaction pathway that produces 3000 kg catechol/day. You conduct some experiments in a laboratory-scale apparatus (called a "pilot plant"). In the experiment, 13.65 gmol benzene/h, 17.75 gmol propylene/h, 12.3 gmol oxygen/h, and 10.4 gmol hydrogen peroxide/h are fed to the pilot plant. The output stream from the plant contains catechol, benzene, propylene, phenol, acetone, and water. Complete a DOF analysis to determine if this problem is completely specified. Calculate raw material requirements and byproduct flows for a 3000-kg/day catechol production rate, assuming the full-scale plant operates exactly the same as the pilot plant.

Solution

We start as always with a flow diagram:

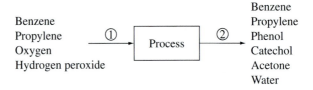

We will calculate the outlet stream from the pilot plant, then scale up the results to the full production scale. Let's start with the DOF analysis.

Number of Independent Variables

	Answer	**Explanation**
Stream variables	10	4 components in input stream 6 components in output stream
System variables	4	4 reactions No accumulation (steady state)
Total	14	

Number of Independent Equations

	Answer	**Explanation**
Specified flows	4	4 inlet flows given
Specified stream compositions	0	
Specified system performance	0	
Material balance equations	10	Includes all compounds, even those that are strictly intermediates
Total	14	

DOF $= 14 - 14 = 0$. The problem is completely specified.

The system is continuous flow and steady state, and there are reactions of known stoichiometry, so we use the steady-state differential mole balance equation, Eq. (3.17b).

$$\sum_{\substack{\text{all } j \text{ out}}} \dot{n}_{ij} = \sum_{\substack{\text{all } j \text{ in}}} \dot{n}_{ij} + \sum_{\substack{\text{all } k \\ \text{reactions}}} \nu_{ik} \dot{\xi}_k$$

Since there is only one stream in and one stream out, the material balance equation (in units of gmol/h) simplifies to:

Benzene	$n_{B2} = 13.65 - \dot{\xi}_1$
Propylene:	$n_{P2} = 17.75 - \dot{\xi}_1$
Cumene:	$0 = \dot{\xi}_1 - \dot{\xi}_2$
Oxygen:	$0 = 12.3 - \dot{\xi}_2$
Cumene HP:	$0 = \dot{\xi}_2 - \dot{\xi}_3$
Phenol:	$\dot{n}_{Ph2} = \dot{\xi}_3 - \dot{\xi}_4$
Acetone:	$n_{A2} = \dot{\xi}_3$
Hydrogen peroxide:	$0 = 10.4 - \dot{\xi}_4$
Catechol:	$n_{C2} = \dot{\xi}_4$
Water:	$n_{W2} = \dot{\xi}_4$

We solve by first identifying the equations with only one unknown, then working our way through the remainder. The extents of reaction are $\dot{\xi}_1 = \dot{\xi}_2 = \dot{\xi}_3 = 12.3$ gmol/h and $\dot{\xi}_4 = 10.4$ gmol/h. So the reactions are limited by the oxidizing reactants oxygen and hydrogen peroxide. (We will discuss limiting reactants further in Chap. 4, and explain why we might want to operate a process with limiting quantities of some reactants.)

Calculating the flows of each compound is straightforward and left for the reader. We then scale up to 3000 kg catechol/day by multiplying each flow by the compound's molar mass and then using a scale factor of 2.622. Results are summarized in convenient table form.

Compound	M_i	Pilot plant flow rate gmol/h		Production plant flow rate kg/day	
		in	out	in	out
Benzene	78	13.65	1.35	2792	276
Propylene	42	17.75	5.45	1955	600
Oxygen	32	12.3		1032	
Acetone	58		12.3		1871
Phenol	94		1.9		468
Hydrogen peroxide	34	10.4		927	
Water	18		10.4		491
Catechol	110		10.4		3000

You can check the solution by seeing if the total mass flow in equals the total mass flow out. (It does.)

3.2.4 General Forms of the Integral Material Balance Equations

In Sec. 3.2.1, we derived differential material balance equations by letting the time interval get infinitesimally small, $\Delta t \to dt$. Equations (3.15) to (3.18) apply for a snapshot—a single instant of time. But what if we want to consider the system over a finite time interval, say between $t = t_0$ and $t = t_f$?

To answer this question, let's return our attention to Fig. 3.2 and Eq. (3.3):

$$\Delta m_{sys} = \dot{m}_1 \Delta t - \dot{m}_2 \Delta t \qquad (3.3)$$

If we define $m_{sys,0}$ as the mass in the system at t_0, and $m_{sys,f}$ as the mass in the system at t_f, then

$$\Delta m_{sys} = m_{sys,f} - m_{sys,0}$$

But, can we similarly replace Δt with $t_f - t_0$ on the right-hand side of Eq. (3.3)? Although we might be tempted to do this, let's consider the possibility that *flow rates might change with time*. This idea is illustrated in Fig. 3.4.

Suppose we start with an initially empty system ($m_{sys,0} = 0$), one input $\dot{m}_1$ that changes over the time interval from $t = t_0$ to $t = t_f$, and no output ($\dot{m}_2 = 0$). Imagine for example that there are several step changes in $\dot{m}_1$, as illustrated in Fig. 3.5a. Or, imagine that $\dot{m}_1$ increases linearly with time, as illustrated in Fig. 3.5c. In either case, the total mass entering the system is simply equal to the area under the

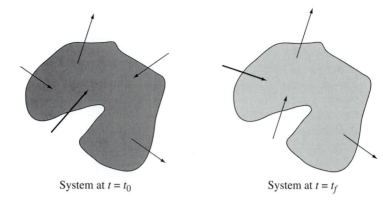

System at $t = t_0$ System at $t = t_f$

Figure 3.4 A system with multiple inputs and outputs, shown at two different times. The system is a three-dimensional enclosed volume, and the system boundary is a two-dimensional surface area. The system is shaded, and the system boundary is shown as a dark line. The darkness of the shading corresponds to the quantity of the material in the system, which changes with time. Material flows in and out are shown as lines with arrows. The thickness of the lines corresponds to the qsuantity of material flow, which changes with time.

curve of $\dot{m}_1$ versus t (Fig. 3.5b and d). Mathematically, this is the integral of $\dot{m}_1$ from $t = t_0$ to $t = t_f$:

$$m_{sys} = \int_{t_0}^{t_f} \dot{m}_1\, dt$$

If both $\dot{m}_1$ and $\dot{m}_2$ (Fig. 3.2) vary over the time interval from $t = t_0$ to $t = t_f$, then,

$$m_{sys,f} - m_{sys,0} = \int_{t_0}^{t_f} \dot{m}_1\, dt - \int_{t_0}^{t_f} \dot{m}_2\, dt$$

We generalize to a system with multiple input and output streams to derive the integral total mass balance equation:

$$m_{sys,f} - m_{sys,0} = \sum_{\text{all } j \text{ in}} \left(\int_{t_0}^{t_f} \dot{m}_j\, dt \right) - \sum_{\text{all } j \text{ out}} \left(\int_{t_0}^{t_f} \dot{m}_j\, dt \right) \qquad (3.19)$$

Example 3.8   **Integral Equation: Blending and Shipping**

At a blending and shipping facility at a large integrated refinery, gasoline from three different processes—called reformer, isomax, and FCC—is pumped into a large storage tank. The processes run continuously, producing 15,400, 18,200, and 10,500 barrels of gasoline per day, respectively. (A barrel is 42 gallons, and gasoline has a

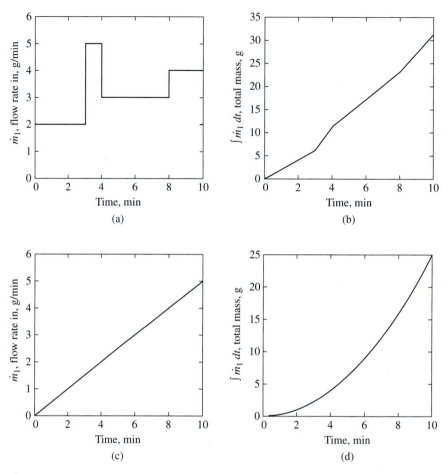

Figure 3.5 (a) Step change in $\dot{m}_1$; (b) integral under the curve of (a); (c) linear increase in $\dot{m}_1$; (d) integral under the curve of (c).

density of about 6.6 lb/gal.) A large tanker comes to port to load up. The tanker holds 60.966 million lb of gasoline, and it takes 54 hours to pump it full. When pumping to the tanker is first started, the storage tank contains 154,000 barrels. The ship's captain is worried that there isn't enough gasoline to fill the tanker, but the supervisor at the blending and shipping facility tells him not to worry. Who's right?

Solution

The flow diagram is shown, with streams numbered.

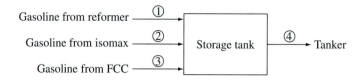

The storage tank is the system. We will do all calculations on a mass basis, which requires us to convert from volumetric flow to mass flow. The conversion factor is

$$\frac{42 \text{ gal}}{\text{barrel}} \times \frac{6.6 \text{ lb}}{\text{gal}} \times \frac{1 \text{ day}}{24 \text{ h}} = 11.55 \frac{\text{lb/h}}{\text{barrel/day}}$$

Since gasoline is the only material in the system, we write a balance on the total mass. The stream variables are:

$$\dot{m}_1 = 15,400 \frac{\text{barrels}}{\text{day}} \times 11.55 \frac{\text{lb/h}}{\text{barrel/day}} = 178,000 \frac{\text{lb}}{\text{h}}$$

$$\dot{m}_2 = 18,200 \frac{\text{barrels}}{\text{day}} \times 11.55 \frac{\text{lb/h}}{\text{barrel/day}} = 210,000 \frac{\text{lb}}{\text{h}}$$

$$\dot{m}_3 = 10,500 \frac{\text{barrels}}{\text{day}} \times 11.55 \frac{\text{lb/h}}{\text{barrel/day}} = 121,000 \frac{\text{lb}}{\text{h}}$$

$$\dot{m}_4 = \frac{60,966,000 \text{ lb}}{54 \text{ h}} = 1,129,000 \frac{\text{lb}}{\text{h}}$$

There are no chemical reactions, no stream composition specifications, and no system performance specifications. Since we are interested in what happens over a specified time interval (54 hours), we use the integral total mass balance equation [Eq. (3.19)].

$$m_{\text{sys},f} - m_{\text{sys},0} = \sum_{\text{all } j \text{ in}} \left(\int_{t_0}^{t_f} \dot{m}_j \, dt \right) - \sum_{\text{all } j \text{ out}} \left(\int_{t_0}^{t_f} \dot{m}_j \, dt \right)$$

Using our known flow rates, we find:

$$m_{\text{sys},f} - m_{\text{sys},0} = \int_{t_0}^{t_f} \dot{m}_1 \, dt + \int_{t_0}^{t_f} \dot{m}_2 \, dt + \int_{t_0}^{t_f} \dot{m}_3 \, dt - \int_{t_0}^{t_f} \dot{m}_4 \, dt$$

$$= \int_{t_0}^{t_f} (\dot{m}_1 + \dot{m}_2 + \dot{m}_3 - \dot{m}_4) \, dt$$

$$= \int_{0}^{54} (178,000 + 210,000 + 121,000 - 1,129,000) \, dt$$

$$m_{\text{sys},f} - m_{\text{sys},0} = -620,000 (54 - 0) = -33,480,000 \text{ lb}$$

Quick Quiz 3.4

If the FCC process was shut down when the tanker was in port, should the ship's captain worry?

With all three processes running, what is the minimum initial quantity in the storage tank to avoid running dry during tanker loading?

We know the quantity of gasoline in the storage tank at the beginning of the pumping operation:

$$m_{\text{sys},0} = 154,000 \text{ barrels} \times \frac{42 \text{ gal}}{\text{barrel}} \times \frac{6.6 \text{ lb}}{\text{gal}} = 42,700,000 \text{ lb}$$

Therefore:

$$m_{sys,f} = -33{,}480{,}000 + 42{,}700{,}000 = 9{,}220{,}000 \text{ lb}$$

After filling up the ship, the storage tank still contains over 9 million lb gasoline, or over 33,000 barrels, so the ship's captain doesn't need to worry.

In Sec. 3.2.2, we derived the differential material balance equations for component mass, total moles, and component moles, from conservation-of-mass principles. For the corresponding integral balance equations, we must account for the possibility that not only flow rates but also reaction rates may vary with time. The corresponding integral material balance equations are

$$m_{i,sys,f} - m_{i,sys,0} = \sum_{\text{all } j \text{ in}} \left(\int_{t_0}^{t_f} \dot{m}_{ij}\, dt \right) - \sum_{\text{all } j \text{ out}} \left(\int_{t_0}^{t_f} \dot{m}_{ij}\, dt \right)$$

$$+ \sum_{\substack{\text{all } k \\ \text{reactions}}} \left(\int_{t_0}^{t_f} M_i \nu_{ik} \dot{\xi}_k\, dt \right) \tag{3.20}$$

$$n_{sys,f} - n_{sys,0} = \sum_{\text{all } j \text{ in}} \left(\int_{t_0}^{t_f} \dot{n}_j\, dt \right) - \sum_{\text{all } j \text{ out}} \left(\int_{t_0}^{t_f} \dot{n}_j\, dt \right)$$

$$+ \sum_{\substack{\text{all } k \\ \text{reactions}}} \sum_{\substack{\text{all } i \\ \text{compounds}}} \left(\int_{t_0}^{t_f} \nu_{ik} \dot{\xi}_k\, dt \right) \tag{3.21}$$

$$n_{i,sys,f} - n_{i,sys,0} = \sum_{\text{all } j \text{ in}} \left(\int_{t_0}^{t_f} \dot{n}_{ij}\, dt \right) - \sum_{\text{all } j \text{ out}} \left(\int_{t_0}^{t_f} \dot{n}_{ij}\, dt \right)$$

$$+ \sum_{\substack{\text{all } k \\ \text{reactions}}} \left(\int_{t_0}^{t_f} \nu_{ik} \dot{\xi}_k\, dt \right) \tag{3.22}$$

⚠ Helpful Hint
Use an integral material balance equation to analyze the performance of batch and semibatch processes over a specified time interval.

Eq. (3.19) to (3.22) are rather fearsome in appearance. In reality, these equations usually simplify considerably. For example, if the system is at steady state, there is no change in the material in the system and the left-hand side equals zero. As another example, if the flow rates are constant, then the integral becomes simply the flow rate multiplied by the time interval ($t_f - t_0$). It is important to recognize that the integral material balance equations allow us to evaluate the *change* in material in the system but not the *absolute* quantity in the system. For that, we need additional information, such as the initial quantity of material in the system.

| **Example 3.9** | **Integral Equation with Unsteady Flow: Jammin' with Cherries** |

At the award-winning Jumpin' Jam Factory, cherry jam is produced by chopping up 100 lb cherries and mixing the chopped cherries all at once with 200 lb sugar in a pot. Then, water is boiled off. As the mixture thickens, the water evaporation rate decreases. Charlie Cherrypit, the engineer at Jumpin' Jam, estimated that the evaporation rate can be modeled as $\dot{m}_{w,evap} = 2.0 - 0.03t$ with t in units of minutes and $\dot{m}_{w,evap}$ in units of lb/min. How long will it take to make 240 lb of jam?

Solution

The flow diagram, with the pot as the system, is shown.

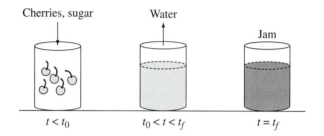

This is a semibatch process: The cherries and sugar are charged all at once to the pot initially, the water vapor is removed continuously, and the jam is collected from the pot at the end. Notice some features of this problem: There is an *initial* charge of material to the system, there is a *specified time interval,* and we want to know the *total mass* left in the system at the end of the time interval. Furthermore, there is *no chemical reaction.* These clues together indicate that the easiest way to solve this problem is to use the integral total mass balance equation:

$$m_{sys,f} - m_{sys,0} = \sum_{\text{all } j \text{ in}}\left(\int_{t_0}^{t_f} \dot{m}_j \, dt \right) - \sum_{\text{all } j \text{ out}}\left(\int_{t_0}^{t_f} \dot{m}_j \, dt \right)$$

From the information given, at $t = t_0$,

$$m_{sys,0} = 100 \text{ lb cherries} + 200 \text{ lb sugar} = 300 \text{ lb}$$

At the end of jam making, there should be 240 lb jam in the pot, or

$$m_{sys,f} = 240 \text{ lb}$$

There is no flow of material into the pot from $t = t_0$ to $t = t_f$, so

$$\sum_{\text{all } j \text{ in}}\left(\int_{t_0}^{t_f} \dot{m}_j \, dt \right) = 0$$

Material leaves the pot in a single output stream of water vapor, with the rate decreasing with time. The total mass leaving the pot from $t = t_0$ to $t = t_f$, is

$$\sum_{\text{all } j \text{ out}} \left(\int_{t_0}^{t_f} \dot{m}_j \, dt \right) = \int_{t_0}^{t_f} \dot{m}_{w,\text{evap}} \, dt = \int_{t_0}^{t_f} (2.0 - 0.03t) \, dt$$

$$= 2(t_f - t_0) - 0.015(t_f - t_0)^2$$

Inserting these expressions into the material balance equation gives

$$240 - 300 = -[2(t_f - t_0) - 0.015(t_f - t_0)^2]$$

This is a quadratic equation, with two solutions: $t_f - t_0 = \Delta t = 45.6$ minutes or 87.7 minutes. Which is right? Only one answer makes physical sense. If $\Delta t = 87.7$ minutes, the water evaporation rate would be negative (-0.063 lb/min), which would mean that water would be entering the pot. The answer of $\Delta t = 45.6$ minutes is reasonable; the water evaporation rate at the end of the jam-making session is $+ 0.063$ lb/min, about 80% less than the evaporation rate initially.

In Table 3.2, the integral equations are compared to the material balance equation you learned in Chapter 2. Compare these equations also to the differential equations in Table 3.1.

Table 3.2   Material Balance Equations: Integral Form

	Accumulation =	Input	−Output	+Generation − Consumption
Total mass	$m_{\text{sys},f} - m_{\text{sys},0} =$	$\displaystyle\sum_{\text{all } j \text{ in}} \left(\int_{t_0}^{t_f} \dot{m}_j \, dt \right)$	$\displaystyle- \sum_{\text{all } j \text{ out}} \left(\int_{t_0}^{t_f} \dot{m}_j \, dt \right)$	
Mass of i	$m_{i,\text{sys},f} - m_{i,\text{sys},0} =$	$\displaystyle\sum_{\text{all } j \text{ in}} \left(\int_{t_0}^{t_f} \dot{m}_{ij} \, dt \right)$	$\displaystyle- \sum_{\text{all } j \text{ out}} \left(\int_{t_0}^{t_f} \dot{m}_{ij} \, dt \right)$	$\displaystyle+ \sum_{\substack{\text{all } k \\ \text{reactions}}} \left(\int_{t_0}^{t_f} M_i \nu_{ik} \dot{\xi}_k \, dt \right)$
Total moles	$n_{\text{sys},f} - n_{\text{sys},0} =$	$\displaystyle\sum_{\text{all } j \text{ in}} \left(\int_{t_0}^{t_f} \dot{n}_j \, dt \right)$	$\displaystyle- \sum_{\text{all } j \text{ out}} \left(\int_{t_0}^{t_f} \dot{n}_j \, dt \right)$	$\displaystyle+ \sum_{\substack{\text{all } k \\ \text{reactions}}} \sum_{\substack{\text{all } i \\ \text{compounds}}} \left(\int_{t_0}^{t_f} \nu_{ik} \dot{\xi}_k \, dt \right)$
Moles of i	$n_{i,\text{sys},f} - n_{i,\text{sys},0} =$	$\displaystyle\sum_{\text{all } j \text{ in}} \left(\int_{t_0}^{t_f} \dot{n}_{ij} \, dt \right)$	$\displaystyle- \sum_{\text{all } j \text{ out}} \left(\int_{t_0}^{t_f} \dot{n}_{ij} \, dt \right)$	$\displaystyle+ \sum_{\substack{\text{all } k \\ \text{reactions}}} \left(\int_{t_0}^{t_f} \nu_{ik} \dot{\xi}_k \, dt \right)$

3.2.5 A Few More Problems

Let's finish this section with two problems that illustrate the use of these material balance equations to complete flow calculations. These problems are a bit more challenging than the ones you have seen already. In working these problems we will use the 10 Easy Steps, but we will not explicitly list each step. See if you can identify each of the steps.

| **Example 3.10** | **Integral Equation with Unsteady Flow and Chemical Reaction: Controlled Drug Release** |

Helpful Hint
If you get stymied, remember accumulation = input − output + (generation − consumption), check for dimensional consistency, and recall the 10 Easy Steps!

Patients with certain diseases are treated by injection of proteins or drugs into their bloodstream. Upon injection there is a sudden increase in the protein or drug concentration in the blood to very high levels, but then the concentration rapidly falls. A steadier blood concentration is often desirable, to reduce toxic side effects and increase therapeutic efficacy. Controlled-release technology reduces the variability in drug concentration in the blood. With controlled release, the protein or drug is encapsulated in a polymer and is released slowly into the bloodstream. This maintains the concentration of drug or protein in the bloodstream at a lower, more constant level.

1. 100 μg of a drug is loaded into a controlled-release capsule and then injected into a patient. The drug is released at a rate of $8e^{-0.1t}$ μg/h, where t is hours after injection. What fraction of the drug has been released after 24 h?
2. Once in the bloodstream, the drug is lost at a rate of 3.1 μg/h as a result of degradation reactions and elimination processes. How does the mass (μg) of drug in the bloodstream vary as a function of time after injection? At what time is the drug concentration the highest?

Solution

1. We choose the capsule as the system. The capsule releases the drug in all directions into the bloodstream.

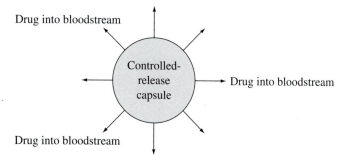

Drug into bloodstream

Drug into bloodstream

Controlled-release capsule

Drug into bloodstream

Drug into bloodstream

We are interested in what happens to the mass of drug D in the capsule over a 24-h time interval, so we use the integral mass balance applied to the drug as the component:

$$m_{D,\text{sys},f} - m_{D,\text{sys},0} = \sum_{\substack{\text{all } j \text{ in}}} \left(\int_{t_0}^{t_f} \dot{m}_{Dj}\, dt \right) - \sum_{\substack{\text{all } j \text{ out}}} \left(\int_{t_0}^{t_f} \dot{m}_{Dj}\, dt \right)$$

$$+ \sum_{\substack{\text{all } k \\ \text{reactions}}} \left(\int_{t_0}^{t_f} M_D \nu_{Dk} \dot{\xi}_k\, dt \right)$$

The initial quantity of drug in the capsule is $m_{D,\text{sys},0} = 100\ \mu g$. There is no new drug entering the capsule, so

$$\sum_{\substack{\text{all } j \text{ in}}} \left(\int_{t_0}^{t_f} \dot{m}_{Dj}\, dt \right) = 0$$

There is no chemical reaction inside the capsule, so

$$\sum_{\substack{\text{all } k \\ \text{reactions}}} \left(\int_{t_0}^{t_f} M_D \nu_{Dk} \dot{\xi}_k\, dt \right) = 0$$

The mass flow rate of drug from the capsule as a function of time is known:

$$\dot{m}_{D,\text{out}} = 8e^{-0.1t}$$

We now evaluate the total mass flow rate of drug from the capsule over the 24-h interval:

$$\sum_{\substack{\text{all } j \text{ out}}} \left(\int_{t_0}^{t_f} \dot{m}_{DJ}\, dt \right) = \int_0^{24} 8e^{-0.1t}\, dt = -\frac{8}{0.1}(e^{-2.4} - e^0) = 72.7\ \mu g$$

These terms are inserted into the integral mass balance equation to yield:

$$m_{D,\text{sys},f} - 100 = -72.7$$

The drug remaining in the capsule after 24 h is

$$m_{D,\text{sys},f} = -72.7 + 100 = 27.3\ \mu g$$

In other words, 27.3% of the drug remains in the capsule and 72.7% has been released 24 hours after injection.

We can use a similar procedure to evaluate drug release profiles at any time t, simply by integrating from 0 to t rather than from 0 to 24 h. Results of these calculations are shown.

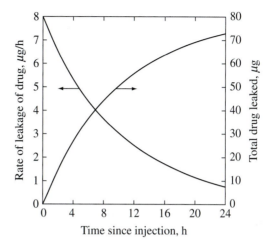

2. Now we are interested in the drug quantity in the bloodstream at any time t after injection. Our system is the blood.

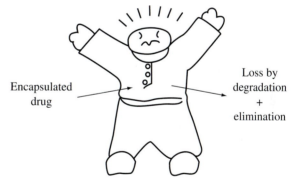

Encapsulated drug

Loss by degradation + elimination

An integral mass balance is still useful, because we are interested in the net accumulation of drug in the system over a finite time interval from $t = 0$ to t. Assuming that there is no drug in the blood initially, $m_{D,sys,0} = 0$, and

$$m_{D,sys,f} = \sum_{\text{all } j \text{ in}} \left(\int_0^t \dot{m}_{Dj} \, dt \right) - \sum_{\text{all } j \text{ out}} \left(\int_0^t \dot{m}_{Dj} \, dt \right) + \sum_{\substack{\text{all } k \\ \text{reactions}}} \left(\int_0^t M_D \nu_{Dk} \dot{\xi}_k \, dt \right)$$

The mass flow rate of drug into the blood equals the mass flow rate of drug out of the capsule, or

$$\sum_{\text{all } j \text{ in}} \left(\int_{t_0}^{t_f} \dot{m}_{Dj} \, dt \right) = \int_0^t 8e^{-0.1t} \, dt = 80 \left(1 - e^{-0.1t} \right)$$

Drug is lost from the blood at a rate of 3.1 μg/h by elimination (a mass flow rate out) and degradation reactions. We don't have enough information to separate these two loss mechanisms, but we can lump them together:

$$\sum_{\substack{\text{all } j \text{ out}}} \left(\int_0^t \dot{m}_{Dj}\, dt \right) + \sum_{\substack{\text{all } k \\ \text{reactions}}} \left(\int_0^t M_D \nu_{Dk}\, \dot{\xi}_k\, dt \right) = \int_0^t 3.1 dt = 3.1t$$

Inserting these expressions into the integral mass balance equation yields

$$m_{D,\text{sys},f} = 80(1 - e^{-0.1t}) - 3.1t$$

A plot of $m_{D,\text{sys},f}$ versus t is illuminating. The rapid initial release of the drug from the capsule causes a rise in the blood concentration. As the release slows, degradation reactions and elimination eventually become faster than the release, and the drug concentration decays. No drug is left 24 h after injection. The maximum concentration is reached at about 9.5 h after injection.

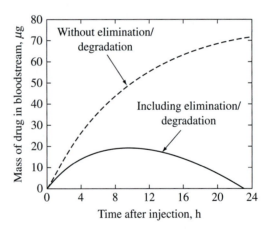

Example 3.11   **Differential Equation with Unsteady Flow and Chemical Reaction: Glucose Utilization in a Fermentor**

Yeast metabolize glucose ($C_6H_{12}O_6$) and make ethanol (C_2H_5OH). Humans have exploited this process for thousands of years to make wine, beer, and other alcoholic beverages. Although the chemical reactions are highly complex, the overall reaction can be written simply as

$$C_6H_{12}O_6 \rightarrow 2C_2H_5OH + 2CO_2$$

Besides making ethanol, yeast grow and reproduce, consuming some of the glucose for maintenance and growth. The rate of glucose consumption depends on the number of yeast in the fermentor as well as the rate of growth of the yeast. So, the glucose consumption rate increases as the yeast multiply.

We start with a 10-L fermentor containing 1000 g glucose and some yeast. The fermentor is operated in semibatch mode. During fermentation, additional glucose is added continuously at a rate of 20 g glucose/h. CO_2 is continuously vented to prevent pressure buildup. No other products or byproducts are removed. The mass rate of glucose consumption $\dot{R}_g$ (g glucose consumed per hour) increases with time as the number of yeast increases:

$$\dot{R}_g = -2.8e^{0.1t}$$

where t is in hours. About 90% of the glucose consumption goes toward ethanol production, with the rest going to support yeast growth.

1. Plot the rate at which glucose in the fermentor changes with time.
2. Plot the grams of glucose and ethanol in the fermentor at any time.
3. Calculate the CO_2 flow rate out of the fermentor as a function of time.
4. How long will it take for the glucose concentration in the fermentor to drop to zero (at which point the fermentation is stopped)? How much ethanol is in the fermentor at that point?

Solution

We start, as always, with a diagram. The fermentor is the chosen system. The fermentor operates in semibatch mode; the glucose and ethanol concentrations inside the vessel change with time, and the CO_2 is continuously removed. All information is given as mass and mass flows, so we'll stick with grams and hours for units.

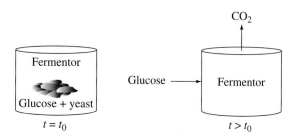

1. We want to know the rate of change of glucose in the fermentor at any specified time, so we'll use a differential equation. We'll use a mass balance, because all information is given in mass units, with glucose as the component. There is no flow of glucose out, so the differential component mass balance simplifies to:

$$\frac{dm_{g,sys}}{dt} = \dot{m}_{g,in} + \dot{R}_g = 20 - 2.8e^{0.1t}$$

where $\dot{m}_{g,in}$ = mass flow of glucose into fermentor.

We'll plot this equation versus time:

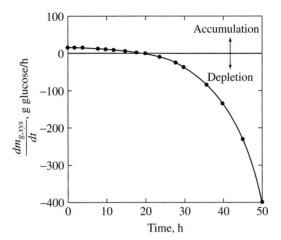

As we see from the graph, initially there is a net accumulation of glucose in the fermentor ($dm_{g,sys}/dt > 0$). At 20 h, however, the rate of consumption of glucose equals the flow rate in ($dm_{g,sys}/dt = 0$); above 20 h, the rate of consumption exceeds the flow rate in and glucose is depleted ($dm_{g,sys}/dt < 0$).

2. To determine the accumulated quantity of glucose in the fermentor at a given time, we need to find an expression for $m_{g,sys}$. We get that by integrating the differential material balance equation:

$$m_{g,sys} = \int(20 - 2.8e^{0.1t})\,dt = 20t - 28e^{0.1t} + C$$

where C is a constant of integration. We know that at $t = 0$, $m_{g,sys} = 1000$ g. From this we find that $C = 1028$ g, and

$$m_{g,sys} = 1028 + 20t - 28e^{0.1t}$$

No ethanol enters or leaves the fermentor, and there is no ethanol in the system at $t = 0$. The differential mass balance equation for ethanol simplifies to:

$$\frac{dm_{e,sys}}{dt} = \dot{R}_e$$

where $m_{e,sys}$ = mass of ethanol in the system and $\dot{R}_e$ = mass rate of reaction of ethanol.

What is $\dot{R}_e$? From the stoichiometry of the glucose-to-ethanol reaction, we know that 2 moles of ethanol are produced per mole of glucose consumed by this reaction, or $\nu_e/\nu_g = -2$. We calculate the molar masses: $M_e = 46$ g/gmol and $M_g = 180$ g/gmol. Finally, from the problem statement we know that, of all the glucose consumed, 90% of it goes toward making ethanol.

Therefore:

$$\dot{R}_e = \frac{\nu_e M_e}{\nu_g M_g}(0.9\dot{R}_g) = -2\left(\frac{46}{180}\right)[0.9(-2.8e^{0.1t})] = 1.29e^{0.1t}$$

We substitute this expression into the ethanol differential mass balance equation, and then integrate (with $m_{e,sys} = 0$ at $t = 0$) to find:

$$m_{e,sys} = 12.9e^{0.1t} - 12.9$$

We plot these two expressions versus time:

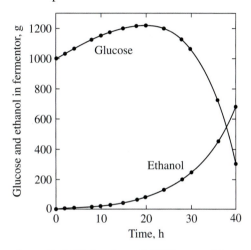

Glucose rises from its initial value of 1000 g, peaking around 20 h into the fermentation before it rapidly drops. Ethanol increases slowly at first, but then at an accelerating rate as the yeast proliferate (until the glucose runs out).

3. CO_2 is generated by reaction along with the ethanol. From the stoichiometry we calculate that

$$\dot{R}_{CO_2}(\text{ethanol}) = \frac{\nu_{CO_2} M_{CO_2}}{\nu_e M_e}\dot{R}_e = \frac{2}{2}\left(\frac{44}{46}\right)\left(1.29e^{0.1t}\right) = 1.23e^{0.1t}$$

Additional CO_2 is generated when glucose is consumed for maintenance and growth of the yeast. We don't have sufficient information to calculate this quantity exactly. However, we can calculate a limiting case where we assume that all the remaining glucose (not consumed in the ethanol production reaction) reacts to CO_2 and H_2O

$$C_6H_{12}O_6 + 6O_2 \rightarrow 6CO_2 + 6H_2O$$

This gives us the maximum amount of CO_2 production possible. Since 10% of the glucose is consumed for maintenance and growth, we calculate that the maximum CO_2 production from this reaction is

$$\dot{R}_{CO_2}(\text{maintenance}) = \frac{\nu_{CO_2} M_{CO_2}}{\nu_g M_g}(0.1\dot{R}_g) = -\frac{6}{1}\left(\frac{44}{180}\right)[0.1(-2.8e^{0.1t})] = 0.41e^{0.1t}$$

The maximum mass rate of reaction of CO_2 is obtained by summing the rate of CO_2 generation from the ethanol reaction and the rate of CO_2 generation from reactions associated with maintenance and growth

$$\sum_{\substack{\text{all } k \\ \text{reactions}}} \dot{R}_{CO_2,k} = 1.23e^{0.1t} + 0.41e^{0.1t} = 1.64e^{0.1t}$$

CO_2 is continuously vented so that it does not build up in the fermentor. The differential mass balance equation for CO_2 is set up and then integrated:

$$\frac{dm_{CO_2,sys}}{dt} = 0 = -\dot{m}_{CO_2,out} + \sum \dot{R}_{CO_2}$$

$$\dot{m}_{CO_2,out} = 1.64e^{0.1t}$$

4. To get the time at which glucose drops to zero in the fermentor, we can solve the integral material balance equation with $m_{g,sys,f} = 0$:

$$m_{g,sys,f} - m_{g,sys,0} = \int_{t_0}^{t_f} \dot{m}_{g,in}\, dt + \int_{t_0}^{t_f} \dot{R}_g\, dt$$

$$0 - 1000 = \int_0^{t_f} 20\, dt + \int_0^{t_f} -2.8 \exp(0.1t)\, dt$$

Evaluating the definite integrals produces:

$$0 - 1000 = 20t_f - 28[e^{0.1t_f} - e^0]$$

$$t_f = 42 \text{ h}$$

The ethanol in the fermentor at 42 h is found, by a similar strategy, to be 847 g.

3.3   Linear Equations and Chemical Reactions (Optional Section)

In Chap. 1, we stoichiometrically balanced chemical reactions and completed generation-consumption analysis. In Chap. 2, we solved many problems requiring simple process flow calculations. In some problems there was only a single process unit, but in other problems we linked together multiple process units into a block flow diagram. We learned how to complete a degree-of-freedom analysis by counting the number of independent variables and independent equations.

In Sec. 3.3 and 3.4, we return to these topics, but with a more mathematical and rigorous approach. In Sec. 3.3, we learn (again) how to balance chemical reactions. We discover a method for determining a set of independent balanced chemical reactions, given only a set of compounds. In Sec. 3.4, we learn how to write linear equations to describe the performance of the four major types of process units on block flow diagrams. Finally, we develop methods for generating linear models of block flow diagrams. To understand these sections, you should have some limited familiarity with linear algebra and matrix manipulations. It would be particularly useful for you to know how to set up and solve matrix equations using your calculator. (This you can do without actually understanding much at all about linear algebra!) For more background, refer to App. A.1.

3.3.1 Linear Equations, Linear Independence, Solution Existence, and Solution Uniqueness

Linear equations are those equations where all the variables are raised to the first power. For example,

$$2^2u + \sqrt{3}v + 4w = 10$$

is a linear equation in variables u, v, and w, whereas

$$2u^2 + 3\sqrt{v} + 4w = 10$$

is linear in w but not in u or v, and so does not qualify as a linear equation.

In chemical process calculations, we are frequently faced with problems involving many linear equations and many variables. To find a solution to a set of linear equations (i.e., find numerical values for all variables that simultaneously satisfy all equations) the number of equations must equal the number of variables. But this is not enough. A crucial feature of a correctly specified set of linear equations is that all equations are linearly **independent**. Equations are linearly independent if a subset of the equations cannot be combined in some way to yield another equation. For example, the set:

$$a + b + c = 10$$

$$a + b = 5$$

$$c = 5$$

has an equal number of equations and variables. But, the equations are *not* linearly independent because the third equation can be derived by multiplying the second equation by −1 and adding the result to the first equation. There are only two independent equations but three variables, and the set of equations is underspecified.

Thus, when faced with finding a solution to a set of linear equations, we ask two questions: (1) Does a unique solution exist? and (2) What is the solution? Sets of linear equations can be solved by the process of substitution and elimination, as we did in earlier chapters. Through this process we incidentally discover whether or not a unique solution exists. Substitution and elimination works well with a few simple linear equations. However, these methods rapidly become tedious and prone to error as the number of equations and variables increases.

Matrix methods provide an alternative strategy for determining solution existence and uniqueness, and for finding the solution, of sets of linear equations. The algebraic manipulations with matrix methods are conceptually no different than those of substitution and elimination. The advantages that matrix methods provide include (1) a compact notation, and (2) a mechanism for solution using calculators or computers. In this section we describe how to set up matrix equations given a set of linear equations. For more details on the mechanics of solving matrix equations, refer to App. A.1.

Suppose we have three equations in three unknowns, such as:

$$u + v + w = 6$$

$$u + 2v + 2w = 11$$

$$2u + 3v - 4w = 3$$

To write this set of three linear equations in matrix notation, we place the coefficients of the variables in a 3×3 matrix, the variables in a column vector, and the numbers on the right-hand side in another column vector:

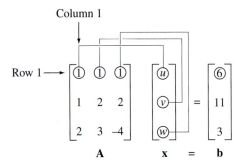

We'll call the matrix of coefficients $\mathbf{A}$. Each entry in $\mathbf{A}$ is identified by row and column number as a_{ij}, where i is the row and j is the column. The column of variables is called the vector $\mathbf{x}$. Each entry in $\mathbf{x}$ is identified as x_j. We'll name the column vector on the right-hand side vector $\mathbf{b}$. Each entry in $\mathbf{b}$ is identified as b_i. To multiple a matrix and a vector, we multiple each *row* in the matrix $\mathbf{A}$ by each entry in the column vector $\mathbf{x}$. The result equals the value in the same *row* of the vector $\mathbf{b}$. In other words, we recover our original equations

$$\sum_{\text{all } j} a_{ij} x_j = b_i,$$

with one equation for each row. The system of equations is written compactly as $\mathbf{Ax} = \mathbf{b}$.

For there to be a unique solution to the equation $\mathbf{Ax} = \mathbf{b}$, the matrix $\mathbf{A}$ must be *square*, that is, the number of rows must equal the number of columns. (Note that the number of rows in $\mathbf{A}$ equals the number of equations, and the number of columns in $\mathbf{A}$ equals the number of variables!) Furthermore, all the equations

Quick Quiz 3.5

What is the numerical value of a_{32} in the matrix equation above?

What is the numerical value of b_1?

must be independent. A straightforward way to check whether equations are independent is to try to reduce the matrix **A** to another matrix **A*** such that *all* entries below the diagonal are zero, and *none* of the entries on the diagonal are zero. Matrix reduction proceeds by repeated multiplication and addition of rows: we multiply row 1 by an appropriate factor and add the result to row 2 such that $a_{21} = 0$, then repeat with row 2, etc. If the equations are linearly independent, you will be able to find **A***. If the equations are not independent, you will end up with a zero in at least one of the entries on the diagonal. **Matrix reduction is available on many scientific calculators (see App. A.1).**

For example, the matrix **A**

$$\begin{bmatrix} 1 & 1 & 1 \\ 1 & 2 & 2 \\ 2 & 3 & -4 \end{bmatrix}$$

is reduced to **A***:

$$\begin{bmatrix} 1 & 0 & 0 \\ 0 & 1 & 0 \\ 0 & 0 & 1 \end{bmatrix}$$

However, for the set of equations

$$a + b + c = 10$$

$$a + b = 5$$

$$c = 5$$

the matrix **A**

$$\begin{bmatrix} 1 & 1 & 1 \\ 1 & 1 & 0 \\ 0 & 0 & 1 \end{bmatrix}$$

is reduced to **A***:

$$\begin{bmatrix} 1 & 1 & 0 \\ 0 & 0 & 1 \\ 0 & 0 & 0 \end{bmatrix}$$

Quick Quiz 3.6

Do the equations
$a + b + c = 2,$
$a + 2b + 2c = 3,$
$2a + 3b - 4c = 4$
form a linearly independent set?

We've got a zero on the diagonal, and no way to get rid of it. Therefore, these equations are not independent (which we knew already for this particular case). **A** is called a *singular* matrix. It corresponds to the case where there are more unknowns than equations (underspecified, DOF > 0). In this case, there will be many possible solutions.

Suppose we've convinced ourselves that the matrix **A** is square and nonsingular. For there to exist *one and only one* solution to **Ax** = **b**, we have another requirement: **b** must be nonhomogeneous, that is, at least one entry in the vector

$\mathbf{b}$ must not equal zero. If $\mathbf{b} = 0$ (meaning that every entry in the $\mathbf{b}$ vector is zero), we have the trivial solution $\mathbf{x} = 0$, but there are many other solutions as well.

Our final task is to find $\mathbf{x}$, where $\mathbf{x} = \mathbf{A}^{-1}\mathbf{b}$, and $\mathbf{A}^{-1}$ is the inverse matrix of $\mathbf{A}$. There is a simple algorithm we can use to solve these systems, called *Gaussian elimination*, if we employ matrix notation. Alternatively, many modern scientific calculators and equation-solving programs easily find the inverse matrix $\mathbf{A}^{-1}$ and can carry out multiplication of $\mathbf{A}^{-1}$ and $\mathbf{b}$ to find $\mathbf{x}$. (See App. A.1). Thus, solving even large sets of linear equations boils down to setting up the equation $\mathbf{Ax} = \mathbf{b}$, checking that $\mathbf{A}$ is square and nonsingular, checking that the vector $\mathbf{b}$ is nonhomogeneous, and then solving for $\mathbf{x}$, using a calculator or computer.

3.3.2 Using Matrices to Balance Chemical Reactions

Recall from Chap. 1 that we balanced chemical reaction equations by invoking the element balance equation:

$$\sum_{all\ i} \varepsilon_{hi} \nu_i = 0 \tag{1.1}$$

where ε_{hi} = the number of atoms of element h in molecule i and ν_i is the (unknown) stoichiometric coefficient. If there are H elements, then H equations must be solved simultaneously. We can use matrices to find the solution: Note the similarity between the element balance equation and

$$\sum_i a_{ij} x_j = b_j$$

Suppose the reaction of interest is oxidation of methane (CH_4) to CO_2 and water, written as an unbalanced reaction as

$$CH_4 + O_2 \rightarrow CO_2 + H_2O$$

There are three elements, so we write three element balance equations (numbering our compounds as $1 = CH_4$, $2 = O_2$, $3 = CO_2$, $4 = H_2O$):

$$C: \quad \nu_1 + \nu_3 = 0$$

$$H: \quad 4\nu_1 + 2\nu_4 = 0$$

$$O: \quad 2\nu_2 + 2\nu_3 + \nu_4 = 0$$

We write this set of equations in matrix form

$$\begin{array}{c} C \\ H \\ O \end{array} \begin{bmatrix} 1 & 0 & 1 & 0 \\ 4 & 0 & 0 & 2 \\ 0 & 2 & 2 & 1 \end{bmatrix} \begin{bmatrix} \nu_1 \\ \nu_2 \\ \nu_3 \\ \nu_4 \end{bmatrix} = \begin{bmatrix} 0 \\ 0 \\ 0 \\ 0 \end{bmatrix}$$

Notice that each column in matrix A represents the chemical formula for one compound. For example, column 1 represents $C_1H_4O_0$, or CH_4. In other words, we can write this matrix from the known chemical formulas for each compound, without bothering to derive the element balance equation!

The matrix is not square: There are four variables but only three equations. Furthermore, the vector **b** is homogeneous—full of zeros. This system of equations has an infinite number of possible solutions. To proceed, we arbitrarily specify one of the stoichiometric coefficients. Let's pick $v_1 = -1$. We'll call CH_4 our *basis* compound, because we choose to base subsequent calculations on its stoichiometric coefficient. We then rewrite the element balance equations so only terms involving the unknowns are on the left-hand side:

$$C: \quad v_3 = 1$$

$$H: \quad 2v_4 = 4$$

$$O: \quad 2v_2 + 2v_3 + v_4 = 0$$

(Of course it would be easy to solve this set of equations, but we continue on for illustration purposes.) We write this new set of three equations in three unknowns in matrix form:

$$\begin{bmatrix} 0 & 1 & 0 \\ 0 & 0 & 2 \\ 2 & 2 & 1 \end{bmatrix} \begin{bmatrix} v_2 \\ v_3 \\ v_4 \end{bmatrix} = \begin{bmatrix} 1 \\ 4 \\ 0 \end{bmatrix}$$

Notice three things. First, we now have a 3 × 3, or *square* matrix. This is a necessary (but not sufficient) condition for finding a unique solution. Second, matrix **A** can be written down by inspection: Each column is simply the chemical formula of the compounds for which the stoichiometric coefficient is not known. Third, vector **b** can be written down by inspection: it is simply the chemical formula for the compound chosen as the basis!

Solution is straightforward; you can solve by substitution and elimination, by using matrix functions on a calculator, or by using equation-solving software. The solution is

$$\begin{bmatrix} v_2 \\ v_3 \\ v_4 \end{bmatrix} = \begin{bmatrix} -2 \\ 1 \\ 2 \end{bmatrix}$$

and the balanced equation is:

$$CH_4 + 2O_2 \rightarrow CO_2 + 2H_2O$$

Let's recap how we use matrices to balance a chemical equation involving *I* compounds and *H* elements:

1. List the elements involved (e.g., C, H, O, N).
2. Choose one of the reactants or products to serve as a basis. Set its stoichiometric coefficient equal to -1 (reactant) or $+1$ (product).
3. List the chemical composition of all other compounds except the basis compound in a column in the matrix. Be sure to list the elements in the order

chosen in step 1, and do not forget the zeros. The matrix will have H rows (corresponding to the H elements) and $I - 1$ columns (I total compounds $- 1$ basis compound).

4. List the unknown stoichiometric coefficients v_j in vector $\mathbf{x}$. The vector will have $I - 1$ entries (I total compounds $- 1$ basis compound). Be sure to list the coefficients in the same order that the compounds were entered into the matrix.

5. List the chemical composition of the basis compound in the $\mathbf{b}$ vector. Be sure to list the elements in the order chosen in step 1, and do not forget the zeros.

6. Find the solution to the matrix equation, using Gaussian elimination, a calculator, or an equation-solving program.

Example 3.12

Balancing Chemical Equations with Matrix Math: Adipic Acid

Cyclohexanol $C_6H_{12}O$ and nitric acid (HNO_3) react to make adipic acid ($C_6H_{10}O_4$), with nitrogen oxide (NO) and water (H_2O) as byproducts. The unbalanced chemical reaction equation is:

$$C_6H_{12}O + HNO_3 \rightarrow C_6H_{10}O_4 + NO + H_2O$$

Find the stoichiometric coefficients using a matrix equation.

Solution

We select $C_6H_{12}O$ as the basis compound, set $v_1 = -1$, and proceed to write by inspection:

$$
\begin{matrix} C \\ H \\ O \\ N \end{matrix}
\begin{bmatrix} 0 & 6 & 0 & 0 \\ 1 & 10 & 0 & 2 \\ 3 & 4 & 1 & 1 \\ 1 & 0 & 1 & 0 \end{bmatrix}
\begin{bmatrix} v_2 \\ v_3 \\ v_4 \\ v_5 \end{bmatrix}
=
\begin{bmatrix} 6 \\ 12 \\ 1 \\ 0 \end{bmatrix}
$$

This problem is solvable by Gaussian elimination, by using matrix function keys on scientific calculators, or by using equation-solving software on a computer.

The solution is

$$
\begin{bmatrix} v_2 \\ v_3 \\ v_4 \\ v_5 \end{bmatrix}
=
\begin{bmatrix} -8/3 \\ 1 \\ 8/3 \\ 7/3 \end{bmatrix}
$$

or, written in conventional format:

$$C_6H_{12}O + \tfrac{8}{3}HNO_3 \rightarrow C_6H_{10}O_4 + \tfrac{8}{3}NO + \tfrac{7}{3}H_2O$$

Quick Quiz 3.7

Refer back to Example 1.1, and write the system of element balance equations for nitric acid synthesis in matrix form. (You don't need to solve).

3.3.3   Using Matrices in Generation-Consumption Analysis

In Chap. 2 you learned how to pull together a set of chemical reactions into a reaction pathway through generation-consumption analysis. One of the steps involved is to balance the reactions such that net generation/consumption of intermediate compounds was zero. You did this by finding values of a multiplying factor χ_k where

$$\sum_{\substack{\text{all } k \\ \text{reactions}}} \nu_{ik}\chi_k = 0 \qquad \text{for intermediates} \tag{1.2}$$

Matrix math can be used to find the correct values of χ_k. These methods are particularly useful for systems of large numbers of reactions, because the matrix equation can be developed by inspection. Matrix methods are even useful in cases where it is not obvious whether the reactions can be combined in such a way that a compound can serve as an intermediate or must be a byproduct!

Our goal is to find an equation $\mathbf{A}\mathbf{x} = \mathbf{b}$, where $\mathbf{A}$ is a matrix containing the stoichiometric coefficients of all compounds that have net-zero generation/consumption, $\mathbf{x}$ is the vector containing the (unknown) multiplying factors χ_k, and $\mathbf{b}$ is a vector containing the stoichiometric coefficients for one reaction chosen as the basis reaction. Then we solve for $\mathbf{x}$, and complete the generation-consumption analysis. Here is the procedure to follow:

1. List all the compounds that appear in any reaction. To write the matrix $\mathbf{A}$, list the stoichiometric coefficients for each reaction in a column, in the order of the compounds in your list. $\mathbf{A}$ will have I rows (one for each compound) and K columns (one for each reaction). There should be at least as many compounds as there are reactions.
2. Scan the rows of $\mathbf{A}$. Cross out any rows that have only a single nonzero entry. These rows correspond to compounds that appear in only a single reaction in the reaction system. Such compounds *cannot* have net-zero generation/consumption and so cannot be intermediates: They must be either a reactant or a product.
3. If there is one fewer row than column in matrix $\mathbf{A}$, go to step 4. If not, scan the remaining rows of $\mathbf{A}$ and identify any compounds that are acceptable as raw materials and/or products. Such compounds may be "acceptable" because they are nontoxic, or because they are cheap raw materials or valuable byproducts. Continue crossing out rows of acceptable compounds until there is one fewer row than column in $\mathbf{A}$ (equivalently, there is one fewer compound than reaction).
4. Choose one of the reactions (one of the columns) to serve as a basis reaction. Let $\mathbf{b} = (-1) \times$ the stoichiometric coefficients of the basis reaction. Delete that column from matrix $\mathbf{A}$.
5. Check that you now have a square coefficient matrix $\mathbf{A}$ with an equal number of rows and columns, a variable vector $\mathbf{x}$ that is the unknown multiplying factors, and a vector $\mathbf{b}$ that describes your basis reaction. Solve for $\mathbf{x}$. Use the solution to complete the generation-consumption analysis.

The procedure sounds more complicated than it is. Example 3.13 illustrates the idea.

Example 3.13

Generation-Consumption Analysis Using Matrix Math: Nitric Acid Synthesis

We want to develop a reaction pathway to make nitric acid (HNO_3) from readily available and cheap raw materials. We think some combination of the following reactions might be useful:

$$O_2 + 2CH_4 \rightarrow 2CO + 4H_2 \quad \text{(R1)}$$

$$CO + H_2O \rightarrow CO_2 + H_2 \quad \text{(R2)}$$

$$N_2 + 3H_2 \rightarrow 2NH_3 \quad \text{(R3)}$$

$$4NH_3 + 5O_2 \rightarrow 4NO + 6H_2O \quad \text{(R4)}$$

$$2NO + O_2 \rightarrow 2NO_2 \quad \text{(R5)}$$

$$3NO_2 + H_2O \rightarrow 2HNO_3 + NO \quad \text{(R6)}$$

Use matrix methods to combine these reactions into a pathway to make nitric acid. Preferably, we'd like to use inexpensive and readily available raw materials like water, methane (CH_4), and oxygen and nitrogen from air, and we want to avoid any net generation of toxic or environmentally damaging compounds such as NO, NO_2, NH_3, and CO.

Solution

1. We list the compounds involved and immediately write down the matrix of stoichiometric coefficients from the balanced chemical reactions:

	R1	R2	R3	R4	R5	R6
O_2	−1	0	0	−5	−1	0
N_2	0	0	−1	0	0	0
CH_4	−2	0	0	0	0	0
H_2O	0	−1	0	6	0	−1
CO	2	−1	0	0	0	0
CO_2	0	1	0	0	0	0
H_2	4	1	−3	0	0	0
NH_3	0	0	2	−4	0	0
NO	0	0	0	4	−2	1
NO_2	0	0	0	0	2	−3
HNO_3	0	0	0	0	0	2

2. Next we scan the list and eliminate any rows (compounds) with just one entry. This includes necessary reactants N_2 and CH_4 and the desired product HNO_3. We also observe that CO_2 *must* be a product of this reaction pathway, because it appears in only one reaction. Our matrix becomes:

	R1	R2	R3	R4	R5	R6
O_2	−1	0	0	−5	−1	0
H_2O	0	−1	0	6	0	−1
CO	2	−1	0	0	0	0
H_2	4	1	−3	0	0	0
NH_3	0	0	2	−4	0	0
NO	0	0	0	4	−2	1
NO_2	0	0	0	0	2	−3

3. We have one more compound than reaction; according to our procedure we need to have one fewer compound than reaction. We look for two materials that are acceptable raw materials or byproducts. O_2 and H_2O fit the bill. We eliminate them from consideration. The remaining five compounds can all be net-zero compounds! The matrix becomes:

	R1	R2	R3	R4	R5	R6
CO	2	−1	0	0	0	0
H_2	4	1	−3	0	0	0
NH_3	0	0	2	−4	0	0
NO	0	0	0	4	−2	1
NO_2	0	0	0	0	2	−3

4. We arbitrarily choose one of the reactions to serve as the basis reaction—let's choose (R1). We create the **b** vector by multiplying the column corresponding to (R1) by −1, and then we delete that column from **A**. The **x** vector is simply the listing of the multiplying factors χ_k for the remaining reactions. We end up with

$$
\begin{bmatrix}
-1 & 0 & 0 & 0 & 0 \\
1 & -3 & 0 & 0 & 0 \\
0 & 2 & -4 & 0 & 0 \\
0 & 0 & 4 & -2 & 1 \\
0 & 0 & 0 & 2 & -3
\end{bmatrix}
\begin{bmatrix}
\chi_2 \\
\chi_3 \\
\chi_4 \\
\chi_5 \\
\chi_6
\end{bmatrix}
=
\begin{bmatrix}
-2 \\
-4 \\
0 \\
0 \\
0
\end{bmatrix}
$$

5. The columns in the matrix **A** correspond to the five remaining reactions, (R2) through (R6). The rows in the matrix correspond to the stoichiometric coefficients of the remaining compounds: CO, H_2, NH_3, NO, and NO_2. These are the compounds

where we want to have no net generation or consumption. We solve, by calculator or by computer, and find the multiplying factors:

$$\mathbf{x} = \begin{bmatrix} 2 \\ 2 \\ 1 \\ 3 \\ 2 \end{bmatrix}$$

Finally, we multiply the stoichiometric coefficients ν_{ik} by the corresponding multiplying factor χ_k to complete the generation-consumption table:

Compound	R1	R2	R3	R4	R5	R6	Net
O_2	-1			-5	-3		-9
N_2			-2				-2
CH_4	-2						-2
H_2O		-2		$+6$		-2	$+2$
CO	$+2$	-2					0
CO_2		$+2$					$+2$
H_2	$+4$	$+2$	-6				0
NH_3			$+4$	-4			0
NO				$+4$	-6	$+2$	0
NO_2					$+6$	-6	0
HNO_3						$+4$	$+4$

The net overall reaction is

$$9O_2 + 2N_2 + 2CH_4 \rightarrow 2H_2O + 2CO_2 + 4HNO_3$$

3.3.4 Using Matrices to Find Linearly Independent Chemical Equations

In Chap. 2 you learned how to determine the degrees of freedom in a problem statement. As part of DOF analysis, the number of independent chemical reactions are counted. If there are just a few reactions this is pretty straightforward. But what if we are faced with many reactions? How do we know the number of independent reactions, and how do we know which they are? A bit of linear algebra is enlightening. In fact, we can start simply with a list of compounds, and *no postulated reactions,* and find a set of independent chemical equations that are stoichiometrically balanced! Since we don't invoke any particular knowledge about chemistry, the reactions may not describe how the compounds really combine and reform—but that is not our quest; we simply want to know enough to complete process flow calculations.

If we have I compounds of known molecular formula composed of H elements, the *maximum* number of independent chemical equations is simply $I - H$. Here are the mechanics of the method to find what those chemical equations are. This method is particularly useful in cases where there are a very large number of compounds—for example, in combustion problems. We illustrate the method as we go because it is easier to do than to explain.

Step 1.   We have I compounds of known molecular formula composed of H elements. First we write a matrix **A** that contains I columns and H rows, where each entry is the number of atoms of element h in molecule i. (In general, **A** is not a square matrix.)

Illustration: We have seven compounds, CH_4, CO, CO_2, O_2, H_2O, H_2, and CH_3OH, composed of three elements, C, H, and O. ($I = 7$ and $H = 3$.) The matrix is

$$
\begin{array}{c c c c c c c c}
 & CH_4 & CO & CO_2 & O_2 & H_2O & H_2 & CH_3OH \\
C & \begin{bmatrix} 1 & 1 & 1 & 0 & 0 & 0 & 1 \\ H & 4 & 0 & 0 & 0 & 2 & 2 & 4 \\ O & 0 & 1 & 2 & 2 & 1 & 0 & 1 \end{bmatrix}
\end{array}
$$

Step 2.   Use matrix reduction to put ones on the diagonal and zeros everywhere else in the first H columns and H rows.

Illustration: The reduced matrix is

$$
\begin{array}{c c c c c c c c}
 & CH_4 & CO & CO_2 & O_2 & H_2O & H_2 & CH_3OH \\
C & 1 & 0 & 0 & 0 & 0.5 & 0.5 & 1 \\
H & 0 & 1 & 0 & -2 & -2 & -1 & -1 \\
O & 0 & 0 & 1 & 2 & 1.5 & 0.5 & 1
\end{array}
$$

Identity matrix   Remaining entries

The first three rows and columns form an *identity matrix.*

Step 3.   Eliminate any rows that are all zeros. The number of independent element balance equations you can write is the number of nonzero rows. Most of the time, the number of independent element balance equations equals the number of elements, but there are rare cases when this is not true.

Illustration: We have all nonzero rows, so there are three independent element balance equations (C, H, O).

Step 4.   Generate a new matrix **A'** as follows: Erase the identity matrix, multiply all the remaining entries by -1, and then put a new identity matrix at the bottom. **A'** should have as many columns as there are independent element balances, and as many rows as there are compounds. For bookkeeping, write the molecular formulas of all compounds in a column, in the same order as they were in the row.

Illustration: When we apply these steps to our example matrix, we get

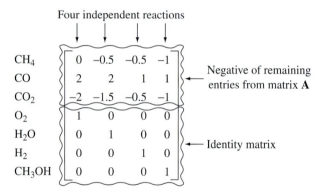

Step 5.   The number of independent chemical equations is the number of columns with two or more nonzero entries. Read off the stoichiometric coefficients of the chemical equations from each column—each coefficient belongs to the compound written in that row.

Illustration: Using the convention that negative stoichiometric coefficients are reactants and positive are products, we get four independent chemical equations:

$$2CO_2 \rightarrow 2CO + O_2$$

$$0.5CH_4 + 1.5CO_2 \rightarrow 2CO + H_2O$$

$$0.5CH_4 + 0.5CO_2 \rightarrow CO + H_2$$

$$CH_4 + CO_2 \rightarrow CO + CH_3OH$$

These are not the *only* possible set of independent equations that you could write. But, we don't care—we just need one such set.

3.4   Linear Models of Process Flow Sheets (Optional Section)

In Chap. 2 we completed process flow calculations on several preliminary block flow diagrams. In those examples, we often made simplifying approximations: that the separation was perfect, for example, or that the reactants were completely consumed. Now we would like to analyze more realistic process flow sheets that contain many interconnected process units, with fewer simplifying approximations.

In this section we'll step back and look at flow sheets in a general sense. What we're after is, in mathematical terms, a **linear model of a process flow sheet.** A process flow sheet such as a block flow diagram is a *visual* representation of process units, process streams, and how they are all connected. A linear model is a *mathematical* representation of that flow sheet.

To develop a linear model of a process flow sheet, we derive a system of linear equations by combining steady-state material balance equations with appropriate system performance specifications for each process unit. These linear equations express the outlet flow from a process unit as a linear function of inlet flows and/or system performance specifications. Matrix methods prove particularly useful in solving a linear model when the inlet flow and system performance are known and we wish to calculate the outlet flows; in this case, the vector **b** contains the known inlet flows and system performance specifications and the vector **x** contains the unknown outlet flows (and extents of reaction, where appropriate).

Not all process flow problems are amenable to linear analysis. Unsteady-state processes, for example, cannot be modeled by linear equations. Only certain kinds of system performance descriptors give rise to linear equations; we will see in Chaps. 4 and 5 that analysis of reactors and separators often includes nonlinear functions. Still, when a process flow problem can be cast as a set of linear equations, there are powerful mathematical tools at our disposal.

The material in this section may seem rather abstract. It needs to be, because we are trying to develop robust methods for analyzing any process flow sheet—in a sense, we are taking away the details that make a process unique, and "abstracting" out the underlying structure. But there is purpose to this approach. If we are able to write linear algebraic equations to describe all the process units in a complex process flow sheet, then mathematical analysis of that flow sheet is simplified. Furthermore, once we set up a model, it is easier to examine the impact of various design decisions on the overall process flow.

3.4.1 Linear Models of Single Process Units

We start by analyzing individual process units. As we learned in Chap. 2, there are only four fundamental kinds of process units in block flow diagrams: *mixers, splitters, reactors,* and *separators.* It is amazing that we can make so many different products by combining these four simple building blocks in myriad ways. This diversity in outcome is achieved through the diversity in chemical reactions, reactor designs, and separation technologies, and through the diversity in how the units are connected.

Let's write general forms of the steady-state component mole balance equation for each of these four kinds of process units. We write all balance equations with the outlet stream flows on the left and all other terms on the right. Then we'll write linear equations describing performance specifications appropriate for that type of process unit. Together, the material balance equations and the performance equations constitute a linear model of a process unit.

Mixers Mixers combine multiple input streams into a single output stream. There is no chemical reaction.

The steady-state component mole balance equation for a mixer is

$$\dot{n}_{i,\,\text{out}} = \sum_{\text{all } j \text{ in}} \dot{n}_{ij} \qquad (3.23)$$

No performance specification for mixers is needed. If there are C components passing through the mixer, there are C independent material balance equations. This set of balance equations is a **linear model** of a mixer.

Example 3.14   **Linear Models of Mixers: Sweet Mix**

Two aqueous solutions are mixed together in a continuous steady-state mixer. One of the solutions (solution A) contains 15 mol% glucose, 12 mol% fructose, and 73 mol% water and is fed to the mixer at 180 kgmol/h. The other solution (solution B) contains 6 mol% glucose, 3 mol% fructose, and 91 mol% water and is fed to the mixer at 250 kgmol/h. What is the flow rate of each component in the stream leaving the mixer?

Solution

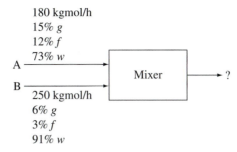

We'll use g to indicate glucose, f for fructose, and w for water. We write three equations because there are three components:

$$\dot{n}_{g,\text{out}} = \sum_{\text{all } j \text{ in}} \dot{n}_{gj} = \dot{n}_{gA} + \dot{n}_{gB} = (0.15 \times 180) + (0.06 \times 250)$$

$$= 42 \,\frac{\text{kgmol glucose}}{\text{h}}$$

$$\dot{n}_{f,\text{out}} = \sum_{\text{all } j \text{ in}} \dot{n}_{fj} = \dot{n}_{fA} + \dot{n}_{fB} = (0.12 \times 180) + (0.03 \times 250)$$

$$= 29 \,\frac{\text{kgmol fructose}}{\text{h}}$$

$$\dot{n}_{w,\text{out}} = \sum_{\text{all } j \text{ in}} \dot{n}_{wj} = \dot{n}_{wA} + \dot{n}_{wB} = (0.73 \times 180) + (0.91 \times 250)$$

$$= 359 \,\frac{\text{kgmol water}}{\text{h}}$$

Written in matrix form, these equations are simply:

$$\begin{bmatrix} 1 & 0 & 0 \\ 0 & 1 & 0 \\ 0 & 0 & 1 \end{bmatrix} \begin{bmatrix} \dot{n}_{g,\text{out}} \\ \dot{n}_{f,\text{out}} \\ \dot{n}_{w,\text{out}} \end{bmatrix} = \begin{bmatrix} 42 \\ 29 \\ 359 \end{bmatrix}$$

The solution is, of course,

$$\begin{bmatrix} \dot{n}_{g,\text{out}} \\ \dot{n}_{f,\text{out}} \\ \dot{n}_{w,\text{out}} \end{bmatrix} = \begin{bmatrix} 42 \\ 29 \\ 359 \end{bmatrix}$$

Notice that the matrix **A** for a mixer is simply the identity matrix! This will always be true for mixers, so the matrix equation can be written down immediately.

Splitters Splitters take a single input stream and divide it into two or more output streams *of identical composition.*

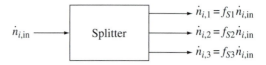

The steady-state mole balance equation is

$$\sum_{\text{all } j \text{ out}} \dot{n}_{ij} = \dot{n}_{i,\text{in}} \tag{3.24}$$

The performance of a splitter is specified most readily by the **fractional split** f_{Sj}:

$$f_{Sj} = \frac{\text{moles of } i \text{ leaving in stream } j}{\text{moles of } i \text{ fed to splitter}} = \frac{\dot{n}_{ij}}{\dot{n}_{i,\text{in}}}$$

f_{Sj} must be the same for all components. *(Why?* This additional constraint is the origin of the "splitter restriction" that you used in DOF analysis in Chap. 2.)
    Rewriting the system performance specification as a linear equation relating output and input streams, we find

$$\dot{n}_{ij} = f_{Sj}\dot{n}_{i,\text{in}} \tag{3.25}$$

If we have S output streams from the splitter, it might seem that we have S fractional splits to be specified. But we have one restriction that must be satisfied: All the fractional splits have to add up to 1, or:

$$\sum_{j=1}^{S} f_{Sj} = 1$$

For each splitter with C components and S outlet streams, we could write $C \times S$ equations like Eq. (3.25) describing splitter performance. But, because the sum of all f_{Sj} must equal 1, only $C \times (S - 1)$ of these system performance equations are independent. These, plus the C material balance equations (Eq. 3.24), constitute a linear model of a splitter. For example, if there are four components and three output streams, there are 4 independent material balance equations and $4 \times (3 - 1)$ or 8 independent system performance equations, for a total of 12 equations. There are a total of 16 stream variables (4 for the input stream and 4×3 for the output streams). If the 4 input component flows are known, then the linear model is completely specified.

Example 3.15	**Linear Model of a Splitter: Sweet Split**

A solution containing 9.8 mol% glucose, 6.6 mol% fructose, and 83.6 mol% water is fed at a rate of 430 kgmol/h to a splitter operated continuously and at steady state. The splitter produces three outlet streams, A, B, and C. Stream A is 25% of the inlet stream ($f_{SA} = 0.25$) and Stream B is 35% of the inlet stream ($f_{SB} = 0.35$). Calculate the flow rates of the output streams.

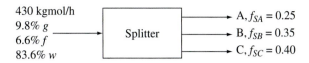

430 kgmol/h
9.8% g
6.6% f
83.6% w

Splitter

A, $f_{SA} = 0.25$
B, $f_{SB} = 0.35$
C, $f_{SC} = 0.40$

Solution

There are three components and three outlet streams, so $C = 3$ and $S = 3$. Therefore there are three material balance equations, plus $3 \times (3 - 1)$ or 6 splitter performance equations, involving 12 stream variables.

Material balance equations:

$$\dot{n}_{gA} + \dot{n}_{gB} + \dot{n}_{gC} = \dot{n}_{g,in}$$
$$\dot{n}_{fA} + \dot{n}_{fB} + \dot{n}_{fC} = \dot{n}_{f,in}$$
$$\dot{n}_{wA} + \dot{n}_{wB} + \dot{n}_{wC} = \dot{n}_{w,in}$$

Splitter performance equations:

$$\dot{n}_{gA} = f_{SA}\dot{n}_{g,in}$$
$$\dot{n}_{fA} = f_{SA}\dot{n}_{f,in}$$
$$\dot{n}_{wA} = f_{SA}\dot{n}_{w,in}$$
$$\dot{n}_{gB} = f_{SB}\dot{n}_{g,in}$$
$$\dot{n}_{fB} = f_{SB}\dot{n}_{f,in}$$
$$\dot{n}_{wA} = f_{SB}\dot{n}_{w,in}$$

Written in matrix form as **Ax = b**, these equations become:

$$
\begin{bmatrix}
1 & 0 & 0 & 1 & 0 & 0 & 1 & 0 & 0 \\
0 & 1 & 0 & 0 & 1 & 0 & 0 & 1 & 0 \\
0 & 0 & 1 & 0 & 0 & 1 & 0 & 0 & 1 \\
1 & 0 & 0 & 0 & 0 & 0 & 0 & 0 & 0 \\
0 & 1 & 0 & 0 & 0 & 0 & 0 & 0 & 0 \\
0 & 0 & 1 & 0 & 0 & 0 & 0 & 0 & 0 \\
0 & 0 & 0 & 1 & 0 & 0 & 0 & 0 & 0 \\
0 & 0 & 0 & 0 & 1 & 0 & 0 & 0 & 0 \\
0 & 0 & 0 & 0 & 0 & 1 & 0 & 0 & 0
\end{bmatrix}
\begin{bmatrix}
\dot{n}_{gA} \\ \dot{n}_{fA} \\ \dot{n}_{wA} \\ \dot{n}_{gB} \\ \dot{n}_{fB} \\ \dot{n}_{wB} \\ \dot{n}_{gC} \\ \dot{n}_{fC} \\ \dot{n}_{wC}
\end{bmatrix}
=
\begin{bmatrix}
\dot{n}_{g,in} \\ \dot{n}_{f,in} \\ \dot{n}_{w,in} \\ f_{SA}\dot{n}_{g,in} \\ f_{SA}\dot{n}_{f,in} \\ f_{SA}\dot{n}_{w,in} \\ f_{SB}\dot{n}_{g,in} \\ f_{SB}\dot{n}_{f,in} \\ f_{SB}\dot{n}_{w,in}
\end{bmatrix}
$$

By matrix reduction of **A**, one can show that this set of equations is independent and well defined. If we now specify the three input streams plus the two fractional splits, we will have nine equations in nine unknowns. The matrix equation for the particular input flows and fractional splits specified in this problem is

$$
\begin{bmatrix}
1 & 0 & 0 & 1 & 0 & 0 & 1 & 0 & 0 \\
0 & 1 & 0 & 0 & 1 & 0 & 0 & 1 & 0 \\
0 & 0 & 1 & 0 & 0 & 1 & 0 & 0 & 1 \\
1 & 0 & 0 & 0 & 0 & 0 & 0 & 0 & 0 \\
0 & 1 & 0 & 0 & 0 & 0 & 0 & 0 & 0 \\
0 & 0 & 1 & 0 & 0 & 0 & 0 & 0 & 0 \\
0 & 0 & 0 & 1 & 0 & 0 & 0 & 0 & 0 \\
0 & 0 & 0 & 0 & 1 & 0 & 0 & 0 & 0 \\
0 & 0 & 0 & 0 & 0 & 1 & 0 & 0 & 0
\end{bmatrix}
\begin{bmatrix}
\dot{n}_{gA} \\
\dot{n}_{fA} \\
\dot{n}_{wA} \\
\dot{n}_{gB} \\
\dot{n}_{fB} \\
\dot{n}_{wB} \\
\dot{n}_{gC} \\
\dot{n}_{fC} \\
\dot{n}_{wC}
\end{bmatrix}
=
\begin{bmatrix}
42.1 \\
28.4 \\
359.5 \\
10.5 \\
7.1 \\
89.9 \\
14.8 \\
9.9 \\
125.8
\end{bmatrix}
$$

The solution is easy to obtain on a calculator with matrix function keys, by solving $\mathbf{x} = \mathbf{A}^{-1}\mathbf{b}$ or

$$
\begin{bmatrix}
\dot{n}_{gA} \\
\dot{n}_{fA} \\
\dot{n}_{wA} \\
\dot{n}_{gB} \\
\dot{n}_{fB} \\
\dot{n}_{wB} \\
\dot{n}_{gC} \\
\dot{n}_{fC} \\
\dot{n}_{wC}
\end{bmatrix}
=
\begin{bmatrix}
10.5 \\
7.1 \\
89.9 \\
14.8 \\
9.9 \\
125.8 \\
16.8 \\
11.4 \\
143.8
\end{bmatrix}
$$

Once the matrix equation is set up, it is straightforward to change input flows or fractional splits and examine the effect on outlet flows.

Notice the general structure of the matrix equation for a splitter. The output flows are all listed in the $\mathbf{x}$ vector. In the $\mathbf{b}$ vector are the input flows and the fractional splits. Notice also the simple pattern of entries in the $\mathbf{A}$ matrix: Can you see how this matrix could be quickly expanded to incorporate many more components and many more output streams?

Reactors convert reactants to products by chemical reaction. An idealized reactor has only one inlet and only one outlet flow.

The steady-state component mole balance equation is

$$\dot{n}_{i,\text{out}} = \dot{n}_{i,\text{in}} + \sum_{\text{all k}} \nu_{ik}\dot{\xi}_k \qquad (3.26)$$

Reactors are so important in chemical processes that we've devoted a whole chapter (Chap. 4) to the topic. All we want right now is a straightforward way to write reactor performance specifications that is useful for linear models. In this chapter we will use **fractional conversion** as the sole measure to specify reactor performance. In Chap. 4 we will discuss fractional conversion in greater detail and also describe other useful measures for specifying reactor performance.

The **fractional conversion** of reactant i, f_{Ci}, is the fraction of the reactant fed to the reactor that gets consumed by chemical reaction, and is defined as

$$f_{Ci} = \frac{\text{moles of } i \text{ consumed by reaction}}{\text{moles of } i \text{ fed to reactor}} = \frac{- \sum_{\substack{\text{all } k \\ \text{reactions}}} \nu_{ik}\dot{\xi}_k}{\dot{n}_{i,\text{in}}}$$

Rearranging, we get a linear equation describing reactor performance:

$$f_{Ci}\dot{n}_{i,\text{in}} = - \sum_{\substack{\text{all } k \\ \text{reactions}}} \nu_{ik}\dot{\xi}_k \qquad (3.27)$$

Note that $0 \le f_{Ci} \le 1$. There are at most the same number of independent fractional conversions as there are reactions. Fractional conversions cannot be defined for products.

If we have a reactor with C components (in or out), and R independent reactions, the linear model contains $C + R$ independent equations: C material balance equations plus R reactor performance equations.

Example 3.16   **Linear Model of a Reactor: Glucose-Fructose Isomerization**

A solution of 9.8 mol% glucose, 6.6 mol% fructose, and 83.6 mol% water is fed to a reactor at a rate of 172.3 kgmol/day. Glucose and fructose are isomers: They both have the same molecular formula, $C_6H_{12}O_6$, but they have different chemical structures, and fructose is much sweeter-tasting than glucose. In the reactor, 53.25% of the glucose is converted to fructose:

$$C_6H_{12}O_6 \text{ (glucose)} \rightarrow C_6H_{12}O_6 \text{ (fructose)}$$

Calculate the reactor output flow rate and composition.

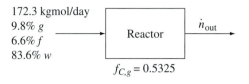

172.3 kgmol/day
9.8% g
6.6% f
83.6% w
$\rightarrow$ Reactor $\xrightarrow{\dot{n}_{\text{out}}}$

$f_{C,g} = 0.5325$

Solution

There are three components and one reaction. We write three material balance equations and one reactor performance equation, putting the unknowns on the left-hand side.

$$\dot{n}_{g,\text{out}} + \xi = \dot{n}_{g,\text{in}}$$
$$\dot{n}_{f,\text{out}} - \xi = \dot{n}_{f,\text{in}}$$
$$\dot{n}_{w,\text{out}} = \dot{n}_{w,\text{in}}$$
$$\xi = f_{C,g}\dot{n}_{g,\text{in}}$$

In matrix form this set of equations, which constitutes a linear model of this reactor, is written

$$\begin{bmatrix} 1 & 0 & 0 & 1 \\ 0 & 1 & 0 & -1 \\ 0 & 0 & 1 & 0 \\ 0 & 0 & 0 & 1 \end{bmatrix} \begin{bmatrix} \dot{n}_{g,\text{out}} \\ \dot{n}_{f,\text{out}} \\ \dot{n}_{w,\text{out}} \\ \dot{\xi} \end{bmatrix} = \begin{bmatrix} \dot{n}_{g,\text{in}} \\ \dot{n}_{f,\text{in}} \\ \dot{n}_{w,\text{in}} \\ f_{C,g}\dot{n}_{g,\text{in}} \end{bmatrix}$$

The output streams and extent of reaction are the four unknown variables, all in the **x** vector. The known fractional conversion of glucose, $f_{C,g} = 0.5325$, and the known input flow rates are all in the *b* vector. Inserting the numerical values into the matrix equation yields

$$\begin{bmatrix} 1 & 0 & 0 & 1 \\ 0 & 1 & 0 & -1 \\ 0 & 0 & 1 & 0 \\ 0 & 0 & 0 & 1 \end{bmatrix} \begin{bmatrix} \dot{n}_{g,\text{out}} \\ \dot{n}_{f,\text{out}} \\ \dot{n}_{w,\text{out}} \\ \dot{\xi} \end{bmatrix} = \begin{bmatrix} 16.9 \\ 11.4 \\ 144 \\ 9.0 \end{bmatrix}$$

The matrix is already in its reduced form, the set of linear equations is independent, and solution is straightforward.

$$\begin{bmatrix} \dot{n}_{g,\text{out}} \\ \dot{n}_{f,\text{out}} \\ \dot{n}_{w,\text{out}} \\ \dot{\xi} \end{bmatrix} = \begin{bmatrix} 7.9 \\ 20.4 \\ 144 \\ 9.0 \end{bmatrix}$$

Notice how we set up the matrix equation in Example 3.16. We placed the reactant output flow at the top of the vector **x**, followed by the product output flow, then any un-reactive compounds and finally the extent of reaction. In the **b** vector, we placed the input flows of the components in the same order, followed by the fractional conversion times the input flow of reactants. Also, notice something very curious about

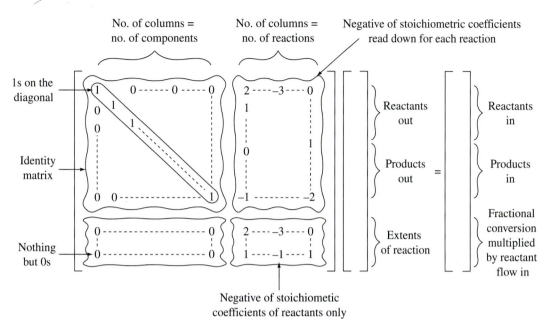

Figure 3.6 The matrix equation $\mathbf{A} = \mathbf{bx}$ for reactors can be generated from this template, without first deriving the linear equations.

the matrix $\mathbf{A}$. We can consider four different parts of the matrix (Fig. 3.6). The top left portion of $\mathbf{A}$ is an identity matrix! It has as many columns and rows as there are components. Below this is a region with nothing but zeros. Now read down the last column in $\mathbf{A}$: The first entries are simply the negative of the stoichiometric coefficients of the components! Finally, the bottom right corner is simply the negative of the stoichiometric coefficient of the reactants. This pattern will hold with much larger sets of equations, involving many compounds and many chemical reactions. Thus, we can write down matrix $\mathbf{A}$ by inspection! This is illustrated in the next example.

Example 3.17   **Linear Model of a Reactor: Multiple Reactions**

A mixture of 34 mol% CH_4, 52 mol% H_2O, 10 mol% O_2, and 4 mol% CO is fed to a reactor at 100 gmol/min. The product stream includes all reactants, plus H_2, CH_3OH, and CO_2. 73% of the methane, 17.3% of the water, 90% of the oxygen and 98% of the CO, are converted to products. What is the output from the reactor?

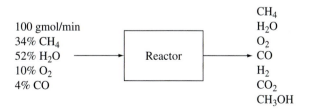

Solution

We examined this set of compounds when we described a method for finding a set of linearly independent chemical reactions given a panel of compounds. From Sec. 3.3.4, we know that there are four independent chemical equations:

$$2CO_2 \rightarrow 2CO + O_2$$
$$0.5CH_4 + 1.5CO_2 \rightarrow 2CO + H_2O$$
$$0.5CH_4 + 0.5CO_2 \rightarrow CO + H_2$$
$$CH_4 + CO_2 \rightarrow CO + CH_3OH$$

Since there are seven components, there are seven material balance equations. Therefore **A** is an 11×11 matrix: seven of the rows correspond to material balance equations and four to reactor performance equations. (Try deriving the equations.) Using the template of Fig. 3.6, we go directly to writing the matrix equation:

$$
\begin{bmatrix}
1 & 0 & 0 & 0 & 0 & 0 & 0 & 0 & 0.5 & 0.5 & 1 \\
0 & 1 & 0 & 0 & 0 & 0 & 0 & 0 & -1 & 0 & 0 \\
0 & 0 & 1 & 0 & 0 & 0 & 0 & -1 & 0 & 0 & 0 \\
0 & 0 & 0 & 1 & 0 & 0 & 0 & -2 & -2 & -1 & -1 \\
0 & 0 & 0 & 0 & 1 & 0 & 0 & 2 & 1.5 & 0.5 & 1 \\
0 & 0 & 0 & 0 & 0 & 1 & 0 & 0 & 0 & -1 & 0 \\
0 & 0 & 0 & 0 & 0 & 0 & 1 & 0 & 0 & 0 & -1 \\
0 & 0 & 0 & 0 & 0 & 0 & 0 & 0 & 0.5 & 0.5 & 1 \\
0 & 0 & 0 & 0 & 0 & 0 & 0 & 0 & -1 & 0 & 0 \\
0 & 0 & 0 & 0 & 0 & 0 & 0 & -1 & 0 & 0 & 0 \\
0 & 0 & 0 & 0 & 0 & 0 & 0 & -2 & -2 & -1 & -1
\end{bmatrix}
\begin{bmatrix}
\dot{n}_{CH_4,out} \\
\dot{n}_{H_2O,out} \\
\dot{n}_{O_2,out} \\
\dot{n}_{CO,out} \\
\dot{n}_{CO_2,out} \\
\dot{n}_{H_2,out} \\
\dot{n}_{CH_3OH,out} \\
\xi_1 \\
\xi_2 \\
\xi_3 \\
\xi_4
\end{bmatrix}
=
\begin{bmatrix}
\dot{n}_{CH_4,in} \\
\dot{n}_{H_2O,in} \\
\dot{n}_{O_2,in} \\
\dot{n}_{CO,in} \\
\dot{n}_{CO_2,in} \\
\dot{n}_{H_2,in} \\
\dot{n}_{CH_3OH,in} \\
f_{C,CH_4}\dot{n}_{CH_4,in} \\
f_{C,H_2O}\dot{n}_{H_2O,in} \\
f_{C,O_2}\dot{n}_{O_2,in} \\
f_{C,CO}\dot{n}_{CO,in}
\end{bmatrix}
$$

Finally, we input numerical values into the **b** vector and solve for **x**:

$$
\begin{bmatrix}
1 & 0 & 0 & 0 & 0 & 0 & 0 & 0 & 0.5 & 0.5 & 1 \\
0 & 1 & 0 & 0 & 0 & 0 & 0 & 0 & -1 & 0 & 0 \\
0 & 0 & 1 & 0 & 0 & 0 & 0 & -1 & 0 & 0 & 0 \\
0 & 0 & 0 & 1 & 0 & 0 & 0 & -2 & -2 & -1 & -1 \\
0 & 0 & 0 & 0 & 1 & 0 & 0 & 2 & 1.5 & 0.5 & 1 \\
0 & 0 & 0 & 0 & 0 & 1 & 0 & 0 & 0 & -1 & 0 \\
0 & 0 & 0 & 0 & 0 & 0 & 1 & 0 & 0 & 0 & -1 \\
0 & 0 & 0 & 0 & 0 & 0 & 0 & 0 & 0.5 & 0.5 & 1 \\
0 & 0 & 0 & 0 & 0 & 0 & 0 & 0 & -1 & 0 & 0 \\
0 & 0 & 0 & 0 & 0 & 0 & 0 & -1 & 0 & 0 & 0 \\
0 & 0 & 0 & 0 & 0 & 0 & 0 & -2 & -2 & -1 & -1
\end{bmatrix}
\begin{bmatrix}
\dot{n}_{CH_4,out} \\
\dot{n}_{H_2O,out} \\
\dot{n}_{O_2,out} \\
\dot{n}_{CO,out} \\
\dot{n}_{CO_2,out} \\
\dot{n}_{H_2,out} \\
\dot{n}_{CH_3OH,out} \\
\xi_1 \\
\xi_2 \\
\xi_3 \\
\xi_4
\end{bmatrix}
=
\begin{bmatrix}
34 \\
52 \\
10 \\
4 \\
0 \\
0 \\
0 \\
24.82 \\
8.996 \\
9 \\
3.52
\end{bmatrix}
$$

Quick Quiz 3.8

Did you notice that
two of the extents of
reaction in Example
3.17 are negative?
How do you interpret
that?

$$
\begin{bmatrix}
\dot{n}_{CH_4,out} \\
\dot{n}_{H_2O,out} \\
\dot{n}_{O_2,out} \\
\dot{n}_{CO,out} \\
\dot{n}_{CO_2,out} \\
\dot{n}_{H_2,out} \\
\dot{n}_{CH_3OH,out} \\
\xi_1 \\
\xi_2 \\
\xi_3 \\
\xi_4
\end{bmatrix}
=
\begin{bmatrix}
9.18 \\
43.0 \\
1 \\
0.08 \\
2.176 \\
5.51 \\
26.56 \\
-9 \\
-8.996 \\
5.51 \\
26.56
\end{bmatrix}
$$

The solution vector lists all component output flows as well as the four extents of
reactions. From the component flows, we calculate that the total reactor output is
87.5 gmol/min total, with 10.5 mol% methane, 49.1 mol% water, 1.1 mol% oxygen,
0.1 mol% CO, 2.5 mol% CO_2, 6.3 mol% H_2 and 30.4 mol% CH_3OH.

Separators always have at least one input stream and at least two output
streams. The output streams differ in composition from each other, and from the
input stream. Separators achieve changes in composition through physical
means, not by chemical reactions. Chap. 5 is devoted to the topic of separation
technologies.

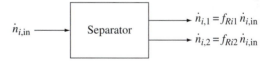

The steady-state differential material balance equation for a separator with
one input stream and two or more output streams is:

$$
\sum_{\text{all } j \text{ out}} \dot{n}_{ij} = \dot{n}_{i,\,in} \tag{3.28}
$$

We build separators to recover desired components from mixtures, and to
produce pure products. For the purpose of building a linear model of a separator,
we will specify separator performance by using **fractional recovery.** (Chapter 5
discusses other measures of separator performance.) **Fractional recovery** f_{Rij} is
the fraction of component i in the inlet stream that is recovered in outlet stream j

$$
f_{Rij} = \frac{\text{moles of } i \text{ leaving in stream } j}{\text{moles of } i \text{ fed to separator}} = \frac{\dot{n}_{ij,out}}{\dot{n}_{i,in}}
$$

(Compare fractional recovery to fractional split; how are they different?) Rearranging, we get a linear performance equation:

$$\dot{n}_{ij,\text{out}} = f_{Rij}\,\dot{n}_{i,\text{in}} \tag{3.29}$$

Suppose we have S outlet streams. We can independently define $S - 1$ fractional recoveries of component i, subject to the constraint that

$$\sum_{j=1}^{S} f_{Rij} = 1$$

We can write similar equations for fractional recovery of all the components. For a separator, the fractional recovery of one component is generally *not* the same as that for any other component: $f_{RAj} \neq f_{RBj} \neq f_{RCj} \neq \cdots$. Thus, for a separator, as many as $C \times (S - 1)$ fractional recoveries can be specified independently.

A linear model of a separator has $C \times S$ independent equations: C material balance equations and $C \times (S - 1)$ separator performance equations. Typically the input stream to the separator and the fractional recoveries are known, and the outputs are then calculated.

Example 3.18   **Linear Model of Separators: Sweet Solutions**

A solution of 9.8 mol% glucose, 6.6 mol% fructose, and 83.6 mol% water is fed to a separator at a rate of 172.3 kgmol/day. Three product streams leave the separator. Stream A contains most of the glucose, stream B contains most of the fructose, and stream C contains most of the water. 94% of the glucose is recovered in stream A, and 4% is recovered in stream B. 85% of the fructose is recovered in stream B, and 10% in stream A. 70% of the water is recovered in stream C, and 15% in each of stream A and stream B. Calculate the flows of all output streams.

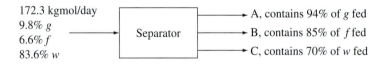

Solution

We write three material balance equations plus six separator performance equations to develop the linear model of this separator. (Try writing down these equations.) By specifying the input flows and the fractional recoveries, we have a completely specified model. To write in matrix form, as usual we put the unknown stream variables in the **x** vector and the known stream and performance variables in the **b** vector:

$$
\begin{bmatrix}
1 & 0 & 0 & 1 & 0 & 0 & 1 & 0 & 0 \\
0 & 1 & 0 & 0 & 1 & 0 & 0 & 1 & 0 \\
0 & 0 & 1 & 0 & 0 & 1 & 0 & 0 & 1 \\
1 & 0 & 0 & 0 & 0 & 0 & 0 & 0 & 0 \\
0 & 1 & 0 & 0 & 0 & 0 & 0 & 0 & 0 \\
0 & 0 & 1 & 0 & 0 & 0 & 0 & 0 & 0 \\
0 & 0 & 0 & 1 & 0 & 0 & 0 & 0 & 0 \\
0 & 0 & 0 & 0 & 1 & 0 & 0 & 0 & 0 \\
0 & 0 & 0 & 0 & 0 & 1 & 0 & 0 & 0
\end{bmatrix}
\begin{bmatrix}
\dot{n}_{gA} \\ \dot{n}_{fA} \\ \dot{n}_{wA} \\ \dot{n}_{gB} \\ \dot{n}_{fB} \\ \dot{n}_{wB} \\ \dot{n}_{gC} \\ \dot{n}_{fC} \\ \dot{n}_{wC}
\end{bmatrix}
=
\begin{bmatrix}
\dot{n}_{g,in} \\ \dot{n}_{f,in} \\ \dot{n}_{w,in} \\ f_{Rg,A}\dot{n}_{g,in} \\ f_{Rf,A}\dot{n}_{f,in} \\ f_{Rw,A}\dot{n}_{w,in} \\ f_{Rg,B}\dot{n}_{g,in} \\ f_{Rf,B}\dot{n}_{f,in} \\ f_{Rw,B}\dot{n}_{w,in}
\end{bmatrix}
$$

We input the known values of stream flow rates and fractional recoveries:

$$
\begin{bmatrix}
1 & 0 & 0 & 1 & 0 & 0 & 1 & 0 & 0 \\
0 & 1 & 0 & 0 & 1 & 0 & 0 & 1 & 0 \\
0 & 0 & 1 & 0 & 0 & 1 & 0 & 0 & 1 \\
1 & 0 & 0 & 0 & 0 & 0 & 0 & 0 & 0 \\
0 & 1 & 0 & 0 & 0 & 0 & 0 & 0 & 0 \\
0 & 0 & 1 & 0 & 0 & 0 & 0 & 0 & 0 \\
0 & 0 & 0 & 1 & 0 & 0 & 0 & 0 & 0 \\
0 & 0 & 0 & 0 & 1 & 0 & 0 & 0 & 0 \\
0 & 0 & 0 & 0 & 0 & 1 & 0 & 0 & 0
\end{bmatrix}
\begin{bmatrix}
\dot{n}_{gA} \\ \dot{n}_{fA} \\ \dot{n}_{wA} \\ \dot{n}_{gB} \\ \dot{n}_{fB} \\ \dot{n}_{wB} \\ \dot{n}_{gC} \\ \dot{n}_{fC} \\ \dot{n}_{wC}
\end{bmatrix}
=
\begin{bmatrix}
16.9 \\ 11.4 \\ 144 \\ 15.9 \\ 1.14 \\ 21.6 \\ 0.68 \\ 9.7 \\ 21.6
\end{bmatrix}
$$

Quick Quiz 3.9

What's the difference between saying "94% of the glucose is recovered in stream A" and "stream A is 94% glucose"?

and readily find the solution:

$$
\begin{bmatrix}
\dot{n}_{gA} \\ \dot{n}_{fA} \\ \dot{n}_{wA} \\ \dot{n}_{gB} \\ \dot{n}_{fB} \\ \dot{n}_{wB} \\ \dot{n}_{gC} \\ \dot{n}_{fC} \\ \dot{n}_{wC}
\end{bmatrix}
=
\begin{bmatrix}
15.9 \\ 1.14 \\ 21.6 \\ 0.68 \\ 9.7 \\ 21.6 \\ 0.32 \\ 0.56 \\ 100.8
\end{bmatrix}
$$

Compare the matrix equation we derived for a separator (Example 3.18) and for a splitter (Example 3.15). The left-hand sides are identical! In matrix **A**, the pattern in the upper rows reflects the distribution of components among the output streams. The lower left portion is simply an identity matrix, and the lower right portion is full of zeros. The **x** vector is simply a listing, in order, of the output flows of each component in each stream. The difference between the splitter

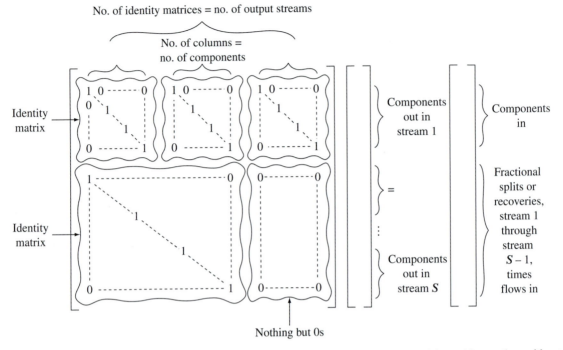

Figure 3.7 The matrix equation $\mathbf{A} = \mathbf{b}x$ for separators and splitters can be generated from this template without explicitly deriving the linear equations.

and separator models lies only in the $\mathbf{b}$ vector, and it is only a minor difference. In both cases, the top entries in the $\mathbf{b}$ vector are the component flows into the process unit. These are followed by the fractional split (for splitters) or the fractional recovery (for separators) multiplied by the component input flows. Following the pattern shown in Fig. 3.7, we can write down the matrix equations for separators and splitters by inspection!

3.4.2 Process Topology

We've just described the four process units, and shown how to define linear models for each that contain independent, correctly specified equations. These linear models relate output flows (and extents of reaction) to input flows, through specification of fractional split, fractional conversion, or fractional recovery.

The other important feature of a process flow sheet is its **topology**—the way in which the process units are connected together, and the direction of flow from one unit to another. In essence this is a map of the process. To obtain a linear model of an entire process, we write linear models for each mixer, splitter, reactor, and separator unit in the process. Then, we superimpose on these unit models the map of the process, simply by appropriately equating output flows from one unit to input flows of another.

There are three basic topologies we'll discuss—*diverging trees, converging trees,* and *loops*. Figure 3.8 illustrates the general shape of these patterns. On any particular flow sheet you will usually see some combination of these patterns, or only a section of one of the patterns. The *diverging tree* is a common pattern for refineries or other similar processes, where a mixture (such as crude oil) is separated into its component parts (such as gasoline, diesel, and asphalt). The *converging tree* is common in synthesis of complex organic chemicals such as polymers or pharmaceuticals, where a single product is produced as a result of multiple chemical reactions involving many reactants. *Loops* are observed often in commodity chemical processes, particularly where low fractional conversions in reactors are advantageous.

Observing the general topological patterns helps in devising strategies for solving process flow calculations for complicated process flow sheets. For diverging trees, it is usually easiest to start at the beginning (feed rate of raw material) and work forward. For converging trees, it is often easiest to start at the end (flow rate out of desired product) and work backward. If necessary, we can use the old trick of changing the basis for calculations, proceeding with all process flow calculations using the new basis, then using a scale factor to switch back to the original basis and get the final answers.

Loops on process flow sheets present challenges. One useful way to attack loops is to use a *tearing algorithm*. This technique is especially useful if we have specifications for fractional conversion, fractional split, and/or fractional recovery for all process units in the loop. We "tear" the loop by choosing a "tear stream" that is internal to the loop. We then follow a key component as it moves around the loop, using the performance specifications to relate one stream to another. We do this until we've looped all the way back to the beginning. By doing this, we derive a new relationship between outlet and inlet streams.

Let's illustrate with a typical loop that contains a mixer, reactor, and separator. There are two reactants, A and B, one product C, and one reaction:

$$A + 0.5B \rightarrow 2C$$

The flow diagram is shown in Fig. 3.9. The fractional conversion of A in the reactor, f_{CA}, the fractional recoveries of A, B, and C in stream 5, f_{Ri}, and the molar flow rates of all components in stream 1, $\dot{n}_{i1}$, are known. From a DOF analysis of the process, we find that there are 14 stream variables and 1 system variable, with 2 specified flows, 4 system performance specifications, and 9 material balance equations. Thus, DOF = $(14 + 1) - (2 + 4 + 9) = 0$ and the system is completely specified.

The linear models for the three process units are written down in matrix form by inspection, using the templates suggested in Fig. 3.6 and 3.7:

Mixer:
$$\begin{bmatrix} 1 & 0 & 0 \\ 0 & 1 & 0 \\ 0 & 0 & 1 \end{bmatrix} \begin{bmatrix} \dot{n}_{A2} \\ \dot{n}_{B2} \\ \dot{n}_{C2} \end{bmatrix} = \begin{bmatrix} \dot{n}_{A1} + \dot{n}_{A5} \\ \dot{n}_{B1} + \dot{n}_{B5} \\ \dot{n}_{C1} + \dot{n}_{C5} \end{bmatrix}$$

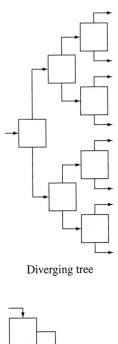

Diverging tree

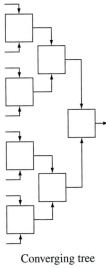

Converging tree

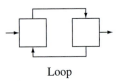

Loop

Figure 3.8 General topological patterns observed in many process flow diagrams.

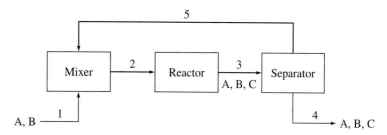

Figure 3.9 Typical block flow diagram with a recycle loop. Streams 2, 3 and 5 are internal to the loop.

Reactor:
$$\begin{bmatrix} 1 & 0 & 0 & 1 \\ 0 & 1 & 0 & 0.5 \\ 0 & 0 & 1 & -2 \\ 0 & 0 & 0 & 1 \end{bmatrix} \begin{bmatrix} \dot{n}_{A3} \\ \dot{n}_{B3} \\ \dot{n}_{C3} \\ \dot{\xi} \end{bmatrix} = \begin{bmatrix} \dot{n}_{A2} \\ \dot{n}_{B2} \\ \dot{n}_{C2} \\ f_{CA}\dot{n}_{A2} \end{bmatrix}$$

Separator:
$$\begin{bmatrix} 1 & 0 & 0 & 1 & 0 & 0 \\ 0 & 1 & 0 & 0 & 1 & 0 \\ 0 & 0 & 1 & 0 & 0 & 1 \\ 1 & 0 & 0 & 0 & 0 & 0 \\ 0 & 1 & 0 & 0 & 0 & 0 \\ 0 & 0 & 1 & 0 & 0 & 0 \end{bmatrix} \begin{bmatrix} \dot{n}_{A5} \\ \dot{n}_{B5} \\ \dot{n}_{C5} \\ \dot{n}_{A4} \\ \dot{n}_{B4} \\ \dot{n}_{C4} \end{bmatrix} = \begin{bmatrix} \dot{n}_{A3} \\ \dot{n}_{B3} \\ \dot{n}_{C3} \\ f_{RA5}\dot{n}_{A3} \\ f_{RB5}\dot{n}_{B3} \\ f_{RC5}\dot{n}_{C3} \end{bmatrix}$$

Without the loop, the equations and solution strategy would be straightforward. But unfortunately there is a loop and we can't just wish it away. With the loop, the **b** vector of the mixer model contains three unknowns: $\dot{n}_{A5}$, $\dot{n}_{B5}$, and $\dot{n}_{C5}$, the component flows in the recycle stream. What's needed is a way to express these three variables in terms of flows and specifications that are known. To demonstrate how to come up with this equation, we'll trace component A around the loop.

We can start at any stream that is internal to the loop; we'll begin the journey at the reactor inlet, stream 2. Now we start marching around the loop. From the reactor performance specification, we relate the flow of component A in stream 3 to the flow of A in stream 2:

$$\dot{n}_{A3} = (1 - f_{CA})\dot{n}_{A2}$$

Continuing on the loop, we relate the flow of A in stream 5 to that in stream 3:

$$\dot{n}_{A5} = f_{RA5}\dot{n}_{A3}$$

Combining these two equations gives the flow of A in stream 5 related to that in stream 2:

$$\dot{n}_{A5} = f_{RA5}(1 - f_{CA})\dot{n}_{A2}$$

Continuing again around the loop, we get back to where we started!

$$\dot{n}_{A2} = \dot{n}_{A1} + \dot{n}_{A5}$$

Combining these two last equations gives

$$\dot{n}_{A5} = f_{RA5}(1 - f_{CA})(\dot{n}_{A1} + \dot{n}_{A5})$$

or, rearranging, we get

$$\dot{n}_{A5} = \left[\frac{f_{RA5}(1 - f_{CA})}{1 - f_{RA5}(1 - f_{CA})} \right] \dot{n}_{A1} = f''_A \dot{n}_{A1}$$

where f''_A indicates the term in brackets. Now we have what we wanted: $\dot{n}_{A5}$ expressed in terms of known system performance specifications and known flows.

To derive an expression for $\dot{n}_{B5}$, we start again at the reactor inlet. We note that

$$\dot{n}_{B3} = \dot{n}_{B2} + \nu_B \dot{\xi} = \dot{n}_{B2} - 0.5\dot{\xi}$$

We do not have a specification on fractional conversion of B—it cannot be independent, because there is only one reaction! But, we can write $\dot{\xi}$ in terms of f_{CA}: $\dot{\xi} = f_{CA}\dot{n}_{A2}$ by definition. Substituting this into the previous equation we find

$$\dot{n}_{B3} = \dot{n}_{B2} + \nu_B \dot{\xi} = \dot{n}_{B2} - 0.5 f_{CA}\dot{n}_{A2}$$

From our previous work, $\dot{n}_{A2} = \dot{n}_{A1} + \dot{n}_{A5} = (1 + f''_A)\dot{n}_{A1}$

Now we continue around the loop until we get back to the starting gate. We find

$$\dot{n}_{B5} = \frac{f_{RB5}(\dot{n}_{B1} - 0.5 f_{CA}\dot{n}_{A1}(1 + f''_A))}{1 - f_{RB5}}$$

Although this is messier than the equation for compound A, still, everything on the right-hand side is known. Similarly, for the product C we find

$$\dot{n}_{C5} = \frac{f_{RC5}(\dot{n}_{C1} + 2 f_{CA}\dot{n}_{A1}(1 + f''_A))}{1 - f_{RC5}}$$

We now can substitute in these expressions for $\dot{n}_{A5}$, $\dot{n}_{B5}$, and $\dot{n}_{C5}$ in the mixer model. This means that everything in vector **b** is known, so we can solve for **x**. Solving for **x** in the mixer model provides the information we need to fill numbers into vector **b** of the reactor model. Likewise, solving for **x** in the reactor model gives the flows we need for vector **b** in the separator model. These connections between vector **x** of one process unit model and vector **b** of the next is essentially a map of the process topology!

Example 3.19   **Linear Models with Multiple Process Units and Recycle: Taking an old Plant out of Mothballs**

A customer is considering your company as a new supplier of an ethylene oxide product. The customer requires that the product contain at least 98 mol% ethylene oxide, and he would like to purchase 1.7 million kgmol product per year. The contract is potentially quite lucrative, and your company could use the business.

Luckily, your company owns an ethylene oxide plant that has been in mothballs for several years, because of low demand for the product. You propose recommissioning the plant now that there is a new customer. Many of the records from the plant were lost, but you are able to dig up an old block flow diagram. The plant produced ethylene oxide by reaction of oxygen and ethylene:

$$2C_2H_4 + O_2 \rightarrow 2C_2H_4O$$

From handwritten notes on yellowed pages you find that the reactor was designed to operate at 6% conversion of ethylene feed, and that the separator was designed to recover 97% of the ethylene oxide as the product stream, along with 98% of the unreacted ethylene and 99.95% of the unreacted oxygen for recycle. From a few old plant operating records, you find that the ethylene flow rate into the plant is set at a maximum of 196 kgmol ethylene/h and the maximum oxygen feed rate is 84.5 kgmol O_2/h.

Can the facility meet the customer's requirements? If not, can you determine which system performance specifications most affect the overall performance of the process, in order to come up with proposals for modifying the plant? Assume that the facility will be operating at maximum feed flow rates.

Solution
Here's the block flow diagram, with streams labeled.

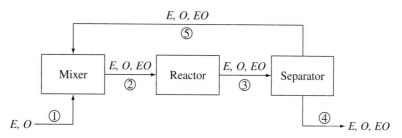

The linear models for each process unit are (in kgmol/h):

Mixer: Three material balance equations

$$\dot{n}_{E2} = 196 + \dot{n}_{E5}$$

$$\dot{n}_{O2} = 84.5 + \dot{n}_{O5}$$

$$\dot{n}_{EO2} = \dot{n}_{EO5}$$

Reactor: Three material balance equations plus one reactor performance specification equation

$$\dot{n}_{E3} = \dot{n}_{E2} - 2\dot{\xi}$$

$$\dot{n}_{O3} = \dot{n}_{O2} - \dot{\xi}$$

$$\dot{n}_{EO3} = \dot{n}_{EO2} + 2\dot{\xi}$$

$$\dot{\xi} = \frac{f_{CE}\dot{n}_{E2}}{2} = 0.03\dot{n}_{E2}$$

Separator: Three material balance equations plus three separator performance specification equations

$$\dot{n}_{E4} + \dot{n}_{E5} = \dot{n}_{E3}$$

$$\dot{n}_{O4} + \dot{n}_{O5} = \dot{n}_{O3}$$

$$\dot{n}_{EO4} + \dot{n}_{EO5} = \dot{n}_{EO3}$$

$$\dot{n}_{E5} = f_{RE5}\dot{n}_{E3} = 0.98\dot{n}_{E3}$$

$$\dot{n}_{O5} = f_{RO5}\dot{n}_{O3} = 0.995\dot{n}_{O3}$$

$$\dot{n}_{EO4} = f_{REO4}\dot{n}_{EO3} = 0.97\dot{n}_{EO3}$$

We work around the loop to derive:

$$\dot{n}_{E5} = \frac{f_{RE5}(1 - f_{CE})}{1 - f_{RE5}(1 - f_{CE})}\dot{n}_{E1} = \frac{0.98 \times 0.94}{1 - (0.98 \times 0.94)}196 = 2291$$

This set of equations constitutes a linear model of the ethylene oxide plant. Now we can proceed to solve this set of equations, either by working through the equations one by one, by using matrix methods, or by going to an equation-solver program. We find $\dot{\xi} = 74.6$ kgmol/h and

	$\dot{n}_{i1}$	$\dot{n}_{i2}$	$\dot{n}_{i3}$	$\dot{n}_{i4}$	$\dot{n}_{i5}$
Ethylene	196	2487	2338	47	2291
Oxygen	84.5	2051	1976	10	1966
Ethylene oxide	0	4.5	154	149	4.5
Total	280	4542	4468	206	4262

Stream 4 is the product stream. Figuring that the plant operates 24 h/day and 350 days/year, 206 kgmol product per hour is equal to 1.73 million kgmol product per year, just barely enough to satisfy our customer. But our product purity is only 72 mol%, well below what our customer demands.

What would be good ways to change the process to get closer to customer requirements? Here are some ideas:

1. Increase fractional conversion of ethylene
2. Improve fractional recovery of ethylene oxide to product
3. Improve fractional recovery of ethylene to recycle
4. Reduce oxygen fresh feed to the mixer

The next step is to figure out which of these changes would have the greatest effect on process operation. Once we have the equations set up, it is easy to explore. If we are able to find changes in operation that might get us closer to our goal, then we can focus efforts on redesigning the equipment to allow such changes.

Idea 1.   *What if we could increase the fractional conversion of ethylene?*

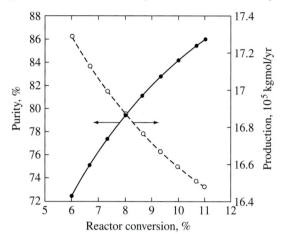

This helps significantly with purity. The maximum conversion we can run at is approximately 11%, as we are limited by the amount of oxygen fresh feed. The maximum purity we can achieve is about 86%. Production rates suffer as purity improves.

Idea 2.   *What if we could improve fractional recovery of ethylene oxide to product?* This doesn't help at all! (Check for yourself!) Any ethylene oxide not recovered just goes around the loop again.

Idea 3.   *What if we could improve fractional recovery of ethylene to recycle?* Just a slight increase from 98 to 99% increases purity quite a bit—to 86.5%. Further improvements don't help, because we run out of oxygen.

Idea 4.   *What if we reduce oxygen fresh feed?*

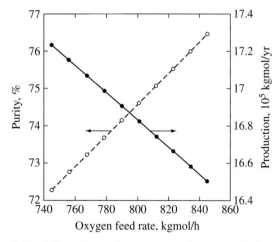

Essentially this reduces the amount of oxygen in the product. Purity increases, but production rate decreases.

Idea 5.   *What if. . .?* Maybe we could combine a couple of changes—what if we could increase oxygen fresh feed and simultaneously improve fractional recoveries in the separator? Perhaps you can come up with some better ideas.

CASE STUDY   Manufacture of Nylon-6,6

Nylon-6,6 was the first completely synthetic polymer to be manufactured and sold on a large scale. Worldwide production of nylons is currently around 8 billion pounds per year. Nylon fibers are used in carpeting, clothing, and tires; nylon resins are used in making injection-molded automotive and electrical components, photographic film, wires and cables.

Nylons are polymers, large molecules (macromolecules) with molar masses in the thousands or even millions, made by repeatedly linking the same basic chemical structure together. Nylons belong to a specific class of polymers called polyamides. (Proteins are also polyamides.) Polyamides contain carbonyl (CO) and amine (NH) groups next to each other. The repeating structure of nylon-6,6 is

Nylon-6,6

The "6,6" simply means that the nylon is built from two different building blocks, or monomers, each containing 6 carbons. The bracket indicates the repeat unit, and the n indicates the number of repeat units. Commercial grades of nylon-6,6 have average molecular weights of about 12,000 to more than 20,000. Since the molecular weight of the repeat unit is 226, nylon-6,6 polymers typically contain about 50 to 100 repeat units.

Nylon-6,6 is built by linking together a dicarboxylic acid, adipic acid (abbreviated AA), and a diamine, hexamethylenediamine (HD):

Adipic acid   Hexamethylenediamine

(Can you pick out these two 6-carbon groups from the nylon-6,6 repeat unit?)

Each of the polymer building blocks is "difunctional"—that is, it has reactive groups at both ends (carboxylic acids for adipic acid and amines for HD). This allows, at least theoretically, unlimited growth by linking the two building blocks end to end. Water (W) is a byproduct. The reaction is called *polycondensation,* and it is an important class of reaction for production of many natural and synthetic polymers:

$$n \text{ HOOC(CH}_2)_4\text{COOH} + n \text{ H}_2\text{N(CH}_2)_6\text{NH}_2 \rightarrow$$

$$\text{HO[OC(CH}_2)_4\text{CONH(CH}_2)_6\text{NH]}_n\text{H} + (2n - 1) \text{ H}_2\text{O} \qquad \text{(R5)}$$

In this case study we'll come up with a preliminary process flow sheet for manufacture of nylon-6,6. This will take several iterations, as we go from the simplest design to a more realistic block flow diagram. Observe that these iterations roughly follow the route we've taken from Chap. 1 through Chap. 3.

First we address the question of raw material source. Adipic acid and hexamethylenediamine are the building blocks for nylon synthesis. Do we want to buy these from outside suppliers, or make them ourselves? Most nylon producers make AA and HD in-house for two reasons: (1) there is essentially no other market for HD, and (2) it is absolutely essential to control the quality of the building blocks in order to get good quality nylon products.

How should we make these two chemicals? First, let's consider HD synthesis. Here is one reaction pathway used commercially. First, butadiene and hydrogen cyanide are combined to make adiponitrile:

$$\text{C}_4\text{H}_6 + 2\text{HCN} \rightarrow \text{NC(CH}_2)_4\text{CN} \qquad \text{(R1)}$$

Then adiponitrile is hydrogenated to make HD:

$$\text{NC(CH}_2)_4\text{CN} + 4\text{H}_2 \rightarrow \text{H}_2\text{N(CH}_2)_6\text{NH}_2 \qquad \text{(R2)}$$

Adipic acid can be synthesized starting with cyclohexane. (In Chap. 1, we described an alternative process for making adipic acid from glucose. That process has not been commercialized yet.) First, cyclohexane is partially oxidized with oxygen to cyclohexanone, with water as a byproduct. (In reality other reactions occur simultaneously; we'll ignore them for simplicity.)

$$\text{C}_6\text{H}_{12} + \text{O}_2 \rightarrow \text{C}_6\text{H}_{10}\text{O} + \text{H}_2\text{O} \qquad \text{(R3)}$$

Then, nitric acid is used as a strong oxidant to produce adipic acid (AA), with nitric oxide (NO) and water as byproducts:

$$\text{C}_6\text{H}_{10}\text{O} + 2\text{HNO}_3 \rightarrow \text{HOOC(CH}_2)_4\text{COOH} + 2\text{NO} + \text{H}_2\text{O} \qquad \text{(R4)}$$

Let's put together a generation-consumption table for production of 1 mole of nylon-6,6 with an average molecular weight of about 15,000 ($n = 66$), using these reaction pathways, and with no net production of AA or HD.

Let's assume that we want to build a process capable of producing 100,000 lb/day nylon-6,6. Raw material requirements and byproduct generation are calculated from the generation-consumption analysis and summarized in Table 3.4.

Table 3.3	Generation-Consumption Table for Nylon-6,6 Manufacture						
Compound	**Abbreviation**	ν_{i1}	ν_{i2}	ν_{i3}	ν_{i4}	ν_{i5}	$\nu_{i,\,net}$
C_4H_6	B	-66					-66
HCN	CN	-132					-132
$NC(CH_2)_4CN$	AN	$+66$	-66				
H_2	H		-264				-264
$H_2N(CH_2)_6NH_2$	HD		$+66$			-66	
C_6H_{12}	CH			-66			-66
O_2	O			-66			-66
$C_6H_{10}O$	CK			$+66$	-66		
H_2O	W			$+66$	$+66$	$+131$	$+263$
HNO_3	NA				-132		-132
$HO_2C(CH_2)_4CO_2H$	AA				$+66$	-66	
NO	NO				$+132$		$+132$
Nylon 6,6	N66					1	1

We've already learned a few things. First, we need large quantities of a strong acid (nitric acid) and of a highly toxic material (hydrogen cyanide). As we develop our process, we need to keep in mind how safety concerns might influence design choices. Second, we'll be generating a lot of wastewater, which will likely be contaminated with organic chemicals. We will need to design systems for handling this wastewater. Third, we'll be generating a lot of NO, something we'll want to minimize because of its negative environmental impact. At this early stage we would complete a preliminary economic analysis by comparing reactant costs and product values, and we might investigate alternative raw materials and reaction schemes.

But let's move forward and make our first attempt at a block flow diagram. We incorporate mixers whenever two or more reactants are needed. We use the heuristics of Chap. 2 (e.g., introduce reactants as late as possible, remove byproducts as soon as possible) to come up with a very preliminary block flow diagram (Fig. 3.10). The process topology is basically that of a converging tree, common for polymer production plants.

Table 3.4 Raw Material and By-Products Flows for Production of 100,000 lb/day Nylon-6,6

Compound	$\nu_{i,\,net}$	M_i lb/lbmol	Flowrate, lb/day (SF = 100,000/14,934)
C_4H_6	−66	54	−23,900
HCN	−132	27	−23,900
H_2	−264	2	−3,500
C_6H_{12}	−66	84	−37,100
O_2	−66	32	−14,100
H_2O	+263	18	+31,700
HNO_3	−132	63	−55,700
NO	+132	30	+26,500
Nylon 6,6	+1	14,934	+100,000

Before we attempt any process flow calculations, let's complete a DOF analysis of the block flow diagram in Fig. 3.10. Let's begin by making a number of simplifying approximations, as discussed in Chap. 2. We'll assume that the reactions go to completion, that the separators work perfectly, and that reactants are fed in stoichiometric ratio.

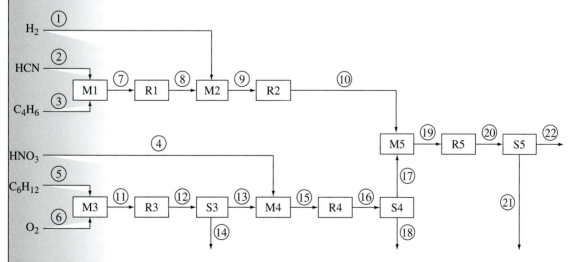

Figure 3.10 Preliminary block flow diagram for nylon-6,6 production. Mixers (M), reactors (R), and separators (S) are indicated, as are stream numbers.

Table 3.5	Enumeration of Stream Variables of Figure 3.10

× indicates that the component is present in that stream, with the simplifying approximations that the reactions go to completion, that the separators work perfectly, and that reactants are fed in stoichiometric ratio.

	1	2	3	4	5	6	7	8	9	10	11	12	13	14	15	16	17	18	19	20	21	22
B			×				×															
CN		×					×															
AN								×	×													
H	×						×															
HD											×							×				
CH					×						×											
O						×					×											
CK												×	×		×							
W												×		×		×		×		×	×	
NA				×											×							
AA																×	×		×			
NO																×		×				
N66																				×		×

Table 3.5 shows how, under these approximations, components are distributed among the process streams. Notice how components appear and disappear on the table—this charts the progression from raw material to product. Table 3.5 shows the count of *stream variables*.

In Table 3.6, we list all process units, and indicate whether the component passes through each unit. This is really a way to enumerate the number of material balance equations.

Now let's complete a DOF analysis:

Count the Number of Variables

Step	Answer	Comment
Independent stream variables	32	The number of ×'s in Table 3.5
Independent system variables	5	5 chemical reactions
Total	37	

Table 3.6   Enumeration of Material Balance Equations of Figure 3.10

× indicates that a material balance equation for that component in that process unit is needed.

	M1	R1	M2	R2	M3	R3	S3	M4	R4	S4	M5	R5	S5
B	×	×											
CN	×	×											
AN		×	×		×								
H			×	×									
HD				×							×	×	
CH					×	×							
O					×	×							
CK						×	×	×	×				
W						×	×		×	×		×	×
NA								×	×				
AA									×	×	×	×	
NO								×	×				
N66												×	×

Count the Number of Equations

Step	Answer	Comment
Specified flows	1	Nylon-6,6 production rate of 100,000 lb/day
Stream composition specifications	0	
System performance specifications	0	
Material balance equations	36	See Table 3.6
Total	37	

The problem is completely specified.

   This analysis tells us exactly what we need to set up a linear model of this system: 36 material balance equations (identified in Table 3.6), written in terms of 5 reaction variables and 32 system variables (identified in Table 3.5), with one known flow rate. We notice that the flow diagram has the general pattern of a converging tree. Thus, starting at the trunk (the product) and working backward is the most straightforward strategy. The nylon-6,6 production rate of 100,000 lb/day is chosen as the basis. We move through the material balance equations, starting with the material balance equation on nylon-6,6 for S5. We leave the detailed calculations as an end-of-chapter exercise.

Whew! We've done a lot of work. But all we really have is the skeleton of a block flow diagram. Still, even at this early stage, the flow diagram aids in raising important design questions such as: What if we use air instead of pure oxygen? Will the nitric acid be pure or in water? What if we eliminate S3 and let S4 do the job? What if we eliminate S3 and S4 and get by with only one separator? Should we add a separator to remove water before R4?

To make a more realistic model of this process, we remove *all* the simplifying approximations about complete conversion of reactants ($f_C = 1.0$), perfect separation ($f_R = 1.0$), and stoichiometric reactant feed ratio. This generalizes the problem. Essentially, this means that *once a component enters the process its*

Table 3.7   Enumeration of Stream Variables of Figure 3.10

× indicates that the component is present in that stream. No simplifying approximations were made.

	1	2	3	4	5	6	7	8	9	10	11	12	13	14	15	16	17	18	19	20	21	22
B			×				×	×	×	×									×	×	×	×
CN		×					×	×	×	×									×	×	×	×
AN								×	×	×									×	×	×	×
H	×								×	×									×	×	×	×
HD										×									×	×	×	×
CH				×							×	×	×	×	×	×	×	×	×	×	×	×
O						×					×	×	×	×	×	×	×	×	×	×	×	×
CK												×	×	×	×	×	×	×	×	×	×	×
W												×	×	×	×	×	×	×	×	×	×	×
NA			×												×	×	×	×	×	×	×	×
AA															×	×	×	×	×	×	×	×
NO															×	×	×	×	×	×	×	×
N66																				×	×	×

Table 3.8 — Enumeration of Material Balance Equations of Figure 3.10

$\times$ indicates that the a material balance equation for that component in that process unit is needed. No simplifying approximations were made.

	M1	R1	M2	R2	M3	R3	S3	M4	R4	S4	M5	R5	S5
B	×	×	×	×							×	×	×
CN	×	×	×	×							×	×	×
AN		×	×	×							×	×	×
H			×	×							×	×	×
HD				×							×	×	×
CH					×	×	×	×	×	×	×	×	×
O					×	×	×	×	×	×	×	×	×
CK						×	×	×	×	×	×	×	×
W						×	×	×	×	×	×	×	×
NA							×	×	×	×	×	×	×
AA								×	×	×	×	×	×
NO								×	×	×	×	×	×
N66												×	×

presence must be accounted for in all downstream streams. This greatly increases the complexity of the problem, but it is actually quite straightforward to trace the components through the process—we simply enter an $\times$ in all cells in Tables 3.7 and 3.8 that are connected downstream of the entry point for that component. We know this information by knowing the process topology.

(Compare Table 3.5 to 3.7, and Table 3.6 to 3.8.) Now let's complete a DOF analysis:

Count the Number of Variables

Step	Answer	Comment
Independent stream variables	111	The number of ×'s in Table 3.7
Independent system variables	5	Still 5 chemical reactions
Total	116	

Count the Number of Equations

Step	Answer	Comment
Specified flows	1	Nylon-6,6 production rate of 100,000 lb/day
Stream composition specifications	0	
System performance specifications	0	See explanation following.
Material balance equations	81	The number of ✕'s in Table 3.6.
Total	82	

DOF = 116 − 82 = 34. Greatly underspecified!

What kind of specifications should we add to make this completely specified? We can use our tables to guide us:

System performance specifications

- *Reactors.* One fractional conversion per reactor per reaction can be specified, for a total of **5**.
- *Separators.* There are $C \times (S - 1)$ independent fractional recoveries per separator. Each separator has 2 outlet streams. S3 has 4 components, S4 has 7, and S5 has 13 components (easily seen from Table 3.8). So, we specify 4 fractional recoveries for S3, 7 for S4, and 13 for S5, or a total of **24**.

Stream composition specifications. Raw materials can be directly identified from examination of Table 3.8. Look for the first entry in each row: Any component that first enters the table in a process unit *other than through generation by chemical reaction* must be a raw material. If there are C raw materials in a process, then there can be $C - 1$ composition specifications. For this process there are 6 raw materials, so **5** specifications are possible.

This adds up to 34 specifications, exactly the number we needed!

This preliminary analysis provides a systematic outline for deriving all the equations that we need to build a linear model of this process flow sheet. Thus, the linear model consists of

- 81 material balance equations
- 1 product flow rate
- 5 reactor performance specifications (fractional conversions)
- 24 separator performance specifications (fractional recoveries)
- 5 stream composition specifications (relative flow of raw materials)

This linear model could be written as a matrix equation using the methods of Sec. 3.4, where the matrix A would have 116 rows and columns. (Engineers use *process simulators* to generate the model equations automatically, given a flow sheet along with a basis and stream composition and system performance specifications.)

Once the hard work is done, we can solve the linear model of this flowsheet for any number of variations in the specifications. Such an effort is an essential feature of process synthesis. Let's illustrate by discussing a few cases.

Case 1: *How does cyclohexane consumption change with reactor fractional conversion?* If all reactors and separators work perfectly, 442.2 lbmol cyclohexane/day is required. We examined several cases where the fractional conversion dropped; we fixed the nylon production rate at 100,000 lb/day and assumed that the fractional recoveries in the separators were still 1.0 and that reactants were fed at stoichiometric ratio.

(a) If fractional conversion in R3 drops to 0.9, cyclohexane consumption rises to 491.3 lbmol/day.

(b) If fractional conversion in R1 drops to 0.9, cyclohexane consumption also rises, to 491.3 lbmol/day! This is surprising at first glance, because cyclohexane is not in the branch with R1. This result is seen because we specified stoichiometric feed ratios of all raw materials. If conversion in one of the branches is higher than that in another, we likely will redesign our process to adjust feed ratios. We can use our model to find the optimum feed ratios as a function of reactor conversion.

(c) If fractional conversion in all the reactors drops to 0.9, cyclohexane consumption rises nearly 40% over the base case, to 606.6 lbmol/day.

(d) A drop in fractional conversion in all reactors to 0.8 is about equivalent to a drop in conversion in just R5 to 0.5—cyclohexane consumption nearly doubles in either case.

Case 2: *Which system performance specifications most affect NO production?* Because of the chemical's adverse contribution to smog, we wish to hold NO production near the minimum possible. In the base case, 884.4 lbmol NO/day is generated.

(a) Reducing fractional conversion in R4 has no effect on NO production rates!

(b) Reducing fractional recovery of adipic acid in S4 to 0.9 increases NO production, to 983 lbmol/day.

(c) Reducing fractional recovery of adipic acid in S4 to 0.9 and reducing fractional conversion in R5 to 0.9 increases NO production even more, to 1092 lbmol/day.

Case 3: *How do separator efficiencies affect the quantity of water in the final product?* If all three separators work perfectly, no water is in the nylon product. We considered four other conditions:

(a) If 99% of the water is removed in each of S3, S4, and S5, the nylon product contains 0.16 wt% water.

(b) If 99% of the water is removed in S4 and S5, but only 90% in S3, the nylon product contains, again, 0.16 wt% water.

(c) If 99% of the water is removed in S3 and S5, but only 90% in S4, the water content of the nylon product increases slightly, to 0.17 wt% water.

(d) If 99% water removal is achieved in S3 and S4, but only 90% in S5, the nylon product is contaminated with 1.6 wt% water.

Clearly, water removal is most important in the last separator. We will want to pay closest attention to careful design of this separator.

Such exploration of the flow sheet model may lead us to changes in design. This analysis is a prelude to further detailed design of each process unit. For example, we might ask: Are there unwanted side reactions in any of the reactors? Is it possible to design a reactor that achieves high fractional conversion, or will separation and recycle be required? What kind of separation technology can achieve the necessary purity? Can these separations be achieved in a single separator, or will multiple pieces of equipment be required? These are the sorts of questions that Chaps. 4 and 5 aim to address.

Summary

- The **material balance equation** derives from the Law of Conservation of Mass.

- **Material balance equations** in differential form are summarized in the following table (i is a component, j is a stream, and k is a chemical reaction).

	Accumulation =	Input	− Output	+ Generation − Consumption
Total mass	$\dfrac{dm_{sys}}{dt} =$	$\displaystyle\sum_{\text{all } j \text{ in}} \dot{m}_j$	$-\displaystyle\sum_{\text{all } j \text{ out}} \dot{m}_j$	
Mass of i	$\dfrac{dm_{i,\,sys}}{dt} =$	$\displaystyle\sum_{\text{all } j \text{ in}} \dot{m}_{ij}$	$-\displaystyle\sum_{\text{all } j \text{ out}} \dot{m}_{ij}$	$+\displaystyle\sum_{\substack{\text{all } k \\ \text{reactions}}} \nu_{ik} M_i \dot{\xi}_k$
Total moles	$\dfrac{dn_{sys}}{dt} =$	$\displaystyle\sum_{\text{all } j \text{ in}} \dot{n}_j$	$-\displaystyle\sum_{\text{all } j \text{ out}} \dot{n}_j$	$+\displaystyle\sum_{\substack{\text{all } k \\ \text{reactions}}} \displaystyle\sum_{\substack{\text{all } i \\ \text{compounds}}} \nu_{ik} \dot{\xi}_k$
Moles of i	$\dfrac{dn_{i,\,sys}}{dt} =$	$\displaystyle\sum_{\text{all } j \text{ in}} \dot{n}_{ij}$	$-\displaystyle\sum_{\text{all } j \text{ out}} \dot{n}_{ij}$	$+\displaystyle\sum_{\substack{\text{all } k \\ \text{reactions}}} \nu_{ik} \dot{\xi}_k$

The differential material balance equation describes a single point in time. At steady state, the accumulation term is set equal to zero. Steady-state continuous-flow processes are analyzed by using the steady-state differential equation.

- **Material balance equations** in integral form are summarized in the following table (i is a component, j is a stream, and k is a chemical reaction).

	Accumulation =	Input	$-$Output	$+$Generation $-$ Consumption
Total mass	$m_{\text{sys},f} - m_{\text{sys},0} =$	$\displaystyle\sum_{\text{all } j \text{ in}} \left(\int_{t_0}^{t_f} \dot{m}_j \, dt \right)$	$- \displaystyle\sum_{\text{all } j \text{ out}} \left(\int_{t_0}^{t_f} \dot{m}_j \, dt \right)$	
Mass of i	$m_{i,\text{sys},f} - m_{i,\text{sys},0} =$	$\displaystyle\sum_{\text{all } j \text{ in}} \left(\int_{t_0}^{t_f} \dot{m}_{ij} \, dt \right)$	$- \displaystyle\sum_{\text{all } j \text{ out}} \left(\int_{t_0}^{t_f} \dot{m}_{ij} \, dt \right)$	$+ \displaystyle\sum_{\substack{\text{all} k \\ \text{reactions}}} \left(\int_{t_0}^{t_f} M_i \nu_{ik} \dot{\xi}_k \, dt \right)$
Total moles	$n_{\text{sys},f} - n_{\text{sys},0} =$	$\displaystyle\sum_{\text{all } j \text{ in}} \left(\int_{t_0}^{t_f} \dot{n}_j \, dt \right)$	$- \displaystyle\sum_{\text{all } j \text{ out}} \left(\int_{t_0}^{t_f} \dot{n}_j \, dt \right)$	$+ \displaystyle\sum_{\substack{\text{all} k \\ \text{reactions}}} \sum_{\substack{\text{all} i \\ \text{compounds}}} \left(\int_{t_0}^{t_f} \nu_{ik} \dot{\xi}_k \, dt \right)$
Moles of i	$n_{i,\text{sys},f} - n_{i,\text{sys},0} =$	$\displaystyle\sum_{\text{all } j \text{ in}} \left(\int_{t_0}^{t_f} \dot{n}_{ij} \, dt \right)$	$- \displaystyle\sum_{\text{all } j \text{ out}} \left(\int_{t_0}^{t_f} \dot{n}_{ij} \, dt \right)$	$+ \displaystyle\sum_{\substack{\text{all } k \\ \text{reactions}}} \left(\int_{t_0}^{t_f} \nu_{ik} \dot{\xi}_k \, dt \right)$

The integral material balance equation describes the system over a defined time interval. Batch systems are usually analyzed using the integral balance.

- **Extent of reaction** $\dot{\xi}$ is a measure of the number of reaction events per unit time. $\dot{\xi}$ relates the rates of consumption and generation of compounds in a chemical reaction. For any reaction k and reactant or product i, $\dot{\xi}_k = \dot{r}_{ik}/\nu_{ik} = \dot{R}_{ik}/\nu_{ik}M_i$, where $\dot{r}_{ik}$ is the molar rate ($\dot{R}_{ik}$ is the mass rate) of reaction of compound i by reaction k and ν_{ik} is the corresponding stoichiometric coefficient (negative for reactants, positive for products).

- **Linear algebra** techniques are used to systematically find stoichiometric coefficients, develop generation-consumption analyses, and identify sets of independent chemical reactions.

- While flow sheets provide visual representations of a process, **linear models** of process flow sheets provide mathematical representations. Each model includes material balance equations and system performance equations. For process flow calculations where input flows and system performance specifications are given, linear models are rapidly derived. Once set up, linear models provide a means for rapid exploration of design variables.

- There are four fundamental types of process units: mixers, splitters, reactors, and separators.
  - Splitter performance is characterized by **fractional split** f_{Sj}

$$f_{Sj} = \frac{\text{moles of } i \text{ leaving in stream } j}{\text{moles of } i \text{ fed to splitter}} = \frac{\dot{n}_{ij}}{\dot{n}_{i,\text{in}}}$$

- Reactor performance is characterized by **fractional conversion** f_{Cj}

$$f_{Ci} = \frac{\text{moles of } i \text{ consumed by reaction}}{\text{moles of } i \text{ fed to reactor}} = \frac{-\sum\limits_{\substack{\text{all } k \\ \text{reactions}}} \nu_{ik}\dot{\xi}_k}{\dot{n}_{i,\,\text{in}}}$$

- Separator performance is characterized by **fractional recovery** f_{Rij}

$$f_{Rij} = \frac{\text{moles of } i \text{ leaving in stream } j}{\text{moles of } i \text{ fed to separator}} = \frac{\dot{n}_{ij}}{\dot{n}_{i,\,\text{in}}}$$

ChemiStory: Of Toothbrushes and Hosiery

*"In olden days a glimpse of stocking
   was looked on as something shocking
   now heaven knows
   anything goes."* Cole Porter

The Roaring 20s was a wild, exciting time in U.S. history—a time of bootleg booze and speakeasies, rising skirts and rising fortunes. The DuPont family was one of several fabulously wealthy families of the time. The DuPont Company started as a gunpowder manufacturer, and had grown to become the major supplier of explosives to the Allied forces in World War I. With the end of WWI and the beginning of the peacetime economic expansion, the company wisely moved from explosives to consumer goods. DuPont illustrated its new consumer focus through their famous motto: "Better Things for Better Living through Chemistry." Using their expertise in cellulose and nitrocellulose chemistry, the company developed and sold a host of new consumer products: cellophane packaging, rayon stockings, lacquers for painting cars bright colors.

Cellulose and nitrocellulose are plant-derived polymers, although at the time little was known about their true nature. Debates raged among European chemists: Were polymers true molecules, albeit very large, or were they aggregates of small molecules held together by some as-yet-unknown noncovalent force? Virtually every well-respected chemist believed the latter—they could not fathom the idea of a molecule with a molecular weight of 100,000, any more than they could imagine "an elephant. . . 1500 ft long and 300 ft high." At a conference held in Europe in 1926, Hermann Staudinger (later awarded the Nobel Prize in Chemistry) stood virtually alone as he argued that polymers were true molecules. A young theoretical organic chemist named Wallace Carothers was one of the small minority who agreed with Staudinger.

Wallace Carothers, born in 1896, had an inauspicious start; he attended his father's secretarial school and studied typing and penmanship. Only later would he study chemistry at the University of Illinois and Harvard. In 1928 Carothers was wooed to DuPont by Charles Stine, a man who believed that corporations should have fundamental research groups for the prestige they would bring to the company. This was a revolutionary idea at the time. DuPont was interested in Carothers because Carothers was interested in polymers. Carothers wanted to prove that Staudinger was right about the molecular nature of polymers, and DuPont seemed to be the place to do it.

Wallace Carothers.
Photo courtesy of DuPont.

In Carothers' first attempts to make polymers, he exploited the well-known chemical reaction between an alcohol and an organic acid to produce an ester. He reasoned that if both the alcohol and the acid were difunctional (that is, possessed two reactive groups, one on each end) he could link them together in an infinite chain, as "poly-esters." This worked, to a point. Polyesters were indeed produced, but their molecular weights were only 5000 to 6000, too short to have any commercial value. Carothers eventually realized that the huge amounts of water produced during the ester reaction might be limiting the extent of polymerization. His group adapted a "molecular still" apparatus to continuously remove water during the condensation reaction. Julian Hill, a chemist in the Carothers lab, set up the molecular still and distilled water from an acid/alcohol reaction mixture. After 12 days, Julian poked the resultant mass with a glass rod. When he pulled the rod back, much to his surprise and delight, along came a long thin filament. By chance, the group had discovered a polymeric material that could be spun into fibers—of great interest for clothing, carpeting, and the like. Although these fibers were strong and pliable, they had one serious drawback as a fabric: they melted at low temperatures, a real problem for ironing clothing. The group attempted to synthesize polyamides, reasoning that polyamides should have higher temperature stability than polyesters, were unable to make anything of commercial interest, and abandoned the project.

The Great Depression of the 1930s changed DuPont. The company laid off workers and cut wages. Charles Stine was promoted and replaced by Elmer Bolton, who was much more interested in applied than fundamental research. Wallace Carothers became deeply depressed and suffered from constant mood swings. Still, his scientific output was prodigious.

Bolton pushed Carothers to work once again on polyamides. In 1934, Donald Coffman in Carothers' group dipped a glass stirring rod into a molten mass made from pentamethylene diamine and sebacic acid and pulled out a fine filament. The product was lustrous, stronger than silk, and impervious to hot water or dry-cleaning solvents. Despite the excitement surrounding Coffman's find, it could not be commercialized—the starting materials were too expensive and the fibers were difficult to spin. The following year, Gerard Berchet (also in Carothers' group) came up with a method of making polyamide fibers from cheaper benzene-derived chemicals—the first nylon-6,6. Significant engineering challenges lay ahead: producing hexamethylenediamine and adipic acid in large quantities and of sufficient purity, controlling the polymer length, melt-spinning a polymer that was insoluble in all common solvents. In the course of the next few years these problems were solved and nylon-6,6 became the first totally synthetic fiber to be sold commercially.

Wallace Carothers had accomplished what he set out to do: He collected irrefutable evidence that supported Staudinger's molecular theory of polymers. In the process, he launched a huge new industry. However, his health, especially his mental health, drastically worsened. He began to doubt his scientific abilities. In February 1936, he surprised everyone by getting married. He was elected to the National Academy of Science, collapsed, and spent months recuperating in the Alps. On April 29, 1937, Wallace checked into a hotel, emptied the contents of a cyanide capsule into a glass of lemon juice, and drank. Seven months after his death, his daughter was born.

Lake County Museum/ Corbis.

Nylon-6,6 was put on the market in 1940; its first commercial use was as toothbrush bristles, to replace the Chinese pig bristles that became unavailable after the Japanese invaded Manchuria. The key to nylon's commercial success, however, was not toothbrushes, but women's stockings: 5 million pairs of nylon stockings went on the market in 1940 and sold out in a single day. During World War II, nylon was diverted to the manufacture of parachutes, tire cords, and tents. It is now used for clothing, carpeting, upholstery, and myriad other items, to the tune of about 1.5 lb for every person on earth.

In the 1960s, Julian Hill, the chemist who first discovered a polyester fiber, and Paul Flory, a prominent polymer physical chemist working at DuPont, began to raise concerns about the huge waste problem brought about by nylon and other plastics, and about the rise of an entire industry based on making single-use throwaway products from cheap petroleum. Increasingly, companies now are exploring the use of agricultural materials to make biodegradable polymers. In a sense, these companies are moving full circle, back to the early days of cellulose-based polymers.

Quick Quiz Answers

3.1 2 gmol/s, 4 gmol/s
3.2 It is simply the net column, scaled to the desired basis.
3.3 $m_{sys,f} - m_{sys,0} = (\dot{m}_1 - \dot{m}_2)t$.
3.4 (a) still left with about 2.7 million lb, so OK; (b) about 33.5 million lb.
3.5 $a_{32} = 3$, $b_1 = 6$.
3.6 Yes.
3.7 With NH_3 as the basis compound,

$$\begin{bmatrix} 0 & 1 & 0 \\ 0 & 1 & 2 \\ 2 & 3 & 1 \end{bmatrix} \begin{bmatrix} \nu_{O_2} \\ \nu_{HNO_3} \\ \nu_{H_2O} \end{bmatrix} = \begin{bmatrix} 1 \\ 3 \\ 0 \end{bmatrix}$$

3.8 Reaction goes the opposite direction as it is written.
3.9 "94% of glucose is recovered" means that, of all the glucose fed to the separator, 94% goes to stream A and the rest goes to other streams. "94% glucose in stream A" means that stream A is a mixture, containing 94% glucose and 6% other materials.

References & Recommended Reading

1. For more on Wallace Carothers and the story of nylon, read "The Nylon Drama" by D. A. Houshell and J. K. Smith Jr., *American Heritage of Invention and Technology,* Fall 1988, pp. 40–55, or *Prometheans in the Lab,* S. B. McGrayne, McGraw-Hill.
2. A good readable introduction to the basics of linear algebra is contained in Chap. 1 of *Linear Algebra and Its Applications,* 3rd ed., by Gilbert Strang, Harcourt, Brace, Jovanovich, San Diego (1988).

Chapter 3 Problems

Hein's Law: Problems worthy of attack prove their worth by hitting back.

Warm-Ups

P3.1 You mix 10 gmol polystyrene (average molecular weight 66,000) with 1000 gmol benzene (C_6H_6). Calculate the weight fraction and mole fraction polystyrene in the mixture.

P3.2 A 5 wt% salt/95 wt% water solution flows into a tank at 15 g/min, where it mixes with some salt already in the tank. A 10 wt% salt/90 wt% water solution flows out of the tank at 15 g/min. What is $\dot{m}_{w,in}$, $\dot{m}_{w,out}$, $\dot{m}_{s,in}$, $\dot{m}_{s,out}$, $\dot{m}_{in}$, $\dot{m}_{out}$, dm_{sys}/dt, $dm_{w,sys}/dt$, and $dm_{s,sys}/dt$?

P3.3 Water flows into a tank. The water flow rate $\dot{m}_w$ is a function of time t, where $\dot{m}_w$ is in kg/h and t is in hours. How much water (kg) is in the tank in 2 hours if $\dot{m}_w = 1 + 2t$ and the tank is initially empty? If $\dot{m}_w = 3e^{2t}$ and the tank initially contains 10 kg water?

P3.4 Given the reaction

$$CH_4 + 2O_2 \rightarrow CO_2 + 2H_2O$$

If $\dot{\xi} = 5$ gmol/min, determine: $\dot{r}_{O_2}$, $\dot{r}_{CO_2}$, $\dot{r}_{H_2O}$, $\dot{r}_{CH_4}$, $\dot{R}_{CH_4}$, $\dot{R}_{O_2}$, $\dot{R}_{CO_2}$, $\dot{R}_{H_2O}$.

P3.5 100 gmol/min hydrogen and 100 gmol/min nitrogen are fed to a reactor at steady state, where they react to ammonia:

$$N_2 + 3H_2 \rightarrow 2NH_3$$

If the ammonia flow rate out of the reactor is 45 gmol/min, what is $\dot{\xi}$? What is the total flow rate out of the reactor (gmol/min)?

P3.6 100 lb/h of a 10 wt% glucose solution is fed to an isomerization reactor, where part of the glucose ($C_6H_{12}O_6$) is converted to its isomer, fructose. (Fructose is sweeter than glucose). The reactor output stream is a 6 wt% glucose solution. What is $\dot{R}_g$? What is $\dot{\xi}$?

P3.7 Are the equations $a + b + c = 2$, $2a + b + c = 3$, and $3a + b + c = 4$ linearly independent?

P3.8 Define fractional split, fractional conversion, and fractional recovery. Draw flow diagrams for a splitter, reactor, and separator. Use the flow diagram to illustrate your definitions.

P3.9 100 lb/h of a 10 wt% glucose/90 wt% water solution is fed to a splitter. The splitter produces two output streams. The glucose flow rate in one of the streams is 2 lb/h glucose. What is the fractional split? What is the flow rate of water in that stream?

P3.10 100 lb/h of a 10 wt% glucose solution is fed to an isomerization reactor, where part of the glucose ($C_6H_{12}O_6$) is converted to its isomer, fructose. (Fructose is sweeter than glucose). The reactor output stream is a 6 wt% glucose solution. What is the fractional conversion?

P3.11 100 lb/h of a 6 wt% glucose/4 wt% fructose solution is fed to a separator. Two product streams are produced: one stream is 50 lb/h of a 9 wt% glucose/1 wt% fructose solution. What is the fractional recovery of glucose in this product stream?

Drills and Skills

P3.12 A burner is fed with 100 gmol/s ethane (C_2H_6) and 400 gmol/s O_2, where the ethane is completely burned to CO_2 and H_2O. The burner operates at steady state. Write the differential forms of the component mass balance equations with the burner as the system and prove that the total mass balance equation is satisfied. Then, write the differential forms of the component mole balance equation and prove that there is no law of conservation of total moles.

For Problems 3.13 to 3.33, complete a DOF analysis before starting calculations. Start with one of the differential or integral material balance equations given in the Chap. 3 Summary and explain why you chose that one.

P3.13 Two aqueous solutions are mixed together in a steady-state mixer. One of the solutions contains 8.3 mol% ethyl acetate, 6.2 mol% ethanol, and water, and is fed to the mixer at 64 kgmol/h. The other solution contains 3.7 mol% ethyl acetate, 2.6 mol% acetic acid, 5.4 mol% ethanol, and water, and is fed to the mixer at 97 kgmol/h. Calculate the flow rate (kgmol/h) and composition (mol%) of all components in the output stream from the mixer.

P3.14 A solution, containing 3.7 mol% ethyl acetate, 2.6 mol% acetic acid, 5.4 mol% ethanol, and water, is fed to a splitter at 97 kgmol/h. The splitter operates at steady-state and has three output streams. 27% of the flow exits in the first stream, 54% exits in the second stream, and the remainder exits in the third stream. Calculate the flow rates (kgmol/h) and composition (mol%) of the three output streams.

P3.15 A solution containing 5.4 mol% ethanol, 8.3 mol% acetic acid and water is fed to a reactor at 97 kgmol/h. In the reactor, operating at steady-state, ethanol and acetic acid react to form ethyl acetate and water:

$$C_2H_5OH + CH_3COOH \rightarrow CH_3COOC_2H_5 + H_2O$$

The molar rate of reaction of ethanol is 4.8 kgmol/h. Calculate the flow rates (kgmol/h) and composition (mol%) of the output stream.

P3.16 A solution containing 6.2 mol% ethanol, 5.4 mol% acetic acid, and water is fed to a separator at 97 kgmol/h. The separator operates at steady state. Three product streams leave the separator. 94% of the ethanol fed to the separator leaves in stream A, and 4% leaves in stream B. 85% of the acetic acid leaves in stream B, and 10% in stream A. 70% of the water fed to the separator leaves in stream C, and 15% in each of stream A and stream B. Calculate the flow rates (kgmol/h) and composition (wt%) of the three product streams.

P3.17 Sugar beet juice, which contains 18 wt% sugar and 82 wt% water, is fed continuously at 100 kg/h to a large vessel. Water is removed by evaporation at a constant rate. If the process produces concentrated sugar beet juice (in the vessel) at 65 wt% sugar, what is the rate of evaporation of water?

P3.18 You're a witch in need of a new magic potion. You've got three flasks, containing the ingredients listed below. You'd like to mix these together in your cauldron, heat the cauldron over a fire to evaporate off excess water, and make 100 g of a liquid potion containing 27 wt% toe of frog, 22 wt% eye of newt, and 11 wt% wool of bat. How many grams from each flask should you add to your cauldron? How many grams of water should you evaporate off? The rate of water evaporation from the cauldron decreases with time, as the potion becomes thicker. If the evaporation rate (in grams per minute) is $30 - 2t$, where t is in minutes, how long will it take to evaporate off the right amount of water?

	Flask A, wt%	Flask B, wt%	Flask C, wt%
Toe of frog	10	0	50
Eye of newt	0	30	0
Wool of bat	40	0	10
Water	50	70	40

P3.19 Milk is pasteurized by heating it rapidly in a pasteurizer to temperatures such that most of the bacteria in the milk are killed. The pasteurizer should kill 99% of the bacteria in the milk. The rate of bacteria killing (bacteria per second) in the pasteurizer is $\dot{r}_{kill} = 10000e^{-0.3t}$ where t is in seconds. How long should milk containing 10^4 bacteria stay in the pasteurizer?

P3.20 A pesticide product containing the active ingredient d-phenothrin is sprayed within passenger aircraft cabins on planes flying international routes to comply with the World Health Organization's International Health Regulations. An airline cabin of 30,000 ft^3 containing 10,000 ppm d-phenothrin is to be flushed with fresh air until the d-phenothrin concentration is reduced to less than 100 ppm. At that concentration of - d-phenothrin the cabin is considered safe. If the flow rate of air into the cabin is 600 ft^3/min, for how many minutes must the cabin be flushed out? Assume that the flushing operation is conducted so that the air in the cabin is well mixed.

P3.21 A 12-oz. mug of coffee contains 200 mg of caffeine. Caffeine is eliminated from the body at a rate of $dm_{c,sys}/dt = -0.116m_{c,sys}$, where $m_{c,sys}$ is the mass of caffeine in the body, $m_{c,sys}$ is in mg and t is in h. After you chug down a mug of coffee, how long will it take for the caffeine in your body to drop to 100 mg? You drink one 12-oz. mug at 6 A.M. and another at 2 P.M. Plot the caffeine content of your body as a function of time for one 24-h interval. If you are having trouble falling asleep at 11 P.M., would it help much to cut out that second mug?

P3.22 Titanium dioxide (TiO_2) is by far the most widely used pigment in white paint. Specifications for the white pigment powder used in paint making require that it contain 70 wt% TiO_2, 5 wt% ZnO, and 25 wt% SiO_2. Each of these powders is received at the paint factory in 50-kg sacks. At 7 A.M., an operator fills a large empty tank, equipped with a mixer, with 28 sacks of TiO_2, 2 sacks of ZnO, and 10 sacks of SiO_2. Then, he starts making paint by continuously drawing off 500 kg/h white powder to mix with the latex paint. At 10 A.M. and again at noon, he adds another 14 sacks of

TiO_2, 1 sack of ZnO, and 5 sacks of SiO_2 to the tank. At 3 P.M. he shuts off the paint-mixing operation and goes home.

Divide the work day into separate time intervals, and apply the material balance equation to each interval. What is the lowest amount of powder in the tank, and when does that occur? Does the tank ever run out? How much pigment powder is left in the tank when the operator leaves work for the day?

P3.23 In Example 3.10, we examined a problem in controlled drug release. Solve the problem again, except add in one complication: The drug inside the controlled release device is known to degrade into an inactive form at a steady rate of 1.1 μg/h. Calculate the mass of drug inside the device at any time, and calculate the total fraction of the initial drug load released in a 24-h period.

P3.24 Formaldehyde (HCHO) is produced by partial oxidation of methanol (CH_3OH). Several side reactions also occur, producing formic acid (HCOOH), CO, CO_2, and H_2O. Suppose 100 kgmol/h methanol and 21.05 kgmol/h O_2 are fed to a reactor, where 40% of the methanol and 95% of the O_2 is converted to products. What must the reactor outlet composition be?

P3.25 A local brewery operation has offered to donate, free of charge, spent grains from its operations, to a heating plant at a college campus. As the heating plant engineer, your job is to see whether it is technically feasible to use this material as fuel. Chemical analysis of the grains gave the following elemental composition: 14.4 wt% C, 6.2 wt% H, 78.8 wt% O, 0.6 wt% S. You placed 1.00 kg of the grains into an evacuated laboratory reactor and heated the material in the absence of air. Only CO, CO_2, SO_2, and H_2O were detected in the exit gases, and no unburned material was left in the reactor at the end of the experiment. How many standard cubic liters of gas were produced? What is the mol% SO_2 in the exit gas? How much oxygen would it take per kg of spent grains to convert all the CO to CO_2?

P3.26 Antibiotics are typically produced by fermentation, where the fermentor is operated under what is called *fed-batch* culture. In an example of fed-batch culture, the fermentor is initially filled with 600 mL fermentation broth, containing 100 g glucose/L and some antibiotic-producing cells. Additional broth (containing 100 g glucose/L) is added continuously to the fermentor at a rate of 200 mL/h. The cells consume the glucose at a rate of 25 g glucose/h. What is the concentration of glucose (g/L) in the fermentor at the end of 6 h? Assume the broth density is 1 g/mL.

P3.27 You are planning to run a fermentor to produce antibiotics from fungi, using a fermentation broth that contains 15 wt% glucose, 6 wt% phosphate, 6 wt% nitrate, various trace nutrients, and water. At 8 A.M. the fermentor is filled with 6000 mL broth and some antibiotic-producing fungi. Over the course of the fermentation, additional broth is pumped into the fermentor at 200 mL/h. The cells consume glucose at a rate of 35 g/h, phosphates at a rate of 13 g/h, and nitrates at a rate of 12 g/h. The fermentation is stopped when the concentration of one of these three nutrients goes to zero (because the cells can no longer grow). Which nutrient—glucose, phosphate, or nitrate—will be depleted first? When is the fermentation stopped? What is the concentration (g/L) of the other two nutrients in the fermentor at the end of the run? Assume broth density is 1 g/mL.

P3.28 A chemical plant has an accidental spill of acrylaldehyde, a volatile liquid. The concentration of acrylaldehyde in the outside air rapidly reaches 10 ppm (parts per million). Acrylaldehyde is extremely toxic: Exposure to concentrations above 4 ppm pose immediate dangers to human health. Operators at the chemical plant work inside a control house, which is located near the spill site. The control house has a volume of 10,000 ft^3 and there are three air exchanges per hour with the outside air (in other words, air flow rate through the control house is 30,000 ft^3/h). The control house normally contains no acrylaldehyde vapors. Assume that the air in the control house is well mixed and that the outside air concentration remains steady at 10 ppm. Calculate how long the operators have to put on protective breathing equipment. (*Hint:* Use the control house as the system.)

P3.29 Immunotoxins are molecules designed to specifically kill cancer cells without damaging healthy noncancerous cells. Immunotoxins must accumulate inside cancer cells before they can be effective. A start-up biotech company has discovered a new immunotoxin. The company is trying to raise venture capital. You are trying to decide if you should invest your millions. The start-up company provides the following data. The immunotoxin enters the typical cancer cell at a rate of 62,000 molecules per hour. Some of the immunotoxin is degraded inside the cancer cell by enzymes. The rate of degradation is estimated to be $2700(1 - e^{-0.3t})$ molecules degraded per hour, where t is time in hours since the cancer cell was first exposed to the immunotoxin. The cancer cell also spits out some of the immunotoxin; the rate of immunotoxin leaving the cancer cell is estimated to be 57,000 molecules per hour. You find independent scientific literature that indicates that, for an immunotoxin to successfully kill a cancer cell, there must be an accumulation of at least 30,000 immunotoxin molecules inside the cell after 8 hours. Should you invest?

P3.30 Natural gas contains about 97 mol% methane (CH_4) and 3 mol% ethane (C_2H_6). Suppose that natural gas is burned with excess air in a home furnace. The furnace is improperly maintained, so that the product gases contain some CO, at a CO: CO_2 ratio of 1:5. All of the natural gas, and 90% of the oxygen in the air, is converted into products. Calculate the composition of the product gases.

P3.31 Fresh fruit juice contains 88 wt% water and 12 wt% solids. A fruit juice processor buys fresh juice every week and makes concentrated juice by evaporating most of the water off. When the evaporator is clean, it removes water at a rate of 1770 lb/day. Over the course of a week, however, the evaporator performance worsens because of fouling. At the end of the week, the evaporator is shut down and cleaned. The plant engineer estimates the evaporation rate decreases by 10% per day. The concentrated juice must be 44 wt% solids. Derive an equation that expresses the fresh feed rate as a function of the day. How much fresh juice should the processor buy per week?

P3.32 A 5000 L tank is filled to capacity with a solution that contains 40 wt% nitric acid in water. The density of this solution is 1.256 g/mL. A small hole develops in the bottom of the tank at a corroded spot, and, as the nitric acid leaks out, the diameter of the hole increases. This causes the flow rate through the hole to increase linearly with time. Assume the flow rate through the leak is initially 5 L/min and is 55 L/min after 10 min. Calculate how many grams of HNO_3 have spilled on the floor in 20 minutes, when you discover the problem.

P3.33 A 15 wt% Na_2SO_4 solution is fed at the rate of 12 lb/min into a mixer that initially holds 100 lb of a 50-50 mixture (by weight) of Na_2SO_4 and water. The exit solution leaves at the rate of 12 lb/min. Assume uniform mixing, which means that the concentration of the exit solution is the same as the concentration in the mixer. What is the total mass in the mixer at the end of 10 minutes? What is the concentration of Na_2SO_4 in the mixer at the end of 10 minutes?

P3.34 Cyclohexanone ($C_6H_{10}O$) and nitric acid (HNO_3) react to form adipic acid ($C_6H_{10}O_4$), with NO and H_2O as byproducts. Use matrix methods to find the correct stoichiometric coefficients for this reaction.

P3.35 Chlorine dioxide is used to bleach pulp in the paper industry. The gas is produced by the following reaction (not balanced):

$$NaClO_3 + H_2SO_4 + CH_3OH \rightarrow ClO_2 + NaHSO_4 + CO_2 + H_2O$$

Using matrix methods, find the stoichiometric coefficients.

P3.36 Redo Example 1.5, but use matrix methods to complete the generation-consumption analysis.

P3.37 Use a matrix method to show whether the following reactions are linearly independent.

$$H_2O + CO \rightarrow CO_2 + H_2$$
$$2H_2 + O_2 \rightarrow 2H_2O$$
$$2CO + 4H_2 \rightarrow 2CH_4 + O_2$$
$$O_2 + H_2 + CO \rightarrow CO_2 + H_2O$$

P3.38 Given the following chemicals: N_2, O_2, NO, NO_2, H_2, H_2O, NH_3 and HNO_3, come up with a set of independent balanced chemical equations. (Automobile exhaust contains these as well as other compounds.)

P3.39 20,000 kgmol ethylene/h and 11,000 kgmol oxygen/h are fed to a reactor operating at steady state. 25% of the ethylene and 90.9% of the oxygen is converted to products. The reactor outlet contains C_2H_4, O_2, CO_2, H_2O, and C_2H_4O. First determine a minimal set of independent balanced chemical equations. Then, use the component mole balance equation to find the composition and flow rate of the reactor outlet stream.

P3.40 Complete the process flow calculations described in Tables 3.5 and 3.6 of the case study.

Scrimmage

P3.41 As part of a start-up process, a liquid solution containing a very hazardous material (compound X) is pumped into an empty feed tank. The cylindrical feed tank is 1 m in diameter and 3 m high. The solution (density = 1000 kg/m³, 0.1 kg X/kg solution) is pumped in at a rate of 40 kg/min. At a point 1 m up the tank wall, a corroded spot gives way, and a leak develops. The leak rate gets worse as the tank fills, with the rate increasing as the square root of the height of the liquid above the leak:

$$\dot{m}_{leak} \text{ (kg/min)} = 4 \times \text{(liquid height in tank} - \text{tank height at leak point)}^{0.5}$$

How much compound X has leaked out, when you walk by the tank 40 minutes after the start of the tank filling process and notice the leak?

P3.42 The oceans already serve as a repository of anthropogenic CO_2 (CO_2 generated from human activity). Deepwater injection of CO_2 is proposed as one way to further reduce atmospheric CO_2 levels. However, since CO_2 is acidic, CO_2 injection lowers the pH, which has potentially adverse effects on marine life. One proposed solution is to build CO_2 sequestration reactors at power plants located on or near the ocean. Stack gases rich in CO_2 (about 10 mol%) would be pumped across a bed of porous limestone (assume 100% $CaCO_3$) that is continuously sprayed with water. The reaction produces calcium bicarbonate $[Ca(HCO_3)_2]$, which is alkaline and highly soluble in water. This would be pumped out continuously into the ocean. You would like to design a process to treat 1 ton CO_2/day, assuming that the reactor would be refilled once per day. How much limestone must be placed in the reactor each day? If the solubility of $Ca(HCO_3)_2$ in water is 5×10^{-3} moles per liter at projected reactor operating conditions, what flow rate of water is required (ton/day)? Capture, transport, and open-ocean injection of CO_2 is estimated to cost from $90 to $180/ton CO_2. Operation of the proposed reactor entails the following costs: $1.45/ton for crushed limestone, $0.04/ton per km for transporting limestone, $0.24/1000 tons per meter for vertical pumping of the required ocean water, and $0.06/ton per km for pipeline transport of CO_2. Derive an equation that relates the cost to N_t, km required to transport limestone; N_p, meters of vertical pumping of water; and N_c, km to transport CO_2. Under what conditions is the proposed reactor economically competitive with simple open-ocean injection? (Reference: www.netl.doe.gov/publications/proceedings/01/carbon%5Fseq/p24.pdf.)

P3.43 Some diseases are treated by injecting proteins into the bloodstream of the patient. The problem with this is that there is a sudden increase in the protein concentration in the blood, and then a rapid fall-off. This means patients require several injections per day. A steadier blood concentration, with fewer injections, could be achieved by using controlled-release technology. Let's look at one example. Protein C is a blood protein important in clotting. Researchers encapsulated protein C in a polymeric particle. The particle is designed to slowly release the encapsulated protein C. 100 "units" of protein C were encapsulated per 100 mg polymeric particles. The researchers placed 100 mg of encapsulated protein C in a beaker containing a blood-like solution, and measured the amount of protein C released as a function of time. Here are some data:

Time, (h)	Total amount of protein C released into beaker, units
0	0
0.34	5
0.56	8
1.0	14
2.0	25
3.0	35

The researchers, who are great polymer synthetic chemists but not too good with engineering, need your help to analyze the data. They want you to determine:

(a) how much protein C is left in the particles as a function of time and

(b) the rate of protein C release (units/h) as a function of time. Then, they want you to use these data to come up with a model equation for how the rate of protein C release depends on the quantity of protein C left in the particle, and use this equation to determine how long it will take for 90% of the protein C to be released. Can you help? Start by calculating $\Delta m_{sys}/\Delta t$ for each time interval, then plotting $\Delta m_{sys}/\Delta t$ versus m_{sys}, and then applying an appropriate material balance equation.

P3.44 As part of the process of producing sugar crystals from sugar cane, raw sugar cane juice is sent to a series of evaporators to remove water. The sugar cane juice, which is 85 wt% water, is fed to the first evaporator at 10,000 lb/h. The concentrated juice out of the last evaporator is 40 wt%

water. First examine a system with two evaporators. Calculate the water evaporated in each evaporator, assuming that the fraction of water in the feed removed in each evaporator is the same. Then, develop a linear model of a flow sheet with N evaporators, assuming that the fractional recovery of water $f_{R,w}$ in each evaporator is the same. Use your model to develop a plot of N versus $f_{R,w}$, letting N vary from 1 to 6.

P3.45 Most pharmaceutical products are complex organic chemicals that are made by multistep synthesis; that is, there are many reactions in series required to convert the raw materials to the desired product. Let's consider how the number of reactors and the fractional conversion per reactor affect the drug production rate. Suppose we feed 1000 kg/day of a reactant to a process requiring multiple reactions. Develop an equation of a flow sheet with N reactors, where the fractional conversion in each reactor is f_C. Use the model to plot the rate of production of the drug product (kg/day) as a function of N and f_C, letting N vary from 1 to 10 and f_C vary from 0.1 to 0.9.

P3.46 Hydrodealkylation is a process in which side chain alkyl groups (like methyl) are removed from aromatics by reaction with hydrogen. A process stream containing 5 mol% benzene (C_6H_6), 20 mol% toluene ($C_6H_5CH_3$), 35 mol% xylene [$C_6H_4(CH_3)_2$] and 40 mol% pseudocumene [$C_6H_3(CH_3)_3$] is fed at a rate of 100 gmol/h to a hydrodealkylation plant at a refinery. This process stream is mixed with 500 gmol H_2/h before being fed to the reactor. The following reactions take place in the reactor:

$$C_6H_5CH_3 + H_2 \rightarrow C_6H_6 + CH_4 \qquad \text{(R1)}$$
$$\text{toluene} \qquad\qquad \text{benzene}$$

$$C_6H_4(CH_3)_2 + H_2 \rightarrow C_6H_5CH_3 + CH_4 \qquad \text{(R2)}$$
$$\text{xylene} \qquad\qquad \text{toluene}$$

$$C_6H_3(CH_3)_3 + H_2 \rightarrow C_6H_4(CH_3)_2 + CH_4 \qquad \text{(R3)}$$
$$\text{pseudocumene} \qquad\qquad \text{xylene}$$

In addition, an unwanted side reaction can occur, in which two benzenes react to form diphenyl:

$$2C_6H_6 \rightarrow C_6H_5C_6H_5 + H_2 \qquad \text{(R4)}$$
$$\text{benzene} \quad \text{diphenyl}$$

Assume that, in the reactor, the following conversions are obtained: 70% conversion of pseudocumene, 17% conversion of xylene, 75% conversion of toluene, and 20% conversion of benzene. First complete a DOF analysis to show that this problem is completely specified. Then set up, and solve, a linear model of the reactor using matrix methods.

P3.47 A waste gas contains 55 mol% DMF (dimethylformamide—a common solvent) in air. A purification unit is available that can remove a fraction of the DMF in the feed *to the unit*. Some of the material leaving the purification unit is recycled back to the inlet. First complete a DOF analysis of this process. Then develop an equation that relates the DMF in the product stream to DMF in the feed through f_S, the fractional split, and f_R, the fractional recovery in the separator. Calculate the purity of the final product, and the composition of the stream fed to the purification unit, as a function of f_S and f_R. Plot your results. What is the required fractional split if $f_R = 0.67$ and the DMF content of the exit gas must be reduced to 10 mol%?

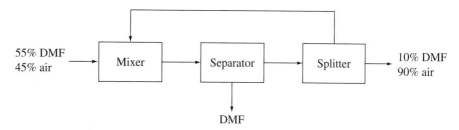

P3.48 In petroleum refining, crude oil is separated into several different mixtures of hydrocarbons. One product stream is the light fraction C_1–C_5 alkanes (methane, ethane, propane, butane, and pentane). Before these hydrocarbons can be sold they are further separated into five different products using a series of separators. Each separator produces two product streams: one called "overhead" and the other called "bottoms."

The light alkane stream that we are in charge of processing has a flow rate of 1000 kgmol/h and contains 10% methane, 30% ethane, 15% propane, 30% butane, and 15% pentane. This stream is sent to Separator 1. 100% of the methane and ethane, and 44.6% of the propane fed to Separator 1 is recovered in an overhead product. All other material is recovered in the bottoms product. The overhead from Separator 1 is sent to Mixer 1, where it is mixed with the overhead from Separator 4. Output from Mixer 1 is sent to Separator 2. 99.5% of the methane fed to Separator 2 is recovered in the overhead product; 99.83% of the ethane and all the propane fed to Separator 2 is recovered in its bottoms product. The bottoms from Separator 2 is fed to Separator 3. 100% of the methane and 99.5% of the ethane fed to Separator 3 is recovered as overhead; 95.8% of the propane fed to Separator 3 is recovered in the bottoms product.

The bottoms from Separator 1 is fed to Separator 4. 96.4% of the propane fed to Separator 4 is recovered as overhead, which is sent to Mixer 1 as mentioned earlier. 100% of the butane and pentane fed to Separator 4 is recovered in the bottoms product, which is sent to Separator 5. 100% of the propane and 99% of the butane sent to Separator 5 is recovered as overhead; 100% of the pentane sent to Separator 5 is recovered as bottoms.

Draw and label a block flow diagram. Is this an example of a diverging or converging tree structure? Then, develop a general linear model of this process at steady state, using material balance equations and fractional recoveries. Solve your model for the specific fractional recoveries stated in the problem. Give the flow rates and compositions of the five product streams.

P3.49 We want to design a system that produces 1000 tons/day of freshwater from saltwater and recovers 30% of the water in the saltwater fed to the system as freshwater. The salt removed from the feed leaves the process as a brine byproduct. (Brine is a concentrated aqueous salt solution.) Explore the following design variations by first setting up a general system of equations, then simplifying by applying the specifications for each of the variations.

(a) The seawater contains 3.5 wt% salt and the remainder water. Compare the seawater feed rate, and the production rate and salt concentration of the briny byproduct, assuming first that concentration of salt in the freshwater product is 0 ppm salt and then assuming it is 10,000 ppm salt.

(b) Now assume that the salt concentration in the freshwater product is fixed at 1000 ppm salt. Examine the effect of using different saltwater feeds. Calculate the saltwater feed rate, the production of brine, and the salt concentration of the brine, as a function of the concentration of salt in the feed, from 1 wt% salt to 10 wt% salt. Plot your results

What affects the process flow calculations more, the amount of salt in the feed or the amount of salt in the product? Why? Would it be a reasonable approximation to assume that the freshwater is pure water, even if there was some salt in it?

Game Day

P3.50 Ethylene glycol [$C_2H_4(OH)_2$, an antifreeze] is produced in two steps. The simplified process flow diagram is sketched below. In the first reactor R-1, ethylene (C_2H_4) is mixed with air (79% N_2, 21% O_2) and oxidized to ethylene oxide (C_2H_4O) in the gas phase:

$$C_2H_4 + \tfrac{1}{2}O_2 \rightarrow C_2H_4O$$

Complete combustion of the ethylene to water and carbon dioxide can also occur, as an undesired side reaction:

$$C_2H_4 + 3O_2 \rightarrow 2CO_2 + 2H_2O$$

All of the O_2 fed to the process is consumed. CO_2, H_2O, and C_2H_4O are separated from unreacted C_2H_4 and N_2 by absorption in column A-1 into cold water. C_2H_4 and N_2 are recycled back to the reactor inlet, with a purge stream taken off. C_2H_4O and CO_2 are then separated from the water in distillation column D-1. The water is discarded, and CO_2 is then removed from C_2H_4O by absorption in column A-2 into triethanolamine (TEA). C_2H_4O is sent to reactor R-2. In the second reactor, ethylene oxide is mixed with liquid water, in which it is very soluble. Ethylene oxide reacts with the water to produce ethylene glycol:

$$C_2H_4O + H_2O \rightarrow C_2H_4(OH)_2$$

Ethylene glycol can react with ethylene oxide to produce diglycol in an unwanted side reaction:

$$C_2H_4O + C_2H_4(OH)_2 \rightarrow (C_2H_4OH)_2$$

Ethylene oxide is very reactive, and 100% conversion of ethylene oxide to products is achieved. Water and diglycol are separated from ethylene glycol in two distillation columns, D-2 and D-3. The fresh ethylene feed rate to the process is 1000 gmol/h. The C_2H_4/O_2 ratio in the fresh feed is 2:1. 50 gmol/h CO_2 is removed in the absorber A-2. The fractional conversion of C_2H_4 to products in R-1 is 0.25. Fresh water is fed to R-2 at a water: C_2H_4O ratio of 5:1. For every 10 mol of ethylene glycol produced, 1 mol of diglycol is produced. Develop a linear model of this flow sheet. Calculate the production rate of C_2H_4O from R-1, the composition and flow rate of the purge gas, the flow rate of the recycle stream to R-1, and the production rate of ethylene glycol. What fraction of the ethylene fed to the process is converted to the desired product?

The process has been operating successfully for about 10 years. Suddenly, the price of ethylene doubles. Propose two process modifications that would improve ethylene utilization. Identify the effects of these proposed modifications on all parts of the plant (i.e., increases/decreases in flows or compositions).

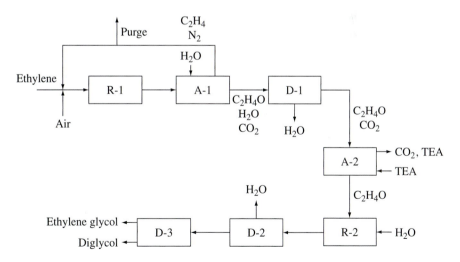

P3.51 Refer to the block flow diagram. A 95 mol% propylene (C_3H_6)/5 mol% propane (C_3H_8) feed is mixed with benzene B (C_6H_6) feed at a molar ratio of 1.2:1 propylene:benzene (P:B). These fresh feeds are mixed with recycled streams, then fed to a reactor. Two reactions occur simultaneously in the reactor, producing the desired product cumene C (C_9H_{12}) and an undesired byproduct diisopropylbenzene D ($C_{12}H_{18}$):

$$C_3H_6 + C_6H_6 \rightarrow C_9H_{12} \tag{R1}$$

$$C_3H_6 + C_9H_{12} \rightarrow C_{12}H_{18} \tag{R2}$$

Under the reaction conditions, propane I is an inert. The reactor effluent is cooled and sent to a separator, where a vapor stream containing propylene and propane is taken off the top and a liquid stream containing benzene, cumene, and diisopropylbenzene is taken off the bottom. A fraction of the vapor stream leaving the separator is purged and the remainder is recycled to be mixed with the incoming feed stream. The liquid stream leaving the reactor is sent to a series of two distillation columns, where benzene is recovered and recycled, and cumene and diisopropylbenzene are separated and sent to storage tanks. The cumene production rate is 25 gmol/s.

Develop a linear model of this flow sheet, where the fractional split and the fractional conversions of propylene and benzene in the reactor are initially unspecified. All separators work perfectly. Then, use your linear model to consider how the flow rate through the reactor (stream 2) is affected by these three performance specifications.

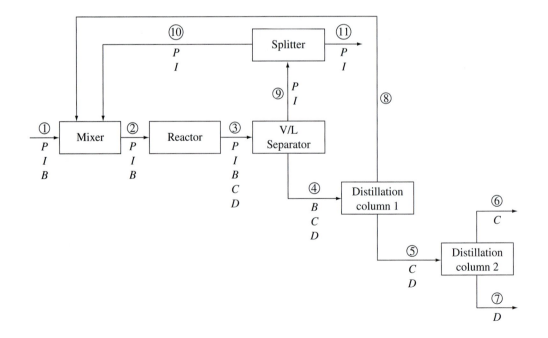

4

Synthesis of Reactor Flow Sheets and Selection of Reactor Process Conditions

In This Chapter

We look more closely inside chemical reactors. We quickly review methods to select chemical reaction pathways that were discussed in Chap. 1. We revisit material balances with chemical reactors, but in a more comprehensive manner. We introduce three ways to think about chemical reactor performance: conversion, yield, and selectivity. We consider the effect of chemical reaction equilibrium and chemical kinetics on reactor performance and discuss ways to design and operate reactors for optimal output. You won't become an expert, but you will gain deeper insight into the inner workings of reactors.

The questions we'll address in this chapter include:

- What are the main classes of industrially important chemical reactions?
- What are the different kinds of chemical reactors?
- How do I describe reactor performance?
- How do I choose the optimum temperature and pressure?
- Why don't all reactors achieve complete conversion of reactants?
- What kinds of approximations are useful for reactor calculations?

Words to Learn

Watch for these words as you read Chapter Four.

Extent of reaction
Limiting reactant
Excess reactant
Conversion
Yield
Selectivity
Recycle
Purge

Chemical reaction equilibrium
Chemical kinetics
Catalyst
Gibbs energy of reaction
Enthalpy of reaction
Gibbs energy of formation
Enthalpy of formation

4.1   Introduction

Chemical reactors are at the heart of any chemical process (Fig. 4.1). Reactors provide the conditions that allow chemical reactions to occur, so raw materials are converted into products. In fact, you might say that the ability to deal with reacting systems is one of the distinct skills that differentiates chemical engineers from other kinds of engineers. The challenge in engineering chemical reactions are many: raw materials and reaction pathways must be selected; reactor shape, size, and operating conditions must be chosen; the reactor flow sheet must be designed; and the chemistry, equipment, and flow sheet must be combined in a way that is safe, economical, and environmentally sound.

4.1.1   Industrially Important Chemical Reactions

It's impossible to list all the myriad chemical reactions that humans employ to convert the raw materials we have to the products we want. Here, we simply list some categories of chemical reactions that are important industrially. Most of these chemical reactions are also important in the natural world—for the functioning of everything from single-cell organisms to ecosystems.

> *Oxidation.* Oxidation was probably the first chemical reaction to be exploited by humans. Complete oxidation of carbon- and hydrogen-containing materials provides heat for cooking and warmth. Oxidizing agents range from oxygen to hydrogen peroxide, widely used for bleaching and disinfecting, to potassium nitrate, used in explosives. Partial oxidation allows introduction of oxygen groups into hydrocarbons derived from fossil fuels and is an important step in production of a huge array of industrial chemicals, including alcohols and organic acids.

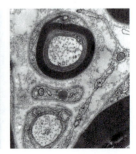

Figure 4.1 Chemical reactors come in all shapes and sizes. In commodity chemical plants multistory reactors in outdoor facilities are common. In a brewery, the fermentors are human size. Mammalian cells are tiny chemical reactors just a few micrometers in diameter.
Left: Courtesy of U.S. Environmental Protection Agency; Middle: © Vol. 5/PhotoDisc/Getty; Right: © The McGraw-Hill Companies Inc./Dr. Dennis Emery Dept. of Zoology and Genetics, Iowa State University.

Hydrogenation and dehydrogenation. These reactions are of utmost importance in organic and inorganic chemistry. Crude oil is hydrogenated to remove sulfur and nitrogen, thus avoiding release of harmful acid gases upon burning of gasoline or other fuels. Dehydrogenation of fats to oils changes the material from solid to liquid as carbon-carbon single bonds are converted to double bonds. Hydrogenation of nitrogen produces ammonia; discovery of this reaction pathway led to huge increases in agricultural output. Either hydrogen gas or stronger reducing agents, like sodium borohydride, are commonly used reactants in hydrogenation processes.

Polymerization. In polymerization reactions, one or two types of small molecules with reactive ends are linked together to form chains that can reach molecular weights in the millions. Rubber, cellulose, starch, proteins, and DNA are all naturally occurring polymers. Nylon, polyester, Teflon, polycarbonate, and other synthetic polymers are ubiquitous in modern life.

Hydrolysis and dehydration. Water is added to compounds in hydrolysis reactions and removed in dehydration reactions. Hydrolysis often leads to the breakdown of larger molecules to smaller—for example, hydrolysis of starch produces sugars—and is a key chemical reaction in biodegradation. Conversely, dehydration is frequently a key step in polymerization reactions.

Halogenation and other substitution reactions. Halogens have strong electron-withdrawing power; chlorine and flourine in particular are added to hydrocarbons to tune their physicochemical properties. Halogenated hydrocarbons include refrigerants like the Freons and polymers such as polyvinylchloride (PVC). They are generally quite resistant to chemical and biological degradation. This resistance to degradation makes them very useful—PVC is popular for underground pipe for example—but also means that these compounds persist for long times in the environment.

Isomerization. Isomers are chemicals with identical molecular formulas but different spatial arrangements of the constituent elements. This spatial arrangement can dramatically alter the properties of isomers. Glucose and fructose are both simple carbohydrates ($C_6H_{12}O_6$), but fructose is much sweeter than glucose. Conversion of glucose to fructose is big business—check out the ingredients listed on a bottle of any nondiet soda. *N*-octane and iso-octane (2,2,4-trimethylpentane) are both alkanes of the same molecular formula (C_8H_{18}), but perform drastically different in automobile engines: A sedan will run like a race car on isooctane but knock and ping on *n*-octane.

Ring opening/ring closing. Cyclic aromatics are mainstays of dyes and pharmaceuticals. Benzene (a cyclic aromatic) and cyclohexane (a cycloalkane) are examples of cyclic compounds that serve as raw materials for the synthesis of a large diversity of compounds. Ring-opening reactions are an important class of reactions leading to polymer synthesis.

4.1.2 Heuristics for Selecting Chemical Reactions

Given the plethora of possible chemical reaction pathways, it is useful to have a few heuristics (rules of thumb) to guide initial selection of raw materials and chemical reaction pathways. Heuristics are guidelines, not laws. Experienced engineers use heuristics to eliminate clearly unsafe or unworkable schemes and to quickly generate a few reasonable choices that can be evaluated in more detail. Some useful heuristics are:

1. Aim to maximize incorporation of reactant atoms into the final product. Choose raw materials that are as close as possible in chemical structure to the final product. Avoid chemical syntheses that use temporary chemical modification of the reactants (e.g., protection/deprotection schemes). Avoid introducing elements that are not incorporated into the final product.
2. Choose reactants to minimize risk of explosions, fires, or release of toxic materials. If use of hazardous materials is unavoidable, design for minimum reactor volume. In syntheses requiring multiple reactors, avoid storage of hazardous intermediates.
3. Use high-purity raw materials to minimize unwanted side reactions. Consider purifying raw materials before introduction into a reactor, if possible.
4. Favor reaction schemes requiring fewer steps.
5. Use a **catalyst** (a material that speeds up the reaction rate) if at all possible.
6. Choose reactions that proceed spontaneously at temperatures and pressures as close to ambient conditions as possible. Temperatures and pressures above ambient are preferable to those below ambient.

The rationale behind many of these heuristics will become clearer as we delve into chemical reactor design and analysis.

4.1.3 A Brief Review: Generation-Consumption Analysis and Atom Economy

Selection of appropriate raw materials and reaction pathways is the first step in design of chemical reactor flow sheets. These ideas were introduced in Chap. 1 and are reviewed here.

A **generation-consumption analysis** is a systematic way to analyze chemical reaction pathways. To complete a generation-consumption analysis, start with a set of balanced chemical reactions, make a table, and

1. List all compounds (reactants or products) in the first column.
2. Using a new column for each chemical reaction, write ν_{ik} for each compound involved in that reaction.
3. Add up the numbers in each row and put the sum in the last column.
4. If there is an unwanted nonzero entry in the last column, find multiplying factors for the reactions involving that species such that the row will sum to zero.

Atom economy is a simple indicator of the efficiency of utilization of raw materials in a given reaction pathway and is easily calculated, once the generation-consumption analysis is complete, as

$$\text{Fractional atom economy} = \frac{\nu_P M_P}{-\sum\limits_{\text{all reactants}} \nu_i M_i} \tag{1.3}$$

where ν_P and M_P are the stoichiometric coefficient and molar mass, respectively, of the product. Generation-consumption analysis and calculation of atom economy tell you the *best* you can do, given the chosen reaction pathway. A real process will never achieve quite as good utilization of raw materials. *All else being equal*, reaction pathways with high atom economy are preferable; these should have fewer waste products and, by making good use of the raw materials, should be more cost-efficient.

Example 4.1

Generation-Consumption and Atom Economy: Improved Synthesis of Ibuprofen

Ibuprofen [2-(p-isobutylphenyl)propionic acid, $C_{13}H_{18}O_2$] is an over-the-counter drug used to treat minor aches and pains. About 30 million lb of the medicine are produced per year. The traditional synthesis of ibuprofen involves six steps, starting with isobutylbenzene ($C_{10}H_{14}$) and acetic anhydride ($C_4H_6O_3$):

$$C_{10}H_{14} + C_4H_6O_3 + AlCl_3 + 6H_2O \rightarrow C_{12}H_{16}O + CH_3COOH$$
$$+ AlCl_3 \cdot 6H_2O \tag{R1a}$$

$$C_{12}H_{16}O + C_4H_7O_2Cl + NaOC_2H_5 \rightarrow C_{16}H_{22}O_3 + C_2H_5OH + NaCl \tag{R1b}$$

$$C_{16}H_{22}O_3 + HCl \rightarrow C_{13}H_{18}O + C_2H_5OOCCl \tag{R1c}$$

$$C_{13}H_{18}O + NH_2OH \rightarrow C_{13}H_{19}ON + H_2O \tag{R1d}$$

$$C_{13}H_{19}ON \rightarrow C_{13}H_{17}N + H_2O \tag{R1e}$$

$$C_{13}H_{17}N + 2H_2O \rightarrow C_{13}H_{18}O_2 + NH_3 \tag{R1f}$$

In the early 1990s, when the patent for ibuprofen expired, a new process was developed and commercialized. The new process involves three reaction steps, again starting with isobutylbenzene and acetic anhydride:

$$C_{10}H_{14} + C_4H_6O_3 \rightarrow C_{12}H_{16}O + CH_3COOH \quad \text{(HF catalyst)} \tag{R2a}$$
$$C_{12}H_{16}O + H_2 \rightarrow C_{12}H_{18}O \quad \text{(nickel catalyst)} \tag{R2b}$$
$$C_{12}H_{18}O + CO \rightarrow C_{13}H_{18}O_2 \quad \text{(palladium catalyst)} \tag{R2c}$$

What is the difference in atom economy between the traditional and the new process?

Did You Know?

Ibuprofen is a widely prescribed pain killer in modern times, but medication for pain has been around for thousands of years. Around 400 B.C., Hippocrates (of the Hippocratic oath for doctors), prescribed chewing on willow leaves for women in labor. He knew what he was talking about—compounds related to salicylic acid (*o*-hydroxybenzoic acid) are plentiful in the roots, bark, and leaves of willow trees as well as other plants such as poplar, birch, beech, and wintergreen. Aspirin, the most widely used painkiller today, is simply acetylated salicylic acid (acetylsalicylic acid, or *o*-acetyoxybenzoic acid). It is made commercially by combining acetic anhydride with salicylic acid. With the introduction of other modern pain killers like ibuprofen and acetaminophen, it was thought that aspirin sales would plummet. However, the discovery that consumption of one "baby" aspirin per day may help prevent heart disease and stroke has breathed new life into a very old drug. Salicycilic acid is seeing use in a number of other applications today—such as in products treating dandruff, acne, and warts.

Solution

First let's complete a generation-consumption analysis and calculate the atom economy of the traditional scheme. For the atom economy calculation, we consider all net reactants but only the desired product.

Compound	ν_{i1a}	ν_{i1b}	ν_{i1c}	ν_{i1d}	ν_{i1e}	ν_{i1f}	$\nu_{i,\text{net}}$	M_i	$\nu_i M_i$
$C_{10}H_{14}$	−1						−1	134	−134
$C_4H_6O_3$	−1						−1	102	−102
$AlCl_3$	−1						−1	133.5	−133.5
H_2O	−6			+1	+1	−2	−6	18	−108
$C_{12}H_{16}O$	+1	−1							
CH_3COOH	+1						+1		
$AlCl_3{^\circ}6H_2O$	+1						+1		
$C_4H_7O_2Cl$		−1					−1	122.5	−122.5
$NaOC_2H_5$		−1					−1	68	−68
$C_{16}H_{22}O_3$		+1	−1						
C_2H_5OH		+1					+1		
$NaCl$		+1					+1		
HCl			−1				−1	36.5	−36.5
$C_{13}H_{18}O$			+1	−1					
C_2H_5OOCCl			+1				+1		
NH_2OH				−1			−1	33	−33
$C_{13}H_{19}ON$				+1	−1				
$C_{13}H_{17}N$					+1	−1			
$C_{13}H_{18}O_2$						+1	+1	206	+206
NH_3						+1	+1		

The fractional atom economy is

$$\frac{\nu_P M_P}{-\sum_{\text{all reactants}} \nu_i M_i} = \frac{206}{134 + 102 + 133.5 + 108 + 122.5 + 68 + 36.5 + 33}$$

$$= 0.28$$

Now let's analyze the new scheme:

Compound	ν_{i2a}	ν_{i2b}	ν_{i2c}	$\nu_{i,\text{net}}$	M_i	$\nu_i M_i$
$C_{10}H_{14}$	-1			-1	134	-134
$C_4H_6O_3$	-1			-1	102	-102
$C_{12}H_{16}O$	$+1$	-1				
CH_3COOH	$+1$			$+1$		
H_2		-1		-1	2	-2
$C_{12}H_{18}O$		$+1$	-1			
CO			-1	-1	28	-28
$C_{13}H_{18}O_2$			$+1$	$+1$	206	$+206$

$$\frac{\nu_P M_P}{-\sum_{\text{all reactants}} \nu_i M_i} = \frac{206}{134 + 102 + 2 + 28} = 0.77$$

This is an incredible improvement. (Evaluate the traditional and new synthesis schemes vis-à-vis the heuristics given in Sec. 4.1.2. Which heuristics are violated by the traditional scheme?)

4.1.4 Reactor Design Variables

In a chemical reactor, chemical reactions take place under controlled conditions. The engineer exercises substantial control in selecting reactor design variables to optimize the performance of the process and the quality of the product. Some of the key choices that must be made include:

Reactor temperature and pressure. Reactor temperature and pressure are manipulated to maximize the conversion of raw material to desired product while reducing or eliminating undesired reactions. A fermentor will usually operate at about 37°C and 1 atm pressure. Reactors processing hydrocarbon gases might operate at 500 to 600°C and 400 bar. Higher temperatures and pressures are feasible, but may require special materials of construction. Some reactions, such as those involving semiconductor materials, are carried out under vacuum. Rigorous control of temperature and pressure are required for optimal performance.

Reactor volume. Chemical reactors differ in size by orders of magnitude. A single yeast cell, for example, is about 1 μm in diameter, yet carries out a huge number of chemical reactions, including the remarkable reactions involved in self-replication. On the other hand, a commodity chemical reactor might process a million tons per year using a single chemical reaction, and could easily be 30 ft high and 15 ft wide. The reactor size is chosen on the basis of the volume of material to be processed as well as the time required for the desired reaction.

Residence time. The *residence time* is the time that the material stays in the reactor. Residence time varies from less than a second to several days, depending on the rate of the reaction. For continuous reactors, residence time is the reactor volume divided by the volumetric flow rate through the reactor. A fast reaction rate means a shorter required residence time, which translates into lower reactor volume and therefore lower equipment cost.

Reactant addition. Most chemical reactions require two or more reactants. The reactants can be mixed and fed to the reactor at exact stoichiometric ratio. This isn't always the best choice; nonstoichiometric feed ratios might be chosen to minimize unwanted byproducts, for example, or to ensure complete conversion of the most expensive reactants, or to control heat release. In semibatch reactors, one of the reactants may be slowly dripped into a pool of the other reactants, for similar reasons.

Catalysts. A **catalyst** is a material that speeds up the rate of a reaction. Speeding up reactions reduces the size of reactors and saves money. Catalysts may allow the reactions to occur at reactor temperatures and/or pressures closer to ambient, thus increasing the safety of a process. A good catalyst will speed up the rate of a desired reaction without affecting the rate of undesired reactions. In this way, more of the desired product and less of undesired byproducts are made. Catalysts are not consumed by reaction, so theoretically they can be used indefinitely. Acids and bases are used quite commonly as catalysts, both in the laboratory and in the plant. In large reactors producing commodity organic chemicals, solid metal catalysts are often preferred. These are expensive, but solid catalysts are easily separated from fluid process streams for recovery and reuse. Enzymes are protein catalysts. The yeast cell requires hundreds of enzymes to control its metabolism and growth. Enzymes are used commercially, especially in the food, pharmaceutical, and biotechnology industries, because they are highly selective and specific catalysts.

Mode of operation. Reactors can operate in batch, semibatch, or continuous mode. Batch reactors have several advantages: The initial capital investment is lower, and operation is more flexible because the same equipment can be used to make many different products. They are preferred for small-volume and specialty chemicals, and are used widely for biochemical, specialty polymer, and prescription pharmaceutical applications. Continuous-flow reactors are cheaper to operate in the long run, and provide greater quality control. They are ubiquitous in the manufacture of large-volume products such as commodity chemicals. Semibatch reactors are specialty reactors, used sometimes for polymer synthesis or fermentation reactions.

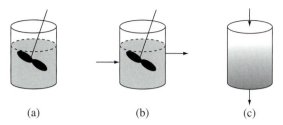

(a)   (b)   (c)

Figure 4.2 Typical flow patterns in chemical reactors include (a) stirred tank batch, (b) stirred tank continuous flow, and (c) plug flow.

Mixing patterns. The extent of mixing inside the reactor is carefully controlled. Batch, semi-batch and continuous reactors can be operated as stirred-tank reactors, with complete mixing of their internal contents. The concentration inside a stirred tank reactor may change with time, but it is the same at every location inside the reactor. Continuous-flow reactors are sometimes designed as plug-flow reactors. In plug-flow reactors, the fluid moves as a "plug" through the reactor, and the concentration changes with distance as the reaction proceeds. There are many other variations in flow pattern that lie in between the stirred-tank and plug-flow reactor (Fig. 4.2).

Throughout the remainder of this chapter, we will discuss how the choice of various reactor design variables affects reactor performance. By knowing how reactor design variables influence reactor performance, we make better engineering choices. But first we will review and discuss reactor process flow calculations.

4.2   Reactor Material Balance Equations

Reactors exist for one purpose: to provide the conditions that allow a desired chemical reaction to occur. At its simplest, a reactor has one input and one output. In this section, we review the use of material balance equations with reacting systems. As you learned in Chap. 3, there are several different forms of the material balance equation. Our primary goal in this section is to discuss which of the material balance equations is most suitable for particular types of problems.

4.2.1   Reactors with Known Reaction Stoichiometry

If the stoichiometric coefficients are known, then the extent-of-reaction concept is very useful in process flow calculations for reacting systems. Recall that reaction rates are characterized by the **extent of reaction** $\dot{\xi}$, which has units of moles/time. $\dot{\xi}$ relates the reaction rates of reactants and products through their stoichiometric coefficients. If $\dot{r}_A$ is the rate of consumption of reactant A and $\dot{r}_B$ is the rate of generation of product B, then $\dot{r}_A$ and $\dot{r}_B$ are related by:

$$\frac{\dot{r}_A}{\nu_A} = \frac{\dot{r}_B}{\nu_B} = \dot{\xi} \tag{3.11}$$

or, in general, the extent of reaction is defined such that:

$$\dot{r}_{ik} = \nu_{ik} \dot{\xi}_k \tag{4.1}$$

where $\dot{r}_{ik}$ is the molar rate of reaction of compound i in reaction k, ν_{ik} is the stoichiometric coefficient for compound i in reaction k, and $\dot{\xi}_k$ is the extent of reaction k.

Assuming that the stoichiometric coefficients are known, then the best choice for the material balance equation depends on the nature of the reactor. Is the reactor continuous-flow, batch, or semibatch? Steady state or unsteady-state? We will consider several variations.

Continuous-flow steady-state reactors are the workhorses of commodity chemical processes. Assuming one input and one output stream, the material balance equation simplifies to

$$\dot{n}_{i,\text{out}} = \dot{n}_{i,\text{in}} + \sum_{\substack{\text{all } k \\ \text{reactions}}} \nu_{ik} \dot{\xi}_k \tag{4.2}$$

$$\dot{n}_{i,\text{in}} \longrightarrow \boxed{\sum \nu_{ik}\dot{\xi}_k} \longrightarrow \dot{n}_{i,\text{out}}$$

Let's consider the continuous-flow steady-state reactor from a degree-of-freedom point of view. If there are R reactants, K reactions, and P products and byproducts of the reactions, then we will have R input stream variables, at most $R + P$ output stream variables (including reactants that are not completely converted to products), and K system variables (extents of reaction). The total number of variables is therefore $2R + P + K$. There are $R + P$ material balance equations, leaving an additional $R + K$ equations required for the system to be correctly specified. Where does this information come from? Most commonly, we have complete information about the input stream (in other words, R known flows), and we have *either* some information about the output stream *or* some information about the reaction rates. Given information about the output stream, then we solve for the extents of reaction; alternatively, given information about the reaction rates, we solve for the output stream. We first apply the material balance equation to the compounds about which most is known, then move on to the material balances on other compounds.

> **! Helpful Hint**
> Don't forget step 1 in the 10 Easy Steps—always draw a diagram!

Example 4.2	**Continuous-Flow Steady-State Reactor with Known Reaction Stoichiometry: Combustion of Natural Gas**

Natural gas, produced over eons by decay of ancient plants, is recovered from wells and used widely as a source of heat and energy. The composition of natural gas varies from well to well. Here are analyses of natural gas from three different sources:

Composition of Natural Gas (mol%)

Component	Rio Arriba, New Mexico	San Juan, New Mexico	Cliffside, Texas
CH_4 (methane)	96.91	77.28	67.0
C_2H_6 (ethane)	1.33	11.18	3.8
C_3H_8 (propane)	0.19	5.83	1.7
C_4H_{10} (butane)	0.05	2.34	0.8
C_5H_{12} (pentane)	0.02	1.18	0.5
CO_2 (carbon dioxide)	0.82	0.80	0.0
N_2 (nitrogen)	0.68	1.39	26.2

Adapted from *Kirk-Othmer Encyclopedia of Chemical Technology*, 1994.

A furnace operating at steady state burns 1.00 MMSCFD (1 million standard cubic feet per day) of natural gas from Rio Arriba, with 23.0 MMSCFD air (assumed to be 79 mol% N_2, 21 mol% O_2). The only detectable compounds in the flue gas are CO_2, H_2O, O_2, and N_2.

Complete combustion of hydrocarbons is described by the general chemical equation

$$C_xH_y + \left(x + \frac{y}{4}\right)O_2 \rightarrow xCO_2 + \frac{y}{2}\,H_2O$$

What is the flow rate (kgmol/h) and composition (mol%) of the flue gas?

Solution
The flow diagram, with the furnace as the system, is

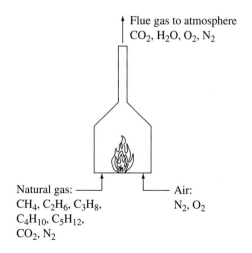

Let's quickly complete a DOF analysis. There are 13 stream variables and 5 system variables (one combustion reaction for each of the hydrocarbons). Flows of all compounds into the furnace are readily calculated from the composition and total flow rate information (9 specified flows/compositions). There are 9 different compounds in the system, so we can write 9 independent material balance equations. Thus, DOF $= (13 + 5) - (9 + 9) = 0$.

Next we need to convert volumetric flow rates to molar flow rates. We'll assume that natural gas and air behave as ideal gases. The volumetric flow rate is reported at standard temperature and pressure: 0°C and 1 atm (Sect. 2.2.4). Therefore, the molar flow rate of natural gas to the furnace is

$$\dot{n} = \frac{P\dot{V}}{RT} = \frac{(1 \text{ atm})(10^6 \text{ ft}^3/\text{day})(1 \text{ day}/24 \text{ h})}{(0.082057 \text{ L atm/gmol K})(1000 \text{ gmol/kgmol})(0.03531467 \text{ ft}^3/\text{L})(273.15 \text{ K})}$$

$$\dot{n} = 52.6 \text{ kgmol/h}$$

An equivalent calculation gives the air flow rate to the furnace as 1210 kgmol/h (956 kgmol/h N_2 and 254 kgmol/h O_2).

Since the system is continuous-flow and steady-state, we use Eq. (4.2). Given the large number of compounds, a table format is convenient for organizing the information. All rates are in kgmol/h.

Component	$\dot{n}_{i,\text{in}}$			$\sum \nu_{ik} \dot{\xi}_k$					$\dot{n}_{i,\text{out}}$
	Natural gas	**Air**	**Total**						
CH_4	50.98		50.98	$-\dot{\xi}_1$					0
C_2H_6	0.70		0.70		$-\dot{\xi}_2$				0
C_3H_8	0.10		0.10			$-\dot{\xi}_3$			0
C_4H_{10}	0.03		0.03				$-\dot{\xi}_4$		0
C_5H_{12}	0.01		0.01					$-\dot{\xi}_5$	0
CO_2	0.43		0.43	$\dot{\xi}_1$	$2\dot{\xi}_2$	$3\dot{\xi}_3$	$4\dot{\xi}_4$	$5\dot{\xi}_5$	$\dot{n}_{CO_2,\text{out}}$
N_2	0.36	956	956.4						$\dot{n}_{N_2,\text{out}}$
O_2		254	254	$-2\dot{\xi}_1$	$-3.5\dot{\xi}_2$	$-5\dot{\xi}_3$	$-6.5\dot{\xi}_4$	$-8\dot{\xi}_5$	$\dot{n}_{O_2,\text{out}}$
H_2O			0	$2\dot{\xi}_1$	$3\dot{\xi}_2$	$4\dot{\xi}_3$	$5\dot{\xi}_4$	$6\dot{\xi}_5$	$\dot{n}_{H_2O,\text{out}}$
Total	52.6	1210	1262.6						

We use the balances on each of the hydrocarbons to determine $\dot{\xi}_k$, then use those numbers to calculate the flowrates of the remaining compounds. We find that the flue gas flow rate is 1263.1 kgmol/h and that it contains 75.7 mol% N_2, 11.8 mol% O_2, 8.3 mol% H_2O and 4.2 mol% CO_2.

Batch reactors are chosen for small-volume specialty chemicals (e.g., prescription drugs, specialty plastics) and sometimes for processes dating back to antiquity (e.g., fermentation of grapes to wine). For batch reactors, the integral mole balance equation is particularly useful when we are interested in the change in the system due to reaction over a defined time interval, from $t = t_0$ to $t = t_f$. If no material is added to or removed from the reactor during that time interval, then the material balance equation simplifies to

$$n_{i,\text{sys},f} = n_{i,\text{sys},0} + \sum_{\substack{\text{all } k \\ \text{reactions}}} \int_{t_0}^{t_f} \nu_{ik} \dot{\xi}_k \, dt \tag{4.3}$$

Generally speaking, with batch reactors we specify the initial contents of the system $(n_{i,\text{sys},0})$. Additionally, we might know the reaction rates and time interval, and solve for the final contents; alternatively, we might know the final contents and solve for the reaction rates.

We may be interested in the rate of change of material inside the batch reactor, in which case the differential material balance equation is used, again with no inputs or outputs:

$$\frac{dn_{i,\text{sys}}}{dt} = \sum_{\substack{\text{all } k \\ \text{reactions}}} \nu_{ik} \dot{\xi}$$

Example 4.3

Batch Reactor with Known Reaction Stoichiometry: Ibuprofen Synthesis

Ibuprofen is synthesized from isobutylbenzene ($C_{10}H_{14}$), acetic anhydride ($C_4H_6O_3$), H_2, and CO in a three-step reaction scheme:

$$C_{10}H_{14} + C_4H_6O_3 \rightarrow C_{12}H_{16}O + CH_3COOH \tag{R1}$$

$$C_{12}H_{16}O + H_2 \rightarrow C_{12}H_{18}O \tag{R2}$$

$$C_{12}H_{18}O + CO \rightarrow C_{13}H_{18}O_2 \tag{R3}$$

The reaction is carried out in a batch reactor. Initially the reactor is charged with 1.4 gmol $C_{10}H_{14}$, 1.4 gmol $C_4H_6O_3$, 3 gmol H_2, and 2 gmol CO. The reactor is brought to reaction temperature and held there for 2.3 h. Then the gases are vented off and the liquid is recovered. From previous work, the reaction rates are known to be constant: $\dot{\xi}_1 = 0.39$ gmol/h, $\dot{\xi}_2 = 0.35$ gmol/h, and $\dot{\xi}_3 = 0.30$ gmol/h. Calculate the quantities of the other compounds in the liquid as well as the amount of H_2 and CO vented.

Solution

We consider this process as a batch reactor followed by a gas-liquid separator.

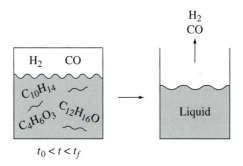

$$t_0 < t < t_f$$

The liquid remaining in the separator after the gases are vented off is the same as the liquid in the batch reactor at $t = t_f$. Therefore we simply analyze the batch reactor part of the process using Eq. (4.3). The integral in Eq. (4.3) is evaluated from $t_0 = 0$ to $t_f = 2.3\,\text{h}$; evaluation is straightforward since the extents of reaction are not functions of time. The results are presented in table form (all units in gmol):

Compound	$n_{i,\text{sys},0}$	$\sum\limits_{\substack{\text{all } k \\ \text{reactions}}} \int\limits_{t_0}^{t_f} \nu_{ik}\dot{\xi}_k \, dt$	$n_{i,\text{sys},f}$
$C_{10}H_{14}$	1.4	$-0.39(2.3)$	0.5
$C_4H_6O_3$	1.4	$-0.39(2.3)$	0.5
$C_{12}H_{16}O$	0	$+0.39(2.3) - (0.35)(2.3)$	0.09
CH_3COOH	0	$+0.39(2.3)$	0.9
H_2	3	$-0.35(2.3)$	2.20
$C_{12}H_{18}O$	0	$+0.35(2.3) - 0.30(2.3)$	0.12
CO	2	$-0.30(2.3)$	1.31
$C_{13}H_{18}O_2$	0	$+0.3(2.3)$	0.69

Semibatch reactors display features of both the continuous-flow and the batch reactor. They usually require more attention from plant operators than their simpler counterparts. They are used only when the semibatch nature provides some important advantage. Probably the most common situation is where the reactor contents are liquid or solid, but one of the reactants or products is a gas. In this case, one might choose to charge the reactor initially with the liquid or solid, and then continuously add the gaseous reactant (or remove the gaseous product). Whether the differential or integral form of the material balance equation is more useful depends on the question to be answered. For analysis of the state of a system at a single point in time, the differential balance is best; the integral balance is appropriate for analysis of processes over a defined time interval. In either case, to analyze semibatch reactors it is usually best to start with the most general form of the material balance equation, Eq. 3.17a or Eq. 3.22, and then simplify as appropriate.

Example 4.4

Semibatch Reactor with Known Reaction Stoichiometry: Ibuprofen Synthesis

The last reaction in the three-step ibuprofen synthesis is

$$C_{12}H_{18}O + CO \rightarrow C_{13}H_{18}O_2$$

Because CO is a highly toxic gas, the reactor is designed to be operated in semibatch mode. Initially, 1.5 gmol $C_{12}H_{18}O$ is charged to the reactor. CO is added at a rate of 0.30 gmol/h. If the reaction rate under these conditions is 0.27 gmol/h, how long should the reactor be operated to produce 1.2 gmol ibuprofen? What is in the reactor at the end of operation?

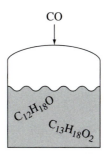

Solution

We are interested in what happens over an interval of time (although we don't yet know what that time interval is!), so we use the integral balance:

$$n_{i,\text{sys},f} = n_{i,\text{sys},0} + \sum_{\substack{\text{all } j \\ \text{in}}} \int_{t_0}^{t_f} \dot{n}_{ij}\, dt - \sum_{\substack{\text{all } j \\ \text{out}}} \int_{t_0}^{t_f} \dot{n}_{ij}\, dt + \sum_{\substack{\text{all } k \\ \text{reactions}}} \int_{t_0}^{t_f} \nu_{ik}\dot{\xi}_k\, dt$$

We start with a balance on ibuprofen ($C_{13}H_{18}O_2$). We know that there is no ibuprofen initially in the reactor ($n_{I,\text{sys},0} = 0$) and that we wish to have 1.2 gmol at the end of the run ($n_{I,\text{sys},f} = 1.2$). From the information given, no ibuprofen enters or leaves the reactor over the time interval of interest, and $\nu_I\dot{\xi} = 0.27$ gmol/h. Inserting these values into the material balance equation gives

$$1.2 = \int_0^t 0.27 \, dt = 0.27t$$

Solving, we find that the reactor run time $t = 4.44$ h.

For $C_{12}H_{18}O$: The integral balance simplifies to

$$\dot{n}_{A,\text{sys},f} = 1.5 - \int_0^{4.44} 0.27 \, dt = 0.13 \, \text{gmol}$$

For CO we need to consider the flow rate in as well as the reaction rate. There is no initial charge of CO to the reactor, and no CO output stream, so the balance equation simplifies to

$$n_{\text{CO},\text{sys},f} = \int_{t_0}^{t_f} n_{\text{CO,in}} \, dt + \int_{t_0}^{t_f} \nu_i \dot{\xi} \, dt$$

$$n_{\text{CO},\text{sys},f} = \int_0^{4.44} 0.30 \, dt - \int_0^{4.44} 0.27 \, dt = 0.13 \, \text{gmol}$$

4.2.2 Reactors with Unknown Reaction Stoichiometry

The extent of reaction concept is very useful when the stoichiometric coefficients are known. But what if we don't know ν_{ik}? And, why wouldn't we know ν_{ik} anyway? There are a couple of common situations: (1) The raw material is highly complex and the molecular formulas are unknown or uncertain. This is true for many natural materials, like wood or coal. (2) The reaction network is highly complex and the reaction products are unknown or uncertain. Combustion and degradation reactions may fall into this category. In these situations, we need a different strategy.

If the elemental composition of the raw material and/or reaction products is known, then we use elements, rather than compounds, as our components. Since elements do not undergo chemical reactions (and we are ignoring nuclear reactions), $\dot{\xi}_k = 0$, and the material balance equations are simplified whether the reactor is steady-state continuous-flow, batch, or semibatch. A complication can arise if some but not all of the molecular formulas of reactants or products are known. In that case, we simply need to convert the moles (or molar flows) of compounds to equivalent moles (or molar flows) of elements:

$$n_h = \varepsilon_{hi} n_i \tag{4.4}$$

where n_h is the moles of element h, n_i is the moles of compound i, and ε_{hi} is the number of atoms of element h in compound i. On a mass basis:

$$m_h = \varepsilon_{hi} \left(\frac{M_h}{M_i} \right) m_i \tag{4.5}$$

where m_i is the mass quantity of compound i and M_h is the atomic mass of the element h.

Even when we know the molecular formulas of the compounds and the chemical reactions, if there are many chemical reactions but only a few elements it may be easier to write balances on elements rather than on compounds.

We illustrate these ideas in the next two examples.

| **Example 4.5** | **Material Balance Equation with Elements: Combustion of Natural Gas** |

Redo Example 4.2, but use elements rather than compounds as components in the material balance equation.

Solution

The feed to the furnace of Example 4.2 contains eight compounds, and there are five combustion reactions to consider. However, there are only four elements: C, H, O, and N. We need only four element balance equations. The steady-state mole balance equation for each element h, where h is C, H, O, or N, simplifies to:

$$\dot{n}_{h,\text{in}} = \dot{n}_{h,\text{out}}$$

Since we know the molar flow rates of the compounds but not of the elements, the big job is to calculate the molar flow rate of each element, using Eq. (4.4) adapted for flows:

$$\dot{n}_{h,\text{in}} = \varepsilon_{hi}\dot{n}_{i,\text{in}}$$

This is readily implemented by using a spreadsheet: Results are shown in tabular form. All flows are given in kgmol/h.

Molar flow,		**C**		**H**		**O**		**N**	
$\dot{n}_{i,\text{in}}$	ε_{Ci}	$\dot{n}_{C,\text{in}}$	ε_{Hi}	$\dot{n}_{H,\text{in}}$	ε_{Oi}	$\dot{n}_{O,\text{in}}$	ε_{Ni}	$\dot{n}_{N,\text{in}}$
CH_4	50.98	1	50.98	4	203.92				
C_2H_6	0.70	2	1.40	6	4.20				
C_3H_8	0.10	3	0.30	8	0.80				
C_4H_{10}	0.03	4	0.12	10	0.30				
C_5H_{12}	0.01	5	0.05	12	0.12				
CO_2	0.43	1	0.43			2	0.86		
N_2	0.36							2	0.72
Total	52.6		53.28		209.34		0.86		0.72

Similar calculations show that 1912 kgmol N/h and 508 kgmol O/h enter in the air stream.

! Helpful Hint
Solve the balance
involving the
fewest number
of unknowns
first.

Now we proceed to evaluate the material balance equation for each element.

N atom balance:

$$\dot{n}_{N,in} = 1912 \frac{\text{kgmol N in air}}{\text{h}} + 0.72 \frac{\text{kgmol N in gas}}{\text{h}} = 1912.7 \frac{\text{kgmol}}{\text{h}} = \dot{n}_{N,out}$$

Since all N leaves as N_2, the N_2 flow rate out with the flue gas is

$$\dot{n}_{N_2,out} = \frac{\dot{n}_{N,out}}{\varepsilon_{N,N_2}} = \frac{1912.7}{2} = 956.4 \frac{\text{kgmol}}{\text{h}}$$

H atom balance:

$$\dot{n}_{H,in} = \frac{209.34 \text{ kgmol}}{\text{h}} = \dot{n}_{H,out}$$

Since all the H leaves as water:

$$\dot{n}_{H_2O,out} = \frac{\dot{n}_{H,out}}{\varepsilon_{H,H_2O}} = \frac{209.34}{2} = 104.67 \frac{\text{kgmol}}{\text{h}}$$

C atom balance:

$$\dot{n}_{C,in} = 53.28 \frac{\text{kgmol}}{\text{h}} = \dot{n}_{C,out}$$

All the C leaves as CO_2:

$$\dot{n}_{CO_2,out} = \frac{\dot{n}_{C,out}}{\varepsilon_{C,CO_2}} = \frac{53.28}{1} = 53.28 \frac{\text{kgmol}}{\text{h}}$$

O atom balance:

$$\dot{n}_{O,in} = \frac{508 \text{ kgmol O in air}}{\text{h}} + \frac{0.86 \text{ kgmol O in gas}}{\text{h}} = 508.86 \frac{\text{kgmol}}{\text{h}} = \dot{n}_{O,out}$$

O leaves in several different forms: as H_2O, CO_2, and O_2. From the H and C balances, we know the flow rates of H_2O and CO_2. (Good thing we left the O balance for last!)

$$\dot{n}_{O,out} = \varepsilon_{O,H_2O}\dot{n}_{H_2O,out} + \varepsilon_{O,CO_2}\dot{n}_{CO_2,out} + \varepsilon_{O,O_2}\dot{n}_{O_2,out}$$

$$508.86\frac{\text{kgmol}}{\text{h}} = 1\left(104.67\frac{\text{kgmol}}{\text{h}}\right) + 2\left(53.28\frac{\text{kgmol}}{\text{h}}\right) + 2\dot{n}_{O_2,out}$$

$$\dot{n}_{O_2,out} = 148.8\frac{\text{kgmol}}{\text{h}}$$

Quick Quiz 4.2

In Example 4.5, what
is the volumetric flow
rate of flue gas, in
MMSCFD?

To summarize:

Total flue gas flow rate $= 956.4 + 104.67 + 53.28 + 148.8 = 1263.1$ kgmol/h

Flue gas contains: 11.8 mol% O_2

4.2 mol% CO_2

8.3 mol% H_2O

75.7 mol% N_2

Another strategy that occasionally works well when stoichiometric coefficients are unknown is to use the mass rate of reaction $\dot{R}_i$. This strategy is applicable, for example, when the component i is a mixture of related species without a unique molecular formula, such as a polymer (where there is a distribution of molecular weights) or protein, where the mixture undergoes a common reaction, such as hydrolysis. Example 4.6 illustrates this approach.

Example 4.6   Mass Rates of Reaction: Microbial Degradation of Soil Contaminants

At an abandoned gasoline station, material from the old underground storage tank has leaked into the ground. After many years, the soil has become contaminated, primarily with aromatics: benzene, toluene, ethylbenzene and xylene (called BTEX).

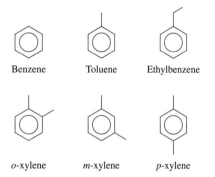

Benzene   Toluene   Ethylbenzene

o-xylene   *m*-xylene   *p*-xylene

Your job is to see if the contaminated soil can be spiked with bacteria that degrade aromatics to less noxious, more volatile compounds. The idea is to turn the soil itself into a batch reactor, and decontaminate the soil *in situ*. The alternative is to dig out the soil and dispose of it as hazardous waste.

You need to determine how long it will take to decontaminate the soil, so you dig up some information from the scientific literature. One study reports that the degradation rate for BTEX is 0.22 µg BTEX per day per gram of soil, microbial. In another study, in which clean soil was spiked with individual aromatic compounds, degradation rates were reported of 6×10^{-5} µmol benzene/g soil/day, 2×10^{-3} µmol toluene/g soil/day, 6×10^{-4} µmol ethylbenzene/g soil/day, and 1.8×10^{-3} µmole xylene/g soil/day.

The contaminated soil contains 8 µg BTEX/g soil. The soil must be decontaminated to 0.1 µg BTEX/g soil. Are the data from the two studies consistent? How long will it take to decontaminate the soil?

Solution

To compare the two studies, we convert the molar rates of degradation from the second study to mass rates of degradation, by multiplying by the molar mass of each component.

$$\dot{R}_b = 6 \times 10^{-5}\frac{\mu\text{mol benzene}}{\text{g soil/day}} \times \frac{78\ \mu\text{g benzene}}{\mu\text{mol benzene}}$$

$$= 0.0047\ \mu\text{g benzene/g soil/day}$$

$$\dot{R}_t = 2 \times 10^{-3}\frac{\mu\text{mol toluene}}{\text{g soil/day}} \times \frac{92\ \mu\text{g toluene}}{\mu\text{mol toluene}}$$

$$= 0.18\ \mu\text{g toluene/g soil/day}$$

$$\dot{R}_{eb} = 6 \times 10^{-4}\frac{\mu\text{mol ethylbenzene}}{\text{g soil/day}} \times \frac{106\ \mu\text{g ethylbenzene}}{\mu\text{mol ethylbenzene}}$$

$$= 0.064\ \mu\text{g ethylbenzene/g soil/day}$$

$$\dot{R}_x = 1.8 \times 10^{-3}\frac{\mu\text{mol xylene}}{\text{g soil/day}} \times \frac{106\ \mu\text{g xylene}}{\mu\text{mol xylene}}$$

$$= 0.19\ \mu\text{g xylene/g soil/day}$$

These rates are reasonably consistent with the rate of $0.22\ \mu$g BTEX per day per gram reported in the first study. The first study is more useful for our purposes: We have a mixture of related compounds and we know the total mass of those compounds, but we don't know the exact composition. The second study does point out, however, that the degradation rates observed will depend on the identity of the contaminants, so we may want to look at a couple of cases.

The soil is a batch reactor, and we are interested in the degradation that occurs over a specified time interval, so we use an integral mass balance equation. Let's choose as a basis 1 g of soil.

$$m_{\text{BTEX,sys},f} - m_{\text{BTEX,sys},0} = \int_{t_0}^{t_f}\dot{R}_{\text{BTEX}}\,dt$$

$$0.1\ \mu\text{g BTEX} - 8\ \mu\text{g BTEX} = \int_0^{t_f} -0.22\frac{\mu\text{g BTEX}}{\text{day}}\,dt = -0.22t_f$$

$$t_f = 36\ \text{days}$$

Using the degradation rate for pure benzene we find something sobering: If the soil were contaminated with pure benzene, the decontamination time would be much longer: about 1580 days!

4.3   Stream Composition and System Performance Specifications for Reactors

Ideally, we choose reactants and a reaction pathway with a high atom economy. Ideally, we design the perfect chemical reactor, providing exactly the right combination of size, residence time, temperature, pressure, catalyst, mixing pattern, and reactant addition to: (1) completely and rapidly convert all the raw materials to useful products by the desired chemical reaction and (2) completely prevent unwanted chemical reactions. The performance of the perfect chemical reactor can be calculated from generation-consumption and atom economy analysis. As you can imagine, the perfect reactor rarely exists.

Figure 4.3 compares an example of a perfect reactor to an example of a real reactor. In the perfect reactor, reactants A and B are fully converted to desired product D by a single reaction:

$$A + B \rightarrow D$$

In the real reactor, reactants A and B, along with contaminant C, are fed at a non-stoichiometric ratio, and are partially converted to desired product D. Along with the desired reaction, a couple of undesired reactions take place:

$$A + C \rightarrow E + F$$

$$A + D \rightarrow G$$

A comparison of the degrees of freedom of the perfect and real reactors is enlightening (Table 4.1). We assume for this analysis that the production rate of D is specified. For the perfect reactor, no additional specifications are required! Indeed, when we completed generation-consumption analysis in Chap. 1 and then scaled up to a desired production rate, we were able to complete all calculations because we implicitly assumed that the reactor was perfect. The real reactor is grossly underspecified, however. By specifying the feed stream composition (e.g., the percent of A, B, and C in the feed), we gain two equations. The remaining three specifications must come from information about the system performance.

Commonly used measures of the performance of a real reactor are: *conversion, selectivity*, and *yield*. These are all related to the extents of reaction $\dot{\xi}_k$. These system performance specifications are used with material balance equations to design and analyze chemical reactors. We will describe each of these measures in turn.

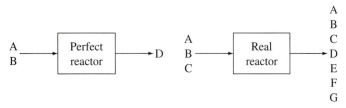

Figure 4.3 In the perfect reactor, reactants A + B are fully converted to desired product D and nothing else. In the real reactor, reactants A and B and contaminant C undergo multiple reactions producing D, E, F, and G.

Table 4.1	DOF Analysis of the Perfect Reactor and the Real Reactor of Fig. 4.3	
	Perfect reactor	**Real reactor**
Stream variables	3	10
System variables	1	3
Total variables	4	13
Specified flows	1	1
Specified stream composition	0	0
Specified system performance	0	0
Material balances	3	7
Total equations	4	8
DOF	0	5

4.3.1 Stream Composition Specification: Excess and Limiting Reactants

In the perfect reactor of Figure 4.3, there is only one reaction and the reactants are fed at exactly the right stoichiometric ratio. But sometimes reactants are fed at nonstoichiometric ratio. (We'll discuss reasons why this is sometimes a good choice later in this chapter.) A reactant fed at less than its stoichiometric ratio (relative to the other reactants) is called the **limiting reactant**. Reactants fed at greater than stoichiometric ratio relative to the limiting reactant are called **excess reactants**. For example, let's say the desired reaction is

$$2A + B \rightarrow D$$

If

$$\frac{\dot{n}_{A,in}}{\dot{n}_{B,in}} > 2 \Rightarrow A \text{ is excess, B is limiting}$$

In general then, reactant A is the excess reactant if

$$\frac{\dot{n}_{A,in}}{\dot{n}_{B,in}} > \frac{\nu_A}{\nu_B} \tag{4.6a}$$

Similarly, reactant A is the limiting reactant if

$$\frac{\dot{n}_{A,in}}{\dot{n}_{B,in}} < \frac{\nu_A}{\nu_B} \tag{4.6b}$$

If there are more than two reactants, than the limiting reactant would be the reactant for which Eq. (4.6b) was true relative to all other reactants.

The *percent excess* reactant indicates the percent by which the feed of the excess reactant exceeds what would be required for stoichiometry. Let's again say that the desired reaction is $2A + B \rightarrow D$. Suppose 100 gmol/min of B were fed to the reactor. The stoichiometric requirement for A would be 200 gmol/min. If 250 gmol/min of A were fed, then A would be fed at $[(250 - 200)/200] \times 100\%$ or 25% excess. In general, then, we define percent excess as

$$\frac{\dot{n}_{A,in} - \left(\dfrac{\nu_A}{\nu_B}\right)\dot{n}_{B,in}}{\left(\dfrac{\nu_A}{\nu_B}\right)\dot{n}_{B,in}} \times 100\% = \% \text{ excess} \qquad (4.7)$$

where A is an excess reactant and B is the limiting reactant.

Percent excess has a special meaning with combustion reactions. Combustion reactions are considered to go to completion if all the C in the fuel is converted to CO_2, all the H is converted to H_2O, and all the S is converted to SO_2. Even if part of the fuel is incompletely combusted, (e.g., some CO is produced), the percent excess is calculated on the assumption of complete combustion.

Example 4.7   **Excess Reactants: A Badly Maintained Furnace**

Natural gas from Rio Arriba is fed to an industrial furnace at 1.00 MMSCFD along with 25% excess air. The flue gas is tested and found to contain both CO and CO_2, at a 1:10 mole ratio. Also in the flue gas is N_2, O_2, and H_2O. Calculate the flow rate and composition (mol%) of the flue gas.

Solution

In Example 4.5, we calculated that Rio Arriba natural gas at 1.00 MMSCFD was equivalent to a molar flow rate to the furnace of: 53.28 kgmol C/h, 209.34 kgmol H/h, 0.86 kgmol O/h and 0.72 kgmol N/h. First we will calculate the stoichiometric amount of oxygen required for complete combustion of these elements. The reactions are

$$C + O_2 \rightarrow CO_2$$

$$4H + O_2 \rightarrow 2H_2O$$

Complete combustion of 53.28 kgmol C/h requires 53.28 kgmol O_2/h. Complete combustion of 209.34 kgmol H/h requires (209.34/4) or 52.34 kgmol O_2/h. The fuel gas itself supplies (0.86/2) or 0.43 kgmol O_2/h (equivalent). Therefore, the total required oxygen flow just to meet stoichiometric demands is $53.28 + 52.34 - 0.43$ or 105.19 kgmol O_2/h. The total oxygen flow to the reactor is therefore the stoichiometric quantity plus 25% more, or $(105.19)(1 + 0.25) = 131.49$ kgmol/h. Since air is (approximately) 21 mol% O_2 and 79 mol% N_2, the nitrogen flow rate from the air is 131.49(0.79/0.21) = 494.7 kgmol/h.

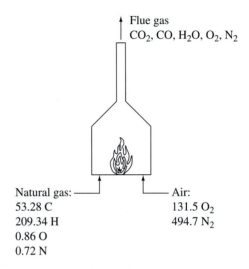

Flue gas
CO_2, CO, H_2O, O_2, N_2

Natural gas: ——
53.28 C
209.34 H
0.86 O
0.72 N

Air:
131.5 O_2
494.7 N_2

We then quickly derive element balance equations. Coupled with the information that the CO_2:CO ratio is 10:1, we solve for the flue gas flow rate and composition.

$$\dot{n}_{N,in} = 2(494.7) + 0.72 = 990.12 \, \frac{kgmol}{h} = \dot{n}_{N,out} = 2\dot{n}_{N_2,out}$$

$$\dot{n}_{N_2,out} = 495 \, \frac{kgmol}{h}$$

$$\dot{n}_{H,in} = 209.34 \, \frac{kgmol}{h} = \dot{n}_{H,out} = 2\dot{n}_{H_2O,out}$$

$$\dot{n}_{H_2O,out} = 104.67 \, \frac{kgmol}{h}$$

$$\dot{n}_{C,in} = 53.28 \, \frac{kgmol}{h} = \dot{n}_{C,out} = \dot{n}_{CO,out} + \dot{n}_{CO_2,out}$$

$$\dot{n}_{CO,out} = 0.1\dot{n}_{CO_2,out}$$

$$\dot{n}_{CO,out} = 4.84 \, \frac{kgmol}{h}, \qquad \dot{n}_{CO_2,out} = 48.44 \, \frac{kgmol}{h}$$

$$\dot{n}_{O,in} = 2(131.49) + 0.86 = 263.84 \, \frac{kgmol}{h}$$

$$= \dot{n}_{O,out} = 104.67 + 4.84 + 2(48.44) + 2\dot{n}_{O_2,out}$$

$$\dot{n}_{O_2,out} = 28.72 \, \frac{kgmol}{h}$$

As a quick check, we know that if the furnace was working well, the oxygen in the flue gas should be exactly the excess oxygen in the air feed, or $(0.25)(105.19) = 26.3$ kgmol/h. Since there is some partial combustion, the O_2 in the flue gas is slightly higher.

The flue gas flow rate is 681.6 kgmol/h; the flue gas composition is 72.6% N_2, 4.2 mol% O_2, 0.7 mol% CO, 7.1 mol% CO_2, and 15.4 mol% H_2O.

4.3.2   System Performance Specification: Fractional Conversion

The fraction of reactant converted to products is one of the most commonly used measures of reactor performance. **Fractional conversion** is defined succinctly in words as

$$\text{Fractional conversion} = \frac{\text{moles of reactant consumed}}{\text{moles of reactant fed}}$$

Let's now define fractional conversion mathematically. (If you read the optional Sec. 3.4 you have already come across this definition.) Suppose reactant i enters a reactor operating at steady-state at flow rate $\dot{n}_{i,\text{in}}$ and leaves at a flow rate $\dot{n}_{i,\text{out}}$. The steady-state mole balance equation for reactant i is

$$\dot{n}_{i,\text{out}} = \dot{n}_{i,\text{in}} + \sum_{\substack{\text{all } k \\ \text{reactions}}} \nu_{ik}\dot{\xi}_k \tag{4.2}$$

Now, subtract $\dot{n}_{i,\text{in}}$ from both sides:

$$\dot{n}_{i,\text{out}} - \dot{n}_{i,\text{in}} = \sum_{\substack{\text{all } k \\ \text{reactions}}} \nu_{ik}\dot{\xi}_k$$

The left-hand side is simply the difference between what comes out of the reactor and what goes in (out − in). The right-hand side is the net reaction rate (generation − consumption). If we divide both sides by the flow rate into the reactor, $\dot{n}_{i,\text{in}}$, and multiply through by (-1), we get the net moles of reactant consumed per mole of reactant fed, or the **fractional conversion** f_{Ci}:

$$f_{Ci} = \frac{\dot{n}_{i,\text{in}} - \dot{n}_{i,\text{out}}}{\dot{n}_{i,\text{in}}} = \frac{-\displaystyle\sum_{\substack{\text{all } k \\ \text{reactions}}} \nu_{ik}\dot{\xi}_k}{\dot{n}_{i,\text{in}}} \tag{4.8a}$$

Rearranging Eq. (4.8a) gives

$$\dot{n}_{i,\text{out}} = (1 - f_{Ci})\dot{n}_{i,\text{in}}$$

$$\dot{n}_{i,\text{in}} \longrightarrow \boxed{\text{Reactor}} \longrightarrow \dot{n}_{i,\text{out}} = (1 - f_{ci})\dot{n}_{i,\text{in}}$$

The fractional conversion in a batch reactor is defined in essentially the same manner:

$$f_{Ci} = \frac{n_{i,\text{sys},0} - n_{i,\text{sys},f}}{n_{i,\text{sys},0}} = \frac{-\sum\limits_{\substack{\text{all } k \\ \text{reactions}}} \int\limits_{t_0}^{t_f} \nu_{ik}\, \xi_k\, dt}{n_{i,\text{sys},0}} \qquad (4.8b)$$

It is always true that $0 \le f_{c,i} \le 1$. For the "perfect" reactor, $f_{c,i} = 1$ for all reactants. *Percent conversion* is simply 100% times the fractional conversion. Conversion is defined only for reactants, never for products. If there is only one reaction and all the reactants are fed to the reactor at exactly the right stoichiometric ratio, then the fractional conversion of one reactant is the same as that of all the others. If reactants are fed at nonstoichiometric ratio, then the fractional conversion of the limiting reactant will be greater than the fractional conversion of any excess reactants.

Example 4.8   **Fractional Conversion: Ammonia Synthesis**

Ammonia (NH_3) is synthesized from nitrogen and hydrogen by the following stoichiometrically balanced reaction:

$$N_2 + 3\,H_2 \rightarrow 2\,NH_3$$

Consider three cases:

Case 1.   Nitrogen and hydrogen are fed to an ammonia synthesis reactor. The nitrogen feed rate is 1000 kgmol/h, the hydrogen feed rate is 3000 kgmol/h, and $\dot{\xi} = 500$ kgmol/h. Calculate the fractional conversion of nitrogen and of hydrogen.

Case 2.   Nitrogen and hydrogen are fed at a 1:5 ratio to an ammonia synthesis reactor. The nitrogen feed rate is 1000 kgmol/h and $\dot{\xi} = 500$ kgmol/h. Identify the limiting reactant. Calculate the fractional conversion of nitrogen and of hydrogen.

Case 3.   Nitrogen and hydrogen are fed at a 1:5 ratio to an ammonia synthesis reactor. The nitrogen feed rate is 1000 kgmol/h, and $\dot{\xi} = 1000$ kgmol/h. Identify the limiting reactant. Calculate the fractional conversion of nitrogen and of hydrogen.

Solution

Case 1.

$$\text{For nitrogen: } f_{C,N_2} = \frac{-\nu_{N_2}\dot{\xi}}{\dot{n}_{N_2,\text{in}}} = \frac{500}{1000} = 0.50$$

$$\text{For hydrogen: } f_{C,H_2} = \frac{-\nu_{H_2}\dot{\xi}}{\dot{n}_{H_2, \text{ in}}} = \frac{(3)500}{3000} = 0.50$$

Nitrogen and hydrogen are fed at stoichiometric ratio (1:3), so the fractional conversion of the 2 reactants is the same.

Case 2.

$$\text{For nitrogen: } f_{C,N_2} = \frac{500}{1000} = 0.50$$

Hydrogen is fed at 5 times the rate of nitrogen, or 5000 kgmol/h. Therefore:

$$f_{C,H_2} = \frac{(3)500}{5000} = 0.30$$

Hydrogen is fed in excess of the stoichiometric ratio. Therefore, hydrogen is the excess reactant, nitrogen is the limiting reactant, and the fractional conversion of the limiting reactant is greater than that of the excess reactant.

Case 3.

$$\text{For nitrogen: } f_{C,N2} = \frac{1000}{1000} = 1.00$$

$$\text{For hydrogen: } f_{C,H_2} = \frac{(3)1000}{5000} = 0.60$$

Hydrogen is the excess reactant and nitrogen is the limiting reactant. 100% of the nitrogen is consumed by reaction, but there is some hydrogen left.

Quick Quiz 4.4

100 kgmol C_2H_4 and 100 kgmol O_2 react to make C_2H_4O. Which reactant is limiting? What is the maximum possible fractional conversion of the excess reactant?

Example 4.9

Effect of Conversion on Reactor Flows: Ammonia Synthesis

Ammonia (NH_3) is synthesized from nitrogen and hydrogen:

$$N_2 + 3H_2 \rightarrow 2NH_3$$

Suppose we want to make 1000 kgmol NH_3/h. Explore the effect of adjusting the fractional conversion of nitrogen on the flow rates in and out of a continuous-flow steady-state reactor. Assume that N_2 and H_2 are fed at stoichiometric ratio.

Solution
The block flow diagram is

The basis is 1000 kgmol NH_3/h leaving the reactor. From the steady-state mole balance equation on ammonia we find the extent of reaction:

$$\dot{n}_{NH_3,out} = 2\,\dot{\xi} = 1000\,\frac{\text{kgmol}}{\text{h}}$$

$$\dot{\xi} = 500\,\frac{\text{kgmol}}{\text{h}}$$

Since we have specified a desired production rate of ammonia, the extent of reaction remains constant as the feed rate of reactants changes. From the definition of fractional conversion:

$$f_{C,H_2} = \frac{-\nu_{H_2}\,\dot{\xi}}{\dot{n}_{H_2,\,in}} = \frac{(3)(500)}{\dot{n}_{H_2,\,in}} = \frac{1500}{\dot{n}_{H_2,in}}$$

$$f_{C,N_2} = \frac{-\nu_{N_2}\,\dot{\xi}}{\dot{n}_{N_2,\,in}} = \frac{(1)(500)}{\dot{n}_{N_2,\,in}} = \frac{500}{\dot{n}_{N_2,in}}$$

Also,

$$\dot{n}_{H_2,out} = (1 - f_{C,H_2})\dot{n}_{H_2,\,in}$$

$$\dot{n}_{N_2,out} = (1 - f_{C,N_2})\dot{n}_{N_2,in}$$

Since the reactants are fed at stoichiometric ratio, $f_{C,N_2} = f_{C,H_2}$.

Now let's use these equations to examine the effect of varying fractional conversion on reactor flows. The calculations are easily carried out on a spreadsheet.

> **! Helpful Hint**
> If, and only if, reactants are fed at stoichiometric ratio, then the fractional conversion is the same for all reactants.

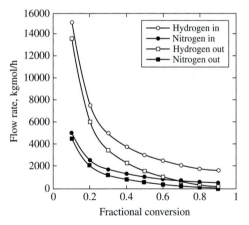

The flow rates in and out of the reactor increase drastically as the fractional conversion decreases.

4.3.3 Fractional Conversion and Its Effect on Reactor Flow Sheet Synthesis: Recycle

If the reactor is operated at low fractional conversion, a large fraction of the reactants passes through the reactor unconverted to products, as we saw in

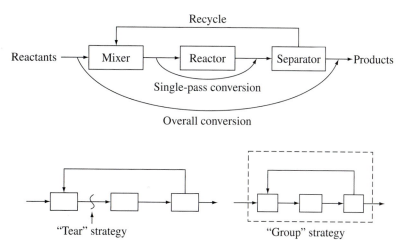

Figure 4.4 When conversion is low, recycle of unreacted reactants is often economical. The flow sheet with recycle requires a mixer and separator as well as a reactor. Analysis of flow sheets with recycle demands new strategies.

Example 4.9. This causes a big problem: We've paid for the reactants and we're not using them! The solution is simple: **Recycle**! Recycle changes the reactor flow sheet: A separator unit must be placed downstream of the reactor, and a process stream must be added to return the unused reactants to a mixer upstream of the reactor.

With recycle, there are two conversions to consider. The conversion based on flows in and out of the *reactor* is called the *single-pass conversion*. The fractional conversion based on flows in and out of the *process* (includes mixer, reactor, separator, and recycle stream) is called the *overall conversion*. (See Fig. 4.4). The overall conversion is always greater than the single-pass conversion.

Completing process flow calculations on flow sheets with recycle can be challenging. Here are two strategies to consider using when you are faced with recycle (Fig. 4.4).

1. "Tear" the loop at the reactor inlet, and write material balance equations around each unit in the loop starting at the "tear" and continuing all the way around until you are back to where you started. Use the equations to relate the flow in the tear stream to the flow in the feed stream.
2. Choose first a system that groups the mixer, reactor, and separator into one block, complete process flow calculations for fresh feed and product rates from this system, then change systems and solve for each process unit separately.

These strategies are illustrated in Example 4.10.

| Example 4.10 | **Low Conversion and Recycle: Ammonia Synthesis** |

We want to produce 1000 kgmol NH_3/h from nitrogen and hydrogen. Derive an equation that relates the nitrogen flow rate into the reactor to the reactor fractional conversion. Assume the reactants are fed at stoichiometric ratio, and that all unreacted reactants are recycled.

Solution
The flow diagram with recycle is shown. We assume that the separator works perfectly—that is, that it completely separates nitrogen and hydrogen from the product ammonia.

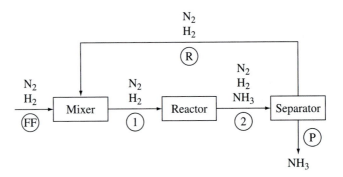

Before we begin our detailed analysis, let's check the degrees of freedom. The process has 10 stream variables and 1 system variable, for a total of 11 variables. There is 1 specified flow (ammonia production rate), 1 stream composition specification (reactants are fed at stoichiometric ratio) and there are 8 material balance equations (2 for the mixer, 3 for the reactor, 3 for the separator). Thus, DOF = 11 − 10 = 1. A system performance specification of the fractional conversion in the reactor would completely specify the process.

To evaluate, let's look at nitrogen flow through the loop, starting and ending with stream 1. With loops, the best strategy often is to tear the loop at the reactor inlet, and follow the material through the loop starting at the tear and continuing all the way around until you are back to where you started. Let's see how this works.

The nitrogen material balance equations for the reactor, separator, and mixer are, respectively:

$$\dot{n}_{N_2,2} = (1 - f_{C,N_2})\dot{n}_{N_2,1}$$

$$\dot{n}_{N_2,R} = \dot{n}_{N_2,2}$$

$$\dot{n}_{N_2,1} = \dot{n}_{N_2,R} + \dot{n}_{N_2,FF}$$

We can combine these three equations to get a simple relationship between the fresh nitrogen feed (stream FF) and the nitrogen feed to the reactor (stream 1):

$$\dot{n}_{N_2,1} = \frac{\dot{n}_{N_2,FF}}{f_{C,N_2}}$$

If we rearrange this equation, we see that the fractional conversion is equal to the ratio of fresh feed rate to reactor feed rate.

Now, if we can determine the nitrogen fresh feed, $\dot{n}_{N_2,FF}$, we can work backward through the loop to calculate all the process flows for a given fractional conversion. To do this, let's think "outside the loop," by changing our choice of system.

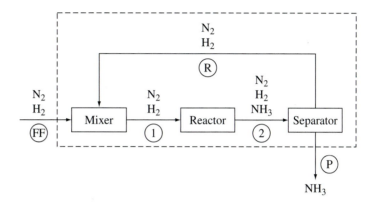

If we treat everything inside the dashed line as the system, the process is straightforward to analyze.

$$\dot{n}_{NH_3,P} = 1000 = \dot{n}_{NH_3,FF} + \nu_{NH_3}\dot{\xi} = 0 + 2\,\dot{\xi}$$

$$\dot{\xi} = 500\,\frac{kgmol}{h}$$

The extent of reaction has not changed from the case with no recycle. That is because the ammonia production rate is still the same, no reactants leave the *process*, and the overall fractional conversion for the process is 1.0!

Now it is simple to complete the process flow calculations for this system:

$$\dot{n}_{N_2,P} = 0 = \dot{n}_{N_2,FF} + \nu_{N_2}\dot{\xi} = \dot{n}_{N_2,FF} - 500$$

$$\dot{n}_{N_2,FF} = 500\,\frac{kgmol}{h}$$

Through similar calculations, we find $\dot{n}_{H_2,FF} = 1500$ kgmol/h. Notice that we never yet used the fractional conversion in the reactor. The fresh feed rate is independent of the single-pass fractional conversion!

Now we can calculate the flow rate into the reactor (stream 1) as a function of fractional conversion, from the equation derived from analysis of the recycle loop:

$$\dot{n}_{N_2,1} = \frac{500}{f_{C,N_2}}$$

This is exactly the same equation that we derived for the reactor inlet flow in the absence of recycle (Example 4.9)! The flows in and out of the *reactor* are the same as those shown in the graph in the solution to Example 4.9. The difference is that the nitrogen and hydrogen flows into and out of the *process* are smaller with recycle than without. The overall fractional conversion is 1.0, even when the single-pass fractional conversion is much lower.

Even with recycle, low single-pass conversion comes with a cost. Low reactor conversions translate into higher reactor flow rates, which translate into larger reactor volumes. This has economic consequences, because larger reactors cost more. Low reactor conversions may have unwanted safety consequences as well, if the reactions involve hazardous materials, because larger-volume reactors have the potential for greater damage in case of accident.

Operation at low fractional conversion with recycle is quite common in the commodity chemicals business, where businesses operate on small profit margins and separation technologies are well developed. Are there cases when we have less than 100% conversion but still don't recycle? Yes. With recycle, we must have a feasible technology for separating unreacted reactants from the products. If the separation is expensive relative to the value of the raw material, and the reactants don't have to otherwise be removed to sell the product, then recycling may not be justifiable. For example, polymerization reactors are usually operated without recycle, and as close to 100% conversion as achievable, because separation of high-molecular-weight molecules is difficult.

4.3.4 Fractional Conversion and Its Effect on Reactor Flow Sheet Synthesis: Recycle and Purge

What if the reactor flow sheet includes recycle but there are contaminants in the raw materials that are not reactants? This might happen, for example, when air is used as the source of oxygen and the nitrogen in the air is inert. In that case, the contaminants enter the process but do not leave; rather, they accumulate in the process and can cause severe problems. One option is simply to not use recycle when the raw materials contain unreactive contaminants. In the pharmaceutical business, for example, recycle is rarely used; the risk of buildup of unwanted and potentially toxic impurities is too great to justify the savings on raw material costs. Another option is to separate the contaminants from either the reactor feed or product stream. But what if separation is impractical? A third option is a compromise between 0% and 100% recycle. This option is implemented by installing a splitter (which can be a simple three-way valve) to split off part of the recycle stream and remove it from the process. This modification to the reactor process flowsheet is called **purge** (Fig. 4.5).

With recycle and purge, the contaminant flow out of the process in the purge stream exactly equals the contaminant flow into the process in the reactant stream (at steady state). The **fractional split** f_{Sj} must be specified, where:

$$f_{Sj} = \frac{\text{moles of } i \text{ leaving in stream } j}{\text{moles of } i \text{ fed}} = \frac{\dot{n}_{ij,\,\text{out}}}{\dot{n}_{i,\,\text{in}}} \tag{4.9}$$

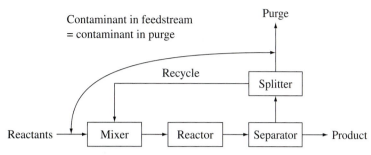

Figure 4.5 When the feed stream to a reactor contains an unreactive contaminant, the contaminant must be removed from the recycle loop to avoid accumulation. If separation is too expensive, a splitter with purge stream is chosen.

! **Helpful Hint**
The fractional split can be defined on the basis of either the purge or the recycle as the output stream. Watch carefully!

where $\dot{n}_{i,\text{in}}$ is the flow rate of component i to the splitter and $\dot{n}_{ij,\text{out}}$ is the flow rate of component i in stream j leaving the splitter.

Rearranging Eq. (4.9) gives:

$$\dot{n}_{ij,\text{out}} = f_{Sj}\dot{n}_{i,\text{in}}$$

Since the stream composition does not change with a splitter, the composition of the purge and recycle streams are the same, or

! **Helpful Hint**
With a Splitter, the composition of all input and output streams is the same.

$$\frac{\dot{n}_{i,\text{recycle}}}{\dot{n}_{\text{recycle}}} = \frac{\dot{n}_{i,\text{purge}}}{\dot{n}_{\text{purge}}} = \frac{\dot{n}_{i,\text{in}}}{\dot{n}_{\text{in}}}$$

Furthermore, the fractional split is the same for all components fed to the splitter.

Example 4.11	**Recycle with Purge: Ammonia Synthesis**

We want to make 1000 kgmol NH_3/h, from nitrogen and hydrogen. We feed nitrogen and hydrogen to a steady-state continuous-flow process at stoichiometric ratio. To save money, we purchase nitrogen that is contaminated with 2 mol% argon. Argon is an inert gas, but it's too expensive to separate argon from nitrogen, although it's easy to separate argon from ammonia. Nitrogen and hydrogen are fed to the process at their stoichiometric ratio. The reactor operates at a single-pass fractional conversion of 0.2. Show why inserting a splitter into the flow sheet is required. Derive an equation that relates the fresh nitrogen feed rate to the fractional split.

Solution
Let's first try to use the flow sheet of Example 4.10, with the exception that argon is in the fresh feed.

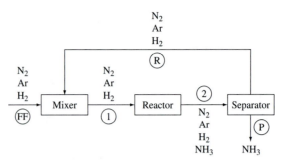

We'll start by writing the material balance equations for argon, starting with the reactor and working around the loop:

$$\dot{n}_{Ar,2} = \dot{n}_{Ar,1} \qquad \text{Reactor}$$

$$\dot{n}_{Ar,R} = \dot{n}_{Ar,2} \qquad \text{Separator}$$

$$\dot{n}_{Ar,1} = \dot{n}_{Ar,R} + \dot{n}_{Ar,FF} \qquad \text{Mixer}$$

If we combine the first two equations, we get

$$\dot{n}_{Ar,1} = \dot{n}_{Ar,R}$$

which contradicts the third equation (unless $\dot{n}_{Ar,FF} = 0$, which we know is not true).

The problem can be clearly seen if we choose our system to be the entire process. Argon is entering in the fresh feed, it's not being consumed by reaction, and it's not leaving the process. Therefore, argon must be accumulating in the system. But that violates the steady-state condition.

Since the recycle loop is the source of the problem, one idea might be to forget about recycling! The problem, of course, is that then we are back to wasting all our unreacted nitrogen and hydrogen.

A second possibility might be to separate argon from the recycle stream and send it on its way. However, the problem statement says that it is too expensive to remove argon from nitrogen, so it'll probably be too expensive to remove argon from a nitrogen/hydrogen mix.

There is a third possibility. What if we bleed off part of the recycle stream, just enough to get rid of the argon?

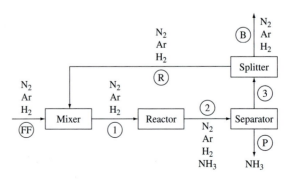

Now, the material balance equations on argon change just a bit. If we work around the loop:

$$\dot{n}_{Ar,2} = \dot{n}_{Ar,1} \qquad \text{Reactor}$$

$$\dot{n}_{Ar,3} = \dot{n}_{Ar,2} \qquad \text{Separator}$$

$$\dot{n}_{Ar,R} + \dot{n}_{Ar,B} = \dot{n}_{Ar,3} \qquad \text{Splitter}$$

$$\dot{n}_{Ar,1} = \dot{n}_{Ar,R} + \dot{n}_{Ar,FF} \qquad \text{Mixer}$$

If we combine the first three balance equations we get

$$\dot{n}_{Ar,R} + \dot{n}_{Ar,B} = \dot{n}_{Ar,1}$$

which, when combined with the last balance equation gives

$$\dot{n}_{Ar,B} = \dot{n}_{Ar,FF}$$

Or, in words, the argon purge flow rate must match the argon feed rate. (We reach the same conclusion if we group the mixer, reactor, separator and splitter into a single system.)

Let's quickly complete the DOF analysis of the flow sheet including purge. There are 20 stream and 1 system variables, for a total of 21 variables. There is 1 specified flow (ammonia production), 14 material balances (3 mixer, 4 reactor, 4 separator, 3 splitter), and 1 system performance specification (fractional conversion in the reactor). The N_2:H_2 ratio and the Ar:N_2 ratio in the fresh feed are both known, counting for 2 stream composition specifications. Finally, we implicitly know 2 additional stream compositions from the splitter (refer back to Sec. 2.5). This gives a total of 20 equations, and DOF = 21 − 20 = 1. Specifying the fractional split

$$f_{S,B} = \frac{\dot{n}_{Ar,B}}{\dot{n}_{Ar,3}} = \frac{\dot{n}_{N_2,B}}{\dot{n}_{N_2,3}} = \frac{\dot{n}_{H_2,B}}{\dot{n}_{H_2,3}}$$

is necessary to completely specify the process.

Let's look at how the flows change as $f_{S,B}$ is varied. Derivation of the material balance equations and performance specifications is left to the reader. The system of equations was solved by allowing $f_{S,B}$ to vary between 0.01 and 0.99.

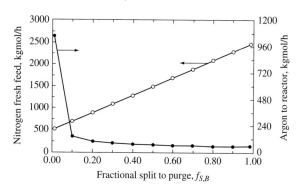

At very low fractional split (low purge, where most of the stream into the splitter is recycled) the nitrogen fresh feed approaches 500 kgmol/h, which is what we observed in the absence of the argon contaminant. However, the argon flow to the reactor is high—almost 1100 kgmol/h. This means that the reactor has to be built large enough to handle a large flow rate of an inert material. In other words, we are paying for extra reactor volume and not doing anything with it! At very high fractional split (high purge, where very little of the stream into the splitter is recycled), the argon flow to the reactor is small but we require almost 2500 kgmol/h fresh nitrogen feed! This is like not having a recycle stream at all. The optimum purge is set by further economic analysis of reactor costs versus raw material costs.

Because raw materials are never 100% pure, a purge stream is frequently required when there is recycle. There are important economic and safety consequences to selecting purge rates: Either we pay more for a larger reactor in order to conserve raw material, or we "waste" raw material in order to reduce the costs of the reactor. If we choose to have a small purge and high recycle, the concentration of contaminants in the reactor feed is high (and the concentration of reactants is low). This could adversely affect the rate of reaction, or might provide a chance for unwanted reactions to occur. If the contaminant is hazardous or toxic, an increase in its concentration presents a safety hazard. If the concentration or type of contaminant in the raw material changes from day to day, a high recycle rate may cause problems with process control. Choosing the optimum recycle and purge rates requires a detailed evaluation of the economic, safety, and environmental impact.

4.3.5 System Performance Specifications: Selectivity and Yield

In the perfect reactor, a single desired chemical reaction takes place in a single reactor. But, in many real reactors, the chemistry is not quite so simple, and more than one reaction may occur. There are two possibilities (Fig. 4.6):

1. *Parallel reactions:* The reactant takes an alternate pathway to form different product(s).
2. *Series reactions:* The desired product reacts further to form another product.

Most of the time, the additional products are undesired, and our design goal is to minimize their production while maximizing production of the desired product.

Fractional conversion of reactants is not sufficient to fully characterize reactor performance in these cases. For example, specification of the fractional conversion of a *reactant* cannot account for conversion of a desired *product* via a series reaction to an undesired product.

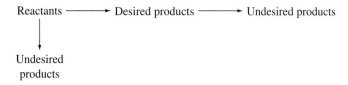

Figure 4.6 Unwanted parallel and series reactions. Reactors provide process conditions so that reactants are converted rapidly to the desired products. Reactants may participate in unwanted side reactions, forming undesired byproducts. Or, the desired product may undergo further chemical reactions to undesired byproducts. Optimally, reactor design variables are chosen to maximize the desired reaction while minimizing all undesired reactions. Realistically, there are often compromises that must be made between high conversion and high selectivity.

We introduce two new means for characterizing reactor performance when multiple reactions occur: **selectivity** and **yield**. In words, they are defined as

$$\text{Fractional selectivity} = \frac{\text{moles of reactant A converted to desired product P}}{\text{moles of reactant A consumed}}$$

$$\text{Fractional yield} = \frac{\text{moles of reactant A converted to desired product P}}{\text{moles of reactant A fed}}$$

(You will sometimes see other definitions of yield and selectivity, but these are the definitions we use in this book.)

We need more mathematical definitions of these two terms. The first question is: Which of the reactants is reactant A? Usually, we will define selectivity and yield on the basis of either the limiting reactant or the most expensive reactant. To be clear, we should always specify the reactant.

Next, let's consider a ratio that is *almost* the same as selectivity and write this ratio with our usual notation:

$$\frac{\text{Moles of desired product P generated}}{\text{Moles of reactant A consumed}} = \frac{\sum \nu_{Pk} \dot{\xi}_k}{-\sum \nu_{Ak} \dot{\xi}_k}$$

where the summation is taken over all k reactions, as usual. Note that the "moles of desired product P generated" is the net generation; the definition includes any reactions that consume P to make undesired byproducts. Now all we need to consider is how "moles of reactant A converted to desired product P" is related to "moles of desired product P generated." There's almost always just one reaction in which reactant A is converted to product P. Let's call that reaction R1. ν_{A1} and ν_{P1} are the stoichiometric coefficients of A and P, respectively, in R1. Then:

$$\frac{\text{Moles of reactant A converted to product P by R1}}{\text{Moles of product P generated by R1}} = -\frac{\nu_{A1}}{\nu_{P1}}$$

Putting these two ratios together gets us what we want: **fractional selectivity** for converting reactant A to product P, denoted as $s_{A \rightarrow P}$

$$s_{A \rightarrow P} = \frac{\nu_{A1}}{\nu_{P1}} \frac{\sum \nu_{Pk} \dot{\xi}_k}{\sum \nu_{Ak} \dot{\xi}_k} \qquad (4.10a)$$

With this definition, $s_{A \rightarrow P}$ is always between 0 and 1. Percent selectivity is simply 100% times fractional selectivity.

Similarly, consider the ratio

$$\frac{\text{Moles of desired product P generated}}{\text{Moles of reactant A fed}} = \frac{\sum \nu_{Pk} \dot{\xi}_k}{\dot{n}_{A,in}}$$

Combining this ratio with the ratio $-\nu_{A1}/\nu_{P1}$ gives the **fractional yield** $y_{A \rightarrow P}$

$$y_{A \rightarrow P} = - \frac{\nu_{A1}}{\nu_{P1}} \frac{\sum \nu_{Pk} \dot{\xi}_k}{\dot{n}_{A,in}} \qquad (4.11a)$$

With this definition, $y_{A \rightarrow P}$ is always between 0 and 1. Percent yield is simply 100% times fractional yield.

The above definitions apply for steady state continuous-flow reactors; similar definitions hold for batch reactors:

$$s_{A \rightarrow P} = \frac{\nu_{A1}}{\nu_{P1}} \frac{\sum \int_{t_0}^{t_f} \nu_{Pk} \xi_k \, dt}{\sum \int_{t_0}^{t_f} \nu_{Ak} \xi_k \, dt} \qquad (4.10b)$$

$$y_{A \rightarrow P} = - \frac{\nu_{A1}}{\nu_{P1}} \frac{\sum \int_{t_0}^{t_f} \nu_{Pk} \xi_k \, dt}{n_{A,sys,0}} \qquad (4.11b)$$

Quick Quiz 4.5

Why is there a negative sign in Eq. (4.11) but not in Eq. (4.10)?

Finally, compare Eq. (4.8), (4.10), and (4.11) for fractional conversion, yield, and selectivity. Notice that

$$y_{A \rightarrow P} = f_{C,A} \times s_{A \rightarrow P}$$

Conversion reports on how much of the reactant has been consumed. *Selectivity* and *yield* tell us how effective we've been at converting that reactant to the desired product. Only two of the three terms are independent.

Equations (4.10) and (4.11) may *look* complicated. We illustrate their use in the following examples.

Example 4.12   **Selectivity and Yield Definitions: Acetaldehyde Synthesis**

To make acetaldehyde (CH_3CHO), ethanol (C_2H_5OH) is partially oxidized using O_2.

$$C_2H_5OH + 0.5O_2 \rightarrow CH_3CHO + H_2O \qquad (R1)$$

Oxidation reactions are not very selective. Some of the ethanol is completely oxidized to carbon dioxide and water:

$$C_2H_5OH + 3O_2 \rightarrow 2CO_2 + 3H_2O \tag{R2}$$

Also, some of the acetaldehyde is partially oxidized to acetic acid, CH_3COOH:

$$2CH_3CHO + O_2 \rightarrow 2CH_3COOH \tag{R3}$$

(These reactions occur both in large chemical plants and when organisms, including humans, consume ethanol.) Derive expressions for conversion, yield, and selectivity.

Solution

Reaction (R1) plus (R2) is an example of parallel reactions, whereas reaction (R1) plus (R3) is an example of series reactions. It is convenient to summarize the net reaction rate for each compound in a table (a kind of generation-consumption table!).

Compound	$\sum \nu_{ik} \dot{\xi}_k$
C_2H_5OH	$-\dot{\xi}_1 - \dot{\xi}_2$
O_2	$-0.5\dot{\xi}_1 - 3\dot{\xi}_2 - \dot{\xi}_3$
CH_3CHO	$\dot{\xi}_1 - 2\dot{\xi}_3$
H_2O	$\dot{\xi}_1 + 3\dot{\xi}_2$
CH_3COOH	$2\dot{\xi}_3$
CO_2	$2\dot{\xi}_2$

The fractional conversion of ethanol is

$$f_{CE} = \frac{-\sum \nu_{Ek}\dot{\xi}_k}{\dot{n}_{E,\,in}} = \frac{\dot{\xi}_1 + \dot{\xi}_2}{\dot{n}_{E,\,in}}$$

Quick Quiz 4.6

If you came up with a way to prevent the complete oxidation of ethanol in Example 4.12, would conversion, selectivity, and yield increase, decrease, or stay the same?

The fractional selectivity and yield of acetaldehyde, with ethanol as the reactant of interest, are

$$s_{E \to Ac} = \frac{\nu_{E1}}{\nu_{Ac1}} \frac{\sum \nu_{Ac,k}\dot{\xi}_k}{\sum \nu_{E,k}\dot{\xi}_k} = \frac{(-1)}{1} \frac{(\dot{\xi}_1 - 2\dot{\xi}_3)}{(-\dot{\xi}_1 - \dot{\xi}_2)} = \frac{\dot{\xi}_1 - 2\dot{\xi}_3}{\dot{\xi}_1 + \dot{\xi}_2}$$

$$y_{E \to Ac} = -\frac{\nu_{E1}}{\nu_{Ac1}} \frac{\sum \nu_{Ac,k}\dot{\xi}_k}{\dot{n}_{E,in}} = -\frac{(-1)}{1} \frac{[\dot{\xi}_1 - 2\dot{\xi}_3]}{\dot{n}_{E,in}} = \frac{\dot{\xi}_1 - 2\dot{\xi}_3}{\dot{n}_{E,in}}$$

Example 4.13	**Using Selectivity in Process Flow Calculations: Acetaldehyde Synthesis**

We want to design and build a plant to produce 1200 kgmol acetaldehyde (CH_3CHO) per hour from ethanol and air. Laboratory data indicate that if we use a new catalyst, and adjust the feed ratio to 5.7 moles ethanol per mole oxygen, we can expect to achieve 25% conversion of ethanol in the reactor, with a selectivity for acetaldehyde of 0.6. The only byproducts are acetic acid and water. Thus, reactions (R1) and (R3) of Example 4.12 are important; reaction (R2) is suppressed, and $\dot{\xi}_2 = 0$.

Assume pure ethanol and air (79 mol% N_2, 21 mol% O_2) are the raw materials available. Determine molar flow rates of all components in and out of the reactor.

Solution

As always, we start with a flow diagram

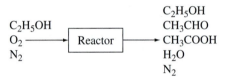

A DOF analysis shows that the problem is completely specified: there are 10 variables (8 stream, 2 system) and 10 equations (5 material balances, 1 specified flow, 2 stream composition specifications, and 2 system performance specifications).

The basis for the design is 1200 kgmol/h acetaldehyde produced, or

$$\dot{n}_{Ac,out} = 1200 \, \frac{kgmol}{h}$$

The mole balance equation for acetaldehyde, assuming steady-state, is

$$\dot{n}_{Ac,out} = 1200 = 0 + (\dot{\xi}_1 - 2\dot{\xi}_3)$$

From the problem statement, the selectivity for acetaldehyde is 0.6:

$$s_{E \to Ac} = 0.6 = \frac{\dot{\xi}_1 - 2\dot{\xi}_3}{\dot{\xi}_1} = \frac{1200}{\dot{\xi}_1}$$

Therefore

$$\dot{\xi}_1 = \frac{1200}{0.6} = 2000 \, \frac{kgmol}{h} \quad \text{and} \quad \dot{\xi}_3 = \frac{2000 - 1200}{2} = 400 \, \frac{kgmol}{h}$$

From the problem statement, the fractional conversion of ethanol is 0.25. Combining this information with the definition of fractional conversion and the above equation gives

$$f_{CE} = 0.25 = \frac{\dot{\xi}_1}{\dot{n}_{E,\,in}} = \frac{2000}{\dot{n}_{E,\,in}} \quad \text{or} \quad \dot{n}_{E,in} = 8000 \, \frac{kgmol}{h}$$

Now, we use the steady-state mole balance with ethanol as the component to get

$$\dot{n}_{E,\text{out}} = \dot{n}_{E,\text{in}} - \dot{\xi}_1 = 8000 - 2000 = 6000 \, \frac{\text{kgmol}}{\text{h}}$$

The O_2 feed rate is

$$\dot{n}_{O,\text{in}} = \frac{\dot{n}_{E,\text{in}}}{5.7} = \frac{8000}{5.7} = 1400 \, \frac{\text{kgmol}}{\text{h}}$$

The mole balance equation for O_2 is

$$\dot{n}_{O,\text{out}} = 0 = \dot{n}_{O,\text{in}} - 0.5 \, \dot{\xi}_1 - \dot{\xi}_3 = 1400 - 0.5(2000) - 400 = 0$$

Finally, the acetic acid mole balance equation is

$$\dot{n}_{AA,\text{out}} = \dot{n}_{AA,\text{in}} + 2\dot{\xi}_3 = 0 + 2(400) = 800 \, \frac{\text{kgmol}}{\text{h}}$$

The results are summarized below in convenient tabular form. All numbers are given as kgmol/h.

	$\dot{n}_{i,\text{in}}$	$\sum \nu_{ik}\,\dot{\xi}_k$	$\dot{n}_{i,\text{out}}$
C_2H_5OH	8000	-2000	6000
O_2	1400	-1400	0
N_2	5300		5300
CH_3CHO		1200	1200
CH_3COOH		800	800
CO_2			0
H_2O		2000	2000

Often, there is a trade-off between high selectivity and high conversion. As a rule of thumb, high selectivity is generally preferred over high conversion. At high selectivity, low single-pass conversion, and high recycle, we achieve a high overall conversion to the desired product, and do not waste raw material making undesired products. If undesired products are made, separation and purification steps are more difficult, complicated, and costly. If the undesired byproducts are toxic or hazardous, we are faced with high waste disposal costs and plant safety concerns. High conversion might be favored over high selectivity if the byproducts are not terribly toxic, the raw materials and/or desired product are toxic, and/or the reactor operates under extreme conditions of temperature or pressure. In these cases, high conversion reduces or eliminates recycle, thus reducing reactor vessel size and decreasing safety concerns.

4.4   Why Reactors Aren't Perfect: Chemical Equilibrium and Chemical Kinetics

We've explained *what* the measures of reactor performance are, but we haven't explained *why*. Why can't we achieve high single-pass conversion? Why can't we achieve high selectivity? There are two major factors that limit reactor performance: chemical reaction equilibria, and chemical reaction kinetics. You could write books on each of these topics (in fact many people have!). Here we can give only a taste of each; you will learn much more in the future.

4.4.1   The Chemical Reaction Equilibrium Constant K_a

Consider a simple reaction of the general type, where reactants A and B are converted to product P and waste product W:

$$2A + B \rightarrow P + 3W$$

But, if A and B can react to form P and W, can P and W react to form A and B?

$$P + 3W \rightarrow 2A + B?$$

Theoretically, at least, a chemical reaction can proceed in either direction. As a practical manner, we engineers want the forward reaction but not the reverse reaction. But all too often the reverse reaction does happen, and in spades. Chemical equilibrium tells us to what extent the reverse reaction must be considered.

When a reacting system reaches **chemical equilibrium**, the concentration of reactants and products do not change with time. But, at a molecular level, the system is still reacting. Thus, at equilibrium, the rates of the forward and reverse reactions are the same. One way to indicate that both forward and reverse reactions are occurring simultaneously is to write

$$2A + B \rightleftarrows P + 3W$$

The concentrations of reactants and products at equilibrium are quantified by a **chemical equilibrium constant K_a**:

$$K_a = \prod_{\text{all } i} a_{i,eq}^{v_i} \tag{4.12}$$

where $a_{i,eq}$ is the *activity* of compound i when the system is *at equilibrium*, v_i is the stoichiometric coefficient for compound i in the reaction, and $\prod$ indicates that the product of all compounds participating in the reaction is calculated.

As a quick example, if the reaction is

$$2A + B \rightleftarrows P + 3W$$

then

$$K_a = (a_{A,\,eq})^{-2}(a_{B,\,eq})^{-1}(a_{P,\,eq})^{+1}(a_{W,\,eq})^{+3} = \frac{a_{P,\,eq}\,a_{W,\,eq}^3}{a_{A,\,eq}^2\,a_{B,\,eq}}$$

You are certainly very familiar with the word *activity*, but perhaps not in a chemical sense. The activity of a compound is related to the chemical energy of that compound in a multicomponent mixture. There is a *lot* more about activity that you will learn in thermodynamics courses. In this textbook, we will *greatly* simplify things and say:

$$a_i \approx \frac{y_i P}{1\ \text{atm}} \qquad \text{if } i \text{ is in the vapor phase}$$

where y_i is the mole fraction of i in the vapor phase, and P is the total pressure (in atm).

$$a_i \approx x_i \qquad \text{if } i \text{ is in the liquid phase}$$

where x_i is the mole fraction of i in the liquid phase.

$$a_i = 1 \qquad \text{if } i \text{ is in the solid phase}$$

There are a number of assumptions that are behind these simplifying equations. For vapor-phase compounds, we are assuming that the vapor behaves as an ideal gas mixture. For liquid-phase compounds, our approximation is reasonably good (1) in dilute solutions of compounds in water or another solvent and (2) in liquid mixtures of two or more compounds that are of similar size and chemical type. For solid-phase compounds, our approximation is good if the solid phase is pure. Many times, even if there are multiple solid-phase compounds, each compound forms its own separate phase, so this approximation applies.

In order to derive an expression for K_a from a chemical reaction equation using Eq. (4.12), the reaction must be stoichiometrically balanced, and we must know the phase of the reactant or product. We will indicate the phase of compounds in a chemical reaction by writing (g) for gas (vapor) phase, (l) for liquid phase, or (s) for solid phase, just after the molecular formula. For example, the reaction between solid barium sulfate and carbon monoxide gas to form solid barium sulfide and carbon dioxide gas is written as:

$$BaSO_4(s) + 4CO(g) \rightleftarrows BaS(s) + 4CO_2(g)$$

Finally, a word about the dimension of K_a. Since the activity coefficient a_i is dimensionless, K_a is also dimensionless. However, for vapor-phase reactions, it is common practice (although not rigorously correct) to write K_a as

$$K_a = \prod_{all\ i}\left(y_{i,eq}P\right)^{\nu_i} = P^{\left(\sum_{all\ i}\nu_i\right)}\prod_{all\ i}\left(y_{i,eq}\right)^{\nu_i}$$

If the number of moles changes due to reaction (e.g., $\Sigma \nu_i \neq 0$) then K_a has pressure units. For example, in the combustion reaction

$$C_2H_6(g) + 3.5O_2(g) \rightleftarrows 2CO_2(g) + 3H_2O(g)$$

$\Sigma \nu_i = 2 + 3 - 1 - 3.5 = 0.5$, and the dimension of K_a would be $P^{0.5}$.

Example 4.14   **Deriving Equations for K_a: Three Cases**

Derive equations for K_a in terms of mole fractions and pressure for the following cases:

Case 1.   The gas-phase synthesis of ammonia:

$$N_2(g) + 3H_2(g) \rightleftarrows 2NH_3(g)$$

Case 2.   The synthesis of aspirin (acetylsalicylic acid) in a dilute aqueous solution from salicylic acid and acetic acid:

$$C_6H_4(OH)COOH(l) + CH_3COOH(l) \rightleftarrows C_6H_4(OCOCH_3)COOH(l) + H_2O(l)$$

Case 3.   The reduction of barium sulfate ore to barium sulfide:

$$BaSO_4(s) + 4CO(g) \rightleftarrows BaS(s) + 4CO_2(g)$$

(g), (l), or (s) indicates that the reactant or product is in the gas, liquid, or solid phase, respectively.

Solution

Case 1.

$$K_a = \frac{a_{NH_3,eq}^2}{a_{N_2,eq} a_{H_2,eq}^3} = \frac{\left(y_{NH_3,eq} P \right)^2}{\left(y_{N_2,eq} P \right) \left(y_{H_2,eq} P \right)^3} = \frac{\left(y_{NH_3,eq} \right)^2}{\left(y_{N_2,eq} \right) \left(y_{H_2,eq} \right)^3} \frac{1}{P^2}$$

Notice that the equation for K_a of gas-phase reactions often includes a pressure term. For this reaction, K_a would be reported in units of atm^{-2}.

Case 2.   Using A for aspirin, SA for salicylic acid, AA for acetic acid, and W for water we write

$$K_a = \frac{(x_{A,eq})(x_{W,eq})}{(x_{SA,eq})(x_{AA,eq})}$$

This is a little tricky: this reaction takes place in water. Therefore, $x_{W,eq}$, the mole fraction of water in the system at equilibrium, includes not just the water produced by reaction but *all* the water in the system. Since the components (other than water) are dilute, we sometimes say that $x_{W,eq} \approx 1$ and

$$K_a \approx \frac{(x_{A,eq})}{(x_{SA,eq})(x_{AA,eq})}$$

Case 3.   Barium sulfate and barium sulfide are both solids, but they don't form a solid solution. Rather they form two separate solid phases. (The two solids may appear to be mixed on a microscopic scale, but they are not mixed on a molecular scale). Therefore, the activity of each of the solids equals 1 and:

$$K_a = \frac{(a_{BaS,eq})\left(a_{CO_2,eq}\right)^4}{(a_{BaS,eq})(a_{CO,eq})^4} = \frac{(1)\left(y_{CO_2,eq}P\right)^4}{(1)(y_{CO,eq}P)^4} = \frac{y^4_{CO_2,eq}}{y^4_{CO,eq}}$$

4.4.2 Calculating K_a

Think about what you are doing right now. You are probably sitting up, reading, breathing, perhaps drinking a soda. Hopefully you are thinking. You are not at your lowest energy state. (If you were, you'd be a dead and decayed heap on the floor.)

There are many ways to describe the energy state of a system. We are interested now in one kind of energy, called the **Gibbs energy**, which we denote as G. When a system is at equilibrium, like our chemically reacting systems, G is at a minimum (but G is *not* zero). If G could become lower by further chemical reaction, then the system would not be chemically equilibrated. G of a chemically reacting system changes as the reaction proceeds to equilibrium. The change in G as the reaction proceeds from reactants to products is called the **Gibbs energy of reaction** and is denoted as $\Delta\hat{G}_r$. (The hat indicates that this is the Gibbs energy change per mole of reaction.)

$\Delta\hat{G}_r$, in essence, describes the thermodynamic driving force for converting reactants to products. It makes sense, then, that $\Delta\hat{G}_r$ is related to the chemical reaction equilibrium constant. This relationship is shown mathematically as

$$K_a = \exp\left(\frac{-\Delta\hat{G}_r}{RT}\right) \tag{4.13}$$

where R is the ideal gas constant and T is the temperature (on an absolute scale).

Equation 4.13 would be a very useful equation if we had access to a table of $\Delta\hat{G}_r$ of every reaction known to humankind. That would be a pretty long table! Fortunately, it is easy to calculate $\Delta\hat{G}_r$ at the *standard state*, typically at 298 K and 1 atm. This value is known as the standard Gibbs energy of reaction and is written with a small "°" to indicate standard state: $\Delta\hat{G}_r^{\circ}$. To calculate $\Delta\hat{G}_r^{\circ}$, we must know the **standard Gibbs energy of formation** $\Delta\widehat{G}^{\circ}_{i,f}$ of the reactants and products. A table of values of $\Delta\widehat{G}^{\circ}_{i,f}$ for some common compounds is included in App. B.

$\Delta\widehat{G}^{\circ}_{i,f}$ is the Gibbs energy change associated with making the compound i from its elements in their natural phase and state of aggregation, at the standard temperature and pressure. The phrase "in their natural phase and state of aggregation" needs a bit of explanation. This phrase is needed because not all elements naturally exist as monoatomic compounds. For example, the natural phase and state of aggregation of oxygen at 298 K and 1 atm is $O_2(g)$, hydrogen is $H_2(g)$, helium is $He(g)$, bromine is $Br_2(l)$, carbon is $C(s)$ and sulfur is $S_8(s)$, where (g) means gas, and (s) means solid. For elements at 298 K, 1 atm, and in their natural phase and state of aggregation, $\Delta\widehat{G}^{\circ}_{i,f} = 0$.

$\Delta\widehat{G}^{\circ}_{r}$ is calculated from $\Delta\widehat{G}^{\circ}_{i,f}$ easily:

$$\Delta\widehat{G}^{\circ}_{r} = \sum \nu_i \Delta\widehat{G}^{\circ}_{i,f} \tag{4.14}$$

where ν_i is the stoichiometric coefficient of species i (negative for reactants, positive for products). It is important to note that $\Delta\widehat{G}^{\circ}_{r}$ depends on the stoichiometric coefficients, so it depends on the way in which the balanced chemical equation is written!

A word about units. $\Delta\widehat{G}_{r}$, $\Delta\widehat{G}^{\circ}_{r}$, $\Delta\widehat{G}^{\circ}_{i,f}$ have units of energy per mole (e.g., kJ/gmol, cal/gmol, or Btu/lbmol). In Eq. (4.13), R must have compatible units. Appropriate values of R include 8.3144 J/gmol K, 1.9872 cal/gmol K, or 1.9872 Btu/lbmol °R. See Chap. 6 for more discussion of energy units.

Determining $\Delta\widehat{G}^{\circ}_{r}$ allows us to use Eq. (4.13) to calculate K_a at 298 K. But most industrial chemical reactors don't operate at 298 K, and unfortunately $\Delta\widehat{G}_{r}$ is a function of T. Are we stuck? Lucky for us, it turns out (we show this without explanation), that to a good approximation, K_a at *any* temperature T can be calculated from:

$$\ln K_{a,T} = -\frac{1}{R}\left[\frac{\Delta\widehat{G}^{\circ}_{r} - \Delta\widehat{H}^{\circ}_{r}}{298} + \frac{\Delta\widehat{H}^{\circ}_{r}}{T}\right] \tag{4.15}$$

$\Delta\widehat{H}^{\circ}_{r}$ is the difference in **enthalpy** of the reactants and products at their standard state, and is called the **standard enthalpy of reaction**. Enthalpy is just another measure of energy that is different from, but related to, Gibbs free energy. We will spend more time on enthalpy in Chap. 6. For now, you just need to know that $\Delta\widehat{H}^{\circ}_{r}$ is calculated in a manner very similar to $\Delta\widehat{G}^{\circ}_{r}$:

$$\Delta\widehat{H}^{\circ}_{r} = \sum \nu_i \Delta\widehat{H}^{\circ}_{i,f} \tag{4.16}$$

where $\Delta\widehat{H}^{\circ}_{i,f}$ is the standard enthalpy of formation of compound i, which we find in the tables in App. B right next to $\Delta\widehat{G}^{\circ}_{i,f}$.

> ⚠ **Helpful Hint**
> The Gibbs energy change is calculated per mole of reaction, not per mole of reactant or per mole of product.

> ⚠ **Helpful Hint**
> Be careful when you use Eq. (4.15); it is easy to make mistakes. Common errors include: forgetting to use an absolute temperature scale (e.g., using °C instead of K), dropping a negative sign, or using inconsistent units.

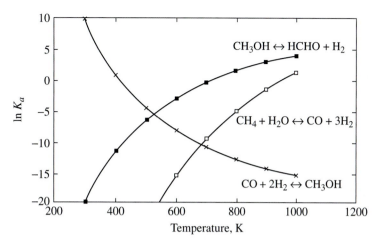

Figure 4.7 The chemical equilibrium constant can change dramatically with temperature. Reactions with $\ln K_a > 0$ are preferred.

K_a varies over many orders of magnitude (Fig. 4.7). If $K_a \gg 1$, then the reaction is considered irreversible, and at equilibrium virtually all the reactants will be converted to products. If $K_a \ll 1$, then the reaction will not "go"; reactions that fall into this category are rarely of industrial significance. The magnitude of K_a may shift dramatically with temperature, depending on the size and sign of $\Delta \hat{H}_r^{\circ}$. As can be seen from Eq. (4.15), if $\Delta \hat{H}_r^{\circ} < 0$, K_a decreases with increasing T, whereas if $\Delta \hat{H}_r^{\circ} > 0$, K_a increases with increasing T. The choice of temperature for a chemical reactor is a key design variable, and is based in part on the desire to have favorable equilibrium.

Example 4.15

Calculating K_a: Ethyl Acetate Synthesis

Ethyl acetate is an industrially important solvent. (Acetates are also partially responsible for the distinctive odor of many fruits.) Ethyl acetate is synthesized on a commercial scale by reacting ethanol (C_2H_5OH) and acetic acid (CH_3COOH) in the liquid phase:

$$C_2H_5OH(l) + CH_3COOH(l) \rightleftarrows CH_3COOC_2H_5(l) + H_2O(l)$$

What is K_a for this reaction at 25°C? at 80°C?

Solution
First we look up $\Delta \hat{G}_{i,f}^{\circ}$ and $\Delta \hat{H}_{i,f}^{\circ}$ for all the reactants and products. These values are for forming the pure compounds in the liquid phase at 25°C and 1 atm.

Species	ν_i	$\Delta \hat{G}_{i,f}^{\circ}$, kJ/gmol	$\Delta \hat{H}_{i,f}^{\circ}$, kJ/gmol
$C_2H_5OH(l)$	-1	-174.7	-277.6
$CH_3COOH(l)$	-1	-392.5	-486.2
$CH_3COOC_2H_5(l)$	$+1$	-318.4	-463.3
$H_2O(l)$	$+1$	-237.2	-285.8

Next we calculate $\Delta \hat{G}_r^{\circ}$ and $\Delta \hat{H}_r^{\circ}$:

$$\Delta \hat{G}_r^{\circ} = \sum \nu_i \Delta \hat{G}_{i,f}^{\circ} = (-1)(-174.7) + (-1)(-392.5) + (-318.4) + (-237.2)$$
$$= +11.6 \text{ kJ/mol}$$

$$\Delta \hat{H}_r^{\circ} = \sum \nu_i \Delta \hat{H}_{i,f}^{\circ} = (-1)(-277.6) + (-1)(-486.2) + (-463.3) + (-285.8)$$
$$= +14.7 \text{ kJ/mol}$$

At 25°C (298 K)

$$K_a = \exp\left(\frac{-\Delta \hat{G}_r}{RT}\right) = \exp\left(\frac{-11,600 \text{ J/gmol}}{(8.3144 \text{ J/gmol K})(298 \text{ K})}\right) = 0.00926$$

To calculate K_a at 80°C (353 K) we use Eq. (4.15):

$$\ln K_{a,353} = -\frac{1}{8.3144 \text{ J/gmol K}} \left[\frac{11,600 - 14,700 \text{ J/gmol}}{298 \text{ K}} \right.$$
$$\left. + \frac{14,700 \text{ J/gmol}}{353 \text{ K}} \right] = -3.757$$

$$K_{a,353} = 0.0233$$

Raising the temperature increases K_a, which means greater conversion of reac- tants to products at equilibrium. This is a reaction we want to run at as high a temperature as possible. We are limited by wanting to keep the mixture in the liquid phase.

Chemical equilibrium is an important consideration in selecting appropriate raw materials and reaction pathways. In Chap. 1 you learned to screen alternative chemical reaction pathways by looking at raw material costs, byproducts, and atom economy. All of these are important criteria, but none consider whether the chemical reaction will actually occur under realistic conditions. As a next step in evaluating the feasibility of reaction pathways, we consider chemical reaction equilibrium. As a rule of thumb, we look for reactions where $K_a \geq 0.1$ to 1.

Lower values may be acceptable only if there are other compelling reasons to prefer a reaction pathway with poor equilibrium.

Equilibrium constants are functions of temperature, so selection of reactor temperature is a part of the design process. Reactor temperatures close to ambient are preferred for safety and energy-cost reasons. For simple commodity chemicals, reactor temperatures up to 500°C are quite reasonable. Even higher temperatures are possible but require special considerations: The upper limit is around 1200°C, which is about the temperature of a hot flame. Manufacture of more complex chemicals, such as polymers or pharmaceuticals, is usually carried out at much lower temperatures (about 25°C to about 100°C), because the compounds degrade at higher temperatures. Unfortunately, the rate of reaction slows considerably as the temperature is lowered (see Sec. 4.4.4).

Example 4.16	**Chemical Equilibrium Considerations in Selection of Reaction Pathway: Safer Routes to Dimethyl Carbonate**

Polycarbonates are transparent impact-resistant polymers used for everything from baby bottles to compact discs to contact lenses. The conventional manufacture of polycarbonates uses phosgene ($COCl_2$) as one of the raw materials, which is a highly toxic chemical. Furthermore, the chlorine is not incorporated into the final product, reducing the atom economy of the conventional reaction pathway and leading to the production of chlorinated byproducts. It might be possible to replace the phosgene-based process with another process involving dimethyl carbonate (DMC). Besides its potential use in polycarbonate manufacture, dimethyl carbonate is of commercial interest because it can be used as a fuel additive, as an electrolyte in lithium-ion batteries, and as a safer raw material in methylating reactions.

However, there is a catch. DMC is now produced from phosgene, so there is no net advantage to replacing phosgene-based processes with DMC-based processes.

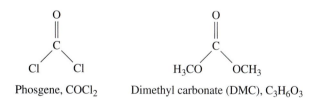

Phosgene, $COCl_2$ Dimethyl carbonate (DMC), $C_3H_6O_3$

The reaction to produce DMC from phosgene is:

$$COCl_2 + 2CH_3OH \rightleftarrows C_3H_6O_3 + 2HCl \qquad (R1)$$

We are in the hunt for alternative reaction pathways for the production of DMC that don't require phosgene. Here are some under consideration:

Oxidative carbonylation of methanol:

$$2CH_3OH + CO + 0.5O_2 \rightleftarrows C_3H_6O_3 + H_2O \qquad (R2)$$

Urea methanolysis:

$$H_2NCONH_2 + 2CH_3OH \rightleftarrows C_3H_6O_3 + 2NH_3 \qquad (R3)$$

Carbonylation of methanol with CO_2:

$$2CH_3OH + CO_2 \rightleftarrows C_3H_6O_3 + H_2O \qquad (R4)$$

Carbonylation of alkyl halide with inorganic carbonate

$$2CH_3I + K_2CO_3 \rightleftarrows C_3H_6O_3 + 2KI \qquad (R5)$$

Which pathway would you pick on the basis of cost? Which on the basis of atom economy? Which on the basis of chemical equilibrium? Look at operating temperatures of 100°C to 500°C as a reasonable range.

Solution

First we collect information about the molar mass of each compound, the cost of each raw material, and the value of the desired product.

Compound	Formula	Molar Mass	$/kg
Phosgene	$COCl_2$	99	1.62
Methanol	CH_3OH	32	0.188
Urea	H_2NCONH_2	60	0.225
Carbon monoxide	CO	28	Negligible
Carbon dioxide	CO_2	44	Negligible
Methyl iodide	CH_3I	142	~8.00
Potassium carbonate	K_2CO_3	138	0.95
Hydrogen chloride	HCl	36.5	
Water	H_2O	18	
Ammonia	NH_3	17	
Potassium iodide	KI	166	
Dimethyl carbonate	$C_3H_6O_3$	90	3.11

To compare raw material costs, we calculate the quantities of raw materials required and the atom economy for production of 1 kg DMC, using the methods of Chap. 1. Without showing the detailed calculations, the results are summarized on the following page:

Reaction	Raw materials required, kg/ kg DMC	Raw material costs, $/kg DMC	Fractional atom economy
R1	0.71 kg CH_3OH 1.1 kg $COCl_2$	1.91	0.55
R2	0.71 kg CH_3OH 0.31 kg CO 0.18 kg O_2	0.134	0.83
R3	0.71 kg CH_3OH 0.66 kg H_2NCONH_2	0.282	0.73
R4	0.71 kg CH_3OH 0.49 kg CO_2	0.134	0.83
R5	3.16 kg CH_3I 1.53 kg K_2CO_3	26.73	0.21

Reactions R2, R3, and R4 are all very attractive from both cost and atom economy points of view. Reaction R2 is less attractive from a safety point of view, since CO is toxic. Reaction R5 is really not worth considering further, because the cost and the atom economy are so unattractive compared to the other alternatives.

Next, we need information regarding the chemical reaction equilibrium constant for each of our reactions. From the data in App. B and Eq. (4.15), we find the following equations that relate K_a to T:

Reaction	$\ln K_{a,T}$ (T in K)
R1	$-8.55 + 15970/T$
R2	$-32.5 + 36000/T$
R3	$-6.77 + 3010/T$
R4	$-22.0 + 1990/T$

Next, we calculate the K_a at 100°C and at 500°C. We prefer $K_a > 1$.

Reaction	K_a at 100°C (373 K)	K_a at 500°C (773 K)
R1	7.5×10^{14}	1.8×10^5
R2	7×10^{27}	1.4×10^6
R3	3.7	0.057
R4	5.6×10^{-8}	3.6×10^{-9}

Whoa. Reaction R4, which was our first choice based on raw materials cost, atom economy, and safety, is unrealistic: Equilibrium lies far toward the left (reactants), and the reaction won't "go" without extraordinary measures. Reaction R3 is also unattractive from an equilibrium point of view. This leaves reaction R2 as the only reasonable alternative to the phosgene-methanol synthesis route. Issues other than chemical reaction equilibrium—side reactions, rates of reaction, ease of separation—now enter the decision-making picture.

4.4.3 Chemical Reaction Equilibrium and Reactor Performance

Many chemical reactors of industrial importance are operated such that the materials leaving the reactor are at chemical equilibrium. Even if equilibrium is not reached, knowing the concentration of reactants and products at equilibrium allows us to calculate the best the reactor can do—the system cannot do better than equilibrium. K_a is useful in chemically reacting systems because it allows us to calculate the fractional conversion *at equilibrium*. For systems with multiple reactions, we can calculate the fractional yield and selectivity as well. Such calculations help us make appropriate choices of reactor design variables such as temperature, pressure, and reactant feed ratio. For reacting systems at equilibrium, *chemical reaction equilibrium relationships coupled with material balance equations allow us to determine the achievable reactor performance.* Our task in this section is to demonstrate strategies for calculating reactor performance, given chemical equilibrium constraints.

Let's consider this problem from a DOF analysis point of view. We'll consider a steady-state continuous-flow reactor in which the exiting stream is at chemical equilibrium. Suppose there are R reactants fed to a reactor, K reactions occurring inside the reactor, and P products generated as a result of these reactions. The process stream into the reactor contains R components, and the stream exiting the reactor contains $R + P$ components. Therefore, there are $2R + P$ stream variables, and K system variables, or $2R + P + K$ variables all together. There are $R + P$ components in the system, so we write $R + P$ material balance equations. The statement that the exiting stream is at equilibrium puts additional constraints on the system: The composition of the exiting stream must satisfy the expression for K_a. If there are K reactions, we write K equations for K_a. We now have counted up $R + P + K$ equations (constraints). Comparing this to the total number of variables, we see that we must provide R additional specifications. Most commonly these are provided by specifying the R inlet flow rates.

For example, suppose we feed reactants A and B at known flow rates to a reactor, where two liquid-phase reactions, R1 and R2, of known stoichiometry occur. Three products: C, D, and E, along with unreacted A and B, exit the reactor. The reactor outlet is at chemical equilibrium, and the numerical values of the equilibrium constants K_1 and K_2 are known.

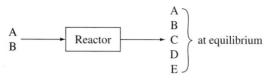

We have nine variables total (seven stream and two system). There are five independent material balance equations and two specified flows (inputs of A and B). The statement that the two chemical reactions are at equilibrium provides two additional system performance specifications! Thus, the system is completely specified.

To solve these problems, we start with a familiar strategy: We write the differential material balance equations for each compound in terms of our stream and system variables.

$$\dot{n}_{A,\text{out}} = \dot{n}_{A,\text{in}} + \nu_{A1}\dot{\xi}_1 + \nu_{A2}\dot{\xi}_2$$

$$\dot{n}_{B,\text{out}} = \dot{n}_{B,\text{in}} + \nu_{B1}\dot{\xi}_1 + \nu_{B2}\dot{\xi}_2$$

$$\dot{n}_{C,\text{out}} = \nu_{C1}\dot{\xi}_1 + \nu_{C2}\dot{\xi}_2$$

$$\dot{n}_{D,\text{out}} = \nu_{D1}\dot{\xi}_1 + \nu_{D2}\dot{\xi}_2$$

$$\dot{n}_{E,\text{out}} = \nu_{E1}\dot{\xi}_1 + \nu_{E2}\dot{\xi}_2$$

Given our problem statement, the unknowns are the five outlet stream flows and the two extents of reaction.

Next we write equations describing the system performance. In this case, these are the chemical equilibrium constants:

$$K_1 = \prod_{\text{all } i}(x_i)^{\nu_{i1}} = (x_A)^{\nu_{A1}}(x_B)^{\nu_{B1}}(x_C)^{\nu_{C1}}(x_D)^{\nu_{D1}}(x_E)^{\nu_{E1}}$$

$$K_2 = \prod_{\text{all } i}(x_i)^{\nu_{i2}} = (x_A)^{\nu_{A2}}(x_B)^{\nu_{B2}}(x_C)^{\nu_{C2}}(x_D)^{\nu_{D2}}(x_E)^{\nu_{E2}}$$

[We use x_i, the mole fraction of i in the exit stream, because this is a liquid-phase reaction. If the component i does not participate in reaction k, then $\nu_{ik} = 0$ and $(x_i)^{\nu_{ik}} = 1$.]

Unfortunately we've introduced several new variables—all the mole fractions. Fortunately, these are easily related to the molar flow rates!

$$x_i = \frac{\dot{n}_{i,\text{out}}}{\displaystyle\sum_{\text{all } i}\dot{n}_{i,\text{out}}} = \frac{\dot{n}_{i,\text{out}}}{\dot{n}_{\text{out}}}$$

We use the material balance equations to write x_i in terms of the (known) stoichiometric coefficients and input flows and the (unknown) extents of reaction. For example,

$$x_B = \frac{\dot{n}_{B,\text{out}}}{\displaystyle\sum_{\text{all } i}\dot{n}_{i,\text{out}}} = \frac{\dot{n}_{B,\text{out}}}{\dot{n}_{\text{out}}} = \frac{\dot{n}_{B,\text{in}} + \nu_{B1}\dot{\xi}_1 + \nu_{B2}\dot{\xi}_2}{\dot{n}_{A,\text{in}} + \dot{n}_{B,\text{in}} + \sum\nu_{i1}\dot{\xi}_1 + \sum\nu_{i2}\dot{\xi}_2}$$

We insert these definitions of x_i into the equations for K_1 and K_2. If $\dot{n}_{i,in}$ and ν_{ik} are known, we have two (messy) equations in two unknowns— $\dot{\xi}_1$ and $\dot{\xi}_2$. Once we solve for $\dot{\xi}_1$ and $\dot{\xi}_2$, we then proceed to calculate conversion, yield, selectivity, outlet flow rates, and so forth.

Example 4.17

Reactor Performance and K_a: Ammonia Synthesis

For the gas-phase synthesis of ammonia, the equilibrium constant is

$$K_a = \frac{\left(y_{NH_3,eq}\right)^2}{\left(y_{N_2,eq}\right)\left(y_{H_2,eq}\right)^3 P^2}$$

$K_a = 6.6 \times 10^5$ atm^{-2} at 298 K. Suppose 1000 kgmol N_2/h and 3000 kgmol H_2/h are fed to a reactor, the reactor operates at 1 atm and 298 K, and the gas leaving the reactor is at chemical equilibrium. What is the fractional conversion of nitrogen to ammonia?

Solution

We always start with a flow diagram:

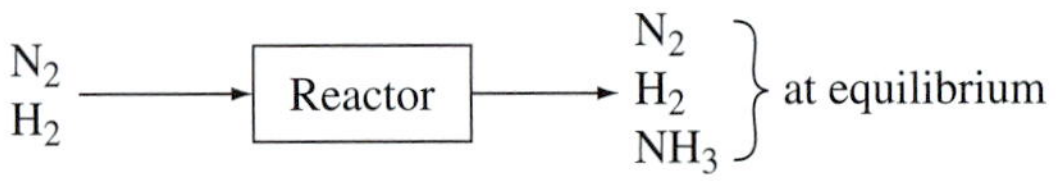

The balanced reaction is

$$N_2 + 3H_2 \rightleftarrows 2NH_3$$

The three material balance equations are

$$\dot{n}_{NH_3,out} = 2\dot{\xi}$$

$$\dot{n}_{N_2,out} = 1000 - \dot{\xi}$$

$$\dot{n}_{H_2,out} = 3000 - 3\dot{\xi}$$

It's useful to calculate the total molar flow out:

$$\dot{n}_{out} = \sum \dot{n}_{i,out} = 2\dot{\xi} + 1000 - \dot{\xi} + 3000 - 3\dot{\xi} = 4000 - 2\dot{\xi}$$

The mole fractions of each component in the effluent stream are

$$y_{NH_3,out} = \frac{\dot{n}_{NH_3,out}}{\dot{n}_{out}} = \frac{2\dot{\xi}}{4000 - 2\dot{\xi}}$$

$$y_{N_2,out} = \frac{\dot{n}_{N_2,out}}{\dot{n}_{out}} = \frac{1000 - \dot{\xi}}{4000 - 2\dot{\xi}}$$

$$y_{H_2,out} = \frac{\dot{n}_{H_2,out}}{\dot{n}_{out}} = \frac{3000 - 3\dot{\xi}}{4000 - 2\dot{\xi}}$$

Since the gas leaving the reactor is at equilibrium, $y_{i,eq} = y_{i,out}$, and we plug these expressions into the equation for K_a:

$$K_a = \frac{\left(y_{NH_3,eq}\right)^2}{\left(y_{N_2,eq}\right)\left(y_{H_2,eq}\right)^3}\frac{1}{P^2}$$

$$6.6 \times 10^5 \text{atm}^{-2} = \frac{(2\dot{\xi})^2(4000 - 2\dot{\xi})^2}{(1000 - \dot{\xi})(3000 - 3\dot{\xi})^3}\frac{1}{(1 \text{ atm})^2}$$

Finally, we want to solve for $\dot{\xi}$. This is a nonlinear equation, and there is more than one value of $\dot{\xi}$ that satisfies the equation *mathematically*. But there is only one value of $\dot{\xi}$ that satisfies *reality*. $\dot{\xi}$ must be nonnegative and $\dot{\xi}$ cannot be any greater than 1000 (why?). It is good practice to look at an equation before solving it, and decide what the upper and lower maximum limits of the solution must be.

We use an equation-solving program, (App. A.2) a programmable calculator, or trial-and-error to find:

$$\dot{\xi} = 970\frac{\text{kgmol}}{\text{h}}$$

The fractional conversion is therefore 0.97. Using the material balance equations, we calculate the molar flows out of the reactor:

$$\dot{n}_{NH_3,out} = 1940\frac{\text{kgmol}}{\text{h}}$$

$$\dot{n}_{N_2,out} = 1000 - 970 = 30\frac{\text{kgmol}}{\text{h}}$$

$$\dot{n}_{H_2,out} = 3000 - 2910 = 90\frac{\text{kgmol}}{\text{h}}$$

Notice carefully the strategy we used to solve the preceding problem.

- We wrote the equilibrium constant in terms of equilibrium mole fractions of reactants and products.
- After defining mole fraction as molar flow of species at equilibrium/total molar flow, we used mole balance equations to express the molar flow at equilibrium (in the reactor outlet) as a function of the inlet molar flow, the reaction rate $\dot{\xi}_k$, and the known stoichiometric coefficients.
- If the inlet flows and reaction stoichiometry are known, then the only unknown is $\dot{\xi}_k$, which can be found from the K_a equation and the known numerical value of K_a.

In the previous example, the reactor temperature and pressure were specified and we calculated the reactor performance. Alternatively, we could specify the desired reactor performance and find the reactor T and P necessary to achieve that specification. Performance of a reactor operating at equilibrium is influenced by reactor *temperature* because K_a is a function of temperature. As a general rule of thumb, K_a decreases with increasing temperature for oxidation, hydrogenation, and hydrolysis reactions. K_a increases with increasing temperature for dehydrogenation and dehydration reactions, and K_a is insensitive to temperature for isomerization reactions. Performance of reactors may be influenced by reactor *pressure*. K_a is not a function of pressure, but equilibrium conversion may be, if the reaction is gas phase and if there is a change in the number of moles with reaction. If the number of moles decreases as the reaction proceeds, conversion increases with increasing pressure and vice versa.

Example 4.18	**Equilibrium Conversion as a Function of *T* and *P*: Ammonia Synthesis**

In the previous example, we observed that the fractional conversion at equilibrium was high at 298 K and 1 atm. Unfortunately, the reaction kinetics are extremely slow at this temperature; equilibrium might not be reached in this lifetime. Temperatures of about 350 to 600°C are necessary for this reaction to come to equilibrium in a reasonable length of time, using modern commercially available catalysts.

Your job is to choose an appropriate ammonia synthesis reactor temperature and pressure, given a target performance specification of 50% single-pass conversion. 1000 kgmol N_2/h and 3000 kgmol H_2/h are fed to the reactor. Assume that the reactor can be designed to achieve equilibrium conversion at the outlet.

Solution

The flow diagram is identical to that in the previous example.

$$N_2 \quad \longrightarrow \quad \boxed{\text{Reactor}} \quad \longrightarrow \quad \left.\begin{array}{l} N_2 \\ H_2 \\ NH_3 \end{array}\right\} \text{ at equilibrium}$$

The balanced reaction is

$$N_2 + 3H_2 \rightleftarrows 2NH_3$$

First we need to find K_a as a function of T.

$$\Delta\hat{H}_r^\circ = \sum \nu_i \Delta\hat{H}_{i,f}^\circ = (-1)(0) + (-3)(0) + (+2)(-46{,}150) = -92{,}300 \text{ J/gmol}$$

$$\Delta G_r^\circ = \sum \nu_i \Delta\hat{G}_{i,f}^\circ = (-1)(0) + (-3)(0) + (+2)(-16{,}600) = -33{,}200 \text{ J/gmol}$$

$$\ln K_a = -\frac{1}{R}\left[\frac{\Delta\hat{G}_r^\circ - \Delta\hat{H}_r^\circ}{298} + \frac{\Delta\hat{H}_r^\circ}{T}\right]$$

$$= -\frac{1}{8.3144}\left[\frac{-33{,}200 + 92{,}300}{298} + \frac{-92{,}300}{T}\right]$$

$$\ln K_a = -23.85 + \frac{11{,}100}{T}$$

(Units in J, gmol, and K.)

In Example 4.17, we obtained an expression for K_a as a function of the extent of reaction that is still valid for this problem:

$$K_a = \frac{(2\dot{\xi})^2(4000 - 2\dot{\xi})^2}{(1000 - \dot{\xi})(3000 - 3\dot{\xi})^3} \frac{1}{P^2}$$

We've specified that the fractional conversion $f_c = 0.5$. From the material balance equation and the definition of fractional conversion, we find that $\dot{\xi} = 500$ kgmol/h. Substituting in this value gives

$$K_a = \frac{[2(500)]^2[4000 - 2(500)]^2}{(1000 - 500)[3000 - 3(500)]^3} \frac{1}{P^2} = \frac{5.333}{P^2}$$

Now we have two equations in K_a, involving two unknowns, T and P. We simply calculate K_a as a function of T from 350 to 600°C (633 to 873 K), then use the calculated K_a to determine P. As an example, we show the calculations at $T = 633$ K:

$$\ln K_a = -23.85 + \frac{11100}{633} = -6.31$$

$$K_a = 1.81 \times 10^{-3} = \frac{5.333}{P^2}, \qquad P = \sqrt{\frac{5.333}{1.81 \times 10^{-3}}} = 54 \text{ atm}$$

These calculations are straightforward to carry out in a spreadsheet. We plot our results as P versus T; the line indicates the reactor process conditions that produce the desired fractional conversion of 0.5. Notice that very high pressures are required; development of the mechanical equipment necessary to work at such high pressures was a crucial innovation needed for commercialization of ammonia synthesis, as was described in Chap. 2's Chemistory.

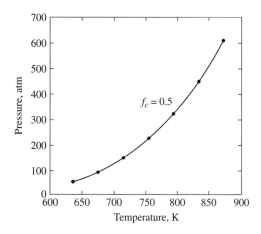

Example 4.19   ### Multiple Chemical Equilibria and Reactor T: NOx Formation.

A significant contributor to airborne pollution is formation of NOx (rhymes with lox, fox, and socks)—compounds like NO, N_2O, and NO_2. One of the main sources of

NOx is combustion in automobiles, industrial furnaces, and other sources. What is the best way to design burners and internal combustion engines so they don't produce much NOx? To begin to answer this question, we ask: How do chemical equilibrium considerations enter into design of engines and burners?

Two reactions of importance are

$$N_2 + O_2 \rightleftarrows 2NO \tag{R1}$$
$$2NO + O_2 \rightleftarrows 2NO_2 \tag{R2}$$

Both reactions occur in the gas phase, and are most important in the hot flame after the fuel has been burned. The equilibrium constants for (R1) and (R2) are K_1 and K_2, respectively, and depend on temperature T as

$$\ln K_1 = +2.97 - \frac{21,700}{T(K)}$$

$$\ln K_2 = -17.5 + \frac{13,700}{T(K)}$$

1. Plot the equilibrium constants K_1 and K_2 as a function of temperature, from 300 K (room temperature) to 2000 K (about as hot as a flame could get).
2. A burner has a postcombustion gas composition of 6.4 mol% O_2, and 73.8 mol% N_2, with the remainder a mix of CO_2 and H_2O. The high flame temperature supports further reaction to form NOx. Calculate the equilibrium composition of NO, NO_2, and NOx (NO + NO_2) as a function of flame temperature. Assume that H_2O and CO_2 are completely inert and the pressure of the flame is 1 atm.

Solution

1. Temperature has opposite effects on the equilibrium constant for reaction R1 versus R2: K_a increases with T for R1 but decreases for R2. R2 is much more favored than R1 except at very high temperatures. Thus, NO formation is favored at high temperatures, NO_2 at low temperatures.

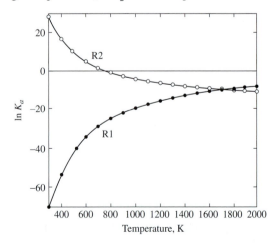

2. We start with a flow diagram. The flame itself acts as a "reactor." We choose a convenient basis of 1000 gmol/h total flow rate into the flame (any basis is fine), and we lump together CO_2 and H_2O as I (for inert).

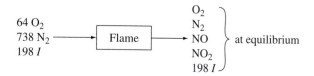

Next we set up the material balance equations, and derive equations for the mole fraction of each compound in the outlet stream as a function of the extents of reaction:

Compound	$\dot{n}_{i,in}$	$\nu_{i1}\dot{\xi}_1 + \nu_{i2}\dot{\xi}_2$	$\dot{n}_{i,out}$	$y_{i,out}$
O_2	64	$-\dot{\xi}_1 - \dot{\xi}_2$	$64 - \dot{\xi}_1 - \dot{\xi}_2$	$\dfrac{64 - \dot{\xi}_1 - \dot{\xi}_2}{1000 - \dot{\xi}_2}$
N_2	738	$-\dot{\xi}_1$	$738 - \dot{\xi}_1$	$\dfrac{738 - \dot{\xi}_1}{1000 - \dot{\xi}_2}$
NO	0	$+2\dot{\xi}_1 - 2\dot{\xi}_2$	$+2\dot{\xi}_1 - 2\dot{\xi}_2$	$\dfrac{+2\dot{\xi}_1 - 2\dot{\xi}_2}{1000 - \dot{\xi}_2}$
NO_2	0	$+2\dot{\xi}_2$	$+2\dot{\xi}_2$	$\dfrac{+2\dot{\xi}_2}{1000 - \dot{\xi}_2}$
I	198	0	198	$\dfrac{198}{1000 - \dot{\xi}_2}$
Total	1000	$-\dot{\xi}_2$	$1000 - \dot{\xi}_2$	1

For this gas-phase reaction, K_1 and K_2 are evaluated as (try deriving these equations yourself!):

$$K_1 = \frac{(y_{NO})^2}{y_{N_2}y_{O_2}} = \frac{(2\dot{\xi}_1 - 2\dot{\xi}_2)^2}{(738 - \dot{\xi}_1)(64 - \dot{\xi}_1 - \dot{\xi}_2)}$$

$$K_2 = \frac{(y_{NO_2})^2}{(y_{NO})^2 y_{O_2}}\left(\frac{1}{P(\text{atm})}\right) = \frac{(2\dot{\xi}_2)^2(1000 - \dot{\xi}_2)}{(2\dot{\xi}_1 - 2\dot{\xi}_2)^2(64 - \dot{\xi}_1 - \dot{\xi}_2)}$$

Given a temperature T, we find K_1 and K_2 from part (1). We then have two equations in two unknowns, $\dot{\xi}_1$ and $\dot{\xi}_2$. We solve simultaneously, noting that there are constraints on what values are physically reasonable, e.g., $\dot{\xi}_1 + \dot{\xi}_2 < 64$, $\dot{\xi}_1 > \dot{\xi}_2$. Then we use the obtained values to calculate the mole fractions of NO and NO_2 from the material balance equation. For example, at $T = 1500\ K$, $K_1 = 1.0 \times 10^{-5}$ and $K_2 = 2.3 \times 10^{-4}$ atm^{-1}. Using these values for K_1 and K_2 in the equations above, we find the solution by using a spreadsheet or an equation solver. (Notice that the equilibrium constants are very low for both reactions. This indicates that the extents of reaction will also be low—close to zero. If we notice this, we can simplify the equations further by noting that $738 - \dot{\xi}_1 \approx 738$, $64 - \dot{\xi}_1 - \dot{\xi}_2 \approx 64$, and $1000 - \dot{\xi}_2 \approx 1000$!). We find $\dot{\xi}_1 = 0.343$ and $\dot{\xi}_2 = 1.3 \times 10^{-3}$ gmol/h. Using these values, we calculate mole fractions: $y_{NO} = 6.8 \times 10^{-4}$ and $y_{NO_2} = 2.6 \times 10^{-6}$.

By repeating at other temperatures, we produce a graph of y_{NO} and y_{NO_2} versus T:

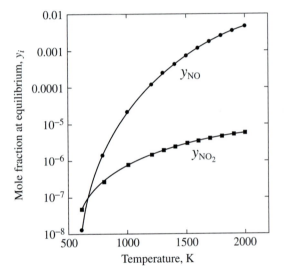

NOx formation increases dramatically with temperature. Industrial boilers and incinerators must operate at a low enough flame temperature to minimize NOx formation while maintaining a high enough temperature to achieve complete combustion. The ratio of NO to NO$_2$ also shifts dramatically with temperature: below 600 K, NO$_2$ is the main source of NOx but at higher temperatures NO becomes the most important contributor to NOx.

Sometimes the desired chemical reaction has a highly unfavorable equilibrium conversion at any reasonable temperature and pressure. We could live with a low single-pass conversion, and simply recycle. Are there other ways to engineer a chemically-reacting system in the face of unfavorable equilibrium? The answer is yes. Here are some ideas to consider:

Adjust the feed ratio. Reactants do not have to be fed at stoichiometric ratio. The fractional conversion of a limiting reactant is higher, and the conversion of an excess reactant is lower, than if reactants are fed at stoichiometric ratio.

Remove one of the products continuously. If a chemical reaction and product separation are judiciously combined in a single piece of equipment, then the equilibrium is driven towards increased conversion.

Couple to a reaction with favorable equilibrium. A reaction that consumes a byproduct of the desired reaction drives the equilibrium toward increased conversion.

We further explore some of these ideas in the case study at the end of this chapter.

4.4.4   Chemical Reaction Kinetics and Reactor Performance (Optional Section)

For some reactions, complete conversion cannot be achieved because of equilibrium constraints. Even when reaction equilibrium is highly favorable, it may be a practical impossibility to achieve equilibrium under industrially-relevant conditions. Why? Because of slow reaction kinetics.

The extent of reaction ξ is a key variable in the material balance equation. In the last section we learned how to calculate ξ if the reaction is at equilibrium. If equilibrium is not attained in an operating reactor, then ξ depends on a large number of factors including:

Factors associated with reaction chemistry:
- Temperature
- Pressure
- Catalyst properties
- Reactant and product concentrations

Factors associated with reactor design
- Reactor flow pattern
- Reactor volume

Chemical kinetics is the science of chemical reaction rates. Kineticists attempt to understand the dependence of the rate of reaction on temperature, pressure, catalyst, and reactant and product concentrations. *Reactor engineering* is the art and technology of designing reactors with the right temperature and pressure control, flow pattern, and residence time to efficiently and safely carry out the desired reactions. Both topics are much too large for us to cover. Here we will just give a small taste of chemical kinetics and reactor engineering.

Chemical kineticists define a reaction rate (which we will call r_i') as the change in concentration of compound *i* per unit time:

$$r_i' = \frac{dc_i}{dt}$$

? Did You Know?

Combustion of wood is an everyday example of a reaction that is highly favored at room temperature from a chemical equilibrium point of view, but (luckily for us) not from a reaction kinetics point of view. Combustion will not occur spontaneously without a spark–a high-temperature source that initiates reaction. Once the reaction begins, enough heat is given off to raise the temperature; this speeds up the reaction kinetics and makes the process self-sustaining. Biodegradation of wood can lead to the same final products as combustion, but uses enzymes as catalysts (supplied by insects, microbes, etc.) rather than temperature to speed up the reaction rate.

where c_i is the molar concentration of i, that is, the moles of i per reactor volume V_R:

$$c_i = \frac{n_i}{V_R}$$

r_i' has dimensions of [moles/volume-time], while $\dot{\xi}$ has dimensions of [moles/time].

In general, r_i' is a complicated function of T, P, catalyst, and reactant and product concentrations. Unfortunately, there is no general expression relating r_i' to these properties. We'll mention two specific cases, for which simple but useful expressions for r_i' are available.

Case 1.   Consider the conversion of reactant A to product P:

$$A \rightarrow P$$

If this reaction is irreversible (e.g., $K_a \gg 1$ and the reverse reaction $P \rightarrow A$ does not happen to any measurable extent), then

$$r_A' = \frac{dc_A}{dt} = -kc_A \tag{4.17}$$

These kinetics are called "first-order irreversible" because the rate depends on the *reactant concentration* raised to the first power, and does not depend on the product concentration. The term k is called a *rate constant* and is always positive. The rate constant generally increases strongly with *temperature*, as described by the Arrhenius expression:

$$k = k_0 \exp\left(-\frac{E_a}{T}\right) \tag{4.18}$$

where k_0 and E_a are experimentally determined parameters, unique to a specific reaction and catalyst, but independent of temperature.

This simple case is surprisingly useful, describing many important reactions such as pasteurization and sterilization.

Case 2.   Heterogeneous reactions are reactions involving two or more phases. Two important heterogeneous reactions include combustion of firewood and catalytic conversion of CO in automobile exhaust to CO_2. Many of these reactions have rate equations of the form

$$r_A' = \frac{dc_A}{dt} = -\frac{kK_{ads}c_A}{1 + K_{ads}c_A}$$

where K_{ads} is an experimentally determined constant that depends on the chemical and physical properties of the catalyst as well as the nature of the reactant and the temperature. Many enzyme-catalyzed reactions are well modeled by a similar equation, in which case this equation becomes the famous Michaelis-Menten equation.

We still have the task of relating r_i' to $\dot{\xi}_i$ in the material balance equation. To do this, we need to turn our attention to the reactor engineering issues: the *reactor volume* V_R and the *reactor mixing* pattern. Reaction engineers generally start by considering two limiting cases—the completely mixed, or stirred-tank, reactor, and the completely unmixed, or plug-flow, reactor (refer back to Fig. 4.2). In the completely mixed reactor, the temperature and concentrations are uniform throughout the reactor. In the plug-flow reactor, the fluid moves as a "plug" through a cylindrical reactor. As a cross-sectional slice moves through the reactor, the concentration and (possibly) the temperature change as the reaction proceeds. Since the reaction rate is a function of concentration and temperature, the reaction rate is a function of position in the reactor. The extent of reaction is calculated by summing up (integrating) the reaction rate in every cross-sectional slice of the reactor. As you can see from these brief explanations, the choice of mixing pattern greatly influences the extent of reaction achieved. You will learn much more about these topics in chemical kinetics and reactor engineering classes.

Just to illustrate the role of the material balance equation in analyzing chemical kinetics and reactor design, we will evaluate one simple case: a batch reactor in which a single irreversible first-order reaction, $A \rightarrow P$, takes place. We assume that the reactor volume does not change with time and that the reactor is charged with $n_{A,\text{sys},0}$ moles of compound A at $t = 0$. Notice that

$$n_{A,\text{sys},0} = c_{A0} V_R$$

where c_{A0} is the initial concentration of A (moles/volume) and V_R is the reactor volume. In the reactor,

$$r_A' = \frac{dc_A}{dt} = -kc_A$$

We rearrange and then integrate from $c_A = c_{A0}$ at $t_0 = 0$ to $c = c_{Af}$ at $t = t_f$:

$$\int_{c_{A0}}^{c_{Af}} \frac{dc_A}{c_A} = -k \int_0^{t_f} dt$$

$$\ln\left(\frac{c_{Af}}{c_{A0}}\right) = -kt_f$$

or

$$c_{Af} = c_{A0} e^{-kt_f}$$

The number of moles of A in the system when the reaction is stopped, at $t = t_f$, is simply

$$n_{A,\text{sys},f} = c_{Af} V_R$$

The integral component mole balance equation for compound A becomes

$$n_{A,\text{sys},f} - n_{A,\text{sys},0} = \int_{t_0}^{t_f} \nu_A \dot{\xi}\, dt = V_R(c_{Af} - c_{A0}) = V_R c_{A0}(e^{-kt_f} - 1)$$

The fractional conversion is a simple equation:

$$f_{CA} = \frac{n_{A,sys,0} - n_{A,sys,f}}{n_{A,sys,0}} = \frac{V_R c_{A0}(1 - e^{-kt_f})}{V_R c_{A0}} = 1 - e^{-kt_f} \qquad (4.19)$$

where we've used the definition of f_{CA}, the fractional conversion of A. t_f is often called the reactor *residence time*. This derivation shows that reactor performance is a function of both reaction kinetics and reactor design.

Example 4.20   **Reaction Kinetics and Reactor Performance: Vegetable Processing**

We need to sterilize cans of vegetables. Each can contains 250 mL and an average of 10,000 spores. For safety and shelf stability, we need to reduce this to an average of 0.1 spores/can, a 99.999% reduction. But, we also want to keep the vegetables tasty.

Spore killing is a first-order irreversible reaction, with a rate constant

$$k = 9 \times 10^{15} \exp\left(-\frac{15000}{T}\right)$$

where T is temperature (K) and k has units of min^{-1}. Loss of flavor is modeled as a first-order irreversible reaction, too, with a rate constant

$$k_f = 9 \times 10^{5} \exp\left(-\frac{5000}{T}\right)$$

where T is in K and k_f has units of min^{-1}. On the basis of consumer taste panels, we decide that a 25% loss of flavor is acceptable.

Is there a way to produce canned vegetables that are both safe to eat and tasty? At what temperature would you heat the cans? For how long?

Solution
The sterilizer is a batch reactor with first-order irreversible kinetics. Before we solve, let's plot the rate constants versus temperature.

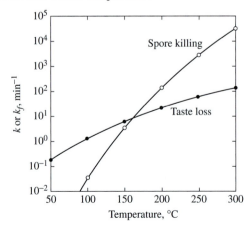

At 100 °C (373 K), k for spore killing is much lower than k_f for flavor loss. However, the spore killing rate increases faster than the flavor loss rate with increasing temperature. Around 160°C, the rates become similar. Therefore, we expect that temperatures higher than 160°C are necessary for sufficient spore killing and minimum flavor loss.

We can use the equation derived for first-order irreversible kinetics in a batch reactor:

Spore killing ($f_{CA} = 0.99999$):

$$f_{CA} = 1 - e^{-kt_f} = 0.99999$$

Rearranging and taking the natural logarithm of each side, we get

$$t_f = -\frac{\ln(1 - 0.99999)}{\left[9x10^{15}\exp\left(\frac{-15000}{T}\right)\right]} = 1.28 \times 10^{-15}\exp\left(\frac{15{,}000}{T}\right)$$

Flavor loss ($f_{CA} = 0.25$):

$$t_f = -\frac{\ln(1 - 0.25)}{\left[9x10^5\exp\left(\frac{-5000}{T}\right)\right]} = 3.20 \times 10^{-7}\exp\left(\frac{5000}{T}\right)$$

We have two equations in two unknowns. We solve to get $T = 517$ K (244°C) and $t_f = 5$ ms. Sterilization and pasteurization are often carried out at high temperature for short times to avoid flavor loss and vitamin degradation.

CASE STUDY   Hydrogen and Methanol

In the 1700s, it was whale oil. In the 1800s, coal was king. And the late 1900s were dominated by liquid petroleum. Some people now forecast that the next century will usher in the hydrogen economy. Hydrogen is a clean-burning fuel, producing nothing but water on oxidation. It is the fuel source of choice for energy-efficient fuel cells that in the future may power everything from portable phones to huge electrical utility plants. Although many have proposed that automobiles and trucks may some day run on hydrogen-fed fuel cells, concern has been expressed about the safety of people driving around with storage tanks of gaseous hydrogen under high pressure. Methanol has been proposed as one possible liquid fuel alternative. The idea is that liquid methanol could be stored in the fuel tank, then converted to hydrogen in situ which could then be fed to the fuel cell. Besides their potential use in fuel cells, hydrogen and methanol are important reactants in atom-economical chemical synthesis routes. But, hydrogen is not a readily available raw material and must be synthesized.

In this case study, we'll examine the synthesis of hydrogen and methanol from methane. Natural gas, an abundant but nonrenewable resource, is currently the major source of methane. In the future it may be possible to generate significant quantities of methane from biomass, using aerobic digesters like the one pictured in Fig. 4.1. We'll identify a number of variables of concern in designing hydrogen and methanol synthesis reactors.

Methane is converted to hydrogen via the steam reforming reaction:

$$CH_4(g) + H_2O(g) \rightleftarrows CO(g) + 3H_2(g) \tag{R1}$$

Let's start by considering the chemical equilibrium constant of this reaction. This will help us identify reactor temperatures and pressures that will give reasonable equilibrium conversions. (Remember—we can never do better than equilibrium conversion.) We'll use M for methane, W for water, CO for carbon monoxide and H for H_2. The reaction takes place in the vapor phase. The chemical equilibrium constant is

$$K_a = \prod_{\text{all } i} (y_i P)^{\nu_i} = \frac{(y_{CO}P)(y_H P)^3}{(y_M P)(y_W P)} = P^2 \frac{(y_{CO})(y_H)^3}{(y_M)(y_W)}$$

K_a for methane-steam reforming is calculated from the Gibbs energy and enthalpy of reaction to be

$$\ln K_a = 25.8 - \frac{24800}{T}$$

where K_a has units of $(\text{atm})^2$ and T is in K.

Now we can figure out how the fractional conversion at equilibrium depends on temperature and pressure. Combining the above equations, and grouping pressure and temperature on one side of the equals sign, gives

$$\frac{\exp\left(25.8 - \dfrac{24800}{T}\right)}{P^2} = \frac{(y_{CO})(y_H)^3}{(y_M)(y_W)}$$

Our first goal is to calculate the equilibrium conversion as a function of reactor T and P. We'll sketch out the flow diagram:

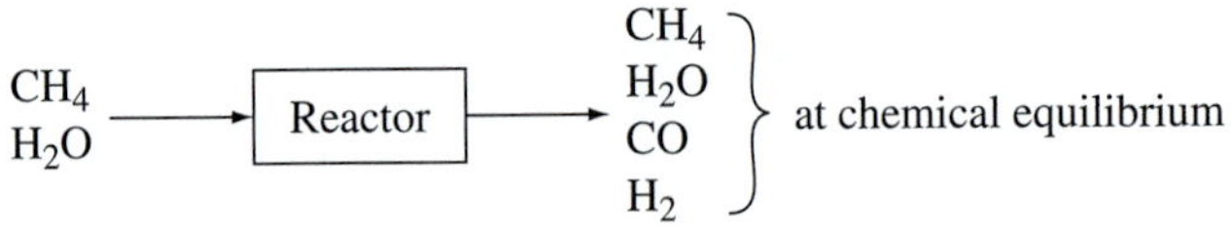

We need to find common ground between the flow diagram and the equilibrium equation so we write mole fractions at equilibrium in terms of component molar flow rates out of the reactor. We can choose any basis we want. Let's choose 1.0 gmol/s CH_4 fed to the reactor. Let's assume for now that the feed ratio of methane:water is 1:1 (the stoichiometric ratio). The steady-state differential material balance reaction for methane M is

$$\dot{n}_{M,\text{out}} = \dot{n}_{M,\text{in}} - \dot{\xi}_1 = 1.0 - \dot{\xi}_1$$

We write similar material balance equations for the other components, sum up the component molar flows out of the reactor to get the total molar flow out, and calculate the mole fraction by dividing the component molar flow by the total material flow. It's convenient to do this in table form;

	$\dot{n}_{i,\text{in}}$	$\dot{n}_{i,\text{out}}$	$y_{i,\text{out}} = \dfrac{\dot{n}_{i,\text{out}}}{\dot{n}_{\text{out}}}$
CH_4	1	$1 - \dot{\xi}_1$	$\dfrac{1 - \dot{\xi}_1}{2 + 2\dot{\xi}_1}$
H_2O	1	$1 - \dot{\xi}_1$	$\dfrac{1 - \dot{\xi}_1}{2 + 2\dot{\xi}_1}$
CO	0	$\dot{\xi}_1$	$\dfrac{\dot{\xi}_1}{2 + 2\dot{\xi}_1}$
H_2	0	$3\dot{\xi}_1$	$\dfrac{3\dot{\xi}_1}{2 + 2\dot{\xi}_1}$
Total	2	$2 + 2\dot{\xi}_1$	1

Now we can write K_a in terms of T, P, and $\dot{\xi}_1$:

$$\frac{K_a}{P^2} = \frac{e^{\left(25.8 - \frac{24,800}{T}\right)}}{P^2} = \frac{(\dot{\xi}_1)(3\dot{\xi}_1)^3}{(1 - \dot{\xi}_1)^2(2 + 2\dot{\xi}_1)^2} \tag{4.20}$$

Quick Quiz 4.8

What is the range of physically reasonable values for $\dot{\xi}_1$?

For a given value of T and P, we can solve for $\dot{\xi}_1$ at equilibrium. This isn't too hard to do with an equation solver. Then, we use the value of $\dot{\xi}_1$ to calculate y_i, $\dot{n}_{i,\text{out}}$, and fractional conversion of methane f_{CM} at equilibrium.

What range of values is reasonable for T and P? From an examination of Eq. (4.20), we see that $\dot{\xi}_1$ increases if (1) the pressure decreases or (2) the temperature increases. From our heuristics, we prefer not to go any lower than ambient pressure. Temperatures above about 600°C (873 K) require specialized materials of construction, but maybe we could push it a bit on this end if needed. With these thoughts in mind, let's examine reactor performance at 1 atm and at 400°C to 800°C. Some results of our calculations are plotted.

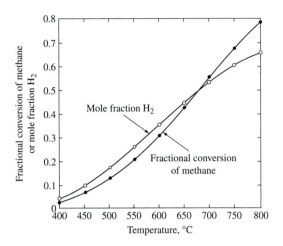

At 600°C, equilibrium conversion is about 30%. This is probably acceptable, since much higher temperatures are frowned on because of materials costs and added safety concerns.

Are there any other ways to change the reactor design to increase equilibrium conversion of methane? Here are a couple of ideas:

1. Increase the amount of water in the reactor to help drive the reaction, making methane the limiting reactant.
2. Install an in situ adsorber: a solid material to which CO selectively sticks, such that the mole fraction CO in the gas phase is always maintained at a low level, e.g., 0.01. (In Chap. 5 we will discuss adsorption in more detail.)

Idea 1.   Increase the water:methane feed ratio above stoichiometric. Let's keep $T = 600$°C, $P = 1$ atm, and $\dot{n}_{M,\text{in}} = 1.0$ and vary the stoichiometric ratio from 1:1 to 9:1. Eq. (4.20) changes somewhat. (See if you can derive the modified equation.) The results are plotted:

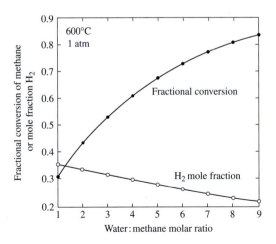

Increasing the water:methane ratio greatly increases methane conversion! We pay a price, though—we are handling a lot of extra steam, and the hydrogen in the product stream is more dilute, which increases separation costs.

Idea 2.   Install an adsorber to remove CO as it is made. The flow diagram changes: our process unit is a combined reactor and separator.

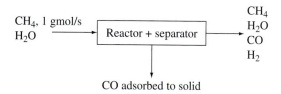

CO adsorbed to solid

For these calculations we'll assume that the adsorber maintains $y_{co} = 0.01$ in the reactor. (In Chap. 5 you will learn more about how absorbers work and how to determine the gas phase composition in the presence of an adsorber.) We'll assume a 1:1 methane:water feed ratio, 600°C, 1 atm. The material balance equations must be adjusted to account for the extra outlet stream:

	$\dot{n}_{i,in}$	$\dot{n}_{i,out}$	$y_{i,out} = \dfrac{\dot{n}_{i,out}}{\dot{n}_{out}}$
CH_4	1	$1 - \dot{\xi}_1$	$\dfrac{1 - \dot{\xi}_1}{2 + 2\dot{\xi}_1 - \dot{n}_{ads}}$
H_2O	1	$1 - \dot{\xi}_1$	$\dfrac{1 - \dot{\xi}_1}{2 + 2\dot{\xi}_1 - \dot{n}_{ads}}$
CO	0	$\dot{\xi}_1 - \dot{n}_{ads}$	0.01
H_2	0	$3\dot{\xi}_1$	$\dfrac{3\dot{\xi}_1}{2 + 2\dot{\xi}_1 - \dot{n}_{ads}}$
Total	2	$2 + 2\dot{\xi}_1 - \dot{n}_{ads}$	1

where $\dot{n}_{ads}$ = flow rate of CO leaving the reactor adsorbed to the solid. With this situation,

$$\frac{K_a}{P^2} = \frac{e^{\left(25.8 - \frac{24800}{T}\right)}}{P^2} = 0.0737 = \frac{(0.01)(3\dot{\xi}_1)^3}{(1 - \dot{\xi}_1)^2(2 + 2\dot{\xi}_1 - \dot{n}_{ads})^2}$$

and we have the specification regarding the separation that

$$y_{CO} = 0.01 = \frac{\dot{\xi}_1 - \dot{n}_{ads}}{2 + 2\dot{\xi}_1 - \dot{n}_{ads}}$$

Solving these two equations simultaneously, we find that adding the adsorber increases conversion from 30% to 65%. Furthermore, the gas leaving the reactor is more enriched in hydrogen ($y_{H_2} = 0.73$, versus 0.35 in the base case), because the byproduct CO is continuously removed. Are there any disadvantages you can think of?

Notice that with both of these ideas we achieved greater fractional conversion, not by violating the law of chemical reaction equilibrium, but by designing around it!

In the analysis so far, we have assumed that only one reaction takes place. In reality, chemistry plays tricks on us all the time. As (bad) luck would have it, most of the time these tricks come in the form of unwanted reactions: Either the reactants convert to other undesired products or the products themselves undergo further reactions. How can we design a reactor to give good fractional conversion *and* good selectivity when there are multiple chemical reactions? We've got a bag of tricks of our own.

We've made CO and H_2 by steam reforming of methane,

$$CH_4(g) + H_2O(g) \rightleftarrows CO(g) + 3H_2(g) \tag{R1}$$

and will want to feed CO and H_2 to another reactor to make methanol,

$$CO(g) + 2H_2(g) \rightleftarrows CH_3OH(g)$$

However, in the steam reforming reactor, CO undergoes a further reaction with steam to make CO_2. This reaction is the famous water-gas shift reaction:

$$CO(g) + H_2O(g) \rightleftarrows CO_2(g) + H_2(g) \tag{R2}$$

The chemical equilibrium constant for the water-gas shift reaction is

$$\ln K_a = -5.1 + \frac{4950}{T}$$

where T is in K.

How should I operate the steam reformer, given these complications and the desire to make methanol? The methanol reaction requires 2 moles H_2 per mole CO, and steam reforming produces 3 moles H_2 per mole CO. So we already have a CO deficit, and the water-gas shift reaction just makes it worse. Are there temperatures and pressures that allow good conversion in steam reforming while suppressing the water-gas shift? Let's start by looking at some trends.

The equilibrium expression for the water-gas shift reaction is

$$K_a = \frac{y_{CD}\, y_H}{y_{CO}\, y_W}$$

where we use *CD* for carbon dioxide. Notice that there is no pressure term! Since the equilibrium conversion for the desired steam reforming reaction R1 decreases with increasing pressure, and the extent of reaction of the undesired reaction R2 is unaffected by pressure, we should design our reactor at as low a pressure as feasible, to favor the desired reaction. Therefore, let's keep the design pressure at 1 atm.

Now, let's look at the temperature dependence by plotting $\ln(K_a)$ for the two reactions R1 and R2:

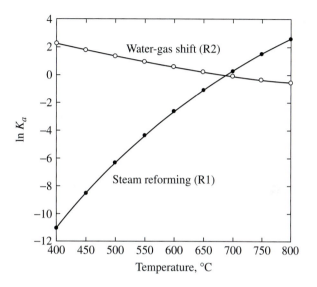

The desired reaction is highly unfavored at low temperature but highly favored at high temperatures. The undesired reaction has the opposite trend. This is good news. We want to operate at reactor temperatures where K_a for the undesired reaction is less than 1, and also less than that of the desired reaction. Our earlier choice of 600°C, made when we considered only one reaction, is too low. Let's pick 725°C, a temperature at which K_a of R1 is greater than that of R2. Although hot, 725°C is still reasonable in terms of materials of construction.

We'll assume that the reactants are fed at stoichiometric ratio (for the desired reaction R1) and use a basis of 1 gmol CH_4/s fed to the reactor.

Let's use our material balance equations to calculate conversion and selectivity. In this analysis, we add in CO_2 as one of the components, and we include the rates of both reactions.

	$\dot{n}_{i,in}$	$\sum \nu_{ik}\dot{\xi}_k$	$\dot{n}_{i,out}$	$y_{i,out} = \dfrac{\dot{n}_{i,out}}{\dot{n}_{out}}$
CH_4	1	$-\dot{\xi}_1$	$1-\dot{\xi}_1$	$\dfrac{1-\dot{\xi}_1}{2+2\dot{\xi}_1}$
H_2O	1	$-\dot{\xi}_1-\dot{\xi}_2$	$1-\dot{\xi}_1-\dot{\xi}_2$	$\dfrac{1-\dot{\xi}_1-\dot{\xi}_2}{2+2\dot{\xi}_1}$
CO	0	$\dot{\xi}_1-\dot{\xi}_2$	$\dot{\xi}_1-\dot{\xi}_2$	$\dfrac{\dot{\xi}_1-\dot{\xi}_2}{2+2\dot{\xi}_1}$
H_2	0	$3\dot{\xi}_1+\dot{\xi}_2$	$3\dot{\xi}_1+\dot{\xi}_2$	$\dfrac{3\dot{\xi}_1+\dot{\xi}_2}{2+2\dot{\xi}_1}$
CO_2	0	$\dot{\xi}_2$	$\dot{\xi}_2$	$\dfrac{\dot{\xi}_2}{2+2\dot{\xi}_1}$
Total	2		$2+2\dot{\xi}_1$	1

Now we have two equations relating mole fractions to the two value of K_a, and we have expressions for mole fraction of all components in terms of $\dot{\xi}_1$ and $\dot{\xi}_2$.

We now substitute the expressions for mole fractions from the table into the equations for K_a and solve. We'll compare to the case where we didn't consider the water-gas shift reaction.

	Water-gas not considered		**Water-gas considered**	
	y_i	$\dot{n}_{i,out}$	y_i	$\dot{n}_{i,out}$
CH_4	0.125	0.38	0.12	0.40
H_2O	0.125	0.38	0.10	0.32
CO	0.19	0.62	0.16	0.53
H_2	0.57	1.86	0.59	1.88
CO_2			0.024	0.077
$\dot{\xi}_1$		0.62		0.60
$\dot{\xi}_2$				0.077

A couple of observations: First, the water-gas shift reaction has had a small effect on the methane conversion. Second, the H_2:CO ratio has changed from 3:1 to 3.6:1. This is bad if our goal is to make methanol, since the stoichiometric ratio for the methanol synthesis reaction is 2:1 H_2:CO. Third, the selectivity, based on methane converted to CO, is $(0.60 - 0.077/0.60) = 0.87$ (compared to 1.0 in the absence of the water-gas shift reaction). Finally, we're making a by-product, CO_2, that will have to be separated out and is of little use.

Are there any other changes we could make? Notice that water is a reactant in both reactions. What happens if we adjust the stoichiometric feed ratio of methane to water? Let's evaluate a couple of different feed ratios: twofold excess water, and twofold excess methane. We'll keep the basis as 1 gmol methane/s fed to the reactor.

	CH$_4$:H$_2$O feed = 1:1		**CH$_4$:H$_2$O feed = 1:2**		**CH$_4$:H$_2$O feed = 1:0.5**	
	y_i	$\dot{n}_{i,\text{out}}$	y_i	$\dot{n}_{i,\text{out}}$	y_i	$\dot{n}_{i,\text{out}}$
CH$_4$	0.12	0.40	0.041	0.19	0.26	0.60
H$_2$O	0.10	0.32	0.22	0.99	0.034	0.079
CO	0.16	0.53	0.13	0.61	0.165	0.37
H$_2$	0.59	1.89	0.57	2.63	0.53	1.22
CO$_2$	0.024	0.076	0.043	0.20	0.009	0.021
$\dot{\xi}_1$		0.60		0.81		0.40
$\dot{\xi}_2$		0.077		0.20		0.021

This has made a big difference! With excess water, we've achieved an amazing 81% conversion of methane. The cost, though, has come in selectivity, which has dropped to 0.75. The H$_2$:CO ratio of 4.4 is worse, and we've generated *more* CO$_2$. The total flow through the reactor has increased, which means a bigger, more expensive, reactor. In contrast, with methane in excess, the methane conversion dropped way down to 40%, but selectivity is better at 0.95, less CO$_2$ is produced, and the H$_2$:CO ratio is lower.

Deciding on the best reactor conditions will require further design analysis. Can unreacted methane be recycled? How difficult is it to remove CO$_2$? Is there anything useful we could do with it? Are there other uses for the excess H$_2$? Are there any other reactions that might occur? Finally, does the reactor operate close to equilibrium, or is conversion and selectivity dominated by kinetic considerations?

Summary

- Three useful measures of reactor system performance are **conversion, selectivity,** and **yield:**

$$\text{Fractional conversion} = \frac{\text{moles of reactant consumed}}{\text{moles of reactant fed}}$$

$$f_{Ci} = \frac{\dot{n}_{i,\text{ in}} - \dot{n}_{i,\text{ out}}}{\dot{n}_{i,\text{ in}}} = \frac{-\displaystyle\sum_{\substack{\text{all } k \\ \text{reactions}}} \nu_{ik}\dot{\xi}_k}{\dot{n}_{i,\text{ in}}}$$

$$\text{Fractional selectivity} = \frac{\text{moles of reactant A converted to desired product P}}{\text{moles of reactant A consumed}}$$

$$s_{A \to P} = \frac{\nu_{A1}}{\nu_{P1}} \frac{\displaystyle\sum_{\substack{\text{all } k \\ \text{reactions}}} \nu_{Pk} \dot{\xi}_k}{\displaystyle\sum_{\substack{\text{all } k \\ \text{reactions}}} \nu_{Ak} \dot{\xi}_k}$$

$$\text{Fractional yield} = \frac{\text{moles of reactant A converted to desired product P}}{\text{moles of reactant A fed}}$$

$$y_{A \to P} = -\frac{\nu_{A1}}{\nu_{P1}} \frac{\displaystyle\sum_{\substack{\text{all } k \\ \text{reactions}}} \nu_{Pk} \dot{\xi}_k}{\dot{n}_{A,\,\text{in}}}$$

- If fractional conversion is low, then reactants are recycled. This requires addition of a separator to the flow sheet. With **recycle,** the overall conversion of raw material to product in the process may be very high even if the single-pass conversion is low. Recycle is more common in commodity-chemical than in specialty-chemical processes. If any inerts are present in the raw material and recycle is used, then **purge** may be required. With purge, a splitter is added after the separator.

- **Chemical reaction equilibrium** limits the maximum achievable conversion and may affect selectivity. The chemical reaction equilibrium constant K_a is

$$K_a = \prod_{\text{all } i} a_i^{\nu_i}$$

where a_i is the activity of species i at equilibrium and ν_i is the stoichiometric coefficient of i. To a first approximation,

$$a_i = \frac{y_i P}{1 \text{ atm}} \qquad \text{for a gas}$$

$$a_i = x_i \qquad \text{for a liquid}$$

$$a_i = 1 \qquad \text{for a solid}$$

K_a is a function of temperature:

$$\ln K_{a,T} = -\frac{1}{R}\left[\frac{\Delta \widehat{G}_r^\circ - \Delta \widehat{H}_r^\circ}{298} + \frac{\Delta \widehat{H}_r^\circ}{T}\right]$$

where

$$\Delta \widehat{G}_r^\circ = \sum_i \nu_i \Delta \widehat{G}_{i,f}^\circ \qquad \text{and} \qquad \Delta \widehat{H}_r^\circ = \sum_i \nu_i \Delta \widehat{H}_{i,f}^\circ$$

By adjusting temperature, pressure, and reactant feed ratio, and by modifying the reactor flow sheet, we can design around some of the limitations imposed by chemical reaction equilibrium.

- Chemical reaction kinetics may limit the conversion and affect selectivity. By adjusting temperature, pressure, reactant feed ratio, catalyst, and reaction time, we can overcome some of the constraints imposed by chemical reaction kinetics.

ChemiStory: Quit Bugging Me!

Not many people like bugs. They are, for most of us, nuisances. More than that, about 1% of the known 1 million species of insects are real pests. Because insects are voracious feeders and disease spreaders, some bugs threaten human life and health. For example, shortly after the Bolshevik Revolution, 25 to 30 million Russians contracted typhus, which is spread by lice; about 3 million died. During World War I, neutral Switzerland suffered severe food shortages because insects consumed much of their grain.

Courtesy of Centers for Disease Control

It was in Switzerland that Paul Hermann Müller was born, in 1899. Paul was a mediocre student who dropped out of school at the age of 17. But he loved chemistry, and kept a lab in his family's home. Paul eventually returned to school, completed his Ph.D. in chemistry in 1925, and went to work for J. R. Geigy. (Geigy later became Ciba-Geigy, which later became the pharmaceutical giant Novartis.) During the early part of Müller's career, the company discovered a mothproofing compound—a chlorinated hydrocarbon that was a stomach poison for moths. This discovery initiated Geigy's expansion into the insecticide business. Prior to this discovery, chemical insect control had been limited to natural compounds like nicotine and rotenone—tropical-plant-based compounds that were expensive and unstable—and arsenic compounds—cheap stomach poisons that worked well against chewing insects, but were also highly toxic to humans and other warm-blooded animals.

In 1935, Müller was assigned the job of finding a better insecticide. He set several criteria for his "ideal" insecticide. It would: kill by contact rather than requiring ingestion; be a broad spectrum pesticide, killing many different kinds of insects; be harmless to fish, plants, and warm-blooded animals; have no odor;

*© Swim Ink 2. LLC/Corbis
Pesticides*

be inexpensive and chemically stable. Müller got to work and started synthesizing and testing compounds. He started with a few chemical structures; if a compound looked promising he would make several related compounds. After 4 years, he had laboriously worked his way through 349 compounds, one by one. (Today, large "libraries" of compounds are synthesized in parallel by combinatorial chemistry, and tested in parallel using high-throughput screening. Hundreds if not thousands of chemicals can be synthesized and tested in weeks.) For his 350th compound, Müller reacted chloral and chlorobenzene with sulfuric acid as a catalyst, and made dichlorodiphenyl trichloroethane, or DDT. He reported that DDT, sprayed in glass cages, killed flies. Even better, the cages remained toxic weeks later. DDT was not degraded by light or oxidation, and had a very low vapor pressure, so it would persist after application. It killed many other insects such as the Colorado potato beetle larvae. It was insoluble in water, suggesting that it would not contaminate the water supplies. Muller had hit his target—he had discovered a cheap, powerful, broad-spectrum contact poison that was very stable.

Geigy patented the use of DDT as an insecticide. It was an instant hit. In 1942, Geigy sold one pound of DDT-laced insecticide per person in Switzerland, and rescued the potato crop. Since Switzerland was a neutral country, Geigy reported their invention to both Allied and Axis countries. Germany was not interested, as they had their own insecticide development program underway. The United States, however, was very interested. After just a few months of testing, DDT was declared safe and effective, and its use skyrocketed. In Naples, Italy, 1.3 million refugees were sprayed with DDT after Allied Forces recaptured that city, thereby preventing a sure outbreak of deadly typhus. Pacific islands were sprayed with DDT to kill mosquitoes before Allied Forces fought to reoccupy the lands. (Earlier in the war, malaria had incapacitated up to 2 out of every 3 soldiers in the mosquito-infested islands.)

In August, 1945, DDT was released for civilian use, and was it ever used! Dairy farmers, apple growers, cattle ranchers and housewives all used DDT to kill any and all insects. In the 1950s, the World Health Organization aggressively pushed for DDT spraying to eradicate malaria. This effort was extraordinarily effective; for example, in Sri Lanka the number of malaria cases dropped from 2.8 million in 1948 to just 17 in 1963! By this time, the

United States was using over 150 million pounds per year of DDT, and a nearly equal quantity of other chlorinated hydrocarbon bug killers.

Among all the enthusiasm for this miracle chemical came worrisome bits of news. Some scientists expressed concern that long-term exposure effects had never been tested. A few stories emerged about fish kills and the development of DDT-resistant insects. As early as 1948, when Paul Müller was awarded the Nobel Prize in Medicine for his discovery, he spoke about his concerns of the effect of DDT on ecosystems, especially with overuse. DDT started showing up in sites distant from where it had been sprayed—in fish-eating birds and cow's milk. In some places, insect-eating birds died for lack of food. In other places, DDT killed more beneficial than damaging insects, upsetting predator-prey relationships. Baby food makers were unable to get fruits and vegetables that were free of DDT residues.

DDT's best features turned out to be its worst. Its broad-spectrum activity meant that entire insect ecosystems were wiped out. Its chemical inertness meant that DDT remained in the environment for months and years after its initial application. Its insolubility in water meant that it was soluble in fats and oils, and accumulated in fatty tissue in fish, birds, and mammals.

Rachel Carson in her laboratory.
© Underwood & Underwood/Corbis

In 1962, Rachel Carson published her book, *Silent Spring*. Analyzing studies on DDT published over the previous 17 years, Carson claimed that DDT overusage was leading inexorably toward a silent spring, bereft of insects and birds. This extraordinary book moved the debate about DDT from the scientists' labs and conference halls to the public's living rooms and town halls. And in 1964, when evidence surfaced that peregrine falcons and bald eagles were dying off because of DDT-caused eggshell thinning, the tide dramatically turned. The public outcry against the once-popular insecticide grew to a loud roar. As one of its first acts, the Environmental Protection Agency (EPA), established in the United States in 1971, banned DDT. Anti-DDT feelings remain strong in the general public: recently a shipment of Zimbabwean tobacco was blocked from entering the United States because it contained traces of DDT. (One wonders whether tobacco or DDT poses the greater risk.)

Thirty years after the ban on DDT, the bald eagles and falcons have returned. Still, because of its incredibly slow degradation rates, it is estimated that about 1 billion pounds of DDT remain in the environment. Farmers have switched to other pesticides. But there have been adverse consequences brought on by the ban on DDT. Some pesticides developed as DDT replacements, such as parathion, have proved to be acutely toxic to humans. The ban on DDT spraying, along with the widespread and rapid development of insecticide-resistant bugs, has led to a rebound in mosquito populations and a stunning increase in mosquito-borne disease. Worldwide, 300 to 500 million people suffer from malaria every year and 1 to 2 million people die from the disease, mostly small children in Africa. Weakness and fevers brought on by the disease cripple the fragile economies of entire villages. The burgeoning mosquito population may be partly responsible for the spread of other pathogens such as West Nile virus. New chemical, biological, and ecological strategies to control the number of mosquitos without upsetting fragile ecosystems are under study. These include use of natural insecticides such as bacillus thurngiensis (BT) to selectively kill larvae, installation of pheromone lure traps, and techniques to encourage mosquito-eating predators such as bats.

Quick Quiz Answers

4.1 $\dot{\xi}_1 = \dot{\xi}_2 = \dot{\xi}_3 = 1.4/2.3 = 0.61$ gmol/h.

4.2   24 MMSCFD.

4.3   Hydrogen is limiting, 20% excess.

4.4 C_2H_4, 50%.

4.5   Because in Eq. (4.10), both ν_{A1} and $\Sigma \nu_{Ak} \dot{\xi}_k$ are negative, so selectivity is positive. In (4.11), ν_{A1} is negative, needs minus sign to make yield positive.

4.6 $\dot{\xi}_2 = 0$, so conversion decreases, selectivity increases, yield stays the same.

4.7 $K_a = y_B y_C / y_A^2$; no.

4.8   Must be between 0 and 1.

References and Recommended Readings

1. *Perry's Chemical Engineers' Handbook* and *Lange's Handbook of Chemistry* have extensive lists of values for Gibbs energy and enthalpy of formation. Appendix B of this text has a limited tabulation.

2. The new synthesis route to ibuprofen won several awards, including the 1993 Kirkpatrick Chemical Engineering Achievement Award and the 1997 Presidential Green Chemistry Challenge Award. The example is described more fully in *Real World Cases in Green Chemistry*, by M. C. Cann and M. E. Connelly (2000). Published by the American Chemical Society.

3. Controversy continues regarding the appropriate use of insecticides like DDT. One recent article published in the popular press is "What the World Needs Now Is DDT," Tina Rosenberg, *New York Times,* April 11, 2004. *Prometheans in the Lab: Chemistry and the Making of the Modern World,* by Sharon Bertsch McGrayne (McGraw-Hill, 2001), has more detail on Paul Müller, Rachel Carson, and the invention of DDT.

Chapter 4 Problems

Warm-Ups

P4.1 Explain why some reactor flow sheets must include purge.

P4.2 Make a list of at least five chemical reactions that you used today. Did the reactions involve a catalyst and/or high temperatures? Would you classify each reaction as reversible or irreversible?

P4.3 1 gmol propane (C_3H_8) is burned with 6 gmol oxygen to form carbon dioxide and water. Which reactant is limiting? What is the percent excess of the other reactant?

P4.4 A burner is designed to operate with 40% excess air for efficient combustion. If we are feeding it 60 L/min methane (CH_4), what air feed rate (in L/min) do you recommend? Air is 2l mol% O_2.

P4.5 100 gmol each of A and B are fed to a reactor where the reaction $2A + B \rightarrow C + D$ takes place. If $f_{CA} = 0.5$, what is f_{CB}?

P4.6 100 gmol each of A and B are fed to a reactor. Two reactions take place: $2A + B \rightarrow C + D$, and $A + C \rightarrow E$. C is the desired product and A is the expensive reactant. If $f_{CA} = 0.60$ and $f_{CB} = 0.25$, what is the yield and selectivity for producing C from A? Check your answer by confirming that yield = conversion $\times$ selectivity.

P4.7 In the manufacture of high-fructose corn syrup (used in enormous quantities in sodas, fruit-flavored drinks, and other beverages), glucose ($C_6H_{12}O_6$) is isomerized to fructose by using an enzyme as catalyst. Despite the fact that the reaction proceeds fairly rapidly, the maximum glucose conversion achievable is slightly less than 50%. Why? What does this tell you about the approximate value of K_a and $\Delta \hat{G}_r^\circ$ for this reaction?

Drills and Skills

P4.8 One tablet of Alka-Seltzer contains 324 mg aspirin ($C_9H_8O_4$), 1904 mg sodium bicarbonate ($NaHCO_3$), and 1000 mg citric acid ($C_3H_5O(COOH)_3$). What's the reaction between the last two ingredients that produces the fizz? (Hint: This reaction also produces a weak base.) Which is the limiting reactant? What volume of gas is produced from one tablet of Alka-Seltzer dissolved in a glass of water? How many milligrams of stomach acid (HCl) can be neutralized by one tablet?

P4.9   In the first step in the manufacture of the over-the-counter pain medication ibuprofen, isobutylbenzene ($C_{10}H_{14}$) and acetic anhydride ($C_4H_6O_3$)
react over a hydrogen fluoride catalyst to make isobutylacetophenone
($C_{12}H_{16}O$), with acetic acid (CH_3COOH) as the byproduct.

$$C_{10}H_{14} + C_4H_6O_3 \rightarrow C_{12}H_{16}O + CH_3COOH$$

1200 gmol isobutylbenzene/h and 1200 gmol acetic anhydride/h are fed to
a continuous-flow steady-state reactor. Chemical analysis of the exit
stream from the reactor reveals that the stream contains 42 mol% acetic
acid. What is the rate of consumption of isobutylbenzene inside the reactor? What is the flow rate of isobutylbenzene out of the reactor? What is
the fractional conversion?

P4.10 Your company is interested in building a plant to make ibuprofen, using
the new catalytic reaction scheme described in Example 4.1. Your job is
to measure the reaction rate with different catalysts, as a first step in
designing a full-scale reactor. In one experiment, you mix 134 g isobutylbenzene (IBB, $C_{10}H_{14}$) with 134 g acetic anhydride (AAn, $C_4H_6O_3$) in a
laboratory-scale batch reactor, adjust the temperature, add some catalyst,
and wait 1 h. At the end of the hour you stop the reaction, collect all the
material in the pot, and send it for chemical analysis. The chemist reports
back that the pot contains 27 g IBB, 52 g acetic anhydride, 121 g isobutylacetophenone (IBA, $C_{12}H_{16}O$), and 68 g acetic acid (AAc, CH_3COOH).
When you see these results you tell the chemist that the analysis is wrong.
He protests. Who's right?

P4.11 Blue light-emitting diodes can be manufactured from gallium nitride
(GaN) by a process called metalorganic chemical vapor deposition
(MOCVD). A 1 cm $\times$ 1 cm chip of Al_2O_3 is placed inside a laboratory-
scale MOCVD reactor to serve as an inert substrate. Trimethyl gallium
$[(CH_3)_3Ga]$ and ammonia (NH_3) are pumped continuously into the reactor, where the following reaction occurs:

$$(CH_3)_3Ga\ (g) + NH_3\ (g) \rightarrow GaN\ (s) + 3\ CH_4\ (g)$$

The solid GaN deposits on the Al_2O_3 chip in an even layer while the
methane gas flows continuously out of the reactor.

   Ammonia and trimethylgallium are fed to the reactor at a steady flow
rate of 10 μmol/h each. Only methane is present in the exit gas. What is
the rate of reaction? If GaN has a density of 6.1 g/cm^3 and a molar mass
of 84 g/gmol, how long will it take to build up a GaN layer 1 μm thick?

P4.12 Ethylene oxide is produced by partial oxidation of ethylene:

$$2\ C_2H_4 + O_2 \rightarrow 2\ C_2H_4O$$

Complete oxidation occurs as an undesired side reaction:

$$C_2H_4 + 3O_2 \rightarrow 2CO_2 + 2H_2O$$

Ethylene, oxygen, and nitrogen are placed in a batch reactor, with an initial composition of 10 mol% C_2H_4, 12 mol% O_2, and 78 mol% N_2. The reactor is heated. After some time, the reactor is cooled so that essentially all the water is condensed. The reactor is opened and the gases are sampled. All the oxygen is gone, and the ethylene concentration is 5.1 mol%. What is the fractional conversion of ethylene, and the yield of ethylene oxide from ethylene?

P4.13 Ethylene glycol ($HOCH_2CH_2OH$), used as an antifreeze, is produced by reacting ethylene oxide with water. A side reaction produces an undesireable dimer, DEG:

$$C_2H_4O + H_2O \rightarrow HOCH_2CH_2OH$$

$$HOCH_2CH_2OH + C_2H_4O \rightarrow HOCH_2CH_2OCH_2CH_2OH$$

The reactor feed is 10 gmol/min ethylene oxide and 30 gmol/min water. Why is water the excess reactant? If the fractional conversion of ethylene oxide is 0.92 and the selectivity is 0.85, what is the reactor outlet composition and the yield?

P4.14 In the case study in Chap. 2, we made diethanolamine from ethylene, oxygen, and ammonia. Now, let's consider just the first reactor, where ethylene oxide is produced. The reaction is

$$2\ C_2H_4 + O_2 \rightarrow 2\ C_2H_4O \tag{R1}$$

There is an unwanted side reaction—combustion of ethylene to CO_2 and water:

$$C_2H_4 + 3\ O_2 \rightarrow 2\ CO_2 + 2\ H_2O \tag{R2}$$

Air (79 mol% N_2, 21 mol% O_2) is fed as the source of oxygen. The C_2H_4:O_2 molar ratio in the feed is 2:1. 25% of the C_2H_4 and all of the oxygen fed to the reactor is converted to products.

We want to produce 2000 kgmol/h ethylene oxide. What's the reactor feed rate and feed composition?

P4.15 100 gmol/min of a solution of 70 mol% ethanol/30 mol% water is fed to a reactor operating at steady state, along with 80 gmol/min of air (79 mol% N_2, 21 mol% O_2). Ethanol (C_2H_5OH) reacts with oxygen to make acetaldehyde (CH_3CHO). Acetaldehyde is further oxidized to acetic acid (CH_3COOH). (Similar reactions take place in your stomach.) Write the two stoichiometrically balanced chemical equations. What is the byproduct of the reactions? What is the limiting reactant? If there is 100% conversion of the limiting reactant and the production rate of acetaldehyde is 25 gmol/min, calculate the fractional conversion of the excess reactant, the yield of acetaldehyde from ethanol, and the composition and flow rate of the reactor effluent stream.

P4.16 "Buckyballs" are large molecular cages made of 60 or 70 carbon atoms. Buckyballs may be useful as molecular carriers of metals such as iron—which are called bucky ferrocenes. In a bucky ferrocene, the iron is sandwiched between cyclopentadienyl rings (C_5H_5, denoted as Cp) attached to the buckyball. One research group made bucky ferrocenes by methylation of 10 g buckyball (C_{60}) to give $C_{60}(CH_3)_5H$ in 95% yield, then heated this compound with $Fe(CpCO)_2$, producing "bucky ferrocene", $Fe[C_{60}(CH_3)_5]Cp$ in 52% yield. What mass of bucky ferrocene was produced by this method? (*J. Amer. Chem. S.* vol. 124, pp. 9354–9355, 2002.)

P4.17 A synthetic fiber plant uses a variety of volatile solvents. An analysis of waste solvent vapors from one plant showed that the gas contained 20 mol% CH_4, 20 mol% CS_2, 10 mol% SO_2 and 50 mol% H_2O. It has been proposed that the gas be combusted and then emitted to the air. Local air pollution regulations limit the SO_2 concentration in stack gases to no more than 2 mol% on a dry basis. What is the minimum percent excess air that must be used to meet these requirements, assuming complete combustion? What is the composition of the combustion gases? Assume air contains 21 mol% oxygen and 79 mol% nitrogen.

P4.18 A local brewery operation has offered to donate, free of charge, spent grains from its operations, to a heating plant at a college campus. As the heating plant engineer, your job is to see whether it is technically feasible to use this material. Chemical analysis of the grains gives the following elemental composition: 14.4 wt% C, 6.2 wt% H, 78.8 wt% O, 0.6 wt% S. You placed 1.00 kg of spent grains in a laboratory reactor, sealed it, and then heated it. Exit gases were continuously vented until no material was left in the reactor. Only CO, CO_2, SO_2 and H_2O were detected in the exit gas. How many standard cubic feet of gas were produced, and what was the gas composition? In the heating plant, the grains would be mixed with air and burned. Per kilogram of grain burned, how much air would you feed? If the maximum SO_2 content is limited to 0.5 mol% for environmental reasons, would you recommend accepting this donation? (Assume the reactor was totally empty before you placed the grains in it.)

P4.19 Waste streams from various processes in a complex chemical facility are combined and burned for heat, because the cost of purifying and recycling the components is prohibitive. On a typical day, the combined waste stream contains 2 mol% acetone $(CH_3)_2CO$, 14 mol% acetic anhydride $(CH_3CO)_2O$, 36 mol% methane CH_4, 20 mol% ethane C_2H_6, 13 mol% N_2, 4 mol% O_2, 8 mol% CO, and 3 mol% H_2S. The waste stream is fed to a furnace along with 20% excess air and combusted. (For combustion reactions, excess air is calculated on the assumption that the only products of combustion are CO_2, H_2O, N_2, and SO_2.) Analysis of the exit flue gas gives a CO_2/CO mass ratio of 4:1. The only other components in the flue gas are N_2, O_2, H_2O, and SO_2. Calculate the composition of the exit flue gas. Assume air is 21 mol% O_2 and 79 mol% N_2.

P4.20 Your neighbors are concerned about the operation of their gas furnace. A handyman came to their door with an offer to check the performance of their furnace at no expense. The handyman explained that if the CO_2

content of the gas leaving the chimney is above 15%, the situation is dangerous to their health, violates city codes, and can cause chimney rot. He carefully took a sample of the gas leaving the chimney and reported that it contained 30% CO_2 on a dry basis (with water not included). He has offered to arrange for the purchase and installation of a safe new high-efficiency furnace at a bargain price. Your neighbors are burning natural gas. What is your estimate of the mole percentage of CO_2 in the chimney gas (dry basis)? What will you tell your neighbors regarding the handyman proposal?

P4.21 Bis-uwiguwi (BUG) is a brand-new and exciting chemical, possessing remarkable insecticide properties, yet biodegradable and nontoxic to birds, fish, or mammals. Your company is interested in manufacturing BUG, using a newly discovered synthesis scheme involving six reaction steps. The scheme is top secret because the patent hasn't been issued yet; all you are told is that it requires several reactants, which are labeled A, B, etc. Each reactor runs under different conditions, so six reactors in series are required. The reaction scheme, along with preliminary estimates of fractional conversions achievable, is given below.

Reaction	Fractional conversion
$A + B \rightarrow C + H_2O$	0.92
$C + HCN \rightarrow D + CO_2$	0.84
$D + 2NaOH \rightarrow E + 2H_2O$	0.97
$E + 2HNO_3 \rightarrow G + H_2O$	0.95
$G + 2F \rightarrow J + 2NaNO_3$	0.97
$J + K \rightarrow BUG + 2H_2O$	0.99

You are in charge of evaluating process economics and waste production for this synthesis scheme. Per 100 kgmol BUG produced, calculate the moles of each reactant required and each waste product generated. Assume all reactants are fed at stoichiometric ratio.

P4.22 You are in charge of designing a chemical process that converts raw material A to product B. The raw material A is contaminated with an inert I. Only 15% conversion of A to B is achieved in the reactor, so the unreacted A (plus I) is separated from the product; some is recycled back and some is purged. One of your jobs as process designer is to determine the optimum recycle:purge ratio (total molar flow rate in recycle stream/total molar flow rate in purge stream). Compare the parameters listed below, and specify whether the parameter value would be *high*, *low*, or *same* at high recycle:purge ratio and at low recycle:purge ratio. Assume the rate of production of B is constant in either case. The first one is done for you.

	High recycle:purge ratio	**Low recycle:purge ratio**
Production rate of B	Same	Same
Fresh feed rate		
% A in purge		
% I in feed to reactor		
Reactor volume required		
Equipment costs		
Raw materials costs		

P4.23 Insulin is a protein hormone that helps to regulate blood glucose. Normally, insulin is produced in the pancreas and secreted into the blood stream as needed. People with some forms of diabetes do not produce sufficient quantities of insulin and must inject insulin daily. Although once made by purification from pig pancreas, now virtually all therapeutic insulin is made by using recombinant DNA technology. *E. coli* bacteria are genetically engineered to produce a fast-acting insulin. The bacteria are grown in a 10.0 L flask, operated in a batch mode. Initially, 20 g of bacteria and 10.0 L broth are loaded into the flask, and the fermentation starts. The mass of *E. coli* in the flask at any time, $m_{b,sys}$, is described by the equation:

$$m_{b,sys} = m_{b,sys,0}e^{0.17t}$$

with t in hours. Insulin production rate is proportional to the mass of bacteria in the flask:

$$\dot{r}_{ins} = 16m_{b,sys}$$

with $\dot{r}_{ins}$ in units of mg insulin/h. The fermentation is stopped when the bacterial mass reaches 500 g. For how long does the fermentation run? What is the total amount of insulin produced?

P4.24 In many drug discovery operations, high-throughput robotic workstations are used. Hundreds or even thousands of reactants are tested in small volume wells. Microwave technology may be used to speed up synthesis rates in the drug discovery business. The basic idea is that by heating solutions of reactants with microwaves rather than with external heaters, there is uniform temperature throughout the sample, which increases the rate of reaction. Uniform heating should also improve selectivity and control of batch production-scale reactors. A Swedish company markets specialized microwave technology for this purpose. It reports for example, the synthesis of a compound in 4 min with 90% yield, compared to 54% yield and 48 h with conventional technology. Use these data to calculate the percentage increase in the net rate of reaction of the desired compounds. (Only professional microwaves are used. Do not attempt the experiment at home; kitchen microwaves cycle and power output is not tightly controlled.)

P4.25 American settlers made soap by boiling potash (a mix of sodium and potassium hydroxides and carbonates left over after burning brush) with animal fat in a pan. Modern large-scale continuous soap-making processes use a very similar chemistry: sodium hydroxide (NaOH) reacts with fatty acids produced from fats (beef tallow and coconut oil are the most common fat sources) in a process called *saponification.*

Fatty acids have the general molecular formula RCOOH, where R is a long hydrocarbon chain. The saponification reaction is

$$RCOOH + NaOH \rightarrow RCOONa + H_2O$$

Beef tallow produces a mix of the following fatty acids:

Fatty acid	Molecular formula	Molecular weight	wt%
Palmitic acid	$CH_3(CH_2)_{14}COOH$	256	32
Stearic acid	$CH_3(CH_2)_{16}COOH$	284	26
Oleic acid	$CH_3(CH_2)_7CH{=}CH(CH_2)_7COOH$	282	42

Inexpensive bars of soap are produced by mixing a 24 wt% NaOH aqueous solution with tallow-derived fatty acids. Inside the saponifier, all the fatty acid is converted to its sodium salt, and the water content is adjusted by adding or removing water. The soap product contains 12 wt% water. If we want to make one metric ton of soap per day, how many kilograms of tallow fatty acid do we need? How much water (kg/day) needs to be added or removed?

P4.26 Hydrodealkylation is a process in which side chain alkyl groups (like methyl) are removed from aromatics by reaction with hydrogen. The following reactions take place in the reactor:

$$\underset{\text{toluene}}{C_6H_5CH_3} + H_2 \rightarrow \underset{\text{benzene}}{C_6H_6} + CH_4 \qquad \text{(R1)}$$

$$\underset{\text{xylene}}{C_6H_4(CH_3)_2} + H_2 \rightarrow \underset{\text{toluene}}{C_6H_5CH_3} + CH_4 \qquad \text{(R2)}$$

$$\underset{\text{pseudocumene}}{C_6H_3(CH_3)_3} + H_2 \rightarrow \underset{\text{xylene}}{C_6H_4(CH_3)_2} + CH_4 \qquad \text{(R3)}$$

In addition, an unwanted side reaction occurs, in which two benzenes react to form diphenyl:

$$2C_6H_6 \rightarrow C_6H_5C_6H_5 + H_2 \qquad\qquad \text{(R4)}$$
$$\text{benzene} \quad \text{diphenyl}$$

A process stream containing 5 mol% benzene (C_6H_6), 20 mol% toluene ($C_6H_5CH_3$), 35 mol% xylene [$C_6H_4(CH_3)_2$], and 40 mol% pseudocumene [$C_6H_3(CH_3)_3$] is fed at a rate of 100 gmol/h to a hydrodealkylation plant. This process stream is mixed with 500 gmol H_2/h before being fed to the reactor. Hydrogen and methane are separated from the remaining compounds (lumped together as aromatics). The aromatics are analyzed and found to contain 28 mol% pseudocumene, 1 mol% diphenyl, 19% benzene, and 1% toluene. Calculate the extents of reaction and fractional conversions of all reactants, the methane production rate, and the mol% methane in the gas stream.

P4.27 Air is mixed with fresh and recycled methanol and fed to a reactor, where formaldehyde (HCHO) is produced by partial oxidation of methanol (CH_3OH). Several side reactions also occur, producing formic acid (HCOOH), CO, CO_2, and H_2O. The reactor inlet stream is 35 mol% CH_3OH, and the fractional conversion of methanol in the reactor is 0.8. The reactor products are sent to a series of separators, which produce pure methanol for recycle, a liquid stream containing HCHO, HCOOH, and H_2O, and an offgas containing 20.2 mol% H_2, 4.8 mol% CO_2, 0.2 mol% CO, 74.5 mol% N_2, and 0.3 mol% O_2. Calculate the moles of formaldehyde produced per mole of fresh methanol fed to the process, assuming steady-state operation.

P4.28 Chlorine dioxide is used to bleach pulp in the paper industry. The gas is produced by the following reaction:

$$6NaClO_3 + 6H_2SO_4 + CH_3OH \rightarrow 6ClO_2 + 6NaHSO_4 + CO_2 + 5H_2O$$

3000 kgmol/h of an equimolar mix of $NaClO_3$ and H_2SO_4 are mixed with 200 kgmol/h CH_3OH in a lead-lined reactor. 90% conversion of limiting reactant is achieved. What is the composition of the reactor outlet?

P4.29 Catalyst performance is usually tested in laboratory or pilot-plant reactors. In one particular run, a supported platinum/tin catalyst was tested for its effectiveness at dehydrogenating light hydrocarbons, particularly isobutane. Data from one day were taken as follows:

Reactor temperature: 602°C
Reactor pressure: 768 torr

Inlet gas flow rate: 51.38 SCCM (standard cubic centimeters per minute)
Outlet gas flow rate: 60.59 SCCM
Inlet gas composition (mole percent): 0.086% propane, 32.9% isobutane, 0.068% *n*-butane, and 66.9% hydrogen

Outlet gas composition (mole percent): 0.35% methane, 0.037% ethane, 0.034% propane, 13.5% isobutane, 13.8% isobutene, 0.20% *n*-butane, 0.0375% cis-butene, 0.045% trans-butene, and 71.6% hydrogen.

Complete balances on carbon and hydrogen to check the reliability of the data. Calculate the fractional conversion of isobutane to products. Calculate the moles of isobutylene per mole of isobutane reacted. If the yield is different from 100%, explain why.

P4.30 100 gmol each of A and B are placed in a reactor where the gas-phase reaction 2A + B → C + D takes place. The reaction reaches equilibrium. Write an expression for K_a in terms of the mole fractions of each reactant and product and the total pressure. Then simplify so that you have an expression for K_a in terms of the fractional conversion of A and the total pressure. Will increasing the reactor pressure increase or decrease the fractional conversion of A?

P4.31 Animal cells use glucose ($C_6H_{12}O_6$) as their predominant energy source. Aerobic metabolism of glucose leads to complete oxidation of glucose to carbon dioxide and water. (*Aerobic* means that the metabolism requires oxygen from air.)

$$C_6H_{12}O_6 + 6O_2 \rightarrow 6CO_2 + 6H_2O$$

What is ΔG_r° of glucose oxidation?

This reaction is coupled to the synthesis of ATP (adenosine triphosphate) from inorganic phosphate and ADP (adenosine diphosphate). ΔG_r° of this reaction is +7.3 kcal/gmol.

$$ADP + H_3PO_4 \rightarrow ATP$$

If 6 moles of ADP are converted to ATP for every mole of oxygen consumed by aerobic glucose, what is the efficiency of conversion of the chemical energy of glucose into chemical energy of the phosophamide bond in ATP?

P4.32 MTBE (methyl tert-butyl ether) was mandated by the U.S. Clean Air Act to be added to gasoline in some urban areas with smog problems. MTBE helps gasoline burn cleanly, thus reducing smog. But, MTBE is highly soluble in water and contaminates groundwater. Dimethyl carbonate (DMC) has been proposed as an alternative to MTBE; reportedly it is even better than MTBE at helping gasoline burn cleanly. The problem is that DMC is made with highly toxic phosgene, and manufacture of DMC generates chlorinated by-products. Many other reaction pathways have been proposed; some were discussed in Example 4.16 and two more are presented here:

Scheme 1: React methanol, CO_2 and ethylene oxide to DMC and formaldehyde (CH_2O)

Scheme 2: React dimethylether (CH_3OCH_3) with water to make methanol, then react methanol with CO_2 to make DMC and water.

Evaluate each scheme considering atom utilization, chemical reaction equilibria, and costs. Are either of these schemes promising compared to those discussed in Example 4.16?

P4.33 A mixture of 30 mol% CO, 65 mol% H_2, and 5 mol% N_2 is fed to a methanol (CH_3OH) synthesis reactor, where the following reaction occurs:

$$CO + 2H_2 \rightleftarrows CH_3OH$$

The reactor is at 200°C and 4925 kPa. The stream leaving the reactor is at equilibrium.

If 100 kmol/h of the feed mixture is fed to the reactor, calculate the flow rates of all species leaving the reactor.

P4.34 Steel (Fe) sits out in the air and slowly rusts. What is the most likely product of oxidation of Fe: FeO, Fe_3O_4, or Fe_2O_3? (Use chemical equilibrium to determine.)

P4.35 Atrazine is a heavily used herbicide, particularly in the Midwest U.S. corn belt. Typically, atrazine is applied to corn fields about 18–30 days after planting, at 0.75 lbs/acre. The herbicide degrades slowly in soil by chemical hydrolysis and microbial activity. If the rate of atrazine degradation in soil is $\dot{R}_{deg} = 0.012m$, where $\dot{R}_{deg}$ is in units of lbs atrazine degraded per acre per day and m is lbs of atrazine per acre at any time t, about how many days will it take for the atrazine in the soil to decay to one tenth of its original value?

Scrimmage

P4.36 Refer back to Example Problems 4.12 and 4.13. You are now trying to build a process to make 1200 kgmol/h acetaldehyde. Two catalysts are under study. One of the catalysts was described in the example problem. With the other catalyst, oxidation of ethanol to carbon dioxide occurs, but none of the acetaldehyde reacts to form acetic acid. If the ethanol:oxygen feed ratio is 8:3, a fractional conversion of 0.25 and a selectivity of 0.6 is achieved. Evaluate this catalyst by calculating flow rates in and out of the reactor. Write a memo describing your results and explain what criteria you would consider in deciding between this catalyst and the one described in the example problem.

P4.37 Methanol (CH_3OH) reacts to form formaldehyde (HCHO) either by decomposition to formaldehyde and hydrogen (H_2) or by oxidation to form formaldehyde and water (H_2O):

$$CH_3OH \rightarrow HCHO + H_2$$
$$CH_3OH + \tfrac{1}{2}O_2 \rightarrow HCHO + H_2O$$

A mixture containing 99 mol% methanol and 1 mol% of an inert contaminant is available as feed to a formaldehyde plant. 1000 kgmol/h of this mixture, along with 200 kgmol/h O_2, are fed to a process as fresh feed. The fresh feed is mixed with a recycle stream and fed to a reactor. The fractional conversion of methanol achieved in the reactor is 25%. All of the oxygen is consumed in the reactor. The reactor output is sent to a separation unit, where formaldehyde, water, and hydrogen are removed, and methanol and the contaminant are recycled to the reactor feed. To control the contaminant level in the reactor, a purge stream is taken off the recycle stream. The maximum contaminant allowed in the recycle stream is 10 moles contaminant/100 moles methanol. Determine (a) the overall fractional conversion of methanol to products, (b) the production rate of formaldehyde, (c) the purge:recycle ratio, and (d) the recycle:fresh methanol feed ratio.

P4.38 Amines are derivatives of ammonia; they are added to shampoo to make it foam, used as a building block in carpet fibers, and cause the stench of rotting fish. Industrially, amines are produced from alcohols and ammonia over solid catalysts, with water as a by-product. Alcohols in turn are produced from alkene hydrocarbons. A synthetic route direct from alkenes to amines would avoid the cost of producing the alchohol intermediate.

Your boss is very enthusiastic about reports (*Science,* vol. 297, p. 1677, 2002) of a new catalyst that might be able to achieve this synthetic route by catalyzing two reactions in the same pot. For example, butene (C_4H_8), CO, H_2, and dimethylamine [$(CH_3)_2NH$] react as shown in (R1) and (R2):

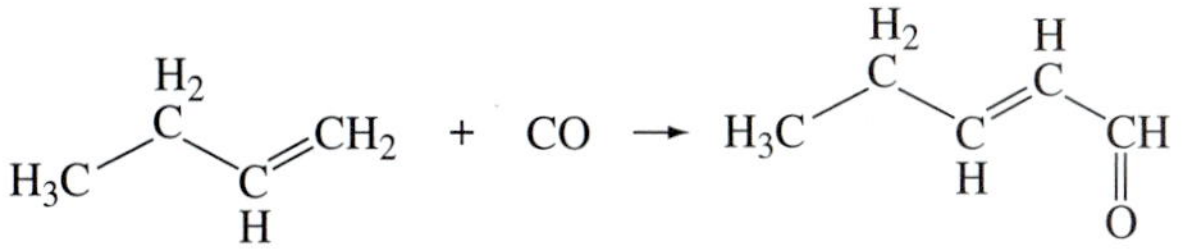

(R1)

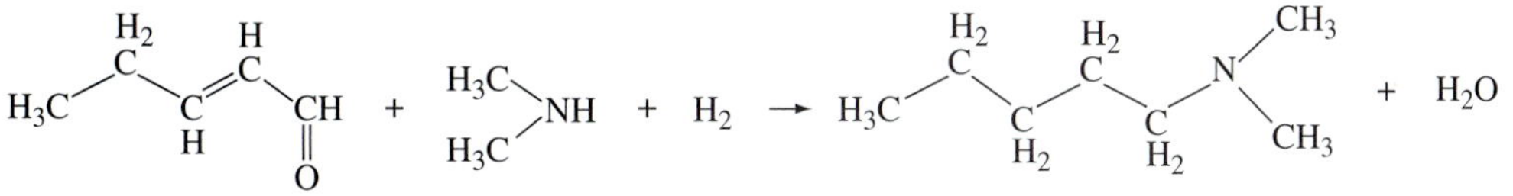

(R2)

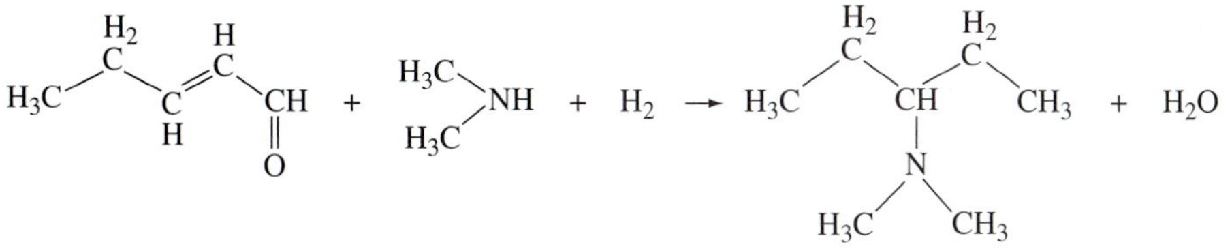

(R3)

(R2) produces what is called a linear amine. Unwanted side reactions also occur, producing branched amines; a typical unwanted reaction is (R3)

There are several minor unwanted reactions that produce other by-products.

Here are some data from experiments carried out in a laboratory-scale batch reactor. CO and H_2 were supplied in great excess. The butene and dimethylamine were dissolved in an organic solvent. In all cases, 15 mmol dimethylamine was added to the reactor initially along with the butene as listed in the table.

Temperature, °C	Moles butene/mole dimethylamine initially	Fractional conversion of dimethylamine, %	Selectivity total amines, %	Selectivity, linear amine, %
100	1.2:1	100	54	52
120	1.2:1	100	77	60
140	1.2:1	100	98	65

For each case, calculate the millimoles of butene, dimethylamine, total amines, and linear amines in the reactor at the end of the experiment. Which operating temperature would you choose? (Consider the entire process flow diagram.) Explain your answer.

P4.39 Hydrogen reacts with iron oxide (Fe_2O_3) to produce metallic iron (Fe), with water as a byproduct. 100% conversion of Fe_2O_3 is achieved, and the metallic iron is easily separated from the hydrogen-water vapor mixture. The water is condensed, and the hydrogen is recycled. The hydrogen source available onsite is contaminated with 1 mol% CO_2, which is inert. Sketch out a process for converting 1 ton/h of Fe_2O_3 to Fe. Assume that the recycle gas:fresh gas feed ratio is 5:1, and the maximum CO_2 concentration allowed in the reactor is 3.5 mol%. Calculate all process flows.

P4.40 A soap manufacturer is considering converting its excess glycerine by-product to hydrogen for use in fuel cells, using a new catalytic process. The process operates the following reaction:

$$C_3H_8O_3 + 3\ H_2O \rightleftarrows 3\ CO_2 + 7\ H_2$$

The reactor operates in the gas phase, at 1.2 atm and 200°C, with a feed ratio of water:glycerine of 5:1 (molar units). The glycerine feed rate in the pilot plant reactor is 150 g/h, and an exit stream hydrogen mole fraction of 0.54 is measured. Find the fractional conversion of glycerine under these conditions.

You are told that the equilibrium constant for this reaction at 200°C is $K_a = 55$ atm⁶. Is the reactor achieving equilibrium? Suggest two ways to do better. What is the mole fraction of H_2 in the exit and the fractional conversion of glycerine if the reactor outlet is at equilibrium?

P4.41 The lactic acid byproduct recovered from cheese plants can be used to make a variety of chemicals. For example, it may be hydrogenated to produce 1,2-propanediol, which is in turn used as a polymer precursor and as a food additive:

$$CH_3CHOHCOOH + 2H_2 \rightleftarrows CH_3CHOHCH_2OH + H_2O$$

The process feed is a vapor stream containg 1:3:2 lactic acid:H_2:inert (mole ratios), at a total pressure of 5 atm and a flow rate of 300 mol/h. The reaction is reversible, with an equilibrium constant $K_a = 3.2$ atm^{-1} at the reactor temperature. What is the fractional conversion of the limiting reactant at equilibrium, and the reactor outlet composition? Will increasing the inert concentration in the feed increase or decrease the equilibrium fractional conversion? First give your answer by using qualitative reasoning, then calculate the conversion at a higher inert concentration and see if your reasoning was right. Would changing the reactor pressure change the conversion? Prove your results by calculating the effect of an increase in pressure by 2 atm.

P4.42 One step in making synthetic detergents is preparation of butanal (also called butyraldehyde) from propene (also called propylene), CO, and hydrogen:

$$CH_3CHCH_2 + CO + H_2 \rightleftarrows CH_3CH_2CH_2CHO$$

The reaction takes place in a gas-phase reactor at 5 atm pressure. Feed flow rate to the reactor is 1200 gmol/h. The reaction is reversible, with an equilibrium constant of 8 atm^{-2} at the operating temperature of 180°C.

First analyze the case where the feed is 20 mol% propylene, 40 mol% CO, and 40 mol% H_2. Determine the fractional conversion of propylene, the butanal production rate, and the percent butanal in reactor effluent. Then, repeat these calculations at several propene concentrations from 10 to 80 mol%. In all cases, CO and H_2 are the remainder of the feed, and are fed at 1:1 mole ratio. Plot your results and examine the trends. What propene feed percent allows for the maximum conversion? What propene feed percent provides maximum butanal production rate? Do you think one of these is optimum, or would you choose a different propene feed percent? Explain your reasoning.

P4.43 The water-gas shift reaction

$$CO + H_2O \rightleftarrows CO_2 + H_2$$

reaches equilibrium quickly. Calculate K_a versus T in the range of 100 to 1000 °C. Assuming that CO and H_2O are fed to a reactor at a stoichiometric ratio, calculate the mole fractions of CO and H_2 in the reactor outlet as a function of temperature in this range, if the reactor outlet is at equilibrium. Plot your results (y_{CO} versus T, y_{H_2} versus T). What reactor temperature would you pick if you wanted the H_2:CO ratio out of the reactor to be 2:1?

P4.44 One method of producing ethanol is the vapor-phase hydration of ethylene

$$C_2H_4 + H_2O \rightleftarrows C_2H_5OH$$

Assume that ethylene and water are fed to a reactor at equimolar ratio. Derive a general expression for the moles of ethanol produced as a function of temperature and pressure at equilibrium. Use this expression to generate plots of ethanol production from 150 to 300°C at 1 atm pressure, and from 1 to 50 atm at 150°C. Comment on your results in terms of optimizing reactor operating conditions.

P4.45 Isobutene (C_4H_8) and isobutene (C_4H_{10}) combine to make an ingredient in high-octane gasoline:

$$C_4H_8 + C_4H_{10} \rightleftarrows C_8H_{18} \text{ (iso-octane)}$$

The reaction equilibrium constant is $K_a = 1.8$ atm^{-1} at the reaction temperature. If the feed to the reactor is 50 gmol/min isobutene and 150 gmol/min isobutane, and the reactor operates at 4 atm pressure, what is the fractional conversion of the limiting reactant at equilibrium? What is the flow rate and composition of the reactor outlet stream if equilbrium is reached? How would you adjust the isobutene:isobutane ratio and/or the reactor pressure to achieve (a) higher fractional conversion of isobutene or (b) higher mole fraction iso-octane in the product stream?

P4.46 Ammonium nitrate is used as a fertilizer but, mixed with a bit of fuel oil, it can explode. Explosive decomposition of ammonium nitrate was the cause of a serious accident in Texas City, Texas, as well as the Oklahoma City bombing. Several decomposition reactions are proposed to occur in an explosion:

$$NH_4NO_3 \rightleftarrows N_2O + H_2O$$
$$NH_4NO_3 \rightleftarrows N_2 + H_2O + O_2$$
$$N_2O \rightleftarrows N_2 + O_2$$

Balance the reactions. Are these three reactions independent?

Pure ammonium nitrate is a solid below 169.6°C. What is the equilibrium mixture at 100°C? 250°C? Why do you think fuel oil is added to ammonium nitrate to make it explosive?

P4.47 Salmonella in raw eggs is a major source of food poisoning. The idea of pasteurizing eggs in the shell to reduce salmonella has been considered, but the trick is to kill the bacteria without cooking the eggs. One of the key chemical processes that occurs when eggs cook is denaturation and coagulation of proteins. At temperatures below 60°C, egg proteins do not significantly denature or coagulate, so one group of researchers proposed that eggs be pasteurized by exposure to hot air at 55°C. (*Food*

Microbiology 1996, vol. 13, pp. 93–101). To evaluate this idea, you inoculate eggs with salmonella bacteria, hold the eggs at 55°C for various time intervals, then measure the number of bacteria remaining in the eggs. (Bacteria are quantified as colony-forming units per milliliter, or cfu/mL.) Use these data to estimate a rate constant k_f, assuming first-order degradation kinetics.

0 min	30 min	90 min	130 min
2.0×10^6 cfu/ml	1.4×10^5 cfu/ml	9.8×10^2 cfu/ml	10 cfu/min

P4.48 Broth used in fermentation processes must be sterilized before use. About 10 L of broth are used in each batch. You've decided to use heat sterilization and need to pick the optimum time and temperature. 10 L of broth contain about 22,000 spores before sterilization. The death rate for the unwanted bacterial spores is

$$\dot{r}_{death} = 3 \times 10^9 e^{-9000/T} \times N_{spores}$$

where N_{spores} is the number of living spores at any time t and $\dot{r}_{death}$ is the number of spores killed per minute.

An expensive nutrient present in the broth at 5 mg/L is destroyed by heat at a rate of

$$\dot{r}_{destruction} = 5e^{-2000/T} \times m_{nutrient}$$

where $m_{nutrient}$ is the milligrams of nutrient at any time t and $\dot{r}_{destruction}$ is the milligrams of nutrient destroyed per minute. T is in K.

Your goal is to produce spore death while minimizing nutrient destruction. First plot the rates of death and destruction at different temperatures from 298 to 500 K. Then determine an appropriate sterilization temperature and time if your goal is 99.999% spore death, and an acceptable level of destruction of nutrient is (a) 50% or (b) 5%.

Suppose the kind of spores contaminating the broth change and now have a death rate of

$$\dot{r}_{death} = 10^{40} e^{-45000/T} \times N_{spores}$$

If you were unaware of this change and continued to operate under the chosen design conditions, what percentage death would you achieve? Would this be a worry?

P4.49 PET, the polymer used to make plastic 2-L soda bottles, is made by polymerization of terephthalic acid (TA; $HOOC\text{-}C_6H_4\text{-}COOH$). In a synthesis of TA carried out in your company, *p*-xylene ($CH_3C_6H_4CH_3$) is dissolved in acetic acid (CH_3COOH) at 10 wt%, and 20% excess air is bubbled through the mixture. The reaction takes place over a proprietary solid catalyst; conversions and selectivities approaching 100% to the desired product is achieved. However, about 10% of the acetic acid is oxidized and the reaction is very energy-intensive, so the search is on to discover a new process. One day Connie Chemist drops by your office with exciting results. She's discovered a process in which hydrogen peroxide (H_2O_2) and *p*-xylene are dissolved in supercritical water using a very dilute, very inexpensive salt as the catalyst. The hydrogen peroxide decomposes to O_2 and H_2O, and the O_2 reacts with the *p*-xylene to form TA. Carried out in a minireactor and using reactants at stoichiometric ratio, Connie produced about 10 g TA per hour, with yields better than 70% and selectivity better than 90%.

Evaluate the new process vis-à-vis the conventional process. First compare the atom economy of the two reaction schemes, assuming no unwanted side reactions. Sketch out simplified block flow diagrams, and complete process flow calculations for each process, using the available data and a basis of 100 kg/day TA production. Would you pursue Connie's idea further? Why or why not?

P4.50 Ammonia (NH_3) and methanol (CH_3OH) react over a catalyst to produce methylamine (CH_3NH_2), which is a useful intermediate in the production of some polymers and pharmaceuticals. Methylamine can react further over the same catalyst to produce dimethylamine ($CH_3)_2NH$, for which there is a limited market as a specialty solvent. Water is a byproduct of both reactions.

You work in a process research lab, and you are testing a new proprietary catalyst for methylamine production. In one experiment, you feed into a laboratory-scale reactor 100 gmol methanol/h and 100 gmol ammonia/h into the reactor. You collect the reactor effluent and send it off for analysis. The results: the effluent contains 20.0 mol% ammonia, 9.4 mol% methanol, 18.6 mol% methylamine, 10.9 mol% dimethylamine, and the remainder water.

Write down the two stoichiometrically balanced equations: (R1) for making methylamine from methanol and ammonia and (R2) for making dimethylamine from methylamine and methanol. Use element balances to see if the laboratory analysis is reasonable. From the experimental data, calculate (a) the fractional conversion of methanol, (b) the yield of methylamine based on methanol, (c) the selectivity for methylamine based on methanol, (d) the reaction rate $\dot{r}_1$ (gmol/h), and (e) the reaction rate $\dot{r}_2$ (gmol/h).

The equilibrium constant for the methylamine production reaction (R1) is $K_1 = 4.0$, and for dimethylamine production from methylamine (R2) is $K_2 = 2.5$. Is the effluent from the laboratory reactor at equilibrium?

Before installing a full-scale reactor, operating parameters must be optimized. Reactor optimization may be based on maximizing different parameters, depending on raw material costs, reactor capital and operating costs, value of product and by-products, difficulty of separation, and other factors. Determine what the optimum ammonia: methanol feed ratio is if you want to maximize (a) concentration of methylamine in the reactor outlet, (b) ratio of methylamine to dimethylamine, (c) yield of methylamine based on ammonia, or (d) yield of methylamine based on methanol. For these calculations, assume the reactor operates at equilibrium. Can you make any general observations?

P4.51 Formaldehyde (HCHO) is produced by partial oxidation of methanol (CH_3OH). Several side reactions also occur, producing formic acid (HCOOH), CO, CO_2, and H_2O.

In one working process, air (21 mol% O_2, 79 mol% N_2) is mixed with fresh and recycled methanol and fed to a reactor. Your job is to evaluate how well this process is working. You take samples at several points in the process and find that (a) the reactor inlet stream is 35 mol% CH_3OH, (b) the recycled methanol is pure, (c) the offgas contains 10.9 mol% H_2, 6.0 mol% CO_2, 0.3 mol% CO, 81.9 mol% N_2, and 0.9 mol% O_2, and (d) the liquid product stream contains 30.3 mol% HCHO, plus HCOOH and H_2O.

Use a DOF analysis to determine if you've got enough stream compositions analyzed to completely describe the process operation. Calculate the flow rate through the reactor, and the production rate for the two valuable products, formaldehyde and formic acid, per 100 moles methanol fed to the process. Write down a set of independent, stoichiometrically balanced chemical equations that completely describe all the reactions taking place in the reactor. Finally, if formaldehyde is a more valuable product than formic acid, how might you consider adjusting the process operation?

P4.52 Ethane (C_2H_6) is cheap and readily available, but not very reactive. Ethane can be dehydrogenated to ethylene (C_2H_4), which is valuable as a raw material for polymer manufacture. If the dehydrogenation goes further, acetylene (C_2H_2) is made. Acetylene is highly reactive and an explosion hazard.

A reactor is fed with ethane and steam at a 1:1 mole ratio. (The steam simply acts as a diluent.) If the reactor runs at 1000 K and 1 atm and the reaction reaches equilibrium, find the yield and selectivity for ethylene. Then, determine yield and selectivity as a function of reactor temperature, from 800 to 1400 K, and plot your results. Finally, consider adjusting the reactor pressure between 0.25 to 4 atm, or the ethane:steam feed ratio between 3:1 to 1:3. What conditions give the best yield? The best selectivity? Considering also safety and separation problems downstream from the reactor, what conditions of T, P, and ethane:steam ratio would you choose?

P4.53 Propylene (C_3H_6) and chlorine (Cl_2) are converted to allyl chloride (C_3H_5Cl) and hydrogen chloride (HCl). Besides allyl chloride, several chlorinated byproducts are generated, of which the major unwanted byproduct is 1,3 dichloropropane ($C_3H_6Cl_2$).

The major reactions are:

$$C_3H_6 + 0.5\,Cl_2 \rightarrow C_3H_5Cl + HCl \tag{R1}$$

$$C_3H_6 + Cl_2 \rightarrow C_3H_5Cl \tag{R2}$$

The reaction rate constants are in units of $\dfrac{\sqrt{\text{gmol}}}{L\,s}$, with T in K:

$$k_1 = 1.4 \times 10^8 e^{-10,000/T}$$
$$k_2 = 1.4 \times 10^{10} e^{-15,000/T}$$

If you were asked about the best reactor temperature, what would you say?
(a) Run the reactor at as high a temperature as possible.
(b) Run the reactor at as low a temperature as possible.
(c) Reactor temperature is irrelevant, do what's convenient.
If you were asked about the best propylene:chlorine molar ratio to feed to the reactor, what would you say?
(a) Run the reactor with excess propylene.
(b) Run the reactor with excess chlorine.
(c) Run the reactor at the stoichiometric ratio of R1.
Finally, if you were asked about the best reactor pressure, what would you say?
(a) Run the reactor with high P.
(b) Run the reactor with low P.
(c) P is irrelevant.
Does the phase of the reactants matter?

P4.54 Hydrogen peroxide solutions are used routinely to bleach cotton fibers before dyeing. A specialty cotton fabric manufacturer produces 50,000 liters/day of spent bleaching water contaminated with 0.3 wt% H_2O_2. You'd like to be able to re-use the water in the dyeing process, but the residual hydrogen peroxide is reactive with the plant-derived dyes, destroying their color. Your job is to design a batch reactor that removes 99% of the hydrogen peroxide so the water can be re-used. You consider two possibilities: (1) spontaneous decomposition of hydrogen peroxide to water and oxygen, or (2) catalytic decomposition of hydrogen peroxide to water and oxygen. From laboratory data, you know that after 24 h at 37°C, 5% of the H_2O_2 decomposes in the absence of a catalyst by a first-order reaction. With the catalyst, the rate of decomposition is 0.0005 gmoles H_2O_2/L-min, and this rate is independent of the H_2O_2 concentration. How long would you have to wait to decompose 99% of the hydrogen peroxide with (1) spontaneous decomposition and (2) catalytic decomposition?

P4.55 Vegetables "breathe," even after they are cut and stored in the refrigerator. For example the rate of respiration of cut broccoli is estimated to be:

$$\dot{r}_{O_2} = \frac{219\, y_{O_2}}{0.014 + y_{O_2}}$$

where $\dot{r}_{O_2}$ is the rate of oxygen uptake by the vegetables, in units of mL O_2/kg broccoli/h and y_{O_2} is the mol fraction oxygen in the gas surrounding the broccoli. If vegetables "breathe" too fast, they spoil quickly. However, if their oxygen supply is completely cut off, they die, emitting foul odors and liquefying in the process.

Packaging films are designed to regulate the oxygen content in packaged fresh vegetables to control respiration rates, thereby increasing the shelf life of vegetables. These films allow some limited transfer of oxygen from the air to the package. In one experiment, 137 grams of cut broccoli are placed in a container (initially containing air) and covered with a low-density polyethylene packaging film. What is the initial rate of oxygen uptake (mL O_2/h)? Some oxygen transfers across the film; at steady state it is found that the mole fraction O_2 in the container is 0.008. What is the rate of transfer of O_2 across the film at steady state?

Game Day

P4.56 Phthalic anhydride ($C_8H_4O_3$—we'll call it PA for short) is widely used to synthesize plasticizers used in vinyl notebooks and auto interiors. The old process to make it uses partial oxidation of naphthalene ($C_{10}H_8$). The process operates at 90% conversion of naphthalene, with 85% selectivity to PA. The by-products are CO_2 and water.

You are studying a newer catalyst that facilitates the production of PA by partial oxidation of o-xylene (C_8H_{10}). In laboratory experiments, 75% conversion of xylene with 65% selectivity to PA was achieved. The undesired byproducts are—you guessed it—CO_2 and water. Perhaps with some tweaking of reactor conditions or catalyst design this could be improved.

Your job is to analyze the economic feasibility of converting to the new process. First, determine the raw material requirements for a reactor producing 2 metric tons/week of PA, using each of the reaction schemes. Find current prices of naphthalene, o-xylene, and PA, and calculate the operating profit or loss, based solely on material costs. Assume air is free and combustion products are worthless. Which process do you recommend? Then, develop new process flow sheets, assuming that the reaction products can be separated. What does the optimum process flow sheet look like? How does this change the process economics, and your thinking on the choice of reaction pathway? Are there any other considerations besides economics that might affect the choice?

P4.57 Steel has an oxide scale (FeO, Fe_2O_3, and Fe_3O_4) that must be removed by "pickling" (treatment with sulfuric acid) prior to further processing. In the pickling vat, H_2SO_4 reacts with the iron oxides to produce iron sulfates and water. "Overpickling," the reaction of H_2SO_4 with Fe to produce H_2 and $FeSO_4$, results in the loss of steel and generation of H_2 bubbles. Other relevant facts:

The rate of reaction slows down with decreasing H_2SO_4 concentration.
The rate of reaction at high H_2SO_4 concentrations is limited by the rate of dissolution of iron oxides into acid.

At temperatures below 100°F, hydrated crystals of iron sulfate ($FeSO_4 \cdot 7H_2O$) start to form. These can be used as flocculating agents in sewage treatment, or in the manufacture of inks, dyes, and pigments.

In an old steel pickling plant operating in a batch mode, the following procedure is used:

(a) Steel is placed into one of four pickling tanks, each of which is vented to the atmosphere. Acid-water vapors escape through the vent.

(b) At the beginning of the process, the pickle tanks are heated by direct steam injection to 160 to 180°F and the acid strength is adjusted to 12 to 14 wt% H_2SO_4.

(c) Steel is left in the tank for 30 min to 3 h. At the end of the pickling process, the acid content is 3 to 5 wt% H_2SO_4 and the Fe content of the liquid is 8 to 12 wt%.

(d) Steel is then removed and placed in a rinse tank and rinsed with high-pressure water. The rinse water is dumped to the sewer.

(e) The waste pickle liquor is hauled away for disposal. The pickle tanks are cleaned out. The sludge on the bottom of the tank, containing FeO and $FeSO_4 \cdot H_2O$, is hauled off to landfill.

Propose three process modifications to reduce waste generation in this plant. Sketch a diagram of your modifications and explain in detail what each modification is and why it would reduce waste generation.

P4.58 Polyvinyl chloride (PVC) is produced by the catalytic polymerization of vinyl chloride and is used extensively to make products like plastic pipe and film. Your assigment is to design a process for making vinyl chloride (C_2H_3Cl). A brief survey of the synthetic chemistry literature unearths the following reactions involving vinyl chloride or similar molecules:

$$C_2H_2 + HCl \rightarrow C_2H_3Cl \qquad (1)$$

$$C_2H_4 + Cl_2 \rightarrow C_2H_4Cl_2 \qquad (2)$$

$$C_2H_4Cl_2 \rightarrow C_2H_3Cl + HCl \qquad (3)$$

$$2\,HCl + \tfrac{1}{2}\,O_2 + C_2H_4 \rightarrow C_2H_4Cl_2 + H_2O \qquad (4)$$

$$C_2H_4Cl_2 + NaOH \rightarrow C_2H_3Cl + H_2O + NaCl \qquad (5)$$

Come up with several different reaction pathways for the production of vinyl chloride by mixing and matching these five reactions. Using the market values below, analyze which of your pathways look most promising.

Ethylene:	$0.27/lb
Dichloroethane:	$0.17/lb
Acetylene:	$1.22/lb
Chlorine:	$0.10/lb
Hydrogen chloride:	$0.72/lb

Sodium hydroxide: $1.13/lb
Vinyl chloride: $0.22/lb

For the reactions listed literature data and pilot plant studies have generated the following information:

Reaction (1) proceeds at 120°C and 5 atm over a catalyst. Essentially 100% conversion can be achieved in a single pass through the reactor.

Reaction (2) proceeds at 95°C and 3 atm pressure over a catalyst. Under these conditions, 90% conversion can be reached .

Reaction (3) proceeds at 400°C and 20 atm over a catalyst. 80% conversion can be achieved in a single pass through the reactor.

Reaction (4) proceeds at 300°C and 5 atm. 70% conversion can be achieved in a single pass through the reactor.

Reaction (5) proceeds at 80°C and 4 atm. Essentially 100% conversion is achievable in a single pass through the reactor.

Dichloroethane is a liquid at the conditions of reactions (2) and (5) and a vapor at the conditions of reactions (3) and (4). Vinyl chloride and ethylene are vapors at the conditions of all four reactions listed.

Also, the following restrictions apply:

(a) No impurities are allowed in vinyl chloride product.

(b) Raw material ethylene contains 2% carbon particles.

(c) Only pure Cl_2 and C_2H_4 are allowed in reaction (2) feed.

(d) Only pure $C_2H_4Cl_2$ is allowed in reaction (3) feed.

(e) No $C_2H_4Cl_2$ or C_2H_3Cl should be in the feed to reaction (4).

(f) No C_2H_4 or HCl should be present in feed to reaction (5).

Devise two or three alternative processing schemes for production of vinyl chloride. Identify the key separations required for each scheme. Consider the tolerance to feed disturbances or plant upsets.

P4.59 In the case study in Chap. 1, we explored a reaction sequence that lead from benzene to catechol through several reaction steps. Now we'll look more closely at the first step in that reaction pathway, which is the reaction of benzene and propylene to form cumene (or isopropylbenzene).

Propylene (C_3H_6) and benzene (C_6H_6) react to form cumene (C_9H_{12}). Unfortunately, a side reaction also occurs, in which diisopropylbenzene is produced. The two reactions are

$$C_3H_6 + C_6H_6 \rightarrow C_9H_{12} \qquad \text{(R1)}$$

$$C_3H_6 + C_9H_{12} \rightarrow C_{12}H_{18} \qquad \text{(R2)}$$

In an existing process in your plant, propylene contaminated with 5 mol% propane (C_3H_8, unreactive under these conditions) is mixed with benzene and fed to a reactor. The reactor operates at 3075 kPa (30.75 bar) and the reaction occurs in the vapor phase. The reactor effluent is partially condensed and sent to a vapor-liquid separator, where all of the unreacted propylene, plus the propane, is separated out as vapor. Part of this stream is purged; the remainder is recycled back to the reactor inlet.

The liquid from the separator contains benzene, cumene, and diisopropylbenzene. This is sent to a series of two distillation columns. In the first distillation column, benzene is taken off the top of the column and recycled back to the reactor inlet. Cumene and diisopropylbenzene are sent to the second distillation column, where the cumene product is taken off the top of the column and stored in a tank for sale. The diisopropylbenzene is burned as fuel.

The reactor has a volume of 8 m³ (8000 L). It is a fluidized bed reactor, containing a solid catalyst that is fluidized by the gas stream flowing through it. To keep the reactor behaving properly, the volumetric flow through the reactor is limited to no more than 600 m³/h, calculated at the reactor temperature and pressure and the total molar flow rate at the reactor inlet. For the purposes of this problem, we will assume that the reactor acts like a well-stirred reactor, meaning that the temperature and concentration inside the reactor are the same everywhere. Under these conditions the performance of the reactor is defined by:

$$\frac{V_R}{\dot{n}_{A,\,in}} = \frac{f_{CA}}{-r'_A}$$

where V_R is the reactor volume, $\dot{n}_{A,\,in}$ is the molar flow rate of the limiting reactant A into the reactor, f_{CA} is the fractional conversion of A, and r'_A is the rate of reaction of A. In this kind of reactor, the rate is calculated at the concentrations in the outlet stream from the reactor.

You have a fantastic new catalyst that you are putting into the reactor. The reaction rate expressions for this new catalyst for (R1) and (R2) are

$$r'_c = k_1 c_p c_b$$

$$r'_d = k_2 c_p c_c$$

where c_p, c_b, c_c and c_d are the molar concentrations (mol/L) of propylene, benzene, cumene, and diisopropylbenzene, respectively. The rate constants are

$$k_1 = 2.8 \times 10^7 e^{-12,530/T}$$

$$k_2 = 2.3 \times 10^9 e^{-17,650/T}$$

where T is in K and the rate constants have units of L/gmol s.

Assume that benzene costs $0.22/kg and the 95% propylene/5% propane mix costs $0.209/kg. The selling price of cumene is $0.46/kg. Diisopropylbenzene is worth $0.20/kg. Assume that the maximum allowable reactor pressure is 30.75 bar and that the benzene:propylene molar ratio fed to the reactor is 1:1. Assume that all the other equipment (distillation columns, pumps, vapor-liquid separator, etc.) can handle any process changes.

Sketch out the process and do a DOF analysis. Analyze the performance of the reactor and the process with the new catalyst. Optimize the process by looking at the influence of reactor temperature, recycle, and purge on overall economics. Calculate the flow rates of all components in all streams. Calculate your earnings in $/day from considering only the values of raw materials and products. Explore how the block flow diagram and the economics would be affected by (a) changing the benzene:propylene molar ratio or (b) replacing the propylene/propane stream with a pure propylene source that costs $0.26/kg.

Write a brief report describing your results. Discuss the key issues in optimizing the process design. Attach the final block flow diagram (with flows shown). Document the work by attaching, as needed, calculations, tables, and/or graphs.

Selection of Separation Technologies and Synthesis of Separation Flow Sheets

In This Chapter

We take a closer look at separations. Virtually all chemical processes require some separation units. For example, the available and affordable raw materials might be impure; the chemical process designer must develop methods for removing contaminants in the raw materials so they can be further processed. Or, the chemical reactor is imperfect; the designer must develop techniques for removing undesired byproducts, recovering reactants for recycle, and purifying the desired products to meet customer requirements. Usually, several separation units must be put together to accomplish all these goals.

The questions you'll be able to answer after finishing this chapter include:

- What are the major kinds of separation technologies used in chemical processes?
- What criteria are used to select the best separation technology for a given problem?
- How do we specify the performance of a separation unit?
- How do we best synthesize a flow sheet containing multiple separations?
- Why don't all separators work perfectly?
- Why is phase equilibrium so important in the design of some separation processes?
- How do we choose the optimum temperature and pressure?

Words to Learn

Watch for these words as you read Chapter 5.

Mechanical separations	Phase equilibrium	Extraction
Rate-based separations	Equilibrium stage	Adsorption
Equilibrium-based separations	Gibbs phase rule	Absorption
Separating agent	Filtration	
Product purity	Centrifugation	
Component recovery	Distillation	
Key component	Crystallization	

5.1 Introduction

Separations are a major part of modern chemical processes, typically accounting for 50% or more of the total capital and operating cost. They provide a remarkable variety of challenging problems for the process engineer to solve, requiring both technical know-how and creative spirit. Fortunately, there are rules that guide the designer in this task. This chapter will discuss some of these rules for choosing appropriate separation methods, will demonstrate how to evaluate the performance of some common separation technologies, and will show how to generate reasonable separation flow sheets. We will often take approximate approaches that get you close enough to the exact answer, and we will liberally use heuristics to guide us.

5.1.1 Physical Property Differences: The Basis for All Separations

Let's say you've just gone grocery shopping and are bagging your own groceries. To make for efficient unpacking, you might put all your frozen foods in one bag, cleaning supplies in another, fruits and vegetables in a third, and canned goods in a fourth. What you've done, of course, is taken the output from your trip through the aisles and separated according to final use. You've exploited differences in physical properties (appearance, temperature, container type) to decide which product goes in which bag.

Similarly, in a chemical process, a mixture of compounds must be separated into the appropriate product stream (or recycle or waste stream). This is accomplished by exploiting *differences in physical properties*. The first step in choosing a separation technology is to gather information about the physical properties of the components to be separated. Then, we ask four questions:

- How do the components to be separated differ in physical properties?
- Is the difference between the components large?
- Can the difference be feasibly exploited?
- Do the components go to the correct product stream?

By answering these four questions, we can choose the best physical property differences on which to base our design.

Example 5.1	**Physical Property Differences: Separating Salt From Sugar**

Go to your kitchen and get a tablespoon of common table salt (NaCl) and another tablespoon of table sugar (sucrose, $C_{12}H_{22}O_{11}$). Mix the salt and sugar together. Now try to separate the mixture into pure salt and pure sugar.

Solution
You observe that salt and sugar have the following physical properties:

Physical property	Salt	Sugar
Appearance	White, crystalline	White, crystalline
Size	A few micrometers	A few micrometers
Taste	Salty	Sweet
Meltability over a stove-top burner	Won't melt	Melts, browns
Solubility in water	Dissolves	Dissolves

Clearly, salt and sugar don't differ much in terms of appearance or size. These properties, then, would not provide a good basis for separating salt from sugar. Appearance and size fail the "large-difference" test.

Salt and sugar differ significantly in taste. Could we design a process that separates salt from sugar on the basis of taste? Potentially, yes, but we would need some method of sensing and discriminating salt from sweet and then of placing the salt or sugar into the appropriate location. (Try to imagine how that might work!) Taste as a basis for a large-scale separation process fails the feasibility test.

Salt and sugar do differ in melting point. Since sugar melts at fairly low temperatures but salt does not, you could design a process that heats the salt/sugar mixture to a low temperature and allows the sugary liquid to drain off. Browning of the sugar is a problem, however.

What about dissolving the salt/sugar mixture in water? Both compounds are soluble in water, but sugar much more so than salt. At certain concentrations, all of the sugar but only some of the salt would dissolve. Salt crystals could then be recovered from the sugar- and salt-containing solution by filtration. This process would produce a pure salt stream, but not a pure sugar stream. Perhaps another solvent could be found that dissolves only one of the compounds but not the other.

Are there other physical or chemical property differences? Salt dissolved in water is charged, but sugar is not. Salt lowers the freezing point of water much more than sugar does. Sugar can be oxidized (by enzymes or heat) to CO_2 and water, whereas salt is fairly unreactive. Can you think of any other differences? Can you think of practical ways to exploit these differences to achieve the desired separation?

5.1.2 Mixtures and Phases

In separation processes, multicomponent mixtures are separated into streams of differing composition. The phase of the streams plays a large role in how separation processes actually work. Therefore, to understand separation processes, you must understand phases and mixtures. Here we'll review a few concepts and definitions. In Section 5.5 we devote ourselves to a detailed study of phases.

According to *Webster's New Collegiate Dictionary,* a **phase** is a "*homogeneous, physically distinct,* and *mechanically separable* portion of matter."

(The emphasis is ours.) Solids, liquids, and vapors are all phases. Supercritical fluids and plasmas are phases that are less commonly encountered.

A phase is homogeneous. Within a phase, the chemical composition and physical properties (e.g., density, viscosity) are uniform. A phase may be a single component, or a phase may be a multicomponent mixture of chemical species, with the species distributed uniformly at the molecular level. Sugar dissolved in water is a multicomponent mixture but a single phase. Emulsions, like oil-and-vinegar salad dressing, or bubbly liquids, like carbonated soda, or suspensions, like muddy water, are *not* single phases, because the components in the mixture are not completely and uniformly distributed at a molecular level.

A phase is physically distinct. Vapors, liquids, and solids differ in some fundamental ways. Vapors are much less dense than liquids. Usually liquids are less dense than solids. (An important exception to this is water, where ice is less dense than the liquid. If this were not true, it'd be impossible to ice-skate on a frozen lake). Vapors are highly compressible (i.e., their density changes a lot with pressure), whereas liquids and solids are almost incompressible. This means that the behavior of vapors is very sensitive to pressure, whereas that of solids or liquids is relatively independent of pressure. Vapors and liquids adopt the shape of their containers, whereas solids retain their shape independent of their container.

Phases are mechanically separable. One phase can be separated from another by using mechanical forces and mechanical devices. For example, diesel fuel spilled in a pond will float on top of the water and can be removed by using a skimmer. Dirt particles in muddy water can be filtered out, producing clarified water.

Comments on terminology. A gaseous phase will be defined as a "vapor" for compounds that, when pure, exist as a condensed phase (liquid or solid) at or near ambient conditions. The word "gas" will typically describe compounds that remain in the gaseous phase at temperatures and pressures near ambient. Thus, steam is a vapor, but air is a gas. A liquid will be defined as a "mixture" if all the components are normally liquid when pure, but as a "solution" when one or more of the components is normally solid when pure. A "fluid" is either a gas or liquid; a "condensed phase" is either a liquid or solid.

Figure 5.1 shows several examples of systems that are multicomponent and/or multiphase. Let's examine each system in turn. (a) Table salt dissolved in water is multicomponent and single phase; this is a solution with two components, NaCl and H_2O. (b) Table salt mixed with water at a concentration that exceeds the salt solubility limit is multicomponent and multiphase: the solution contains two components, NaCl and H_2O, while the solid is pure NaCl. (c) A glass of ice cubes in water is an example of a single-component multiphase system; H_2O is the only component in both the solid and liquid phases. (d) With a pot of boiling water we have a single-component liquid phase while the vapor phase is multicomponent,

Figure 5.1 Examples of systems that are multicomponent and/or multiphase. See text for further discussion. (a) Table salt dissolved in water, (b) salt dissolved in water with visible salt crystals, (c) ice cubes in water (d) boiling water (e) turkey drippings, (f) carbonated soda (g) pebbles in seawater (h) gold ore.

containing both air (O_2, N_2, argon, etc.) and H_2O. (e) Drippings collected from a roasted turkey and poured into a cup will separate into two liquid phases, both of which are multicomponent: the oil phase on top and the water-based phase below. (f) Carbonated soda is multiphase: the bubbles are vapor phase and pure CO_2, while the liquid phase is a multicomponent mixture of water, flavorings, sugars or sugar substitutes, and dissolved CO_2. (g) A bucket of sea and sand contains a multicomponent solution phase and one or more solid phases. (h) The gold particles in an ore rock are not mixed homogeneously with the other inorganic compounds at a molecular phase, so this is an example of multiple solid phases.

5.1.3 Classification of Separation Technologies

Separation technologies can be divided into three categories based on their operating mechanism: **mechanical, rate-based,** and **equilibrium-based.** Table 5.1 summarizes the key differences between mechanical, rate-based, and equilibrium-based separation processes.

Table 5.1	Classification of Separation Technologies		
Technology	**Input**	**Output**	**Basis for separation**
Mechanical	Two phases	Two phases	Differences in size or density
Rate-based	One phase	One phase	Differences in rate of transport through a medium
Equilibrium-based	One phase	Two phases	Differences in composition of two phases at equilibrium

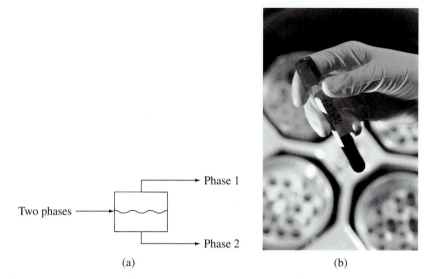

Figure 5.2 In a mechanical separator, a mixture of two (or more) phases is divided into product streams of different phases. For example, a centrifuge separates blood into a fraction containing blood cells and a fraction containing plasma fluid. *Photo © Vol. 72 PhotoDisc/Getty.*

In **mechanical separation** processes (Fig. 5.2), the feed contains two phases (e.g., suspended solids in liquid, solid particles in gas, or two immiscible fluids), and differences in size or density are exploited to separate the two phases from each other. Some of these processes are energy-intensive, like centrifugation; others are not, like sedimentation. See Table 5.2 for some examples of mechanical separation technologies.

Rate-based separation technologies rely on differences in the rate of transport through a medium of the components to be separated. Most often, the medium is a porous solid, and the feed and product streams are all the same phase. If you've done any thin-layer chromatography or gel electrophoresis experiments in a chemistry or biochemistry laboratory, then you are familiar with rate-based processes. The compounds are applied to one side of the paper or gel. They move through the paper or the gel at different rates, and can be collected one at a time as they emerge from the opposite side. If you waited forever, all the compounds would move all the way across the paper or gel, and there would be no separation. Rate-based separation technologies are very important on an analytical scale; commercially they are most commonly employed in the biotechnology industry where very high purities are required, in water desalination, and in isotope concentration. A few examples are listed in Table 5.3 and illustrated in Fig. 5.3.

In **equilibrium-based separation** processes (Fig. 5.4), the feed is a multi-component mixture but a single phase (e.g., solid, liquid, or gas). Within the process, a second phase is generated. *The compositions of the two phases are different.* The two phases are the two products. Generation of the second phase does

> **Helpful Hint**
> Rate-based separations are sometimes compared to shoppers in a mall: Some shoppers make a beeline to purchase only one item and exit quickly, while other shoppers sample every store and spend all day.

Table 5.2		Some Mechanical Separation Technologies		
Technology	**Feed phases**	**Physical property difference**	**How it works**	**Examples**
Filtration	Solid and fluid	Size	Mixture is pumped across a porous barrier such as a membrane; solids are retained while most fluid passes through	Removal of yeast from beer Removal of particulates at an engine air intake
Sedimentation	Solid and liquid	Density	Suspended solid particles are partially separated from liquid by gravity settling	Removal of sludge from municipal wastewater
Flotation	Solid and liquid Two immiscible liquids	Density	Less-dense solid or liquid droplets collect and rise to surface	Removal of contaminants from metal ores Recovery of spilled crude oil in a harbor
Expression	Solid and liquid	Size	Wet solids are compressed, allowing liquid to escape	Recovery of sugar cane juice from chopped cane
Centrifugation	Liquid and vapor Solid and fluid Two immiscible liquids	Density	Mixture is spun rapidly; centrifugal force causes denser phase or solids to migrate outwards	Purification of viruses from cell culture fluid Separation of cream from milk

not happen spontaneously. Rather, it requires an input of a **separating agent.** The separating agent can be energy, or it can be an added material. Intelligent choice of the separating agent is crucial for the success of equilibrium-based separations.

Many of the most popular large-scale equilibrium-based separations add or remove energy (e.g., heat or cool) to produce a change in temperature and generate a second phase. Some of these technologies are listed in Table 5.4. Sometimes,

Table 5.3	A Few Rate-Based Separation Technologies			
Technology	**Feed**	**Physical property difference**	**How it works**	**Examples**
Size-exclusion chromato-graphy	Macromolecules dissolved in solvent	Size	Mixture is injected onto a column containing porous beads, then solvent is pumped continuously over column; larger molecules can't enter the pores and elute quickly while smaller molecules enter the pores and take longer to exit from the column	Purification of proteins
Microfiltration/ Ultrafiltration	Solutes dissolved in solvent	Size	Solution is pumped at high pressure across a membrane with micron- to nanometer-sized pores. Some of the solvent (e.g., water) passes through the membrane, but all the solutes are rejected	Ultracleaning of water for electronics manufacture
Reverse osmosis	Solutes dissolved in solvent	Size	Solution is pumped at high pressure across a membrane with very small pores. Some of the solvent (e.g., water) passes through the membrane, but all the solutes are rejected	Production of fresh water from seawater
Gel electro-phoresis	Macromolecules dissolved in solvent	Size, charge	Mixture is injected onto the top of a thin gel slab or tube which is placed in an electric field; molecules migrate at different velocities through the gel	Separation of DNA fragments

changing the temperature is not feasible. For example, some materials chemically decompose before they are sufficiently hot to vaporize, so distillation is not an option. Or materials will condense only at exorbitantly cold temperatures. In these cases, the separating agent is an added material: The necessary second phase is generated by adding a material that is a different phase from the feed. Table 5.5 lists some of the most important equilibrium-based separation technologies that work in this manner.

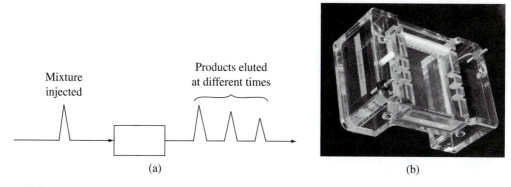

(a)   (b)

Figure 5.3 In a rate-based separation, components in a mixture travel at different rates through a medium. For example, in gel electrophoresis, protein or DNA fragments migrate through a porous polymer gel in response to an applied electric field, with the smallest molecules travelling the fastest. *Photo © The McGraw-Hill Companies. Inc./Auburn University Photographic Services.*

5.1.4 Heuristics for Selecting and Sequencing Separation Technologies

Given the plethora of possible technologies available for separation, it is useful to have a few heuristics to guide initial selection of feasible technologies. Heuristics are guidelines, not hard-and-fast rules. Experienced engineers use heuristics wisely, to eliminate clearly unworkable schemes, and to quickly generate a few reasonable choices that can then be evaluated in more detail. Some useful heuristics are:

1. If the feed is already two phases, use a mechanical separation technology.
2. If the feed is a single phase, first consider equilibrium-based separation technologies, especially for products manufactured in large quantities.

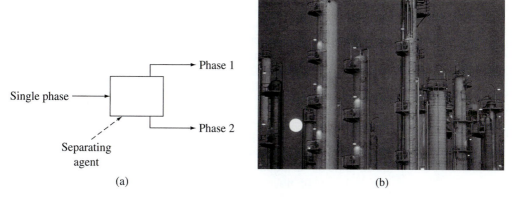

(a)   (b)

Figure 5.4 In an equilibrium-based separation, a separating agent, which could be either a material or energy, is used to convert a single-phase feed into two phases that differ in composition. For example, in distillation, energy is used to partially vaporize hydrocarbons and separate them based on differences in volatility. *Photo © Vol. 13/Corbis.*

Table 5.4   Some Equilibrium-Based Separation Technologies That Use Energy as a Separating Agent

Technology	Feed phase/ Product phases	Physical property difference	How it works	Examples
Evaporation	Liquid/liquid and vapor	Vapor pressure (boiling point)	Liquid mixture is heated until some of the material vaporizes	Separation of butane from asphalt
Condensation	Vapor/liquid and vapor	Vapor pressure (boiling point)	Vapor mixture is cooled until some of the material condenses	Recovery of volatile organic compounds from fume hood exhaust
Distillation	Liquid or vapor/ liquid and vapor	Vapor pressure (boiling point)	Mixture is fed into a multistage column, where repeated evaporation and condensation occurs	Separation of crude oil into gasoline, jet fuel, diesel, etc.
Crystalliza-tion	Liquid/solid and liquid	Solubility at low temperatures (melting point)	Solution is cooled until solubility limit is exceeded	Purification of aspirin
Drying	Solution or suspension/solid and vapor	Vapor pressure	Feed is heated to volatilize solvent, leaving behind nonvolatile solid	Drying of photographic film

3. Consider rate-based separation technologies for small-volume, high-value-added products that demand high purity.

4. For equilibrium-based separations, consider differences in (a) boiling point, (b) melting point, (c) solubility in common solvents, and (d) binding to solid surfaces, in that order. Differences of 10°C or less in boiling point can be effectively exploited. Larger differences in melting point, solubility, or binding are usually necessary.

5. Operate at temperatures and pressures as close to ambient as possible, but prefer temperatures and pressures above ambient rather than below.

6. Avoid adding foreign materials if possible. If a foreign material is added, avoid toxic or hazardous materials and remove it as soon as possible.

7. For recovery of trace quantities, use separation methods where the cost increases with the quantity of material to be recovered, not the quantity of the stream to be processed.

8. For removal of small quantities of contaminants that do not need to be recovered, consider using destructive chemical reactions rather than physical separations.

Table 5.5 Some Equilibrium-Based Separation Technologies That Use an Added Material as a Separating Agent

Technology	Feed phase	Physical property difference	How it works	Examples
Absorption	Gas	Gas solubility in added solvent	Gas mixture is contacted with solvent; one of the components in the gas is more soluble in the solvent	Separation of CO_2 from H_2 by adding ethanolamine-water solvent
Adsorption	Fluid (gas or liquid)	Affinity for solid surface	Fluid is contacted with a solid material; one of the components in the mixture sticks to the solid	Removal of colored impurities from corn syrup using charcoal
Leaching	Solid	Solubility of solid components in added solvent	Solid contains both soluble and insoluble components; soluble components dissolve in added solvent	Recovery of caffeine from coffee beans
Extraction	Liquid	Distribution between two immiscible fluids	An immiscible solvent is contacted with the feed; solute in the fluid preferentially partitions into the added solvent	Purification of antibiotics from fermentation broth

These ideas are illustrated in a few examples. The rationale behind many of these heuristics will become clearer as we delve further into design and analysis of separation processes.

Example 5.2   **Selection of Separation Technology: Separating Benzene from Toluene**

In Chap. 1, we saw how useful benzene is as a feedstock for a number of chemical processes. Benzene is purified from a refinery stream that contains a closely related compound, toluene. The *CRC Handbook of Chemistry and Physics* has this information regarding benzene and toluene:

	Boiling point, °C	**Melting point, °C**	**Soluble in**
Benzene	80.1	5.5	Ethanol, diethyl ether, acetone
Toluene	110.6	−95	Ethanol, diethyl ether, acetone, benzene

What might be a good method for separating a liquid mixture of 50% benzene and 50% toluene into two pure products?

Benzene (C_6H_6)   Toluene (C_7H_8)

Solution

The feed is a single liquid phase, and benzene is a high-volume, relatively low-value product. Therefore, by heuristics (1), (2), and (3), an equilibrium-based separation process is the best. Benzene and toluene differ in a number of ways. Boiling points and melting points are both quite distinct, so either could be exploited (heuristic (4)). Heuristic (5) would favor exploiting differences in boiling point, to keep the process just above ambient conditions. There is no need to go to a process requiring a solvent (indeed, these chemically-similar compounds are soluble in the same solvents), so no foreign material is added (heuristic (6)). Benzene concentration is not low, so heuristic (7) does not apply. *Distillation* is the separation method of choice, because it is an equilibrium-based separation technology that exploits differences in vapor pressure, which are related to differences in boiling point. (Why distillation and not evaporation? We will learn more later in this Chapter.)

Example 5.3

Selection of Separation Technology: Cleaning up Off-Gas from a Printing Press

Printers use solvents in printing presses that end up in the exhaust air leaving the pressrooms. The exhaust air from one pressroom contains 100 ppm isopropanol vapor. The isopropanol cannot be released to the atmosphere because of air quality restrictions. What's the best way to clean up the exhaust air before it is released to the outside?

Solution

First we look up physical property data for isopropanol and air in the *CRC Handbook of Chemistry and Physics* or in App. B:

	Boiling point, °C	Melting point, °C	Soluble in
Isopropanol	82.3	−89.5	Ethanol, acetone, chloroform
Oxygen	−218.9	−182.9	Slightly soluble in ethanol
Nitrogen	−210.0	−195.8	Slightly soluble in ethanol

Isopropanol has a much higher boiling point than does air. Since the feed is a vapor, condensation looks like a good choice. [Isopropanol at low concentration in air is a vapor at room temperature (20 to 25°C), even though pure isopropanol is a liquid at that temperature. How can that be? What temperature does the air need to be cooled to for condensation of the isopropanol? We'll address these questions in a later section.] The melting points are all so low that we would have to operate well below ambient temperature, violating heuristic 5.

But, notice that the isopropanol is present at very low concentrations. To condense a small quantity of isopropanol would require cooling a large amount of air. In accordance with heuristic 7, we might recommend the use of adsorption, absorption, or solvent extraction. Activated carbon is a great adsorbent for many organic compounds and might work well in this application to adsorb the isopropanol from the air. The system could operate at room temperature, and the quantity of activated carbon required would be small because the quantity of isopropanol is small. In a separate step, the isopropanol could be removed from the carbon, and the carbon regenerated and reused.

There is still another, less obvious, solution. The amount of isopropanol may be so small that it's not worth recovering. Instead, why not just burn it? Isopropanol and oxygen can react to CO_2 and water, both of which can be released to the atmosphere. In other words, we can sometimes use chemical reactions to solve separation problems (heuristic 8).

What do we do if we have a multicomponent mixture that must be separated into three, four, or more products? If there are N products desired, then there may be as many as $N - 1$ separation units. The designer has to choose not only the best technology for each individual separator, but also the best sequence in which to place the separators. This is a difficult task, but a few heuristics help make it easier. The underlying basis of many of the heuristics is simple: Separation costs increase as the volume of material to be processed increases, and as the two components to be separated become more similar to each other. Other heuristics come from the need to economize on energy utilization.

Here are some useful, simple heuristics.

1. Remove hazardous, corrosive materials early.
2. Separate out the components present in the greatest quantity first.
3. Save difficult separations for last.

4. Divide streams into equal parts.
5. Avoid recombining components that have been separated.
6. Meet all product specifications, but do not overpurify.
7. When possible, use stream splitting and stream blending to reduce the separation load.
8. In distillation trains, remove the most volatile component first.

As you develop flow diagrams requiring multiple separation units, keep these simple rules of thumb in mind.

Example 5.4   **Sequencing of Separation Technologies: Aromatics and Acid**

Given the process stream described below, devise what you think is the best scheme for separating it into three essentially pure product streams (toluene, *m*-xylene, and *p*-xylene), and a waste sulfuric acid stream. Indicate the types of separation technologies you would use and the sequence of separation steps.

Component	mol% in feed	Boiling point (°C)	Melting point (°C)	Soluble in water?	Soluble in benzene?
Toluene	51	110.6	−95	No	Yes
p-xylene	25	138.4	13.2	No	Yes
m-xylene	24	139.1	−47.2	No	Yes
Sulfuric acid	Trace	330	10.5	Yes	No

Solution

First, consider the physical properties. All compounds are liquids at room temperature. Sulfuric acid is chemically quite distinct, differing tremendously from the other components in boiling point and in solubility. It is also corrosive. Toluene and the xylenes are all chemically similar. Toluene differs from the xylenes in boiling and melting points. The boiling point is above room temperature whereas the melting point is below; by the "prefer above-ambient temperature" heuristic, separations based on boiling point differences are preferable. This leads to *distillation* as the technology of choice for producing the toluene product. The xylene isomers differ very little in boiling point or solubility, but differ significantly in melting point. We probably have no choice but to take advantage of the large melting point difference, even though it requires slightly colder-than-ambient temperatures. Thus, *crystallization* is the method of choice for separation of one xylene from the other. Since sulfuric acid is present in trace quantities only, separation that scales with the amount of acid and not the amount of the other streams makes sense. We can use *liquid-liquid extraction* with water, since the sulfuric acid is soluble in water, but water and toluene/xylene are mutually insoluble.

Second, consider sequencing. Sulfuric acid is hazardous, so it's best to remove that early. Toluene is present in the largest quantity, and a separation between toluene and the xylenes leads to a pretty even split. It makes sense to complete this step next. The separation of *p*-xylene from *m*-xylene is likely the most expensive, so this should come last.

The proposed separation scheme is shown.

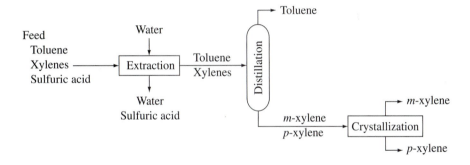

Quick Quiz 5.1

In Example 5.4 we used water as a solvent to remove sulfuric acid. Why not use benzene as a solvent to remove the aromatics instead?

5.2   Separator Material Balance Equations

At their simplest, separation units take in one feed and produce two products that differ in composition. You already have some experience in using material balances around separation units and in calculating process flows. Our primary goal in this section is to review aspects of material balance equations that are particularly important for analysis of separators.

Since by our definition there is no chemical reaction in a separation unit, the choice of mass versus mole units is simply one of convenience. The choice is typically based on whether stream composition specifications and/or any physical property information are given in mass or molar units. (As an aside, there are situations in which a chemical reaction is used to solve a separation problem, as suggested in Example 5.3, and there are situations in which chemical reactions and separations are carried out in the same equipment, as indicated in the case study of Chap. 4. In these situations, we must analyze the process as a combination of reactor and separator. Our discussion here presumes the absence of chemical reaction.)

Since separation units produce two (or more) products that differ in composition, it is often more convenient to write material balance equations on a given component as the mole fraction of that component times the total molar flow rate, rather than as the molar flow rate of the component. We will generally use the convention that

z_i = mole (or mass) fraction of i in the input stream or in a
multiphase mixture

y_i = mole (or mass) fraction of i in a vapor phase

x_i = mole (or mass) fraction of i in a liquid or solid phase, or
in an output stream if the phase is unknown.

For example, the molar flow rate of component i in the feed to a separator is

$$\dot{n}_{i,\text{in}} = z_i \dot{n}_{\text{in}}$$

The most appropriate form of the material balance equation depends on the mode of operation (Fig. 5.5).

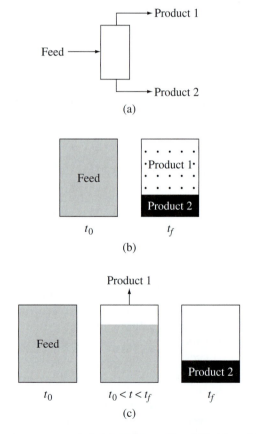

Figure 5.5 Separators may operate in (a) continuous-flow (b) batch, or (c) semi-batch modes. There are several other semi-batch modes not shown.

Continuous-flow steady-state separators are the workhorses of commodity chemical plants. For a steady-state continuous-flow separator, the differential form of the material balance equation is the most useful. The differential component mole balance equation for steady-state continuous-flow separators with a single input simplifies to

$$\sum_{\text{all } j \text{ out}} \dot{n}_{i,j} = \dot{n}_{j,\text{in}} \qquad (5.1a)$$

or, written in terms of mole fractions and total molar flows,

$$\sum_{\text{all } j \text{ out}} x_{ij} \dot{n}_j = z_i \dot{n}_{\text{in}} \qquad (5.1b)$$

The equivalent expressions in mass units (using mass rather than mole fractions!) are

$$\sum_{\text{all } j \text{ out}} \dot{m}_{i,j} = \dot{m}_{i,\text{in}} \qquad (5.1c)$$

$$\sum_{\text{all } j \text{ out}} x_{ij} \dot{m}_j = z_i \dot{m}_{\text{in}} \qquad (5.1d)$$

Batch separators are commonly used at laboratory scale, sometimes used for manufacture of low-volume products especially when solids are involved, but infrequently used at large commodity-chemical scale. In a batch separator, we need to distinguish between the moles (or mass) in the different products at the end of the separation. We will do so by using $n_{P,\text{sys},f}$ to denote the moles of product P in the system at $t = t_f$. Written in terms of mole fractions and total moles, the integral balance for a batch separator is

$$\sum_{\substack{\text{all } P \\ \text{products}}} x_{iP} n_{P,\text{sys},f} = z_i n_{\text{sys},0} \qquad (5.2a)$$

or, in mass units (using mass rather than mole fractions!),

$$\sum_{\substack{\text{all } P \\ \text{products}}} x_{iP} m_{P,\text{sys},f} = z_i m_{\text{sys},0} \qquad (5.2b)$$

Semibatch separators combine features of both batch and continuous-flow separators. They might be used for small-throughput processes, or where one of the components to be separated is present in very low quantities. These separators usually require more attention from plant operators than steady-state continuous-flow separators. One example of a semibatch operation is particle filtration: Fluid with suspended particles is pumped continuously across a filter at a steady rate, but particles accumulate on the filter. Eventually the system must be shut down and the particles cleaned out. A laboratory distillation apparatus may be operated in semibatch mode; a larger-scale version of this is common in the pharmaceutical industry. In this apparatus, a vessel is filled initially with material, and heat is applied. Vapors are continuously removed overhead; the volume in the vessel decreases with time, and nonvolatile materials become concentrated in the vessel. Whether the differential or integral form of the material balance equation is more useful depends on the question to be answered. For analysis at a single point in time, the differential balance is best:

$$\frac{dn_{i,\text{sys}}}{dt} = \dot{n}_{i,\text{in}} - \sum_{\substack{\text{all } j \\ \text{out}}} \dot{n}_{ij} \tag{5.3a}$$

For analysis of system behavior averaged over a specified time interval, the integral balance is best:

$$\sum_{\substack{\text{all } P \\ \text{products}}} n_{iP,\text{sys},f} - n_{i,\text{sys},0} = \int_{t_0}^{t_f} \dot{n}_{\text{in}}\, dt - \sum_{\text{all } j \text{ out}} \int_{t_0}^{t_f} \dot{n}_{ij}\, dt \tag{5.3b}$$

Equivalently, in mass units:

$$\frac{dm_{i,\text{sys}}}{dt} = \dot{m}_{i,\text{in}} - \sum_{\substack{\text{all } j \\ \text{out}}} \dot{m}_{ij} \tag{5.3c}$$

$$\sum_{\substack{\text{all } P \\ \text{products}}} m_{iP,\text{sys},f} - m_{i,\text{sys},0} = \int_{t_0}^{t_f} \dot{m}_{\text{in}}\, dt - \sum_{\text{all } j \text{ out}} \int_{t_0}^{t_f} \dot{m}_{ij}\, dt \tag{5.3d}$$

Example 5.5   **Semibatch Mechanical Separation: Filtration of Beer Solids**

Raw beer (density = 1.04 g/mL) contains 0.5 wt% solids. The solids must be removed before the beer is bottled. Filtration is chosen as the separation technology. Raw beer at 800 L/h is filtered through a basket filter. The process must be shut down and the basket cleaned out after 1000 kg of solids are deposited in the filter. What is the rate of deposition of solids in the basket? How long can the filtration system run until the basket must be cleaned out?

Solution
This is a mechanical separation. The feed contains two phases, liquid and solid, and filtration separates the two phases into a liquid product and a solids by-product. The system is the basket filter, in which the filtered solids accumulate over time. The flow diagram is shown.

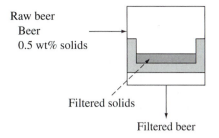

Raw beer
Beer
0.5 wt% solids

Filtered solids

Filtered beer

We'll choose two composite materials—beer (B) and solids (S)—as our components. We need to convert from a volumetric to a mass flow rate. Assuming that the specific gravity of both raw and filtered beer is the same, we calculate:

$$\dot{m}_{in} = 800 \, \frac{L}{h} \times \frac{1000 \, mL}{L} \times \frac{1.04 \, g}{mL} \times \frac{1 \, kg}{1000 \, g} = 832 \frac{kg}{h}$$

$$\dot{m}_{S,in} = 832 \, \frac{kg}{h} \times 0.005 = 4 \, \frac{kg}{h}$$

This is a semibatch operation. The liquid beer is pumped continuously through the system, but the solids are pumped in and then accumulate inside the system, with maximum accumulation set at 1000 kg solids. To determine over what interval of time this quantity of solids accumulates, we use the integral material balance equation.

$$m_{S,sys,f} - m_{S,sys,0} = \int_{t_0}^{t_f} \dot{m}_{S,in} \, dt$$

$$1000 \, kg = \int_{0}^{t_f} 4 \, \frac{kg}{h} \, dt$$

$$t_f = 250 \, h$$

The filter should be shut down for cleaning about every 250 hours.

Example 5.6   ## Rate-Based Separation: Membranes for Kidney Dialysis

Our kidneys are separation devices, removing urea and other waste products from blood. Patients with kidney failure must go on dialysis, unless and until their kidneys can be repaired or replaced. In dialysis, membranes are used that allow passage of some low-molecular-weight solutes (like urea) from the patient's blood into fluid that can be thrown away, but do not allow passage of high-molecular-weight solutes (like proteins) that should stay with the patient. Dialysis is a rate-based separation process, because in practice both urea and proteins pass across the membrane, but the rate of urea passage is much greater than that of proteins.

Your job is to evaluate the performance of several dozen membranes provided by various manufacturers, for possible use in a new kidney dialysis machine that your company is manufacturing. You quickly build a small test device, consisting of a stirred tank with a membrane holder to hold the test membrane. Above the membrane flows dialysis fluid. The system is designed so the volume of fluid in the stirred tank below the membrane does not change. The experiment is simple: You load a sample of urea-containing plasma (blood with the cells removed and anticoagulant added) into the small tank, insert the test membrane in the holder, pump dialysis fluid across the membrane, then measure the concentration of urea in the stirred tank over time.

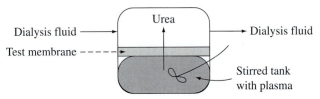

The design goal is for the urea concentration in the plasma sample to drop to 3% of its original concentration in 3 hours. However, you've got a lot of membranes to test, and not much time. You'd like to be able to measure the urea concentration after 30 minutes, and then predict whether or not the membrane meets the design goal. You do know one useful thing: The mass flow rate at which urea (U) passes through the membrane $\dot{m}_{U,\text{out}}$ decreases linearly with a decrease in the mass of urea in the tank $m_{U,\text{sys}}$:

$$\dot{m}_{U,\text{out}} = \beta m_{U,\text{sys}}$$

where β is a constant that characterizes the membrane's performance.

What is the maximum percent urea that should be left in the tank after 30 minutes, to ensure that the design goal is met?

Solution

We'll choose as our system the tank containing the plasma, and urea as our component. This separation operates in a semibatch mode.

Let's start with the differential material balance equation, Eq. (5.3c), written for urea. No urea enters the system; so:

$$\frac{dm_{U,\text{sys}}}{dt} = -\dot{m}_{U,\text{out}} = -\beta m_{U,\text{sys}}$$

Rearranging and integrating from $t = 0$ to t, we find

$$\int_{m_{U,\text{sys},0}}^{m_{U,\text{sys}}} \frac{dm_{U,\text{sys}}}{m_{U,\text{sys}}} = -\beta \int_0^t dt$$

or

$$\ln\left(\frac{m_{U,\text{sys}}}{m_{U,\text{sys},0}}\right) = -\beta t$$

which is our design equation.

Quick Quiz 5.2

If the design goal of Example 5.6 changed and the membrane had to remove 90% of the urea in 1 h, what would be the minimum β required?

At the end of 3 hours ($t = 3$ h), the design goal is to drop the urea concentration to 3% of its initial value:

$$m_{U,sys,f} = 0.03 \times m_{U,sys,0}$$

Inserting these values into our design equation, we solve for the value of β that meets the design goal:

$$\ln\left(\frac{0.03m_{U,sys,0}}{m_{U,sys,0}}\right) = \ln(0.03) = -\beta(3)$$

$$\beta = 1.17 \text{ h}^{-1}$$

So, if the membrane has a value of $\beta = 1.17$ h^{-1} (or greater) it will meet design criteria. What decrease in urea will be achieved with such a membrane after 30 minutes (0.5 h)? We return to our design equation:

$$\ln\left(\frac{m_{U,sys,0.5h}}{m_{U,sys,0}}\right) = -1.17(0.5) = -0.585$$

$$\frac{m_{U,sys,0.5h}}{m_{U,sys,0}} = 0.56$$

To meet the design goal, the urea in the tank must decrease to at least 56% of the initial quantity in the first 30 minutes. It is interesting to plot how the urea quantity in the tank changes with time.

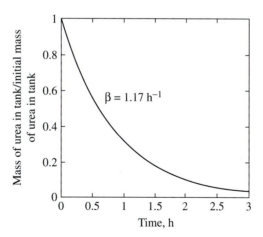

Because the mass flow rate out decreases with time, almost half of the removal is accomplished in the first 30 minutes, and it takes another 2.5 h to remove the remainder.

5.3 Stream Composition and System Performance Specifications for Separators

In a perfect separator, the products are pure, and each component in the feed ends up completely in the appropriate product stream. The perfect separator is rare indeed. More realistically, the products are not pure, and there is not complete recovery of the desired component in the appropriate product stream.

In Fig. 5.6, a perfect and a real separator are compared. (We consider only separators with one input and two outputs.) In the perfect separator (Fig. 5.6a), the feed is a mixture of two components B and C, product 1 is pure B, and product 2 is pure C. More realistically, the feed is a mixture of not only the two components B and C that are to be separated, but also contaminants A, D, and E (Fig. 5.6b). Although B becomes relatively more concentrated in product 1 and C becomes relatively more concentrated in product 2, the two products contain all five components. To simplify the analysis of a real separator somewhat, we can identify "key" and "nonkey" components. Since the separator is designed to separate B from C, these become the **key components.** The **nonkey components,** A, D, and E, are just along for the ride. We then assume that the key components distribute into both product streams, albeit with more B going to product 1 and more C going to product 2. The nonkey components are assumed to be nondistributing, that is, they exit the separator in only one product stream (Fig. 5.6c). Which product stream depends on the physical properties of the nonkey components and the basis for the separation chosen. For example, suppose the basis for separation is a difference in molecular size. If size increases in the order A < B < C < D < E, then all of A, most of B and a bit of C might exit in product 1, whereas a bit of B, most of C, and all of D and E might exit in product 2.

A comparison of the degrees of freedom of the perfect versus real separators shown in Table 5.6 is enlightening. We assume steady-state operation, and that the flow rate and composition of the feed stream are known.

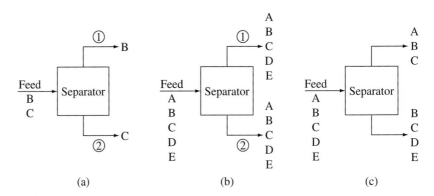

Figure 5.6 (a) A perfect separator. (b) A real separator. Component B is preferentially recovered in product 1, while component C is preferentially recovered in product 2. (c) A real separator, assuming "nonkey" components do not distribute.

	Perfect separator	Real separator	Real separator, non-distributing nonkeys
Table 5.6 DOF Analysis of a Perfect Separator and the Real Separators of Fig. 5.6			
Stream variables	4	15	12
System variables	0	0	0
Total variables	4	15	12
Specified flows	1	1	1
Specified stream composition	1	4	4
Specified system performance	0	0	0
Material balances	2	5	5
Total equations	4	10	10
DOF	0	5	2

Quick Quiz 5.3

Suppose the real separator of Figure 5.6b has only two components, B and C. What is the DOF?

The perfect separator is completely specified. The real separator requires five additional specifications, but only two additional specifications if we make the simplifying approximation that the nonkeys are nondistributing. These specifications are typically provided in one of three ways: product purity, component recovery, or separation factor. We'll describe each one in turn.

Product purity is a stream composition specification. We define **fractional purity** of product stream j as

$$\text{Fractional purity} = \frac{\text{moles (or mass) of desired component } i \text{ in product stream } j}{\text{total moles (or mass) of product stream } j}$$

For steady-state continuous-flow separators, a mathematical definition using molar units is

$$x_{ij} = \frac{\dot{n}_{ij}}{\dot{n}_j} \tag{5.4}$$

where x_{ij} is the mole fraction of the desired component i in product j. (Purity may also be specified in mass units, or for batch separators.) Percent purity is simply fractional purity $\times$ 100. Product purity specifications set the minimum acceptable content of a component in a product stream. Product purity specifications are typically set by the customer. The process design goal is to operate at purities as close as possible to the minimum acceptable to the customer, because that generally minimizes costs. It is always true that $0 \leq x_{ij} \leq 1.0$. Since

$$\sum_{\text{all } i} x_{ij} = 1$$

for each stream j, with N components in the product we can specify at most $N - 1$ product purities.

Component recovery is a system performance specification that relates the output of the process unit to the input. We define **fractional component recovery** as

$$\text{Fractional recovery} = \frac{\text{moles (or mass) of component } i \text{ in product } j}{\text{moles (or mass) of component } i \text{ in feed}}$$

In mathematical terms, for steady-state continuous-flow separators, we define fractional recovery f_{Rij} in molar units as

$$f_{Rij} = \frac{\dot{n}_{ij}}{\dot{n}_{iF}} = \frac{x_{ij}\dot{n}_j}{z_{iF}\dot{n}_F} \tag{5.5}$$

where the F subscript indicates the feed, or input, stream. (Recovery may also be specified in mass units, or for batch separators.) Percent recovery is simply fractional recovery $\times$ 100. According to this definition, it is always true that $0 \le f_{Rij} \le 1.0$. Since

> **Helpful Hint**
> Purity calculated in mole units is numerically different than purity calculated in mass units. However, component recovery is numerically the same for either mass or mole units.

$$\sum_{\text{all } j \text{ out}} f_{Rij} = 1$$

for each component i, if there are only two products then there can be only one independent recovery specification per component. The process design goal is usually to get the highest recovery possible, because higher recovery means more product sold per raw material fed.

Separation factor is a third way to characterize the performance of a separation unit. The separation factor gives a measure of how well we have separated the two key components from each other, and can be thought of as a measure of the selectivity of the separation process. The separation factor is generally defined only in terms of the key components, not the nonkey components. If we have two components, B and C, to separate, and two products, 1 and 2, then the separation factor α_{BC} is

$$\text{Separation factor} = \left(\frac{\text{moles (or mass) of B in product 1}}{\text{moles (or mass) of C in product 1}}\right)$$
$$\times \left(\frac{\text{moles (or mass) of C in product 2}}{\text{moles (or mass) of B in product 2}}\right)$$

or, for steady-state continuous-flow separations using mole units:

$$\alpha_{BC} = \frac{\dot{n}_{B1}}{\dot{n}_{C1}} \frac{\dot{n}_{C2}}{\dot{n}_{B2}} \tag{5.6a}$$

(The separation factor may also be specified in mass units, or for batch separators.) The separation factor is related to purity or recovery specifications; equivalent expressions are derived by combining Eq. (5.6) with Eq (5.4) or (5.5):

$$\alpha_{BC} = \frac{x_{B1}}{x_{C1}} \frac{x_{C2}}{x_{B2}} \tag{5.6b}$$

$$\alpha_{BC} = \frac{f_{RB1}}{f_{RB2}} \frac{f_{RC2}}{f_{RC1}} = \frac{f_{RB1}}{(1 - f_{RB1})} \frac{f_{RC2}}{(1 - f_{RC2})} \tag{5.6c}$$

Quick Quiz 5.4

What is the purity, recovery, and separation factor of a 'perfect' separator?

Is α_{BC} numerically the same or different if defined in mass versus mole units?

By convention, we define B and C such that always $1 \le \alpha_{BC} < \infty$.

Let's return to our real separator of Fig. 5.6c. From Table 5.6, we know we need two additional independent specifications. Frequently used sets of specifications include (1) key component purities in each product (e.g., x_{B1} and x_{C2}), (2) key component purity and recovery in one product (e.g., x_{B1} and f_{RB1}), and (3) key component recoveries in each product (e.g., f_{RB1} and f_{RC2}).

In the next few examples, we clarify the definition of separator performance specifications. Then we show how separator performance specifications are coupled with material balance equations and feed composition specifications to design and analyze separation units.

Example 5.7

Defining Separator Performance Specifications: Separating Benzene from Toluene

In Example 5.2, we chose distillation as a means to separate benzene from toluene. In a distillation column, the feed enters more or less near the middle of the column and two product streams are removed. One product stream, the "distillate," leaves the top of the column and is enriched in the more volatile (higher vapor pressure, lower boiling point) material. The other stream, called the "bottoms" (guess where it exits the column!), is enriched in the less volatile component. Since benzene is more volatile than toluene, in this example the distillate is benzene-rich and the bottoms product is toluene-rich.

The feed to a distillation column contains 60 wt% benzene and 40 wt% toluene. The feed rate is 100 g/s, and the column is operated in the steady-state continuous-flow mode. The distillate stream is 57 g/s benzene and 1.2 g/s toluene. Calculate (a) purity of distillate, (b) purity of bottoms, (c) fractional benzene recovery in the distillate, (d) fractional toluene recovery in the bottoms, and (e) the separation factor.

Solution

We start with a flow diagram:

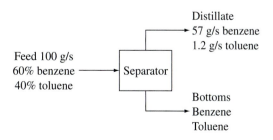

All information is given in units of g/s and mass fraction, so we will stick with those units. We'll use b and t subscripts for our components benzene and toluene, and F, D, and B subscripts for our feed stream and distillate and bottoms product streams, respectively.

(a) Distillate purity $= x_{bD} = \dfrac{\dot{m}_{bD}}{\dot{m}_D} = \dfrac{\dot{m}_{bD}}{\dot{m}_{bD} + \dot{m}_{tD}} = \dfrac{57}{57 + 1.2} = 0.98 \dfrac{\text{g benzene}}{\text{g distillate}}$

(b) To calculate the purity of the bottoms, we first need to find the flow rates of each component in that stream from material balances. For benzene,

$$\dot{m}_{bF} = z_{bF}\dot{m}_F = \dot{m}_{bD} + \dot{m}_{bB}$$

$$(0.6)100 = 57 + \dot{m}_{bB}$$

$$\dot{m}_{bB} = 3 \text{ g/s}$$

Similarly, for toluene,

$$\dot{m}_{tF} = z_{tF}\dot{m}_F = \dot{m}_{tD} + \dot{m}_{tB}$$

$$(0.4)100 = 1.2 + \dot{m}_{tB}$$

$$\dot{m}_{tB} = 38.8 \text{ g/s}$$

Therefore, the bottoms purity is

$$x_{tB} = \dfrac{\dot{m}_{tB}}{\dot{m}_B} = \dfrac{38.8}{38.8 + 3} = 0.93 \dfrac{\text{g toluene}}{\text{g bottoms}}$$

The purity is based on the component that is enriched in that product stream.

With the material balances solved, the remaining calculations are straightforward.

(c) Fractional recovery of benzene in the distillate:

$$f_{RbD} = \dfrac{\dot{m}_{bD}}{\dot{m}_{bF}} = \dfrac{57 \text{ g/s}}{60 \text{ g/s}} = 0.95$$

(d) Fractional recovery of toluene in the bottoms:

$$f_{RtB} = \dfrac{\dot{m}_{tB}}{\dot{m}_{tF}} = \dfrac{38.8 \text{ g/s}}{40 \text{ g/s}} = 0.97$$

(e) Separation factor:

$$\alpha_{bt} = \dfrac{\dot{m}_{bD}\,\dot{m}_{tB}}{\dot{m}_{tD}\,\dot{m}_{bB}} = \dfrac{57}{1.2} \times \dfrac{38.8}{3} = 614$$

Example 5.8   **Purity and Recovery Specifications in Process Flow Calculations: Separating Benzene and Toluene**

Feed (100 g/s) containing 60 wt% benzene and 40 wt% toluene is sent to a distillation column. The distillation column recovers 95% of the benzene in the distillate product, which is 98 wt% pure benzene. Calculate the flow rates and compositions of the distillate and bottoms products.

Solution

The flow diagram is shown. There are 6 stream variables, 1 specified flow rate, 2 stream compositions specified, 1 performance specification (95% recovery of benzene in the distillate) and 2 material balance equations. Therefore, $DOF = 6 - (1 + 2 + 1 + 2) = 0$ and the problem is completely specified.

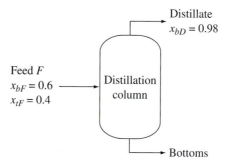

Streams will be designated as F (feed), D (distillate), and B (bottoms). Benzene and toluene will be designated by subscripts b and t, respectively. All flow rates are in units of g/s.

The distillate product purity is specified to be 98 wt% benzene.

$$x_{bD} = 0.98 = \frac{\dot{m}_{bD}}{\dot{m}_D}$$

The fractional recovery is specified: 95% of the benzene fed to the distillation column must be recovered in the distillate product, or

$$0.95 = \frac{\dot{m}_{bD}}{\dot{m}_{bF}} = \frac{x_{bD}\dot{m}_D}{z_{bF}\dot{m}_F}$$

▄Helpful Hint
Since stream composition information is often given in separator flow calculation problems, and since there is no reaction to worry about, writing a material balance equation first on total mass is often the fastest way to a solution.

Substituting in the known basis, feed composition specifications, and product purity specifications, we get:

$$0.95 = \frac{0.98\dot{m}_D}{0.6(100)}$$

$$\dot{m}_D = 58\ \frac{\text{g}}{\text{s}}$$

We have two components (benzene and toluene) and can write two material balance equations. We're going to use a little trick here—we will write material balance equations on benzene and on total mass. (Why? Because we have already solved for the total mass flow in the distillate.)

The material balance equation on total mass is:

$$\dot{m}_F = \dot{m}_D + \dot{m}_B$$

$$\dot{m}_B = 100 - 58 = 42\,\frac{g}{s}$$

The material balance equation for benzene is

$$z_{bF}\dot{m}_F = x_{bD}\dot{m}_D + x_{bB}\dot{m}_B$$

$$0.6(100) = 0.98(58) + x_{b,B}(42)$$

$$x_{bB} = 0.075$$

Since the mass fractions in each stream must sum to 1:

$$x_{tD} = 0.02$$

$$x_{tB} = 0.925$$

As a check, we write the toluene material balance:

$$z_{tF}\dot{m}_F = x_{tD}\dot{m}_D + x_{tB}\dot{m}_B$$

$$0.4(100) = 0.02(58) + 0.925(42)$$

$$40 = 40$$

The separation factor is

$$\alpha_{bt} = \frac{x_{bD}}{x_{tD}}\frac{x_{tB}}{x_{bB}} = \frac{0.98}{0.02}\frac{0.925}{0.075} = 604$$

Example 5.9

Fractional Recovery in Rate-Based Separations: Membranes for Kidney Dialysis

In Example 5.6, you evaluated a test apparatus for selecting membranes appropriate for kidney dialysis, using as a design criteria 97% removal of urea in the plasma sample in 3 hours.

Let's suppose you completed your experiments and identified two membranes that you want to investigate further. For FlowTru membranes, $\beta = 1.1\ \text{h}^{-1}$, while for DiaFlo membranes, $\beta = 1.7\ \text{h}^{-1}$. A secondary design goal is to minimize loss of protein through the membrane; specifically, at least 95% of the protein in the plasma should remain. You repeat the experiment, but measure protein concentration in the plasma sample. The flow rate of protein through the membrane decreases with the mass of protein in the tank (just as for urea), and you define a new parameter β_P, where

$$\dot{m}_{P,\text{out}} = \beta_P m_{P,\text{sys}}$$

where the subscript P refers to protein. From the data you calculate that for FlowTru membranes, $\beta_P = 0.016$ h^{-1}, whereas for DiaFlo membranes $\beta_P = 0.05$ h^{-1}.

Which membrane, FlowTru or DiaFlo, would you choose?

Solution

This is a rate-based separation. The goal is to separate urea and protein, using differences in the rate of transport of the two through the membrane. If the separation worked perfectly, all of the urea and none of the protein would be recovered in the dialysis fluid product, and none of the urea but all of the protein would be recovered in the treated plasma product. However, the separation is not perfect.

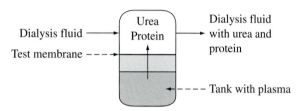

In Example 5.6 we derived a design equation that should apply to either urea or protein:

$$\ln\left(\frac{m_{U,sys}}{m_{U,sys,0}}\right) = -\beta t$$

$$\ln\left(\frac{m_{P,sys}}{m_{P,sys,0}}\right) = -\beta_P t$$

If we consider the dialysis fluid exiting the apparatus as the urea-enriched product 1, then the fractional urea recovery in that product f_{RU1} at any time t is

$$f_{RU1} = 1 - \frac{m_{U,sys}}{m_{U,sys,0}} = 1 - e^{-\beta t}$$

The treated plasma is the protein-enriched product 2, so the fractional recovery of protein in the plasma product f_{RP2} is

$$f_{RP2} = \frac{m_{P,sys}}{m_{P,sys,0}} = e^{-\beta_P t}$$

We'll use these equations to evaluate the membranes.

For FlowTru, after 3 h,

$$f_{RU1} = 1 - e^{-\beta t} = 1 - e^{-(1.1h^{-1})(3h)} = 0.963$$

or 96.3% of the urea is removed. The protein remaining in the tank is

$$f_{RP2} = e^{-0.016(3)} = 0.953$$

For DiaFlo, similar calculations give 99.4% removal of urea but only 86% retention of protein.

Neither membrane quite meets both specifications. What if we adjust the operating times? Let's use the equations for f_{RU1} and f_{RP2}, and plot fractional recoveries versus time, to get insight into the characteristics of these systems.

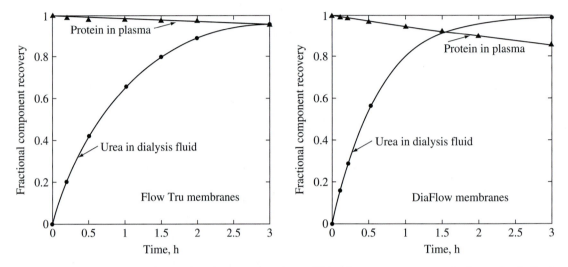

Increasing the operating time will lead to a greater removal of urea. Let's fix the urea removal at the design specification of 97% and calculate the time required to achieve that. For FlowTru

$$f_{RU1} = 0.97 = 1 - e^{-1.1t}$$

$$t = 3.2 \text{ h}$$

At this operating time, protein remaining in plasma is

$$f_{RP2} = e^{-0.016(3.2)} = 0.950$$

For DiaFlo, similar calculations show that 97% removal of urea is reached in 2.1 h, at which point the percent protein remaining in the plasma is 90%.

So neither membrane fulfills all the design requirements. The choice depends on how hard and fast the specifications are. The advantage of a shorter operating time for DiaFlo to the patient is significant, and may overcome the concern about protein loss.

5.3.1 Recycling in Separation Flow Sheets

We used recycling to advantage in chemical reactors, to overcome low reactor conversion. Is recycle useful for separations? How does recycle affect recovery and purity?

A separation flow sheet that includes recycle is shown in Fig. 5.7. We need a splitter in Fig 5.7 because we can't recycle all of product 2 from the separator

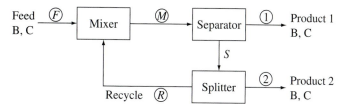

Figure 5.7 Separator flow sheet with recycle.

back to the feed. Why? Because the flow sheet would then produce only product 1, and at steady state product 1 would have to be exactly the same as the feed, which would have gained us nothing! This mental exercise helps us to qualitatively see what happens when we join recycle with separation. At high recycle, product 1 becomes more and more like the feed in composition. The recovery of component B in product 1 increases as recycle increases, but the purity decreases. The opposite trends are observed with product 2—the recovery of component C decreases, but the purity increases.

Let's put this qualitative reasoning on a more quantitative basis. We define single-pass separator recoveries: f_{RB1} and f_{RCS} are the fractional recovery of component B in stream 1 and the fractional recovery of component C in stream S, respectively, *based on the input stream M to the separator.* We define overall fractional recoveries: f_{RB} and f_{RC} are the fractional recoveries of component B in product 1 and component C in product 2, respectively *based on the input stream F to the process.* Finally, we define a fractional split f_s as the fraction of the splitter feed S that is recycled to the mixer ($f_s = \dot{n}_R/\dot{n}_S$).

From these definitions of fractional recoveries and the material balance equations, we find

$$f_{RB} = \frac{\dot{n}_{B1}}{\dot{n}_{BF}} = \frac{f_{RB1}}{1 - f_s(1 - f_{RB1})}$$

$$f_{RC} = \frac{\dot{n}_{C2}}{\dot{n}_{CF}} = \frac{f_{RCS}(1 - f_s)}{1 - f_s f_{RCS}}$$

The left-hand side of these equations is the overall fractional recovery—based on feed to the process and products from the process—while the right-hand side includes only performance specifications for the individual units on the flow sheet. (Don't try to memorize these equations—they can be derived easily from analysis of the flowsheet.)

Product purities are

$$x_{B1} = \frac{\dot{n}_{B1}}{\dot{n}_{B1} + \dot{n}_{C1}} = \frac{f_{RB}\dot{n}_{BF}}{f_{RB}\dot{n}_{BF} + (1 - f_{RC})\dot{n}_{CF}}$$

$$x_{C2} = \frac{\dot{n}_{C2}}{\dot{n}_{B2} + \dot{n}_{C2}} = \frac{f_{RC}\dot{n}_{CF}}{f_{RC}\dot{n}_{CF} + (1 - f_{RB})\dot{n}_{BF}}$$

We use these equations to explore how overall recoveries and purities change with recycle in the following example.

Example 5.10   **Separation with Recycle: Separating Sugar Isomers**

Fructose and glucose are isomers—with the same molecular formula ($C_6H_{12}O_6$), but different molecular structures. They are both simple sugars, but gram for gram fructose tastes much sweeter than glucose and is therefore in high demand. Fructose is naturally present in fruit, but is made commercially on a very large scale by hydrolysis of corn-starch to glucose, followed by enzymatically catalyzed isomeration of glucose to fructose. High-fructose corn syrup is a mix of mainly glucose and fructose in water that is widely used in sodas, juice-flavored beverages, and many other sweetened foods.

Because of chemical reaction equilibrium constraints, the fractional conversion of glucose to fructose in the isomerization reactor is less than 50%. A reactor produces a mix, at 500 kg/h, containing 8 wt% fructose and 12 wt% glucose in water. The mix is fed to a separator. 90% of the fructose fed to the separator is recovered in product I and 90% of the glucose is recovered in product II. (The water splits such that the total sugar concentration remains 20 wt% in both products.)

(a) Calculate the flow rates and purities of the two products.
(b) Fructose is a more valuable product than glucose. Consider ways to use recycle to adjust recovery of fructose and/or purity of the fructose-enriched product.

Solution
(a) The flow diagram is shown (why is it OK to ignore the water?)

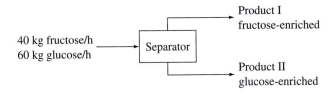

Let's start by writing material balance equations. This is a steady-state, continuous-flow process. We'll use F for fructose and G for glucose.

$$\dot{m}_{F,\text{feed}} = 40\ \frac{\text{kg}}{\text{h}} = \dot{m}_{F,\text{I}} + \dot{m}_{F,\text{II}}$$

$$\dot{m}_{G,\text{feed}} = 60\ \frac{\text{kg}}{\text{h}} = \dot{m}_{G,\text{I}} + \dot{m}_{G,\text{II}}$$

The separator performance is specified through component recoveries:

$$\dot{m}_{F,\text{I}} = f_{RF,\text{I}}\dot{m}_{F,\text{feed}} = 0.90(40) = 36\ \text{kg/h}$$

$$\dot{m}_{G,\text{II}} = f_{RG,\text{II}}\dot{m}_{G,\text{feed}} = 0.90(60) = 54\ \text{kg/h}$$

Also considering the material balance equations, we find that 86% of sugar in product 1 is fructose, and the total sugar flow rate is 42 kg/h. In product II 93% of the sugar is glucose, and the sugar flow rate is 58 kg/h.

(b) Now we want to look at the effect of recycle. One thing we notice is that we are losing 10% of the fructose in product II. Can we use recycle to increase recovery of the fructose in product I? Here is the proposed scheme.

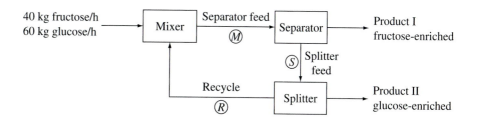

Now we have more material balance equations:

$$\text{Mixer:} \qquad \dot{m}_{F,\text{feed}} + \dot{m}_{FR} = 40 + \dot{m}_{FR} = \dot{m}_{FM}$$

$$\dot{m}_{G,\text{feed}} + \dot{m}_{GR} = 60 + \dot{m}_{GR} = \dot{m}_{GM}$$

$$\text{Separator:} \qquad \dot{m}_{FM} = \dot{m}_{FS} + \dot{m}_{F,I}$$

$$\dot{m}_{GM} = \dot{m}_{GS} + \dot{m}_{G,I}$$

$$\text{Splitter:} \qquad \dot{m}_{FS} = \dot{m}_{FR} + \dot{m}_{F,II}$$

$$\dot{m}_{GS} = \dot{m}_{GR} + \dot{m}_{G,II}$$

If the separator performance remains the same, then:

$$0.90 = \frac{\dot{m}_{F,I}}{\dot{m}_{FM}}$$

$$0.90 = \frac{\dot{m}_{GS}}{\dot{m}_{GM}}$$

The fractional split is a design variable that has not yet been specified:

$$f_S = \frac{\dot{m}_{FR}}{\dot{m}_{FS}} = \frac{\dot{m}_{GR}}{\dot{m}_{GS}}$$

Now we can solve the system of equations as a function of fractional split. (It's a good idea to solve for $f_S = 0.0$, the base case of no recycle, as a check.) Let's plot fractional recovery of fructose and purity of product I as a function of f_S.

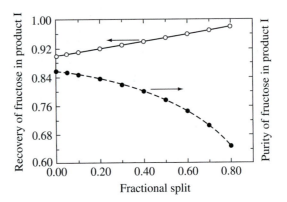

As the recovery of fructose in product I increases, the purity decreases. If a lower-purity product is acceptable, then recycle is a good idea. What would happen if we changed the splitter position?

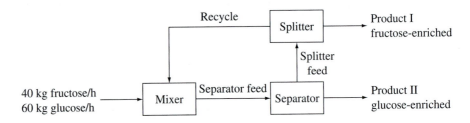

Briefly, the fructose recovery in product I decreases, but the purity increases. (We'll leave the details for you to calculate.)

5.4   Why Separators Aren't Perfect: Entrainment and Equilibrium

Why aren't most separation process units perfect? What factors prevent the separator from achieving 100% recovery or 100% purity? In this section we'll first discuss the role of incomplete mechanical separation, called *entrainment*, on separation performance. Then we'll discuss how phase equilibrium places limits on component recovery and product purity in separations, somewhat akin to the way in which chemical reaction equilibrium places constraints on conversion and yield in reactors.

5.4.1   Entrainment: Incomplete Mechanical Separation

Recall that in a mechanical separation, a two-phase feed is separated into two phases, which are the two products. The phases are separated on the basis of

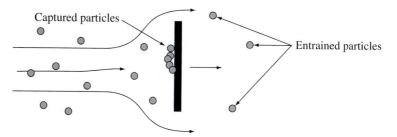

Figure 5.8 Entrainment in mechanical separation. The particles are carried in the rapidly-flowing fluid stream. As the stream impinges on the plate, the particles tend to collect on the plate. However a few renegade particles are swept along with the fluid and manage to escape the collector plate. Separation is incomplete.

differences in density or size. In a "perfect" mechanical separator, the separation into two phases is complete. Real separators often suffer from incomplete mechanical separation of the two phases, or **entrainment.** Consider, for example, particles or liquid droplets suspended in a vapor stream. If this two-phase stream flows rapidly past a plate, the particles or droplets will tend to collect on the plate. Some of the particles or droplets, however, will remain "caught up" or entrained in the vapor stream, and the separation will be incomplete (Fig. 5.8). Entrainment is even more of a problem in liquid-solid separations. Think, for example, of scooping up sand and seawater in a bucket. The sand rapidly sediments, and the seawater can be easily poured off as a single phase. Still, the sand is wet—the seawater becomes entrained in the solid phase.

Entrainment hurts the performance of a separator: Entrainment decreases component recovery in the product that is a single phase, and decreases purity in the product with the entrained phase. Since equilibrium-based separations require generation and then separation of two phases, entrainment is a concern in both mechanical separations and in equilibrium-based separations. Analysis of entrainment is simplified if one reasonable approximation is invoked: *The composition of the entrained material is the same as that of its bulk phase.* This idea is illustrated in the next example.

Example 5.11	**Accounting for Entrainment: Coffee Making**

Roasted coffee beans contain a complex mixture of chemical species, some of which are soluble in water (e.g., caffeine) and some of which are not. Typically, beans contain about 60 wt% soluble components and 40 wt% insoluble components. To make coffee, ground roasted coffee beans are contacted with hot water. Soluble components are leached out of the beans by the hot water, making an aromatic, addictive brown liquid.

Suppose 65 grams of coffee beans are contacted with 1800 g (about 8 cups) of hot water. After 10 minutes, most of the soluble components in the beans are leached into the liquid—the liquid phase contains 33 g solubles plus water while the ground beans contain 6 g solubles and all the insolubles. (Not all the solubles in the beans actually dissolved.) The resulting solid-liquid mixture is poured into a coffee filter and allowed to drain out. Most of the liquid passes through the filter and goes into the coffeepot. Some liquid is entrained with the coffee grounds captured by the filter; specifically, 2.5 grams of liquid solution is entrained per gram of dry solid material. How much coffee is in the pot? What is the percent solubles in the coffee?

Solution

Coffee brewing is an example of a solvent leaching process, in which a soluble component is removed from a solid by addition of a solvent. Leaching is very common in food processing; other examples include tea brewing and extraction of fish and vegetable oils. Leaching is an equilibrium-based separation: A second phase (hot water) is added to the solids feed, and soluble components in the solids are transferred to the added phase. After leaching, the solids and liquids are separated by filtration, a mechanical separation technology.

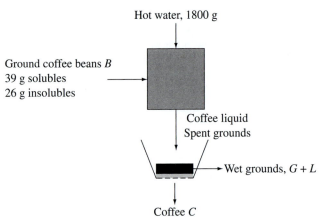

In this problem we focus on the operation of the filter. There are three components: solubles s, insolubles i, and water w. (Solubles and insolubles are composite materials—mixtures that can be treated as single components because they behave identically in the process unit.)

The feed to the filter is a two-phase mixture. The liquid phase contains 33 g solubles plus all the water (1800 g). The solid phase is 6 g solubles plus 26 g insolubles. The two product streams are liquid coffee C and wet grounds. The wet grounds contain both dry ground solids G and entrained liquid L. It is advantageous to consider the two phases in the product stream, because the entrained liquid L has the same composition as the liquid coffee product C. What is this composition? The coffee product and the entrained liquid in the wet grounds have the same composition as the liquid phase in the feed!

$$x_{sC} = x_{sL} = \frac{33}{1800 + 33} = 0.018 \ \frac{\text{g solubles}}{\text{g solution}}$$

By the same reasoning, the water mass fraction in the coffee and in the entrained liquid is 0.982 (g water/g solution).

The solid phase in the feed to the filter contains 6 g solubles and all the insolubles originally in the beans, which is (0.4)65 or 26 g insolubles. Since the dry solid grounds G contain all of this material,

$$m_G = 6 + 26 = 32 \text{ g}$$

Treating the filter as a batch separator, the integral material balance equation is

$$m_{i,sys,0} = \sum_{\substack{\text{all } P \\ \text{products}}} m_{iP,sys,f}$$

With all quantities written in grams, the solubles material balance equation is

$$39 = x_{sC}m_C + x_{sL}m_L + x_{sG}m_G = 0.018(m_C + m_L) + 6$$
$$33 = 0.018(m_C + m_L)$$

Finally, we know that 2.5 g liquid is entrained per g of dry grounds, or

$$m_L = 2.5m_G = 2.5(32) = 80 \text{ g entrained liquid}$$

Inserting this value into the solubles material balance equation yields

$$m_C = 1753 \text{ g coffee}$$

Quick Quiz 5.5

Calculate the mass fraction of insolubles in the coffee grounds without entrainment and then with entrainment.

The coffee contains 0.018(1753) or 31.6 g coffee solubles. 33 g out of the 39 g solubles in the beans actually dissolved in the water. Thus, in the absence of entrainment, the recovery of solubles in the liquid is (33/39) or 0.85. The recovery of solubles in the coffee product has been decreased by entrainment:

$$f_{RsC} = \frac{31.6}{39} = 0.81$$

If we considered the coffee grounds as a product enriched in insolubles, the purity of this product would be adversely affected by entrainment.

5.4.2 Phase Equilibrium and the Equilibrium Stage

With equilibrium-based separation technologies, the feed is a single-phase multicomponent mixture. Within the separation unit, a second phase is generated (by addition of a separating agent). Importantly, *the compositions of the two phases are different.* The difference in composition of the two phases is the basis for achieving the separation. Unfortunately, we cannot arbitrarily decide the compositions of the two phases; rather, thermodynamics is in charge.

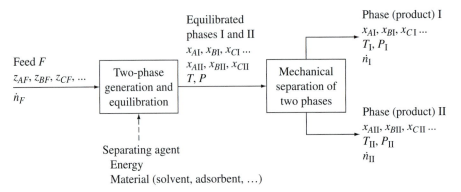

Figure 5.9 Schematic of an equilibrium stage. The stage includes generation of a second phase by addition of a separating agent, equilibration of the two phases, and complete mechanical separation of the phases.

!
Helpful Hint
Study Fig. 5.9 carefully! A clear understanding of the concept of an equilibrium stage is required for success in design and analysis of equilibrium-based separations.

Phase equilibrium constraints limit the achievable component recovery and product purity in separators, just as chemical reaction equilibrium constraints limit the achievable fractional conversion in a reactor. Just as with reactors, there are some tricks we can play to get around these limitations. The topic of phase equilibrium and its effect on separations is important enough to devote several sections and many pages and examples to it. Even so, in this limited space you will only scratch the surface of the topic. We will introduce the fundamental concept in this section. In Sec. 5.5 we discuss phase equilibria in some depth. In Sec. 5.6, we show how your newfound knowledge of phase equilibria is put to good use in the design and analysis of equilibrium-based separations.

An **equilibrium stage** is the fundamental concept for understanding how phase equilibrium affects performance of equilibrium-based separations. A general schematic of an equilibrium stage is shown in Fig. 5.9.

An equilibrium stage is a concept, not necessarily a real piece of equipment. The feed F contains a mixture of components A, B, C, . . . to be separated. The mole or mass fraction of each component in the feed is $z_{AF}, z_{BF}, z_{CF}, \ldots$. In the equilibrium stage, the separating agent (which could be energy or material, see Tables 5.4 and 5.5) is added. The temperature and/or pressure may change. Addition of the separating agent generates two phases. The phases reach equilibrium within the stage, and are then mechanically separated. We assume that the mechanical separation is perfect in an equilibrium stage. Two separate products—the two phases—leave the stage.

The two phases leaving an equilibrium stage are by definition at thermodynamic phase equilibrium. This means the following:

- The *temperatures* of both phases are the same: $T_I = T_{II}$
- The *pressures* of both phases are the same: $P_I = P_{II}$
- The *chemical potentials* of all species in both phases are the same: *????*

Chemical potential is a very useful thermodynamic property that describes something about the Gibbs energy of that component in the mixture. There is a very specific mathematical description of chemical potential but we will say *nothing* more about it except this: even though the chemical potentials of each component in both phases are equal, *the compositions of the two phases are not equal but they are related to each other.* (Your future study of thermodynamics will teach you quite a bit about chemical potential.) Importantly, knowledge of the phase equilibrium behavior of the mixture of interest provides the relationship between the composition of the two phases.

Let's look at a simple separation problem in an equilibrium stage, where the feed F contains two components B and C and there are two product streams, I and II. There are six stream variables in total. Since there are two components, we can write two material balance equations. If the flow rate and composition of the feed are known, we are left with $(6 - 4) = 2$ degrees of freedom. Previously we observed that the required information may be supplied by specification of product purity or component recovery. In an equilibrium stage, the required information is provided by phase equilibrium relationships.

Material balance equations around the equilibrium stage take the form of

$$z_{BF}\dot{n}_F = x_{BI}\dot{n}_I + x_{BII}\dot{n}_{II}$$

$$z_{CF}\dot{n}_F = x_{CI}\dot{n}_I + x_{CII}\dot{n}_{II}$$

> **Helpful Hint**
> Sometimes students make the mistake of saying $x_{BI} + x_{BII} = 1.0$. This is *not true!* Take care to avoid this common error.

where z_{BF}, z_{CF}, and $\dot{n}_F$ are known, $x_{BI} + x_{CI} = 1.0$, and $x_{BII} + x_{CII} = 1.0$.

From knowledge of the phase equilibrium behavior of the components in our system, we relate the mole fraction of each component in one phase to that in the other phase by some (as yet unknown) function f:

$$x_{BI} = f(x_{BII})$$

$$x_{CI} = f(x_{CII})$$

These two equations provide the last bit of information needed to completely specify our separator! (If there are N components equilibrated between the two phases, then we require N such equations.) The information to find the function f is available in various forms: tables, figures, or equations. Section 5.5 is devoted entirely to learning about these phase equilibrium relationships. Phase equilibrium relationships describe the composition of the two phases *leaving* the equilibrium stage, *if and only if* there are two phases present and *if and only if* the two phases are at equilibrium with each other. For equilibrium-based separations, *phase equilibrium relationships coupled with material balance equations allow us to determine the achievable product purities and component recoveries.*

5.5   An Exhausting (but Not Exhaustive) Look at Phase Equilibrium

In Sec. 5.5 we discuss phase equilibrium in some depth. We begin, in Sec. 5.5.1, with a brief introduction to the Gibbs phase rule. In Sec. 5.5.2, we review single-component phase equilibrium to provide needed background. We begin our real work on multicomponent phase equilibrium in Sec. 5.5.3, where we show how various kinds of information allow us to find the needed function relating compositions in two phases at equilibrium. Then, in Sec. 5.6, we use this new knowledge to show how to design and analyze the performance of equilibrium-based separation units.

You may work through this material in one of two ways: First, you can read all about different kinds of phase equilibria information and then learn how to apply this information to related separation technologies. This is how the material is laid out. Or you can skip from a description of phase equilibrium directly to the related separation technology, moving back and forth through the text. The chart below indicates the relevant sections.

Phase behavior	**Section**	**Separation technology**	**Section**
How to find the phase equilibrium relationship		*How to use the phase equilibrium relationship*	
Two-component liquid-solid	5.5.3.1	Crystallization	5.6.1
Two-component vapor-liquid	5.5.3.2, 5.5.3.4	Evaporation, condensation, equilibrium flash	5.6.2
		Distillation	5.6.3
Multicomponent gas-liquid	5.5.3.3, 5.5.3.4	Absorption	5.6.4, 5.6.5
Multicomponent liquid-liquid	5.5.3.5	Extraction	5.6.4, 5.6.5
Multicomponent fluid-solid	5.5.3.6	Adsorption	5.6.4

5.5.1 The Gibbs Phase Rule

The variables that describe a system at equilibrium include temperature, pressure, and mole (or mass) fraction of each component in each phase. How many of these variables are independent? The answer to this question lies in a very simple but useful rule called the **Gibbs phase rule:**

$$F = C + 2 - \Pi \tag{5.7}$$

where

Π = the number of phases present
C = the number of components present
F = the number of variables that can be independently set

Let's look first at a single-component, single-phase system such as pure liquid water, so $C = 1$ and $\Pi = 1$. Therefore, $F = 1 + 2 - 1 = 2$. We can independently specify two variables, e.g., temperature T and pressure P of the system. In other words, there are multiple combinations of T and P at which liquid water exists. (The Gibbs phase rule does *not* tell us whether, at a given T and P, water *will* be liquid—only that there are multiple values of T and P at which water *can* be a liquid.)

Now, suppose our system is pure water, present as both vapor and liquid, and the system is at equilibrium. Then $C = 1$ and $\Pi = 2$. Therefore $F = 1 + 2 - 2 = 1$. This means that we can specify one additional condition, either temperature T or pressure P, but not both. If T is specified, there is *one and only one P* at which pure water can exist as both vapor and liquid phases simultaneously. If P is specified, there is *one and only one T* at which pure water can exist as both vapor and liquid phases simultaneously.

Suppose we have pure water present simultaneously as vapor, liquid, and solid. Then $F = 1 + 2 - 3 = 0$. There are 0 independent variables. There is only one T and P at which this is possible, the triple point, and Nature has set these values for us.

Now let's see how the Gibbs phase rule works for multicomponent systems. Imagine we have a two-component two-phase system such as an ethanol-water mixture, present as both vapor and liquid simultaneously. For this case, $C = 2$, $\Pi = 2$, and $F = 2 + 2 - 2 = 2$. Two independent variables must be specified to completely specify the system. There are four possible independent variables: T, P, x_E (mole fraction of ethanol in the liquid phase) and y_E (mole fraction of ethanol in the vapor phase). Picking any two of these completely defines the system. For example, if we specify T and y_E, there is only one possible P and only one possible x_E. The Gibbs phase rule does not ensure us that there are two phases present at T and y_E, only that if there are two phases present, at specified T and y_E we cannot independently specify anything else about the system. To put a more positive spin on the story, if T and y_E are specified, then we can determine P and x_E if we simply possess the correct phase equilibrium relationship. (Of course, we would also know y_W and x_W, because $y_E + y_W = 1.0$ and $x_E + x_W = 1.0$.) The Gibbs phase rule is handy to keep in mind as we learn about phase equilibrium.

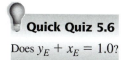

Quick Quiz 5.6

Does $y_E + x_E = 1.0$?

5.5.2 Single-Component Phase Equilibrium

Let's consider first a single-component system: water. Water is a liquid, vapor, or solid, depending on the temperature T and pressure P. You probably already know, for example, that water changes from liquid to vapor at 100°C and 760 mmHg. We call this the **normal boiling point** T_b of water. You probably also know that water changes from solid to liquid at 0°C and 760 mmHg—the **normal melting point** T_m of water. You may know that water coexists as vapor, liquid, and solid at a single temperature and pressure, 0.01°C and 4.58 mmHg, which is the **triple point** of water. There is one other point you should know about—the **critical point.** The critical point for water is 1.67×10^5 mmHg (**critical pressure** P_c) and 374°C (**critical temperature** T_c). If the temperature and pressure are both above the critical point, then the material becomes a supercritical fluid, which is neither liquid nor vapor. All of this information is very succinctly presented on a **P-T phase diagram.** The *P-T* phase diagram for pure water is shown in Fig. 5.10.

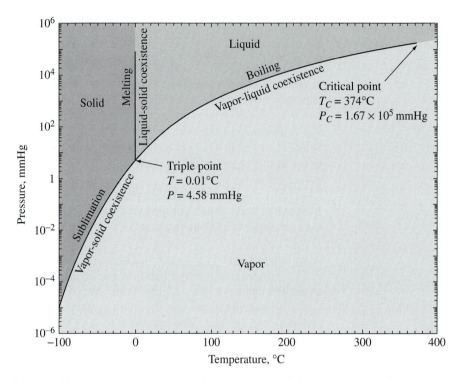

Figure 5.10 Pressure-temperature diagram for pure H_2O. Based on data from *Perry's Chemical Engineers' Handbook,* 6th edition, McGraw-Hill.

Let's examine Fig. 5.10 closely. First find the triple point. There are three phases at the triple point, so by the Gibbs phase rule, $F = 1 + 2 - 3 = 0$. Once we know we are at the triple point of water, we know we *must* be at 0.01°C and 4.58 mmHg; we have no flexibility about picking the temperature or the pressure! Next, identify the solid lines, which are called **coexistence curves**. They show all the combinations of P and T at which two phases (liquid and vapor, liquid and solid, or solid and vapor) may coexist. Notice that these curves make sense in light of the Gibbs phase rule: If two phases are present, then there is only one degree of freedom, e.g., if P is specified then T is known from the coexistence curve. The normal boiling point of water lies on the vapor-liquid coexistence curve, while the normal melting point lies on the liquid-solid coexistence curve, at $P = 760$ mmHg. Both the liquid-vapor and solid-vapor coexistence curves show a strong pressure dependence, indicating that the temperature at which a condensed phase will vaporize increases significantly with an increase in pressure. In contrast, the liquid-solid coexistence curve is nearly independent of pressure, indicating that the melting point temperature is nearly constant.

The pressure corresponding to a specific T on the liquid-vapor or solid-vapor coexistence curve is called the **saturation pressure P^{sat}**. Vapor at T and P lying on the coexistence curve is called **saturated vapor**. Vapor at T and P lying below the vapor-liquid coexistence curve of Fig. 5.10 is called **superheated**. Liquid at T and P lying on the coexistence curve is called **saturated liquid**. Liquid at temperatures above the vapor-liquid coexistence curve in Fig. 5.10 is called **subcooled**. As an example, at 100°C and 760 mmHg, water is either a saturated liquid or a saturated vapor (or a mixture of both). At 100°C and 10,000 mmHg, water is a subcooled liquid. At 100°C and 0.01 mmHg, water is a superheated vapor. H_2O at −50°C and 10,000 mmHg is a solid. These points are marked on the diagram, as are a few additional conditions. The diagram also shows how to evaluate changes in phase with changing P at constant T, or with changing T at constant P.

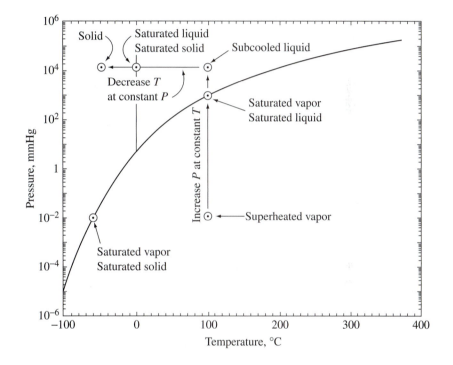

P-T diagrams of the type shown in Fig. 5.10 are available in handbooks for many common substances. Sometimes the information is shown in tabular form instead, which is more convenient if precision is required.

But what if we can't find a P-T diagram for a specific compound? For many compounds, values for T_m, T_b, T_c, and P_c are available. App. B, *Perry's Chemical Engineers' Handbook*, *CRC Handbook of Chemistry and Physics*, and *Lange's Handbook of Chemistry* are all good sources of data. At a minimum, these data provide a few points on a P-T diagram. Furthermore, it is reasonable to approximate T_m as independent of P. What we have left to find is a relationship between P^{sat} and T, which would allow us to plot out the solid-vapor and liquid-vapor coexistence curves.

Helpful Hint
For a pure com-
ponent at liquid-
vapor or solid-
vapor equilib-
rium, $P = P^{sat}$.

Quick Quiz 5.8

From the Antoine
equation, what's the
saturation pressure for
H_2O at 100°C?

There are several useful *model equations* available to describe P^{sat} as a function of T. One of the simplest and most widely used is the **Antoine equation.**

$$\log_{10} P^{sat} = A - \frac{B}{T + C} \tag{5.8}$$

where A, B, and C are empirically-determined constants for a specific material.
Refer to App. B for Antoine constants for many common chemicals.

Illustration: The saturation pressure of water, according to Antoine's equation
coefficients in App. B, is

$$\log_{10} P^{sat}(\text{mmHg}) = 8.10765 - \frac{1750.286}{T(°C) + 235.0} \quad \text{from 0°C to 60°C}$$

$$\log_{10} P^{sat}(\text{mmHg}) = 7.96681 - \frac{1668.21}{T(°C) + 228.0} \quad \text{from 60°C to 150°C}$$

We use this equation to calculate P^{sat} versus T from 0°C to 150°C. Compare the
plot to Fig. 5.10; notice that we have just constructed the vapor-liquid coexistence curve for this temperature range!

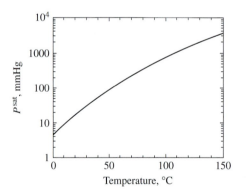

5.5.3 Multicomponent Phase Equilibrium

Everything in Sec. 5.5.2 applies for systems that contain one component. But, if
we are interested in separations problems, then we have at least two components!
This means that we have to understand a bit about **multicomponent phase equilibrium**. This is a complicated topic but we can understand a few basics well
enough to enable preliminary work on designing separation processes.

The phase behavior of multicomponent systems involves not only T and P,
but also the mole or mass fractions of the components. This information is
harder to show graphically than a single-component P-T diagram. In this section,
we describe diagrams, tables, and model equations for multicomponent phase
equilibria that will prove useful to us.

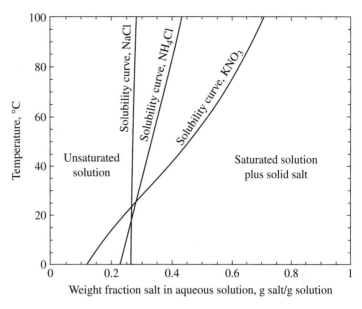

Figure 5.11 Solubility curves for KNO_3, NH_4Cl and NaCl. Data taken from *Perry's Chemical Engineers' Handbook,* 6th edition, McGraw-Hill.

Multicomponent phase equilibrium data are available in sources such as *Perry's Chemical Engineers' Handbook* and many thermodynamics textbooks. Some data are collected for your convenience in App. B of this text. As you read the following sections, you may want to scan through the appendix to familiarize yourself with the information presented there.

5.5.3.1 Two-Component Liquid-Solid Phase Equilibrium   Common
table salt (NaCl) dissolves easily in water. If too much salt is added, a second, solid phase appears. The solid phase is pure NaCl. When both solid and liquid phases coexist and are at equilibrium, the liquid is called a **saturated solution**, and the concentration of NaCl in the solution is at its solubility limit.

The solubility of most solids is a function of T, but is essentially independent of P. Therefore, information on the solubility of salts in water as a function of T is compactly shown in ***T-x* diagrams** like the one in Fig. 5.11. (See also data in App. B.) There are two phases represented on the diagram: a liquid solution, containing both salt and water, and a solid, which is pure salt. The solid line gives the weight fraction salt in a saturated liquid solution (x_{AL}) as a function of temperature. Unsaturated salt solutions—where the salt concentration is below the solubility limit—lie to the left of the curve. The line $x_A = 1$ (the right-hand side vertical axis) corresponds to pure salt.

To determine the phase behavior of an equilibrated salt-water mixture at a given temperature T and mass fraction salt z_A, one first finds the point corresponding to

z_A and T on the T-x diagram. If the point lies to the left of the solubility curve, the salt concentration is below the solubility limit and the mixture is a single-phase unsaturated solution. If the point lies to the right of the solubility curve, then the mixture splits into two phases. One phase is a saturated liquid solution, with the salt concentration at the solubility limit. To determine x_{AL} of the saturated liquid solution, we read across at constant T to the solubility curve. The other phase is a solid—pure salt ($x_{AS} = 1$). To determine the relative quantity of the solution and the solid phase, we invoke the *inverse lever rule:*

$$\frac{m_S}{m_L} = \frac{z_A - x_{AL}}{x_{AS} - z_A} \tag{5.9}$$

where m_S is the mass of the solid phase and m_L is the mass of the liquid solution. The inverse lever rule is simply a restatement of the law of conservation of mass. (Try to prove this to yourself by deriving Eq. (5.9) from material balance equations!)

Illustration: *Reading Solubility Graphs*
A mixture of 50 wt% KNO_3 in water ($z_A = 0.5$) at 30°C separates into two phases: a saturated solution of 31 wt% KNO_3 ($x_{AL} = 0.31$) and a solid phase of 100% KNO_3 ($x_{AS} = 1.0$). The ratio of the masses of solid and liquid phases is

$$\frac{m_S}{m_L} = \frac{z_A - x_{AL}}{x_{AS} - z_A} = \frac{0.5 - 0.31}{1 - 0.5} = 0.38$$

or 72.5% of the mixture is in the liquid phase and 27.5% is solid.
       If the mixture is heated above 57°C, the solid phase disappears and the mixture is a single liquid phase.

Quick Quiz 5.9

A mixture of 50 wt% KNO_3 at 90°C is cooled to 20°C. What happens?

At 60°C, is NaCl or KNO_3 more soluble in water?

What about at 10°C?

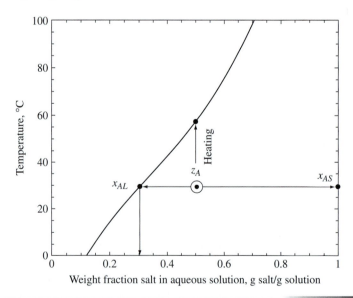

Weight fraction salt in aqueous solution, g salt/g solution

For some mixtures, the liquid-solid phase diagram is a bit more complicated. Consider the benzene (C_6H_6)–napththalene ($C_{10}H_8$) system. T_m of pure naphthalene is 80.2°C and that of pure benzene is 5.5°C. What happens when we mix liquid naphthalene and benzene? Because they are chemically similar, naphthalene dissolves readily in benzene, and vice versa—benzene dissolves readily in naphthalene.

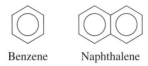

Benzene   Naphthalene

The *T-x* phase diagram for mixtures of benzene and naphthalene is shown in Fig. 5.12. (See also App. B.) There are four regions on the diagram. The solid curves are liquid-solid coexistence curves. Above these curves lies a region in which the mixture forms a single liquid solution. Below these curves, the mixture forms a saturated liquid solution and either solid naphthalene or solid benzene, as indicated on the diagram. The far right and left ends of these curves correspond to T_m of pure naphthalene and of pure benzene, respectively. Surprisingly, at 0.14 mol fraction naphthalene, the mixture will remain liquid down to −4°C, lower than the melting points of either benzene or naphthalene alone! This point is called a eutectic point and is a common feature of solid-liquid phase equilibrium when the two components are chemically similar. Below the eutectic temperature, all benzene-naphthalene mixtures are in the solid phase.

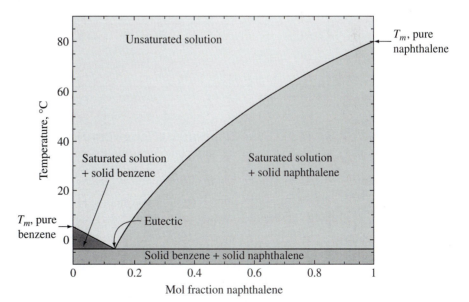

Figure 5.12 Liquid-solid phase equilibrium for benzene-naphthalene mixtures.

Illustration: *Reading Solid-Liquid T-x Graphs*

6 gmol naphthalene (N) and 4 gmol benzene (B) initially at 80°C are mixed
($z_N = 0.6$) and form a single liquid phase. The mixture is then slowly cooled to
20°C. At 54°C the mixture reaches the solubility limit and as cooling continues
the mixture separates into a saturated liquid solution of 27 mol% naphthalene/73
mol% benzene ($x_{NL} = 0.27$) and a solid phase of pure naphthalene ($x_{NS} = 1.0$).
The ratio of solid and liquid phases is

$$\frac{n_S}{n_L} = \frac{z_N - x_{NL}}{x_{NS} - z_N} = \frac{0.6 - 0.27}{1 - 0.6} = 0.825$$

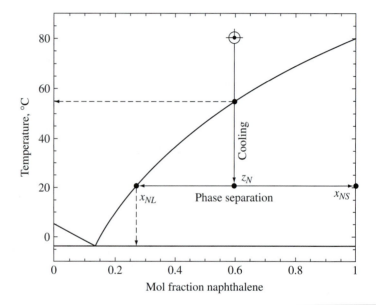

Mol fraction naphthalene

5.5.3.2 Two-Component Vapor-Liquid Phase Equilibrium   Two-
component vapor-liquid phase equilibrium differs from solid-liquid phase equilibrium because the composition of the phases is sensitive to both temperature
and pressure. Graphing the data presents a challenge. Generally, we solve this
problem by either (1) holding P constant and plotting the data as a T-y-x diagram
or (2) holding T constant and plotting the data as a P-y-x diagram. Both are useful representations; we will focus on T-y-x diagrams with constant P.

Figure 5.13 is a T-y-x diagram for ethanol and water, with $P = 1$ atm. (For
the same data presented in table form, see App. B.) The single-component
boiling point temperatures are shown in Fig. 5.13; for pure water ($y_W = x_W = z_W = 1.0$, so $y_E = x_E = z_E = 0.0$), the boiling point at 1 atm is 100°C, and for

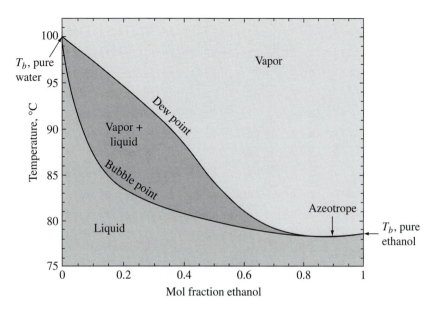

Figure 5.13 *T-x-y* diagram for ethanol-water mixtures at $P = 1$ atm. Drawn from data taken from *Perry's Chemical Engineers' Handbook,* 7th edition, McGraw-Hill.

pure ethanol ($y_E = x_E = z_E = 1.0$), the boiling point temperature at 1 atm is 78.3°C. The two solid curves in Fig. 5.13 form an envelope that encompasses mixture compositions and temperatures at which vapor and liquid phases coexist. The top curve, called the **dew point** curve, shows the mole fraction ethanol in the saturated vapor phase, y_E, at the corresponding dew point temperature. The bottom curve, labeled the **bubble point** curve, shows the mole fraction of ethanol in the saturated liquid phase, x_E, at the corresponding bubble point temperature.

!
Helpful Hint
A mixture at its dew point temperature is all vapor except for a tiny drop of liquid "dew". A mixture at its bubble point temperature is all liquid except for a tiny "bubble" of vapor.

Illustration: *Reading a Vapor-Liquid T-y-x Diagram*
A mixture of 40 mol% ethanol/60 mol% water at 77°C ($z_E = 0.4$, $T = 77$°C, $P = 1$ atm) is a subcooled liquid. We raise the temperature and cross the bubble point curve at 80.5°C. The liquid is now saturated and the first bubble of vapor emerges. At the bubble point, the ethanol mole fraction in the saturated liquid phase is the same as the mixture mole fraction ($x_E = z_E = 0.4$). The composition of the vapor is different from that of the liquid, because the more volatile component (ethanol) is concentrated in the vapor. We read the bubble (vapor) mole fraction at 80.5°C by drawing a line horizontally across the two-phase region until we intersect the dew point curve, at $y_E = 0.61$.

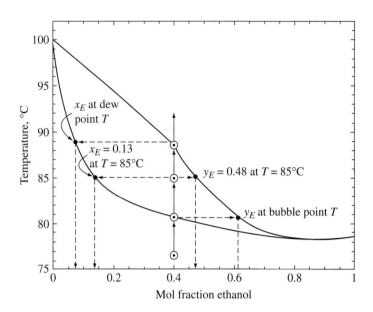

Helpful Hint
By simply measuring the lengths of the two line segments $z_E - x_E$ and $y_E - z_E$ with a ruler and taking the ratio, you can determine the ratio of vapor to liquid.

Now let's raise the temperature of the mixture to 85°C ($z_E = 0.4$, $T = 85°C$). This point clearly lies in the two-phase region. The mixture spontaneously separates into saturated liquid and saturated vapor phases. We determine the compositions of the liquid and vapor phases that are in equilibrium with each other by drawing a horizontal line in both directions until it intersects with the dew point and bubble point curves. The saturated liquid composition lies on the bubble point curve at $x_E = 0.13$, $T = 85°C$. The saturated vapor composition lies on the dew point curve at $y_E = 0.48$, $T = 85°C$. The relative quantity of vapor and liquid is calculated from the inverse lever rule:

$$\frac{n_V}{n_L} = \frac{z_E - x_E}{y_E - z_E} = \frac{0.4 - 0.13}{0.48 - 0.4} = 3.375$$

Now, let's warm the mixture to the dew point temperature. At 88.5°C, the mixture is a saturated vapor ($y_E = 0.4$) in equilibrium with a drop of liquid ($x_E = 0.08$). Finally, if the temperature of the mixture is raised further, we move into the superheated vapor region.

Quick Quiz 5.11

What are the dew point and bubble point temperatures of a 20 mol% ethanol/80 mol% water mixture?

A 20 mol% ethanol/80 mol% water mixture is heated to 92°C. Is there one phase or two? If two, what is the mole fraction ethanol in each phase?

Look at Fig. 5.13 again. Notice that at $z_E = 0.89$, the dew point and bubble point curves intersect, and the vapor and liquid compositions are the same. The temperature at this mole fraction (78.15°C) is lower than the normal boiling point of pure ethanol and that of pure water. This point is called an **azeotrope**. The ethanol-water system is an example of a minimum-boiling azeotrope. Some two-component mixtures form a maximum-boiling azeotrope, where a temperature greater than the boiling points of either pure component is required to generate

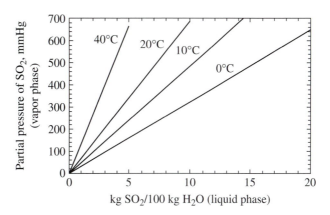

Figure 5.14 Gas-liquid phase equilibrium for SO_2-H_2O mixtures, Based on data from *Perry's Chemical Engineers Handbook,* 6th edition, McGraw-Hill.

a vapor phase. The existence of an azeotrope is a consequence of strong interactions between the two different components in the liquid phase.

5.5.3.3 Multicomponent Gas-Liquid Phase Equilibrium   Let's

move on to systems where we have a multicomponent gas mixture in contact with a liquid phase. We distinguish between "gas" and "vapor" in this way: a vapor may condense at commonly encountered temperatures and pressures but a gas does not. For example, acetaldehyde (CH_3CHO) would be called a vapor at room temperature and pressure, since its normal boiling point is 20.3°C, but oxygen (O_2) is called a gas, since its normal boiling point is –182.9°C. The distinction between gas and vapor is not hard and fast, but still useful.

In gas-liquid equilibrium, one (or more) component is sparingly soluble in the liquid phase. The solubility of the gas in the liquid phase depends on both T and P. Equilibrium data are usefully presented graphically, as shown in Fig. 5.14 for SO_2 dissolved in water. (See App. B for additional gas-liquid equilibrium data.) The gas-phase composition is given as **partial pressure p_i,** which for ideal gas mixtures is simply

$$p_i = y_i P \tag{5.10}$$

where y_i is the mole fraction of compound i in the gas phase. p_{SO_2} is plotted on the y axis in Fig. 5.14. The quantity of gas dissolved in the water is given as kg SO_2/kg water in the liquid phase. The solid lines are *isotherms*—lines of constant temperature. Points on the lines give the relationship between gas and liquid phase compositions at the specified T.

Illustration: *Reading Gas Solubility Graphs*
(1) A gas mixture at 1320 mmHg and 10°C contains 9 mol% SO_2 and 91 mol% N_2. The partial pressure $p_{SO_2} = y_{SO_2} P = 0.09(1320) = 120$ mmHg. The point (p_{SO_2}

$= 120$ mmHg, $T = 10°C$) is circled. The gas mixture is in equilibrium with water. The solubility of SO_2 in the water is determined from the x axis as about 2.8 kg SO_2 per 100 kg H_2O, corresponding to a mass fraction of SO_2 of 0.027 and mole fraction $x_{SO_2} = 0.0079$.

(2) An aqueous solution containing 10 kg SO_2 dissolved in 100 kg H_2O is equilibrated with air at $10°C$ and 760 mmHg. The partial pressure p_{SO_2} is read from the diagram to be 470 mmHg. From this we calculate the mole fraction of SO_2 in the gas phase:

$$y_{SO_2} = \frac{p_{SO_2}}{P} = \frac{470 \text{ mmHg}}{760 \text{ mmHg}} = 0.62$$

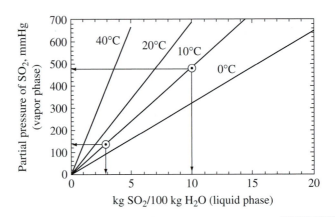

5.5.3.4 Model Equations for Vapor-Liquid and Gas-Liquid Phase Equilibrium

Graphs and tables of the type we've just described are very useful. Sometimes, though, it would be useful to express phase equilibrium information in the form of equations, because we want to use a calculator or computer to solve a problem. Or, perhaps reliable data at the exact temperature and pressure of interest are not available. In these cases, we turn to *model equations*. Here we present two simple but useful equations for modeling phase equilibrium behavior in special cases: Raoult's law and Henry's law.

For vapor-liquid equilibrium, when the molecules in the mixture are similar in size and chemical structure, and don't interact strongly with each other, we relate vapor and liquid mole fractions by **Raoult's law:**

$$y_i = \left(\frac{P_i^{\text{sat}}}{P}\right) x_i \tag{5.11}$$

where P_i^{sat} is the saturation pressure of pure component i at the temperature of the mixture. Raoult's law is not a law at all, just a very useful model equation. The law works well with mixtures such as hexane and heptane, or benzene and styrene. It does not work well to predict vapor-liquid equilibrium curves for ethanol and water (where there are strong hydrogen-bonding interactions), polystyrene in benzene (where the molecular sizes are very different), or DNA in water (where there are strong electrostatic interactions and the molecular sizes are very different). *Equation (5.11) applies only when both vapor and liquid phases coexist and are in equilibrium.*

We can use Raoult's law in many ways. If we have saturation pressure data for component i or we have an equation such as the Antoine equation, Eq. (5.8), given T we can find P_i^{sat}. Then, if we specify P also we can use Raoult's law to calculate the mole fractions in the vapor and liquid phases, subject to two constraints:

$$\sum_i x_i = 1 \quad \text{and} \quad \sum_i y_i = 1$$

It's particularly simple to calculate dew point and bubble point pressures at a given mixture composition and temperature. First, we note that

$$\sum_i y_i P = P \sum_i y_i = P$$

Applying Raoult's law to all components and summing together, we find that at vapor-liquid equilibrium

$$\sum_i y_i P = \sum_i x_i P_i^{sat}$$

At the bubble point, $z_i = x_i$. Combining this idea with the two previous equations we find:

$$P = \sum_i x_i P_i^{sat} = \sum_i z_i P_i^{sat} \tag{5.12}$$

Helpful Hint
At the bubble point, $z_i = x_i$ and y_i is found from phase equilibrium. At the dew point, $z_i = y_i$ and x_i is found from phase equilibrium.

Given T we find P_i^{sat} for all compounds. Then, given the bubble point mole fractions, we use Eq. (5.12) to find the bubble point pressure P at the given T and mixture composition. A similar analysis leads to an equation to calculate the dew point pressure:

$$\frac{1}{P} = \sum_i \frac{y_i}{P_i^{sat}} = \sum_i \frac{z_i}{P_i^{sat}} \tag{5.13}$$

By repeating dew point and bubble point pressure calculations at different mixture compositions, we can map out an entire P-y-x diagram. Also, we can invert the problem: Given P and a mixture composition, we can calculate dew point and bubble point temperatures and generate a T-y-x diagram. The calculations are a bit more involved, as illustrated in Example 5.12.

Example 5.12   **Using Raoult's Law: Dew Point and Bubble Point Temperatures of Hexane-Heptane Mixtures**

What are the dew point and bubble point temperatures of a mixture of 40 mol% n-hexane (C_6H_{14})/60 mol% n-heptane (C_7H_{16}) at 2 atm pressure?

Solution
We'll use the subscript 6 to indicate hexane and 7 to indicate heptane. First we'll find Antoine's constants so we can calculate saturation pressures. From App. B, we find

$$\log_{10}(P_6^{sat}) = 6.87601 - \frac{1171.17}{T + 224.41}$$

$$\log_{10}(P_7^{sat}) = 6.89677 - \frac{1264.90}{T + 216.54}$$

where T is in °C and P_i^{sat} is in units of mmHg. We quickly convert the total pressure P from 2 atm to 1520 mmHg, to make the units consistent.

At the dew point, the mixture composition is the same as the liquid composition, or $y_6 = 0.4$, $y_7 = 0.6$. Applying Raoult's law [Eq. (5.11)] we find

$$0.4 = \frac{10^{6.87601 - [1171.17/(T+224.41)]}}{1520} x_6$$

$$0.6 = \frac{10^{6.89677 - [1264.90/(T+216.54)]}}{1520} x_7$$

We also know that $x_6 + x_7 = 1$. Combining these three equations and rearranging gives

$$\frac{0.4(1520)}{10^{6.87601 - [1171.17/(T+224.41)]}} + \frac{0.6(1520)}{10^{6.89677 - [1264.90/(T+216.54)]}} = 1.0$$

We can now solve to find $T = 114°C$. (T should be somewhere between the boiling point temperature for pure hexane and that for pure heptane at 2 atm pressure.) We also calculate that $x_6 = 0.234$ and $x_7 = 0.776$ at the dew point of this mixture.

Bubble point temperature calculations proceed in exactly the same way, except now $x_6 = 0.4$, $x_7 = 0.6$, and $y_6 + y_7 = 1$. You should get $T = 108.5°C$ at the bubble point. (Does it make sense that the bubble point temperature is lower than the dew point temperature?)

When low concentrations of noncondensable gases dissolve in a liquid and there are no strong interactions between the liquid and the dissolved gas, the solubility of gases in the liquid can be modeled by using **Henry's law:**

$$y_i = \left(\frac{H_i}{P}\right)x_i \tag{5.14}$$

H_i is called, cleverly, the Henry's law constant, and is a function of T as well as the identity of the dissolved gas and the liquid. Henry's law constants for several gases dissolved in water are given in App. B. Henry's law is not a law at all, just a useful model equation in some situations. The law works well for gases such as N_2 or CO_2 dissolved in water, but is a poor model for gases such as NH_3 or SO_2 in water except at very low concentrations. *Equation 5.14 applies only when gas and liquid phases co-exist and are in equilibrium.*

Illustration: *Using Henry's Law*

A gas at 10°C and 20 atm, containing 98 mol% N_2 and 2 mol% CO_2, is in equilibrium with water. Since both N_2 and CO_2 are noncondensable at this temperature and pressure and do not interact strongly with water, we use Henry's law, with the constants obtained from App. B.

$$y_{CO_2} = 0.02 = \frac{H_{CO_2}}{P} x_{CO_2} = \frac{1040 \text{ atm}}{20 \text{ atm}} x_{CO_2}$$

$$x_{CO_2} = 0.0004$$

$$y_{N_2} = 0.98 = \frac{H_{N_2}}{P} x_{N_2} = \frac{66800 \text{ atm}}{20 \text{ atm}} x_{N_2}$$

$$x_{N_2} = 0.0003$$

A higher Henry's law constant indicates a lower solubility. Despite the high concentration of nitrogen in the gas phase, it is much less soluble in water than is CO_2. Notice too that the mole fractions of dissolved gas in the liquid phase are very low, which is the case when the Henry's law equation is a reliable model of gas-liquid phase equilibrium. Also, remember, that x_{H_2O} is close to 1.0.

Quick Quiz 5.12

A gas mixture of N_2 and CO_2 at 10°C and 20 atm is in equilibrium with water. Would you use Raoult's law or Henry's law to calculate the mole fraction of water in the gas?

5.5.3.5 Multicomponent Liquid-Liquid Phase Equilibrium

Perhaps you have made oil and vinegar salad dressing and observed that the oil and vinegar tend to separate into two distinct layers, with each layer a separate liquid phase. Indeed, it is quite common for organic and aqueous liquids to phase-separate, much to the dismay of salad-dressing manufacturers. Yet this tendency for organic and aqueous liquids to separate can also be gainfully exploited by separation process designers; if a third component is present, then that component may prefer to be in one or the other phase.

Let's consider a three-component system containing an organic solvent S, water W, and a third component A. There are two possible cases.

Case 1: W and S are completely *immiscible* (insoluble) in each other. Two liquid phases form, aqueous phase I and organic phase II, and all of W is in phase I and all of S is in phase II. Component A then *partitions* or *distributes* between the two immiscible phases. The mass (or mole) fraction of A in each phase is a function of temperature T and the identity of the organic material. Pressure effects are, as usual in liquid and solid systems, negligible. Phase equilibrium is described in terms of a **distribution coefficient K_D**:

$$x_{AII} = K_D x_{AI} \tag{5.15}$$

where x_{AI} is the mass (or mole) fraction of component A in phase I, and x_{AII} is the mass (or mole) fraction of component A in phase II. K_D is a function of T and the identity of the partitioning compound as well as the two solvents. App. B contains K_D data for several solutes partitioning between various organic solvents and water. Equation (5.15) is usually adequate only when component A is very dilute.

Illustration: *Using Distribution Coefficients*

A solution of 1.2 wt% acetic acid in water, $x_{AI} = 0.012$, is in equilibrium with methyl acetate. Methyl acetate and water form two immiscible liquid phases. From App. B, $K_D = 1.273$ for acetic acid partitioning between methyl acetate and water. Therefore:

$$x_{AII} = 1.273(0.012) = 0.0153$$

Quick Quiz 5.13

From K_d data available in App. B, is acetic acid more soluble in benzene or in water?

Case 2. *W* and *S* are partially immiscible in the presence of *A*. In this case, *W* and *S* still tend to separate into two liquid phases, but a bit of *W* dissolves in *S* and vice versa. Indeed, their mutual solubility usually increases in the presence of the solute A. This makes sense if we think about the heuristic *like dissolves like*. If solute A is soluble in solvent *S*, then A must be "like" *S* (in a chemical sense). If solute A is also soluble in solvent *W*, then A must be like *W*. But for *W* and *S* to be mutually immiscible, *W* must be most definitely *not* like *S*. How can that be, if A is like both *W* and *S*? An example of this is the water–acetic acid–methylisobutylketone (MIBK) system.

| Water | Acetic acid | MIBK |

Acetic acid is very soluble in both water and MIBK, but water and MIBK are mutually insoluble (or nearly so). You may be able to see from the chemical structures why this might be true. A mixture of acetic acid-water-MIBK separates into two liquid phases if the acetic acid content is low, but forms a single liquid phase at higher acetic acid content. When two separate liquid phases exist, all three components are present in both phases.

Since there are three components present, graphical data on liquid-liquid systems must include the mass fraction of two of these components. It is hard to

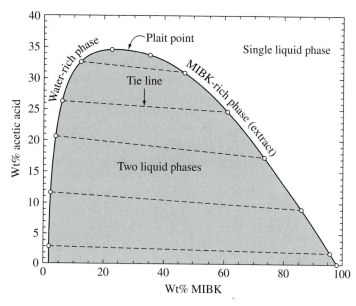

Figure 5.15 MIBK-acetic acid-water liquid-liquid phase diagram at 25°C. Drawn from data in *Perry's Chemical Engineers' Handbook,* 6th edition.

include the temperature effect as well and keep the graph two-dimensional, so data are commonly presented at a single temperature!

A liquid-liquid phase diagram for the three-component water–acetic acid–MIBK system is shown in Fig. 5.15. (These data are also included in App. B.) The weight percent acetic acid is plotted versus the weight percent MIBK; the weight percent water can be calculated by difference (wt% water = 100 − wt% MIBK − wt% acetic acid.) Notice that there is a dome-like curve drawn on the graph. Any point that lies under the dome is in the two-phase region—at these compositions the mixture spontaneously splits into two liquid phases. The dashed lines under the dome are *tie lines*; points on the curve connected by the tie line show the compositions of the two liquid phases at equilibrium. The MIBK (organic) phase composition is the rightmost point on the tie line, while the water-rich (aqueous) phase composition is the leftmost point. The organic phase is commonly called the *extract* and the aqueous phase is called the *raffinate*. At any mixture composition that falls outside the dome, the mixture forms a single liquid phase. The point near the top of the dome is called the plait point; it is the mixture composition that separates the single-phase from the two-phase region.

Illustration: Reading Liquid-Liquid Phase Diagrams
A mixture of 57 wt% MIBK, 10 wt% acetic acid, 33 wt% water at 25°C falls under the dome so it splits into two liquid phases. The point lies exactly on the

Is a mixture of 38% acetic acid, 22% water, and 40% MIBK one phase or two?

I have a mixture of acetic acid, water and MIBK that has separated into two liquid phases at equilibrium. One of the phases is 61 wt% MIBK. What is the composition of the two phases?

tie line. We follow the tie line in both directions to find the compositions of the two phases. The extract phase is 86 wt% MIBK, 9 wt% acetic acid, and (by difference) 5 wt% water. The raffinate phase is 2.5 wt% MIBK, 11.5 wt% acetic acid, and 86 wt% water.

If a mixture composition does not fall exactly on a tie line, then one must visually interpolate (or "eyeball") the position of the tie line. The slope of the tie line should be in between the slope of the two adjacent tie lines. For example, the mixture of 30 wt% MIBK, 5 wt% acetic acid, 65 wt% water splits into two phases: The raffinate phase is about 2 wt% MIBK, 6 wt% acetic acid, and 92 wt% water, while the extract phase is about 93 wt% MIBK, 3 wt% acetic acid, and 4 wt% water.

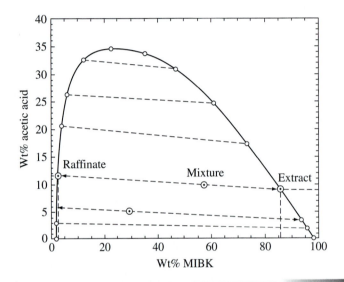

5.5.3.6 Multicomponent Fluid-Solid Phase Equilibrium   Now let's move to a case where one of the components is present only in the solid phase and not in the fluid phase. For example, a liquid mixture of phenol and water is in equilibrium with activated carbon. The activated carbon is only in the solid phase. Some of the phenol sticks (*adsorbs*) to the solid carbon surface. In other words, the phenol *partitions* or *distributes* between the liquid and the solid surface.

Figure 5.16 depicts the phase equilibrium between an antibody in solution and the antibody adsorbed to a special solid called an immunosorbent, at 25°C. (Antibodies are large proteins that bind tightly and specifically to other proteins or cells. They are used in health care and biotechnology products, primarily for diagnostic purposes but occasionally as drugs.) Figure 5.16 is an example of an *adsorption isotherm* and is very easy to use. If the concentration of the antibody in solution is known, then the quantity of adsorbed antibody per gram immunosorbent is read directly from the graph.

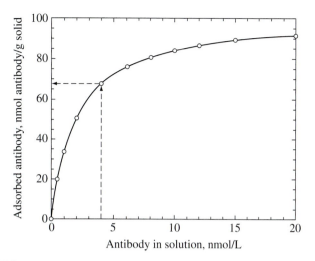

Figure 5.16 Adsorption isotherm for solid-fluid phase equilibrium in an antibody-immunosorbent system.

Illustration: *Reading Adsorption Isotherms*

A solution contains 4 nmol monoclonal antibody per liter and the solution is in equilibrium with the immunosorbent. The amount of antibody on the immunosorbent is determined from Fig. 5.16 to be about 67 nmol/g solid. (*Note:* 1 nmol $= 10^{-9}$ gmol.)

5.6   Equilibrium-Based Separations

Now that we've learned how to find phase equilibrium relationships, we return to the main theme of this chapter. We are ready to predict the performance of an equilibrium-based separator, choose operating conditions, compare different kinds of separations—in short, we can now start designing separators!

Recall that the equilibrium stage is the basis for design and analysis of equilibrium-based separations. In an equilibrium stage, a feed is brought to a two-phase mixture by addition of a separating agent; the phases are equilibrated and then separated into two products. (You may want to briefly review Sec. 5.4.2 before proceeding.) To analyze the equilibrium stage, we combine material balance equations with phase equilibrium information. Phase equilibrium information tells us the composition of the two equilibrated phases but nothing about the quantity of each phase. Material balance equations tell us the quantity of each phase. We write one material balance equation for each component, as usual. We write one phase equilibrium relationship for each component *that is present in both phases*.

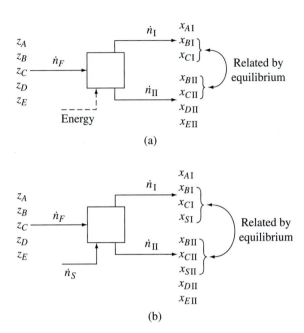

Figure 5.17 Representative equilibrium stage separators. (a) An energy separating agent changes the temperature and/or pressure of the feed. Phase equilibrium data are used to relate x_{BI} to x_{BII} and x_{CI} to x_{CII}. (b) A solvent S is a material separating agent that is mixed with the feed. Phase equilibrium data are used to relate mole fractions of the 3 components that appear in both phases.

Let's illustrate by examining the situations depicted in Fig. 5.17. An equilibrium stage with an energy separating agent is shown in Fig. 5.17a. The feed is a mixture of five components, and the equilibrium stage is designed to separate component B from component C. Components B and C are the key components and they appear in both phases, while components A, D, and E are nondistributing components and exit in only a single stream. There are 12 independent stream variables. If the feed flow rate and feed composition are known, these 5 specifications plus the 5 material balance equations yield DOF $= 12 - (5 + 5) = 2$. Specifying that the streams exiting the stage are at equilibrium at a known temperature and pressure provides us with the last two bits of required information to completely specify the system: the two phase equilibrium relationships between the two components, B and C, that are present in both phases. Figure 5.17b depicts an equilibrium stage with a material separating agent, solvent S. There are now 15 stream variables and 6 material balance equations. If the feed flow rate, feed composition, and solvent flow rate are known, then DOF $= 15 - (6 + 6) = 3$. Specifying that the streams exiting the stage are at equilibrium at a known temperature and pressure provides us with the last three bits of required information to completely specify the system: the phase equilibrium relationships between the three components, B, C, and S, that are present in both phases.

In this section we discuss several common separation technologies. Observe carefully how, in each case, we combine material balance equations with appropriate phase equilibrium information. We start with separation technologies that rely on energy as a separating agent: crystallization, evaporation, condensation, and distillation. Then we discuss separation technologies in which a material is added to act as the separating agent: absorption, adsorption, and extraction. Be forewarned that what we cover is just the tip of the iceberg! The strategies you learn are applicable to a wider variety of separation processes than just those we can illustrate here.

5.6.1   Crystallization

Crystallization is a common separation process in the commodity inorganic industry (e.g., salts), in food processing (e.g., sugars), and in pharmaceutical manufacturing (e.g., chiral drugs). The aim is to take advantage of the differences in composition of liquid and solid phases in equilibrium. Generally, the desired product is a pure solid. There are two ways to crystallize out a solid product from a liquid solution by adjusting temperature: (1) by cooling the solution, if the solubility of the desired product decreases with temperature, or (2) by evaporation of the volatile component, if the crystallizable material is nonvolatile. In either case, the addition or removal of energy is required to shift the equilibrium toward the two-phase region.

Example 5.13

Process Flow Calculations with Liquid-Solid Equilibrium Data: Potassium Nitrate Crystallization

100 g of a 50 wt% KNO_3 aqueous solution at 80°C is cooled to 20°C. What is the wt% KNO_3 of the liquid? How much solid salt is produced? What is the fractional recovery of KNO_3 into the solid salt product?

Solution

As always, we start by drawing and labeling a flow diagram. We'll use K and W to denote our two components KNO_3 and water, with

z_i = mass fraction i in feed
x_{iL} = mass fraction i in the liquid solution,
x_{iS} = mass fraction i in the solid,
m_F = initial mass (g) in system
m_L = total mass (g) of liquid phase in system at equilibrium
m_S = total mass (g) of solid phase in system at equilibrium

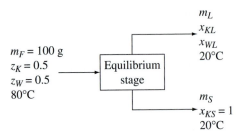

The initial mass and stream compositions are known from the problem statement:

$$m_F = 100 \text{ g}$$

$$z_W m_F = 0.5(100) = 50 \text{ g H}_2\text{O}$$

$$z_K m_F = 0.5(100) = 50 \text{ g KNO}_3$$

This is a batch process; the integral material balance equations for the two components simplify to

$$x_{WS} m_S + x_{WL} m_L = 50$$

$$x_{KS} m_S + x_{KL} m_L = 50$$

We need phase equilibrium information to find the composition of the solid and liquid phases at equilibrium, so we turn to Fig. 5.11. At 80°C and $z_K = 0.5$, the solution is unsaturated. To follow what happens to the solution as it cools, we move vertically down at a constant mol fraction $z_K = 0.5$. A solid phase starts to form at 55°C. As the system cools further, the solubility decreases. At 20°C, the saturated liquid is 24 wt% KNO$_3$, and the solid phase is pure salt. Thus, $x_{KL} = 0.24$, $x_{WL} = 0.76$, $x_{KS} = 1.0$, and $x_{WS} = 0.0$.

Inserting these values into the material balance equations yields

$$0.76 m_L = 50$$

$$m_S + 0.24 m_L = 50$$

from which we find

$$m_L = 66 \text{ g}$$

$$m_S = 34 \text{ g}$$

The fractional recovery f_R of KNO$_3$ in the solid phase is

$$f_R = \frac{x_{KS} m_S}{z_K m_F} = \frac{34}{50} = 0.68$$

Thus, one equilibrium stage has produced pure (100 wt%) KNO$_3$ salt but has recovered only 68% of the KNO$_3$ fed.

The solid and liquid phases produced during crystallization need to be separated out by mechanical means. This is often accomplished by filtration. Entrainment of solution in the filter cake is a common problem, as we observed in Example 5.11. The performance of crystallization processes is affected by entrainment, as illustrated in the following example.

Example 5.14   **Entrainment Effects in Equilibrium-Based Separations: Separation of Benzene and Naphthalene by Crystallization**

1000 kg/h of a liquid solution of 60 wt% naphthalene/40 wt% benzene at 80°C is cooled to 10°C. The resulting solid and liquid phases are separated in a rotary drum filter. The process operates at steady state. Your job is to determine how well the drum filter is performing. You determine that the filtrate liquid flow rate is 505 kg/h. How much entrainment (kg entrained liquid/kg solids) does the filter leave? By how much does any entrainment change the recovery of naphthalene in the filter cake product or the purity of the product?

Solution
The flow diagram is shown.

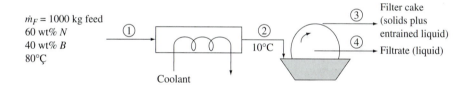

Solid-liquid equilibrium data for the benzene-naphthalene system are available (Fig. 5.12). However, the data are given as mole fractions, whereas the information in the problem statement is in mass units, so we need to convert to consistent units:

$$\dot{n}_F = \sum \frac{\dot{m}_{iF}}{M_{iF}} = \frac{600 \text{ kg/h naphthalene}}{128 \text{ kg/kgmol}} + \frac{400 \text{ kg/h benzene}}{78 \text{ kg/kgmol}} = 9.82 \text{ kgmol/h}$$

$$z_N = \frac{600/128}{9.82} = 0.478$$

$$z_B = \frac{400/78}{9.82} = 0.522$$

We will analyze each unit as a separate system, starting with the crystallizer. A liquid solution enters the crystallizer (stream 1) and a multiphase mix of liquid and

solid leaves (stream 2). The material balance equations for naphthalene (N) and benzene (B) are

$$0.478(9.82) = 4.69 = x_{NL}\dot{n}_{L2} + x_{NS}\dot{n}_{S2}$$

$$0.522(9.82) = 5.13 = x_{BL}\dot{n}_{L2} + x_{BS}\dot{n}_{S2}$$

where $\dot{n}_{L2}$ and $\dot{n}_{S2}$ are the liquid and solid flow rates in stream 2, respectively, and
x_{iL} and x_{iS} are the liquid and solid mole fractions of component i at equilibrium.

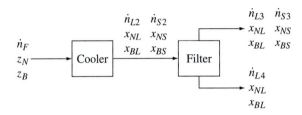

Now we need phase equilibrium information to determine the mole fractions of
naphthalene and benzene in each phase. Consulting Fig. 5.12, we learn that a mixture of 48 mol% naphthalene/52 mol% benzene at 10°C produces a saturated liquid
at 20 mol% naphthalene ($x_{NL} = 0.2$, $x_{BL} = 0.8$) and a pure naphthalene solid ($x_{NS} =$
1.0). Therefore:

$$4.69 = 0.2\dot{n}_{L2} + 1.0\dot{n}_{S2}$$

$$5.13 = 0.8\dot{n}_{L2}$$

Solving, we get

$$\dot{n}_{L2} = 6.41 \text{ kgmol/h}$$

$$\dot{n}_{S2} = 3.41 \text{ kgmol/h}$$

Now we analyze the performance of the filter. If the rotary drum filter yielded a perfect mechanical separation, the fractional recovery of naphthalene as solid would be

$$f_{RNS} = \frac{x_{NS}\dot{n}_{S2}}{z_N\dot{n}_F} = \frac{3.41}{0.478(9.82)} = 0.726 \qquad \text{(perfect filter)}$$

with a product purity of 100%.

The measured filtrate flow is 505 kg/h. Since we know that the saturated liquid
solution is 20 mol% naphthalene/80 mol% benzene, we can now convert this mass
flow rate to a molar flow rate:

$$\dot{n}_{L4} = \frac{\dot{m}_{L4}}{x_{NL}M_N + x_{BL}M_B} = \frac{505 \text{ kg/h}}{0.2(128 \text{ kg/kgmol}) + 0.8(78 \text{ kg/kgmol})}$$

$$= 5.74 \text{ kgmol/h}$$

If the filter worked perfectly, the filtrate (stream 4) would contain all the liquid in the stream leaving the cooler, or 6.41 kgmol/h. The difference between the perfect and the actual flow rate equals the entrained liquid in the filter cake:

$$\dot{n}_{L3} = \dot{n}_{L2} - \dot{n}_{L4} = 6.41 - 5.74 = 0.67 \text{ kgmol/h}$$

Assuming that all the solids are in the filter cake, then the filter entrainment on a mass basis is

$$\frac{\dot{n}_{L3}(x_{NL}M_N + x_{BL}M_B)}{\dot{n}_{S3}M_N} = \frac{0.67[0.2(128) + 0.8(78)]}{3.41(128)} = 0.135 \text{ kg solution/kg solid}$$

The fractional recovery is

$$f_{RNS} = \frac{x_{NS}\dot{n}_{S3} + x_{NL}\dot{n}_{L3}}{z_N \dot{n}_F} = \frac{1(3.41) + 0.2(0.67)}{0.478(9.82)} = 0.755 \qquad \text{(real filter)}$$

The mole fraction of naphthalene in the filter cake (solids + entrained liquid) is

$$\frac{x_{NL}\dot{n}_{L3} + x_{NS}\dot{n}_{S3}}{\dot{n}_{L3} + \dot{n}_{S3}} = \frac{0.2(0.67) + 1(3.41)}{0.67 + 3.41} = 0.87$$

Entrainment has slightly increased the fractional recovery but decreased the purity of the filter cake from 100% to 87 mol% naphthalene.

5.6.2 Evaporation, Condensation, and Equilibrium Flash

Evaporation, condensation, and **equilibrium flash** are related: Each of these separation processes produces vapor and liquid streams in equilibrium with each other. The term "evaporation" is used most often when a liquid feed is partially or completely vaporized. "Condensation" is used commonly when a vapor feed is partially or completely condensed. "Equilibrium flash" is conventionally reserved for processes where the components in the mixture have fairly similar volatilities. A flash drum is just a vessel in which a liquid stream is "flashed"—that is, heated until some of it is vaporized and the liquid and vapor phases reach equilibrium. Then the phases are separated into two products. Evaporation, condensation, and equilibrium flash are all designed and analyzed in similar fashion: by combining vapor-liquid phase equilibrium and material balance equations.

| **Example 5.15** | **Process Flow Calculations with Raoult's Law: Dehumidification of Air by Condensation** |

Wet air at 27°C and 760 mmHg contains 2.6 mol% water vapor. We need to feed the air at 1000 kgmol/h into a reactor, but the air must contain at most 0.6 mol% water vapor. You propose to remove the water by condensation. Calculate the flow rate of dry air leaving the separator. Determine the appropriate operating temperature. Finally, determine the relative humidity of the wet air. (Relative humidity, or relative saturation, is the ratio of the actual water vapor content of the air to the water vapor content if the air were saturated with water.) You may assume that air is completely insoluble in water.

Solution

Water and air differ greatly in their relative volatility. Let's try cooling the air to condense out part of the water:

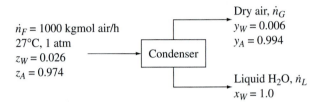

This is a steady-state continuous-flow separator. We will treat dry (water-free) air A as a single composite material rather than consider O_2, N_2, argon, etc. as separate components (Why is this OK?). Water W is the second component. The gas stream leaving the condenser contains both components whereas the liquid stream is presumed to contain only water. The steady-state material balance equations for air and water are

$$z_A \dot{n}_F = 0.974(1000) = 974 \text{ kgmol/h} = y_A \dot{n}_G = 0.994 \dot{n}_G$$

$$z_W \dot{n}_F = 0.026(1000) = 26 \text{ kgmol/h} = y_W \dot{n}_G + x_W \dot{n}_L = 0.006 \dot{n}_G + 1.0 \dot{n}_L$$

where $\dot{n}_G$ and $\dot{n}_L$ are the molar flow rates (kgmol/h) of the gas and liquid products, respectively. We solve to find

$$\dot{n}_G = 980 \frac{\text{kgmol}}{\text{h}}$$

$$\dot{n}_L = 20 \frac{\text{kgmol}}{\text{h}}$$

Now we need to find the operating temperature at which $y_W = 0.006$. Let's model vapor-liquid equilibrium of water using Raoult's law:

$$y_W = 0.006 = \frac{P_W^{\text{sat}}}{P} x_W$$

! **Helpful Hint**
If a component appears in only one phase, we do not need a phase equilibrium relationship for that component.

Since air is assumed to be insoluble in water, the liquid is pure water, $x_W = 1.0$. From the problem statement, $P = 760$ mmHg. Therefore:

$$P_W^{\text{sat}} = \frac{y_W P}{x_W} = \frac{0.006(760)}{1} = 4.56 \text{ mmHg}$$

Saturation pressure is a function of temperature. We can use either tabulated data or Antoine's equation to determine the temperature at which $P_w^{sat} = 4.56$ mmHg. We'll use Antoine's equation (App. B):

$$\log_{10} P_W^{sat} = \log_{10}(4.56 \text{ mmHg}) = 8.10765 - \frac{1750.286}{T(^\circ C) + 235}$$

$$T = 0^\circ$$

We need to cool the air down to 0°C to condense out sufficient water. We'll need to be concerned about ice formation.

Finally, to calculate the relative humidity of the wet air feed, we first need to find the mole fraction of water vapor in the air if it were saturated. At 27°C,

$$\log_{10} P_W^{sat} = 8.10765 - \frac{1750.286}{27 + 235} = 1.427$$

$$P_W^{sat} = 10^{1.427} = 26.74 \text{ mmHg}$$

From Raoult's law, the saturation composition is

$$y_w = \frac{P_w^{sat}}{P} x_w = \frac{26.74}{760}(1) = 0.035$$

3.5 mol% is the maximum water vapor composition of the air at 27°C; this corresponds to 100% relative humidity. The actual relative humidity is

$$\frac{0.026}{0.035} \times 100\% = 74\%$$

Example 5.16   **Process Flow Calculations with Raoult's Law: Equilibrium Flash of a Hexane/Heptane Mixture**

A mixture of 40 mol% n-hexane (C_6H_{14})/60 mol% n-heptane (C_7H_{16}) is fed to a flash drum operating at 1520 mmHg pressure. The feed rate is 100 gmol/s. The mixture is heated to 111°C, and vapor and liquid products are removed continuously. Calculate the vapor and liquid flow rates and compositions. What are the fractional recoveries of hexane in the vapor and of heptane in the liquid?

Solution:
The flow diagram is

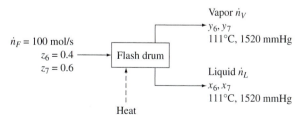

Helpful Hint

Write a phase equilibrium relationship for every component that is present in both phases at equilibrium.

(We use 6 and 7 subscripts to distinguish hexane and heptane, respectively.) The steady-state material balance equations are

$$0.4(100) = y_6 \dot{n}_V + x_6 \dot{n}_L$$

$$0.6(100) = y_7 \dot{n}_V + x_7 \dot{n}_L$$

We use Raoult's law to model the phase equilibrium behavior of both hexane and heptane.

$$y_6 = \frac{P_6^{\text{sat}}}{P} x_6$$

$$y_7 = \frac{P_7^{\text{sat}}}{P} x_7$$

From Antoine's equation (App. B) we find saturation pressures of hexane and heptane at 111°C:

$$\log_{10}(P_6^{\text{sat}}) = 6.87601 - \frac{1171.17}{111 + 224.41} = 3.3843$$

$$P_6^{\text{sat}} = 10^{3.384} = 2422 \text{ mmHg}$$

$$\log_{10}(P_7^{\text{sat}}) = 6.89677 - \frac{1264.90}{111 + 216.54} = 3.035$$

$$P_7^{\text{sat}} = 10^{3.035} = 1084 \text{ mmHg}$$

Inserting these values, plus the total pressure (1520 mmHg) into Raoult's law yields

$$y_6 = \frac{2422}{1520} x_6 = 1.593 x_6$$

$$y_7 = \frac{1084}{1520} x_7 = 0.713 x_7$$

We also know that

Quick Quiz 5.16

What's the separation factor for the hexane-heptane flash of Example 5.16? What's the ratio of vapor pressures of hexane and heptane? Is it a coincidence that your answers to these questions are the same?

$$x_6 + x_7 = 1$$

$$y_6 + y_7 = 1$$

(*Note:* $x_6 + y_6 \neq 1$!)
Combining, we get

$$1 = 1.593 x_6 + 0.713 x_7 = 1.593 x_6 + 0.713(1 - x_6)$$

We now work our way back through these equations to find

$$x_6 = 0.326, \, x_7 = 0.674$$

$$y_6 = 0.519, \, y_7 = 0.481$$

$$\dot{n}_V = 38.5 \text{ gmol/s}, \, \dot{n}_L = 61.5 \text{ gmol/s}$$

The fractional recoveries are

$$f_{R6} = \frac{y_6 \dot{n}_V}{z_6 \dot{n}_F} = \frac{0.519(38.5)}{0.4(100)} = 0.50$$

$$f_{R7} = \frac{x_7 \dot{n}_L}{z_7 \dot{n}_F} = \frac{0.674(61.5)}{0.6(100)} = 0.69$$

The products are not particularly pure, and the fractional recoveries are less than impressive. Is there a better way? Stay tuned!

Example 5.17   **Vapor-Liquid Separations with Nonideal Solutions: Equilibrium Flash Separation of Ethanol-Water Mixture**

A liquid stream containing 40 mol% ethanol and 60 mol% water is fed to a flash drum operating at 1 atm. The feed rate is 100 gmol/s. Within the stage the temperature is raised to 85°C. What are the flow rates and compositions of the vapor and liquid streams leaving the stage? Assume steady state operation.

Solution
The flow diagram is shown.

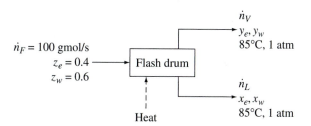

The material balance equations on ethanol and water simplify to

$$0.4(100) = y_e \dot{n}_V + x_e \dot{n}_L$$

$$0.6(100) = y_w \dot{n}_V + x_w \dot{n}_L$$

Now we need to determine the composition of the vapor and liquid phases. This requires phase equilibrium information. Luckily, we've got Fig. 5.13. First, check that you are in the two-phase vapor-liquid coexistence region. (You are.) Second, use Fig. 5.13 to determine that the ethanol vapor and liquid mole fractions at 85°C are

$$x_e = 0.13$$

$$y_e = 0.48$$

$$x_w = 0.87$$

$$y_w = 0.52$$

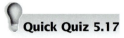

Quick Quiz 5.17

Why mustn't we use Raoult's law for Example 5.17?

Now we insert the phase equilibrium information into the material balance equations. Solving the equations simultaneously yields

$$\dot{n}_V = 77\frac{\text{gmol}}{\text{s}}$$

$$\dot{n}_L = 23\frac{\text{gmol}}{\text{s}}$$

5.6.3   Distillation (Optional)

Let's look back at the equilibrium flash separation of the hexane-heptane mixture in Example 5.16. The vapor product was only 52% pure hexane, and the recovery of hexane in the vapor product was only 50%. Is there a way to get higher purity and component recovery?

Luckily, yes. The basic idea is this: Take the liquid product from the equilibrium flash, heat it up to partially vaporize, and separate into vapor and liquid. The liquid product from the second flash will be more concentrated in heptane. Similarly, take the vapor product from the first flash, cool to partially condense, and separate into vapor and liquid. The vapor product from this process will be more concentrated in hexane. Repeat equilibrium flashes over and over again.

Distillation is the separation technology of multiple equilibrium flashes. In distillation columns, the feed is introduced roughly in the middle of the column, while products come out the top and bottom (Fig. 5.18). The product from the top of the column is called the *distillate* and contains preferentially the more volatile (lower boiling point) components. The product from the bottom of the column is called, appropriately, *bottoms*, and it contains preferentially the less volatile (higher boiling point) components. A reboiler (heater) is added at the bottom, to

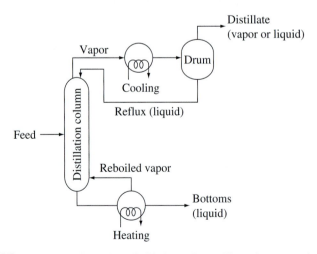

Figure 5.18 Schematic of a typical distillation column. The column contains many equilibrium stages.

provide vapor to flow up the column, and a condenser (cooler) is added at the top, to provide liquid to flow down the column. Essentially, energy is put into the column at the bottom and taken out at the top, and this drives the separation.

The column pressure is nearly constant, but the temperature decreases as we move up the column. The top of the column operates close to the dew point temperature of the distillate (at the column pressure), and the bottom of the column operates close to the bubble point of the bottoms product.

The distillation column is essentially a stack of trays, each of which functions as an equilibrium stage. Distillation is very useful because it is relatively easy to build lots of stages. It is not uncommon to have distillation columns with 20 or 30 stages, or even more. That means that we often can economically separate compounds with similar boiling point temperature (e.g., a few degrees apart) using distillation. There is an easy method to estimate the total number of stages required for a given separation. The equation, known as the **Fenske equation,** is

$$N_{min} = \frac{\log\left(\dfrac{y_{Ad}x_{Bb}}{y_{Bd}x_{Ab}}\right)}{\log\left(\dfrac{P_A^{sat}}{P_B^{sat}}\right)} \qquad (5.16)$$

where A is the more volatile compound (which will end up primarily in the distillate d) and B is the less volatile compound (which will end up primarily in the bottoms b). N_{min} is the minimum number of stages required; the actual number of stages is typically about 20% to 100% more than N_{min}. Notice that the numerator in the Fenske equation is simply the log of the separation factor α_{AB} for the entire column. The denominator is simply the log of the separation factor for a single equilibrium stage, if the mixture obeys Raoult's law, because at equilibrium

$$\frac{P_A^{sat}}{P_B^{sat}} = \frac{y_A P/x_A}{y_B P/x_B} = \left(\frac{y_A}{y_B}\frac{x_B}{x_A}\right)$$

Thus, we can think of the Fenske equation in this way: The total number of stages required is the ratio of the total separation required (the numerator) divided by the separation achieved per stage (the denominator). The saturation pressures are calculated at the average column temperature, which is the mean of the dew point of the distillate and the bubble point of the bottoms, or

$$T_{ave} = \frac{T_{dp,\ distillate} + T_{bp,bottoms}}{2}$$

If the feed is a multicomponent mixture, Eq. (5.16) can still be used: components A and B are the key components (see Fig. 5.6). Estimating the number of stages required is an early step in estimating the cost of building a distillation column.

Example 5.18	The Power of Multistaging: Distillation versus Equilibrium Flash for Hexane/Heptane Separation

In Example 5.16, you analyzed the separation of a mix of 40 mol% n-hexane and 60 mol% n-heptane using a single-stage flash drum, operated at 2 atm and 111°C. The result was not impressive: The vapor (38.5% of the feed) was only 51.9 mol% n-hexane and the liquid was only 67.4 mol% n-heptane.

Your customer demands 99 mol% pure n-hexane and 99.5 mol% pure n-heptane. You decide to design a distillation column to meet customer specifications, and you choose an operating pressure of 1520 mmHg. Estimate the temperature at the top and bottom of the column. About how many stages should the column have?

Solution

The temperature at the top of the column is approximately the dewpoint temperature of the distillate (99% n-hexane) at the column operating pressure. We use Raoult's law to calculate the dew point temperature; using a strategy similar to that used for Example 5.12.

$$x_6 + x_7 = 1 = \frac{y_6 P}{P_6^{sat}} + \frac{y_7 P}{P_7^{sat}} = 1520 \left(\frac{0.99}{P_6^{sat}} + \frac{0.01}{P_7^{sat}} \right)$$

with the saturation pressures described by Antoine's equations

$$\log_{10} P_6^{sat} = 6.87601 - \frac{1171.17}{T(°C) + 224.41}$$

$$\log_{10} P_7^{sat} = 6.89677 - \frac{1264.90}{T(°C) + 216.54}$$

We combine these equations and solve for T; the dew point temperature of the distillate is 93.1°C.

The bottoms bubble point temperature is found from the specification that the bottoms product contain 99.5 mol% n-heptane:

$$y_6 + y_7 = 1 = \frac{P_6^{sat}}{P} x_6 + \frac{P_7^{sat}}{P} x_7 = \frac{0.005 P_6^{sat} + 0.995 P_7^{sat}}{1520}$$

Combining with the Antoine equations and solving for T, we find that the bubble point temperature of the bottoms is 123.7°C.

Now we use the Fenske equation to calculate the minimum number of stages. The average column temperature is $(T_{dp} + T_{bp})/2 = 108.4°C$. The saturation pressures of hexane and heptane at 108.4°C, calculated from the Antoine equation, are 2275 and 1009 mmHg, respectively. Therefore,

$$N_{min} = \frac{\log\left(\frac{y_{6d}}{y_{7d}} \frac{x_{7b}}{x_{6b}}\right)}{\log\left(\frac{P_6^{sat}}{P_7^{sat}}\right)} = \frac{\log\left(\frac{0.99}{0.01} \frac{0.995}{0.005}\right)}{\log\left(\frac{2275}{1009}\right)} = \frac{\log(19{,}700)}{\log(2.255)} = 12$$

A real column typically has about 1.2 to 2 times the minimum number of stages; for this case that yields an estimate of approximately 15 to 25 required stages.

Once we have the equations set up, we can look at all kinds of variations. For example, it is interesting to see how the minimum number of stages changes with the product purity requirements. This is important, because the cost of building the column increases with increasing number of stages. The graph below shows the results of a calculation where we varied the distillate and bottoms purity from 80% to 99.99%.

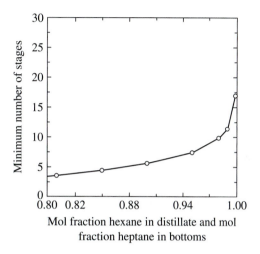

Mol fraction hexane in distillate and mol fraction heptane in bottoms

This graph shows that getting the last bit of improvement in purity is very expensive! It makes sense to give customers a product with the lowest purity that they will accept.

5.6.4 Absorption, Adsorption, and Extraction

In the previous sections, we described some very common equilibrium-based separation technologies: crystallization, equilibrium flash, and distillation. In all example problems, the temperature of the feed was adjusted so that two phases differing in composition were produced.

Sometimes, however, temperature (or pressure) manipulations cannot be exploited to achieve desired separations, either for economic or technical reasons. For example, many pharmaceutical compounds degrade before they reach the high temperatures that would be needed for vaporization. As another example, the very low temperatures required to condense some gases can make these processes expensive, or difficult to implement at low production quantities. As a third example, removal of a trace nonvolatile contaminant from a polluted water stream by evaporation of all the water is much less desirable than a method that specifically plucks out the contaminant and lets the water through undisturbed.

In cases like these, separation technologies employing material separating agents are used. The feed is a single-phase multicomponent mixture, and a foreign

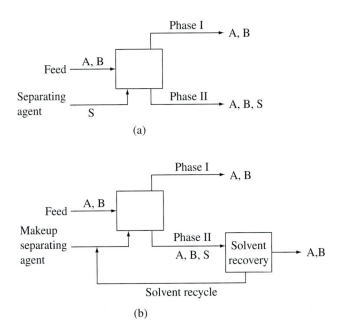

Figure 5.19 Typical equilibrium-based separation with material separating agent. A second separation unit as shown in (b) is frequently required for recovery and reuse of the separating agent.

material that differs in phase from the feed is added (Fig. 5.19a). One of the components in the feed mixture preferentially partitions into the second, added, phase, while the other feed components do not.

One of the key advantages of separation technologies that employ material separating agents is that we have the whole chemical world at our disposal. We can select a material (or mixture of materials) that is best tuned for the exact application at hand. This wide choice of chemistry confers the possibility of high specificity—with luck and skill, we can very efficiently perform the desired separation.

One of the key disadvantages of separation technologies that employ material separating agents is that almost always the separating agent must be removed and recovered if the process is to be technically and economically feasible. If the desired component partitions into the added phase, then it must be recovered from that phase. This means that we have not one separation problem but two—the original separation, plus the separation of the desired product from the separating agent (Fig. 5.19b). Selection of the best separating agent for a given application requires consideration of both facets.

We'll discuss three related separation technologies in this section. Analysis of many other separation technologies with material separating agents, such as leaching or flocculation, proceeds in exactly the same manner.

In **absorption,** the feed is a gas mixture. It is brought into contact with a liquid solvent. One of the components of the gas preferentially dissolves in the liquid.

Gas and liquid phases are separated after equilibration; the gas phase is depleted in the soluble component and enriched in the insoluble component. **Stripping** is the companion separation process to absorption. The feed is a liquid containing a dissolved gas; the feed is contacted with a gas phase and the solute in the liquid phase transfers to the gas phase.

In **adsorption,** the feed is a fluid (vapor or liquid) mixture. The feed is brought into contact with a solid. One of the components of the fluid preferentially adsorbs (sticks to) the solid. Fluid and solid phases are separated after equilibration; the solid phase is enriched in the adsorbing component.

In **solvent extraction,** the feed is a liquid mixture. It is brought into contact with a liquid solvent that is not miscible with the feed, so that there are two liquid phases present. One of the components in the feed phase preferentially partitions into the solvent phase.

Absorption, adsorption, and solvent extraction are used frequently to remove small quantities of a contaminant in a fluid stream that contains many components. This practice follows the heuristic: *For recovery of trace quantities, use separation methods where the cost increases with the quantity of material to be recovered, not the quantity of the stream to be processed.* Furthermore, because adsorbents and solvents can be tailor-made for specific applications, very high separation factors can be obtained. In the following examples, we consider each of these technologies in turn.

Absorption   The success of an absorption process depends on the choice of the solvent added. Ideally, the solvent is nontoxic, cheap, nonflammable, environmentally benign, noncorrosive, stable, nonfoaming, nonvolatile, and not very viscous. Water is a good default choice. Example 5.19 illustrates how to combine material balance equations with gas-liquid phase equilibrium information to design an absorber. Gas-liquid phase equilibrium data may be available in graphical form; or, for some cases, a model equation like Henry's law is appropriate.

Example 5.19

Process Flow Calculations Using Gas-Liquid Equilibrium Data: Cleaning up Dirty Air by Absorption

A gas stream is contaminated with SO_2 (300 kg air, 63 kg SO_2.) The concentration of SO_2 must be reduced before the gas can be released to the atmosphere. To remove SO_2, the gas is contacted with water at 0°C and 21,000 mmHg (27.6 atm), the streams are equilibrated, then the gas and water streams are separated. If 95% of the SO_2 in the gas must be removed, how much water is required? You may assume that the air is completely insoluble in water, and that the water is nonvolatile.

Solution
As always, we start with a labeled flow diagram. We'll work in units of kg, and use the subscripts a, s, and w to indicate air, SO_2 and water, respectively.

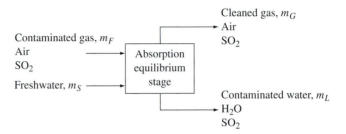

The fractional recovery of SO_2 in the liquid product is specified as 0.95, therefore:

$$f_{RsL} = 0.95 = \frac{m_{sL}}{m_{sF}} = \frac{m_{sL}}{63}$$

$$m_{sL} = 60 \text{ kg } SO_2$$

By material balance, 3 kg SO_2 remain in the cleaned gas.

$$m_{sG} = m_{sF} - m_{sL} = 63 - 60 = 3 \text{ kg } SO_2$$

The cleaned gas and the contaminated water streams *leaving* the absorber are in equilibrium with each other. Thus, we need phase equilibrium data, conveniently provided by Fig. 5.14. In order to use this data, we need to convert the mass fraction of SO_2 in the cleaned gas to a partial pressure:

$$p_s = y_s P = \left[\frac{3 \text{ kg } SO_2 \left(\frac{\text{kgmol } SO_2}{64 \text{ kg } SO_2} \right)}{3 \text{ kg } SO_2 \left(\frac{\text{kgmol } SO_2}{64 \text{ kg } SO_2} \right) + 300 \text{ kg air} \left(\frac{\text{kgmol air}}{28.8 \text{ kg air}} \right)} \right] (21{,}000 \text{ mmHg})$$

$$= 94 \text{ mmHg}$$

From Fig. 5.14, at 0°C, $p_s = 94$ mmHg corresponds to about 3.4 kg SO_2 per 100 kg water in the liquid phase at saturation. Since the contaminated water stream needs to carry 60 kg SO_2,

$$m_{wL} = \left(\frac{100 \text{ kg water}}{3.4 \text{ kg } SO_2} \right) 60 \text{ kg } SO_2 = 1765 \text{ kg water}$$

By material balance on water, the solvent requirement is calculated:

$$m_S = m_{wL} = 1765 \text{ kg water}$$

Adsorption   The water softener in the basement and the carbon filter on the kitchen faucet are two very widely used adsorption devices. Both are designed for applications where a material present at very low concentration (e.g., calcium and magnesium ions, or organic contaminants) must be removed from another material used in high volume (water). Because the surface of solid adsorbents can be chemically modified, adsorption systems can be highly selective in recovering a single component at low concentrations from complex multicomponent fluids. The

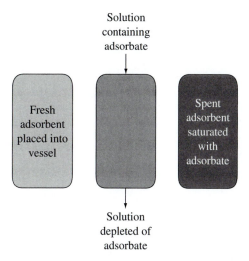

Figure 5.20 Adsorption processes are frequently run in semibatch mode. The solid adsorbent is loaded into the vessel initially. Then feed solution is pumped through the adsorbent bed, and the adsorbate sticks to the adsorbent. After some time, the adsorbent is completely saturated with adsorbate. The spent adsorbent is replaced with fresh adsorbent, and the process resumes. Spent adsorbent is generally regenerated for reuse.

component that sticks to the solid adsorbent is called the *adsorbate*. Adsorbers are usually run in either batch or semibatch mode, as shown in Fig. 5.20, because solids are hard to pump around. The design and analysis of adsorbers as separators is analyzed in the same manner as absorbers—by combining phase equilibrium data and material balance equations.

Example 5.20	**Process Flow Calculations Using Adsorption Isotherms: Monoclonal Antibody Purification**

Hybridoma cells are special mammalian cells that produce monoclonal antibodies, large proteins that bind very tightly and very specifically to other proteins or cells. Monoclonal antibodies are used in both health care and biotechnology products such as pregnancy tests and cancer treatment. The purity required for these products is extremely high. The cells secrete monoclonal antibodies at low concentrations into a culture media (an aqueous solution of vitamins, proteins, growth factors, sugars, amino acids, etc.). Monoclonal antibodies, like most proteins, are very temperature sensitive and are degraded in most organic solvents or at pH extremes. These facts make distillation, crystallization, and solvent extraction infeasible separation technologies. Monoclonal antibodies have the very useful property that they bind very strongly to a solid immunosorbent. Other components in the cell media do not adsorb to the immunosorbent, resulting in high purity of the adsorbed antibody product.

Suppose we've made 5 L of culture media containing monoclonal antibodies at a concentration of 0.5 μmol/L (5×10^{-7} M). We load 34.9 g of immunosorbent beads into a vessel, mix with the culture media, allow the phases to equilibrate, and then remove everything and mechanically separate the beads from the liquid phase. The adsorption equilibrium isotherm is shown in Fig. 5.16. What is the fractional recovery of the antibodies in the feed onto the beads?

Solution

The separation is carried out in batch mode.

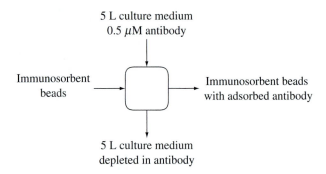

The vessel serves as the system, and the antibody (Ab) is our component. The integral material balance equation is simply

$$n_{Ab,L} + n_{Ab,S} = n_{Ab,F}$$

The moles of antibody initially in the system is simply the antibody concentration times the volume of culture medium, or

$$n_{Ab,F} = c_{Ab,F}V_L = 0.5 \frac{\mu mol}{L} \times 5\,L = 2.5\,\mu mol = 2500\,nmol$$

where $c_{Ab,F}$ is the concentration of antibody in the feed and V_L is the total volume of liquid. At equilibrium, this quantity of antibody is distributed between the liquid and solid (bead) phases. To fix the distribution between the phases, we refer to Fig. 5.16. The data on this figure are given as concentration of Ab in solution (nmol/L) and nmol Ab per g immunosorbent. Since the antibodies are present at a very dilute concentration, the total volume of liquid media does not change upon equilibration with the solid adsorbent. Therefore:

$$n_{Ab,L} = c_{Ab,L}V_L = c_{Ab,L}(5\,L)$$

where $c_{Ab,L}$ is the concentration (nmol/L) of antibody in the liquid at equilibrium. In the solid phase,

$$n_{Ab,S} = q_{Ab,S}M_S = q_{Ab,S}(34.9\,g)$$

$q_{Ab,S}$ is the concentration (nmol/g) of antibody adsorbed to the beads at equilibrium and M_s is the total mass (g) of immunosorbent beads. Combining these definitions with the material balance equation, we find:

$$5c_{Ab,L} + 34.9q_{Ab,S} = 2500\,nmol$$

Helpful Hint
In an equilibrium stage the two *product* streams are in equilibrium with each other; avoid the temptation to use the feed concentration with phase equilibrium data!

$c_{Ab,L}$ and $q_{Ab,S}$ are related as shown in Fig. 5.16. Finding the right values is a trial-and-error procedure. We guess a value of $c_{Ab,L}$, find $q_{Ab,s}$ from Fig. 5.16, then check whether the material balance equation is satisfied. (Be careful here—$c_{Ab,L}$, not $c_{Ab,F}$, is the concentration needed in Fig. 5.16!)

For our first iteration, we guess $c_{Ab,L} = 20$ nmol/L. At this concentration in the fluid phase, $q_{Ab,S} = 90$ nmol/g. These values give

$$5(20 \text{ nmol/L}) + 34.9(90 \text{ nmol/g}) = 3240 \text{ nmol} \;>\; 2500 \text{ nmol}$$

This guess was too high. A few more iterations get us to the correct answer: $c_{Ab,L} =$ 5 nmol/L, and $q_{Ab,S} = 71$ nmol/g, because

$$5(5 \text{ nmol/L}) + 34.9(71 \text{ nmol/g}) = 2500 \text{ nmol}$$

The fractional recovery of antibody on the solid phase is

$$f_R = \frac{n_{Ab,S}}{n_{Ab,F}} = \frac{34.9(71 \text{ nmol/g})}{2500 \text{ nmol}} = 0.99$$

Desorption of the antibody from the immunosorbent must be carried out in a separate step. This is frequently done by changing the pH or salt content of the solvent, or by adding other solutes.

Extraction In solvent extraction, a solvent is mixed with a liquid feed. Two liquid phases form, are brought to phase equilibrium, then are separated according to differences in density (Fig. 5.21). The phases are commonly called the *raffinate* and the *extract*. To analyze and design solvent extraction processes, we combine liquid-liquid phase equilibrium information with material balance equations. (This should sound very familiar!)

Discovering the right solvent is the key to successful solvent extraction processes. The perfect solvent is nontoxic, cheap, nonflammable, environmentally benign, and compatible with downstream uses of the product. The perfect solvent has a very different density than the feed, so that the two liquid phases can be easily mechanically separated. The perfect solvent is completely immiscible with the feed phase, but has infinite solubility for the recovered solute. Finally, the perfect solvent can be easily separated from the solute, so that the solvent can be reused and the solute recovered. As you can imagine, there is no perfect solvent!

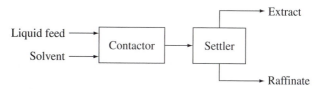

Figure 5.21 Solvent extraction. The extract is "solvent-rich" (contains most or all of the solvent) and the raffinate is "solvent-poor" (contains little to none of the solvent).

Example 5.21

Process Flow Calculations Using Liquid-Liquid Distribution Coefficients: Cleanup of Wastewater Stream by Solvent Extraction

A wastewater stream (10,000 kg/h) is contaminated with 6 wt% acetic acid. The acetic acid content must be reduced to 0.5 wt% before the water can be discharged. Two solvents are under consideration: methyl acetate and 1-heptadecanol. The distribution coefficient K_D (weight fraction acetic acid in solvent phase/weight fraction acetic acid in water phase) for methyl acetate is 1.273 and that for 1-heptadecanol is 0.312. How much of each solvent is required if the extraction is done in a single equilibrium stage? Which solvent would you choose? You can assume that methyl acetate and 1-heptadecanol are both completely insoluble in water.

Solution

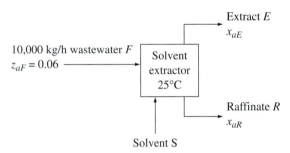

We'll work in units of kg/h. Let the subscripts S = solvent in, F = aqueous feed in, E = extract (solvent-rich stream out), and R = raffinate (water-rich stream out). There are three components: water (w), acetic acid (a), and solvent (s). Streams E and R are in equilibrium; acetic acid distributes between both liquid phases, but the solvent and the water are completely immiscible. The basis ($\dot{m}_F = 10{,}000$ kg/h) and two stream compositions (the acetic acid content of the wastewater, $z_{aF} = 0.06$, and the acetic acid content in the raffinate, $x_{aR} = 0.005$) are specified. The three steady-state material balance equations are

$$z_{aF}\dot{m}_F = x_{aE}\dot{m}_E + x_{aR}\dot{m}_R$$

$$0.06(10000) = 600 = x_{aE}\dot{m}_E + 0.005\dot{m}_R$$

The phase equilibrium relationship is

$$x_{aE} = K_D \cdot x_{aR}$$

Which when combined with the material balance equation yields

$$600 = 0.005 K_D \dot{m}_E + 0.005 \dot{m}_R$$

where $K_D = 1.273$ for methyl acetate and 0.312 for 1-heptadecanol. Additionally, we have material balance equations on solvent and water:

$$\dot{m}_S = x_{sE}\dot{m}_E$$

$$z_{wF}\dot{m}_F = x_{wR}\dot{m}_R = 0.94(10000) = 9400 \text{ kg/h}$$

!**Helpful Hint**
If a component is present in only one phase, do not try to find a phase equilibrium relationship for that component!

Because the solvent and water are totally immiscible, each is present in just one phase and there is no phase equilibrium relationship to be written! Since the mass fractions *in each phase* must add up to 1:

$$x_{wR} = 1 - x_{aR} = 1 - 0.005 = 0.995$$

$$x_{sE} = 1 - x_{aE} = 1 - 0.005K_D$$

From these equations we find

$$\dot{m}_R = \frac{9400}{0.995} = 9447 \text{ kg/h}$$

$$\left.\begin{array}{l} \dot{m}_E = 86{,}845 \text{ kg/h} \\ \dot{m}_S = 86{,}290 \text{ kg/h} \end{array}\right\} \text{ with methyl acetate}$$

$$\left.\begin{array}{l} \dot{m}_E = 354{,}335 \text{ kg/h} \\ \dot{m}_S = 353{,}780 \text{ kg/h} \end{array}\right\} \text{ with heptadecanol}$$

The extract contains 0.6 wt% acetic acid if the solvent is methyl acetate and 0.2 wt% if the solvent is heptadecanol. (Note that the ratio of K_D's of the two solvents under consideration is close to the ratio of the solvent flows required.)

Looks like methyl acetate is the winner. But the solvent requirement is high; recycle would be required to make this process economical. We need to consider whether separation of methyl acetate from acetic acid is readily achievable before finalizing our choice of solvent.

Another question to consider: Why didn't we choose to separate the acetic acid from water by equilibrium flash rather than solvent extraction? The reason is that the water, which is in great excess, has a higher vapor pressure than acetic acid. This means that most of the water would need to go to the vapor phase, which brings with it a very large energy cost.

Example 5.22   **Process Flow Calculations Using Triangular Phase Diagrams: Separating Acetic Acid from Water**

100 g of a 45 wt% acetic acid/55 wt% water solution at 25°C is contacted with 100 g of MIBK. Does the mixture form one or two liquid phases? If two phases are formed, what is the composition of each phase? How much of each phase is formed? If all the MIBK is removed from the extract, what is the product purity and component recovery?

Solution

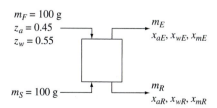

This is a batch process; the solution and solvent are added initially, and the extract and raffinate are removed at the end of the equilibration period. There are three components, as in Example 5.21, but this time we cannot assume that all the MIBK leaves in the extract, nor that all the water leaves in the raffinate.

A total of 200 g is added to the extractor: 100 g MIBK (50 wt%), 45 g acetic acid (22.5 wt%), and 55 g water (27.5 wt%). Now we need phase equilibrium data, conveniently provided by Fig. 5.15.

To find the equilibrium compositions of extract and raffinate, we first locate the point on Fig. 5.15 corresponding to the mixture composition. This point lies below the dome, so it is in the two-phase region—two liquid phases do form. The mixture composition lies in between two tie lines, so we must imagine a third tie line that falls between these two and goes through the mixture composition point. Estimating where this imaginary tie line would intersect with the solid curve, we find that the raffinate contains about 24 wt% acetic acid ($x_{aR} = 0.24$), 71 wt% water ($x_{wR} = 0.71$), and 5 wt% MIBK ($x_{mR} = 0.05$), while the extract contains about 21 wt% acetic acid ($x_{aE} = 0.21$), 12 wt% water ($x_{wE} = 0.12$) and 67 wt% MIBK ($x_{mE} = 0.67$).

There are three material balance equations; we insert the known variables:

$$m_S = x_{mE}m_E + x_{mR}m_R \qquad\qquad 100 = 0.67m_E + 0.05m_R$$

$$x_{wF}m_F = x_{wE}m_E + x_{wR}m_R \qquad\qquad 0.55(100) = 0.12m_E + 0.71m_R$$

$$x_{aF}m_F = x_{aE}m_E + x_{aR}m_R \qquad\qquad 0.45(100) = 0.21m_E + 0.24m_R$$

(Only two of these are independent!) We solve the first two to get $m_E = 145$ g and $m_R = 55$ g. (If you use a different pair of equations you might get a slightly different answer. This is simply because of inaccuracies in reading compositions from the graph.)

The fractional recovery of acetic acid in the extract is

$$f_{RaE} = \frac{x_{aE}m_E}{z_a m_F} = \frac{0.21(145)}{0.45(100)} = 0.68$$

The purity of the extract, on a solvent-free basis, is

$$\frac{x_{aE}m_E}{x_{aE}m_E + x_{wE}m_E} = \frac{0.21(145)}{0.21(145) + 0.12(145)} = 0.64$$

(The solvent-free basis assumes that all the MIBK is removed from the extract.) Even though the mass fraction of acetic acid in the extract is slightly lower than that in the feed, the mass fraction of acetic acid *on a solvent-free basis* is higher, because the water mass fraction in the extract is very low.

5.6.5 Multistaged Separations Using Material Separating Agents (Optional)

Let's look back at the example problems from the previous section. Although some processes (e.g., antibody adsorption) produced a product of high purity and achieved high recovery, other processes (e.g., acetic acid extraction into

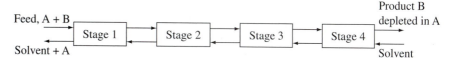

Figure 5.22 A four-stage countercurrent extraction process. Each stage functions as a separate equilibrium stage. The composition of the streams leaving each stage are related by phase equilibrium.

MIBK) were pretty marginal in both purity and recovery. In still other processes (e.g., absorption of SO_2 into water), the quantity of solvent required was quite high. Is there a way to get higher purity and component recovery? Is there a way to reduce solvent requirements?

There is. The basic idea is this: Break the separation problem into little steps, and take a lot of little steps. Each step corresponds to a single equilibrium stage. This concept is called **multistaging**. Distillation (Sec. 5.6.3) is an example of a multistage separation, where energy is used as the separating agent. With separations requiring addition of a second material (i.e., absorption, extraction), multistaging (almost) always requires less solvent or adsorbent than does a single-stage operation. Multistaging (almost) always improves the product purity and/or component recovery.

The idea is illustrated in Fig. 5.22, for a four-stage solvent extraction. In the figure, the feed enters on the left and flows to the right. The solvent enters from the right and flows to the left. This is called **countercurrent flow**.

In each stage, two streams are mixed, equilibrated, and then separated into two product phases. Streams leaving *each stage* are equilibrated with each other. The streams leaving the *process* are *not* in equilibrium. For example, in Fig. 5.22, the feed, containing a significant amount of A, is contacted with a solvent stream (extract) leaving stage 2. This solvent has already picked up some A from earlier stages. The extract leaving stage 1 is in equilibrium with the raffinate leaving stage 1. (Recall that extract is the solvent-rich phase and raffinate is the solvent-poor stage.) The raffinate leaving stage 1, partially depleted in A, passes to stage 2, where it mixes with extract from stage 3. The raffinate leaving stage 2 is in equilibrium with the extract leaving stage 2. Notice that the raffinate entering stage 4 is significantly depleted of A, and it is contacted with the "cleanest" solvent. The "dirtiest" feed is contacted with the "dirtiest" solvent in stage 1. In other words, the average amount of A decreases from left to right, in either the raffinate or the extract streams.

To calculate the performance of staged processes, we proceed more or less as usual. Material balance equations are written around each stage, as are phase equilibrium relationships for streams leaving each stage. If there are 2 components in the feed plus one component in the solvent, and each component appears at equilibrium in both phases, we have 3 material balance equations and 3 phase-equilibrium relationships *per stage*, or 12 material balance equations and 12 phase-equilibrium relationships all together for all 4 stages. The equations and relationships are connected, because the equilibrated stream leaving one stage becomes the feed to the next. Design or analysis of the separation unit then requires specifications regarding feed rate and quality, product quality, and/or unit performance.

Analyzing multistage processes gets tedious pretty fast. For some limited (but important) cases, simple expressions describe the relationship between number of stages, solvent flow, and performance of the cascade. In the case where the solute is dilute and the phase equilibrium relationship is linear, the Kremser-Souder-Brown (KSB) equation is a useful approximation:

$$f_{RiS} = \frac{\left(\dfrac{\dot{m}_S K_i}{\dot{m}_F}\right)^{N+1} - \left(\dfrac{\dot{m}_S K_i}{\dot{m}_F}\right)}{\left(\dfrac{\dot{m}_S K_i}{\dot{m}_F}\right)^{N+1} - 1} \tag{5.17}$$

where f_{RiS} is the fraction of i recovered in the added phase, $\dot{m}_S$ is the flow rate of the added separating agent, $\dot{m}_F$ is the feed flow rate, N is the number of stages, and K_i is the constant describing the equilibrium between the two phases. For absorption where the gas-liquid equilibrium is described by Henry's law:

$$K_i = \frac{x_i}{y_i} = \frac{P}{H_i}$$

For solvent extraction where the two phases are totally immiscible and liquid-liquid equilibrium is described by a distribution coefficient K_D:

$$K_i = \frac{x_{iE}}{x_{iR}} = K_D$$

You can see how simple and useful the KSB equation is. For example, given the desired component recovery, you can look at the relationship between the number of stages required and the solvent flow rate. This allows optimization of the cost of the process, because the capital (investment) cost is related mainly to the number of stages, whereas the operating cost is related to the solvent requirements.

| Example 5.23 | **The Power of Multistaging: Recovery of Acetic Acid from Wastewater** |

In Example 5.21, we examined the treatment of a wastewater stream (10,000 kg/h) contaminated with 6 wt% acetic acid. The acetic acid content must be reduced to 0.5 wt% before the water can be discharged. With methyl acetate and a single equilibrium stage, the solvent flow required was more than 86,000 kg/h. Now, let's look at using a countercurrent two-stage process to do the same job. By how much does the solvent quantity change? Is there any advantage to having more than two stages? Methyl acetate and water can be considered totally immiscible.

Solution

The process is a multistage steady-state continuous-flow solvent extraction.

Stream numbering is shown on the flow diagram; w, a, and m are used to indicate water, acetic acid, and methyl acetate, respectively. From the problem statement, $\dot{m}_{wF} = 9400$ kg/h and $\dot{m}_{aF} = 600$ kg/h. If methyl acetate and water are completely immiscible in each other,

$$\dot{m}_{wF} = \dot{m}_{wR1} = \dot{m}_{wR2} = 9400 \text{ kg/h},$$

$$\dot{m}_{mS} = \dot{m}_{mE2} = \dot{m}_{mE1}$$

From the specification on product purity, the raffinate stream leaving the process must contain only 0.5 wt% acetic acid:

$$x_{aR2} = 0.005 = \frac{\dot{m}_{aR2}}{\dot{m}_{aR2} + \dot{m}_{wR2}} = \frac{\dot{m}_{aR2}}{\dot{m}_{aR2} + 9400}$$

from which we find

$$\dot{m}_{aR2} = 47 \text{ kg/h}$$

From a material balance equation around the entire process, we find that

$$\dot{m}_{aE1} = \dot{m}_{aF} - \dot{m}_{aR2} = 600 - 47 = 553 \text{ kg/h}$$

We now proceed to write material balance equations for acetic acid around each stage, along with the phase equilibrium relationships, using the distribution coefficient $K_D = 1.273$ given in Example 5.21.

Stage 1:

$$600 + \dot{m}_{aE2} = \dot{m}_{aR1} + 553$$

$$x_{aE1} = K_D x_{aR1}, \quad \Rightarrow \quad \frac{553}{553 + \dot{m}_{mE1}} = 1.273 \left(\frac{\dot{m}_{aR1}}{\dot{m}_{aR1} + 9400} \right)$$

Stage 2:

$$\dot{m}_{aR1} = \dot{m}_{aR2} + \dot{m}_{aE2} = 47 + \dot{m}_{aE2}$$

$$x_{aE2} = K_D x_{aR2} \quad \Rightarrow \quad \frac{\dot{m}_{aE2}}{\dot{m}_{aE2} + \dot{m}_{mE2}} = 1.273(0.005) = 0.0063$$

We solve this system of equations to find:

$$\dot{m}_{aR1} = 186 \text{ kg/h}$$

$$\dot{m}_{aE2} = 139 \text{ kg/h}$$

$$\dot{m}_{mS} = 21{,}800 \text{ kg/h}$$

!**Helpful Hint**
In a multistage separation, the phase equilibrium relationships apply to the compositions of the two streams leaving *each stage.*

Compare this result to the solvent requirement calculated in Example 5.21. Changing the design from one stage to two drops our solvent requirement by a factor of 4! Also notice that the acetic acid content of the water between stage 1 and stage 2 is intermediate between the acetic acid content of the wastewater feed and the cleaned water product.

The number of equations increases with three, four, or more stages. That's when the KSB equation comes in handy. Suppose we wish to know the solvent flow required if there are five stages. We use the KSB equation with $N = 5$:

$$f_{RaS} = \frac{553}{600} = 0.922 = \frac{\left(\dfrac{1.273\dot{m}_S}{\dot{m}_F}\right)^6 - \left(\dfrac{1.273\dot{m}_S}{\dot{m}_F}\right)}{\left(\dfrac{1.273\dot{m}_S}{\dot{m}_F}\right)^6 - 1}$$

and we solve to find

$$\frac{\dot{m}_S}{\dot{m}_F} = 1.023$$

The solvent requirement for five stages is about half that required with two stages. We can use the KSB equation to estimate solvent requirements given any number of stages.

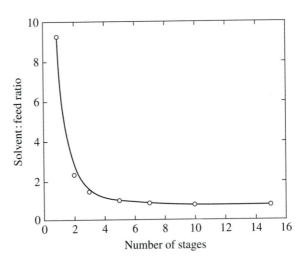

Most of the reduction in solvent usage occurs with the first few stages. There is a balance between the increased capital cost of more stages and the decreased operating cost of less solvent; typically a few stages (2 to 10) is the economic optimum.

Figure 5.22 may leave you with the impression that, in multistaged separation processes, each stage is carried out in a physically distinct vessel. This is sometimes true. However, many times it is better to build up than out, and stages are stacked. The stack of stages is inside a single tall column. If you get a chance to see a chemical process plant, the tall columns are most likely staged separation processes such as distillation or absorption towers. The height of the column gives you some idea of the number of stages. As a rule of thumb, a stage is between 1 and 2 feet high.

CASE STUDY   Scrubbing Sour Gas

Coal was once an important energy source in Europe and the United States, but the difficulty and mess of handling solid fuels has led to a switch to liquid and gas fuels. Still, there are large reservoirs of coal (and its cousin, oil shale) that could be tapped if the price of crude oil skyrockets. Processes to liquefy or gasify coal have been developed that convert the solid coal to a more convenient form for use in modern homes and buildings. Unfortunately, coal contains sulfurous compounds that are converted during gasification to hydrogen sulfide (H_2S). H_2S and other volatile sulfur-containing compounds have a sour "rotten egg" smell . Besides the nasty smell, H_2S is an extremely toxic material. If H_2S-containing gases are burned as fuel, SO_x (pronounced "socks," a mixture of SO_2 and SO_3) is generated. Atmospheric release of SO_x is strictly restricted because SO_x reacts with moisture in the air to make acid rain. Therefore, either the H_2S must be removed before the "sour" gas is burned for energy, or the SO_x must be removed from the flue gases. In this case study, we'll investigate processes for scrubbing sour gas before burning it. These processes produce "sweet" gas, low in H_2S.

In our case study, we've got 1000 kgmol/h of a sour gas, produced by coal gasification, to clean up. The composition of the gas varies somewhat, depending on the coal source and the reaction conditions; the gas may contain CH_4, H_2, and CO_2, as well as up to 2 mol% H_2S. The gas is at 20 atm and 100°C. The H_2S content of the gas must be reduced to 0.04 mol%. We'd also like to remove the CO_2 if possible, because it has no heating value, but not the CH_4 or H_2.

Our first question is: What's the best separation technology to use? Let's recall our heuristics for selecting separation technologies. We have a single gas phase, so mechanical separations are out. This is a high-volume process, so equilibrium-based separations are likely the best choice. The boiling point of H_2S is –60°C and that of CH_4 is –161°C. At these low temperatures, any separation technology that requires cooling and condensing would be prohibitively expensive. Furthermore, our job is to remove a relatively low-concentration contaminant from a process stream. Given our heuristics, gas absorption looks like a promising technology. H_2S is much more soluble in water than are methane and hydrogen, making absorption into water an attractive choice.

At what temperature and pressure should we operate? According to the heuristics, we want to work as close to ambient temperature and pressure as possible. On the other hand, gas solubility generally increases with decreasing temperature and increasing pressure, so we want to operate at as low a temperature and as high a pressure as possible. Since liquid water is our solvent of choice at this preliminary design stage, we are restricted to temperatures above 0°C. Let's choose 10°C as our operating temperature. Since the gas is already at 20 atm, let's choose that as our operating pressure.

The composition of the sour gas varies from day to day. Let's assume for now that the sour gas contains nothing but 2 mol% H_2S and 98 mol% H_2; later we'll examine how changes in the composition affect process operation. We will *not* make the simplifying assumption that the methane is completely insoluble in water, nor that the water is completely nonvolatile. We'll use s, h, and w to indicate the three components, H_2S, H_2 and H_2O, respectively, and use z_i, x_i, and y_i for mole fractions in the feed, liquid product, and gas product streams, respectively. We'll indicate the total flow rates using subscripts of F for feed, S for solvent (fresh water), L for liquid leaving the stage, and G for the sweet gas leaving the stage. All units are in kgmol/h.

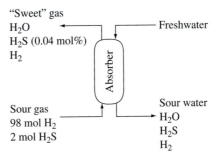

Like any problem involving an equilibrium stage, we combine phase equilibrium data with material balance equations and any stream composition or system performance specifications.

We need to find data, or use model equations, to describe the phase equilibrium for all three components in our system. Since H_2S and H_2 are both "noncondensable" gases, we'll use Henry's law to model their solubility in water. Henry's law is not appropriate for water; rather, we'll model water vapor-liquid equilibrium with Raoult's law. In App. B we find that the Henry's law constants at 10°C are 63,600 and 367 atm for H_2 and H_2S, respectively. From the Antoine equation, we calculate that the saturation pressure of water at 10°C is 0.012 atm. Thus we have three phase equilibrium equations, one for each component, relating the gas and liquid compositions at equilibrium:

$$y_s = \frac{H_s}{P}x_s = \frac{367}{20}x_s$$

$$y_h = \frac{H_h}{P}x_h = \frac{63,600}{20}x_h$$

$$y_w = \frac{P_w^{sat}}{P}x_w = \frac{0.012}{20}x_w$$

This is a steady-state continuous-flow system, with three independent material balance equations:

$$z_s \dot{n}_F = 0.02(1000) = x_s \dot{n}_L + y_s \dot{n}_G$$

$$z_h \dot{n}_F = 0.98(1000) = x_h \dot{n}_L + y_h \dot{n}_G$$

$$\dot{n}_S = x_w \dot{n}_L + y_w \dot{n}_G$$

We also know that

$$x_s + x_h + x_w = 1$$

$$y_s + y_h + y_w = 1$$

Finally, the purity specification dictates that the absorber produce a sweet gas with no more than 0.04 mol% H_2S, or

$$y_s = 0.0004$$

This set of equations is solvable. We proceed by first using the purity specification along with Henry's law for H_2S to find

$$x_s = 2.2 \times 10^{-5}$$

The rest of the equations need to be solved simultaneously for an exact solution. Solution yields two stunning results. First, the required water flow rate is just under 905,000 kgmol/h (to treat 1000 kgmol/h gas!). Second, the sour water contains 0.031 mol% H_2. This doesn't sound like much hydrogen, but when multiplied by the huge liquid flow rate it adds up to a loss of nearly 30% of the H_2 fed to the system!

The amount of water required is incredible. At higher temperatures, the gas solubility decreases and the water requirements increase even further. To see how much, let's look at an operating temperature of 50°C. The Henry's law constants are 76,500 atm and 884 atm for H_2 and H_2S, respectively, at 50°C, and the saturation pressure of water is 0.12 atm. Solving the equations with these new physical properties gives us the disheartening news that the required water flow rate is now over 2 million kgmol/h, and hydrogen losses total 58%! What kind of pressure could cut our water flow rate? We can use the same equations, but let P vary, and find that to reduce the water flow rate to 10,000 kgmol/h (still 10 times the gas feed rate) requires a pressure of over 1700 atm! This is unreasonable.

Clearly, this design isn't too promising. And, there's another complication: The sour gas stream composition varies from day to day. We need a robust technology that will adjust to these variations. Suppose that some days the sour gas contains 2 mol% H_2S, but also 3 mol% CO_2 and 18 mol% CH_4. If we did build an absorber operating at 10°C and 20 atm, with a water flow rate of 905,000 kgmol/h, how would changes in gas composition affect its operation? CO_2 and CH_4 can be considered noncondensable gases at this temperature and pressure, so Henry's law will be used to model their phase equilibrium behavior. At 10°C,

$H_{CO_2} = 1040$ atm and $H_{CH_4} = 29,700$ atm. We have two more phase equilibrium equations:

$$y_c = \frac{H_{CO_2}}{P} x_c = \frac{1040}{20} x_c$$

$$y_m = \frac{H_{CH_4}}{P} x_m = \frac{29,700}{20} x_m$$

and two more material balance equations:

$$0.03(1000) = x_c \dot{n}_L + y_c \dot{n}_G$$

$$0.18(1000) = x_m \dot{n}_L + y_m \dot{n}_G$$

which are combined with the modified constraints:

$$x_s + x_h + x_w + x_c + x_m = 1$$

$$y_s + y_h + y_w + y_c + y_m = 1$$

We set the fresh water flow rate equal to 905,000 kgmol/h and solve the system of equations:

	Sour gas	Sweet gas	Fresh solvent	Sour liquid
Total, kgmol/h	1000	620	905,000	905,380
Mol%:				
H_2S	2	0.04	0	0.0022
H_2	77	85.1	0	0.027
CO_2	3	0.17	0	0.0033
CH_4	18	14.6	0	0.0098
H_2O	0	0.05	100	99.96

We find that we can still meet our sweet gas purity specification for H_2S (good), but additionally we've removed about 50% of the methane (bad) and about 99% of the CO_2 (good) from the gas stream. The sweet gas production stream drops to 620 kgmol/h (from 700 kgmol/h with just hydrogen and H_2S in the sour gas). This is because methane is more water-soluble than hydrogen. Thus, the process operation could tolerate some changes in gas composition and still meet the purity specification, but at the expense of the sweet gas production rate.

The huge quantity of water required, the huge quantity of sour water produced (which will have to be treated), and the significant economic cost due to the loss of these fuels provide incentive to consider alternatives. What are some possibilities?

Idea 1.　Use multiple stages. Let's do an initial evaluation, using the KSB equation:

$$f_{RsL} = \frac{\left(\dfrac{\dot{n}_S K_s}{\dot{n}_F}\right)^{N+1} - \left(\dfrac{\dot{n}_S K_s}{\dot{n}_F}\right)}{\left(\dfrac{\dot{n}_S K_s}{\dot{n}_F}\right)^{N+1} - 1}$$

All the assumptions that go into the KSB equation aren't exactly true for our case. (For instance, the flow rates aren't constant). But, for a back-of-the-proverbial-envelope calculation, let's set as a design goal that we recover 98.6% of the H_2S fed in the liquid product (about what we achieved in the single-stage design), and let's suppose that our engineering experience dictates that a realistic number of stages is 10 ($N = 10$). For our system, at 10°C and 20 atm, $K_s = x_s/y_s = P/H_s = 0.0545$. Inserting these values into the KSB equation yields

$$0.986 = \frac{\left(\dfrac{0.0545\dot{m}_S}{\dot{m}_F}\right)^{1} - \left(\dfrac{0.0545\dot{m}_S}{\dot{m}_F}\right)}{\left(\dfrac{0.0545\dot{m}_S}{\dot{m}_F}\right)^{11} - 1}$$

We find

$$\frac{\dot{m}_S}{\dot{m}_F} = 24.4$$

or the required water flow rate is reduced to about 24,400 kgmol/h, which is a big improvement over 905,000 kgmol/h—but is still a lot of water!

Idea 2.　*Use a better solvent.* Water is certainly cheap and nontoxic, but there's got to be a better solvent. By tuning the chemistry of the solvent to match the chemistry of the gas, we should be able to greatly enhance selectivity and capacity of the solvent for the gas. A little study of chemistry reveals that H_2S and CO_2 are both acidic gases, but H_2 and CH_4 are not. Can we take advantage of this difference in physical properties? Perhaps a solvent that is basic will have a greater preferential solubility for acidic gases. With a bit of research, we find that a much better solvent for H_2S and CO_2 is a solution of monoethanolamine (MEA, $NH_2C_2H_4OH$) or diethanolamine [DEA, $NH(C_2H_4OH)_2$] in water. Aqueous MEA or DEA solutions are basic (because of the amine group) and therefore will complex with acidic gases, greatly increasing their solubilities. H_2, the gas we do not want to remove, is not acidic, so its solubility in MEA or DEA solutions should not be much different than in water.

In cases such as H_2S dissolving into aqueous MEA solutions, the strong acid-base interactions mean that Henry's law is not reliable. We find the following equilibrium data for H_2S and 5 mol% MEA aqueous solution at 40°C.

Partial pressure H_2S (mmHg)	1	10	100	200	500	700
Moles H_2S per mole MEA	0.128	0.374	0.802	0.890	0.959	0.980

Let's pick an operating pressure of 20 atm and an operating temperature of 40°C (because we have the data at that temperature). The three material balance equations remain unchanged, as does the performance specification on fractional recovery of H_2S.

We still require that $y_s = 0.0004$. At an operating pressure of 20 atm (15,200 mmHg), the partial pressure of H_2S in the sweet gas is

$$p_s = y_s P = 0.0004(15200) = 6.08 \text{ mmHg}$$

The sweet gas stream is in equilibrium with the sour liquid stream, so we can interpolate with the data in the table to estimate that there must be about 0.27 moles H_2S per mole MEA in the sour liquid.

To simplify the problem a bit, let's assume that all the CH_4 and H_2 exit with the gas stream, and all the MEA solution exits with the liquid stream. This is a much better assumption when the solvent is MEA than when the solvent is water; MEA was chosen just because it is highly selective for acid gases like H_2S. In this case we can solve the material balance equation on H_2S to find that there must be 19.6 kgmol H_2S/h exiting with the sour liquid. Combining this result with the equilibrium data, we calculate that the MEA flow rate is (19.6/0.27) = 72.6 kgmol/h MEA. The solution described in the table is 5 mol% MEA in water, meaning that this process requires a solvent flow rate of 1450 kgmol/h. Compare MEA solvent flow to that for pure water. By considering the chemistry of solutions, we've reduced the amount of solvent required by more than 600-fold! This translates into enormous savings in equipment and operating costs.

Idea 3. *Use a better solvent and use multistaging.* If two heads are better than one, are two ideas better than one? In this case, yes. Our ideas are complementary; using the MEA solvent in a multistage absorber will further reduce solvent requirements. We cannot use the KSB equation, because Henry's law does not accurately model the phase equilibrium data. Rather, we need to complete calculations for all stages—in each stage, combining a material balance equation with the phase equilibrium data. We leave this for an end-of-chapter problem.

We aren't done yet. We've produced a sweet gas stream, but what do we do with the sour liquid? The solvent is too expensive to simply throw away. Rather, we'd like to clean up the solvent and recycle it back to the absorber. We are confronted with the classic problem that accompanies any separation solution involving the use of a material separating agent: we solve one separation problem only to create another. What is the best choice for a separation technology for our new problem? One idea is to use stripping, in which a stripping gas is contacted with the sour liquid and strips out the H_2S, leaving clean solvent behind. To complete our design, we will need to choose a stripping gas (nitrogen, perhaps), determine the optimal operating temperature and pressure, and calculate the correct number of stages. This leaves us now with another H_2S-containing gas stream, and we still need to decide what to do with that! The classic choice is to react the H_2S to SO_2, then make H_2SO_4 for sale. Another idea may be to find a way to allow in-situ reaction of H_2S, right in the solvent, to SO_2, or perhaps to another product like elemental sulfur. We could use the ideas described in this and previous chapters to push forward toward a complete design of an integrated process for removing the hydrogen sulfide contaminant from a gas stream and converting it to a useful product.

Summary

- Separations account for 50% or more of the total capital and operating costs of a typical chemical process facility. There is enormous diversity in the choice of separation technologies, but they can be handily classified as one of three kinds: (a) **mechanical**, (b) **rate-based**, and (c) **equilibrium-based**. Separations work by exploiting differences in physical and/or chemical properties of the species to be separated. Engineers use heuristics to guide their choice of technology.

- Three useful measures of the performance of a separation process are **purity**, **component recovery**, and **separation factor**:

$$\text{Fractional purity} = \frac{\text{quantity of desired component } i \text{ in product stream } j}{\text{quantity of product stream } j}$$

$$x_{ij} = \frac{\dot{m}_{ij}}{\dot{m}_j}$$

$$\text{Fractional recovery} = \frac{\text{quantity of desired component } i \text{ in product } j}{\text{quantity of desired component } i \text{ in feed}}$$

$$f_{Rij} = \frac{\dot{m}_{ij}}{\dot{m}_{i,feed}}$$

$$\text{Separation factor} = \alpha_{AB} = \frac{x_{A1}}{x_{A2}} \frac{x_{B2}}{x_{B1}}$$

- Separators are not "perfect" for two reasons: (1) **entrainment**, or incomplete mechanical separation of the two phases and (2) **phase equilibrium** limitations.

- An **equilibrium stage** is the basic unit of equilibrium-based separations processes. In an equilibrium stage, a multicomponent mixture is brought to conditions such that two phases are generated. This requires addition of a separating agent—energy or material (or both).

- When two phases of a multicomponent mixture are at equilibrium, the temperature and pressure are the same in both phases but the compositions are not the same. Graphs, tables, or model equations are used to describe the compositional relationship between the two phases. Phase equilibrium relationships are combined with material balance equations to describe the performance of an equilibrium stage.

- Two useful equations for modeling phase equilibrium are:

$$\textbf{Raoult's law: } y_i = \frac{P_i^{sat}}{P} x_i$$

$$\textbf{Henry's law: } y_i = \frac{H_i}{P} x_i$$

  Raoult's law is used to model vapor-liquid equilibrium when the components are similar in size and chemical nature, and are condensable at the system temperature and pressure. Henry's law is used to model gas-liquid equilibrium when the component is noncondensable at the system temperature and pressure and is sparingly soluble in the liquid phase.

- By using **multiple stages** and countercurrent flow, much higher product purity and/or component recovery are achieved, and solvent or energy costs are reduced.

ChemiStory: How Sweet It Is

Sugarcane is a perennial grass, native to tropical southern Asia. After Christopher Columbus brought sugarcane to the New World, the European colonial powers Spain, England, and France rapidly established sugarcane plantations on the tropical Caribbean islands and produced molasses—a brown, unrefined sugar syrup. Ships transported the molasses to New England, where it was made into rum, the rum was shipped to slave traders in Africa, and then the ships returned to the islands with new slaves to work the plantations.

When slaves revolted on French Caribbean islands, French plantation owners fled to New Orleans. Louisiana's rich soils and extensive waterways

Norbert Rillieux
Courtesy of Louisiana State Museum.

proved conducive to establishment of a sugar industry, and by the early 1800s, sugarcane was the dominant crop grown in southern Louisiana. A great deal of slave labor and fuel was required to produce sugar. The cane was first crushed at a mill to release the juices, then water was evaporated in the sugarhouse. Working in the "Jamaica train," slaves ladled boiling sugar juice from one steaming open kettle to another. The work was hot, dirty, and dangerous. When the sucrose concentration in the syrup was finally high enough to crystallize, the juice was cooled, allowing further crystallization. The juice-crystal mixture was stored in a barrel with a perforated bottom, allowing the molasses to run out and leaving behind brown sugar crystals. This raw sugar was shipped north for further refining into white sugar by re-crystallization.

In 1806, just 3 years after the Louisiana Purchase, Norbert Rillieux was born in New Orleans. He was the son of a wealthy white cotton merchant and his mixed-race mistress—a "quadroon," or one-fourth black and three-fourths white. He grew up a free man of color—educated, well-to-do, with the right to own land and slaves but not to vote or marry whites. During the 1820s, as sugar became king in Louisiana, Norbert was sent to Paris for his higher education and became interested in mechanics and thermodynamics. (Rather ironically, it was more common for southern free men of color than for whites

Sugar manufacture in Antigua, West Indies. Drawing by William Clark, 1823.
Courtesy of the John Carter Brown Library at Brown University

to be educated at European universities, which were far more advanced than American schools at that time.) For a gifted student such as Rillieux, Paris in the 1820s was the place to be. The Industrial Revolution was underway. Engine efficiency, and the relationship between heat and work, were hot topics. The Parisian Sadi Carnot published pioneering studies of steam engines in 1822–1824 and conducted work leading to formulation of the second law of thermodynamics. (Carnot's work preceded by about 30 years that of James Joule, whose work led to formulation of the first law of thermodynamics— the energy balance.) Rillieux became especially interested in latent heat—the energy required to convert liquid to vapor.

At that time, much of European sugar derived from the sugar beet. French scientists and engineers had worked extensively to establish the science underpinning sugar processing and to use scientific reasoning as the basis for technology development. This was a far cry from the empirical and tradition-bound methods of Louisiana sugarmakers. Evaporation of water from sugar beet juice required an enormous amount of energy, and the French were attempting to develop methods to use the energy in the steam emanating from the boiling sugar juice. Rillieux became interested in this problem; he was familiar with the Jamaica train method of transferring sugarcane juice from one kettle to the next during the boil-up process. His idea was to set up a cycle, where steam evaporated from one pot would give up its latent heat to provide the energy for evaporation from the next pot. From his study of thermodynamics, Rillieux knew that heat flows only from hot to cold, but he also knew that the boiling temperature dropped with a drop in pressure. He figured he could build a series of three enclosed containers, each operating at greater vacuum than the previous. The syrup would boil at progressively lower temperatures, and the steam from one container would be used to heat the next.

The idea was great on paper, but Rillieux needed to prove it would work by building a prototype. Unfortunately, the French economy was sputtering by this time, and he could not find funding or a manufacturer to test his idea. In contrast, on the other side of the Atlantic, the sugar business was undergoing explosive growth. In 1833, Rillieux left Paris and returned to Louisiana, as the chief engineer at a sugar refinery owned by the wealthy Edmond Forstall. It was probably a difficult decision, since he was returning to a land where slavery was still legal. What Rillieux found upon his return was that Louisiana sugar production technology was still in the dark ages. Sugar produced by the Jamaica train was dark, heavy, and dirty, but federal tariffs protected the processors from competition. Sugar syrup evaporation took a lot of energy and labor, and the local swamps were stripped bare of timber.

Forstall hired Rillieux to solve the problems of poor sugar quality, high energy use, and high labor costs. Rillieux worked for 10 years, first for Forstall and then alone, perfecting his triple-effect evaporator design, filing patents, and building prototypes. These failed to work reliably, because the equipment was home-made. He needed a professional machinery company to build the equipment to tight specifications, and he needed financial backing. Rillieux's chance came when he met the millionaire planter Judah Benjamin, the first openly Jewish U.S. senator. Finally, Rillieux was able to get a professionally built apparatus. The system had 3 stages and was fueled entirely on discarded dried cane (called bagasse). It was installed on Benjamin's plantation in 1843 and worked extraordinarily well. Profits went up 70%. To top it off, the sugar was a much higher quality, as good as any produced by secondary refiners in the north. In fact, Benjamin's beautiful white sugar crystals were prize winners.

But the situation in the South was deteriorating in the years leading up to the Civil War. Legal rights that free men of color had enjoyed for years were curtailed. Even as he traveled around the state installing his now-heralded invention, Rillieux was forced to stay in slave quarters. Patent examiners challenged his legal right to file patents as a nonwhite. Increasingly frustrated, he moved back to France in the 1860s, married a young French woman, took up the study of Egyptian hieroglyphs, and abandoned his engineering and science work.

Rillieux's invention may have had the greatest impact in central Europe, where farmers eagerly adopted new technology, and both land and labor were scarce resources. By 1888, about 150 Rillieux evaporators were installed in Germany, Austria, and Russia. By 1900, German agriculture was transformed; by building on a strong scientific and technological basis the country began exporting food products that led to expansion of the German economy. Rillieux's multi-stage evaporator is still in use worldwide, adapted to a slew of energy-intensive processing industries.

Quick Quiz Answers

5.1 Because the sulfuric acid is present in small quantities but the aromatics in large; liquid-liquid extraction scales with the quantity of material to be recovered, so we would need a lot of benzene. Additionally, benzene is more difficult to separate from the other aromatics and is more expensive than water.

5.2 $\beta = 2.3\ \mathrm{h}^{-1}$.

5.3 DOF = 2. (6 stream variables, 1 specified flow, 1 specified composition, 2 material balances)

5.4 Purity = 1.0, recovery = 1.0, separation factor approaches infinity. Same.

5.5 0.8125 without entrainment, 0.78 with entrainment.

5.6 *No!*

5.7 Liquid, about 1000 mmHg, unknown.

5.8 760 mm Hg.

5.9   Some salt precipitates out, at 60°C KNO_3 is more soluble, at 10°C NaCl is more soluble.

5.10 Two phases. About 10 mol% naphthalene. Pure solid benzene.

5.11 About 95°C, about 83°C, two phases, about 6% ethanol in liquid, about 30% ethanol in vapor.

5.12 Raoult's law.

5.13 Water.

5.14 One phase; 61% MIBK, 24.5% acetic acid, 14.5% water; 6% MIBK, 26% acetic acid, 68% water.

5.15 26.8.

5.16 2.23, 2.23, no!

5.17 Because ethanol and water hydrogen bond, so Raoult's law doesn't work well.

References and Recommended Readings

1. *Perry's Chemical Engineers' Handbook* has several sections describing separation technologies in detail and includes a fairly extensive list of phase equilibrium data. App. B of this text has a limited tabulation.

2. Physical property data useful for initial selection of appropriate separation technologies are available in reference books such as the *CRC Handbook of Chemistry and Physics, Lange's Handbook of Chemistry,* or *Physical and Thermodynamic Properties of Pure Chemicals* (DIPPR database), published by Taylor and Francis in 1999. The Knovel Engineering and Scientific Online Reference is also a useful source.

3. For more on the life and times of Norbert Rillieux, see *Prometheans in the Lab,* by S. B. McGrayne.

Chapter 5 Problems

Warm-Ups

P5.1 Match up each separation technology (left) with the physical property difference exploited (right).

crystallization	difference in solubility in two immiscible liquids
adsorption	difference in freezing point
liquid-liquid extraction	difference in binding to solid
distillation	difference in size
membrane filtration	difference in volatility
absorption	difference in solubility of gas in a liquid

P5.2 Briefly explain the similarities and differences between flash vaporization, condensation, and distillation. Briefly explain the similarities and differences between adsorption, absorption, and solvent extraction.

P5.3 Use the Antoine equation to calculate the boiling point temperature of water at 1 atm pressure and at 2.5 atm.

P5.4 What is the pressure of saturated steam at 150 psig? (Refer to Fig. 5.10.)

P5.5 Acetone boils at 1 atm at 56°C, and has a saturation pressure of 0.25 atm at 20°C. Sketch the expected trend for its saturation pressure from 0-100°C.

P5.6 A saturated liquid solution of benzene and naphthalene is equilibrated at 45°C and 1 atm. What is the mol% benzene in solution? If the solution is cooled slightly such that a solid phase forms, will the solid be pure benzene, pure naphthalene, or a mixture?

P5.7 A 30 mol% ethanol/70 mol% water mixture is equilibrated at 1 atm pressure and 88°C. List the vapor and liquid compositions. Use the inverse lever rule to determine the percent of the mixture that is vaporized. (Refer to Fig. 5.13.)

P5.8 Refer to Fig. 5.15. Suppose you have a mixture containing 20 wt% acetic acid, 60 wt% water and the remainder MIBK in a glass flask sitting on a lab bench. Should you see one phase or two? If two, what are the compositions of the two phases? How does the Gibbs phase rule apply to this situation?

P5.9 A gas mixture containing 9 mol% SO_2 with the remainder nitrogen is equilibrated with water. If the mixture is equilibrated at 20°C and 760 mm Hg, what is the mol fraction of SO_2 in the liquid phase? At 10°C and 3040 mm Hg?

P5.10 For the following situations, state whether you think the phase equilibrium relationship would be adequately described by Raoult's law, Henry's Law, or neither. Briefly state your reasoning.

A mixture of xylene isomers that partially crystallizes.
Dissolved CO_2 concentration in a bottle containing carbonated soda.
Water content of room air, with no standing puddles in the room.
A mixture of ethylbenzene and styrene at 1 atm pressure and 140°C that splits into vapor and liquid phases.
A mixture of vinegar (acetic acid) and water at a temperature and pressure such that liquid and vapor phases form.

P5.11 Water has a normal boiling point at 1 atm of 100°C. There is water vapor and no liquid water in a classroom at 22°C. Explain this apparent discrepancy. What is the maximum mole fraction water vapor in the air at 22°C?

P5.12 In the summer, larger fish tend to leave the warmer shallow waters near the shoreline and collect in cooler, deeper waters. Use the Henry's law constant for oxygen dissolved in water to explain this phenomenon.

P5.13 100 mL hexane, 100 mL water, and 1 mL of an oil-based colored dye are mixed gently in a cup, then allowed to sit. Two layers form. Is the colored dye in the top or bottom layer? Explain.

P5.14 "Sour" gas, 1% H_2S in CH_4, can be "sweetened" by absorption of H_2S into liquid water. The absorption is carried out at 1 atm pressure and 25°C. Unfortunately, the amount of water required is enormous. Name two ways to reduce the water requirement.

Drills and Skills

P5.15 You are faced with solving several different separation problems, as listed below. For each problem, choose the best separation technology from this list: distillation, sedimentation, flash vaporization, condensation, absorption, filtration, leaching, crystallization, solvent extraction, adsorption. Write the name of the chosen technology in the table. State whether it is a strictly mechanical separation, or an equilibrium-based separation.

Separation problem	Best separation technology
Recovery of antibiotics from fermentation broth	
Removal of isopropanol vapor from air	
Recovery of limestone sludge from saline solution	
Recovery of soybean oil from soybeans	
Removal of colored impurities from high fructose corn syrup	
Separation of methane from digested manure	
Separation of CO_2 and H_2	
Separation of ethylbenzene and styrene	
Removal of yeast from beer	
Recovery of potassium nitrate from aqueous solution	

P5.16 Given the process stream described in the following table, devise what you think is the best flow diagram for separating it into four essentially pure product streams. Indicate the separation technology you would choose and the sequence of separation steps. Briefly explain your reasoning. A quantitative answer is not required.

Compound	Mol% in feed	Normal boiling point (°C)	Normal melting point (°C)	Soluble in water?	Soluble in benzene
Naphthalene	12	218	80.2	yes	yes
Ethylene glycol	18	197	−11.5	yes	no
Ethylbenzene	32	136.2	−95	no	yes
Styrene (vinylbenzene)	38	145.2	−30.6	no	yes

P5.17 A liquid bromine (Br_2) stream contains 2% chlorine (Cl_2) and 0.02% chloroform ($CHCl_3$) as contaminants. The contaminants must be removed before the bromine can be used for further fine chemical manufacturing. The problem is that the boiling point of chloroform is very similar to that of bromine. (Refer to App. B.) However, the following reaction takes place at 250°C over a catalyst:

$$\tfrac{3}{2}\,Br_2 + CHCl_3 \;\rightarrow\; CHBr_3 + \tfrac{3}{2}\,Cl_2$$

Sketch out a process flow sheet for removing the chlorine contaminants from the liquid bromine. Write a paragraph justifying your design.

P5.18 100 gmol/min of a gas stream containing 30 mol% ethane (C_2H_6) and 70 mol% methane (CH_4) is fed to a distillation column, where it is separated into an overhead product containing 90 mol% methane and a bottoms product containing 98 mol% ethane. Calculate the overhead and bottoms flow rates and the fractional recoveries of methane and ethane in their corresponding product streams.

P5.19 A 30 wt% Na_2CO_3 aqueous solution is fed at 10,000 lb/h to an evaporator, where 40% of the water is removed. This produces a two-phase stream containing crystals of pure Na_2CO_3 and a solution of 17.7 wt% Na_2CO_3. This stream is fed to a rotary drum filter, which produces two products: a filter cake and a filtrate solution. The filter cake entrains 1 lb solution per 3.5 lb crystals. Calculate (a) the filter cake production rate, (b) the purity of the filter cake, and (c) the fractional recovery of Na_2CO_3 in the filter cake.

P5.20 A gas mixture containing 72 mol% CH_4, 13 mol% CO_2, 12 mol% H_2S, and 3 mol% COS is to be purified in an absorber by contacting the gas with a liquid solvent. The gas is fed at 3200 gmol/h and the gas feed rate/solvent feed rate ratio is 3:1. The solvent absorbs 97.2% of the H_2S in the gas feed stream. The COS concentration in the exiting gas stream is 0.3 mol%. CH_4 and CO_2 are not absorbed in the solvent at all, and no solvent leaves with the gas. First complete a DOF analysis and show that the problem is completely specified. Calculate the flow rate and composition of the exit gas and the concentration of H_2S and COS in the exit liquid (solvent) stream.

P5.21 Popcorn is to be dried with hot air as shown in the flow sheet. The gas stream recycle flow rate is 4 times the hot air feed flow rate. The desired popcorn production rate is 50 kg/hr. What feed rate of hot air is needed if the exit air is to be at 15 volume% water? Give your answer as a volumetric flowrate (liters/h). Model air as an ideal gas. Assume the air temperature is 80°C and the pressure is 1 atm. Find the moisture content of the air entering the dryer.

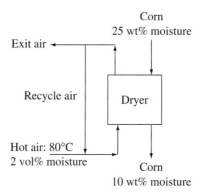

P5.22 Whey from cheese contains a number of proteins that may have specific uses when purified. For example, glycomacropeptide is the only known natural protein that contains no phenylalanine, and is therefore a protein source that could be consumed by people with the disease phenylketonuria (PKU). We are interested in developing a process to separate glycomacropeptide from whey. In one experiment, whey containing 1.2 g/L glycomacropeptide and 0.8 g/L beta-lactoglobulin (along with other proteins, lactose, and salts) is fed to a separator at 150 mL/min for 30 minutes. 89% of the glycomacropeptide and 24% of the beta-lactoglobulin are adsorbed to an ion exchange resin in the separator, while all the other constituents of whey pass through the separator. Then the whey feed is shut down and a buffer containing 0.25 M NaCl is pumped through the separator at a rate of 150 mL/min for 10 minutes. All the adsorbed proteins are eluted into the buffer. The buffer is then dried down to evaporate off water.

Draw a flow diagram. To analyze the process, consider three separate time intervals: (a) first 30 minutes, during adsorption to the ion exchanger, (b) second 10 minutes, during elution, and (c) the drying phase. Write material balance equations for glycomacropeptide, beta-lactoglobulin, and NaCl during each of these time intervals. What is the purity (wt%) and quantity (g) of the glycomacropeptide product? What is the separation factor, if the separation is between glycomacropeptide and beta-lactoglobulin?

P5.23 100 kg/h of a 30 wt% KNO_3 aqueous solution is cooled to 5°C. The solid and liquid phases are separated by filtration. The filter cake entrains 1 kg liquid per 19 kg solid. Calculate the purity of the filter cake and the fractional recovery of KNO_3 in the filter cake. Calculate the separation factor. (Refer to Fig. 5.11 or to App. B for solid-liquid equilibrium data.)

Installation of a recycle loop, that recycles 50% of the filtrate back to the feed, is under consideration. How does the recycle change the purity of the product, the fractional recovery of KNO_3, and the separation factor for the process?

P5.24 Sodium sulfate forms three crystalline solids: anhydrous (Na_2SO_4) and decahydrate ($Na_2SO_4°10H_2O$). Solubility data are given in App. B. Use the data to generate a *T-x* phase diagram. A 30 wt% aqueous Na_2SO_4 solution is fed to a cooling-type crystallizer at 50°C. To what temperature must the solution be cooled so that crystallization will begin (assuming that equilibrium is maintained)? Will the crystals be decahydrate, heptahydrate or anhydrous? To what temperature will the mixture have to be cooled to crystallize 50% of the Na_2SO_4? If I prefer to evaporate off water at 50°C rather than cooling the salt solution, how much water must I remove to just begin to see a crystalline phase at equilibrium?

P5.25 Use the Antoine equation along with constants given in App. B, to calculate saturation pressures of the *n*-alkanes (*n*-pentane, *n*-hexane, *n*-heptane, and *n*-octane), the *n*-alcohols (methanol, ethanol, *n*-propanol, and *n*-butanol) and water at 80°C. Plot the saturation pressure versus the molecular weight of the compound. Discuss any trends you see, and rationalize these trends on the basis of the chemical structures.

P5.26 Vapor-liquid phase equilibrium data for methanol and water at 101.3 kPa are tabulated. The data are from J. G. Dunlop's 1948 M.S. thesis at the Brooklyn Polytechnic Institute. Plot the data to generate a *T-y-x* diagram (similar to Fig. 5.13). Then refer to your plot to answer the following questions.

(a) What is the boiling point of methanol at 101.3 kPa? What is the boiling point of water at 101.3 kPa? What are the dew point and bubble point temperatures for a mixture containing 50 mol% methanol/50 mol% water at 101.3 kPa? If a mixture of 50 mol% methanol/50 mol% water is at 101.3 kPa and 95°C, is it a liquid, a vapor, or both?

(b) A liquid containing 20 mol% methanol/80 mol% water is heated to 89.3°C. The pressure is 101.3 kPa. Use these data to determine the fraction of the feed that is vaporized, and the composition of the vapor and liquid phases. Calculate the fractional recovery of methanol in the vapor stream and the separation factor.

T, °C	64.5	65.0	66.0	67.5	69.3	71.2	73.1	75.3	78.0	81.7	84.4	87.7	89.3	91.2	93.5	96.4	100.0
y_{meth}	1.00	0.979	0.958	0.915	0.870	0.825	0.779	0.729	0.665	0.579	0.517	0.418	0.365	0.304	0.230	0.134	0.00
x_{meth}	1.00	0.950	0.900	0.800	0.700	0.600	0.500	0.400	0.300	0.200	0.150	0.100	0.080	0.060	0.040	0.020	0.00

P5.27 Styrene is the building block for polystyrene, a polymer that enjoys ubiquitous use in products such as throwaway coffee cups and packaging "peanuts." Styrene ($C_6H_5C_2H_3$) is made by dehydrogenation of ethylbenzene ($C_6H_5C_2H_5$). Since reactor conversion is much less than 100%, a separation between styrene and ethylbenzene is needed. Using Antoine equations to calculate vapor pressures and Raoult's law, calculate
   (a) the boiling point temperature of pure styrene,
   (b) the boiling point temperature for pure ethylbenzene, and
   (c) the dew point and bubble point temperatures for a 50:50 (mol:mol) mix of styrene-ethylbenzene system at 2 bar.

P5.28 Automobile antifreeze and coolant is a mixture of water and ethylene glycol ($HOCH_2CH_2OH$). The Antoine equation coefficients for both compounds can be found in App. B.
   (a) For proper performance in the summer, the mixture must begin to vaporize (at 1 atmosphere pressure) no lower than 260°F. What minimum mole percent of ethylene glycol do you recommend? What is the corresponding weight percent?
   (b) A typical thermostatic pressure cap can withstand radiator pressurization to 15 psig. At what temperature will your radiator coolant mixture from part (a) begin to form vapor at this pressure?

P5.29 You've got 100 gmol of a liquid mixture containing 40 mol% benzene and 60 mol% toluene at 10 atm (7600 mmHg) pressure. Use the Antoine equation and Raoult's law to answer the following questions: (a) What's the boiling point of pure benzene at 10 atm pressure? (b) If you heat the mixture to 202°C (with pressure at 7600 mm Hg), a fraction of the mixture is vaporized. What is the mol fraction benzene in the vapor phase? How much vapor is produced?

P5.30 100 gmol/h liquid methyl ethyl ketone and 100 gmol/h of nitrogen are fed to a flash tank, from which vapor and liquid streams can be removed. Assume that the flash tank is at 80°C and 3050 mmHg. Calculate the vapor and liquid flow rates and the composition of the vapor stream. Suppose the tank pressure is 3050 mmHg and only vapor is produced. What is the minimum tank temperature? Suppose the tank temperature is 80°C and only vapor is produced. What is the maximum tank pressure?

P5.31 A gas stream contains 40 mol% isopropanol in air and flows at 240 gmol/min, 70°C, and 1 atm. You are to use a condenser operating at a lower temperature to liquefy and recover isopropanol for reuse in the process.
   (a) The present condenser uses a chilled water coil to cool the gas stream to 25°C and produces a liquid condensate stream that is in equilibrium with the exiting gas stream. Sketch a process flow sheet showing this condenser. Find all stream flow rates and compositions and determine the percent recovery of isopropanol as condensate.

(b) We wish to change the condenser operating temperature to recover 98% of the isopropanol, to cut materials costs and produce a gas stream that can be vented or used elsewhere. With this new specification, find the required operating temperature for the condensor.

P5.32 100 gmol/min of a gas stream containing 30 mol% ethane (C_2H_6) and 70 mol% methane (CH_4) is fed to a distillation column, where it is separated into an overhead product containing 90 mol% methane and a bottoms product containing 98 mol% ethane. A customer wants to purchase a product that contains 97% methane. To produce this, the overhead from the column described above is sent to a second column. The overhead from the second column is the desired product, and the bottoms product, which has a flow rate of 100 gmol/min, is recycled back to the first column. Calculate the flow rate of the final product and the composition of the recycled stream. Calculate the fractional recoveries of methane and ethane in their corresponding product streams.

P5.33 A mixture of 30 mol% ethanol and 70 mol% water is charged to a tank, which is at 1 atm pressure. Use Fig. 5.13 to determine:
(a) the dew point temperature of the mixture,
(b) the composition of the liquid droplet at the dew point temperature,
(c) the bubble point temperature, and
(d) the composition of the vapor bubble at the bubble point temperature.

P5.34 A mixture of 30 mol% ethanol and 70 mol% water is charged to an equilibrium flash tank, which is at 1 atm pressure. To what temperature should the tank be adjusted to produce exactly 50% vapor and 50% liquid? What are the liquid and vapor phase compositions at this temperature? Calculate the fractional recoveries of ethanol and water, and the separation factor.

P5.35 You have a mixture of 30 mol% benzene (C_6H_6)/70 mol% toluene (C_7H_8) that you'd like to separate into two products: a benzene-rich product and a toluene-rich product. If the separation is carried out in a single-stage flash drum at 130°C, what is the minimum allowable operating pressure?

The customer demands that the toluene-rich product be at least 98% pure. Your boss requires that at least 96% of the toluene fed to the separation process be recovered in the toluene-rich product. You decide to design a distillation column. What is the minimum number of stages required? You may assume the average column operating temperature is 130°C.

P5.36 MEA (monoethanolamine) solution is a popular solvent used in absorption of "sour" gases like H_2S from natural gas and other process gas streams. Gas-liquid equilibrium data for the H_2S-MEA system are summarized in the following table. Assuming a total pressure of 760 mmHg, convert the data at 40°C to y_{H_2S} versus x_{H_2S}. (Recall the definition of partial pressure: $p_{H_2S} = y_{H_2S}P$, and note that x_{H_2S} = moles of H_2S per mole of liquid solution). Plot y_{H_2S} versus x_{H_2S}. If the data fall on a straight line, the H_2S-MEA system obeys Henry's law [Eq. (5.14)]. Does it?

Partial pressure H$_2$S (mm Hg)	Moles H$_2$S per mole MEA, 40°C	Moles H$_2$S per mole MEA, 100°C
1	0.128	0.029
10	0.374	0.091
100	0.802	0.279
200	0.890	0.374
500	0.959	0.536
700	0.980	0.607

P5.37 A natural gas stream (100 kmol/h) contains 98 mol% methane (CH$_4$) and 2 mol% hydrogen sulfide (H$_2$S). H$_2$S is a highly toxic gas. The concentration of H$_2$S in the natural gas stream must be reduced to 0.01 mol% H$_2$S before the natural gas can be used as a fuel or chemical feedstock. A process engineer recommends that the separation be carried out in a single-stage absorber, with water as the solvent. The absorber operates at 5 atm and 40°C.
(a) Estimate the required water flow rate (kmol/h).
(b) What is the required flow rate if the temperature is dropped to 10°C and the pressure is raised to 40 atm?

P5.38 We managed to spill 100 lb diethylamine (DEA) into a pond that contains 50,000 lb water. We need to remove 99% of the spilled DEA before we can release the pondwater to a local river. You've decided to look into single-stage solvent extraction for this clean-up job. Consider the solvents listed in App. B. Which solvent would you choose, and why? How much solvent would you need?

P5.39 1000 kg of a 25 wt% acetic acid/75 wt% water solution is mixed with 1000 kg methyl isobutyl ketone (MIBK). After equilibration, how many kilograms of extract (MIBK-rich phase) and raffinate (water-rich phase) are produced? What is the composition (wt%) of each phase? (Refer to Fig. 5.15.)

P5.40 Acetic acid is to be extracted from water in a single mixer-settler unit using pure methyl isobutyl ketone (MIBK) as the solvent. The feed contains 25 wt% acetic acid in water, and the water-rich product stream must contain no more than 6 wt% acetic acid. Calculate the kg solvent/kg feed and the compositions and flow rates of the two streams leaving the process. (Refer to Fig. 5.15.)

P5.41 Trace organic contaminants in waste water readily adsorb onto a solid called *activated carbon,* which is not very different from charcoal briquets. Equilibrium data for adsorption of phenol onto activated carbon is shown. One liter of wastewater containing 1 g phenol is contacted with 5 g activated carbon at 20°C. At equilibrium, what fraction of the phenol is adsorbed?

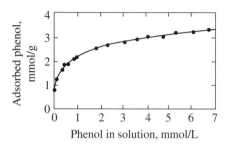

P5.42 10 L of water are contaminated with 200 mmol phenol. The water must be cleaned up to contain no more than 5 mmol phenol before it can be discharged. This will be accomplished by pumping the solution into a vessel loaded with fresh activated carbon, waiting for the phases to equilibrate, then removing the liquid solution. How much activated carbon should be loaded into the vessel? Assume an operating temperature of 20°C. Use the equilibrium data given with Prob. P5.41.

P5.43 Monoclonal antibodies (MAb) are proteins made by growing special cells called *hybridomas* in a complex nutrient broth. We need to develop a process for purifying MAb from the broth. We have decided to use adsorption onto immunosorbent beads. The broth initially contains 300 nmol/L MAb. We want to adsorb 99% of the MAb initially in the broth to the beads. If we have 2 L of broth, how many grams of beads are needed? Refer to Fig. 5.16 for phase equilibrium data.

Scrimmage

P5.44 Try some simple experiments at home with sugar and salt to figure out ways to separate a mixture of the two materials. You might want to look at heating, using (safe) solvents or freezing, for example (or be more creative!). Make some measurements and observations. Write a report on what you did, what you found, and how you would use that information to design a separation process.

P5.45 Look in *Kirk-Othmer Encyclopedia of Chemical Technology* or a similar reference book and identify the one or two most important separation technologies in the following industries:
(a) food manufacture,
(b) petroleum refining,
(c) pharmaceutical manufacturing,

(d) photographic film manufacture,

(e) water purification.

Discuss why these are the most important, given the heuristics for choosing appropriate separation technology.

P5.46 Complete the table below. Categorize each technology as mechanical, rate-based, or equilibrium-based, specify the separating agent (e.g., heat, solvent), the product phases (e.g., liquid + solid), and the physical property exploited. If you are unfamiliar with a particular technology, look up information in *Perry's Chemical Engineers' Handbook* or other sources. A few of the entries have been completed for you.

Technology	Category	Feed phase(s)	Product phase(s)	Property exploited	Separating agent
Drying of solids		Moist solid			
Adsorption					
Electrophoresis		Liquid	Liquid	Differences in ion mobility in an electric field	
Distillation	Equilibrium				
Crystallization					
Absorption					
Settling					
Reverse osmosis					Pressure gradient

P5.47 The average Wisconsin cheese plant makes 300,000 lb/day of whey as a byproduct. Dumping the whey into the nearest river is not a great solution for disposal of this waste product. It'd be much better to develop ways to make products from whey.

Cheese whey contains about 93.4 wt% water, 0.9 wt% protein, 5 wt% lactose ($C_{12}H_{22}O_{11}$, milk sugar), 0.2 wt% lactic acid (CH$_3$CHOH-COOH), and 0.5 wt% inorganic salts. Here is some information about each of these components.

Proteins: A mixture of proteins with molecular weights from 15,000 to 150,000. Highly soluble in water. Will precipitate as a gel if concentrated to 50 to 60 wt%. Valuable as animal feed, or as an ingredient in processed food.

Lactose: Molecular weight = 342. Soluble in water to about 0.1 g/mL. In large concentrations has an unwelcome laxative effect in mammals. Of some value as animal feed, in fermentation broth, or as a raw material for polymer production.

> *Lactic acid:* Molecular weight = 90. Very soluble in water. Fairly high mobility in electric field. A colorless viscous liquid that is highly soluble in ethanol and ether.
>
> *Mineral salts:* primarily calcium and sodium salts, with molecular weight of 50 to 100. Highly soluble in water as ions. High mobility in an electric field. Salts limit palatability of feedstuffs.

Your job is to develop a preliminary flow diagram for making two valuable products—a high-protein, low-salt dry solid and a high-lactose, low-salt dry solid—from cheese whey. Sketch out what you think is the best block flow diagram. Write one or two paragraphs describing your design and justifying why you think your design is best. Complete preliminary process flow calculations if possible. You might want to look up information in *Perry's Chemical Engineers' Handbook* or elsewhere on various separation technologies.

P5.48 Benzene is chlorinated to produce chlorobenzene (the desired product) and HCl:

$$C_6H_6 + Cl_2 \rightarrow C_6H_5Cl + HCl$$

The Cl_2 concentration in the feed is kept below 10 mol% to prevent unwanted additional chlorination of the monochlorobenzene. Essentially all of the chlorine reacts under the reactor conditions.

The byproduct HCl is converted back to Cl_2 by oxidation:

$$4HCl + O_2 \rightarrow 2Cl_2 + 2H_2O$$

The reaction is reversible; at the reactor conditions 60% of the reactants are converted to products.

Physical property data for these compounds are given in App. B. You are given the following additional information:

No O_2 or H_2O is allowed in the feed to the first reactor.
Cl_2 dissolved in water will preferentially partition into carbon tetrachloride, with a distribution coefficient (moles Cl_2 per liter carbon tetrachloride/moles Cl_2 per liter water) equal to 5.0.
The solubility of Cl_2 in water is 2.5 g Cl_2/liter water.
The solubility of HCl in water is 380 g HCl/liter water.

Sketch out a process flow sheet for production of chlorobenzene. Show the components in each stream on your flow sheet. Indicate the basis for separation in all cases (for example, difference in volatility or solubility) You do *not* have to do any calculations.

P5.49 Propylene (C_3H_6) and chlorine (Cl_2) are converted to allyl chloride (C_3H_5Cl) and hydrogen chloride (HCl). Besides allyl chloride, several chlorinated byproducts are generated, of which the major unwanted byproduct is 1,3 dichloropropane ($C_3H_6Cl_2$).

The major reactions are:

$$C_3H_6 + Cl_2 \rightarrow C_3H_5Cl + HCl \tag{R1}$$

$$C_3H_6 + Cl_2 \rightarrow C_3H_6Cl_2 \tag{R2}$$

Given the reactor effluent information described below, devise what you think is the best block flow diagram for producing pure allyl chloride, while recycling unreacted propylene and chlorine back to the reactor feed. Indicate the types of separation technologies you would use and the sequence of separation steps. Briefly explain your reasoning. A quantitative answer is not required.

Component	Relative amount, weight basis	Normal boiling point, °C	Solubility in water, wt%
1,3 dichloropropane	1.8	112	Insoluble
Acrolein chloride	0.2	84	Insoluble
Allyl chloride	9.3	50	0.33
Chlorine	3	−34	1.46
Propylene	105	−48	0.89
Hydrogen chloride	93	−85	72

P5.50 Reconsider Prob. 5.38. Once you've got the diethylamine in the solvent, you'd like to recover the diethylamine and the solvent as (nearly) pure products for reuse. Describe how you would separate diethylamine from the solvent of your choice. (You will have to look up physical property data.) Does the solvent recovery problem affect your choice of solvent for extraction? Is solvent extraction the best technology to use for recovery of the diethylamine, or can you suggest something else?

P5.51 Sulfur dioxide can be manufactured by direct oxidation of sulfur:

$$S + O_2 \rightarrow SO_2$$

The reaction generates a lot of heat, so in order to maintain a manageable temperature in the reactor (which is basically a burner), sufficient amounts of cool, inert gas must dilute the oxygen before it is fed to the reactor. The inert:oxygen gas ratio should be at least 7:3. Oxygen and sulfur are to be fed to the reactor at stoichiometric raio, using air (79% nitrogen, 21% oxygen) as the source of oxygen. All of the oxygen and all of the sulfur fed to the reactor should be consumed. Two possible process schemes are suggested. Scheme A: Use the nitrogen in air as the inert gas. Feed air and sulfur to the burner, then separate the nitrogen from the sulfur dioxide. Scheme B: Separate the nitrogen from the oxygen in the air, feed pure oxygen and sulfur to the burner, cool some of the product sulfur dioxide and recycle it to the reactor as the inert gas.

Sketch out process flow diagrams for the two alternative schemes described above. Using a production rate of 100 gmol/min sulfur dioxide and assuming perfect separations, calculate the composition and flow rates of all streams in your diagram. Which process scheme do you recommend? Consider in particular the difficulty of any separations and the tolerance of the process to sudden changes in feed flow rate or quality.

P5.52 Acetaldehyde (C_2H_4O) is produced by partial oxidation of ethane (C_2H_6) over a catalyst:

$$C_2H_6 + O_2 \rightarrow C_2H_4O + H_2O$$

A number of side reactions also occur, the most important of which are:

$$C_2H_6 + 3.5O_2 \rightarrow 2CO_2 + 3H_2O$$
$$C_2H_6 + 1.5O_2 \rightarrow CH_3OH + CO + H_2O$$

In a process to produce acetaldehyde, ethane at 6000 gmol/h is mixed with 30,952 gmol/h air. The fresh feed is mixed with a recycle stream, then fed to a reactor. The ethane:oxygen ratio in the reactor feed is maintained at 6:1. The reactor outlet stream is fed to gas-liquid Separator 1, where N_2, CO, CO_2, and C_2H_6 are taken off the top and recycled. Part of the recycle stream is split off and sent to a flare to be burned. This purge stream is analyzed for composition: It contains 10% C_2H_6, no O_2, and the CO_2:CO ratio is 2:1. The bottoms stream from Separator 1 is sent to Distillation Column 2, where acetaldehyde and methanol (CH_3OH) are separated from water. Acetaldehyde is further separated from methanol in Distillation Column 3. A simplified process flow diagram is shown. You may assume that air is 79 mol% N_2, 21 mol% O_2, and that the products from the distillation columns are essentially pure, with only trace contaminants.

Which is the limiting reactant, and why was it chosen? Why isn't all of the overhead from separator 1 recycled to the reactor inlet? Given the heuristics you learned, do you think this is the best sequence of separation tasks? Explain why or why not.

Complete a DOF analysis, and show that the process is correctly specified. Calculate the composition and flow rate of the purge gas, and the fractional yield, selectivity, and conversion of acetaldehyde from ethane for the overall process.

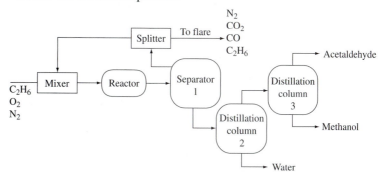

P5.53 In a process for making cellulose acetate, an aqueous acetic acid waste stream (30% acetic acid, 0.2% sulfuric acid, and water) is produced. (All compositions are mass percent.) The acetic acid is too valuable to be thrown away but its concentration is too low to be useful. Therefore, a solvent extraction process, using ether as the solvent, was developed to purify and concentrate the acetic acid.

The process is described as follows. The aqueous acetic acid waste stream is fed to a multistage extraction column, along with the solvent diethyl ether, which is contaminated with a small bit of water. The ether-rich phase leaving the top of the extraction column contains 24% acetic acid, water, and ether. This is fed to a solvent recovery distillation column. The distillate from this column contains 98.8% ether and 1.2% water and is recycled back to the extraction column. The bottoms from the solvent recovery column contains 60% acetic acid and 40% water, and is fed to an acid finishing distillation column. The bottoms from the acid finishing column contains 99% acetic acid with the remainder water; this concentrated acetic acid stream is the desired product. 67.5% of the acetic acid fed to the acid finishing column is recovered as product. The distillate from the acid finishing column, which is dilute acetic acid in water, is recycled back and mixed with the fresh feed. There is 1 lb acetic acid recycled for every 2.3 lb acetic acid in the fresh feed.

The water-rich raffinate stream leaving the bottom of the extraction column contains 7% ether, acetic acid, water, and sulfuric acid. This stream is fed to an ether-stripping distillation column. The distillate from the ether-stripping column contains 98.8% ether and 1.2% water; this is recycled back to the extraction column. The bottoms contains 0.1% ether, acetic acid, water, and sulfuric acid and is discarded. To make up for the loss of ether in this bottoms column, fresh ether solvent (contaminated with 1% water) is mixed with the other recycled ether streams and fed to the extraction column.

(a) What is cellulose acetate, and what is it used for?

(b) Evaluate the choice of separation technologies and the sequence of separations, in light of the heuristics. You may want to look up relevant physical properties of the components in this process. Why is solvent extraction used as the first step rather than distillation?

(c) Draw and label a simplified process flow diagram. For each of the four columns in the process, identify the two key components that are being separated. Complete DOF analysis and show that the process is correctly specified, except for the choice of a basis. Calculate the flows and compositions of all streams, assuming that the feed rate of the aqueous acetic acid waste stream is 1000 lb/h. Summarize the flows and compositions on a table accompanying your flow sheet. What fraction of the acetic acid fed to the process is recovered in the concentrated product?

P5.54 The following feed stream is to be separated by a series of distillation columns into four products:

Species	Flow rate, gmol/h
Pentane	4000
Benzene	1000
Toluene	1000
Orthoxylene	6000

Product 1: 98% pentane, no orthoxylene or toluene
Product 2: 90% benzene, 4% toluene, no orthoxylene
Product 3: 90% toluene, 2% benzene, no pentane
Product 4: 99% orthoxylene, no pentane or benzene

Your supervisor proposes the following design. Do you think this is the most economical sequence? Why or why not? If not, propose an alternative design. For the best design, calculate the product flow rates and purities and the fractional recoveries of each species in the appropriate product.

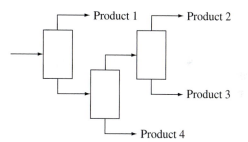

P5.55 A batch thickener is simply a cylindrical tank with an opening at the bottom, used sometimes for sedimentation operations. Initially the bottom opening is closed, and the tank is filled with a slurry. (A slurry is just a mixture of liquid solution and suspended solid particles.) The material is allowed to settle, and then after some time clear liquid is drawn off the top of the tank and a thickened sludge is pulled out the bottom opening.

One liter of a slurry containing 2 g NaCl and 230 g limestone (mostly $CaCO_3$) is poured into a small cylindrical glass tank. The tank is equipped with a bottoms drawoff. The slurry comes 36 cm up the side of the tank. After 8 h settling time, an interface between sludge and clear liquid is apparent, 10 cm up the side of the tank. (The top of the clear liquid is still 36 cm up.) The sludge is carefully drained out the bottom and then the clear liquid is removed from the top. What is the quantity (in total liters and in g NaCl and limestone) and composition (in g/L NaCl and limestone) of the two product streams? The solubility of limestone in water is 0.015 g/L, the solubility of NaCl in water is 360 g/L, and the density of limestone can be taken as 2.7 g/cm^3.

There are three components in the feed stream and two product streams. Is this process designed to separate limestone from NaCl, limestone from water, NaCl from water, or . . . ? What is the product purity and product recovery? What is the separation factor?

P5.56 1000 kg/h of 30 wt% KNO_3 in water at 90°C must be processed to produce dry KNO_3 crystals. I've got the following equipment available:

> One vacuum evaporator, which can operate at 90°C maximum and 50 wt% salts maximum
> One cooler, which can cool to a minimum 5°C
> One rotary drum filter, which entrains 1 kg solution per 19 kg crystalline solids
> One drum dryer, which has a maximum drying capacity of 15 kg water removed/h

My boss says I've got to build a process, using only these pieces of equipment, that produces at least 245 kg/h pure KNO_3 crystals. Not only that, I've got to get a flow sheet ready for his perusal within an hour, that'll convince him the process is workable. If I don't deliver, I'm fired. If I do deliver, I get a huge bonus.

Should I dust off my resume, or buy that sports car I've always wanted?

P5.57 In a process to synthesize the pain reliever acetylsalicylic acid (abbreviated as ASA, also known as aspirin), the effluent stream from the final reactor contains 11 wt% ASA and 2 wt% sodium acetate in water. The final dry product is to be obtained by drying, crystallizing, or some combination. The product purity, measured as weight percent ASA in the final dry powder, is a key value.

(a) If the water is evaporated off, what product purity results?

(b) The sodium acetate can be considered to be infinitely soluble in water, so an alternative process is suggested. Part of the water is removed by evaporation, then the concentrated solution is cooled to crystallize some pure ASA. The filter cake, which contains 1 kg entrained solution for each 4 kg of ASA crystals, is then dried. At the crystallizer/filter temperature, the solubility of ASA in water is 35 wt%. If enough water is evaporated so that half of the ASA fed is recovered as dry product, what is the ASA product final purity? How much water is removed per 100 kg of reactor effluent?

(c) Obtain a general equation that relates product purity to fractional recovery. Plot your expression. What purity do you predict at 90% ASA recovery? At 10% ASA recovery?

(d) Suggest changes to this process that could increase yield, purity, or both.

P5.58 A 30 wt% Na_2CO_3 solution is fed at 10,000 lb/h to an evaporator/crystallizer system shown in the figure. The filter cake contains 3.5 lb crystals per lb entrained solution, and the entrained and recycled solution both contain 17.7 lb Na_2CO_3 per 100 lb of solution. What is the production rate of crystals? How much water is removed in the evaporator? If 40% of the water fed to the evaporator is evaporated, what is the ratio of recycled solution to fresh feed? What is the purity (% Na_2CO_3) of the product stream?

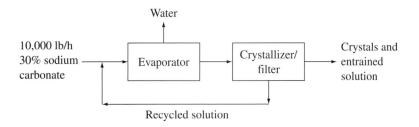

You'd like to improve the performance of the process. The evaporator has a maximum capacity of 7500 lb water evaporated per hour. The concentration of Na_2CO_3 is fixed at its solubility limit of 1.17 lb/100 lb solution. (a) Assuming that the entrainment remains the same, will changing the recycle ratio improve product purity and/or recovery? Explain. (b) Suggest at least one other process modification and analyze how your proposal will affect product purity and/or recovery.

P5.59 You are in charge of designing a process to separate 1000 lb/h of a 60 wt% benzene/40 wt% naphthalene, at 80°C and 1 atm pressure, into 99% pure benzene and 96% pure naphthalene products. Some physical property data are given in App. B, and in Fig. 5.12.

Formation of tarry polymers becomes a problem if the mixture is kept at temperatures much above 150°C.

You have the following idle equipment available on site:

A cooler with a minimum allowable operating temperature of 12°C.
A rotary drum filter that entrains 1 lb solution per 10 lb solids in the filter cake.
A distillation column with 4 equilibrium stages and a minimum allowable operating pressure of 1 atm.
A mixer-settler unit that could be used for single-stage solvent extraction.
A vacuum evaporator with a minimum operating pressure of 300 mmHg and a maximum operating temperature of 250°C.

Sketch out a process flow diagram. Indicate which (if any) of the available pieces of equipment you would use. To the extent possible, determine temperatures, pressures, flow rates, and any other design information. Write a brief memo explaining your design and justifying your choices.

P5.60 1000 kg/h of 10 wt% Na_2SO_4 in water at 20°C must be processed to produce anhydrous Na_2SO_4 crystals. I have available the following pieces of equipment: a vacuum evaporator, which can operate at a maximum of 70°C and maximum 60 wt% Na_2SO_4; a cooler, which can cool to a minimum 5°C; a rotary drum filter, which entrains 1 kg solution per 19 kg crystalline solids; and a drum dryer, which can remove all residual water. The drum dryer is fairly expensive to operate compared to the other items. Put these pieces together (you do not have to use all of them) to design a process that produces the desired product. Sketch out a preliminary process flow diagram to showcase your chosen design. To the extent possible, specify operating temperatures, flow rates, and compositions of all streams. Calculate product recovery. Briefly explain the rationale for your design. (Refer to App. B, for phase equilibrium data.)

P5.61 10,000 lb/h of a 20 wt% K_2SO_4 solution is fed to a evaporator/crystallizer process (see figure). The entire process operates at 40°C. Enough water is removed in the evaporator to produce a 40 wt% K_2SO_4 solution. Crystals form in the crystallization unit and are separated from the solution by filtration and removed as product. 1 lb of solution is entrained with every 10 lb of crystals. The remaining solution is recycled to the evaporator. Solubility data are given in App. B.

(a) Calculate the crystal production rate, the water evaporation rate, and the recycle rate.

(b) An engineer proposes to replace the evaporator with a unit that simply cools the feed down to 0°C. (without removing water) to initiate crystallization. What do you think about this idea?

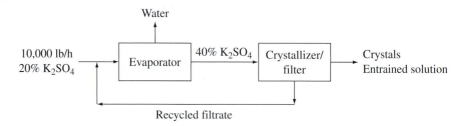

Recycled filtrate

(c) During operation, the capacity of the evaporator slowly decreases because of fouling of the heat-exchange tubes in the evaporator. The plant is shut down for cleaning when the evaporator capacity drops below 6000 lb water evaporated per minute. All other specifications remain the same. Plot how the fresh feed rate, the production rate of solid K_2SO_4, and the ratio kg recycle/kg fresh feed changes as the evaporator capacity changes. Explain your results.

(d) The plant superintendent wants to maintain the fresh feed rate constant at the initial value throughout the time that the plant operates and insists that this can be done by adjusting the recycle ratio and the fraction of water fed to the evaporator that is evaporated. (The concentration of the solution leaving the crystallizer and the ratio of crystals to solution in the filter cake cannot be changed.) Write a brief memo to the superintendent describing your response to his idea.

P5.62 The xylene isomer *p*-xylene is a starting material for the production of polyester fibers. *m*-xylene is blended into gasoline, and is less valuable than *p*-xylene. The volatilities of the two xylenes are quite similar, but their melting points are different, so crystallization is proposed as a means to separate them. Liquid-solid phase equilibrium data are given in App. B.

A mixture of 30 mol% *p*-xylene and 70 mol% *m*-xylene is fed at a rate of 1000 gmol/h to a heat exchanger, where the stream is cooled to −50°F. The stream is next sent to a crystallizer/filter unit, where crystals are separated from the liquid solution. Some solution is entrained with the crystals at 1 mol solution per 19 mol crystals. What is the production rate and purity of the crystals leaving the process? What is the composition and flow rate of the solution leaving the filter? What is the fractional recovery of *p*-xylene in the *p*-xylene-rich product?

A colleague proposes the following idea: Why not send the liquid solution to an isomerization reactor, where some of the *m*-xylene fed to the reactor is converted to *p*-xylene. (See diagram.) The product from the reactor, which is 30 mol% *p*-xylene, is then mixed with the fresh feed. What is the fractional recovery and purity of *p*-xylene in the *p*-xylene-rich product? What do you think about this idea?

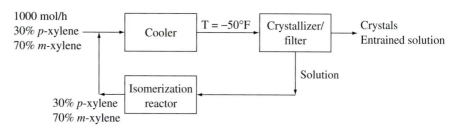

P5.63 At a plant that manufactures styrene, the reactor outlet (100 kgmol/h, 760 mmHg, and 500°C) contains 50 mol% H_2, 30 mol% ethylbenzene (C_8H_{10}), and 20 mol% styrene (C_8H_8). The reactor outlet is cooled and partially condensed, then separated into vapor and liquid streams in a flash drum, which operates as a single equilibrium stage. The hydrogen-rich vapor stream is sent to a desulfurization process elsewhere in the facility. To keep the operators happy at the desulfurization process, the hydrogen-rich stream must contain no more than 0.2 mol% styrene. To what temperature should the reactor outlet be cooled? How much ethylbenzene is in the hydrogen-rich stream?

P5.64 In Prob. 5.26 you examined some methanol-water vapor-liquid equilibrium data. Now, assume that this system can be modeled by using Raoult's law, with the Antoine equation for calculating saturation pressure.

Generate a T-y-x diagram at 1 atm based on these assumptions. (One way to do this is to calculate dew point and bubble point temperatures at each methanol mole fraction listed in the table.) Plot your calculated T-y-x diagram alongside the data. Is Raoult's law a good model for the methanol-water system?

P5.65 300 gmol/s CO, 300 gmol/s H_2, and 100 gmol/s CO_2 are fed to a reactor operating at 200°C and 4000 kPa. Two reactions occur:

$$CO + 2H_2 \rightleftarrows CH_3OH$$

$$CO_2 + H_2 \rightleftarrows CO + H_2O$$

The reactor effluent is at chemical reaction and phase equilibrium. Is the effluent gas phase, liquid phase, or a mix of gases and liquids? (Hint: assume it is gas phase and check the dewpoint.)

P5.66 Humid air enriched with oxygen is needed for a fermentation process. To prepare the warm humid air, room air is mixed with pure dry oxygen and liquid water in a special humidifying chamber, which also heats the air. Room air is at $T = 20°C$, $P = 1$ atm (absolute), and a relative humidity h_r of 20%. (Relative humidity is defined as $h_r (\%) = 100 \, y_{H_2O} P / P_{H_2O}^{sat}$, where y_{H_2O} is the mole fraction of water in the air.

It can be assumed that the N_2/O_2 (molar) ratio in the room air is 3.76 (moles N_2/mol O_2). The humid air leaving the humidifying chamber contains 40 mol% O_2 ($y_{O_2} = 0.40$), has a relative humidity of 100% and is at 37°C and 1 atm pressure. The total volumetric flow of the humid air is 2000 ft³/min. Calculate the mass flow rate (in kg/min) and the volumetric flow rate (in ft³/min) of air, liquid water, and pure dry oxygen into the humidifying chamber.

P5.67 A mixture of 40 mol% isopentane and 60 mol% n-pentane is to be separated by distillation into nearly pure products. 99.5% of the isopentane is to be recovered in the distillate product and 98% of the n-pentane is to be recovered in the bottoms product. Raoult's law is a good approximation of the vapor-liquid equilibrium behavior.

(a) First consider a single equilibrium stage. If the stage operates at 350 kPa, what operating temperature would you pick if you wanted equal-sized vapor and liquid flows? At this temperature, what is the fractional recovery of each component?

(b) Estimate the minimum number of stages in a distillation column needed to achieve the specified recoveries of isopentane in the distillate and n-pentane in the bottoms. Assume the average column temperature is 60°C.

(c) If the distillation column operates at 350 kPa, what (roughly) is the temperature of the the distillate stream leaving the column?

P5.68 We want to separate methanol from benzene, using an equilibrium flash drum. The process feed is 20 mol% methanol, flowing at 5 gmol/min total. We want the vapor stream leaving the drum to be 50 mol%

methanol and 50 mol% benzene, and to contain 90% of the methanol fed. The flash drum operates at 1 atm. Calculate the molar flow rates and compositions of the vapor and liquid streams leaving the drum. Can the desired separation be accomplished in a single equilibrium stage? If yes, determine the appropriate operating temperature. If not, determine the operating temperature that:

(a) meets purity specification but not recovery specification,
(b) meets recovery specification but not purity specification. (Vapor-liquid equilibrium data are given in App. B.)

P5.69 You have a mixture of four isomers of C_8H_{10}: 30% ethylbenzene, 20% *m*-xylene, 35% *o*-xylene, and 15% *p*-xylene. You want to design a sequence of distillation columns that produces the following four product streams:

Product 1: 98% benzene, no *p*-xylene or *o*-xylene
Product 2: 95% toluene, 3% benzene, 1% each *p*-xylene and *o*-xylene
Product 3: 90% *p*-xylene, 8% *o*-xylene, no benzene
Product 4: 95% *o*-xylene, no benzene or toluene

Let's assume that the cost of a distillation column varies roughly as:

$$\$ = C_1 \sqrt{Q}\, N$$

where Q is the molar feed rate, N is the number of stages, and C_1 is a cost factor. (*Note:* This is a gross oversimplification!)

Look at three different ways to sequence the distillation columns for this separation problem. Estimate the temperature of the distillate and bottoms streams for each column, and calculate the minimum number of stages. Then estimate the actual number of stages and the total costs for the columns in all three of your sequencing schemes. Is there any one sequence that is most preferable? How does your answer fit in with the heuristics you learned?

P5.70 Suppose that you want to separate benzene and toluene using a continuous equilibrium flash tank. The pressure and temperature in the tank can be adjusted to any desired values.

Derive an equation that relates the mole fraction of benzene in the liquid stream x_b to the total pressure P and the saturation pressures of benzene and toluene P_b^{sat} and P_t^{sat}. Then, using the Antoine equation, write an equation that relates x_b to P and T. (You may assume that Raoult's law applies.) Starting with steady-state material balance equations, derive an equation that relates the ratio $\dot{n}_V/\dot{n}_L$ to z_b, y_b and x_b. (*Hint*: How is the equation you derived related to the inverse lever rule?)

Assume that the feed rate to the tank is 100 gmol/h and that the pressure is 1 atm. Using the equations you derived, calculate $\dot{n}_V$, $\dot{n}_L$, y_b and x_b for the following three cases: (i) $z_b = 0.2$, (ii) $z_b = 0.5$, (iii) $z_b = 0.8$. For each case, complete calculations assuming that the tank temperature is (a) at its minimum, (b) at its maximum, and (c) halfway between its minimum and maximum. What do these minima and maxima correspond to?

P5.71 A natural gas stream (1000 kmol/h) contains 90 mol% methane (CH_4) and 10 mol% hydrogen sulfide (H_2S). H_2S is a highly toxic gas. The concentration of H_2S in the natural gas stream must be reduced to 0.1 mol% H_2S before the gas can be used as a fuel or chemical feedstock. Purification is achieved by gas absorption with MEA/water solution. The MEA/water solution is regenerated by stripping the H_2S out with a N_2 stream flowing into the stripper at 650 kmol/h. (See diagram.) CH_4 and N_2 are insoluble in MEA/water. Solubility data for H_2S is given below.

Assume that both absorber and stripper behave as single equilibrium stages. The pressure must be between 760 mmHg and 10,000 mmHg, and the temperature must be between 40°C and 100°C. Pick appropriate operating temperatures and pressures for the absorber and stripper (they can be different). Calculate the flow rate of MEA (kmol/h) required. If the N_2 flow rate was decreased and no other flow rate changed, what would be the consequences for the performance of the absorber? Explain.

Partial pressure H_2S, mmHg	Moles H_2S per mole MEA, 40°C	Moles H_2S per mole MEA, 100°C
1	0.128	0.029
10	0.374	0.091
100	0.802	0.279
200	0.890	0.374
500	0.959	0.536
700	0.980	0.607

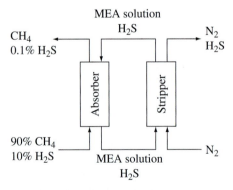

P5.72 The main active ingredients in dishwashing detergents are surfactants, or surface-active molecules, which alter oil-water interfaces. These molecules have a part that is oil soluble, which dissolves oil and grease, and a part that is water soluble, which partitions into the water phase, thereby solubilizing the oil and grease. Detergents differ from soaps because detergents are synthesized from hydrocarbons, while soaps are made from natural fatty acids.

In one process to manufacture a surfactant, a straight-chain hydrocarbon $C_{10}H_{22}$ (decane) is chlorinated in a photocatalytic step to produce monochlorodecane ($C_{10}H_{21}Cl$, MCD):

$$C_{10}H_{22} + Cl_2 \rightarrow C_{10}H_{21}Cl + HCl$$

An undesired reaction also occurs, where MCD is chlorinated to dichlorodecane ($C_{10}H_{20}C_{12}$, DCD):

$$C_{10}H_{21}Cl + Cl_2 \rightarrow C_{10}H_{20}Cl_2 + HCl$$

The following data are available:

Compound	Mol. weight	$T_m °C$	$T_b °C$	Solubility in water, g/L
MCD	177	-50	215	Insoluble
DCD	211	-40	241	Insoluble
Decane	142	-30	174	Insoluble
Cl_2	71	-101	-34	2.5
HCl	36.5	-111	-85	380

The product from the reactor is fed to a high-pressure separator, where chlorine and HCl are taken off as gases, and liquid products are sent to a distillation unit. The liquid product contains 40 mol% decane, 50 mol% monochlorodecane, and 10 mol% dichlorodecane, and enters the distillation unit at a flow rate of 1000 gmol/min. The distillation unit contains two distillation columns. The distillation unit is to be designed to recover 99% of the monochlorodecane in the feed to the unit in a monochlorodecane-rich product stream. The monochlorodecane product should be at least 95% pure. The dichlorodecane product should contain no decane. The decane product is recycled back to the reactor; to avoid excessive chlorination the decane product should contain no dichlorodecane and no more than 2 mol% monochlorodecane.

Explain why distillation was chosen for the separation of the decanes. Propose two different configurations for the two columns in the distillation unit. Identify the key and nonkey components in each. Assume that 97% of the decane fed to the distillation unit is recovered for recycle. Use a DOF analysis to determine whether the problem is completely specified. Calculate the flow rates and compositions of streams (as many as possible) for the two different configurations and summarize your results in table form along with the two block flow diagrams. Is one configuration preferable? If so, why?

P5.73 A gas stream (100 gmol/s, 40°C, 26 atm) contains 95 mol% air and 5 mol% CO_2. The CO_2 content of the air must be reduced to 0.5 mol%. You have decided to carry out the separation using absorption with water in a single equilibrium stage. Assume the absorber operates at the gas feed temperature and pressure.

(a) Assume that the water is nonvolatile and the air is insoluble in water. Calculate the amount of water required (gmol/s).

(b) You assumed that the water is nonvolatile and the air is insoluble in water. Show whether these assumptions are reasonable, using Raoult's or Henry's law, as appropriate.

(c) Let's say we want to design a better separation. One possibility is to adjust the temperature and pressure. Plot the water usage as a function of temperature, from 0°C to 50°C, and of pressure, from 1 atm to 100 atm. What T and P would you choose? (You may continue to assume that the water is nonvolatile and the air is insoluble in water.)

(d) At the chosen T and P from part (c), explore whether adding more stages will significantly decrease the water requirements. First, calculate the water flow for two stages by combining material balance equations and Henry's law for each of the two stages. Then, calculate the water flow using the KSB equation as a function of number of stages, varying from 1 to 10 stages. Comment on your results. (You may continue to assume that water is nonvolatile and air is insoluble.)

P5.74 Your job is to design an absorber that removes 98% of the H_2S in a sour gas stream (2 mol% H_2S and 98 mol% H_2), using water as the absorbent. The operating conditions are set at 10°C and 20 atm. Generate a plot showing how water flow rate decreases with increasing number of countercurrent stages, up to 10 stages. Solve in two ways: (1) by setting up material balances and phase equilibrium relationships for each stage and using equation-solving software, and (2) by using the KSB equation. Compare the answers; if they are not the same, explain why. For this problem you may assume that H_2 is insoluble in water and that H_2O is nonvolatile. Discuss the economic trade-offs between increased number of stages versus increased water flow.

P5.75 An absorber treats a gas stream containing 22.5 gmol/min methane and 2.5 gmol/min carbon dioxide with an aqueous stream with proprietary additives that is flowing at 50 gmol/min. The process is currently operated in a large spray tower, and you may treat the *exiting* liquid stream as being in equilibrium with the *entering* gas stream. The present process recovers 80% of the CO_2 fed to the aqueous stream.

(a) Determine a value for the equilibrium distribution coefficient between the aqueous and gas streams (moles CO_2/mole liquid/moles CO_2/mole gas).

(b) We'd like to improve the separation performance of the absorber. Frannie proposes using a new superadditive that increases the CO_2 solubility in the aqueous phase (increases the distribution coefficient) by 10%, that can be mixed with the incoming liquid stream. What is % recovery of CO_2 if the new additive is used? Zooey suggests staying with the original aqueous solution, but splitting the liquid feed into two equal streams and operating two separators in a cross-current mode, as shown in the figure. You may still assume that, in each separator, the exiting liquid is in equilibrium with the entering gas. What percent recovery CO_2 is achieved with this process change? Do you prefer Frannie's idea or Zooey's, and why?

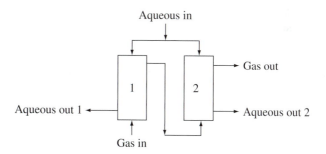

P5.76 Carbon dioxide must be recovered as a pure gas stream from a 25 mol% CO_2/75 mol% N_2 stream. The feed rate of this stream is 100 kgmol/h, and the design goal is to recover 85% of the CO_2. You've decided to use absorption as the separation method, and are considering an aqueous solution of 15.3 wt% monoethanolamine (MEA), as the solvent. Since the solvent is expensive, your design must include a separation unit to remove the CO_2 as a pure stream, allowing recycle of the solvent. To achieve separation of CO_2 from the solvent, you plan to simply adjust the temperature and pressure of the CO_2-laden MEA solution leaving the absorber in a separate unit called a stripper. For preliminary design calculations, you may assume that the absorber and the stripper each act as an equilibrium stage. Generate a block flow diagram for a process that produces pure CO_2 plus a nitrogen-rich waste gas stream, and recycles the solvent. Pick reasonable operating pressures and temperatures for the absorber and the stripper, and calculate the required flow rate of MEA solution (kgmol/h) at your chosen values. (*Hint*: There will be CO_2 in the MEA solution entering the absorber!) Assume the pressure allowed is 1 to 20 atm. Equilibrium data are given.

Solubility of CO_2 in 15.3 wt% MEA		
Partial pressure CO_2 (mm Hg)	**moles CO_2/mole solution, 40°C**	**moles CO_2/mole solution, 100°C**
1	0.383	0.096
10	0.471	0.194
100	0.576	0.347
200	0.614	0.393
760	0.705	0.489
1000	0.727	0.509

P5.77 Benzoic acid can be extracted from a dilute water solution by solvent extraction with benzene in a single equilibrium stage. The feed rate to the unit is 10,000 kg/h, and the feed solution contains 0.02 kg benzoic acid/kg water. Benzene is slightly soluble in water, so the water-rich raffinate stream will contain 0.07 kg benzene per kg water. The benzene-rich extract stream can be assumed to contain no water. Benzoic acid will be present in both raffinate and extract, and will distribute between them as:

$$K_D = 4 = \frac{\text{kg benzoic acid/kg benzene}}{\text{kg benzoic acid/(kg water + kg benzene)}}$$

Suppose that benzoic acid extracted into the benzene is worth \$1.35/kg and that fresh benzene costs \$0.03/kg. Calculate the net profit as a function of benzene flow rate and select the optimum benzene flow rate.

P5.78 In the case study, we considered two ways to reduce solvent requirements: use a multistage absorber with water as the solvent, or use MEA solvent in a single-stage absorber. Now consider combining these two ideas. Determine the flow rate of MEA solvent required if a two-stage counter-current absorber is built. You cannot use the KSB equation, because Henry's law does not accurately model the phase equilibrium data.

P5.79 Penicillin and other antibiotics are often recovered from aqueous fermentation broths by extraction with amyl acetate (also famous as banana flavoring). The distribution coefficient $K_D = 9.0$, where $y = K_D x$, with $y =$ concentration of penicillin in amyl acetate (g/L) and $x =$ concentration of penicillin in aqueous stream (g/L). Both solvents are insoluble in the other, and the concentrations are low enough that density or volume changes are negligible. A typical fermentation batch product is 5000 L of aqueous solution, including 2 g/L penicillin and many undesired compounds that are fortunately not soluble in amyl acetate. Our goal is to recover 95% of the penicillin.

(a) If the separation is done in a single-step batch process, how many liters of amyl acetate will be needed?

(b) We've got on hand an unused 7000-L tank. Our idea is to put the 5000-L broth in the tank, add solvent to fill, equilibrate, then remove the solvent (leaving the broth in the tank). How many times would we need to fill/remove solvent to meet our recovery requirement? Why not drain the broth and leave the solvent instead?

P5.80 Alcohol is removed from a mixture of 40 wt% alcohol/60 wt% ketone by washing with water. The wash water used is contaminated with 4 wt% alcohol. The water and ketone are mutually insoluble, and the distribution coefficient

$K_D = 0.25$, where ketone is solvent phase I and water is solvent phase II.

Suppose that you have 500 kg/h of the alcohol-ketone mixture to process, and 300 kg/h of the wash water is available. Compare the fraction of alcohol removed in three different contacting schemes

(a) Single-stage contacting.

(b) Crosscurrent contacting using three stages, where the wash water is equally split between the three stages.

(c) Countercurrent contacting using three stages, where the fresh feed is introduced into stage 1 and all of the fresh wash water is introduced into stage 3.

P5.81 You are the chief engineer for a top-secret crash program to manufacture large quantities of a new antibiotic "Picro." Picro is made by fermentation of a fungus. The fermentation broth contains 0.001053 g Picro /cm³ broth, along with water (by far the main ingredient), sugars, acids, and other byproducts.

An engineering co-op, Ima Smartee, has discovered a remarkable solvent, code-named "Elixir." Picro is highly soluble in Elixir, but sugars, acids, and other by-products are not. Furthermore, Elixir is completely immiscible with water, and is nonvolatile and nontoxic. Ms. Smartee conducted a number of experiments in the lab and reported the following data.

Experiment 1: The distribution coefficient K_D for Picro distributing between Elixer (solvent phase I) and water (solvent phase II) is 45,000.

Experiment 2: Pure Elixir has a melting point temperature of 5°C. Pure Picro has a melting point temperature of 80°C. A mixture of Picro and Elixer has a eutectic temperature of −4°C and a eutectic composition of 20 wt% Picro and 80 wt% Elixir. The liquid-solid phase equilibrium diagram for Picro and Elixir can be sketched based just on these data.

Before she returned to school, Ima left a preliminary flow diagram for a process to purify Picro from fermentation broth. She proposed using two pieces of unused equipment already at your facility: a mixer/settler unit that functions as a single-stage solvent extractor, and a crystallization/filter unit with a minimum operating temperature of 10°C that (remarkably) produces a solid product with no entrained solution. In Ima's process, the broth along with Elixer is fed to the mixer/settler, then the extract from the mixer/settler is fed to the crystallization/filter unit. Ima's proposed process handled a flow rate of 10,000 cm³/day fermentation broth at 37°C. The broth has a density of 1 g/cm³. Ima claims the process reduces the Picro concentration in the broth to 0.00001053 g Picro/cm³ broth. But Ima did not leave any record of process flow calculations.

(a) Calculate the quantity of Elixer fed to the mixer/settler unit.
(b) Calculate the quantity of Picro crystals (g/day) produces. What is the fractional recovery of Picro?
(c) Suggest ways to improve the fractional recovery of Picro. Sketch out any proposed process modifications, and briefly explain your reasoning.

Game Day

P5.82 Ethylene glycol [$C_2H_4(OH)_2$, an antifreeze] is produced in two steps. The simplified process flow diagram is sketched.

In the first reactor R-1, ethylene (C_2H_4) is mixed with air (79% N_2, 21% O_2) and oxidized to ethylene oxide (C_2H_4O) in the gas phase over a silver catalyst:

$$C_2H_4 \text{ (g)} + \tfrac{1}{2} O_2 \text{ (g)} \rightarrow C_2H_4O \text{ (g)}$$

Complete combustion of the ethylene to water and carbon dioxide can also occur, as an undesired side reaction:

$$C_2H_4 \text{ (g)} + 3O_2 \text{ (g)} \rightarrow 2CO_2 + 2H_2O \text{ (g)}$$

All of the O_2 fed to the process is consumed. CO_2, H_2O, and C_2H_4O are separated from unreacted C_2H_4 and N_2 by absorption in column A-1 into cold water. C_2H_4, O_2, and N_2 are recycled back to the reactor inlet, with a purge stream taken off. C_2H_4O and CO_2 are then separated from the water in distillation column D-1. The water is discarded, and CO_2 is then removed from C_2H_4O by absorption in column A-2 into triethanolamine (TEA). C_2H_4O is sent to reactor R-2.

In the second reaction, ethylene oxide is mixed with liquid water, in which it is very soluble. Ethylene oxide reacts with the water to produce ethylene glycol:

$$C_2H_4O \text{ (g)} + H_2O \text{ (l)} \rightarrow C_2H_4(OH)_2 \text{ (l)}$$

Ethylene glycol can react with ethylene oxide to produce diglycol in an unwanted side reaction:

$$C_2H_4O \text{ (g)} + C_2H_4(OH)_2 \text{ (l)} \rightarrow (C_2H_4OH)_2O \text{ (l)}$$

Ethylene oxide is very reactive, and 100% conversion of ethylene oxide to products is achieved. Water and diglycol are separated from ethylene glycol in two distillation columns D-2 and D-3.

Identify each separation unit. Specify the main purpose of the separation unit and the separation technology used. Explain why that separation technology is or is not the best choice for that separation problem. Simplify the flow diagram to show what it would look like if the unwanted reactions did not occur.

The fresh ethylene feed rate to the process is 1000 gmol/h. The C_2H_4/O_2 ratio in the fresh feed is 2:1. 50 gmol/h CO_2 is removed in the absorber A-2. The fractional conversion of C_2H_4 to products in R-1 is 0.25. Calculate the production rate of C_2H_4O from R-1, the composition and flow rate of the purge gas, and the flow rate of the recycle stream to R-1.

Freshwater is fed to R-2 at a water:C_2H_4O ratio of 5:1. For every 10 mol of ethylene glycol produced, 1 mol of diglycol is produced. Calculate the production rate of ethylene glycol. Calculate the overall conversion of ethylene to ethylene glycol for the entire process.

The process has been operating successfully for about 10 years. Suddenly, the price of ethylene doubles. Propose two process modifications that would improve ethylene utilization. Identify the effects of these proposed modifications on all parts of the plant (i.e., increases/decreases in flows or compositions). Explain your reasoning.

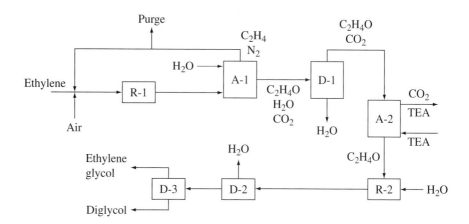

P5.83 The most important route for preparing styrene involves alkylation of benzene with ethylene to produce ethylbenzene followed by the dehydrogenation of ethylbenzene to styrene. Your company has an existing styrene production plant; the following information is available on reactor performance.

Alkylation reactor. The following reactions take place in the alkylation reactor over a solid acid alkylation catalyst at 95°C and 0.3 atm.

$$C_6H_6 + CH_2CH_2 \rightarrow C_6H_5CH_2CH_3 \qquad\qquad \text{(R1)}$$

$$C_6H_6 + 2CH_2CH_2 \rightarrow C_6H_4(CH_2CH_3)_2 \qquad\qquad \text{(R2)}$$

$$C_6H_6 + C_6H_4(CH_2CH_3)_2 \rightleftarrows C_6H_5CH_2CH_3 \qquad\qquad \text{(R3)}$$

Reactions (R1) and (R2) are irreversible and reaction (R3) is equilibrated. To minimize the formation of diethylbenzene, the ethylene to benzene ratio in the reactor feed is controlled to 0.58 moles ethylene/mole benzene. The reactor effluent contains an equilibrated gaseous mixture with the following composition:

Benzene: 65 mol%
Ethylbenzene: 30 mol%
Diethylbenzene: 5 mol%

Dehydrogenation reactor. Ethylbenzene with a purity of 99.5 mol% is mixed with superheated steam (steam temperature 710°C) at a steam-ethylbenzene molar ratio of 15 to achieve a reactor inlet temperature of 630°C. The following reactions take place over an iron dehydrogenation catalyst at 1.0 atm.

$$C_6H_5CH_2CH_3 \rightarrow C_6H_5CHCH_2 + H_2 \qquad\qquad \text{(R4)}$$

$$C_6H_5CH_2CH_3 + H_2 \rightarrow C_6H_6 + C_2H_6 \qquad\qquad \text{(R5)}$$

$$C_6H_5CH_2CH_3 + H_2 \rightarrow C_6H_5CH_3 + CH_4 \qquad\qquad \text{(R6)}$$

$$2C_6H_5CH_2CH_3 \rightarrow C_6H_5CHCHC_6H_5 + C_2H_6 + H_2 \qquad\qquad \text{(R7)}$$

The reactor effluent composition is determined to be

Hydrogen   5.70 mol%
Methane   0.25 mol%
Ethane   0.39 mol%
Benzene   0.30 mol%
Water   84.70 mol%
Toluene   0.25 mol%
Ethylbenzene   2.64 mol%
Styrene   5.64 mol%
Stilbene   0.13 mol%

Demand for polystyrene is increasing and therefore your company plans to build a new styrene production plant. Plans call for producing 130 kgmol/h styrene in the new plant. The new plant will use the same reactor conditions as the existing plant.

Your job is to develop a process for separating the dehydrogenation reactor effluent into three products: a hydrogen-rich stream that goes to fuel, a styrene-rich product stream, and an ethylbenzene-rich stream that is recycled back to the reactor inlet. The styrene product stream must be at least 98% pure. The process must recover at least 95% of the styrene in the reactor outlet. Styrene tends to spontaneously polymerize at temperatures above 100°C. Develop the preliminary design of the separation section of the new plant. Choose appropriate separation technologies. Specify operating temperatures and pressures of each process unit. Determine flow rates and compositions of all major streams.

Process Energy Calculations and Synthesis of Safe and Efficient Energy Flow Sheets

In This Chapter

We take a look at the energy requirements of chemical processes. You've seen, for example, that reactor temperature and pressure must be manipulated to get fast reaction rates or favorable reaction equilibrium, or that the temperature and pressure of a separation unit must be adjusted to achieve efficient separation. You've seen that raw materials and products must be moved into and out of various reactors, mixers, and separators. All these operations involve energy. In this chapter, we learn how to calculate the energy of a system. We define terms like work and heat. We introduce the energy balance equation, and learn how to use this equation to solve practical problems.

The questions we address in Chap. 6 include:

- What natural resources provide our energy needs?
- What are the major types of energy-related process equipment?
- What are the different forms of energy that a system may have?
- What are work and heat?
- What is the energy balance equation?
- How do we complete process energy balance calculations?
- How do we design chemical processes that are energy efficient?

Words to Learn

Watch for these words as you read Chapter 6.

Potential energy	Heat capacity	Enthalpy of solution
Kinetic energy	Enthalpy of vaporization	Enthalpy of reaction
Internal energy	Enthalpy of melting	Work
Enthalpy	Enthalpy of formation	Heat
Reference state	Enthalpy of combustion	Energy balance equation
State function	Enthalpy of mixing	

6.1   Introduction

Energy is a familiar word to us all. Government officials talk about energy policy. We guzzle caffeinated drinks to get energy. We see the Energizer Bunny® that keeps going and going and. . . .

Chemical processes and products require energy as well. We've seen that energy is needed to drive separation processes like distillation and crystallization. We've observed that reactor conversion depends on temperature, which must be controlled by energy supply or withdrawal. Devices such as artificial kidneys require energy to pump fluids around. Designing chemical processes that work and that are economical requires careful consideration of energy needs.

Averaged over all chemical industries, energy costs contribute about 8% to the total expense of product manufacture. Energy costs are very important in determining the profitability of commodity products (e.g., food), but are less important for specialty products (e.g., drugs). As an example, with nitrogen-based fertilizers the energy costs are about 70% of the total manufacturing costs. Even for those processes and products where energy costs are not important, control of energy needs is crucial for the safety and reliability of the process. For example, sterilization of cell culture broths requires heating of the broth to a sufficiently high temperature for a sufficiently long time. As another example, heat removal is essential for safe operation when a reaction is highly exothermic.

6.1.1   Energy Sources

Just as the earth's resources provide the raw materials needed for making products, the earth's resources supply energy. In the United States, combustible fuels (mostly fossil fuels) serve as the major source of raw materials for energy (Fig. 6.1).

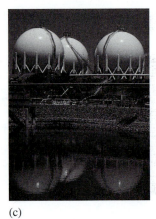

(a)   (b)   (c)

Figure 6.1 Fossil fuels constitute the largest energy resource, although the contribution from renewable energy resources is growing. (a) Large coal pile stacked at a power plant, (b) offshore oil platform, (c) natural gas storage tanks. (a) © *Vol. 160/Corbis;* (b) © *Vol. 57/PhotoDisc/Getty;* (c) © *Vol. 160/Corbis.*

Solid fuels: This category includes coal, waste paper, wood, bagasse from sugarcane—anything solid that burns. These fuels are often relatively inexpensive, but may require special burners and/or solids handling facilities, and typically leave an ash residue, which is a waste disposal problem. (Your great/grandparents probably shoveled coal into the family furnace, or cut logs for the woodstove. Ask your great/grandparents or the local power plant operator about "clinkers.") Solid fuels were once the dominant fuel source; now in the U.S. solid fuels supply about 25% of our energy needs.

Liquid fuels: In the mid-1800s, whale oil was an important liquid fuel source. Starting in the 1920s in the United States, and around 1950 in Europe, crude oil (petroleum) became the primary source of liquid fuels. Ethanol and biodiesel, produced from renewable agricultural resources, are becoming increasingly important alternative liquid fuels. For transportation, the lighter (more volatile) and cleaner liquids are used—gasoline, jet fuel, and diesel. In industrial or home-heating furnaces, less volatile (heavier) oils are burned. About 35 to 40% of U.S. energy needs are supplied by crude oil.

Gas fuels: Natural gas (which is mostly methane) is relatively inexpensive, widely available, and clean-burning. No wonder it is a popular fuel, both in homes and in industry. Until fairly recently, natural gas produced at crude oil wells was simply flared, because the necessary collection and distribution systems were not in place. Natural gas now supplies about 25% of the U.S. energy demand.

Current information about fuel costs and consumption is available from the U.S. Department of Energy. Fuel costs depend on the type of fuel as well as on the type of buyer; residential, commercial, and industrial consumers pay different prices. Here are some approximate fuel costs that a large industrial user might pay. Note that fuel is priced based on its energy content. Be aware that costs fluctuate significantly with time and geographic location, and are very sensitive to global political forces; these prices should be used only for very rough estimates.

Fuel type	$/million kJ
Natural gas	3.60
Light oil	5.50
Heavy oil	3.00
Coal	1.60
Wood/waste	1.50

The carbon:hydrogen ratio decreases as we move from solid fossil fuels (about 4:1 C:H or higher) to liquid fuels (about 1:1 to 1:2 C:H) to natural gas (about 1:4 C:H). This ratio is important, since a higher C:H ratio means greater CO_2 emissions per unit of energy generated. There is a great deal of current interest in moving toward a "hydrogen economy," where hydrogen (with zero CO_2 emissions) becomes a major energy source. Unfortunately, H_2 is not a readily available raw material. Currently, H_2 is manufactured commercially from fossil fuels, primarily methane, a process that generates as much CO_2 as direct combustion of methane. Therefore, although the use of hydrogen as a fuel reduces CO_2 emissions locally, it does not reduce CO_2 emissions globally. Ongoing research is directed towards developing processes for making H_2 from renewable energy resources.

The fuels compare as follows:

Fuel	Energy content: kJ produced per g fuel combusted	Emissions: g CO_2 generated per kJ produced	Energy density: kJ produced/cm³ fuel at 25°C and 1 atm
Coal (anthracite)	31.5	0.112	27.4
Gasoline (C_8H_{18})	48.0	0.064	33.6
Natural gas (methane)	55.6	0.0495	0.0364
Hydrogen	142.9	0.0	0.0117

Coal, which has a high energy per volume but the lowest energy per mass, is inexpensive, but is messy and difficult to handle. Furthermore, sulfur and nitrogen compounds in coal release acid gases upon combustion. Coal is used as a fuel source primarily at large stationary power plants where the cheaper price of the fuel justifies the high capital investment in solids handling and pollution control equipment. Gasoline has a high energy density and is cleaner and easier to use, making it a more valuable fuel especially for applications like transportation, where ease of distribution is important. It produces less CO_2 than coal. Per gram, natural gas is a great fuel, and CO_2 emissions are reduced by one-fourth compared to gasoline. But it has a very low energy density. The gas can be compressed, but the pressure must increase several hundred-fold to approach the energy density of liquid fuels like gasoline. Hydrogen has a very high energy content on a mass basis but is poor on a volume basis.

Sun, wind, flowing water, and nuclear reactions can all be harnessed for energy generation. Their contribution to the world energy picture is still small compared to fossil fuel combustion, but is growing. Nuclear, hydroelectric, and renewable (solar, wind, geothermal, and biomass) sources contributed about 8%, 3% and less than 1% of total U.S. energy needs, respectively.

6.1.2 Energy Distribution: Electricity, Heating Fluids, and Cooling Fluids

Electricity is not a harnessable energy source but a means of distributing energy. It is clean, generally reliable, and a convenient source of energy for doing work. (Certainly it is easier to plug a hair dryer into a wall outlet rather than use an open flame to dry your hair!) Most large municipal and commercial power plants produce electricity by burning fossil fuels; the energy of combustion produces steam, and the steam drives a turbine that drives a generator. Electricity is also generated with nuclear reactors, at dams, with windmill farms, and with solar panels. Small diesel-powered generators are less efficient but useful when a portable source of electricity is required and as backups when power lines are down.

Heating and cooling fluids are used throughout a process. When a process fluid must be heated, one could conceivably burn fuel right at the site. However, it is cheaper, more convenient, and safer to generate hot fluids at a central location, and then distribute that hot fluid to all process units requiring thermal energy. (This is no different than the system used in most modern homes, where a furnace generates hot air or steam by burning fuel, then the hot air or steam is distributed to all the rooms in the house. Compare this to an old home where there is a fireplace in every room as the sole source of heat.) Far and away, *steam* is the hot fluid of choice in large chemical processing facilities. Typical steam temperatures range from 250 to 500°F (120 to 260°C). *Hot water* is quite useful (and very cheap) as a heat source in the 30 to 90°C range. Commercial fluids such as *Dowtherm* are useful in the 200 to 400°C range.

There are times when process streams need to be cooled; for example, to crystallize a product from a liquid mixture, to cool an exothermic reactor, or to keep a computer chip from overheating. *Air* and *water* are commonly used as cold fluids for cooling purposes. Fans may blow cold air across hot surfaces. Cold water from nearby rivers, lakes, or oceans may serve as a heat sink. Large manufacturing facilities often generate their own *cooling water* in cooling towers. Cooling water is useful for cooling streams down to about 25 to 30°C. To cool process streams to temperatures below ambient, *refrigerants* are required. Refrigerant fluids include halocarbons (fluorine- and/or chlorine-containing organics), hydrocarbons (e.g., propane) and inorganics (e.g., ammonia). The choice of refrigerant fluid depends on a number of issues including: the operating temperature, toxicity, and flammability. Freon-12 (CCl_2F_2) is a widely used refrigeration fluid that is being phased out because of evidence that chlorinated hydrocarbons contribute to ozone depletion. These fluids are being replaced by fluorinated hydrocarbons. Ammonia is widely used as a refrigerant in food manufacture; it has excellent thermodynamic properties and is relatively benign environmentally, but because of toxicity concerns ammonia can be employed only where there are not many people.

Costs for electricity, heating fluids, and cooling fluids are based on the cost of the fuel used to generate them as well as costs of distribution. Prices fluctuate with fossil fuel costs, and vary widely depending on location, size and nature of facility, and other factors. Some rough guidelines for current prices are listed.

	$/million kJ	Heating fluids	$/million kJ	Cooling fluids	$/million kJ
Electricity	13–25	High pressure steam, 260°C	5–6	Cooling water, 25°C	0.20
		Medium pressure steam, 180°C	3–4	Chilled water, 5°C	20
		Low pressure steam, 120°C	1–3	Refrigerant, −20°C	32
				Refrigerant, −50°C	60

6.1.3 Energy Transfer Equipment

As we move from block flow diagrams to process flow diagrams, energy-related equipment must be included. Besides mixers, splitters, reactors and separators, we now need to consider the process units necessary to change pressure or temperature, to apply or extract work, to provide or remove heat. We'll call these units energy transfer equipment. Figure 6.2 shows equipment icons used in flow diagrams.

Equipment for Work Transfer and Pressure Changes   *Pumps* increase the pressure of liquids, generally with little to no temperature change. Pumps are relatively cheap and inexpensive to operate. Work input required to drive a pump is usually supplied by electric motors, although larger pumps may be driven by steam or gas turbines. *Compressors* increase the pressure of gases. The temperature of a gas increases as its pressure is increased; therefore, larger compressors are equipped with inter-stage coolers. They are relatively expensive to purchase, maintain, and operate. Compressors require a work input, usually supplied by electric motors, steam turbines, or gas turbines. Blowers or fans are used rather than compressors when only a small increase in pressure is needed.

   Turbines decrease the pressure of gases and produce rotational or reciprocating motion. If directly attached to a generator, a turbine will produce electricity.

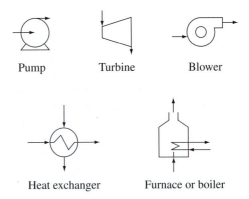

Figure 6.2 Representative icons for energy transfer equipment used in process flow diagrams.

Alternatively, turbines may be directly coupled to pumps or compressors, providing the work necessary to drive them. *Expansion valves* reduce the pressure of gases but don't get work out of the system. They are much cheaper than turbines. It's not always worth it to purchase and install a turbine, just to recover a small amount of energy from pressure reduction.

Equipment for Heat Transfer and Temperature Changes   There are two modes by which heat can be transferred from a heat source to a process material, or from a process material to a heat sink: direct and indirect.

In *direct injection,* hot (or cold) materials are mixed directly with the process material, to quickly increase (or decrease) the process temperature. Direct injection is advantageous in that the temperature change is rapid. However, there is a disadvantage—the heating (or cooling) material is mixed with the process material. Only materials that are compatible with the process stream can be used. Hot (or cold) solids can be used in a direct injection mode as well. Native Americans heated water in baskets, which could not have withstood direct fire, by first heating rocks in a fire, then throwing the rocks into the basket of water. Most of us have thrown a few ice cubes in a drink to quickly cool it down—another example of direct injection.

The indirect mode of heat transfer is the more common mode in the chemical process industry. In indirect heating, a hot fluid such as steam is indirectly contacted with a process fluid to be heated, by passing both fluids past a heat-conducting solid. The key advantage is that process streams are not mixed. The disadvantage is that the rate of heat transfer is not as fast as in direct injection. This heat contact is carried out in *heat exchangers.* Shell-and-tube heat exchangers are the workhorse of the chemical process industry. They consist of a bundle of tubes placed inside a large cylindrical shell, with one fluid pumped through the tubes and the other pumped through the shell. The tubes provide the heat contacting surface area. There are several other kinds of heat exchangers for specialized uses. Double-pipe exchangers are simply a pipe within a pipe—they are used for low flow rates and high temperature changes. Plate-type heat exchangers are useful when there are suspended solids or viscous fluids, since pressure drop is low. Scraped-surface heat exchangers have a rotating element with scraper blades, especially useful for heat transfer with crystallization, heat transfer with severe fouling of surfaces, or highly viscous fluids. (A home ice-cream maker is a scraped-surface heat exchanger.) In air-cooled heat exchangers, fans blow air across a bundle of finned tubes—the fins greatly increase the surface area available for heat transfer. Heating or cooling directly through the walls of the reactor or separator may be provided by fluids pumped through a thin jacket that surrounds the process vessel.

In *furnaces,* fuel and air are combined in the firebox. Solid fuels are shoveled into the bottom of furnaces, whereas fluid fuels are fed to burners (akin to burners on a gas stove) located at the bottom or sides of the firebox. Ignition of the fuel-air mixture initiates the combustion reaction. Hot combustion gases heat the process material. In other words, a chemical reaction occurs that converts the chemical energy stored in the fuel into thermal energy. Process materials can reach temperatures as high as 1650°F (900°C), or even higher. In direct-fired furnaces, the flames and combustion products are in direct contact with the material (limestone kiln, for example, or marshmallows in a campfire). With indirect-fired heaters, flames and combustion products are separated from the process material by metallic or refractory walls (process fluids flowing through furnace tubes, for example, or potions in a witch's cauldron). In a standard indirect-fired heater used in a large-scale chemical process, the process materials pass through tubes lining the walls of the furnace. The hottest temperatures are found near the lower levels of the furnace, close to the flames. As the process materials heat up, the temperature of the combustion gases decreases. The combustion gases rise up the furnace, giving off heat to the process materials, and then exit through the flue (the chimney stack). *Boilers* are simply furnaces with water in the tubes; heat from the combustion gases boils the water and generates steam.

Perry's Chemical Engineers' Handbook has informative sketches of energy transfer equipment as well as more detailed descriptions of their operation.

6.1.4 A Brief Review of Energy-Related Dimensions and Units

Before we delve into process energy calculations, let's review energy-related units. Back in Chap. 2, we stated that "There are only a few base dimensions; the ones we concern ourselves with in this book are: mass M, length L, time t, thermodynamic temperature T, and amount of substance N." Then we went on to describe a large number of different units, useful for measuring quantities such as temperature, volume, or mass flow rate. There are just a few more units needed so we can discuss force, energy and power.

Force has dimension of $[M{\cdot}L/t^2]$. Commonly used units for force include the newton (N) and the dyne. The English system uses pound-force (lb_f) or poundals as units of force. Note that lb_f $[M{\cdot}L/t^2]$ and lb $[M]$ have different dimension.

Use the conversion factors below to convert between different units of force.

$$1 \text{ N} = 1 \text{ kg m/s}^2 = 0.10197 \text{ kg}_f = 10^5 \text{ dynes} = 0.22481 \text{ lb}_f = 7.233 \text{ poundals}$$

$$1 \text{ lb}_f = 32.174 \text{ lb ft/s}^2 = 0.45359 \text{ kg}_f = 4.4482 \text{ N} = 4.482 \times 10^5 \text{ dynes}$$
$$= 32.174 \text{ poundals}$$

Illustration:

$$18.0 \text{ N} \left(\frac{0.22481 \text{ lb}_f}{\text{N}} \right) = 4.05 \text{ lb}_f$$

Weight is the force exerted on an object by gravitational attraction. Thus, a person's weight on Earth is different than his weight on Jupiter, but his mass is the same on either planet. Weight W and mass m are related by

$$W = mg$$

where g is the acceleration due to gravity. On Earth at sea level,

$$g = 9.8066 \, \frac{m}{s^2} = 980.66 \, \frac{cm}{s^2} = 32.174 \, \frac{ft}{s^2}$$

Illustration: An object has a mass of 1 kg. Its weight on Earth is

$$W = mg = 1 \text{ kg} \times 9.8066 \text{ m/s}^2 = \left(9.8066 \text{ kg m/s}^2\right) \times \frac{1 \text{ N}}{\text{kg m/s}^2}$$

$$= 9.8066 \text{ N} = 1 \text{ kg}_f$$

An object has a mass of 1 lb. Its weight on Earth is

$$W = mg = 1 \text{ lb} \times 32.174 \text{ ft/s}^2 = (32.174 \text{ lb ft/s}^2)$$

$$\times \frac{1 \text{ lb}_f}{32.174 \text{ lb ft/s}^2} = 1 \text{ lb}_f = 32.174 \text{ poundals}$$

Although lb_f and lb have different dimension, at sea level on earth they are numerically equivalent.

Energy has dimension of force $\times$ length, or equivalently, $[M{\cdot}L^2/t^2]$. The chemical industry employs a bewildering assortment of energy units—kWh (kilowatt-hour), Btu (British thermal unit), J (joule), kcal (kilocalorie, what is popularly called "a calorie" by dieters), EFOB (equivalent fuel oil barrel—the energy obtained from burning a standard 42-gal barrel of fuel oil). Use the conversion factors below to convert between different units of energy.

$$1 \text{ J} = 1 \text{ N m} = 1 \text{ kg m}^2/\text{s}^2 = 10^7 \text{ ergs} = 0.2389 \text{ cal} = 0.737562 \text{ ft lb}_r = 9.47817$$
$$\times 10^{-4} \text{ Btu} = 2.7778 \times 10^{-7} \text{ kWh} = 0.009869 \text{ L atm} = 0.0003485 \text{ ft}^3 \text{ atm}$$

$$1 \text{ kcal} = 1000 \text{ cal} = 4.184 \text{ kJ} = 3086 \text{ ft lb}_r = 3.966 \text{ Btu}$$

$$1 \text{ Btu} = 1.055056 \text{ kJ} = 1055.056 \text{ J} = 252 \text{ cal} = 778 \text{ ft lb}_f = 3.929 \times 10^{-4} \text{ hp h}$$

Illustration:

$$50 \text{ Btu} \left(\frac{778 \text{ ft lb}_f}{\text{Btu}} \right) = 38{,}900 \text{ ft lb}_f$$

Power has dimension of energy/time, or $[M{\cdot}L^2/t^3]$. Use the conversion factors below to convert between different units of power.

$$1 \text{ kW} = 1 \text{ kJ/s} = 860.4 \text{ kcal/h} = 3412 \text{ Btu/h} = 737.6 \text{ ft lb}_f/\text{s} = 1.3405 \text{ hp}$$

$$1 \text{ hp} = 745.7 \text{ J/s} = 641.88 \text{ kcal/h} = 2545 \text{ Btu/h} = 550 \text{ ft lb}_f/\text{s} = 0.7457 \text{ kW}$$

Illustration:

$$\left(\frac{42\ \text{Btu}}{\text{s}}\right)\left(\frac{3600\ \text{s}}{\text{h}}\right)\left(\frac{1\ \text{hp}}{2545\ \text{Btu/h}}\right) = 59\ \text{hp}$$

6.2   Process Energy Calculations: The Basics

To incorporate energy needs into the design of a chemical process, we need to complete process energy calculations, in much the same way as we have already learned to do process flow calculations. The remainder of this chapter is devoted to teaching you the strategies and skills needed for completing these calculations. In Sec. 6.2 we introduce the energy balance equation and discuss some fundamental definitions and properties of energy. In Secs. 6.3 and 6.4, we discuss the different kinds of energy that a system or a stream may possess, the ways in which energy can be transferred across system boundaries, and the methods used for quantifying energy. Then in Sec. 6.5, we return to the energy balance equation, now written in more useful forms. Finally, in Secs. 6.6 and 6.7, we develop strategies for using the energy balance equation and energy specifications in process energy calculations, and we show how process energy calculations contribute to the design of safe and efficient processes.

6.2.1   The Energy Balance Equation

You learned in Chaps. 2 and 3 that a system is a volume enclosed by a surface. At any time t the system possesses mass. Material can enter or leave the system by crossing the system boundaries. Mass cannot be generated or consumed (in the absence of a nuclear reaction), so the material balance equation for total mass, as you learned in Chap. 2, is

$$\text{Accumulation} = \text{In} - \text{Out}$$

In mathematical terms, the differential mass balance equation is (Chap. 3):

$$\frac{dm_{\text{sys}}}{dt} = \sum_{\text{all } j \text{ in}} \dot{m}_j - \sum_{\text{all } j \text{ out}} \dot{m}_j \tag{3.15a}$$

and the integral mass balance equation is

$$m_{\text{sys},f} - m_{\text{sys},0} = \sum_{\text{all } j \text{ in}} \left(\int_{t_0}^{t_f} \dot{m}_j\, dt \right) - \sum_{\text{all } j \text{ out}} \left(\int_{t_0}^{t_f} \dot{m}_j\, dt \right) \tag{3.19}$$

Just like mass, energy is conserved. (The law of conservation of energy is sometimes called the First Law of Thermodynamics, even though it was discovered *after* the Second Law of Thermodynamics.) Just like total mass, energy can't be generated or consumed (without a nuclear reaction). In words, the energy balance equation is very familiar:

$$\text{Accumulation} = \text{In} - \text{Out}$$

In mathematical terms, the energy balance equation is also very familiar; the differential form is

$$\frac{dE_{sys}}{dt} = \sum_{\text{all } j \text{ in}} \dot{E}_j - \sum_{\text{all } j \text{ out}} \dot{E}_j \tag{6.1}$$

and the integral form is

$$E_{sys,f} - E_{sys,0} = \sum_{\text{all } j \text{ in}} \left(\int_{t_0}^{t_f} \dot{E}_j \, dt \right) - \sum_{\text{all } j \text{ out}} \left(\int_{t_0}^{t_f} \dot{E}_j \, dt \right) \tag{6.2}$$

where E_{sys} is the total system energy, and $\dot{E}$ indicates an energy flow in or out of the system.

Quick Quiz 6.1

What is the dimension of each term in Eq. (6.1)? in Eq. (6.2)?

In many ways the energy balance equation is simpler than the material balance equation: there are no generation or consumption terms; there is only a total energy balance equation, not a component energy balance equation; we don't write different equations for mass vs. mole units. However, in other ways the energy balance equation can be harder to understand. Quantification of the mass of a system, or mass flows in and out of a system, as required for the material balance equation, is fairly intuitive. But, what do we mean by energy and energy flows? How do we quantify system energy and energy flows? Before we delve further into the use of the energy balance equation to solve process problems, we need to take a long digression to answer these questions.

6.2.2 System Energy, Energy Flows, Specific Energy

Imagine a pot of room-temperature water placed over a fire, and a cup of boiling hot milk poured into the pot (Fig. 6.3). If the pot and its contents are a system, then clearly the system possesses energy, and the system energy changes with time as a result of two different kinds of energy flows: heat from the fire

Figure 6.3 The pot and its contents are the system. Energy is transferred into the system in two ways: (1) hot liquid poured into the pot carries energy into the system through material transfer; (2) heat from the fire transfers energy to the pot without material transfer.

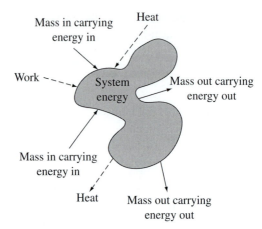

Figure 6.4 A system with multiple energy inputs and outputs. The system is a three-dimensional volume enclosed by system boundaries. Energy can enter or leave the system through material flows in and out of the system (denoted as solid lines), or as energy flows in the absence of material flow, such as heat or work (denoted as dashed lines). The system as shown is an open system.

and hot milk. *These two kinds of energy flows are fundamentally different:* heat from the fire is energy flow without material flow, while the hot milk is energy flow because of material flow.

In general, energy flows via two mechanisms across system boundaries: (1) Energy can be carried in or out of the system by material moving in or out, because the material possesses energy. (2) Energy can move in or out in the absence of material transfer, through **heat** or **work.** An **open system** is one in which material streams cross system boundaries. In a **closed system**, no material crosses system boundaries, but energy may still cross boundaries in the form of heat and/or work. (See Fig. 6.4.)

E_{sys} (system energy) has dimension of energy $[M \cdot L^2/t^2]$, and $\dot{E}_j$ (energy flow) has dimension of energy per time, or power $[M \cdot L^2/t^3]$. Sometimes it is more convenient to speak of the system energy in terms of its energy per unit mass (or per unit mole). Energy per unit mass is called *specific energy* and has dimension of $[L^2/t^2]$, while energy per unit mole is called *molar energy* and has dimension of $[M \cdot L^2/N \cdot t^2]$. (In practice, people do not always make this rigorous distinction between specific and molar energies, and will call something specific energy when they mean molar energy. We will sometimes be similarly sloppy, just to avoid the clumsiness of having to say "specific or molar" energy when we are describing something that is true for both specific and for molar energies.) We will denote either specific or molar energy with a "hat", as $\hat{E}$; the meaning should be apparent from a test of dimensional consistency. By definition,

$$\hat{E}_{sys} = \frac{E_{sys}}{m_{sys}} \qquad \text{or} \qquad \hat{E}_{sys} = \frac{E_{sys}}{n_{sys}} \tag{6.3}$$

Similarly, the specific or molar energy of a stream flowing in or out of the system is defined as

$$\hat{E}_j = \frac{\dot{E}_j}{\dot{m}_j} \quad \text{or} \quad \hat{E}_j = \frac{\dot{E}_j}{\dot{n}_j} \tag{6.4}$$

where $\dot{m}_j$ and $\dot{n}_j$ are the total mass or molar flow rates, respectively, of stream j.

There are a couple of important characteristics of $\hat{E}$.

1. *$\hat{E}$ is a function of the state of the system or stream*, and not the path by which the system or stream got to that state. Hence, $\hat{E}$ is called a **state function.** The state of a system or stream is defined by knowing the following properties:
   - Velocity
   - Position
   - Pressure
   - Temperature
   - Phase
   - Composition
2. *$\hat{E}$ is not an absolute quantity.* Its numerical value depends on the chosen **reference state** of a system. By definition, $\hat{E} = 0$ at the reference state. The reference state is identified by its:
   - Velocity
   - Position
   - Pressure
   - Temperature
   - Phase
   - Composition

Helpful Hint
The energy of the system is always relative to the reference state of the system.

6.3 Putting a Number on Energy: Energy Data and Model Equations

If energy is not an absolute quantity, how can we quantify it? There are two ways:

Method 1: Find data on $\hat{E}$ of the material of interest at the state of interest. The numerical value of $\hat{E}$ depends on the reference state for that data.

Method 2: Define a reference state and set $\hat{E} = 0$ at this state. Find a model equation that describes how $\hat{E}$ depends on the state (velocity, position, pressure, temperature, phase and composition) relative to the reference state. Use the model equation to calculate the difference in energy from the reference state to the actual state.

Sometimes we are interested in the *difference* in energy of two systems or streams that are in different states. Or, we want to calculate the *change* in energy of a system as the system changes from one state to another state. Again, we use one of these two methods:

Method 1: Find data on $\hat{E}$ of the material of interest at state 1 and at state 2. *Be sure that the reference state is the same for both data points.* Subtract $\hat{E}$ of state 1 from $\hat{E}$ of state 2.

Method 2: Find model equations that describe how $\hat{E}$ depends on the property that is changing (velocity, position, pressure, temperature, phase and/or composition). Use the model equations to calculate the change in $\hat{E}$ as the system moves from state 1 to state 2.

In Sec. 6.3, we describe the forms of energy that are important in chemical processes and demonstrate how to quantify each form of energy, using data and/or model equations as appropriate.

6.3.1   Two Forms of Energy: Kinetic and Potential

Two forms of energy that you are probably already familiar with are kinetic and potential energy. Kinetic and potential energy are functions of velocity and position, respectively: the first two items on our list of state properties.

A system or stream possesses **kinetic energy** E_k as a result of its mass m and its *velocity v:*

$$E_k = \frac{mv^2}{2} \tag{6.5}$$

A system or stream possesses **potential energy** E_p as a result of its *position in a force field*. A gravitational force field is the only one that will concern us in this text. E_p is a function of the height h of the system above a reference surface:

$$E_p = mgh \tag{6.6}$$

where g is the acceleration due to gravity.

The value of E_k and E_p depends on the reference state of a system. E_k is known relative to the velocity of the reference state, and E_p is known relative to the position of the reference state in the force field. For example, suppose an airplane is the system. A pilot announces the velocity and height of an airplane *relative* to the earth. The reference state is the earth's surface. The airplane has both kinetic and potential energy if the earth's surface is the reference state. Does the pilot have kinetic and potential energy? It depends on the choice of reference state. If the earth's surface is the reference state, the answer is yes. If the pilot's seat is the reference state, the answer is no.

Similarly, the flow of kinetic energy due to a stream j of mass flow $\dot{m}_j$ and velocity v_j, or the flow of potential energy due to a stream j at height h_j, is

$$\dot{E}_{k,j} = \frac{\dot{m}_j v_j^{\,2}}{2}$$

$$\dot{E}_{p,j} = \dot{m}_j g h_j$$

Example 6.1   ### Kinetic and Potential Energy: Toddler Troubles

Two 33-pound toddlers are playing in their backyard. One toddler is running full-speed at 2 miles per hour (2.9 ft/s) into a brick garden wall. Her twin sister is perched, rather wobbily, on top of a 10-foot-high ladder. Which toddler should Mom rescue first?

Solution

Quick Quiz 6.2

Suppose our toddler on the ladder falls to the ground. What is her change in potential energy? Now suppose she first jumps high into the air, then falls to the ground. Is her change in potential energy the same or different than if she doesn't jump first?

The toddler is the system, and the earth's surface is the reference state. The running toddler possesses kinetic energy (relative to the earth's surface) of

$$E_k = \frac{mv^2}{2} = \frac{33 \text{ lb} \times (2.9 \text{ ft/s})^2}{2} \times \frac{1 \text{ lb}_f}{32.174 \text{ lb ft/s}^2} = 4.3 \text{ ft lb}_f$$

The perching toddler possesses potential energy (relative to the ground) of

$$E_p = mgh = 33 \text{ lb} \times 32.174 \text{ ft/s}^2 \times 10 \text{ ft} \times \frac{1 \text{ lb}_f}{32.174 \text{ lb ft/s}^2} = 330 \text{ ft lb}_f$$

So the perching toddler could do herself more damage than the running one.

Calculating the change in kinetic energy of a system as it undergoes a change in velocity from v_1 to v_2 is straightforward (assuming no change in system mass):

$$E_{k2} - E_{k1} = \frac{m}{2}\left(v_2^2 - v_1^2\right)$$

Similarly, calculation of the change in potential energy as the system height changes from h_1 to h_2 (at constant mass) is

$$E_{p2} - E_{p1} = mg(h_2 - h_1)$$

It is important to note that (1) you must use the same reference state for both state 1 and state 2 and (2) the change in energy is independent of the reference state chosen!

Example 6.2   **Change in Potential Energy: Snow Melt**

Melting snow from the top of a 3100-m-high mountain runs down the mountainside to a lake at 1200 m elevation. What is the change in potential energy (kJ) of a ton of snow melting?

Solution

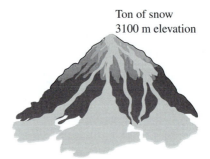

Ton of snow
3100 m elevation

Lake, 1200 m elevation

The system is 1 ton (909 kg) of snow. Suppose we choose sea level as the reference state. Initially, the snow is at $h_1 = 3100$ m. At the end, the snow is at $h_2 = 1200$ m. Therefore, the change in potential energy is

$$E_{p2} - E_{p1} = mg(h_2 - h_1) = 909 \text{ kg} \times 9.8066 \text{ m/s}^2$$

$$\times (1200 - 3100 \text{ m})\left(\frac{1 \text{ kJ}}{1000 \text{ kg m}^2/\text{s}^2}\right) = -16{,}900 \text{ kJ}$$

Example 6.3   **Change in Kinetic Energy of a Stream: Thomas Edison or Rube Goldberg?**

An inventor reports a remarkable new Energy Booster device. 120 cm³/s water flows into the device through a 4.0-cm-diameter pipe. Water at 120 cm³/s flows out of the device through a 2.5-cm-diameter pipe. The inventor claims that the device increases the kinetic energy of the water. Does it? By how much? Assume the density of water is 1.0 g/cm³.

Solution

4 cm

Energy Booster
Patent pending

2.5 cm

To calculate the specific kinetic energy of the water, we need the velocity. At the inlet:

$$v_{in} = \frac{\text{volumetric flow rate in}}{\text{cross-sectional area of pipe}} = \frac{120 \text{ cm}^3/\text{s}}{(\pi/4)(4.0\text{cm})^2} = 9.6 \text{ cm/s}$$

A similar calculation gives us $v_{out} = 24.4$ cm/s.

The flow rates in and out are the same. Volumetric flows are converted to mass flows by multiplying by the density:

$$\dot{m} = 120 \frac{\text{cm}^3}{\text{s}} \times 1.0 \frac{\text{g}}{\text{cm}^3} = 120 \frac{\text{g}}{\text{s}}$$

Therefore

$$\dot{E}_{k,out} - \dot{E}_{k,in} = \frac{\dot{m}}{2}\left(v_{out}^2 - v_{in}^2\right)$$

$$= \frac{120 \text{ g/s}}{2}\left[\left(24.4 \frac{\text{cm}}{\text{s}}\right)^2 - \left(9.6 \frac{\text{cm}}{\text{s}}\right)^2\right]\left(\frac{1 \text{ J}}{10^7 \text{g cm}^2/\text{s}^2}\right) = 0.0030 \text{ J/s}$$

Since the difference between the outlet and inlet kinetic energy is positive, the device does increase the kinetic energy flow. (But how? Hmmm . . .)

6.3.2 A Third Kind of Energy and a Convenience Function: Internal Energy and Enthalpy

A third kind of energy that is important in chemical processes is called **internal energy.** A system possesses internal energy because of its *molecular activity.* Internal energy includes energy stored in molecules in the form of covalent chemical bonds, noncovalent intermolecular forces (such as the forces that hold solids or liquids together), and thermal motion. Internal energy excludes any system energy due to external forces; that is, it excludes energy attributed to the system's velocity or its position in a force field. The specific internal energy $\hat{U}$ is a function of the last four state properties on our list:

- Pressure P
- Temperature T
- Phase ϕ
- Compositions x_i

(*Side note:* Internal energy is often expressed as a function of volume rather than pressure, which turns out to be more convenient in analyzing systems where the volume is constant and the pressure is not.)

A property related to internal energy is called **enthalpy H.** Specific enthalpy $\hat{H}$ is related to internal energy:

$$\hat{H} = \hat{U} + P\hat{V} \tag{6.7}$$

Enthalpy is sometimes called a convenience function, because it is very convenient for many energy calculations, as you will see. You will learn a great deal more about internal energy and enthalpy when you study thermodynamics. In this book we simply demonstrate how we use $\hat{U}$ and $\hat{H}$ in process energy calculations.

Just like kinetic and potential energy, $\hat{U}$ and $\hat{H}$ are defined *relative to some reference state*. With potential energy, the reference state is a position in a gravitational field (e.g., the earth's surface). With kinetic energy, the reference state is a velocity. With $\hat{U}$ and $\hat{H}$, the reference state is a *state of matter*. Specifying the reference state for internal energy and enthalpy requires that *pressure* (or specific volume), *temperature, phase,* and *composition* of the reference state are specified.

In the next sections we discuss in some depth how we find numerical values for $\hat{U}$ and $\hat{H}$ given P, T, ϕ, and x_i. The best way is to use actual data. Alternatively, we turn to model equations, but these are not quite as simple for $\hat{U}$ and $\hat{H}$ as they were for E_k or E_p.

Quick Quiz 6.3

Show that pressure $\times$ volume (PV) has the same dimension as energy.

6.3.3 Using Tables and Graphs to Find $\hat{U}$ and $\hat{H}$

There are a few pure compounds, like H_2O, that are so important that extensive tables and graphs of $\hat{U}$ and $\hat{H}$ as a function of pressure, temperature, and phase are available in resources such as *Perry's Chemical Engineering Handbook*. Table 6.1 is a truncated example of such a table, called a "steam table." A more complete version appears in App. B. For Table 6.1, the reference state is $P =$ 0.006116 bar, $T = 0.01°C$, $\phi =$ liquid, $x_w = 1$. At these conditions (the triple point of water), $\hat{U} = 0$ by definition.

Table 6.1 is simple to use. Pressure P is listed in the first column, with the saturation temperature (temperature at which water can coexist as liquid and vapor) in parentheses. Specific enthalpy $\hat{H}$ (kJ/kg), specific internal energy $\hat{U}$ (kJ/kg), and specific volume $\hat{V}$ (m³/kg) are listed in the next columns for saturated liquid water and saturated steam. The remaining columns contain data for subcooled liquid water and superheated steam at given values of T. (Refer back to Chap. 5 if you've forgotten the meaning of saturated, subcooled, and superheated.) At values of T and/or P that are not shown on the table, $\hat{H}$, $\hat{U}$, or $\hat{V}$ are found by linear interpolation. (Refer to App. A.3 for more on this technique.)

Table 6.1 Specific Enthalpy $\hat{H}$ (kJ/kg), Specific Energy $\hat{U}$ (kJ/kg), and Specific Volume $\hat{V}$ (m³/kg) of water and steam. Reference state is liquid water at its triple point, $T = 0.01°C$, $P = 0.006116$ bar

P, bar (T^{sat}, °C)		Sat'd liquid	Sat'd vapor	Temperature (°C)						
				50	100	150	200	250	300	350
0.006116 (0.01)	$\hat{H}$	0.00	2500.9	2594.5	2688.6	2783.7	2880.0	2977.8	3077.0	3177.7
	$\hat{U}$	0.00	2374.9	2445.4	2516.4	2588.4	2661.7	2736.3	2812.5	2890.1
	$\hat{V}$	0.00100	206.55	244.45	282.30	320.14	357.98	395.81	433.64	470.69
0.1 (45.806)	$\hat{H}$	191.81	2583.9	2592.0	2687.5	2783.1	2879.6	2977.5	3076.8	3177.6
	$\hat{U}$	191.80	2437.2	2443.3	2515.5	2587.9	2661.4	2736.1	2812.3	2890.0
	$\hat{V}$	0.00101	14.670	14.867	17.197	19.514	21.826	24.137	26.446	28.755
1.0 (99.606)	$\hat{H}$	417.50	2674.9	209.46	2675.8	2776.6	2875.5	2974.5	3074.6	3175.8
	$\hat{U}$	417.40	2505.6	209.36	2506.2	2583.0	2658.2	2733.9	2810.7	2888.7
	$\hat{V}$	0.00104	1.6939	0.00101	1.6959	1.9367	2.1725	2.4062	2.6389	2.8710
5.0 (151.83)	$\hat{H}$	640.09	2748.1	209.80	419.51	632.24	2855.9	2961.1	3064.6	3168.1
	$\hat{U}$	639.54	2560.7	209.30	418.99	631.69	2643.3	2723.8	2803.3	2883.0
	$\hat{V}$	0.00109	0.37481	0.00101	0.00104	0.00109	0.4250	0.4744	0.5226	0.57016
10.0 (179.88)	$\hat{H}$	762.52	2777.1	210.19	419.84	632.5	2828.3	2943.1	3051.6	3158.2
	$\hat{U}$	761.39	2582.7	209.18	418.80	631.41	2622.2	2710.4	2793.6	2875.7
	$\hat{V}$	0.00113	0.1944	0.00101	0.00104	0.00109	0.2060	0.2328	0.2580	0.2825
20.0 (212.38)	$\hat{H}$	908.5	2798.3	211.06	420.59	633.12	852.45	2903.2	3024.2	3137.7
	$\hat{U}$	906.14	2599.1	209.03	418.51	630.94	850.14	2680.2	2773.2	2860.5
	$\hat{V}$	0.00118	0.0996	0.00101	0.00104	0.00109	0.00116	0.1115	0.1255	0.1386

(Continued)

Table 6.1 (Continued)

40.0 (250.35)	$\hat{H}$	1087.5	2800.8	212.78	422.10	634.36	853.27	1085.8	2961.7	3093.3
	$\hat{U}$	1082.5	2601.7	208.74	417.93	630.01	848.65	1080.8	2726.2	2827.4
	$\hat{V}$	0.00125	0.04978	0.00101	0.00104	0.00109	0.00115	0.00125	0.0589	0.0665
60.0 (275.58)	$\hat{H}$	1213.9	2784.6	214.50	423.60	635.61	854.09	1085.7	2885.5	3043.9
	$\hat{U}$	1206.0	2589.9	208.44	417.36	629.08	847.18	1078.2	2668.4	2790.4
	$\hat{V}$	0.00132	0.03245	0.00101	0.00104	0.00109	0.00115	0.00125	0.0362	0.0423
100.0 (311.00)	$\hat{H}$	1408.1	2725.5	217.94	426.62	638.11	855.8	1085.8	1343.3	2924.0
	$\hat{U}$	1393.5	2545.2	207.86	416.23	627.27	844.31	1073.4	1329.4	2699.6
	$\hat{V}$	0.00145	0.0180	0.00101	0.00104	0.00108	0.00115	0.00124	0.00140	0.0224
150.0 (342.16)	$\hat{H}$	1610.2	2610.7	222.23	430.39	641.27	857.99	1086.1	1338.3	2693.1
	$\hat{U}$	1585.3	2455.6	207.15	414.85	625.05	840.84	1067.6	1317.6	2520.9
	$\hat{V}$	0.00166	0.01034	0.00101	0.00104	0.00108	0.00114	0.00123	0.00138	0.0115
200.0 (365.75)	$\hat{H}$	1827.2	2412.3	226.51	434.17	644.45	860.27	1086.7	1334.4	1646.0
	$\hat{U}$	1786.4	2295.0	206.44	413.50	622.89	837.49	1062.2	1307.1	1612.7
	$\hat{V}$	0.00204	0.00586	0.00100	0.00103	0.00108	0.00114	0.00123	0.00136	0.00166
220.64 (373.95)	$\hat{H}$	2084.3	2084.3	228.28	435.73	645.77	861.23	1087.0	1333.0	1635.6
	$\hat{U}$	2015.7	2015.7	206.16	412.95	622.01	836.14	1060.0	1303.1	1599.6
	$\hat{V}$	0.00311	0.00311	0.00100	0.00103	0.00108	0.00114	0.00122	0.00135	0.00163

Source: E.W. Lemmon, M.O. McLinden, and D.G. Friend, "Thermophysical Properties of Fluid Systems" in NIST Chemistry WebBook, NIST Standard Reference Database Number 69, Eds. P.J. Linstorm and W.G. Mallard, June 2005, National Institute of Standards and Technology, Gaithersburg MD, 20899 (http://webbook.nist.gov).

Quick Quiz 6.4

What is the temperature and specific enthalpy for saturated steam at 5 bar?

Is H_2O at 300°C and 100 bar subcooled liquid, saturated liquid, saturated steam, or superheated steam?

Illustration: Liquid water at 1 bar and 99.6°C is saturated, with $\hat{H} = 417.5$ kJ/kg and $\hat{U} = 417.4$ kJ/kg.

Water at 200°C and 10 bar is superheated vapor, with $\hat{H} = 2828.3$ kJ/kg and $\hat{U} = 2622.2$ kJ/kg.

$\hat{H}$ for steam at 1 bar and 127°C is not listed on the table. At 1 bar and 100°C, $\hat{H} = 2675.8$ kJ/kg while at 1 bar and 150°C, $\hat{H} = 2776.6$ kJ/kg. By linear interpolation we find that

$$\hat{H} = \left(\frac{127 - 100}{150 - 100}\right)\left(2776.6 - 2675.8\right) + 2675.8 = 2730.2 \; \frac{\text{kJ}}{\text{kg}}.$$

Plotting the data in Table 6.1 is a useful exercise, as it illustrates how enthalpy and internal energy change with pressure, temperature and phase (Fig. 6.5.) There are a few trends from Fig. 6.5 worth particular notice. First, there is a large increase in $\hat{H}$ and $\hat{U}$ with a change in phase from liquid to vapor. (The temperature at which the phase change occurs is that at which $P^{\text{sat}} = P$.) Second, $\hat{H}$ and $\hat{U}$ increase linearly with an increase in temperature at constant phase. Third, $\hat{H}$ and $\hat{U}$ are nearly identical in the liquid phase at identical temperature and pressure, but $\hat{H}$ is slightly greater than $\hat{U}$ in the vapor phase. Fourth, $\hat{H}$ and $\hat{U}$ are not strong functions of pressure, especially at low pressure and in the liquid phase. These trends prove enlightening when we develop model equations for $\hat{H}$ and $\hat{U}$, in Sec. 6.3.4.

Use of steam tables is illustrated in the following examples.

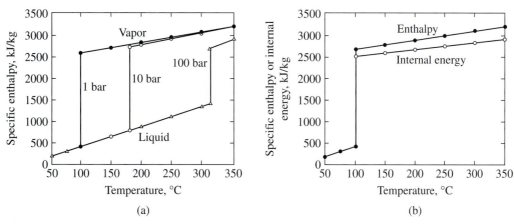

Figure 6.5 $\hat{H}$ and $\hat{U}$ of pure water. (a) Specific enthalpy $\hat{H}$ of H_2O as a function of P, T, and ϕ. (b) Specific enthalpy $\hat{H}$ and internal energy $\hat{U}$ of H_2O at 1 bar.

Example 6.4   **Using Steam Tables to Find $\hat{H}$: Several Cases**

1. What is the specific enthalpy of
   (a) H_2O at 150 bar and 50°C?
   (b) H_2O at 150 bar and 300°C?
   (c) H_2O at 0.1 bar and 50°C?
   (d) H_2O at 0.1 bar and 300°C?
   (e) Saturated liquid H_2O at 1 bar?
   (f) Saturated vapor H_2O at 1 bar?
2. What is the change in specific enthalpy associated with
   (a) Heating H_2O at 150 bar from 50°C to 300°C,
   (b) Heating H_2O at 0.1 bar from 50°C to 300°C,
   (c) Heating saturated liquid H_2O at 1 bar to saturated vapor?

Solution

1. These questions can be answered by consulting Table 6.1 and/or App. B. We'll summarize:

Conditions	P, bar	T, °C	ϕ	$\hat{H}$, kJ/kg
a	150	50	Liquid	222.23
b	150	300	Liquid	1338.3
c	0.1	50	Vapor	2592.0
d	0.1	300	Vapor	3076.8
e	1	99.6	Saturated liquid	417.5
f	1	99.6	Saturated vapor	2674.9

All of these specific enthalpies are based on the reference state of $\hat{H} = 0$ for liquid water at the triple point.

2. We'll indicate the change in enthalpy in going from state 1 to state 2 as $\hat{H}_2 - \hat{H}_1$.

   (a) $\hat{H}_2 - \hat{H}_1 = 1338.3 - 222.23 = 1116.1$ kJ/kg
   (b) $\hat{H}_2 - \hat{H}_1 = 3076.8 - 2592.0 = 484.8$ kJ/kg
   (b) $\hat{H}_2 - \hat{H}_1 = 2674.9 - 417.5 = 2257.4$ kJ/kg

   The largest increase in enthalpy occurs with the change in phase, at constant temperature and pressure [case (c)]. Increasing the temperature of liquid water by 150°C [case (a)] produces more than twice the enthalpy change of increasing the temperature of water vapor by the same amount [case (b)].

Example 6.5   **Using Steam Tables: Pumping Water, Compressing Steam**

Calculate the change in enthalpy flow associated with:

1. Pumping 1.5 kg/s water at 60°C from 5 bar to 100 bar, and
2. Compressing 1.5 kg/s steam from 5 bar and 200°C to 100 bar and 350°C

Use Table 6.1 and App. B for enthalpy data.

Solution

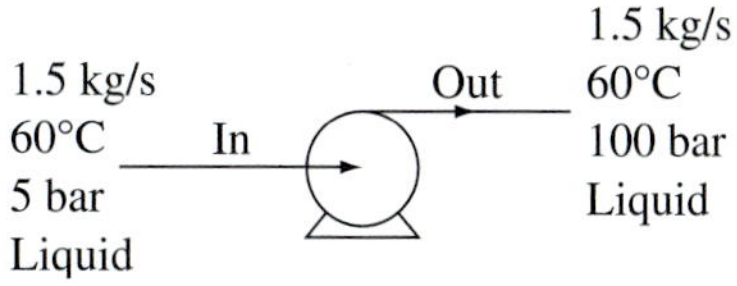

Case 1.   First we find the specific enthalpy at the two conditions. The table gives no data at 60°C, so we must linearly interpolate, using the data at 50°C and 100°C.

At $P = 5$ bar, $T = 50°C$, $\phi = $ liquid:

$$\hat{H} = 209.80 \text{ kJ/kg}$$

At $P = 5$ bar, $T = 100°C$, $\phi = $ liquid:

$$\hat{H} = 419.51 \text{ kJ/kg}$$

By linear interpolation,

At $P = 5$ bar, $T = 60°C$, $\phi = $ liquid,

$$\hat{H} = \frac{(60 - 50)}{(100 - 50)}(419.51 - 209.80) + 209.80 = 251.74 \text{ kJ/kg}$$

Using a similar method, we find that $\hat{H} = 259.67$ kJ/kg at $P = 100$ bar, $T = 60°C$, $\phi = $ liquid.

$$\hat{H}_{out} - \hat{H}_{in} = 259.67 - 251.74 = 7.93 \text{ kJ/kg}$$

To get the change in enthalpy per unit time, we multiply by the mass flow rate:

$$\dot{H}_{out} - \dot{H}_{in} = \dot{m}(\hat{H}_{out} - \hat{H}_{in}) = 1.5 \text{ kg/s} (7.93 \text{ kJ/kg}) = 11.9 \text{ kJ/s}$$

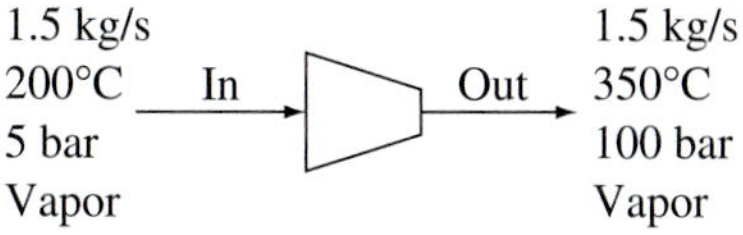

Case 2.   From the steam tables:

At $P = 5$ bar, $T = 200°C$, $\phi = $ vapor:

$$\hat{H} = 2855.9 \text{ kJ/kg}$$

At P = 100 bar, T = 350°C, ϕ = vapor:

$$\hat{H} = 2924.0 \text{ kJ/kg}$$

Therefore

$$\hat{H}_{out} - \hat{H}_{in} = 2924.0 - 2855.9 = 68.1 \text{ kJ/kg}$$

$$\dot{H}_{out} - \dot{H}_{in} = \dot{m}(\hat{H}_{out} - \hat{H}_{in}) = 1.5 \text{ kg/s}\,(68.1 \text{ kJ/kg}) = 102.2 \text{ kJ/s}$$

What if a system possesses kinetic, potential *and* internal energy? The total energy of the system is simply the sum of these forms of energy:

$$E = E_k + E_p + U \tag{6.8a}$$

and any change in system energy as it moves from one state (state 1) to another (state 2) is simply the sum of changes in kinetic, potential and internal energy:

$$E_2 - E_1 = (E_{k,2} - E_{k,1}) + (E_{p,2} - E_{p,1}) + (U_2 - U_1) \tag{6.8b}$$

Changes in internal energy tend to be very large compared to changes in kinetic and potential energy. This point is illustrated in the next example.

Example 6.6   **Comparing Kinetic, Potential, and Internal Energy: Frequent Flyer**

A jug containing 10 kg of water, at 1 bar pressure and 75°C, is placed in an airplane while the plane is on the ground. The airplane takes off, and an hour later the airplane is cruising at 550 mph at an altitude of 30,000 ft. Meanwhile, the water in the jug cools off to 50°C. What is the change in energy of the water?

Solution

The system is the water in the jug. We'll use "1" to indicate the initial state, when the plane is on the ground, and "2" to indicate the final state, when the plane has reached its cruising altitude. We'll choose the earth surface as the reference state for E_k and E_p, and the triple point of water as the reference for U.

State "1"   State "2"

In state 1, since the jug's velocity and position is the same as that of the reference state:

$$E_{k,1} = E_{p,1} = 0,$$

From linear interpolation of the data in Table 6.1, we find $\hat{U} = 313.4$ kJ/kg at 1 bar and 75°C, and therefore

$$U_1 = m\hat{U}_1 = (10 \text{ kg})(313.4 \text{ kJ/kg}) = 3134 \text{ kJ}$$

When the jug is in flight (state 2),

$$E_{k,2} = \frac{1}{2}mv^2 = \frac{1}{2}(10 \text{ kg})\left[\left(550 \frac{\text{miles}}{\text{h}}\right)\left(1609 \frac{\text{m}}{\text{mile}}\right)\left(\frac{\text{h}}{3600 \text{ s}}\right)\right]^2$$

$$\times \left(\frac{1 \text{ kJ}}{1000 \text{ kg m}^2/\text{s}^2}\right) = 302 \text{ kJ}$$

$$E_{p,2} = mgh = (10 \text{ kg})(9.8066 \text{ m/s}^2)(30{,}000 \text{ ft})(0.305 \text{ m/ft})$$

$$\times \left(\frac{1 \text{ kJ}}{1000 \text{ kg m}^2/\text{s}^2}\right) = 897 \text{ kJ}$$

and, from Table 6.1, $\hat{U} = 209.36$ kJ/kg at 1 bar and 50°C, and

$$U_2 = m\hat{U}_2 = (10 \text{ kg})(209.36 \text{ kJ/kg}) = 2094 \text{ kJ}$$

The change in kinetic energy is

$$E_{k,2} - E_{k,1} = 302 \text{ kJ} - 0 \text{ kJ} = +302 \text{ kJ}$$

The change in potential energy is:

$$E_{p,2} - E_{p,1} = 897 \text{ kJ} - 0 \text{ kJ} = +897 \text{ kJ}$$

The change in internal energy is

$$U_2 - U_1 = 2094 \text{ kJ} - 3134 \text{ kJ} = -1040 \text{ kJ}$$

The change in total system energy is

$$E_2 - E_1 = (E_{k,2} - E_{k,1}) + (E_{p,2} - E_{p,1}) + (U_2 - U_1)$$
$$= 302 + 897 + (-1040) = 159 \text{ kJ}$$

A mere 25°C drop in temperature decreases the internal energy of the system almost enough to compensate for the increase in kinetic and potential energy due to the 30,000-ft climb and increase in velocity of 550 mph.

Besides pressure, temperature, and phase, $\hat{U}$ and $\hat{H}$ depend on composition. This means, of course, that $\hat{U}$ and $\hat{H}$ of water are different from $\hat{U}$ and $\hat{H}$ of ammonia, but it also means that $\hat{U}$ and $\hat{H}$ of an ammonia-water mixture are different from that of either pure component, and depend on the concentration of ammonia in the solution. One example of this is shown in Fig. 6.6, where $\hat{H}$ of

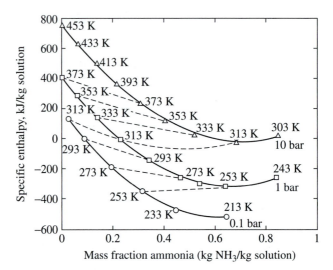

Figure 6.6 Enthalpy-composition diagram for saturated liquid ammonia-water solutions. Reference states are pure liquid water at 273 K and pure liquid ammonia at 196 K at 1 bar. The solid lines are isobars—lines of constant pressure. The dotted lines are isotherms—lines of constant temperature.
Source: Data extracted from a figure in Perry's Chemical Engineers' Handbook (McGraw-Hill).

an ammonia-water mixture is plotted as a function of pressure, temperature, phase, and composition. Use of this *enthalpy-composition* graph is illustrated in the next example.

Example 6.7	**Using Enthalpy-Composition Graphs: Ammonia-Water Mixtures**

Use Fig. 6.6 to determine

1. $\hat{H}$ of liquid water at its melting point
2. $\hat{H}$ of a saturated ammonia solution at 1 bar and 293 K
3. $\hat{H}$ of a saturated solution of ammonia and water at 34 wt% ammonia and 10 bar
4. The change in $\hat{H}$ associated with mixing 1 kg liquid water at 1 bar and 373 K with 1 kg of an 80 wt% saturated ammonia solution at 1 bar, and bringing the liquid mixture to its saturation temperature at 1 bar.

Solution

1. Pure water melts at 0°C (273 K). Since this is the reference state for Fig. 6.6, $\hat{H} = 0$. ($\hat{H} = 0$ also for liquid ammonia at its melting point of 196 K; these are the reference states *for this graph*).

2. Reading directly from Fig. 6.6, saturated ammonia solution at 1 bar and 293 K contains about 34 wt% ammonia, with $\hat{H} = -140$ kJ/kg solution.
3. We start at 0.34 wt fraction ammonia and draw a line straight up until we intersect the curve at 10 bar. Reading across to the x axis, we see that for this solution $\hat{H} = +200$ kJ/kg. The temperature of the saturated liquid solution is about 370 K.
4. We must use Fig. 6.6 for not only the mixtures but also the pure compound.
   For water at $P = 1$ bar, $T = 373$ K, ϕ = liquid, $x_w = 1$, $x_a = 0.0$ (0.0 mass fraction ammonia): $\hat{H}_w = 405$ kJ/kg.
   For saturated liquid ammonia solution at $P = 1$ bar, ϕ = saturated liquid, $x_a = 0.8$ (and T about 245 K): $\hat{H}_s = -260$ kJ/kg.
   Therefore,

$$H_1 = m_w\hat{H}_w + m_s\hat{H}_s = \left(1 \text{ kg} \times 405 \frac{\text{kJ}}{\text{kg}}\right) + \left[1 \text{ kg} \times \left(-260 \frac{\text{kJ}}{\text{kg}}\right)\right] = 145 \text{ kJ}$$

By material balance, the mixture is 40 wt% ammonia ($x_a = 0.4$). From the graph, at $P = 1$ bar, ϕ = saturated liquid, $x_a = 0.4$ (and $T = 288$ K). $\hat{H}_s = -200$ kJ/kg. Therefore,

$$H_2 = 2 \text{ kg} \times \left(-200 \frac{\text{kJ}}{\text{kg}}\right) = -400 \text{ kJ}$$

and

$$H_2 - H_1 = -400 \text{ kJ} - 145 \text{ kJ} = -545 \text{ kJ}$$

6.3.4 Using Model Equations to Find $\hat{U}$ and $\hat{H}$

We've seen from the tables and graphs in Sec. 6.3.3 that $\hat{U}$ and $\hat{H}$ are functions of four state properties:

- Pressure P
- Temperature T
- Phase ϕ
- Composition x_i

!Helpful Hint
When using model equations to calculate U or H, always change just one property at a time.

Unfortunately, more often than not we don't have tables or charts that give us data on enthalpy or internal energy of the compound or mixture of interest. Fortunately, there are model equations that allow us to calculate the change in $\hat{U}$ or $\hat{H}$ with a change in one property (P, T, phase, or composition), *while holding all other properties constant*. We just need to know what the model equations are, and how to use them. In this section we describe model equations that allow us to calculate changes in $\hat{U}$ and $\hat{H}$ with a change in pressure, temperature, phase, and/or composition.

Change in $\hat{U}$ and $\hat{H}$ with Change in P   For *ideal gases* undergoing a pressure change from P_1 to P_2, with T, ϕ, and x_i staying constant:

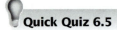

Quick Quiz 6.5

Use Table 6.1 to look at the change in enthalpy and internal energy of H_2O with changes in pressure, at constant temperature and phase. Does steam obey Eq (6.9)? Does liquid water obey Eq. (6.10)?

$$\hat{U}_2 - \hat{U}_1 = 0 \tag{6.9}$$

$$\hat{H}_2 - \hat{H}_1 = 0$$

Equation (6.9) is exact for ideal gases and a very good approximation for "real" gases (gases that do not obey the ideal gas law), unless the pressure change is very large.

For *solids and liquids*,

$$\hat{U}_2 - \hat{U}_1 = 0$$
$$\hat{H}_2 - \hat{H}_1 \approx \hat{V}(P_2 - P_1) \tag{6.10}$$

where $\hat{V}$ is the specific volume ($\hat{V} = 1/\rho$, where ρ is the density).

Illustration: The specific volume of liquid water at 50°C is 1.01×10^{-3} m³/kg. The enthalpy change associated with increasing the pressure of liquid water at 50°C from 1 bar to 10 bars is

$$\hat{H}_2 - \hat{H}_1 \approx \hat{V}(P_2 - P_1) = 1.01 \times 10^{-3} \frac{\text{m}^3}{\text{kg}}$$

$$\times (10 \text{ bar} - 1 \text{ bar}) \times \frac{10^5 \text{ N/m}^2}{\text{bar}} \times \frac{1 \text{ kJ}}{1000 \text{ N m}} = 0.91 \frac{\text{kJ}}{\text{kg}}$$

Change in $\hat{U}$ and $\hat{H}$ with Change in T   If the temperature of the system increases, then $\hat{U}$ and $\hat{H}$ increase. This can be understood physically as an increase in the random thermal motion of the molecules. These are called *sensible* heat effects (as opposed to silly heat effects?) because we can *sense* (feel) changes in temperature.

The change in $\hat{H}$ is proportional to the change in T (at constant P, ϕ, and x_i). The proportionality constant is called the constant-pressure **heat capacity** C_p.

If C_p is independent of temperature, then, with pressure, phase, and composition constant,

$$\hat{H}_2 - \hat{H}_1 = C_p (T_2 - T_1) \tag{6.11a}$$

The change in $\hat{U}$ with change in temperature is determined at constant volume, phase, and composition, and is proportional to the **constant-volume heat capacity** C_v. If C_v is independent of temperature, then, with volume, phase, and composition constant,

$$\hat{U}_2 - \hat{U}_1 = C_v (T_2 - T_1) \tag{6.11b}$$

Values of C_p and C_v for many compounds, and for some mixtures, are available in handbooks (and in App. B).

Equations (6.11a) and (6.11b) are reasonable approximations for most solids and liquids, and for gases as long as the temperature change is not too large. For greater accuracy, especially with gases undergoing large temperature changes, C_p and C_v are functions of temperature, and we say:

$$\hat{H}_2 - \hat{H}_1 = \int_{T_1}^{T_2} C_p \, dT \qquad (6.11c)$$

$$\hat{U}_2 - \hat{U}_1 = \int_{T_1}^{T_2} C_v \, dT \qquad (6.11d)$$

For the most accurate work, tabulated expressions for C_p are used for calculations. These usually take the form of polynomial expressions that can easily be integrated. Polynomial expressions for C_p of some compounds are included in App. B; more extensive tables are found in many reference books.

You have probably noticed the similarity between expressions for $\hat{U}$ and $\hat{H}$ with changing T. So, what's the difference? The difference is in whether pressure or volume is kept constant. Imagine a gas-filled balloon. If the gas is heated, then the volume of the balloon expands. The pressure inside the balloon stays constant. For this system, we'd calculate the *constant-pressure* change, $\hat{H}_2 - \hat{H}_1$. Now imagine a gas-filled steel cylinder. If the gas is heated, the volume can't increase, so the pressure goes up. For this system, we'd calculate the *constant-volume* change, $\hat{U}_2 - \hat{U}_1$.

C_p and C_v, like $\hat{U}$ and $\hat{H}$, are related:

$$C_p \approx C_v \qquad \text{for liquids and solids}$$

$$C_p \approx C_v + R \qquad \text{for gases}$$

where R is the ideal gas constant.

◢Helpful Hint
As rules of thumb,

1. C_p of liquid water $\approx$ 4 J/g °C, 1 cal/g °C, 1 Btu/lb °F.
2. C_p of organic liquids $\approx$ 2 J/g °C, 0.5 cal/g °C, 0.5 Btu/lb °F.
3. C_p of gases at 25°C $\approx$ 1 J/g °C, 0.25 cal/g °C, 0.25 Btu/lb °F.

Illustration: The constant-volume heat capacity C_v for liquid water is 4.19 kJ/kg °C. The specific internal energy change associated with increasing the temperature of liquid water at 1 bar from 50°C to 60°C is

$$\hat{U}_2 - \hat{U}_1 = C_v (T_2 - T_1) = 4.19 \, \frac{\text{kJ}}{\text{kg °C}} \times (60°\text{C} - 50°\text{C}) = 41.9 \, \frac{\text{kJ}}{\text{kg}}$$

Illustration: The heat capacity C_p of CO_2 is a function of temperature:

$$C_p \, (\text{J/gmol °C}) = 19.8 + 0.07344T - 5.602 \times 10^{-5}T^2 + 1.715 \times 10^{-8}T^3$$

with T in K. The enthalpy change associated with increasing the temperature of CO_2 gas at 200 kPa from 50°C to 200°C (323K to 473K) is

$$\hat{H}_2 - \hat{H}_1 = \int_{T_1}^{T_2} C_p\, dT = \int_{323}^{473} (19.8 + 0.07344T$$

Use the rules of thumb in the helpful hints to estimate the enthalpy change in this illustration.

$$- 5.602 \times 10^{-5}T^2 + 1.715 \times 10^{-8}T^3)\, dT$$

$$= 19.8(473 - 323) + \frac{0.07344}{2}(473^2 - 323^2)$$

$$- \frac{5.602 \times 10^{-5}}{3}(473^3 - 323^3) + \frac{1.715 \times 10^{-8}}{4}(473^4 - 323^4)$$

$$= 6175 \frac{J}{gmol}$$

Change in $\hat{U}$ and $\hat{H}$ with Change in ϕ   As we observed in Fig. 6.5, $\hat{U}$ or $\hat{H}$ of liquid water is less than $\hat{U}$ or $\hat{H}$ of water vapor. Energy is required to break the intermolecular forces that hold a liquid together, allowing the molecules to escape into the vapor. Similarly, you need to add energy to melt ice into water. Thus, a change in phase, at constant temperature and pressure, must involve a change in internal energy and enthalpy.

The enthalpy change due to a change in phase from liquid (state 1) to vapor (state 2) is called the **enthalpy of vaporization, $\Delta\hat{H}_v$**:

$$\hat{H}_2 - \hat{H}_1 = \Delta\hat{H}_v \qquad (6.12a)$$

Here are some rules of thumb for phase changes:

1. $\Delta\hat{H}_v$ of water ≈ 1000 Btu/lb or 2000 kJ/kg.

2. $\Delta\hat{H}_v$ for organic chemicals ≈ 500 kJ/kg or about 250 Btu/lb.

3. $\Delta\hat{H}_v$ ≈ 2- to 6-fold larger than $\Delta\hat{H}_m$.

The enthalpy change due to a change in phase from solid (state 1) to liquid (state 2) is called the **enthalpy of melting, $\Delta\hat{H}_m$**:

$$\hat{H}_2 - \hat{H}_1 = \Delta\hat{H}_m \qquad (6.12b)$$

Sometimes the enthalpy change associated with a phase change is called the heat of vaporization, the heat of melting, or simply **latent heat**. The enthalpies of condensation (change from vapor to liquid) and of fusion (change from liquid to solid) are simply the negative of the vaporization and melting enthalpies.

Enthalpy and internal energy of phase changes are related:

$$\Delta\hat{U}_v \approx \Delta\hat{H}_v - RT_b \qquad (6.12c)$$

$$\Delta\hat{U}_m \approx \Delta\hat{H}_m \qquad (6.12d)$$

where T_b is the boiling point. Tabulated data of $\Delta\hat{H}_v$ and $\Delta\hat{H}_m$ for selected compounds are available in App. B; much more data are readily at hand by consulting *Perry's Chemical Engineers' Handbook* and similar references. Usually, $\Delta\hat{H}_v$ and $\Delta\hat{H}_m$ are listed only at the normal boiling or melting point of the material of interest. (The normal boiling or melting point is the temperature at which the material boils or melts at 1 atm pressure.) This presents a problem—what do we do if our system undergoes a phase change at a pressure other than atmospheric? We'll get to that in a moment, after we discuss the effect of composition on enthalpy and internal energy.

Illustration: From App. B, we learn that acetic acid melts at 16.6°C, and the enthalpy change associated with the phase change from solid to liquid is 11.535 kJ/gmol, or 192 kJ/kg. At 1 atm pressure, pure acetic acid undergoes a phase from liquid to vapor at 118.3°C. The enthalpy change accompanying this phase change is 23.7 kJ/gml, or 395 kJ/kg.

Change in $\hat{U}$ and $\hat{H}$ Due to Change in x_i: Mixing and Solution

Mixing oil and vinegar, dissolving sodium chloride into water, crystallizing a protein from solution: These are ubiquitous processes, all of which involve a change in composition. Mixing processes involve a change in chemical composition of the system. Dissolution and crystallization processes involve a simultaneous change in phase and chemical composition.

Try mixing ethanol and water—notice that the mixture feels cooler. What happens if you dilute a 37 wt% solution of hydrochloric acid with water? The solution gets hot. In either case, there is a significant **enthalpy of mixing, $\Delta\hat{H}_{mix}$** (often called the heat of mixing), where:

$$\hat{H}_{mix} - \sum x_i \hat{H}_{i,pure} = \Delta\hat{H}_{mix} \tag{6.13a}$$

$\Delta\hat{H}_{mix}$ for mixing ethanol and water is positive; for dilution of aqueous HCl with water, $\Delta\hat{H}_{mix}$ is negative. For many mixtures, we can safely assume that $\Delta\hat{H}_{mix} \approx 0$. Vapors and gases fall into this category, as do many liquid mixtures of chemically similar components—hexane and heptane, for instance. Any time there are strong noncovalent attractive or repulsive interactions between two components (due to charge or hydrogen bonding, for example) then you likely cannot ignore $\Delta\hat{H}_{mix}$. (These are the same renegade mixtures that won't obey Raoult's law.)

Perhaps in chemistry lab you've dissolved sodium hydroxide pellets in water. Did you notice that the solution became warm? In this case, there is a phase change, as the solid goes into solution, *and* a composition change, due to mixing of the dissolving NaOH with water. The enthalpy change associated with this combined phase and composition change is known as the **enthalpy of solution, $\Delta\hat{H}_{soln}$** (often called the heat of solution), where

$$\hat{H}_{soln} - (x_{iL}\hat{H}_{iL} + x_{iS}\hat{H}_{iS}) = \Delta\hat{H}_{soln} \tag{6.13b}$$

where $\hat{H}_{soln}$ is the enthalpy of the solution, x_{iL} and x_{iS} are the mole (or mass) fractions of the solvent and solute components, respectively, and $\hat{H}_{iL}$ and $\hat{H}_{iS}$ are the specific enthalpies of the solvent component as a pure liquid and the solute component as a pure solid, respectively. The enthalpy of crystallization $\Delta \hat{H}_{cry}$, the enthalpy change associated with crystallization of a solid from a solution, is simply the negative of $\Delta \hat{H}_{soln}$. $\Delta \hat{H}_{soln}$ can be either positive or negative.

$\Delta \hat{H}_{mix}$ and $\Delta \hat{H}_{soln}$ vary with the dilution factor—the moles of diluting solvent added per mole of compound. App. B contains values for $\Delta \hat{H}_{mix}$ and $\Delta \hat{H}_{soln}$ for selected mixtures and solutions. More extensive data are available in *Perry's Chemical Engineers' Handbook* and similar references.

Illustration: What is the enthalpy change associated with dissolving 4 g NaOH pellets in 10 L of water at 18°C? 10 L water is 10,000 g or 555.55 gmol H_2O. 4 g NaOH equals 0.1 gmol NaOH. So the dilution is about 5555 gmol water/gmol NaOH. In the absence of more accurate information, we'll assume this is "infinite dilution." From App. B, at this dilution $\Delta \hat{H}_{soln} = -42.59$ kJ/gmol. The units are important—this is the enthalpy change associated with dissolving NaOH in water, *per gmol NaOH*. The total enthalpy change is therefore (0.1 gmol NaOH)(-42.59 kJ/gmol NaOH) $= -4.259$ kJ.

Change in $\hat{U}$ and $\hat{H}$ with Change in x_i: Chemical Reaction   When chemicals react, the energy of a system can change a lot. The situation you are probably most familiar with is combustion—a chemical reaction where carbon and hydrogen are oxidized to CO_2 and water. Harnessing the energy change associated with this reaction means we can cook our food, heat our homes, and drive our cars.

The enthalpy change that accompanies a chemical reaction is called the **enthalpy of reaction,** or the heat of reaction. By convention, data are generally available at standard T and P, typically 25°C (298 K) and 1 atm pressure. The **standard enthalpy of reaction** is denoted as $\Delta \hat{H}_r^{\circ}$ and is the enthalpy change per mole of reaction at standard T and P. There are two useful ways to calculate $\Delta \hat{H}_r^{\circ}$. We'll illustrate both.

Method 1. From enthalpy of formation:

$$\Delta \hat{H}_r^{\circ} = \sum_{\text{all } i} \nu_i \Delta \hat{H}_{f,i}^{\circ} \tag{6.14a}$$

where the standard **enthalpy of formation** $\Delta \hat{H}_{f,i}^{\circ}$ is the enthalpy change associated with formation of one mole of compound i at 25°C and 1 atm from its elements in their natural phase and state of aggregation. This last point, about "natural state of aggregation," is needed because not all elements naturally exist as monoatomic compounds. For example, the natural phase and state of aggregation (at 25°C and 1 atm) of oxygen is O_2 (g), hydrogen is H_2 (g), helium is He (g), bromine is Br_2 (l), and carbon is C (s), where (g) means gas, (l) means liquid, and (s) means solid, $\Delta \hat{H}_f^{\circ}$

data are available in App. B. and many handbooks. By definition, $\Delta\hat{H}^\circ_f = 0$ for elements in their "natural phase and state of aggregation."

Method 2. From enthalpy of combustion:

$$\Delta\hat{H}^\circ_r = -\sum_{\text{all } i} \nu_i\Delta\hat{H}^\circ_{c,i} \tag{6.14b}$$

where the standard **enthalpy of combustion** $\Delta\hat{H}^\circ_{c,i}$ is the enthalpy change associated with the reaction of oxygen with compound i, per mole of that compound, at 25°C and 1 atm, to generate specific products of reaction: CO_2 (g), H_2O (g), SO_2 (g), N_2 (g), and Cl_2 (g)—and nothing else. $\Delta\hat{H}^\circ_c$ values are tabulated for some compounds in App. B; more extensive lists are available in *Perry's Chemical Engineers' Handbook* and similar references. $\Delta\hat{H}^\circ_{c,i} = 0$ for CO_2 (g), H_2O (g), SO_2 (g), O_2 (g), N_2 (g) and Cl_2 (g). Why? Because these are the defined products of combustion that serve as the reference state.

The two methods for calculating $\Delta\hat{H}^\circ_r$ are illustrated in Fig. 6.7. Use of these methods is demonstrated in the following illustrations. Method 1 ($\Delta\hat{H}^\circ_f$) is applicable for any reaction. Method 2 ($\Delta\hat{H}^\circ_c$) works only for reactions involving molecular species containing no elements except C, H, O, N, S, and Cl—which covers a lot of important chemistry!

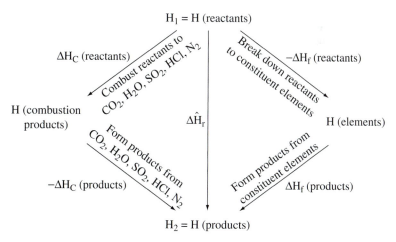

Figure 6.7 The enthalpy change taking the straight route from reactants to products is $\Delta\hat{H}_r$, the enthalpy of reaction. This can be calculated by taking either the left-side route (combustion of reactants to combustion products plus reformation of combustion products to desired products) or the right-side route (breakdown of reactants to elements plus recombination of elements to desired products).

Illustration: Liquid methanol (CH_3OH) is produced by partial oxidation of methane:

$$CH_4(g) + 0.5O_2(g) \rightarrow CH_3OH(l)$$

$\Delta \hat{H}_r^\circ$ (kJ/gmol) is calculated from $\Delta \hat{H}_f^\circ$ of each compound as listed in App. B, using Eq. (6.14a).

$$\Delta \hat{H}_r^\circ = \sum_{\text{all } i} \nu_i \Delta \hat{H}_{f,i}^\circ = (-1)\Delta \hat{H}_{f,CH_4}^\circ + (-0.5)\Delta \hat{H}_{f,O_2}^\circ + (1)\Delta \hat{H}_{f,CH_3OH(l)}^\circ$$

$$= (-1)(-74.52) + (-0.5)(0) + (1)(-238.66) = -164.14 \text{ kJ/gmol}$$

Alternatively, $\Delta \hat{H}_r^\circ$ (kJ/gmol) is calculated from $\Delta \hat{H}_c^\circ$ of each compound as listed in App. B, using Eq. (6.14b):

$$\Delta \hat{H}_r^\circ = -\sum_{\text{all } i} \nu_i \Delta \hat{H}_{c,i}^\circ = (1)\Delta \hat{H}_{c,CH_4}^\circ + (0.5)\Delta \hat{H}_{c,O_2}^\circ + (-1)\Delta \hat{H}_{c,CH_3OH(l)}^\circ$$

$$= (-802.6) + (0) + (-1)(-638.46) = -164.14 \text{ kJ/gmol}$$

Illustration: $\Delta \hat{H}_{c,CH_4(g)}^\circ = \Delta \hat{H}_r^\circ$ for the reaction:

$$CH_4(g) + 2O_2(g) \rightarrow CO_2(g) + 2H_2O(l)$$

Therefore $\Delta \hat{H}_{c,CH_4}^\circ(g)$ can be calculated from $\Delta \hat{H}_f^\circ$ (App. B) and Eq. 6.14(a):

$$\Delta \hat{H}_r^\circ = \sum_{\text{all } i} \nu_i \Delta \hat{H}_{f,i}^\circ = (-1)\Delta \hat{H}_{f,CH_4}^\circ + (-2)\Delta \hat{H}_{f,O_2}^\circ$$

$$+ (+1)\Delta \hat{H}_{f,CO_2}^\circ + (+2)\Delta \hat{H}_{f,H_2O(l)}^\circ$$

$$\Delta \hat{H}_r^\circ = (-1)(-74.52) + (-2)(0) + (+1)(-393.5)$$

$$+ (+2)(-285.84) = -890.67 \text{ kJ/gmol}$$

$\Delta \hat{H}_{c,CH_4(g)}^\circ = -802.6$ kJ/gmol (see App. B) but this is for combustion of methane with water vapor as product. With liquid water as product, $\Delta \hat{H}_{c,CH_4(g)}^\circ = -802.6 - 44.0(2) = -890.6$ kJ/gmol, which agrees with the result of our $\Delta \hat{H}_r^\circ$ calculation. It should. $\Delta \hat{H}_f^\circ$ values are not measured directly, but are determined by measuring $\Delta \hat{H}_c^\circ$ in an apparatus called a calorimeter!

Here are a couple of other points to keep in mind regarding $\Delta \hat{H}_r^\circ$:

1. $\Delta \hat{H}_{f,i}^\circ$ and $\Delta \hat{H}_{c,i}^\circ$ are functions of the phase of compound i. Always indicate the phase of the reactants and products when writing the balanced chemical reaction. If $\Delta \hat{H}_{f,i}^\circ$ or $\Delta \hat{H}_{c,i}^\circ$ data for two different phases are given, the difference is simply the $\Delta \hat{H}$ of phase change at 25°C.

Illustration: For H_2O (g), $\Delta\hat{H}°_f = -241.83$ kJ/gmol while for H_2O (l), $\Delta\hat{H}°_f = -285.84$ kJ/gmol .

The difference is the enthalpy of vaporization of water at 25°C:

$$\Delta\hat{H}_v = -241.83 - (-285.84) = 44.01 \text{ kJ/gmol}$$

2. **Exothermic** reactions are defined as those where $\Delta\hat{H}°_r < 0$. Combustion is a highly exothermic reaction. Oxidation (addition of O_2), hydrogenation (addition of H_2), and hydrolysis (addition of H_2O) reactions are virtually always exothermic. **Endothermic** reactions are those where $\Delta\hat{H}°_r > 0$. Dehydrogenation and dehydration are typically endothermic. With isomerization reactions, usually $\Delta\hat{H}°_r \approx 0$.

<table>
<tr><td>❗</td></tr>
</table>

Helpful Hint
It is easy to make a sign error in calculating $\Delta\hat{H}_r$, so use your knowledge of chemistry to check your calculations.

3. For chemical reactions, $\Delta\hat{H}°_r$ is calculated *per mole of reaction*. Be sure to write out the stoichiometrically balanced reaction explicitly and keep clearly in mind the basis for your calculations. The total enthalpy change associated with a reaction equals $\Delta\hat{H}°_r$ multiplied by the extent of reaction ξ.

Illustration: For the reaction:

$$N_2(g) + 3H_2(g) \rightarrow 2NH_3(g)$$

$$\Delta\hat{H}°_r = \sum \nu_i \Delta\hat{H}°_{f,i} = (-1)\Delta\hat{H}°_{f,N_2} + (-3)\Delta\hat{H}°_{f,H_2} + (+2)\Delta\hat{H}°_{f,NH_3}$$

$$= (-1)(0) + (-3)(0) + (+2)(-46.15) = -92.3 \text{ kJ/gmol}$$

If 100 gmol N_2 is fed along with an excess of H_2 to a reactor and 60% of the N_2 is converted to ammonia, then $\xi = 0.6(100) = 60$ gmol and the total enthalpy change equals $\xi\Delta\hat{H}°_r = 60$ gmol $(-92.3$ kJ/gmol$) = -5538$ kJ (at 25°C.)

For the reaction:

$$0.5N_2(g) + 1.5H_2(g) \rightarrow NH_3(g)$$

$$\Delta\hat{H}°_r = \sum \nu_i \Delta\hat{H}°_{f,i} = (-0.5)\Delta\hat{H}°_{f,N_2} + (-1.5)\Delta\hat{H}°_{f,H_2} + (+1)\Delta\hat{H}°_{f,NH_3}$$

$$= (-0.5)(0) + (-1.5)(0) + (+1)(-46.15) = -46.15 \text{ kJ/gmol}$$

If 100 gmol N_2 is fed along with an excess of H_2 to a reactor and 60% of the N_2 is converted to ammonia, then $\xi = 0.6(100)/0.5 = 120$ gmol and the total enthalpy change equals $\xi\Delta\hat{H}°_r = 120$ gmol $(-46.15$ kJ/gmol$) = -5538$ kJ (at 25°C).

What about $\Delta\hat{U}°_r$? We go back to the general relationship between internal energy and enthalpy:

$$\Delta H_r = \Delta U_r + \Delta(PV)_r$$

where $\Delta(PV)_r$ equals the change in (pressure $\times$ volume) with reaction. For reactions taking place in the *liquid* or *solid* phase, the change in pressure and volume with chemical reaction is usually small, so $\Delta(PV)_r \approx 0$. Therefore, $\Delta U_r \approx \Delta H_r$ and $\Delta \hat{U}_r^\circ \approx \Delta \hat{H}_r^\circ$.

For an ideal gas, $PV = nRT$. For reactions taking place in the *gas* phase, we use the ideal gas equation to find that $\Delta(PV)_r = \Delta(nRT)_r$. Since we are considering changes in enthalpy due only to reaction, the temperature T is constant. Therefore we say that $\Delta(PV)_r = RT\Delta n_r$, where $\Delta n_r = n_{\text{products}} - n_{\text{reactants}}$. Thus, $\Delta U_r \approx \Delta H_r - RT\Delta n_r$ for gas-phase reactions.

Change in $\hat{U}$ and $\hat{H}$ with Change in *P, T,* and ϕ
What if:

the pressure *and* the temperature of a system change?

a solid is heated to its melting point, melted, and then the liquid is heated further?

a gas is vaporized at high pressure, at a temperature other than its normal boiling point?

What you've learned so far is not quite enough to handle these situations. We need to explain one very important and useful point about the character of $\hat{U}$ and $\hat{H}$: they are **state functions,** which is very lucky for us. A state function is any function that depends only on the state of matter, and not the path for getting there. The basic concept is illustrated in Fig. 6.8.

The state function nature of $\hat{U}$ and $\hat{H}$ has enormous practical importance. Essentially, this fact means that we can calculate $\hat{U}$ or $\hat{H}$ at *any* specified condition of T, P, ϕ, using the following procedure:

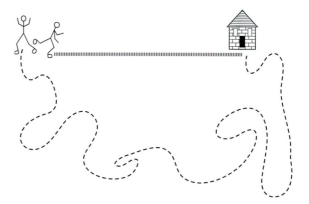

Figure 6.8 Fran races to school, taking the most direct path. Jo meanders through the woods. Both Fran and Jo start and stop at the same locations. The *net change in Fran's and Jo's location* is the same—it is a state function, depending only on the start and stop locations. The time it takes for Fran and Jo to travel to school, or the number of miles each walks, is not a state function; it is a path function.

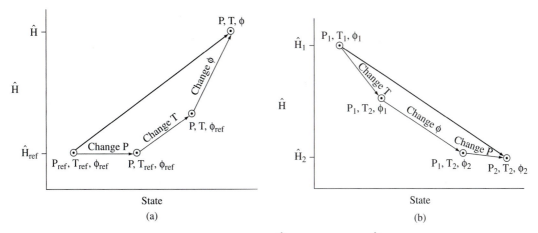

Figure 6.9 (a) Representative pathway for calculating $\hat{H}$ at (P, T, ϕ) given $\hat{H}_{ref}$ at $(P_{ref}, T_{ref}, \phi_{ref})$. In using model equations it is important to change only one property at a time. Calculations can be carried out in any order. (b) Representative pathway for calculating $\hat{H}_2 - \hat{H}_1$, where P, T, and ϕ all change from state 1 to state 2.

1. Pick a convenient reference state of P_{ref}, T_{ref}, ϕ_{ref}.
2. Set $\hat{U}$ or $\hat{H}$ at the reference condition equal to zero.
3. Calculate the change in $\hat{U}$ or $\hat{H}$ as the pressure changes from P_{ref} to P.
4. Calculate the change in $\hat{U}$ or $\hat{H}$ as the temperature changes from T_{ref} to T.
5. Calculate the change in $\hat{U}$ or $\hat{H}$ as the phase changes from ϕ_{ref} to ϕ.

We can calculate $\hat{U}$ or $\hat{H}$ (steps 3, 4, 5) in any convenient order!

6. Add up the results from steps 3, 4, and 5.

We can use a similar strategy to calculate $\hat{U}_2 - \hat{U}_1$ or $\hat{H}_2 - \hat{H}_1$, as the system moves from one specified condition of T_1, P_1, ϕ_1, to a different specified condition of T_2, P_2, ϕ_2. These ideas are illustrated in Fig. 6.9.

There can be a lot to keep track of in doing these sorts of calculations, so it is really useful to have a good bookkeeping scheme. It is a good idea to (a) sketch a diagram similar to Fig. 6.9 or make a table showing the total change to be calculated and the one-step-at-a-time pathway, (b) collect all data needed to evaluate each step in the pathway, (c) write down the data or evaluate the equation to calculate the change in enthalpy for each step, and (d) sum.

Helpful Hint
Change only one variable at each step in the pathway, choose the pathway that is most compatible with the available data, keep organized!

Be aware that sometimes we have to take a rather circuitous route, because that is the only path with the data available. (See Fig. 6.10.) For example, what if we are interested in knowing the enthalpy change associated with vaporization of a compound at 10 atm, but we have data for $\Delta\hat{H}_v$ only at the normal boiling point and 1 atm? Luckily, enthalpy is a state function! We generate a circuitous route such as: Decrease the pressure from 10 atm to 1 atm, decrease the temperature to the normal boiling point, change the phase at the normal boiling point, then raise the temperature back up, then raise the pressure back up. The net change in enthalpy calculated by summing up all these changes is the enthalpy change we want.

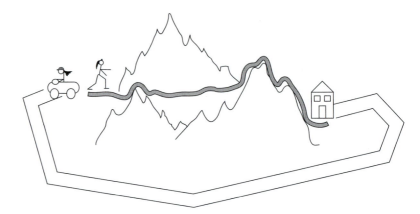

Figure 6.10 Fran and Jo are both traveling to a great little restaurant. Fran walks, so she can take the direct footpath through the mountains. This path is not available, though, to Jo, because she is driving. The only route available to her is the highway. By necessity, Jo takes a longer, more circuitous route than Fran, but both Fran and Jo leave from and arrive at the same locations.

Example 6.8	**Enthalpy Calculations: Enthalpy of Vaporization of Water at High Pressure**

What is the enthalpy change associated with vaporizing water at 10 bar? At its normal boiling point (100°C at 1.01325 bar), $\Delta \hat{H}_v$ of water = 2259 kJ/kg. C_p of liquid water = 4.19 kJ/kg °C. C_p of water vapor = 1.9 kJ/kg °C. The specific volume of liquid water = 0.00113 m³/kg at 179.9°C.

Solution

We want $\Delta \hat{H}_v$ at 10 bar, at which pressure water boils at 179.9°C. We construct a pathway where we change only one parameter at a time, and we utilize the T and P at which we have data:

Change P: from (10 bar, 179.9°C, liquid) to (1.01325 bar, 179.9°C, liquid).

$$\hat{V}(\Delta P) = \frac{0.00113 \text{ m}^3}{\text{kg}} (1.01325 - 10 \text{ bar}) \frac{100 \text{ kJ/m}^3}{\text{bar}} = -1 \text{ kJ/kg}$$

Change T: from (1.01325 bar, 179.9°C, liquid) to (1.01325 bar, 100°C, liquid).

$$C_{p,H_2O,l}(\Delta T) = \frac{4.19 \text{ kJ}}{\text{kg °C}} (100°C - 179.9°C) = -335 \text{ kJ/kg}$$

Change ϕ: from (1.01325 bar, 100°C, liquid) to (1.01325 bar, 100°C, vapor).

$$\Delta \hat{H}_v^o = +2259 \text{ kJ/kg}$$

Change T: from (1.01325 bar, 100°C, vapor) to (1.01325 bar, 179.9°C, vapor).

$$C_{p,H_2O(v)}(\Delta T) = \frac{1.9 \text{ kJ}}{\text{kg °C}} (179.9°C - 100°C) = +152 \text{ kJ/kg}$$

Change P: from (1.01325 bar, 179.9°C, vapor) to (10 bar, 179.9°C, vapor).

$$\hat{V}(\Delta P) = 0$$

The net result of these 5 steps is (10 bar, 179.9°C, liquid) → (10 bar, 179.9°C, vapor). The enthalpy change associated with this change in state is simply the enthalpy of vaporization at 10 bar, which we calculate by summing the enthalpy changes of each step:

$$\Delta \hat{H} = \Delta \hat{H}_v(10 \text{ bar}) = -1 + -335 + 2259 + 152 + 0 = 2075 \text{ kJ/kg}$$

At the higher temperature the latent heat is about 9% lower than at the normal boiling point. This is typical—because C_p for vapors is less than for liquids, as the temperature increases the latent heat will decrease. The change is not huge, and for very rough calculations you might be able to get away with ignoring the temperature effect on the latent heat.

Quick Quiz 6.7

What is $\Delta \hat{H}_v$ at 10 bar from steam tables, and why is it different than what we calculated?

At critical T and P, what is $\Delta \hat{H}_v$?

Change in $\hat{U}$ or $\hat{H}$ with Change in P, T, ϕ and x_i

You've learned that changes in enthalpy are associated with changes in composition due to mixing, solution, or reaction, and you've learned how to find or calculate the specific enthalpy (or internal energy) change associated with these processes, at standard conditions (typically 25°C and 1 atm). However, most chemical reactors and mixers of commercial interest do not operate at 25°C and 1 atm. What do we do if we want to know $\Delta \hat{H}_r$ at 500°C and 10 atm? What if we want to calculate the enthalpy change associated with dissolving NaOH in water at 50°C? We take advantage of the fact that enthalpy is a state function, of course! We construct a pathway that takes us from the reference state to the desired state (or from state 1 to state 2), considering the nature of the data available. This is no different than what we did in the previous section, except for one point: With a change in chemical composition, it is often better to calculate U or H rather than $\hat{U}$ or $\hat{H}$. This avoids any confusion about the basis on which $\hat{U}$ or $\hat{H}$ are calculated that might otherwise arise. For example, the enthalpy change associated with reaction is calculated *per mole of reaction* while the enthalpy change associated with changing the temperature of reactants is calculated *per mole of reactants*. Similarly, the enthalpy change associated with dissolution is calculated *per mole of solute* while the enthalpy change associated with changing the temperature of the solution is calculated *per mole of solution*.

Example 6.9

Enthalpy Calculations: Enthalpy of Reaction at High Temperature

Methane (CH_4) and oxgyen (O_2) are completely combusted to CO_2 and H_2O vapor at 500°C and 1 atm. What is the enthalpy change associated with this reaction per gmol methane?

$$CH_4(g) + 2O_2(g) \rightarrow CO_2(g) + 2H_2O(g)$$

Solution

First we collect relevant data from App. B:

C_p, kJ/gmol°C, with T in °C

CH_4 (g):	$19.25 + 0.05213T + 1.197 \times 10^{-5} T^2 - 1.132 \times 10^{-8} T^3$
O_2 (g):	$28.11 - 3.68 \times 10^{-6} T + 1.746 \times 10^{-5} T^2 - 1.065 \times 10^{-8} T^3$
CO_2 (g):	$19.80 + 0.07344T - 5.602 \times 10^{-5} T^2 + 1.715 \times 10^{-8} T^3$
H_2O (g):	$32.24 + 0.001924T + 1.055 \times 10^{-5} T^2 - 3.596 \times 10^{-9} T^3$
H_2O (l):	$72.43 + 0.0104T$

$$\Delta \hat{H}^{\circ}_{v,H_2O} = 40.65 \text{ kJ/gmol at } 100°C.$$

$$\Delta \hat{H}^{\circ}_{c,CH_4(g)} = -890.6 \text{ kJ/gmol at } 25°C, \text{ with } H_2O \text{ (l) as product}$$

Then we construct a pathway consistent with the available data, evaluate the enthalpy change at each step, then sum.

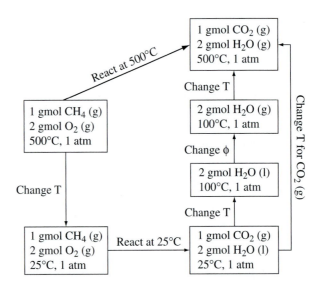

Change T: 1 gmol CH_4 (g) and 2 gmol O_2 (g) from 500°C to 25°C (detailed calculations are left for the reader).

$$\int_{773}^{298} \left[n_{CH_4(g)} C_{p,CH_4(g)} + n_{O_2(g)} C_{p,O_2(g)} \right] dT = -53.1 \text{ kJ}$$

React: 1 gmol CH_4 (g) and 2 gmol O_2 (g) to 1 gmol CO_2 (g) and 2 gmol H_2O (l) at 25°C.

$$\xi \Delta \hat{H}^{\circ}_r = \xi \Delta \hat{H}^{\circ}_{c,CH_4(g)} = (1 \text{ gmol})(-890.6 \text{ kJ/gmol}) = -890.6 \text{ kJ}$$

Change T: 1 gmol CO_2 (g) from 25°C to 500°C.

$$\int_{298}^{773} n_{CO_2(g)} C_{p,CO_2(g)} \, dT = +21.1 \text{ kJ}$$

Change T: 2 gmol H_2O (l) from 25°C to 100°C.

$$\int_{298}^{373} n_{H_2O(l)} C_{p,H_2O(l)} dT = +11.4 \text{ kJ}$$

Change ϕ: 2 gmol H_2O from liquid to vapor at 100°C.

$$n_{H_2O} \Delta \hat{H}^\circ_v = (2 \text{ gmol})(40.65 \text{ kJ/gmol}) = +81.3 \text{ kJ}$$

Change T: 2 gmol H_2O (g) from 100°C to 500°C.

$$\int_{373}^{773} n_{H_2O(v)} C_{p,H_2O(v)} \, dT = +29.0 \text{ kJ}$$

Summing up gives the total enthalpy change:

$$H_2 - H_1 = -53.1 - 890.6 + 21.1 + 11.4 + 81.31 + 29.0 = -800.9 \text{ kJ}$$

Since the net result of our six-step process is to combust 1 gmol CH_4 plus 2 gmol O_2 to 1 gmol CO_2 and 2 gmol H_2O at 500°C in the vapor phase, $\Delta \hat{H}_r = -800.9$ kJ/gmol for methane combustion at 500°C.

6.3.5 Minisummary

This section has covered a lot of information about finding $\hat{U}$ and $\hat{H}$. Here is a summary of what we've learned.

You can look it up. For common compounds or mixtures, tabulated values are available and you can just read from the graph or table. Be sure you know what the reference conditions are. If you use more than one source of information, *check to be sure that the reference states are the same—* if not, you will get into trouble.

You can calculate it. Figure out a pathway to get from state 1 to state 2 (or from reference state to system state) by changing only one parameter (temperature, pressure, phase, or composition) at a time. When setting up the pathway, consider what physical property data are available. Calculate the change in $\hat{U}$ or $\hat{H}$ for each step of the pathway, then add up all the steps. The table summarizes how to complete these calculations.

Change in	$\Delta\hat{H}$	$\Delta\hat{U}$	Comments
Pressure	~ 0 (gas) ~$\hat{V}\Delta P$ (solid or liquid)	~ 0	Usually can be neglected unless pressure change is large.
Temperature	$\int C_p\,dT$	$\int C_v\,dT$	Polynomial expressions for C_p available. $C_p \approx C_v + R$ (gases) $C_p \approx C_v$ (solids and liquids)
Phase	$\Delta\hat{H}_v$ (liquid to vapor) $\Delta\hat{H}_m$ (solid to liquid)	$\Delta\hat{U}_v \approx \Delta\hat{H}_v - RT_b$ $\Delta\hat{U}_m \approx \Delta\hat{H}_m$	Data often available only at normal boiling point T_b and normal melting point T_m.
Composition due to reaction	$\Delta\hat{H}_r^\circ = \sum_i \nu_i \Delta\hat{H}_{f,i}^\circ$ $\Delta\hat{H}_r^\circ = -\sum_i \nu_i \Delta\hat{H}_{c,i}^\circ$	$\Delta\hat{U}_r \approx \Delta\hat{H}_r - RT\Delta n_r$ (gas) $\Delta\hat{U}_r \approx \Delta\hat{H}_r$ (solid or liquid)	Standard conditions are 25°C and 1 atm. Watch basis: per mole of reaction, not per mole of reactant or product.
Composition due to mixing or solution	$\Delta\hat{H}_{soln}$ $\Delta\hat{H}_{mix}$	$\Delta\hat{U}_{soln} \approx \Delta\hat{H}_{soln}$ $\Delta\hat{U}_{mix} \approx \Delta\hat{H}_{mix}$	Important for nonideal mixtures: acids, bases, alcohols, polymers. Watch basis: per mole of solute, not per mole of solution.

6.4   Energy Flows: Heat and Work

Kinetic, potential, and internal energy (E_k, E_p, and U) are kinds of energy that a system may *possess*. These are all state functions. We are also interested in how energy flows into or out of a system. Energy *transfers* in and out of a system by material transferring in and out of the system, because the material stream possesses kinetic, potential and internal energy. But energy can transfer in and out of a system in the absence of any material transfer! This is accomplished by two means: heat and work.   Heat and work are such commonplace words that it is easy to forget that, in the context of process energy, they have very specific and definite meanings.

Heat   flow is energy flow across system boundaries due to a *difference in temperature* between the system and the surroundings, *in the absence of material flow*. For example, refer back to Fig. 6.3: The temperature of the fire under the pot is greater than the temperature of the water in the pot, so there is energy flow *from* the surroundings *to* the system in the form of heat. The energy added to the system by pouring the hot milk into the pot is *not* heat; it is energy transfer due to transfer of a material that possesses energy.

We denote the rate of heat transfer as $\dot{Q}$ and the total heat transferred over a specified time interval as Q, where

$$Q = \int \dot{Q}\, dt \qquad (6.15)$$

Q and $\dot{Q}$ *are positive if heat flows from the surroundings to the system,* and negative if heat flows from the system to the surroundings. If a system is perfectly thermally insulated, then no heat will be transferred between system and surroundings, *even if there is a temperature difference.* Such a system is called **adiabatic**.

Heat is always transferred *from hot to cold.* The rate of heat transfer depends on the size of the temperature difference, on the insulating properties of the material that separates the system from its surroundings, and on the area of the surface that separates the system from its surroundings. For example, one quart of water in a copper pot set over an open flame will heat up much faster than the same quart of water in a small pot made of brick tucked into some dying embers.

Work is the flow of mechanical energy across system boundaries due to driving forces other than temperature. We denote the rate of work transfer as $\dot{W}$ and the total work transferred over a finite time interval as W, with

$$W = \int \dot{W}\, dt \qquad (6.16)$$

We define *work as positive if there is mechanical energy flow from the surroundings to the system*, and negative if there is mechanical energy flow from the system to the surroundings.

For chemical processes, there are two important kinds of work: "shaft work" W_s and "flow work" W_f, where

$$\begin{aligned} W &= W_s + W_f \\ \dot{W} &= \dot{W}_s + \dot{W}_f \end{aligned} \qquad (6.17)$$

Shaft work W_s is work that is done on the system or by the system that involves rotating or reciprocating equipment such as pumps, compressors, turbines, or mixing blades. (The name arises because the equipment has a shaft.) Flow work W_f is work associated with the flow of material into a system pushing on the material already in the system (or material leaving a system getting pushed out by the material in the system). Flow work is equal to the pressure at the point of entry times the volume of the packet of material, so it is sometimes called PV work. If there are multiple streams entering and leaving the system, with each stream at pressure P_j we write:

$$\dot{W}_f = \sum_{\text{all } j \text{ in}} P_j \dot{V}_j - \sum_{\text{all } j \text{ out}} P_j \dot{V}_j \qquad (6.18)$$

where $\dot{V}_j$ is the volumetric flow rate of stream j.

Figure 6.11 illustrates these two different kinds of work.

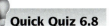

Quick Quiz 6.8

What is the dimension of $\dot{Q}$?

A cup of boiling water is poured into a thermos. If the thermos is the system, is $\dot{Q}$ positive, negative, or zero?

A teakettle full of boiling water is set down on a tray. If the teakettle is the system, is $\dot{Q}$ positive, negative, or zero?

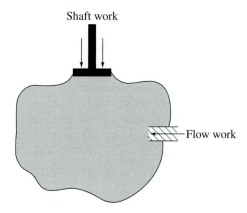

Figure 6.11 Two kinds of work. The system boundary is defined as the solid line. An external force is applied so that the piston is pushing on the system boundary; the piston is doing shaft work on the system. Some material (denoted by the diagonal lines) is entering the system; to do so it must push on the material already in the system. The work done by the material entering the system to push on the material in the system is called flow work.

6.5   The Energy Balance Equation—Again

Back in Sec. 6.2, we introduced the **energy balance equation** in words:

$$\text{Accumulation} = \text{In} - \text{Out}$$

as a differential equation:

$$\frac{dE_{\text{sys}}}{dt} = \sum_{\text{all } j \text{ in}} \dot{E}_j - \sum_{\text{all } j \text{ out}} \dot{E}_j \tag{6.1}$$

and as an integral equation:

$$E_{\text{sys},f} - E_{\text{sys},0} = \sum_{\text{all } j \text{ in}} \left(\int_{t_0}^{t_f} \dot{E}_j \, dt \right) - \sum_{\text{all } j \text{ out}} \left(\int_{t_0}^{t_f} \dot{E}_j \, dt \right) \tag{6.2}$$

where E_{sys} is the total system energy and $\dot{E}$ indicates an energy flow in or out of the system.

Now we are *finally* ready to write E_{sys} and $\dot{E}$ in more useful form. As you've already learned, the energy of a system includes the system's kinetic, potential, and internal energy, or

$$E_{\text{sys}} = E_{k,\text{sys}} + E_{p,\text{sys}} + U_{\text{sys}} \tag{6.12}$$

There are two different kinds of energy flows $\dot{E}_j$ in and out of the system. If material enters or leaves the system, energy enters and exits the system along with the material. Each stream entering a system carries with it kinetic, potential, and internal energy. Energy flow $\dot{E}_j$ due to material flow in stream j is therefore equal to the sum of the material flow rate times the specific kinetic, potential, and internal energy of the material:

$$\dot{E}_j = \dot{m}_j \left(\hat{E}_{k,j} + \hat{E}_{p,j} + \hat{U}_j \right)$$

Equivalently, since $\hat{H} = \hat{U} + P\hat{V}$,

$$\dot{E}_j = \dot{m}_j \left(\hat{E}_{k,j} + \hat{E}_{p,j} + \hat{H}_j - P\hat{V}_j \right) \tag{6.19}$$

Since $\dot{Q}$ and $\dot{W}$ are energy flows across system boundaries in the absence of material flow, these flows must also be included in the energy balance equation. We substitute Eqs. (6.12), (6.17), (6.18) and (6.19) into Eq. (6.1) and find:

$$\frac{d(E_{k,\text{sys}} + E_{p,\text{sys}} + U_{\text{sys}})}{dt} = \sum_{\text{all } j \text{ in}} \dot{m}_j \left(\hat{E}_{k,j} + \hat{E}_{p,j} + \hat{H}_j - P\hat{V}_j \right)$$

$$- \sum_{\text{all } j \text{ out}} \dot{m}_j \left(\hat{E}_{k,j} + \hat{E}_{p,j} + \hat{H}_j - P\hat{V}_j \right) + \sum_j \dot{Q}_j$$

$$+ \sum_{\text{all } j \text{ in}} \dot{m}_j P\hat{V}_j - \sum_{\text{all } j \text{ out}} \dot{m}_j P\hat{V}_j + \sum_j \dot{W}_{s,j}$$

Notice that the PV terms cancel, yielding a useful form of the differential energy balance equation:

$$\frac{d(E_{k,\text{sys}} + E_{p,\text{sys}} + U_{\text{sys}})}{dt} = \sum_{\text{all } j \text{ in}} \dot{m}_j \left(\hat{E}_{k,j} + \hat{E}_{p,j} + \hat{H}_j \right) - \sum_{\text{all } j \text{ out}} \dot{m}_j \left(\hat{E}_{k,j} + \hat{E}_{p,j} + \hat{H}_j \right)$$

$$+ \sum_j \dot{Q}_j + \sum_j \dot{W}_{s,j} \tag{6.20}$$

There is an integral energy balance equation counterpart to Eq. (6.20):

$$(E_{k,\text{sys}} + E_{p,\text{sys}} + U_{\text{sys}})_f - (E_{k,\text{sys}} + E_{p,\text{sys}} + U_{\text{sys}})_0 = \left[E_{k,\text{in}} - E_{k,\text{out}} \right]$$

$$+ \left[E_{p,\text{in}} - E_{p,\text{out}} \right] + \left[H_{\text{in}} - H_{\text{out}} \right] + Q + W_s \tag{6.21}$$

where we've used the notation

$$M = \sum_j \left(\int_{t_0}^{t_f} \dot{M}_j \, dt \right)$$

where M is any energy form, and $\dot{M}_j$ is an energy flow in stream j.

Finally, we point out that molar rather than mass flow rates are used in Eqs. (6.20) and (6.21) if molar rather than specific energies are given.

Although these equations look complex, we are usually able to simplify them significantly for a given problem. Here are three cases that arise frequently:

Steady-state continuous-flow process with negligible differences in temperature, pressure, phase and composition between inlet and outlet streams (common

Quick Quiz 6.9

In Eq. (6.20), identify the terms describing Accumulation, (In − Out) due to material flow, and (In − Out) due to energy flow in the absence of material flow.

What is the dimension of each term in Eq. (6.20)?

Quick Quiz 6.10

In Eq. (6.21), identify the terms describing Accumulation, (In − Out) due to material flow, and (In − Out) due to energy flow in the absence of material flow.

What is the dimension of each term in Eq. (6.21)?

situation when mechanical equipment and fluid flow are the important parts of the process):

$$\sum_{\text{all } j \text{ out}} \dot{m}_j(\hat{E}_{k,j} + \hat{E}_{p,j}) - \sum_{\text{all } j \text{ in}} \dot{m}_j(\hat{E}_{k,j} + \hat{E}_{p,j}) = \sum_j \dot{Q}_j + \sum_j \dot{W}_{s,j}$$

Steady-state continuous-flow process where kinetic and potential energy differences between inlet and outlet streams are negligible (common situation when reactors, separators, or heat exchangers are the important parts of the process):

$$\sum_{\text{all } j \text{ out}} \dot{m}_j\hat{H}_j - \sum_{\text{all } j \text{ in}} \dot{m}_j\hat{H}_j = \sum_j \dot{Q}_j + \sum_j \dot{W}_{s,j}$$

Batch process where kinetic and potential energy changes in the system are negligible:

$$U_{\text{sys},f} - U_{\text{sys},0} = Q + W_s$$

6.6   Process Energy Calculations

6.6.1   A Systematic Procedure for Process Energy Calculations

Now we wish to pull together all these strands—energy flows, heat, work, and energy balance equations—to complete process energy calculations. A systematic approach is the only approach that will reliably ensure successful and accurate completion of process flow calculations. (Sound familiar?) Here is a highly recommended procedure to follow.

Process Energy Calculations in 12 Easy Steps

Step 1.   Draw a *flow diagram.*

Step 2.   Define a *system.*

Step 3.   Choose *components* and define *stream variables* for all material and energy streams entering or leaving the system.

Step 4.   Convert all numerical information into *consistent units* of mass or moles and energy.

Step 5.   Define a *basis.*

Step 6.   Set up process flow calculations, and solve if possible.

Step 7.   If the system is open, write down model equations and/or find data for the energy flows due to material flows ($\dot{E}_k$, $\dot{E}_P$, $\dot{H}$) in and out of the system.

Step 8.   Identify any *heat* or *work* flows into or out of the system, and write down equations for $\dot{Q}$ and $\dot{W}_s$.

Step 9.   Decide if the *system energy changes*. If so, write down model equations and/or find data for $E_{k,sys}$, $E_{p,sys}$, and/or $U_{sys.}$

Step 10.   Write the *energy balance equation* using the stream and system variables you've defined.

Step 11.   *Solve* the equations.

Step 12.   *Check* your solutions.

6.6.2 Helpful Hints for Process Energy Calculations

Here are some Helpful Hints that should assist you as you apply the 12 Easy Steps to solve process energy calculations. Scan the list quickly now, and refer back to it if you get stuck while working a problem.

Step 1. Draw a diagram.
- Indicate material flows with solid lines, and work or heat flows with dashed lines.
- Indicate the direction of work or heat flows, if known.
- Indicate changes in elevation schematically.

Step 2. Choose a system.
- Systems can be either open (with material flowing in and out) or closed.

Step 4. Convert all numerical information into consistent units.
- Use caution with units of specific enthalpy—the data may be reported as energy per unit mass or energy per unit mole.

Step 6. Set up equations for process flow calculations, and solve if possible.
- Sometimes you will need to solve process flow and process energy equations simultaneously.

Step 7. Write equations or find data for the energy of all material streams in and out of the system.
- If you choose the state of one of the streams as the reference state, then the specific energy of that stream equals zero.
- For many chemical processes involving changes in temperature, phase, or chemical composition, kinetic and potential energy contributions are negligible compared to enthalpy contributions.
- The effect of pressure on enthalpy is usually much less than the effect of temperature, phase, or composition.

Step 8. Identify any heat or work flows into or out of the system, and write down equations for Q and W_s.
- Heat flows because of temperature differences between system and surroundings, but only if the system is not perfectly insulated.
- A piece of equipment with a shaft (pump, compressor, etc.) is a good indication that you need a W_s term.
- Pay close attention to the sign (positive or negative) of work and heat terms.

Step 9.   If system energy changes, write down equations for $E_{k,sys}$, $E_{p,sys}$, and U_{sys}.
- Use the same reference state as in step 7.
- Kinetic or potential energy changes if the system's velocity or elevation changes. Internal energy changes if the system's pressure, temperature, phase or composition changes.
- For many chemical processes, changes in kinetic and potential energy are negligible compared to changes in internal energy.
- The effect of pressure on internal energy is usually much less than the effect of temperature, phase, or composition.

Step 10.   Write the energy balance equation.
- Use the differential energy balance equation for a specific point in time, and the integral energy balance equation for a time interval.

Step 11.   Solve the equations.
- Don't forget the equations you developed for process flow calculations.

Step 12.   Check your solutions.
- Pay close attention to the sign (negative or positive) of calculated work and heat.

6.6.3   A Plethora of Problems

Each of the following problems illustrates different features of process energy calculations. We strongly recommend that, for each example, you try to solve the problem first by yourself before looking at the solution. If you have difficulty, study the solution, then cover it up and try to work the problem by yourself. If you still have trouble, be sure you follow the 12 Easy Steps and check the helpful hints. The examples progress generally from simple to more difficult.

Example 6.10

Potential Energy into Work: Water over the Dam

Water falls 1000 m from a tranquil mountain lake to a placid pool below. If the water falls at 1 kg/s and turns the blades of a turbine, how much work can be extracted?

Solution

Step 1.   Here's the diagram, drawn pictorially and schematically:

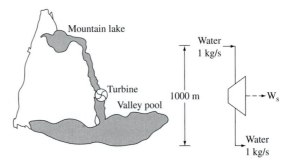

Step 2. Let's choose as a system everything that is in the falling stream, from the lake to the pool, and including the material passing through the turbine. This is an open system.

Step 3. There are only two material stream variables: $\dot{m}_{in}$ (water flow into the system), and $\dot{m}_{out}$ (water flow out of the system). The specific energies associated with these streams are $\hat{E}_{k,in}$, $\hat{E}_{k,out}$, $\hat{E}_{p,in}$, $\hat{E}_{p,out}$, $\hat{H}_{in}$, and $\hat{H}_{out}$. Additional energy stream variables are $\dot{W}_s$ (shaft work of turbine) and $\dot{Q}$ (heat flow between system and surroundings)

Steps 4 and 5.   We'll work in SI units (kg, kJ, s) throughout; all information is already given in these units. The basis is $\dot{m}_{in} = 1$ kg water/s.

Step 6. The material balance equation is very simple: $\dot{m}_{out} = \dot{m}_{in} = 1$ kg water/s.

Step 7. The reference state is chosen to be the water in the valley pool. Because the outlet stream is at the reference state, $\hat{E}_{k,out} = \hat{E}_{p,out} = \hat{H}_{out} = 0$.

The water in the mountain lake is quiescent (zero velocity), just like that of the valley pool, or

$$\hat{E}_{k,in} = \hat{E}_{k,out} = 0$$

The water is pure and liquid, so the composition and phase at the inlet and outlet are the same. Since the water is exposed to the air at both the inlet and outlet, it's also fair to say that the temperature and pressure at the inlet and outlet are the same. This means that the specific enthalpy of the water at the inlet and the outlet is the same, or

$$\hat{H}_{in} = \hat{H}_{out} = 0$$

The potential energy of the water in the mountain lake, with the valley pool as reference ($\hat{E}_{p,out} = 0$), is

$$\hat{E}_{p,in} = gh = (9.8066 \text{ m/s}^2)(1000 \text{ m})\left(\frac{1 \text{ kJ}}{1000 \text{ kg m}^2/\text{s}^2}\right) = 9.8 \text{ kJ/kg}$$

Step 8. The air and water temperatures are roughly the same (do you think this is a reasonable approximation?), so without any temperature difference between system and surroundings,

$$\dot{Q} = 0$$

There is only one work term, $\dot{W}_s$, for the turbine. This is what we want to solve for.

Steps 9 to 12.   This is a continuous-flow steady-state process with only one input and one output stream, no difference in enthalpy or kinetic energy between input and output, and no heat transfer. The energy balance equation reduces to:

$$\dot{m}_{out} \hat{E}_{p,out} - \dot{m}_{in} \hat{E}_{p,in} = \dot{W}_s$$

Helpful Hint
If system and surroundings are at the same *T*, then $\dot{Q}$ must equal zero.

Inserting the numerical values from steps 6 and 7, we find this further simplifies to

$$-9.8 \text{ kJ/s} = \dot{W}_s$$

Since work is extracted from the system by the turbine, the negative sign makes sense. This is about enough power to operate a couple of electric push lawn mowers.

Example 6.11

Integral Energy Balance with a Closed System: Unplugging the Frozen Pipes

You're back home in Calgary after a wonderful winter vacation in Cancun and find that the water in your pipes is frozen solid. You estimate that there is a 2-lb chunk of ice in one of the pipes. Is it feasible to try to melt the ice chunk with a hairdryer?

Solution

Steps 1 to 3. The system is the chunk of ice. This is a closed system, with no material entering or leaving the system, so no material stream variables are required. There is one energy flow into the system from the hair dryer: the only stream variable needed is $\dot{Q}$.

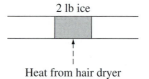

2 lb ice

Heat from hair dryer

Steps 4 to 8. The mass of the system is $m_{sys} = 2$ lb. We check our hair dryer—it's rated at 1.5 kW, or, converting to British units, about 5000 Btu/h. Since heat is added from the surroundings to the system, $\dot{Q} = +5000$ Btu/h. There is no work term.

Step 9. There is no change in kinetic or potential energy of the ice chunk since its position and velocity do not change. There is a change in phase, so the internal energy of the system changes. We aren't given any information about the ice temperature or pressure, but, since changes in phase often contribute more to changes in internal energy than do small changes in temperature, we'll just calculate the change in internal energy as though the only change in the state of the system is the change in phase. From the data in App. B we find that for water $\Delta \hat{H}_m = 6.008$ kJ/gmol $= 144$ Btu/lb. Since $\Delta \hat{H}_m \approx \Delta \hat{U}_m$,

$$U_{sys,f} - U_{sys,0} = m_{sys} \, \Delta \hat{U}_m = 2 \text{ lb} \left(144 \frac{\text{Btu}}{\text{lb}} \right) = 288 \text{ Btu}$$

Steps 10 to 12. To determine if the hair dryer strategy is reasonable, we'll estimate the total time it takes to melt the ice chunk. Since we are interested in what happens over an interval of time, we'll choose the integral energy balance equation. Given our analysis, this equation simplifies to

$$U_{sys,f} - U_{sys,0} = Q$$

or

$$U_{sys,f} - U_{sys,0} = 288 = \int_{t_0}^{t_f} \dot{Q} \, dt = \int_0^{t_f} 5000 \, dt = 5000t_f$$

Solving, we find

$$t_f = \frac{288 \text{ Btu}}{5000 \text{ Btu/h}} = 0.058 \text{ h, or about 3.5 minutes}$$

Our answer puts a lower limit on the actual time required to melt the ice chunk, because we've neglected the time needed to heat the ice to its melting temperature or to heat up the pipe, and we've made the assumption that all of the heat from the hair dryer goes to melt the ice. Since the purpose of this calculation was simply to determine whether using the hair dryer to melt the ice was a reasonable strategy, these approximations are justifiable.

> **! Helpful Hint**
> Use the integral energy balance equation to analyze a process over a specified time interval.

Example 6.12   **Temperature Change with Dissolution: Caustic Tank Safety**

Anaerobic fermentation of sugars produces organic acids such as lactic and acetic acid. These acids reduce the pH of the fermentor to unacceptably low levels unless base is added. You are in charge of designing a small system to make an aqueous NaOH (sodium hydroxide, or caustic soda) solution, to be used in a fermentor to maintain pH control. In your design, solid NaOH pellets are added to a tank containing 10,000 g water at a rate of 40 g/min, over a 10-min interval. The tank is continually stirred with a 0.2-hp mixer. The water in the tank is initially at 50°C, while the NaOH pellets are initially at 25°C.

   If the tank temperature exceeds 70°C at any time, safety regulations require that the tank be insulated to protect workers from possible burns. Will you need to insulate the tank?

Solution

Steps 1 to 4.   Here's a diagram of the mixing tank.

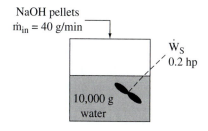

We'll choose the tank as our system and analyze as a semibatch process, with material entering and accumulating inside the tank, but no material leaving (at least, not during the mixing process). Initially, the tank contains 10,000 g water, so $m_{sys,0} = 10,000$. There is only a single material stream variable: $\dot{m}_{in} = 40$ g/min. We'll work with units of grams, joules, and minutes; the only unit conversion necessary is with the mixer: 0.2 hp = 8950 J/min.

Steps 5 and 6.   An integral material balance equation on this semi-batch system yields

$$m_{sys,f} = m_{sys,0} + \int_{t_0}^{t_f} \dot{m}_{in}\, dt = 10,000 + \int_0^{10} 40\, dt = 10,400 \text{ g solution}$$

Step 7.   For the reference state of the NaOH stream, we'll choose the pure solid at 25°C and 1 atm. Therefore, the inlet stream's specific enthalpy $\hat{H}_{NaOH(s)} = 0$. We neglect potential and kinetic energy contributions (because the inlet doesn't change its position and the velocity of the stream is unknown and probably small), so $\hat{E}_k \approx \hat{E}_p \approx 0$.

Step 8.   There are no heat inputs to the system. There is likely some heat loss through the tank walls (because the tank contents are warmer than the surroundings). But the maximum possible temperature will be reached if there is no heat loss. We'd like to solve for the worst-case scenario, so we state that $\dot{Q} = 0$. There is a constant work input from the mixer: $\dot{W}_s = +8950$ J/min, and

$$W_s = \int_{t_0}^{t_f} \dot{W}_s\, dt = 8950 \frac{J}{min} (10 \text{ min}) = 89,500 \text{ J} = 89.5 \text{ kJ}$$

Step 9.   Now we look at the mixing tank. The velocity and position don't change, so

$$E_{k,sys,f} - E_{k,sys,0} = E_{p,sys,f} - E_{p,sys,0} = 0$$

The internal energy of the system does change, because there is a change in temperature, composition, and mass of the system. To evaluate U_{sys}, we need reference states as well as data and/or model equations.

We've already chosen pure NaOH(s) at 25°C and 1 atm as one reference state; for water we'll choose the pure liquid at 25°C and 1 atm.

The temperature and the composition of the system change from t_0 to t_f. We need to find data and/or model equations to describe the change in U with a change in T and x_i; the necessary data are C_v of water and aqueous NaOH solution, and $\Delta\hat{H}_{soln}$ for dissolving NaOH in water. We head to our handbooks and dig out the following information:

1. The enthalpy of solution of NaOH depends on dilution. At the end of the process we are at a dilution of 55 gmol water/gmol NaOH (10,000 g water/400 g NaOH). For this dilution, $\Delta\hat{H}_{soln} = -966$ J/g NaOH, with a reference state of pure solid NaOH and pure liquid water at 25°C.

2. $C_v = 4.19$ J/g °C for water, and $C_v = 4.0$ J/g °C for an aqueous NaOH solution at 1 gmol NaOH/55 gmol water.

The last step is to evaluate $U_{sys,f}$ and $U_{sys,0}$. At $t = t_0$, the tank contains 10,000 g liquid water at 50°C and 1 atm. The only difference between this state and the reference state (pure liquid water at 25°C and 1 atm) is the temperature. From the mini-summary in Sec. 6.3 we know that with a change in temperature

$$\hat{U}_2 - \hat{U}_1 = C_v(T_2 - T_1)$$

Therefore

$$U_{sys,0} - U_{ref} = m_{sys,0}\left(\hat{U}_{sys,0} - \hat{U}_{ref}\right) = m_{sys,0}C_v(T_0 - T_{ref})$$

$$= 10{,}000 \text{ g}\left(4.19\frac{\text{J}}{\text{g°C}}\right)(50°C - 25°C)\left(\frac{1 \text{ kJ}}{1000 \text{ J}}\right) = 1050 \text{ kJ}$$

At time $t = t_f$, both the temperature and the composition of the system differ from the reference state. We construct a two-step pathway from the reference state to the final state:

1. Dissolve 400 g NaOH in 10,000 g water at 25°C and 1 atm.
2. Increase T of 10,400 g solution from 25°C to T_f (°C).

Changes in U for this pathway are calculated as

$$U_{sys,f} - U_{ref} = m_{NaOH}\Delta\hat{U}_{soln} + m_{sys,f}C_v(T_f - T_{ref})$$

$$= 400 \text{ g NaOH } (-966 \text{ J/g NaOH})$$

$$+ \; 10{,}400 \text{ g solution } (4.0 \text{ J/g °C})(T_f - 25°C)$$

$$= -386{,}400 + 41{,}600\,(T_f - 25°C) \text{ J}$$

$$= -386.4 + 41.6\,(T_f - 25°C) \text{ kJ}$$

! Helpful Hint
Choosing the state of one of the process streams as the reference state can simplify a problem, because then the energy of that stream equals zero.

Steps 10 to 12.   Since we are interested in the temperature change over a defined interval of time, we use the integral energy balance equation. We have already found that kinetic and potential energy terms are negligible, and that there is no heat input. By judicious choice of reference state, we've zeroed out any enthalpic contributions from the input stream. Therefore, the integral energy balance equation simplifies to

$$U_{sys,f} - U_{sys,0} = W_s$$

We substitute numerical values into the equation and solve for T_f:

$$-386.4 + 41.6\,(T_f - 25) - 1050 \text{ kJ} = 89.5 \text{ kJ}$$

$$T_f = 62°C$$

The tank temperature increases to 62°C, not enough to require protective insulation.

Example 6.13	**Simultaneous Energy and Material Balances: Mel and Dan's Lemonade Stand**

Mel and Dan run a lemonade stand. They've noticed that some customers like their lemonade barely cool, while others prefer their drinks ice-cold. Always looking for a new way to please their customers (and increase their business), Mel comes up with the brilliant idea of lemonade-to-order. Mel says he will ask customers for their temperature preference (cool: 60°F, cold: 45°F, or icy: 35°F), and then fine-tune the lemonade temperature by adding just the right number of ice cubes to the cup on the spot. Dan wonders how Mel can know how many ice cubes to add, but Mel claims to have come up with a simple mathematical formula that calculates the number of ice cubes to add, given the size of the lemonade (8 oz., 12 oz., or 16 oz.) and the desired temperature.

The lemonade at Mel and Dan's, before adding the ice cubes, is the same as the air temperature, which tends to be about 85°F. The ice cubes are 1 in³ in volume and are kept just below their freezing point temperature. The lemonade is sold to the customer in insulated cups. Always quality-conscious, Mel refuses to sell the customer any cup whose contents are not thermally equilibrated. Dan finds the following notes scribbled on a napkin: $\Delta \hat{H}^{\circ}_{m,\text{ice}} = 330$ J/g $\approx \Delta \hat{U}^{\circ}_{m,\text{ice}}$, $C_{p,\text{water}} \approx C_{v,\text{water}} = 2.3$ J/g °F, density of water $= 1$ g/cm³ $= 29.6$ g/fl. oz. $= 16.4$ g/in³.

Mel won't divulge his secret formula—can you figure it out?

Solution

Steps 1 to 5.   Our system is the contents of the cup. We'll model this system as a semibatch process: the system initially contains lemonade but no ice, over some defined time interval ice is added, and we are interested in the state of the system after the addition of the ice. We'll use units of g for mass, J for energy, °F for temperature. Let $N =$ the number of ice cubes, and $m_{\text{ice}} =$ the mass of each cube. Each ice cube is 1 in³ in volume; the density of ice is about 1.0 g/cm³ or about 16.4 g/in³, so $m_{\text{ice}} = 16.4$ g/cube. The total mass of ice added to the cup $= Nm_{ice} = 16.4N$. Three different size cups are sold at Mel and Dan's; we'll denote the initial mass of lemonade in the cup as $m_{sys,0}$. If we assume that the density of lemonade is the same as that of water, then

$$m_{sys,0} = \frac{29.6 \text{ g}}{\text{fl. oz.}} \times V_{cup} = 29.6 \, V_{cup}$$

where V_{cup} is the volume of the lemonade in the cup, in fluid ounces (fl. oz.).

!

Helpful Hint

You sometimes need the energy balance equation in order to solve the material balance equation.

Step 6.   From an integral material balance on this semibatch process we find

$$m_{sys,f} - m_{sys,0} = \int_{t_0}^{t_f} \dot{m} \, dt$$

$$m_{sys,f} = m_{sys,0} + Nm_{ice} = 29.6 V_{cup} + 16.4N$$

We can't go any further with the material balance, so we bravely press on.

Steps 7 and 8.   Let's choose as reference state pure liquid water at its melting temperature of 32°F. We'll model the lemonade as pure water, because we don't have any data on the thermophysical properties of lemonade. There are no kinetic or potential energy contributions to speak of. The only difference between the state of the inlet stream and that of the reference is the phase—we change from liquid to solid as we move from the reference to the stream state. Therefore, the specific enthalpy of the ice added to the cup is $\hat{H}_{ice,in} = -\Delta \hat{H}°_{m,ice} = -330$ J/g. Since the cup is insulated, $Q = 0$. There's no mechanical equipment, so $W_s = 0$.

Step 9.   The internal energy of the system changes over time. Initially, the cup contents are liquid lemonade at 85°F and 1 atm. Given that the reference state (where $\hat{U}_{ref} = 0$) is pure liquid water at 32°F and 1 atm, the only difference between $\hat{U}_{sys,0}$ and $\hat{U}_{ref}$ is due to the difference in temperature:

$$\hat{U}_{sys,0} = \hat{U}_{sys,0} - \hat{U}_{ref} = C_v(T_0 - T_{ref}) = 2.3 \, \frac{J}{g \, °F} (85°F - 32°F) = 122 \frac{J}{g}$$

At the end, the contents are cool: 60°F, cold: 45°F, or icy: 35°F. Mel waits until the system is thermally equilibrated; since the temperatures are all above the melting temperature, all the added ice is melted. Thus, relative to the chosen reference state (liquid water, 32°F, 1 atm), we need to consider only the effect of temperature on the specific internal energy of the final state:

$$\hat{U}_{sys,f} = \hat{U}_{sys,f} - \hat{U}_{ref} = C_v(T_f - T_{ref}) = 2.3 \, \frac{J}{g \, °F} (T_f - 32°F)$$

where $T_f = 60°F$, 45°F or 35°F, depending on our product.

Steps 10 to 12. We are interested in a specific interval of time, so we select the integral energy balance equation. For this situation, the equation simplifies to

$$U_{sys,f} - U_{sys,0} = \int_{t_0}^{t_f} \dot{m}_{in} \hat{H}_{in} \, dt$$

or

$$m_{sys,f} \hat{U}_{sys,f} - m_{sys,0} \hat{U}_{sys,0} = N m_{ice} \hat{H}_{ice,in}$$

Substituting in for known values, and using our results from material balance calculations, we get

$$(29.6 V_{cup} + 16.4 N)[2.3(T_f - 32)] - 29.6 V_{cup}(122) = 16.4 N(-330)$$

Rearranging, we get an expression for the number of ice cubes N to be added as a function of the size of the lemonade, V_{cup}, and the desired final temperature T_f:

$$N = \frac{3611.2 - 88.8(T_f - 32)}{5412 - 37.72(T_f - 32)} V_{cup}$$

We plot N as a function of T_f for 8-oz., 12-oz., and 16-oz. cups. Each solid line indicates a different size and each dotted line indicates a different temperature. We'd need to add only 2 ice cubes for a cool 8-oz. cup, but 10 ice cubes for an icy 16-oz. serving.

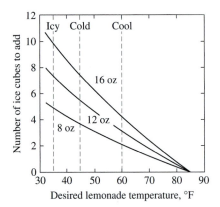

Notice that zero ice cubes are required for lemonade at 85°F—this is a good check of our solution. Also notice that the number of ice cubes is directly proportional to the lemonade volume, which makes sense. One other item we might want to check is this: What percent increase is there in the volume of liquid in the cup after the ice has melted?

Should Mel and Dan worry that customers will complain about thin and watery lemonade? Answering this question is left for the marketing experts.

Example 6.14   ### Energy Balance with Equilibrium Flash: Separation of Hexane and Heptane

A liquid mixture of 40 mol% n-hexane (C_6H_{14})/60 mol% n-heptane (C_7H_{16}) at 25°C is pumped to a flash drum operating at 1520 mmHg pressure. The feed rate is 100 gmol/s. The mixture is heated to 111°C, and vapor and liquid products are removed continuously. How much heat must be added to the flash drum? Data:

$$C_p = 143 \text{ J/gmol °C for hexane vapor and 189 J/gmol °C for liquid hexane}$$

$$C_v = 166 \text{ J/gmol °C for heptane vapor and 212 J/gmol °C for liquid heptane}$$

$$\Delta \hat{H}_{vap} = 28{,}900 \text{ J/gmol at 69°C for hexane and 31,800 at 98.4°C for heptane}$$

Solution

This problem might seem familiar—you completed the process flow calculations for this equilibrium flash in Example 5.16!

Steps 1 to 6.   We'll use the same flow diagram, stream variables, units, and basis as we did for Example 5.16, except we will show the heat flow into the drum on the diagram. We'll use J for energy units. The material balance calculations were completed in Example 5.16; we'll show those results here. (Now is a good time to review how those calculations were done.)

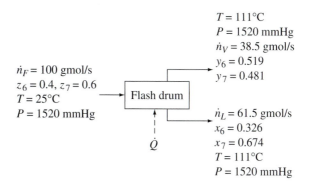

Steps 7 to 9.   The flash drum is an open system, with heat but no work inputs. Since the drum operates at steady state there is no change in system energy. There are changes in temperature, phase, and composition between input and output material flows that must be accounted for in evaluating the enthalpies of the input and output streams; any kinetic or potential energy differences between input and output streams are negligible.

Let's evaluate the enthalpies of each stream given a reference state of pure hexane and pure heptane at $T = 25°C$, $P = 1520$ mmHg and $\phi = $ liquid.

The feed stream is at the reference T, P, and ϕ, but it is a mixture rather than pure components. Since hexane and heptane are chemically similar, we can safely say that $\Delta \hat{H}_{mix} = 0$. Therefore

$$\hat{H}_F = \hat{H}_{ref} = 0$$

The liquid stream leaving the flash drum is at the same P and ϕ as the reference state but at a different T. Therefore,

$$\hat{H}_L - \hat{H}_{ref} = \hat{H}_L = \sum x_i \int_{T_{ref}}^{T} C_{pi} \, dT = [0.326(189)$$
$$+ 0.674(212)](111 - 25) = 17{,}590 \text{ J/gmol}$$

This is the enthalpy of the liquid mixture; we again assume $\Delta \hat{H}_{mix} = 0$.

The vapor stream leaving the flash drum is at the same P as the reference state but at a different T and in a different phase. We have enthalpy of vaporization data at the normal boiling points T_b of hexane and heptane. To use this data, we construct the following path from reference state to vapor state:

1. Increase T of components from 25°C to T_b at 1520 mmHg and in liquid phase
2. Change phase from liquid to vapor at T_b and 1520 mmHg
3. Increase T of components from T_b to 111°C at 1520 mmHg and in vapor phase
4. Mix components (assumed negligible)

Notice once again that this is not the path that the components actually take—there is not a sudden change from liquid to vapor phase at the normal boiling point when the components are in a mixture and at a higher pressure! But, since enthalpy is a state function, our imagined pathway is fine as long as the start and end states are correct.

We'll calculate the enthalpy change for each component first, then mix:

$$\hat{H}_{6,V} = (189)(69 - 25) + 28900 + (143)(111 - 69) = 43{,}220 \text{ J/gmol}$$
$$\hat{H}_{7,V} = (212)(98.4 - 25) + 31800 + (166)(111 - 98.4) = 49{,}450 \text{ J/gmol}$$
$$\hat{H}_V = 0.519(43{,}220) + 0.481(49{,}450) = 46{,}220 \text{ J/gmol}$$

($\hat{H}_{6,V}$ is J/gmol hexane, $\hat{H}_{7,V}$ is J/gmol heptane, $\hat{H}_V$ is J/gmol of the mixture.)

Steps 10 to 12.   The steady-state differential energy balance equation simplifies to

$$\dot{n}_V H_V + \dot{n}_L H_L - \dot{n}_F H_F = \dot{Q}$$

Inserting the values we've calculated gives

$$38.5(46{,}220) + 61.5(17{,}590) - 100(0) = 2{,}861{,}000 \text{ J/s} = 2861 \text{ kJ/s} = \dot{Q}$$

The answer should be positive, as heat is added to the tank.

| **Example 6.15** | **Energy Balance with Chemical Reaction: Adiabatic Flame Temperature** |

Liquid sulfur at 130°C and pure oxygen at 25°C are mixed at stoichiometric ratio and fed to a burner, where complete oxidation of sulfur to SO_2 takes place. The burner is perfectly insulated, the pressure is 1 atm, and the system operates at steady state. What is the temperature of the exiting stream? Data:

$$\Delta \hat{H}_r^\circ = -296.8 \text{ kJ/gmol for the reaction S (s)} + O_2(g) \rightarrow SO_2(g)$$

$$\Delta \hat{H}_m = 1.73 \text{ kJ/gmol for S at } 114°C$$

$$C_p = 32 \text{ J/gmol °C for S (l)}$$

$$C_p = 23.2 \text{ J/gmol °C for S (s)}$$

$$C_p = 39.9 \text{ J/gmol °C for } SO_2(g)$$

$$C_p = 29.3 \text{ J/gmol °C for } O_2(g)$$

Solution

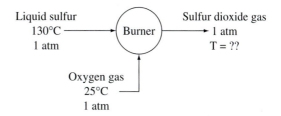

Steps 1 to 6.   We'll choose the burner as the system. We'll work in units of J, gmol, and °C. No basis is given, so we'll arbitrarily choose one:

$$\dot{n}_{S,in} = 1 \frac{\text{gmol}}{\text{s}}$$

Since the sulfur and oxygen are fed at stoichiometric ratio:

$$\dot{n}_{O_2,in} = 1 \frac{\text{gmol}}{\text{s}}$$

Given the reaction stoichiometry and that 100% of the S fed is converted to product:

$$\dot{\xi} = 1 \frac{\text{gmol}}{\text{s}}$$

From a material balance:

$$\dot{n}_{SO_2,out} = \dot{\xi} = 1 \frac{\text{gmol}}{\text{s}}$$

Step 7.   Now we consider energy flows due to material flow in and out of the system. There is negligible change in position or velocity, so no kinetic or potential energy terms need to be considered. The enthalpy of the two entering streams and the one leaving stream must be evaluated. Let's choose the pure elements at 25°C, 1 atm, and in their normal phase and state of aggregation as the reference state: S (s) and O_2 (g). Since the feed oxygen gas is at the temperature, pressure, phase, and composition of the reference state,

$$\hat{H}_{O_2,in} = 0$$

The sulfur feed stream is at the pressure (1 atm) and composition (pure S) of the reference state, but not its temperature or phase. Thus we must account for changing both the phase and the temperature of this stream compared to the reference state. We are given data on the enthalpy change with melting at 114°C. Therefore, we construct a path from the reference state to the actual state of the inlet stream:

1. Heat sulfur from 25°C to 114°C at constant ϕ (solid) and P (1 atm)
2. Melt sulfur from solid to liquid at constant T (114°C) and P (1 atm)
3. Heat sulfur from 114°C to 130°C at constant ϕ (liquid) and P (1 atm)

The calculation associated with this pathway is

$$\hat{H}_{in} - \hat{H}_{ref} = \left(23.2 \frac{J}{gmol \ °C}\right)(114 - 25°C) + 1730 \frac{J}{gmol}$$

$$+ \left(32\frac{J}{gmol \ °C}\right)(130 - 114°C)$$

$$\hat{H}_{in} = 2065 + 1730 + 512 = 4307 \ J/gmol$$

Finally, we calculate $\hat{H}_{SO_2,out}$ at 1 atm pressure and unknown temperature T_{out}. The reference state must be the same as that for the inlet stream: the pure elements at 25°C, 1 atm, and normal phase and state of aggregation. We need to construct a path from the reference state to the actual state of the exiting stream. Given that we have the enthalpy of reaction at 25°C, a workable path includes the following steps:

1. React 1 gmol S(s) and 1 gmol O_2(g) at 25°C and 1 atm to 1 gmol SO_2 (g)
3. Increase the temperature of 1 gmol SO_2 (g) from 25°C to T_{out}

We have to take extra care because we have a change in moles in the reaction.

$$H_{out,SO_2} - H_{ref} = (\Delta \hat{H}_r) \ \dot{\xi} + C_{p,SO_2(g)} (T_{out} - T_{ref}) \dot{n}_{SO_2,out}$$

$$H_{out,SO_2} = -296,800 + 39.9 (T_{out} - 25°C) \ J/s$$

Steps 8 and 9.   The burner is perfectly insulated, so $\dot{Q} = 0$ and there are no pumps, compressors, or turbines, so $\dot{W}_s = 0$.

Steps 10 to 12.   Since we are interested in the behavior of a steady-state continuous-flow system, we use the differential energy balance equation. The energy balance equation reduces to a deceptively simple equation:

$$\sum_{\text{all } j \text{ out}} \dot{n}_j \hat{H}_j = \sum_{\text{all } j \text{ in}} \dot{n}_j \hat{H}_j$$

Quick Quiz 6.11

If air rather than O_2 was fed to the burner of Example 6.15, would the exit gas temperature be higher, lower, or the same?

or

$$\dot{n}_{SO_2,\text{out}} \hat{H}_{SO_2,\text{out}} = \dot{n}_{O_2,\text{in}} \hat{H}_{O_2,\text{in}} + \dot{n}_{S,\text{in}} \hat{H}_{S,\text{in}}$$

We substitute in our numerical evaluations to find

$$-296{,}800 + 39.9\,(T_{\text{out}} - 25) = 4307$$

Solving, we find that $T_{\text{out}} > 7500°C$! Ouch!!

Example 6.15 is an example of an *adiabatic flame temperature* calculation. This is the temperature reached for combustion of any material under adiabatic conditions—with no heat loss to the surroundings. This is highly unrealistic, because there will be heat loss at a flame by radiation, conduction, and convection. However, the calculation is still useful because it gives an upper limit on the achievable temperature.

Example 6.16	**Energy Balances with Multiple Reactions: Synthesis of Acetaldehyde**

We want to design and build a reactor that produces 1200 kgmol acetaldehyde (CH_3CHO) per hour from ethanol and air. Two reactions are important at the reactor conditions; (R1) produces our desired product while (R3) produces an acetic acid byproduct.

$$C_2H_5OH + 0.5O_2 \rightarrow CH_3CHO + H_2O \tag{R1}$$

$$2CH_3CHO + O_2 \rightarrow 2CH_3COOH \tag{R3}$$

Laboratory data indicate that if we use a new catalyst, adjust the feed ratio to 5.7 moles ethanol per mole oxygen, and operate the reactor at 300°C and 1.5 atm pressure, we can expect to achieve 25% conversion of ethanol in the reactor, with a selectivity for acetaldehyde of 0.6. The preliminary block flow diagram calls for mixing pure ethanol vapor and air (79 mol% N_2, 21 mol% O_2) at 300°C and 1.5 atm and feeding the mixture to the reactor. The reactor effluent is also at 1.5 atm and all vapor. To maintain a constant reactor temperature of 300°C, will heat need to be added or removed? If so, state how much.

Solution

Steps 1 to 6:   In Example 4.13 we calculated the extents of reaction and reactor flow rates for this design. Let's draw our flow diagram, showing the results of the process flow calculations and adding the possibility of a heat flow in or out of the reactor.

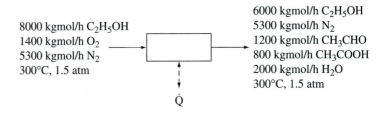

8000 kgmol/h C₂H₅OH
1400 kgmol/h O₂
5300 kgmol/h N₂
300°C, 1.5 atm

6000 kgmol/h C₂H₅OH
5300 kgmol/h N₂
1200 kgmol/h CH₃CHO
800 kgmol/h CH₃COOH
2000 kgmol/h H₂O
300°C, 1.5 atm

$\dot{Q}$

In Example 4.13, we calculated that $\dot{\xi}_1 = 2000$ kgmol/h and $\dot{\xi}_3 = 400$ kgmol/h. These values come in handy in our energy calculations. We'll work with kJ for energy units.

Steps 7 to 9:   The reactor operates at steady state, so there are no changes in the system energy. There is a heat term but no work. Kinetic and potential energy differences between inlet and outlet streams are negligible. T, P, and ϕ of the input and output streams are identical; the only change is in composition due to reaction. The reactions actually occur at 300°C but App. B has $\Delta \hat{H}^\circ_f$ data only at 25°C. Therefore we construct an imaginary pathway:

1. Decrease pressure of reactor inlet stream from 1.5 atm to 1 atm at 300°C and in vapor phase
2. Cool reactants from 300°C to 25°C at 1 atm in the vapor phase
3. React at 25°C, 1 atm, in vapor phase
4. Increase temperature of reactor effluent from 25°C to 300°C at 1 atm in vapor phase
5. Increase pressure of reactor effluent from 1 to 1.5 atm at 300°C and in vapor phase.

We now evaluate the enthalpy change associated with each step in the path. The change in enthalpy due to change in pressure is zero for ideal gases, so we need to calculate only the temperature and reaction effects. Since the nitrogen does not react (and it does not change T, P, or ϕ), no calculations involving nitrogen are required.

It's time to look up some data in App. B and in handbooks. C_p and $\Delta \hat{H}^\circ_f$ are

	C_p, J/gmol °C (approx.)	$\Delta \hat{H}^\circ_f$, kJ/gmol
O₂ (g)	29.3	0
C₂H₅OH (g)	65.5	−234.95
CH₃CHO (g)	54.7	−166.2
CH₃COOH (g)	66.5	−432.8
H₂O (g)	33.6	−241.83

We calculate the standard enthalpy of reaction for (R1) and (R3):

$$\Delta \hat{H}^\circ_{r1} = \sum \nu_{i1} \Delta \hat{H}^\circ_{f,i} = (-1)(-234.95) + (-0.5)(0)$$

$$+ (1)(-166.2) + (1)(-241.83) = -173.1 \text{ kJ/gmol}$$

$$\Delta \hat{H}^\circ_{r3} = \sum \nu_{i3} \Delta \hat{H}^\circ_{f,i} = (-2)(-166.2) + (-1)(0)$$

$$+ (2)(-432.8) = -533.2 \text{ kJ/gmol}$$

Since both reactions are oxidations, it is reasonable that the reactions are both exothermic.

The enthalpy pathway we constructed is expressed mathematically as

> ! **Helpful Hint**
> Multiply $\Delta \hat{H}_r$ by the extent of reaction $\dot{\xi}$ to get the total enthalpy change due to reaction.

$$\sum_{\text{input}} \dot{n}_i C_{pi}(T_{\text{ref}} - T_{\text{reactor}}) + \dot{\xi}_1 \Delta \hat{H}^\circ_{r1} + \dot{\xi}_3 \Delta \hat{H}^\circ_{r3} + \sum_{\text{output}} \dot{n}_i C_{pi}(T_{\text{reactor}} - T_{\text{ref}})$$

Substituting in numerical values (being careful to convert C_p from J to kJ) gives

$$\dot{H}_{\text{out}} - \dot{H}_{\text{in}} = [8000(66.5) + 1400(29.3)](300 - 25)\left(\frac{1}{1000}\right)$$

$$+ 2000(-173.1) + 400(-533.2) + [6000(65.5) + 1200(54.7)$$

$$+ 800(66.5) + 2000(33.6)](25 - 300)\left(\frac{1}{1000}\right)$$

$$\dot{H}_{\text{out}} - \dot{H}_{\text{in}} = 157{,}580 - 559{,}480 - 159{,}240 = -561{,}140 \text{ kJ/h}$$

Steps 10 to 12:   The steady-state differential energy balance equation is

$$\dot{H}_{\text{out}} - \dot{H}_{\text{in}} = \dot{Q} = -561{,}140 \text{ kJ/h}$$

561,000 kJ/h heat must be removed to maintain the reactor temperature at 300°C.

Example 6.17   **Unsteady-State Heat Loss: Cooling a Batch of Sterilized Broth**

A 10,000-L batch of fermentation broth has been sterilized by holding it under pressure at 121°C for 26 min. Now it must cool to 37°C before it can be used. The batch is cooled by placing it in a large room where the temperature is set at 35°C. The rate of cooling $\dot{Q}$ is proportional to the temperature difference between the broth and the room:

$$\dot{Q} = 3760(T - T_{\text{room}})$$

where T is given in °C and $\dot{Q}$ is in kJ/min. What is the total amount of heat that must be removed? Plot the temperature of the broth as a function of time. How long will it take to cool the broth? Assume that the density of the broth is 1.05 kg/liter and its heat capacity $C_v = 4$ kJ/kg °C.

Solution

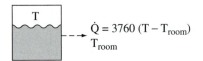

Steps 1 to 6:   The system is simply the vat of broth. We'll work in units of kg for mass, kJ for energy, °C for temperature, and minutes for time. The mass of the system doesn't change with time, so $m_{sys,0} = m_{sys,f} = m_{sys}$ = 10,000 L (1.05 kg/L) = 10,500 kg. This quantity of broth serves as our basis.

Steps 7 to 9:   There are no material streams in or out of the system, so no evaluation of stream energies is required. There are no pieces of mechanical equipment, so $\dot{W}_s = 0$. The kinetic and potential energies of the system do not change over time, but the internal energy of the system does change because of the change in temperature. We will pick as the reference state the liquid broth at room temperature (35°C). (Any other temperature is equally valid as a reference temperature. Other good choices include the desired temperature, 37°C, or the initial temperature, 121°C). With this as a reference state, the initial state (hot broth) differs from the reference state only in temperature, so

$$\hat{U}_{sys,0} - \hat{U}_{ref} = \hat{U}_{sys,0} = C_v(T_0 - T_{ref}) = 4\,\frac{kJ}{kg\,°C}\,(121°C - 35°C) = 344\,\frac{kJ}{kg}$$

The final state (the cooled broth at 37°C) differs from the reference state only in temperature, so

$$\hat{U}_{sys,f} - \hat{U}_{ref} = \hat{U}_{sys,f} = C_v(T_f - T_{ref}) = 4\,\frac{kJ}{kg\,°C}\,(37°C - 35°C) = 8\,\frac{kJ}{kg}$$

Steps 10 to 12:   First we are interested in determining the total heat removed over a specific interval of time. Therefore we want an integral energy balance equation. This equation simplifies quite readily down to

$$U_{sys,f} - U_{sys,0} = m_{sys}(\hat{U}_{sys,f} - \hat{U}_{sys,0}) = \int \dot{Q}\,dt = Q$$

Now substituting in information we have gathered, we get

$$(10{,}500\text{ kg})\left(8\frac{kJ}{kg} - 344\frac{kJ}{kg}\right) = Q = -3{,}528{,}000\text{ kJ}$$

We've answered the first question, the total amount of heat to be removed. (The negative sign indicates heat transfer from system to surroundings.) The second question asks us to determine the temperature T at a specific time t. In other words, we need an equation that relates T to t. We return to the differential energy balance equation. Because there is no material flow in and out, no work, and no kinetic or potential energy changes, the equation simplifies to:

$$\frac{dU_{sys}}{dt} = \dot{Q}$$

Since the *only* property that changes between the reference state and the actual state is the temperature (e.g., pressure, phase, composition stay the same), then at any time

$$\hat{U}_{sys} - \hat{U}_{ref} = \hat{U}_{sys} = C_v(T - T_{ref})$$

and

$$\frac{dU_{sys}}{dt} = \frac{d}{dt}[m_{sys}C_v(T - T_{ref})] = m_{sys}C_v\frac{dT}{dt}$$

(since m_{sys}, C_v, and T_{ref} are independent of t). Substituting this equation along with the equation for $\dot{Q}$ into the differential energy balance equation produces

$$m_{sys}C_v\frac{dT}{dt} = -3760\,(T - T_{room})$$

Substituting in known numerical values for m_{sys}, C_v, and T and rearranging gives

$$\frac{dT}{T - 35} = -\frac{3760}{(10,500)(4)}\,dt = -0.0895\,dt$$

Now we integrate both sides from initial conditions ($t = 0$, $T = T_0 = 121°C$) to any later time t and temperature T:

$$\int_{121°C}^{T} \frac{dT}{T - 35°C} = -\int_0^t 0.0895\,dt$$

$$\ln\left[\frac{T - 35}{121 - 35}\right] = -0.0895t$$

$$T = 35 + 86e^{-0.0895t}$$

where T is in °C and t is in minutes.

We plot the function:

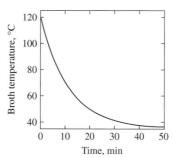

The temperature drops rapidly at first, but then the rate of cooling slows down as the broth temperature approaches the room temperature.

From the equation we derived, we find that $T = 37°C$ at $t = 42$ minutes. It takes longer to cool down the broth than to sterilize. How could you shorten this time?

6.7   A Process Energy Sampler (Optional Section)

There are so many important and interesting applications for process energy calculations that we cannot possibly cover all topics here. Instead, we provide a sampler of topics, to illustrate the power and variety of process energy calculations. Each of the topics is independent, so you may select only one or a few to read, and you may read them in any order. In each example problem we will not explicitly use the 12 Easy Steps, but they are implicit in our solution strategies.

6.7.1   Work and the Engineering Bernoulli Equation

Suppose your job is to purchase a pump. The pump's job is to move a liquid solution at a specified flow rate through a pipe and up to a tank on a hill. The tank is under pressure. How do you calculate the size of the motor required? This and similar questions are answered with a variation of the energy balance called the engineering Bernoulli equation.

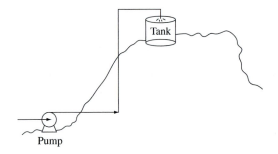

We choose as our system the pump and all the piping from the inlet of the pump to the outlet at the tank, as shown in the sketch. We write the differential energy balance equation, assuming steady-state operation:

$$0 = (\dot{E}_{k,\text{in}} - \dot{E}_{k,\text{out}}) + (\dot{E}_{p,\text{in}} - \dot{E}_{p,\text{out}}) + (\dot{H}_{\text{in}} - \dot{H}_{\text{out}}) + \dot{W}_s + \dot{Q}$$

The mass flow in equals the mass flow out at steady state, $\dot{m}_{\text{in}} = \dot{m}_{\text{out}} = \dot{m}$. By substituting in $\dot{E}_k = \frac{1}{2}\dot{m}v^2$ and $\dot{E}_p = \dot{m}gh$ and rearranging slightly, we find

$$\dot{W}_s = \frac{1}{2}\dot{m}(v_{\text{out}}^2 - v_{\text{in}}^2) + \dot{m}g(h_{\text{out}} - h_{\text{in}}) + \dot{m}(\hat{H}_{\text{out}} - \hat{H}_{\text{in}}) - \dot{Q}$$

What about $\hat{H}_{\text{out}} - \hat{H}_{\text{in}}$? The pressure of the liquid at the inlet is different than the pressure at the outlet, so there will be an enthalpy change due to the change in pressure, or $\hat{H}_{out} - \hat{H}_{in} = \hat{V}(P_{out} - P_{in})$. The phase and composition of the fluid have not changed. Does the temperature change? Is there a heat term?

To consider these questions, we must realize that as fluid flows through pipes, around corners and through valves, some energy is dissipated as friction. (This is sometimes called viscous dissipation.) If the pipe is well-insulated, this could result in an increase in the fluid temperature. (Similarly, your hands warm up if you rub them together rapidly.) If the pipe is poorly insulated, the energy dissipated as friction is lost as heat to the surroundings. The extent of dissipation depends on factors such as the pipe diameter, the roughness of the pipe surface, the number of elbows, the number and kind of valves, the viscosity of the fluid, and the fluid velocity. We lump all of the energy dissipated into a single term $\dot{F}_s$, which includes the energy that goes to heat up the fluid and the energy lost to the surroundings. $\dot{F}_s$ equals the rate of energy dissipation due to friction and will always increase the shaft work required.

With these considerations, the energy balance equation becomes:

$$\dot{W}_s = \frac{1}{2}\dot{m}\left(v_{out}^2 - v_{in}^2\right) + \dot{m}g\left(h_{out} - h_{in}\right) + \dot{m}\hat{V}\left(P_{out} - P_{in}\right) + \dot{F}_s \quad (6.22)$$

Equation (6.22) is sometimes called the engineering Bernoulli equation or the mechanical energy balance equation. It is useful for calculating work requirements at pumps. Because pumps are not 100% efficient–not all the power output of the motor is converted to shaft work detected by the fluid–a correction for pump efficiency η is made:

$$\dot{W}_s = \eta\dot{W}_{pump} \quad (6.23)$$

Pump efficiency η is typically approximately 0.5 to 0.8.

Example 6.18   **The Engineering Bernoulli Equation: Sizing a Pump**

You need to purchase a pump to move 69 gal/min of a liquid solution with a density of 115 lb/ft^3. At the pump inlet, the pipe diameter is 3 in and the liquid pressure is 18 psia. At the pump outlet, the pipe diameter is 2 in. The liquid is pumped 50 ft uphill and discharged into a tank under 40 psia nitrogen pressure. Suppose the pump efficiency is 65%, and the frictional loss is calculated to be 10.0 ft lb$_f$/lb. What pump size (in horsepower) should you purchase?

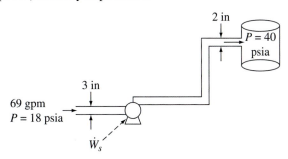

Solution

The engineering Bernoulli equation is applicable. We simply need to calculate each term in the equation in consistent units, substitute into the equation, and solve.

First we convert the volumetric flow rate to a mass flow rate:

$$\dot{m} = 69\frac{\text{gal}}{\text{min}} \times \frac{1 \text{ ft}^3}{7.4805 \text{ gal}} \times \frac{115 \text{ lb}}{\text{ft}^3} \times \frac{\text{min}}{60 \text{ s}} = 17.7 \frac{\text{lb}}{\text{s}}$$

Let's calculate the velocities.

$$v_{\text{in}} = \frac{\text{volumetric flow rate}}{\text{pipe cross-sectional area}} = \frac{69\frac{\text{gal}}{\text{min}} \times \frac{1 \text{ ft}^3}{7.4805 \text{ gal}} \times \frac{\text{min}}{60 \text{ s}}}{\pi(1.5 \text{ in})^2 \times \frac{1 \text{ ft}^2}{144 \text{ in}^2}} = 3.13 \frac{\text{ft}}{\text{s}}$$

Similarly, $v_{\text{out}} = 7.05$ ft/s.
Therefore

$$\dot{E}_{k,\text{out}} - \dot{E}_{k,\text{in}} = \frac{\dot{m}}{2}(v_{\text{out}}^2 - v_{\text{in}}^2)$$

$$= \frac{17.7 \text{ lb/s}}{2}\left(\frac{1 \text{ lb}_\text{f}}{32.174 \text{ lb ft/s}^2}\right)\left[\left(7.05\frac{\text{ft}}{\text{s}}\right)^2 - \left(3.13\frac{\text{ft}}{\text{s}}\right)^2\right] = 11 \frac{\text{ft lb}_\text{f}}{\text{s}}$$

For potential energy:

$$\dot{E}_{p,\text{out}} - \dot{E}_{p,\text{in}} = \dot{m}g(h_{\text{out}} - h_{\text{in}}) = 17.7\frac{\text{lb}}{\text{s}}\left(\frac{1 \text{ lb}_\text{f}}{\text{lb}}\right)(50 \text{ ft} - 0 \text{ ft}) = 885 \frac{\text{ft lb}_\text{f}}{\text{s}}$$

For the contribution from the pressure change:

$$\dot{m}\hat{V}(P_{\text{out}} - P_{\text{in}}) = 17.7\frac{\text{lb}}{\text{s}}\left(\frac{\text{ft}^3}{115 \text{ lb}}\right)\left(40\frac{\text{lb}_\text{f}}{\text{in}^2} - 18\frac{\text{lb}_\text{f}}{\text{in}^2}\right)\left(\frac{144 \text{ in}^2}{\text{ft}^2}\right) = 488 \frac{\text{ft lb}_\text{f}}{\text{s}}$$

The energy dissipated by friction is

$$\dot{F}_s = 10.0\frac{\text{ft lb}_\text{f}}{\text{lb}}\left(17.7\frac{\text{lb}}{\text{s}}\right) = 177 \frac{\text{ft lb}_\text{f}}{\text{s}}$$

Substituting into Eq. (6.22) we find

$$\dot{W}_s = 11 + 885 + 488 + 177 = 1561 \frac{\text{ft lb}_\text{f}}{\text{s}} \qquad \text{or 2.8 hp}$$

To calculate the work at the pump we use the known efficiency of 65%

$$\dot{W}_{\text{pump}} = \frac{\dot{W}_s}{\eta} = \frac{2.8}{0.65} = 4.3 \text{ hp}$$

Since the pump efficiency is 65%, delivering 2.8 hp to the fluid requires a 4.3 hp motor.

6.7.2 Heat and the Synthesis of Heat Exchange Networks

Throughout a chemical process plant, there are hot streams to be cooled and cold streams to be heated. By bringing these streams into indirect contact, heat can be transferred between the hot and cold streams. Myriad kinds of heat exchangers have been designed to facilitate this indirect contact.

Recall that heat is defined as energy flow due to temperature difference in the absence of material flow. The rate of heat transfer $\dot{Q}$ in heat exchangers increases with the contact area between the two streams and the size of the temperature difference. The performance of a heat exchanger can be modeled by a simple equation:

$$\dot{Q} = U_h A (\Delta T)_{LM} \tag{6.24}$$

where U_h is called a heat transfer coefficient and $(\Delta T)_{LM}$ is an average temperature difference between the hot stream and the cold stream. U_h depends on a number of factors including the process streams being heated and cooled, the heat exchanger material of construction, the fluid velocity, and other factors. Accurate calculation of U_h is beyond our purposes, but rough values are:

Nature of process stream	U_h, Btu/h °F ft^2	U_h, J/s °C m^2
Vapor	2–20	10–100
Liquid	10–1000	50–5000
Boiling or condensing	200–20,000	1000–100,000

Calculation of $(\Delta T)_{LM}$ requires knowledge of the configuration of the heat exchanger. For countercurrent heat exchange (Fig. 6.12),

$$(\Delta T)_{LM} = \frac{(T_{hot,in} - T_{cold,out}) - (T_{hot,out} - T_{cold,in})}{\ln\left(\dfrac{T_{hot,in} - T_{cold,out}}{T_{hot,out} - T_{cold,in}}\right)} \tag{6.25}$$

Use of Eqs. (6.24) and (6.25) is illustrated in the following example.

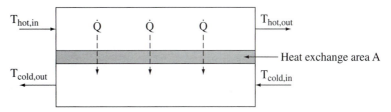

Figure 6.12 Schematic of a countercurrent heat exchanger. Heat is transferred across the heat exchange surface from the hot stream to the cold stream. At any point in the exchanger the hot stream must be warmer than the cold stream. Since the hot stream is being cooled and the cold stream is being heated, $T_{hot,\,in} \geq T_{hot,\,out}$ and $T_{cold,\,out} \geq T_{cold,\,in}$

| **Example 6.19** | **Heat Exchanger Sizing: Steam Heating of Methanol Vapor** |

Formaldehyde (HCHO) is produced by dehydrogenation of methanol (CH_3OH) over a silver catalyst. Before being fed to the reactor, methanol vapor (1000 lbmol/h) at 175°F and 1 atm pressure is heated to 300 °F in a heat exchanger. Saturated steam is available as a heat source at three pressures: 700 psig, 150 psig, and 15 psig.

First calculate the flow rate of steam required and the area of the countercurrent heat exchanger, with high pressure steam (700 psig) as the heat source. Assume the overall heat transfer coefficient $U_h = 1000$ Btu/h °F ft^2 and the heat capacity of methanol vapor is 12 Btu/lbmol °F. Also assume no heat loss from the exchanger to the surroundings.

Next consider medium or low pressure steam as a replacement for the high pressure steam. Given the steam costs in Sec. 6.1, which steam source would you choose?

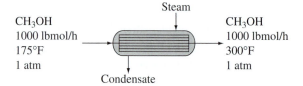

$$
\begin{array}{ccc}
 & \text{Steam} & \\
\text{CH}_3\text{OH} & \downarrow & \text{CH}_3\text{OH} \\
\text{1000 lbmol/h} & & \text{1000 lbmol/h} \\
\text{175°F} & & \text{300°F} \\
\text{1 atm} & & \text{1 atm} \\
 & \uparrow & \\
 & \text{Condensate} &
\end{array}
$$

Solution

First we will determine the amount of heat that must be transferred to the methanol vapor. We sketched a shell-and-tube heat exchanger, with the methanol flowing through the tubes and the steam in the shell. The condensate (condensed steam) leaves at the bottom of the exchanger shell. The system chosen is the methanol vapor passing through the tubes of the heat exchanger.

This is a steady-state system, with

$$\dot{n}_{in} = \dot{n}_{out} = 1000 \text{ lbmol/h}$$

There is no work term, and we neglect kinetic and potential energy contributions. The steady-state energy balance equation simplifies to

$$\dot{n}_{in}(\hat{H}_{out} - \hat{H}_{in}) = \dot{Q}$$

The phase and pressure of the methanol stream remain constant, but the temperature increases from 175°F to 300°F. Therefore,

$$\hat{H}_{out} - \hat{H}_{in} = \int_{T_{in}}^{T_{out}} C_p \, dT = (12 \text{ Btu/lbmol °F})(300 - 175°F) = 1500 \text{ Btu/lbmol}$$

and

$$\dot{n}_{in}(\hat{H}_{out} - \hat{H}_{in}) = \dot{Q} = (1000 \text{ lbmol/h})(1500 \text{ Btu/lbmol}) = 1.5 \times 10^6 \text{ Btu/h}$$

1.5 million Btu/h must be supplied to the system (the methanol vapor) from the surroundings (the steam).

Next we calculate the quantity of steam required. Now the system chosen is the steam flowing through the heat exchanger. (Do you see why we have to use the steam as our system, in order to figure out the steam flow rate?) The steady-state energy balance is

$$\dot{m}_{steam}(\hat{H}_{condensate} - \hat{H}_{sat'd\ steam}) = \dot{Q}$$

From the first part of the solution, we know that $\dot{Q} = -1.5 \times 10^6$ Btu/h. (Why is this now a negative number?) The only change in enthalpy between the steam entering and the condensate leaving is the enthalpy of condensation. This we can find directly from steam tables: $\Delta\hat{H}_v$ (700 psig) = 710 Btu/lb. Therefore,

$$\dot{m}_{steam}(\hat{H}_{condensate} - \hat{H}_{sat'd\ steam}) = \dot{m}_{steam}(-\Delta\hat{H}_v) = \dot{Q}$$

$$\dot{m}_{steam} = \frac{-1.5 \times 10^6\ \text{Btu/h}}{-710\ \text{Btu/lb}} = 2110\ \text{lb/h}$$

From the prices given in Sec. 6.1, high-pressure steam costs about \$5 to 6/million Btu. (Roughly, 1 Btu = 1 kJ.) Therefore, the operating costs are about \$7.50/h to \$9/h.

Next we calculate the heat exchanger area. From steam tables we learn that the temperature of saturated steam at 700 psig is 503°F. We calculate $(\Delta T)_{LM}$ using Eq. (6.25) and setting $T_{hot,in} = T_{hot,out} = 503°F$ (because the saturated steam condenses to saturated liquid at the same pressure and temperature), $T_{cold,in} = 175°F$, and $T_{cold,out} = 300°F$.

$$(\Delta T)_{LM} = \frac{(503 - 300) - (503 - 175)}{\ln\left(\dfrac{503 - 300}{503 - 175}\right)} = 260°F$$

Plugging this value into Eq. (6.24), along with $U_h = 1000$ Btu/h °F ft², gives the heat exchange area required:

$$A = \frac{\dot{Q}}{U_h(\Delta T)_{LM}} = \frac{1.5 \times 10^6\ \text{Btu/h}}{(1000\ \text{Btu/h °F ft}^2)(260°F)} = 5.8\ \text{ft}^2$$

Now, let's consider the alternative of using medium- or low-pressure steam. Saturated steam at 150 psig is 358°F., and saturated steam at 15 psig is only 250°F. Medium-pressure steam is less expensive than the high-pressure variety, but the lower steam temperature means that $(\Delta T)_{LM}$ is smaller, which means the heat exchanger area must increase. We'd love to be able to use the low-pressure steam since it's so cheap, but we can't because it's not hot enough to heat methanol to 300°F. One possibility is to use low-pressure steam in one heat exchanger, heating the methanol from 175°F to, say, 230°F., then use medium-pressure steam to heat the methanol from 230°F to 300°F.

Let's compare three alternative designs: Design I uses high-pressure steam, Design II uses medium-pressure steam, and Design III uses low-pressure steam to raise the methanol temperature to 230°F, then medium-pressure steam to raise the temperature to 300°F. Design I has been worked out already. Calculations for Design II and III proceed in a similar manner. Results are summarized in the table below.

Design	Steam P, psig	Steam T, °F	ΔH_{vap}, Btu/lb	Steam flow, lb/h	Steam cost, $/h	ΔT_{LM}, °F	Area, ft^2
I	700	505	710	2120	7.50-9	260	5.7
II	150	365	860	1745	4.5-6	116	12.9
III	15	250	950	695	0.6-2	41.6	15.9
	150	365	860	975	2.5-3.4	95.8	8.8

The average temperature difference is much greater for the high-pressure steam. The extra operating cost entailed in using high-pressure steam is compensated for by the lower capital cost (smaller required surface area) of the heat exchanger. The choice between Designs I and II will depend on a more detailed cost analysis. Design III looks particularly unpromising. Savings in steam cost compared to Design II are very small, and the size of the heat exchangers really balloons.

In Example 6.19 we analyzed a single heat exchange problem, but in a modern complex processing facility there could be dozens of streams that need to be heated and dozens to be cooled. We could use a heat exchanger for each stream, and use steam to heat all the streams-to-be-heated and refrigerants to cool all the streams-to-be-cooled. But, can we be more efficient? Sure. If a hot process stream needs to be cooled, why not exchange heat with a cold process stream that needs to be heated? Savings in both steam and cooling fluids are realized.

The difficulty lies in the sheer number of possible ways to combine many cooling and heating needs. How do we synthesize an efficient network of heat exchangers? There are very detailed, computerized ways of designing heat exchange networks that we will not discuss. Here are just a few heuristics that provide useful qualitative guidance.

1. Exchange heat between the hottest stream to be cooled and the warmest stream to be heated.
2. Exchange heat between the coldest stream to be heated and the coldest stream to be cooled.
3. If heat needs to be added to the network, add it at the lowest possible temperature.
4. If heat needs to be removed from the network, remove it at the highest possible temperature.

Did You Know?

Nature used countercurrent heat exchange between warm and cold fluids long before any modern chemical processes were designed. For example, warm-blooded animals such as whales have extensive contact between blood vessels carrying warm blood to the peripheral appendages and other vessels carrying cool blood from the periphery back to the heart. In the cold ocean waters, this built-in energy efficiency is crucial for survival.

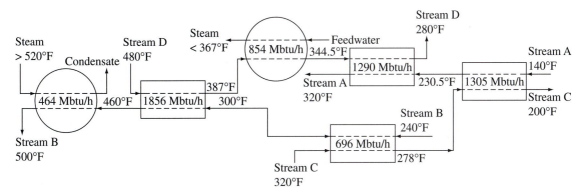

Figure 6.13 Heat exchanger network. Exchangers requiring utilities (steam or cooling water) are drawn as circles; exchangers with process streams only are drawn as oblongs. Hot streams to be cooled are arranged from left to right, while cold streams to be heated flow right to left. The "duty" of each exchanger, in thousand Btu/h (MBtu/h), is shown.

Rules 3 and 4 make sense if you remember that heat sources are cheapest at lower temperatures (as long as it is still above ambient). Any time you put heat into a system, you want to use the cheapest (lowest-temperature) source available. Any time you take heat out of the network, you want to take it out in its most valuable (highest-temperature) form. The same rules apply to cooling to temperatures below ambient, although the logic is a bit different. To take heat out of such a system, you need to use a refrigerant. The closer the refrigerant temperature is to ambient, the less costly is the cooling.

An example of a simple heat exchange network is shown in Fig. 6.13. The network handles four process streams, requires 464,000 Btu/h from expensive high-pressure (high-temperature) steam, but also makes 854,000 Btu/h of useful medium-pressure steam. Examine the network. Are the heuristics followed? Can you come up with a different way to pair the process streams?

	T_{in}	T_{out}	$\dot{m}$	C_p	$\dot{Q} = \dot{H}_{out} - \dot{H}_{in} = \dot{m}C_p(T_{out} - T_{in})$
Stream	°F	°F	lb/h	Btu/lb °F	1000 Btu/h
A	140	320	20,600	0.7	2595.6
B	240	500	23,200	0.5	3016
C	320	200	27,800	0.6	−2001.6
D	480	280	25,000	0.8	−4000

6.7.3   Energy Conversion Processes

Energy is released when fuels are combusted. In this section, we'll briefly discuss three ways we might use that released energy:

1. To convert the energy of reaction to heat, in a furnace.

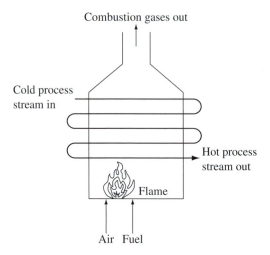

Figure 6.14 Schematic of a typical indirect-fired furnace. For a boiler, the cold process stream is replaced by water, and the hot process stream is steam.

2. To convert the energy of reaction to work, in a heat engine.
3. To convert the energy of reaction to work, in a fuel cell.

Converting Reaction Energy to Heat: Furnaces   In a furnace, whether it be in a house or a large chemical facility, fuel and air are mixed and ignited (Fig. 6.14). The combustion reaction has a large and negative enthalpy of reaction. Hot combustion gases are brought into indirect contact with process fluids. Heat is transferred from the combustion gases to the fluid. Cooled combustion gases leave through a vent or stack and are sometimes referred to as flue gases or stack gases. The heated process fluids are used in various ways, perhaps to warm a home or to feed a reactor.

Ideally, all of the enthalpy of combustion is transferred to the process fluid. However, there is a practical limit on the minimum temperature of the flue gas. If water vapor in the exiting combustion gases condenses in the flue, corrosion problems ensue. Water condensation is avoided by keeping the flue gas temperature above the gas' dewpoint temperature. Typically, flue gas temperature is anywhere from 50 to 175°C (120 to 350°F), and 80 to 90% of the heating value of the fuel is captured in the process fluids.

Example 6.20   **Converting Reaction Energy to Heat: Furnace Efficiency**

What is the maximum efficiency (actual heat recovered/heat recovered in a perfect furnace) of a furnace that burns methane (CH_4) with 25 mol% excess air, if water in the exit flue gases must not condense? Ambient temperature is 25°C.

You may use the following properties:

$$\Delta \hat{H}^{\circ}_{c,CH_4} = -890.6 \text{ kJ/gmol } [\text{with } H_2O \text{ (l) and } CO_2 \text{ (g) as products}]$$

$$\Delta \hat{H}_{v,H_2O} = 44.01 \text{ kJ at } 25 \text{ °C}$$

$$C_{p,H_2O(g)} = 33.6 \text{ J/mol °C}$$

$$C_{p,O_2(g)} = 29.3 \text{ J/mol °C}$$

$$C_{p,N_2(g)} = 29.1 \text{ J/mol °C}$$

$$C_{p,CO_2(g)} = 37 \text{ J/mol °C}$$

Solution

The perfect furnace:

In the "perfect furnace," air and methane enter at 25°C, methane is completely com-busted to CO_2 and H_2O, and combustion products leave at 25°C, with all water con-densed and all O_2 consumed.

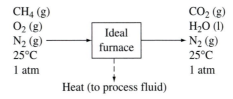

We'll choose the furnace, including the firebox contents but not the process fluids, as the system. An arbitrary basis is chosen of CH_4 flow to the burner of 1 gmol/s. Assuming steady-state operation, 25% excess air at 79 mol% N_2/21 mol% O_2, and complete conversion of CH_4 to products via the combustion reaction,

$$CH_4 (g) + 2O_2 (g) \rightarrow CO_2 (g) + 2H_2O (l)$$

we readily complete process flow calculations to find $\dot{\xi} = 1$ gmol/s and the fol-lowing flow rates:

	In, gmol/s	Out, gmol/s
CH_4	1	0
O_2	2.5	0.5
N_2	9.4	9.4
CO_2	0	1
H_2O	0	2
Total	12.9	12.9

There is no work term, and we'll neglect any changes in kinetic or potential energy. The differential energy balance equation simplifies to:

$$\dot{n}_{out}\hat{H}_{out} - \dot{n}_{in}\hat{H}_{in} = \dot{Q}$$

T and P of the input and output streams are the same, but the composition and phase have changed because of chemical reaction. So, the difference in enthalpy between the output and input streams is entirely due to combustion. Furthermore, since the input and output streams are both at standard T and P,

$$\dot{n}_{out}\hat{H}_{out} - \dot{n}_{in}\hat{H}_{in} = \dot{\xi}\Delta\hat{H}_{c}^{0}$$

$\Delta\hat{H}_{c}^{\circ} = -890.6$ kJ/gmol for combustion of methane to CO_2 (g) and H_2O (l). So,

$$\dot{\xi}\,\Delta\hat{H}_{c}^{0} = \left(1\,\frac{gmol}{s}\right)\left(-890.6\,\frac{kJ}{gmol}\right) = -890.6\,\frac{kJ}{s} = \dot{Q}$$

890.6 kJ/s can be transferred from the combustion gases in the furnace to a process stream in the ideal case.

The real furnace:

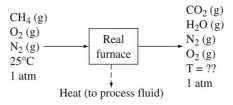

The process flow calculations are identical to those for the ideal furnace. The temperature of the flue gas is greater than 25°C because of the necessity of avoiding condensation. The minimum flue gas temperature is the dew point temperature of the flue gas, because that is the temperature at which the first drop of liquid forms. Raoult's law comes into play:

$$y_w = \frac{x_w P_w^{sat}}{P}$$

The only condensable component of the flue gas is water, so $x_w = 1.0$. The mole fraction of water in the exiting gases, calculated from the molar flows, is

$$y_w = \frac{\dot{n}_w}{\dot{n}} = \frac{2}{12.9} = 0.155$$

Since the gases exit at $P = 1$ atm,

$$P_w^{sat} = \frac{y_w P}{x_w} = \frac{0.155(1\text{ atm})}{1.0} = 0.155\text{ atm}$$

To find T at which $P_w^{sat} = 0.155$ atm, we turn to Antoine's equation for water:

$$\log_{10}P_w^{sat}(\text{mmHg}) = 8.10765 - \frac{1750.286}{T(°C) + 235.0}$$

We substitute in for P_w^{sat} (118 mmHg) and solve to find

$$T = 55°C$$

This is the *minimum* allowable temperature; we add a 10°C safety margin and set the exit temperature of the combustion gases at 65°C.

The energy balance equation for the real furnace is the same as that for the ideal furnace:

$$\dot{n}_{\text{out}}\hat{H}_{\text{out}} - \dot{n}_{\text{in}}\hat{H}_{\text{in}} = \dot{Q}$$

The difference is that we need to account for reaction, phase change, and temperature change in the enthalpy terms.

We construct the following path to take us from inlet to outlet conditions, taking advantage of the available data:

Change composition: React 1 gmol/s CH_4 (g) and 2 gmol/s O_2 (g) at 25°C and 1 atm to produce 1 gmol/s CO_2 (g) and 2 gmol/s H_2O (l).
Change phase: Vaporize 2 gmol/s H_2O (l) to 2 gmol/s H_2O (g) at 25°C and 1 atm.
Change T: Heat the outlet stream of 1 gmol/s CO_2 (g), 0.5 gmol/s O_2 (g), 9.4 gmol/s N_2, 1 gmol/s CO_2, and 2 gmol/s H_2O (g) from 25°C to 65°C.

Evaluating the enthalpy change for each step and then summing gives

Helpful Hint
Enthalpy changes due to reaction and phase change are often given as kJ/mol while heat capacities are often given as J/mol°C. Convert to all kJ or all J to avoid errors!

$$\dot{n}_{\text{out}}\hat{H}_{\text{out}} - \dot{n}_{\text{in}}\hat{H}_{\text{in}} = \xi\Delta\hat{H}_c° + \dot{n}_w\Delta\hat{H}_v + \sum_{\text{all } i \text{ out}} n_i \int_{T_{\text{in}}}^{T_{\text{out}}} C_{pi}\,dT$$

$$= (1)(-890.6) + (2)(44.01)$$

$$+ \frac{[(1)(37) + (0.5)(29.3) + (9.4)(29.1) + (2)(33.6)]}{1000}(65 - 25)$$

$$= -890.6 + 88.01 + 15.70 = -786.9 \text{ kJ/s}$$

The amount of heat that can be transferred to a process material is 786.9 kJ/s. Compare to the 890.6 kJ/s of the ideal furnace, and we see that the real furnace efficiency is about 88%. This is a reasonable estimate of the efficiency of standard modern furnaces. (Can you think of any ways to improve the efficiency?)

Converting Reaction Energy into Work: Heat Engines   We've just seen how chemical energy released from combustion is converted to heat in furnaces and boilers. But, what if we want to do work—electrical or mechanical energy—instead of heat? (We plug our vacuum cleaners and power tools into an electrical outlet, not into a furnace!) The solution is to build a *heat engine,* to produce work from heat.

A heat engine is a cyclical operation, as shown in Fig. 6.15. There is a "working fluid," often water/steam, which circulates through the heat engine. Let's consider the working fluid as our system, and describe each step in the cycle.

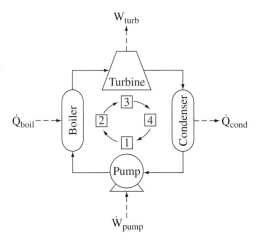

Figure 6.15 Schematic of a cyclic heat engine. Solid lines indicate circulating fluid; dashed lines indicate work or heat inputs/outputs.

Step 1.   The pressure of liquid water is increased using a pump. $\dot{W}_{pump} > 0$.

Step 2.   The high-pressure water is fed to a boiler. Fuel is burned and the combustion gases heat the water to produce high-pressure steam. $\dot{Q}_{boil} > 0$.

Step 3.   The high-pressure steam is fed to a turbine. The steam forces the turbine blades to move, which rotates a shaft. The steam leaves the turbine at a lower pressure. To avoid damage to the turbine blades, the steam must not condense. $\dot{W}_{turb} < 0$. $|\dot{W}_{turb}| > |\dot{W}_{pump}|$.

Step 4.   The low-pressure steam is cooled and condensed in the condenser. $\dot{Q}_{cond} < 0$. $|\dot{Q}_{boil}| > |\dot{Q}_{cond}|$. The condensate is then recycled to the pump, and the cycle continues.

With a cyclic heat engine, there is a net input of heat and a net output of work. Heat is added at a high temperature (in the boiler) and discharged at a low temperature (in the condenser). This means that "expensive" heat is added to the cycle and "cheap" (or worthless) heat is removed. Some work input, at the pump, is required, but this work input is much less than the work output at the turbine. The work output at the turbine might be used directly to drive a pump or compressor, or it might be coupled to a generator to produce electricity. Modern municipal and industrial power plants operate as heat engines that work fundamentally as described in Fig. 6.15. (Large power plants are often located near rivers or lakes so that the water provides cooling in the condenser. The heat released into the river or lake can significantly change the local aquatic flora and fauna.)

Example 6.21   **Converting Reaction Energy to Work: Heat Engine Analysis**

1000 kg/h water at 75°C and 1 bar is fed to a pump, where the pressure of the water is increased to 20 bar. The high-pressure water is fed to the boiler to produce steam at 250°C and 20 bar. The steam is fed to a turbine. The exhaust steam exits the turbine as saturated steam at 1 bar. The exhaust steam is cooled further, leaves the condenser as a subcooled liquid at 75°C, and the cycle continues.

Calculate (a) the work produced by the turbine, (b) the heat input in the boiler, (c) the amount of natural gas required in the boiler, if the boiler efficiency is 85%, (d) the heat removed in the condenser, and (e) the work required by the pump. Assume that the turbine and pump are 100% efficient. If the turbine generates electricity and the pump has an electric motor, what is the net electricity cost per kW, considering only operating costs?

Solution

Steam tables are consulted to find enthalpies at the given T, P, and ϕ:

Stream	Temperature, °C	Pressure, bar	Phase	Enthalpy, kJ/kg
Feed to pump	75	1	Liquid	314.3
Feed to boiler	75	20	Liquid	315.8
Feed to turbine	250	20	Vapor	2903.2
Exhaust from turbine	99.6	1	Vapor	2674.9

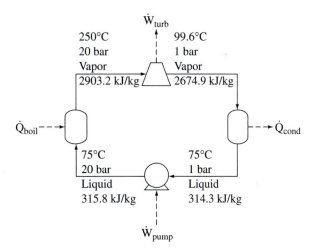

Now it is simple to complete energy balances around each piece of equipment.

Pump: $\dot{W}_s = \dot{m}(\hat{H}_{out} - \hat{H}_{in}) = (1000 \text{ kg/h})(315.8 - 314.3 \text{ kJ/kg})$

$\dot{W}_s = 1500 \text{ kJ/h} = 0.42 \text{ kW}$

Boiler: $\dot{Q} = \dot{m}(\hat{H}_{out} - \hat{H}_{in}) = (1000 \text{ kg/h})(2903.2 - 315.8 \text{ kJ/kg})$

$\dot{Q} = 2,587,400 \text{ kJ/h} = 718.7 \text{ kW}$

Turbine: $\quad \dot{W}_s = \dot{m}(\hat{H}_{out} - \hat{H}_{in}) = (1000 \text{ kg/h})(2675.9 - 2903.2 \text{ kJ/kg})$

$\qquad \dot{W}_s = -227{,}300 \text{ kJ/h} = -63.1 \text{ kW}$

Condenser: $\quad \dot{Q} = \dot{m}(\hat{H}_{out} - \hat{H}_{in}) = (1000 \text{ kg/h})(314.3 - 2674.9 \text{ kJ/kg})$

$\qquad \dot{Q} = -2{,}360{,}600 \text{ kJ/h} = -656 \text{ kW}$

Notice indeed that more heat is put into the boiler (718.7 kW) than taken out in the condenser (−656 kW), and that the work input to the pump (0.4 kW) is much less than the work output in the turbine (−63.1 kW). The efficiency of this heat engine (defined as net work output/heat input at the boiler) is very low:

$$\eta = \frac{-\dot{W}_{net}}{\dot{Q}_{boiler}} = \frac{63.1 - 0.42}{718.7} = 0.087$$

Heat in the boiler is supplied by combustion of natural gas. An 85% efficient boiler would require the equivalent of just over 3 million kJ/h of methane. At $3.60/million kJ for natural gas (see Sec. 6.1), we'd be spending about $11/h for fuel to run the boiler. The net electricity production is 62.9 – 0.4 = 62.5 kW. Thus, the operating cost to produce the electricity for this cycle is about 17 cents per kWh, or about $47/million kJ. (Although modern power plants achieve much better efficiency than this cycle, this example illustrates why electricity is so much more expensive than fossil fuels. Recall from Sec. 6.1 that electricity is valued at about $13 to 25/million kJ.)

Refrigerators work like heat engines in reverse. Work input is greater than work output, and heat input is less than heat output. Refrigerators are also called heat pumps—because they "pump" heat from low temperature sources to high temperature sinks. This doesn't happen spontaneously; work must be added to the cycle. Like the heat engine, the working fluid circulates through the system, and changes from vapor to liquid and back again. Common working fluids include ammonia, the freons, and propane. A refrigeration cycle is illustrated in Fig. 6.16.

Step 1. Saturated liquid refrigerant is passed through an expansion valve to "flash" the working fluid to a lower pressure. This causes the temperature to drop, and some of the liquid may be vaporized. Expansion valves are usually used rather than turbines because the capital cost is so much less. $\dot{W}_{exp} = 0$ for expansion valves.

Step 2. The low-pressure cold fluid is contacted with the process fluid to be cooled in the evaporator (e.g., refrigerator box) and is vaporized to its saturation temperature. $\dot{Q}_{evap} > 0$.

Step 3. The saturated vapor is compressed to high pressure. Compression increases the temperature of the vapor. For compressors to work well, the fluid must remain a vapor during the entire compression cycle. $\dot{W}_{comp} > 0$.

Step 4. The vapor from the compressor is condensed to liquid at high temperature (e.g., kitchen temperature). $\dot{Q}_{cond} < 0$. $|\dot{Q}_{cond}| > |\dot{Q}_{evap}|$.

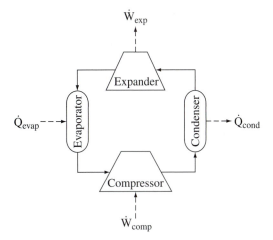

Figure 6.16 Schematic of a refrigeration cycle. Solid lines indicate circulating fluid, dashed lines indicate work and heat inputs and outputs. In a home refrigerator, the evaporator is inside the box and picks up heat. The fluid inside the evaporator must be colder than the temperature of the box. The condenser is outside the box and discharges heat to the kitchen. The fluid in the condenser must be warmer than the kitchen. The turbine is replaced with a simple expansion valve in small refrigeration units and no work is produed.

(Why is it a bad idea to cool off the kitchen by leaving the refrigerator door open?)

Converting Energy of Reaction to Work: Fuel Cells   In a conventional power plant operating as a heat engine, chemical energy from a combustion reaction is used to heat boiler water to steam, and then the steam drives a turbine, which generates electricity. Inefficiencies in converting chemical energy to heat and then converting heat to work motivate a search for technologies that allow direct conversion of chemical energy to work. *Fuel cells* provide the means for doing exactly that. Fuel cells have been proposed for all kinds of

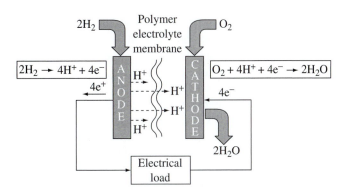

Figure 6.17 Schematic of a typical fuel cell with hydrogen as fuel.

applications, from power plants to car engines to batteries for laptop computers and cell phones.

In a fuel cell, hydrogen (or other fuel) is fed to the anode compartment and air is fed to the cathode compartment. Two electrochemical half-reactions occur at each electrode, with generation of an electric current from the chemical reaction:

$$H_2 \rightarrow 2H^+ + 2e^-$$

$$0.5\,O_2 + 2H^+ + 2e^- \rightarrow H_2O$$

The overall reaction is simply:

$$H_2\,(g) + 0.5O_2\,(g) \rightarrow H_2O\,(l)$$

Hydrogen is the perfect fuel for fuel cells; since the byproduct of oxidation is water, fuel cells are very clean and don't contribute to greenhouse gas production. They are also very quiet. However, hydrogen is not naturally available (unlike natural gas and other fossil fuels). Instead, hydrogen for fuel cells is typically generated by reforming natural gas, a process that generates as much CO_2 as if the natural gas were combusted. Efforts are underway to make hydrogen from renewable agricultural resources such as glucose.

Example 6.22	### Converting Reaction Energy to Work: Hydrogen Fuel Cells

We're evaluating different fuel cell materials for use in a 200 kW emergency power generator. Two types of fuel cells are under consideration: PEFC and SOFC.

In the PEFC (polymer electrolyte fuel cell), a polymeric ion exchange membrane serves as the electrolyte. The only liquid present is water. The maximum operating temperature is limited to about 100°C. Advantages include absence of corrosive liquid, simple fabrication, and long life. One major disadvantage is the high cost of the polymeric membrane. Because the operating temperature is limited to less than 110°C, the hydrogen gas can contain at most a few ppm CO, to avoid damage to the Pt electrodes.

In the SOFC (solid oxide fuel cell), a metal oxide such as yttrium oxide–stabilized zirconia serves as a solid electrolyte deposited as a thin layer between solid electrodes. These materials operate at very high temperatures (1000°C), but are difficult to manufacture.

Assume that air (79 mol% N_2, 21 mol% O_2) is the source of oxygen and that oxygen is fed at its stoichiometric ratio. The air and hydrogen are fed to the fuel cell at 25°C. In a typical fuel cell, about 90% of the hydrogen fed reacts to water; the remainder leaves the fuel cell unconverted. Assume that the exhaust stream is at the operating temperature of the fuel cell. A search of the literature indicates that the theoretical heat losses are a function of the operating temperature of the fuel cell, and can be estimated as $0.045T$ kJ per gmol H_2 reacted, where T is the operating temperature in K.

What is the hydrogen feed requirement for the PEFC vs. the SOFC? The following data may be used:

$$\Delta \hat{H}^{\circ}_{f,H_2O(v)} = -241.8 \text{ kJ/mol}$$

$$C_{p,H_2O(v)} = 33.6 \text{ J/mol } ^{\circ}C$$

$$C_{p,O_2(g)} = 29.3 \text{ J/mol } ^{\circ}C$$

$$C_{p,N_2(g)} = 29.1 \text{ J/mol } ^{\circ}C$$

$$C_{p,H_2(g)} = 29.1 \text{ J/mol } ^{\circ}C$$

Solution

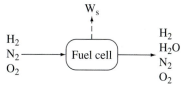

The stoichiometrically balanced reaction is

$$H_2 + 0.5O_2 \rightarrow H_2O$$

Material balance equations are summarized below, using as a basis H_2 fed at 1 gmol/s. (We will calculate the work output for this basis, then scale up or down to the desired work output of 200 kW.)

	$\dot{n}_{in}$, gmol/s	$\nu_i \dot{\xi}$	$\dot{n}_{out}$, gmol/s
H_2	1	-0.9	0.1
O_2	0.5	-0.45	0.05
N_2	1.88	0	1.88
H_2O	0	$+0.9$	0.9

The operating temperature of the fuel cell is 100°C for the PEFC and 1000°C for the SOFC. The stream leaving the fuel cell will be at the operating temperature. Fuel cells operate near atmospheric pressure.

The fuel cell produces some heat, so the energy balance must include any heat losses. We'll neglect kinetic and potential energy changes and assume steady-state operation. The energy balance equation simplifies to

$$\dot{n}_{out}\hat{H}_{out} - \dot{n}_{in}\hat{H}_{in} = \dot{W}_s + \dot{Q}$$

Let's consider each term in the balance equation. First, we found an expression for heat loss as a function of operating temperature and gmols H_2 reacted. Since the PEFC operates at 100°C:

$$\dot{Q}_{PEFC} \approx -0.045 T \dot{\xi} = -0.045(373)(0.9) = -15 \frac{kJ}{s}$$

and for the SOFC, which operates at 1000°C:

$$\dot{Q}_{SOFC} \approx -0.045(1273)(0.9) = -51.6 \frac{kJ}{s}$$

To evaluate the enthalpy terms in the energy balance equation, we pick as the reference state the mix of gases entering the fuel cell at 25°C and the inlet pressure (about 1 atm). Thus, $\hat{H}_{in} = 0$. Now we need to calculate the enthalpy of the outlet stream. We construct a two-step pathway to move from inlet to outlet conditions:

Change composition: React 0.9 gmol/s H_2 (g) and 0.45 gmol/s O_2 (g) to 0.9 gmol/s H_2O (v) at 25°C and 1 atm.

Change T of outlet gases: Heat 0.1 gmol/s H_2 (g), 0.05 gmol/s O_2 (g), 1.88 gmol/s N_2 (g), and 0.9 gmol/s H_2O (v) from 25°C to the outlet temperature.

For the reaction

$$H_2\,(g) + 0.5 O_2\,(g) \rightarrow H_2O\,(v)$$

we calculate

$$\Delta \hat{H}^{\circ}_r = \sum \nu_i \Delta \hat{H}^{\circ}_{f,i} = (-1)(0) + (-0.5)(0) + (1)\left(-241.8 \frac{kJ}{gmol}\right)$$

$$= -241.8 \frac{kJ}{gmol}$$

The enthalpy change due to reaction

$$\dot{\xi} \Delta \hat{H}^0_r = 0.9\,(-241.8) = -217.6\ kJ/s$$

The enthalpy change for increasing T of outlet gases is

$$\sum_{out} \dot{n}_i C_{pi}(T_{out} - T_{ref}) = [(0.1)(29.1) + (0.05)(29.3) + (1.88)(29.1)$$

$$+ (0.9)(33.6)](T_{out} - 25°C)$$

$$= 89.3(T_{out} - 25°C)\ J/s = 0.0893(T_{out} - 25°C)\ kJ/s$$

$T_{out} = 100°C$ for the PEFC and 1000°C for the SOFC. Therefore,

$$\sum n_i C_{pi}(T_{out} - T_{ref}) = 0.0893(100 - 25°C) = 6.7 \frac{kJ}{s} \qquad \text{for PEFC}$$

$$\sum n_i C_{pi}(T_{out} - T_{ref}) = 0.0893(1000 - 25°C) = 87.1 \frac{kJ}{s} \qquad \text{for SOFC}$$

Summing the enthalpy changes from both steps gives

$$\dot{H}_{out} = -217.6 + 6.7 = -210.9 \frac{kJ}{s} \qquad \text{for PEFC}$$

$$\dot{H}_{out} = -217.6 + 87.1 = -130.5 \frac{kJ}{s} \qquad \text{for SOFC}$$

Finally, we can calculate the theoretical work output from the energy balance:

For PEFC: $\dot{W}_s = \dot{H}_{out} - \dot{H}_{in} - \dot{Q} = -210.9 + 15 = -195.9 \frac{kJ}{s} = -195.9 \text{ kW}$

For SOFC: $\dot{W}_s = \dot{H}_{out} - \dot{H}_{in} - \dot{Q} = -130.5 + 51.6 = -78.9 \frac{kJ}{s} = -78.9 \text{ kW}$

These values are the theoretical work outputs assuming 1 gmol/s hydrogen fed. In a typical fuel cell, the actual work output is about 65% of the theoretical work output for PEFC and about 70% for SOFC. Therefore, the PEFC produces about 127 kW per gmol hydrogen fed, and the SOFC produces about 55 kW per gmol hydrogen. Scaling up to 200 kW, the PEFC requires about 1.6 gmol hydrogen/s and the SOFC requires about 3.6 gmol hydrogen/s.

Can you come up with a way to make the SOFC work better? (*Hint*: what will you do with that hot exit gas?)

6.7.4 Chemical Energy and Chemical Safety: Explosions

Explosions and fires are unavoidable hazards of our modern chemistry-dependent life, but careful design and operation will reduce the risk of explosion or fire, and mitigate the consequences of an accident if one does happen. (Hazards and risks are different: A hazard is something with the potential to cause injury or damage, while risk is the likelihood that the hazard will be realized.)

Fires require three conditions: (1) a combustible material, (2) an oxidizing agent (e.g., oxygen in air), and (3) an ignition source. Remove any one of these and you remove the possibility of a fire. Blanketing a storage tank with nitrogen and using spark-proof electrical devices are safety measures taken to reduce the risk of fire.

Explosions, which are rapid and sudden releases of stored chemical energy, are arguably more devastating than fires. Explosions occur when a compound rapidly decomposes into more stable compounds; unlike fires, explosions do not require an oxidizing agent. Indeed, many times fires are the consequence of an explosion that releases combustible materials into the air. Early in the design process, it behooves us to establish the explosion potential of a compound. On the basis of this determination, we might choose to abandon the process, find an alternative chemical reaction pathway that is less hazardous, or install safeguards against explosions and their aftermath.

Five classes of compounds are considered to pose particular explosion and/or fire hazards: hydrocarbons, oxygen-containing organic compounds, nitrates and nitro compounds, other nitrogen-containing compounds, and organic peroxides. Not all compounds that fall into these categories are serious hazards. Methods for rapidly estimating the explosive potential of a particular compound have been developed. We will introduce one method here. Essentially, this method relies on calculating, or estimating, two measures of the energy released during reaction, and then comparing those numbers to compounds of known explosive danger:

1. Calculate $\Delta \hat{H}_r^\circ$ for the decomposition of the compound of interest to stable products.
2. Calculate the adiabatic reaction temperature of the decomposition.
3. If $-\Delta \hat{H}_r^\circ$ is 5 kJ/g or larger, and if the adiabatic reaction temperature is 1000 K or greater, the compound is likely an explosion hazard.

Mixtures can be analyzed by using a similar procedure. For example, ammonium nitrate is a chemical that is widely available in farming regions because it is a useful fertilizer. By itself, it has a low explosive hazard, but its adiabatic reaction temperature is high. When combined with a small amount of fuel oil, ammonium nitrate becomes a powerful, oxygen-balanced explosive. This mixture is used commercially as a blasting agent and has also been used violently, for example, in the Oklahoma City bombing.

Estimating $\Delta \hat{H}_r^\circ$ requires that we first decide what the decomposition reaction is! We want to find the reaction mixture that produces the most stable products, because this will give us the largest (negative) $\Delta \hat{H}_r^\circ$. For compounds containing C, H, O, and/or N, the decomposition products will be some combination of C, CO, CO_2, H_2, H_2O, and N_2. There is generally more than one possible stoichiometrically balanced chemical equation for the decomposition of the compound of interest to these products (or a subset). By trial and error, we find the chemical reaction that produces the most negative $\Delta \hat{H}_r^\circ$. These calculations require that we have information on $\Delta \hat{H}_f$ of all compounds in the reaction. Note that these calculations apply at standard temperature and pressure (25°C and 1 atm).

Next we calculate the adiabatic reaction temperature. (See Example 6.15.) The energy balance equation is simple: $\dot{H}_{in} = \dot{H}_{out}$. Now, however, there is one minor complication: We don't know what the "out" composition is! To take care of this, we specify that the outlet mixture is at chemical equilibrium. Now, if we know (or can calculate) the composition at chemical equilibrium we proceed. But the equilibrium composition depends on temperature, which we don't know! So the job requires that we derive an equation for the equilibrium composition as a function of temperature, and then find T that satisfies both the energy balance equation and the equilibrium expression. In other words, we must combine material balances, an energy balance, and chemical reaction equilibria to find the answer.

These methods are illustrated in the following example.

| Example 6.23 | **Estimating Explosive Potential: Trinitrotoluene** |

Is TNT (trinitrotoluene, $C_7H_5(NO_2)_3$ an explosion hazard? Data: (we are content with approximate C_p values because we are after an order-of-magnitude analysis, not a precise calculation.)

Compound	$\Delta \hat{H}_f$, kJ/gmol	$\Delta \hat{G}_f^\circ$, kJ/gmol	C_p, J/gmol °C
$C_7H_5(NO_2)_3$ (s)	-65.5	—	
C (s) (as graphite)	0	0	9
CO (g)	-110.6	-137.3	29
CO_2 (g)	-393.5	-394.4	43
H_2 (g)	0	0	29
H_2O (g)	-241.8	-228.6	34
N_2 (g)	0	0	29

Solution

We need to first determine the most likely products of explosive decomposition of TNT. For compounds containing only C, H, N, and O, the products to consider are C, CO, CO_2, H_2, H_2O, and N_2. We have four elements and seven compounds altogether, so there is a maximum of three independent chemical reactions (Chap. 3). We are interested only in reactions where the stoichiometric coefficients are positive for all compounds except TNT. We find a set of three independent balanced reactions, and calculate the enthalpy of reaction for each.

$$C_7H_5(NO_2)_3 \rightarrow 5.25\ C + 1.75\ CO_2 + 2.5\ H_2O + 1.5\ N_2$$

$$\Delta \hat{H}_r^\circ = -1223\ \frac{kJ}{gmol\ TNT}$$

$$C_7H_5(NO_2)_3 \rightarrow 4\ C + 3\ CO_2 + 2.5\ H_2 + 1.5\ N_2 \qquad \Delta \hat{H}_r^\circ = -1109\ \frac{kJ}{gmol\ TNT}$$

$$C_7H_5(NO_2)_3 \rightarrow 3.5\ C + 3.5\ CO + 2.5\ H_2O + 1.5\ N_2 \qquad \Delta \hat{H}_r^\circ = -925\ \frac{kJ}{gmol\ TNT}$$

The molar mass of TNT is 227 g/gmol. For the first reaction, which has the most negative enthalpy of reaction, we calculate $\Delta \hat{H}_r^\circ = -1223$ kJ/gmol TNT = -5.4 kJ/g TNT. This is slightly greater (more negative) than the first criteria for explosiveness, that $-\Delta \hat{H}_r^\circ > 5$ kJ/g. Thus, we have our first indication that TNT is explosive.

Next, we check the adiabatic reaction temperature. Recall that the adiabatic reaction temperature is the outlet temperature reached, assuming that the system is perfectly insulated. We must allow for the presence of all compounds in the exit stream, not just the products of the first reaction listed.

To make this problem more manageable we will set up an imaginary process. We imagine two events: First, 1 gmol/s TNT decomposes to C, CO_2, H_2O, and N_2 at room temperature, releasing some heat; second, the heat is returned and the gases re-equilibrate at some unknown temperature:

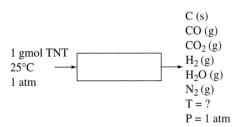

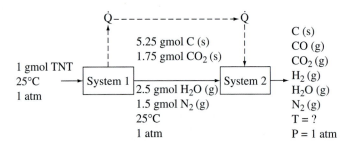

Notice that the net effect of our imaginary systems is exactly the same as the adiabatic system. Notice further that system 1 corresponds to the reaction scenario where we have already calculated $\Delta \hat{H}°_r = -1223$ kJ/gmol TNT.

Applying an energy balance around system 1, we find:

$$\dot{H}_{out} - \dot{H}_{in} = \dot{Q}_1 = -1223 \text{ kJ/s}$$

because the difference in enthalpy between the outlet and inlet streams is simply the standard enthalpy of reaction,

The heat removed in system 1, $\dot{Q}_1$, is equal (but opposite in sign) to the heat added to system 2, $\dot{Q}_2$. If there were no chemical reactions, it would be a simple problem to calculate the increase in temperature of the material leaving system 2 due to the addition of heat. However, the adiabatic reaction temperature calculation requires that the material leave at chemical equilibrium. Since in system 2 there are six compounds involving four elements, there are two independent chemical reactions. Two additional compounds, CO and H_2, are potentially in the exit stream, so we add two chemical reactions that could produce them:

$$C + CO_2 \rightleftarrows 2\,CO \tag{R1}$$

$$H_2O + CO \rightleftarrows CO_2 + H_2 \tag{R2}$$

(There are other possible sets of chemical reactions that are equally valid. As long as the reactions are independent, it doesn't matter which set of reactions is written!)

Let's recap what constraints must be satisfied simultaneously on system 2: material balance equations, the energy balance equation, and the chemical equilibria equations! The material balance equations give us (all in units of gmol/s):

$$\dot{n}_{C,out} = 5.25 - \dot{\xi}_1$$

$$\dot{n}_{CO_2,out} = 1.75 - \dot{\xi}_1 + \dot{\xi}_2$$

$$\dot{n}_{CO,out} = 2\dot{\xi}_1 - \dot{\xi}_2$$

$$\dot{n}_{H_2O,out} = 2.5 - \dot{\xi}_2$$

$$\dot{n}_{H_2,out} = \dot{\xi}_2$$

$$\dot{n}_{N_2,out} = 1.5$$

Now we apply the additional constraint that the outlet from system 2 is at equilibrium. Noting that C is a solid, that $P = 1$ atm, that all other compounds are gases, and that the total molar flow of *gases* in the exit stream $= 5.75 + \dot{\xi}_1$, we use expressions for the equilibrium constants for R1 and R2 to derive:

$$K_1 = \Pi \, (a_i)^{\nu_{i1}} = \frac{y_{CO}^2 P^2}{a_C y_{CO_2} P} = \frac{(2\dot{\xi}_1 - \dot{\xi}_2)^2}{(1.75 - \dot{\xi}_1 + \dot{\xi}_2)(5.75 + \dot{\xi}_1)}$$

$$K_2 = \Pi \, (a_i)^{\nu_{i2}} = \frac{y_{CO_2} y_{H_2}}{y_{CO} y_{H_2O}} = \frac{(1.75 - \dot{\xi}_1 + \dot{\xi}_2)(\dot{\xi}_2)}{(2\dot{\xi}_1 - \dot{\xi}_2)(2.5 - \dot{\xi}_2)}$$

We also know that K_1 and K_2 are functions of temperature from the van't Hoff equation (Chap. 4):

$$\ln K_1 = -\frac{1}{R}\left[\frac{\Delta \hat{G}^\circ_{r1} - \Delta \hat{H}^\circ_{r1}}{298} + \frac{\Delta \hat{H}^\circ_{r1}}{T}\right]$$

$$\ln K_2 = -\frac{1}{R}\left[\frac{\Delta \hat{G}^\circ_{r2} - \Delta \hat{H}^\circ_{r2}}{298} + \frac{\Delta \hat{H}^\circ_{r2}}{T}\right]$$

From the data given in the problem statement, we calculate

$$\Delta \hat{H}^\circ_{r1} = 170.4 \text{ kJ/gmol}$$

$$\Delta \hat{G}^\circ_{r1} = 119.8 \text{ kJ/gmol}$$

$$\Delta \hat{H}^\circ_{r2} = -39.2 \text{ kJ/gmol}$$

$$\Delta \hat{G}^\circ_{r2} = -28.5 \text{ kJ/gmol}$$

Inserting these values into the van't Hoff equations and then combining with the expressions for K_1 and K_2, we get two equations in three variables:

$$\frac{(2\dot{\xi}_1 - \dot{\xi}_2)^2}{(1.75 - \dot{\xi}_1 + \dot{\xi}_2)(5.75 + \dot{\xi}_1)} = \exp\left(20.4 - \frac{20495}{T}\right)$$

$$\frac{(1.75 - \dot{\xi}_1 + \dot{\xi}_2)(\dot{\xi}_2)}{(2\dot{\xi}_1 - \dot{\xi}_2)(2.5 - \dot{\xi}_1)} = \exp\left(-4.56 + \frac{4715}{T}\right)$$

The last equation we need is the energy balance:

$$\dot{H}_{out} - \dot{H}_{in} = \dot{Q}_2 = 1223 \text{ kJ/s}$$

$\dot{Q}_2 = +1223$ kJ/s, from solution of the energy balance around system 1. We design a pathway to move from the inlet to the outlet conditions:

1. React by (R1) and (R2) at 25°C to the equilibrium composition.
2. Increase the temperature of the equilibrium mixture from 25°C to T.

This yields the following equation:

$$\dot{\xi}_1 \Delta \hat{H}^{\circ}_{r1} + \dot{\xi}_2 \Delta \hat{H}^{\circ}_{r2} + \sum_{\text{all } i} \dot{n}_{i,\text{out}} C_{p,i} (T - 298K) = \dot{Q}$$

Now, substituting in known numerical values, we get:

$$\dot{\xi}_1(170.4) + \dot{\xi}_2(-39.2) + [(5.25 - \dot{\xi}_1)9 + (1.75 - \dot{\xi}_1 + \dot{\xi}_2)43 + (2\dot{\xi}_1 - \dot{\xi}_2)29$$

$$+ (2.5 - \dot{\xi}_2)34 + 29\dot{\xi}_2 + 44]\left(\frac{1}{1000}\right)(T - 298 \text{ K}) = 1223 \text{ kJ/s}$$

We have three equations (the two chemical equilibrium equations plus the energy balance equation) in three variables ($\dot{\xi}_1$, $\dot{\xi}_2$, and T). In theory, at least, we should be able to find a solution. But, with highly nonlinear equations such as we have here, finding a physically reasonable solution can be tricky. It is worthwhile to first examine the equations and spot any important trends. First, we note that at $T = 1000$ K (the break-even point for determining if a compound is an explosion hazard), K_1 and K_2 are both approximately equal to 1. If TNT is a serious explosive hazard, the adiabatic reaction temperature should be significantly greater than 1000 K. As T increases above 1000 K, K_1 becomes larger but K_2 becomes smaller. We can calculate $\dot{\xi}_1$ and $\dot{\xi}_2$ as a function of T by guessing a T, then simultaneously solving the two chemical equilibrium equations at that T. These calculations reveal that, at $T = 1000$ K, $\dot{\xi}_1 = 2.05$ and $\dot{\xi}_2 = 1.36$, but as T increase above 1000 K, $\dot{\xi}_1$ and $\dot{\xi}_2$ both reach asymptotic values: $\dot{\xi}_1 \rightarrow 2.5$ and $\dot{\xi}_2 \rightarrow 0.75$. If we insert these asymptotic extents of reaction into the energy balance equation, we quickly calculate that the adiabatic reaction temperature reaches about 3300 K! This result justifies our use of the asymptotic values of $\dot{\xi}_1$ and $\dot{\xi}_2$ in the energy balance equation. Clearly, TNT is an explosion hazard!

CASE STUDY   Energy Management in a Chemical Reactor

Hydrogen cyanide (HCN) is a colorless, poisonous liquid with a "bitter almond" odor. It is present naturally in a number of plant products at low concentrations. Years ago, hydrogen cyanide was sprayed on orange trees as a fumigant. Now its major use is as a basic chemical building block, especially for making nylon, acrylic plastics, and herbicides.

HCN is made from ammonia (NH_3) and methane (CH_4):

$$CH_4 (g) + NH_3 (g) \rightarrow HCN (g) + 3 H_2 (g)$$

Commercially, the reaction is carried out over a platinum catalyst at about 1100°C and 2 atm pressure. Suppose we try to carry this endothermic reaction out by heating up a stoichiometric mix of methane and ammonia to 1100°C, and then feeding the mixture to an insulated reactor. We've been advised that the reactor temperature must not go below 900°C, or the reaction rate slows too much, so we'll set the outlet temperature to 900°C.

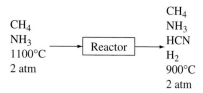

We'll set an arbitrary basis of 1 gmol/s NH_3 fed to the reactor. Analyzing the steady-state material balance equation gives

$$\dot{n}_{CH_4,in} = 1$$

$$\dot{n}_{NH_3,in} = 1$$

$$\dot{n}_{CH_4,out} = 1 - \dot{\xi}_1$$

$$\dot{n}_{NH_3,out} = 1 - \dot{\xi}_1$$

$$\dot{n}_{HCN,out} = +\dot{\xi}_1$$

$$\dot{n}_{H_2,out} = 3\dot{\xi}_1$$

(all in units of gmol/s). There is no work term, and the reactor is well insulated. We will assume that the insulation is good enough that the reactor operates adiabatically, so there is negligible heat transfer between the reactor and its surroundings. At steady state, the energy balance becomes simply:

$$\dot{H}_{out} = \dot{H}_{in}$$

We need to collect some data so we can evaluate the enthalpy change. We'll use the following values:

$$C_p[CH_4\,(g)] = 36 \text{ J/gmol °C}$$

$$C_p[NH_3\,(g)] = 36 \text{ J/gmol °C}$$

$$C_p[HCN\,(g)] = 36 \text{ J/gmol °C}$$

$$C_p[H_2\,(g)] = 29 \text{ J/gmol °C}$$

$$\Delta\hat{H}_r^\circ = +255.8 \text{ kJ/gmol } (\textit{highly} \text{ endothermic!})$$

We construct a pathway to move from the inlet conditions to the outlet conditions, and evaluate the enthalpy change associated with the pathway. This analysis is summarized in table form. (The phase remains vapor throughout.)

	T, °C	P, (atm)	Flow rate, gmol/s				Equation	kJ/s
			CH_4	NH_3	HCN	H_2		
Inlet stream	1100	2	1	1	0	0		
Cool to standard T	25	2	1	1	0	0	$\sum \dot{n}_i \int C_{p,i}\, dT$	-77.4
Lower to standard P	25	1	1	1	0	0		~0
React	25	1	$1 - \dot{\xi}_1$	$1 - \dot{\xi}_1$	$\dot{\xi}_1$	$3\dot{\xi}_1$	$\dot{\xi}_1 \Delta \hat{H}_{r1}$	$+255.8 \dot{\xi}_1$
Increase P	25	2	$1 - \dot{\xi}_1$	$1 - \dot{\xi}_1$	$\dot{\xi}_1$	$3\dot{\xi}_1$		~0
Heat to outlet T	900	2	$1 - \dot{\xi}_1$	$1 - \dot{\xi}_1$	$\dot{\xi}_1$	$3\dot{\xi}_1$	$\sum \dot{n}_i \int C_{p,i}\, dT$	$63.0 + 44.6 \dot{\xi}_1$
Sum								$-14.4 + 300.4 \dot{\xi}_1$

Now, we use this result in the energy balance:

$$0 = \dot{H}_{out} - \dot{H}_{in} = -14.4 + 300.4\,\dot{\xi}_1$$

$$\dot{\xi}_1 = 0.048\, \frac{\text{gmol}}{\text{s}}$$

Less than 5% of the raw materials are converted into products. Why? The problem is that the reaction is highly endothermic. The temperature drops as the reaction proceeds. We quickly bump into the 900°C limit on reactor outlet temperature.

We'd like to get much higher conversion. What's the best way to do this?

Idea 1:   We heat the inlet material in a heat exchanger hot enough to allow complete conversion in the reactor and still maintain a reactor outlet temperature of 900°C. Using an energy balance as above, we calculate that this would require an inlet temperature greater than 5200°C! The maximum temperature for stainless steel alloys is about 1100°C. Ceramics are good up to about 1450°C. [This alternative doesn't look too promising.]

Idea 2:   We encase the reactor in steam jacketing, or even embed tubes inside the reactor, and flow steam through the tubes to supply heat directly to the reactor volume. The maximum steam temperature available is about 550°C, not nearly hot enough.

Idea 3:   We coat the inside of furnace tubes with the platinum catalyst and place the tubes in a furnace. The raw materials flow through the tubes and react; fuel is burned in the furnace to provide the required heat for the reaction. Temperatures up to 1100°C are feasible with this method. Essentially, we are using an exothermic reaction—combustion of fuel in the furnace—to provide the heat needed for the endothermic reaction.

Idea 4:   We combine an exothermic reaction with the desired endothermic reaction, right inside the reactor! This idea is called *reaction combination*. If the two reactions are compatible, this is a viable solution.

Let's explore *idea 4* in our HCN reactor. Oxidation is a highly exothermic reaction. It may be possible to use an oxidative reaction to supply the energy needed to drive the desired reaction. Methane, one of the reactants, is easily oxidized:

$$CH_4 + 2O_2 \rightarrow CO_2 + 2H_2O$$

By controlled feeding of oxygen to the reactor, we might be able to burn just enough methane to release enough energy to balance the demands of the endothermic reaction. Let's say we prefer to maintain the reactor temperature at a constant 1100°C. How much extra methane must be fed to the reactor? How much oxygen?

To answer these questions, we first complete process flow calculations, where we assume the same basis of 1 gmol/s NH_3 fed to the reactor, complete conversion of ammonia, methane and oxygen, and two reactions:

	$\dot{n}_{in}$	$\dot{n}_{out}$
CH_4	$\dot{n}_{CH_4, in}$	$\dot{n}_{CH_4, in} - \dot{\xi}_1 - \dot{\xi}_2 = 0$
NH_3	1	$1 - \dot{\xi}_1 = 0$
O_2	$\dot{n}_{O_2, in}$	$\dot{n}_{O_2, in} - 2\dot{\xi}_2 = 0$
HCN		$\dot{\xi}_1$
H_2		$3\dot{\xi}_1$
CO_2		$\dot{\xi}_2$
H_2O		$2\dot{\xi}_2$

From these material balance equations, we easily calculate that $\dot{\xi}_1 = 1$, and $\dot{\xi}_2 = \frac{1}{2}\dot{n}_{O_2,in} = \dot{n}_{CH_4,in} - 1$.

Now we turn back to the energy balance equation. The reactor is adiabatic and steady-state:

$$\dot{H}_{in} = \dot{H}_{out}$$

We need a couple of additional pieces of data:

$$C_p[O_2\,(g)] = 29 \text{ J/gmol °C}$$
$$C_p[CO_2\,(g)] = 37 \text{ J/gmol °C}$$
$$C_p[H_2O\,(g)] = 34 \text{ J/gmol °C}$$
$$\Delta\hat{H}^\circ_{r2} = -802.6 \text{ kJ/gmol (\textit{highly} exothermic!)}$$

Let's redo the enthalpy calculations:

	T, °C	P, atm	Flow rate, gmol/s							Equation	kJ/s
			CH_4	NH_3	O_2	CO_2	H_2O	HCN	H_2		
Inlet stream	1100	2	$1+\dot{\xi}_2$	1	$2\dot{\xi}_2$	0	0	0	0		
Cool to standard T	25	2	$1+\dot{\xi}_2$	1	$2\dot{\xi}_2$	0	0	0	0	$\sum \dot{n}_i \int C_{p,i}\,dT$	$-77.4 - 101.1\dot{\xi}_2$
Lower to standard P	25	1	$1+\dot{\xi}_2$	1	$2\dot{\xi}_2$	0	0	0	0		~0
React by (R1)	25	1	$\dot{\xi}_2$	0	$2\dot{\xi}_2$	0	0	1	3	$\dot{\xi}_1\Delta\hat{H}_{r1}$	$+255.8$
React by (R2)	25	1	0	0	0	$\dot{\xi}_2$	$2\dot{\xi}_2$	1	3	$\dot{\xi}_2\Delta\hat{H}_{r2}$	$-802.6\dot{\xi}_2$
Increase P	25	2	0	0	0	$\dot{\xi}_2$	$2\dot{\xi}_2$	1	3		~0
Heat to outlet T	1100	2	0	0	0	$\dot{\xi}_2$	$2\dot{\xi}_2$	1	3	$\sum \dot{n}_i \int C_{p,i}\,dT$	$132.2 + 104.3\dot{\xi}_2$
Sum											$310.6 - 799.4\dot{\xi}_2$

Now, we use this in the energy balance:

$$0 = \dot{H}_{out} - \dot{H}_{in} = 310.6 - 799.4\dot{\xi}_2$$

$$\dot{\xi}_2 = 0.39 \frac{gmol}{s}$$

So, by feeding about 39% more CH_4 along with some oxygen, the reactor temperature could be maintained at 1100°C and all the ammonia could be converted to products. Can you think of any problems with this approach? (*Hint:* what happens downstream of the reactor?)

A similar plan might call for controlled oxidation of the hydrogen, a product of the desired reaction, to H_2O. Whether methane or hydrogen oxidation happens depends on the design of the reactor and the catalyst choice. Both processes are used commercially, as is idea 3.

Summary

- The energy of a system (or stream) is the sum of its **kinetic energy** E_k, **potential energy** E_p, and **internal energy** U. E_k varies with system velocity v, E_p depends on the height h of the system in a gravitational field. U varies with the system's pressure P, temperature T, phase ϕ, and chemical composition x_i. **Enthalpy** H is a convenient function for characterizing the energy of a stream entering or leaving a system, where $H = U + PV$.

- Energy is a **state function**—the energy of a system is a function of the state (v, h, P, T, ϕ, x_i) of the system and not the path by which that state was reached. Because energy is a state function, we can calculate the energy of a system simply by knowing its state. E_k, E_p, U and H are always calculated relative to some defined **reference state**.

- There are two methods to determine $\hat{U}$ and $\hat{H}$

  1. For a few common materials, data in the form of graphs and tables are available.
  2. For most other materials, we use physical property data such as **heat capacity** C_p, **enthalpy of phase change** (e.g., $\Delta\hat{H}_v$ or $\Delta\hat{H}_m$), **enthalpy of solution** $\Delta\hat{H}_{soln}$, and **standard enthalpy of formation** $\Delta\hat{H}_f^\circ$. Starting from a well-defined reference state, we construct a pathway in which one state variable (P, T, ϕ, or x_i) is changed at a time, we use model equations to calculate the change in $\hat{U}$ or $\hat{H}$ due to that change in variable, then we sum up all changes.

Change in	$\Delta\hat{H}$	$\Delta\hat{U}$	Comments
Pressure	~ 0 (gas) $\sim \hat{V}(P_2 - P_1)$ (solids or liquids)	~ 0	Usually can be neglected unless pressure change is very large.
Temperature	$\int C_p\, dT$	$\int C_v\, dT$	Polynomial expressions for C_p available. $C_p \approx C_v + R$ (gases) $C_p \approx C_v$ (solids and liquids)
Phase	$\Delta\hat{H}_v$ (liquid to vapor) $\Delta\hat{H}_m$ (solid to liquid)	$\Delta\hat{U}_v \approx \Delta\hat{H}_v - RT_b$ $\Delta\hat{U}_m \approx \Delta\hat{H}_m$	Data often available only at normal boiling point T_b and normal melting point T_m.
Composition due to reaction	$\Delta\hat{H}_r^\circ = \sum_i \nu_i \Delta\hat{H}_{f,i}^\circ$ $\Delta\hat{H}_r^\circ = -\sum_i \nu_i \Delta\hat{H}_{c,i}^\circ$	$\Delta\hat{U}_r \approx \Delta\hat{H}_r - RT\Delta n_r$ (gas) $\Delta\hat{U}_r \approx \Delta\hat{H}_r$ (solids or liquids)	Standard conditions are 25°C and 1 atm. Per mole of reaction.
Composition due to mixing or solution	$\Delta\hat{H}_{soln}$ $\Delta\hat{H}_{mix}$	$\Delta\hat{U}_{soln} \approx \Delta\hat{H}_{soln}$ $\Delta\hat{U}_{mix} \approx \Delta\hat{H}_{mix}$	Important for nonideal mixtures—acids, bases, for example.

- **Work** and **heat** are energy flows in the absence of material flows. Heat $\dot{Q}$ is energy flow due to differences in temperature between the system and surroundings. Work $\dot{W}$ is mechanical energy flow. **Shaft work** $\dot{W}_s$ is work due to rotating or reciprocating equipment.

- The First Law of Thermodynamics states that energy is conserved. This principle is expressed quantitatively in the **energy balance equation.**
  The differential energy balance equation is

$$\frac{d(E_{k,\text{sys}} + E_{p,\text{sys}} + U_{\text{sys}})}{dt}$$

$$= \sum_{\text{all } j \text{ in}} \dot{m}_j \left(\hat{E}_{k,j} + \hat{E}_{p,j} + \hat{H}_j\right) - \sum_{\text{all } j \text{ out}} \dot{m}_j \left(\hat{E}_{k,j} + \hat{E}_{p,j} + \hat{H}_j\right) + \sum_j \dot{Q}_j + \sum_j \dot{W}_{s,j}$$

The integral energy balance equation is

$$\left(E_{k,\text{sys}} + E_{p,\text{sys}} + U_{\text{sys}}\right)_f - \left(E_{k,\text{sys}} + E_{p,\text{sys}} + U_{\text{sys}}\right)_0$$

$$= [E_{k,\text{in}} - E_{k,\text{out}}] + [E_{p,\text{in}} - E_{p,\text{out}}] + [H_{\text{in}} - H_{\text{out}}] + Q + W_s$$

with $M = \sum_j \left(\int_{t_0}^{t_f} \dot{M}_j\, dt\right)$, where M is any energy form.

Both differential and energy balance equations can be written in molar as well as than mass units.

- Successful use of the energy balance equation to solve process energy problems requires a systematic approach such as the 12 Easy Steps.
    1. Draw a diagram.
    2. Define a system.
    3. Choose components and define stream variables.
    4. Check units for consistency.
    5. Define a basis.
    6. Set up process flow calculations, and solve if possible.
    7. Find data or write model equations for energy of process streams.
    8. Identify heat and work flows.
    9. Find data or write model equations for changes in system energy.
    10. Write and simplify the energy balance equation.
    11. Solve equations.
    12. Check solution.

 ## ChemiStory: Get the Lead Out!

Thomas Midgley, born May 18, 1889, rumored to be a descendant of James Watt, was a remarkable inventor. Counted among his many important contributions to modern technology were his discovery that elm-tree juice worked better than spit for curveball pitchers and his invention of the Screwdriver (orange juice and vodka). Midgley graduated from Cornell University with a degree in mechanical engineering. He was a mediocre student, uninterested in chemistry. His first job was at National Cash Register (NCR) in Dayton, Ohio, one of the glamorous companies of the day. After only a year at NCR, Midgely joined his father's tire company, which collapsed in 1916. With little in the way of attractive job prospects, and a wife and two small children to support, he joined the fledgling firm of Charles Kettering. The company, Dayton Engineering Laboratories, was a two-bit operation—with a fake door on the front of an old warehouse. The labs were prolific, however; Kettering invented the electrical ignition, which replaced the heavy hand cranks on "horseless carriages" and made it possible for ordinary men and women to start up an automobile engine. Midgely too was an enthusiastic inventor, bubbling over with ideas—most screwy, but some absolutely ingenious. For instance, it was his idea to put an ice cube in a package of Birdseye frozen vegetables, to prove that the package had never thawed.

Automobile engines of the early 1900s suffered from knock—that "ping" heard whenever an internal combustion engine was under strain. The problem was more than just an annoyance—engines would suddenly lose power, and could even be destroyed. Internal combustion engine design was a balancing act between better performance and knock avoidance. Higher compression

Thomas Midgley
© *Corbis*

ratios (higher pressures of the air and fuel mix inside the cylinders) meant more efficient engine operation but increased the knock. Gasoline from some petroleum sources did not have as much of a knock problem but this was expensive high-octane gasoline, naturally high in aromatics like benzene and toluene. The American auto industry was interested in making powerful engines that could run on cheap, low-octane gas. General Motors contracted with Kettering's company to see what it could come up with. Midgley had no knowledge of petroleum chemistry or internal combustion. But neither did anyone else.

Midgley set out to first understand what knock was. He jury-rigged together a high-speed camera by wrapping film around a tomato can and spinning it on roofing nails. By filming the operating engine through a quartz window cut into the cylinder wall, Midgley discovered that knock was due to premature explosion of the fuel in the cylinder. Kettering guessed that adding a dye to the gasoline would make the fuel ahead of the flame front absorb radiant energy better and vaporize sooner. So Midgley decided to try whatever he could find in the stockroom. The only thing available was iodine-a dark-colored liquid. Midgley added the iodine to gasoline, and lo and behold, no more knock! He went around town and found every oil-soluble dark dye he could find. Nothing worked. Then he tried iodine again—this time as a colorless preparation—and, no knock!

But iodine was corrosive, smelly, and expensive. Still, the experiment convinced Midgley that chemistry could solve the knocking problem. He screened, one by one, over 30,000 compounds, and finally discovered aniline—a highly aromatic compound derived from coal tar. Aniline was much cheaper than iodine, and worked better, but it was still highly toxic, corrosive, and stinky. In 1921 he discovered selenium and tellurium compounds that were antiknock compounds, but these also stank so bad that Midgley's wife banished him to the basement for 7 months.

About this time, Midgley learned about the periodic table, which had been developed in 1913. He noticed that almost all the antiknock compounds he had discovered were bunched together in the lower right-hand corner of the table. He predicted that lead might work. An oil-soluble form of lead was found in tetraethyl lead, $Pb(C_2H_5)_4$, a toxic and rare compound described by a German chemist 75 years earlier. The material proved to be cheap and extraordinarily effective. "Leaded" gasoline worked like magic to improve engine performance, increase mileage, and stop ping. A joint venture with Dow Chemical Company was started to market tetraethyl lead, called Ethyl-Dow, and Midgley became vice president. (Later the company was renamed Ethyl Corporation.)

Lead was a known neurotoxin, and it was understood by the 1930s that lead accumulated in the body and poisoned workers that were exposed to it through their occupation—house painters, for instance. Ethyl Corporation decided to call its leaded gasoline "ethyl," to avoid the adverse connotations of the word "lead". Despite costing about 15% more than regular gasoline, ethyl was a big hit with the motoring public. But in 1924, six workers at ethyl manufacturing facilities died from lead poisoning. A few workers went insane, or suffered from hallucinations. Local newspapers called the lead fumes "looney gas." Midgley counteracted the bad press by going on the offensive. When asked if tetraethyl lead was hazardous, he deliberately poured a bottle of the material on his hands, and held the bottle to his nose. He indicated that the only action required was close supervision of workers—laying them off if they showed signs of poisoning, and firing them if they spilled any lead on themselves. In 1925, the Surgeon General held a public hearing on the topic. Industry representatives argued that the deaths were simply accidents, and that fear must not be allowed to destroy the chance to use this wonder chemical. Several independent scientists went on the attack, claiming that the studies purporting to show the safety of ethyl were seriously flawed. But they were outgunned—and outfinanced. Those fighting the automotive and the petrochemical industry were cast as a ragtag bunch of feminists, socialists, and pacifists.

©Bettmanri/Corbis

American drivers responded warmly to the "Put a Tiger in your Tank" ads, promoting high-octane leaded gasolines. The rapid improvements in gasoline engines led to an American love affair for muscle cars—and to miles of newly paved roads and highways, sharp drops in funding for public transportation, and suburban sprawl. By 1960, nearly 90% of all gasoline sold in the United States contained tetraethyl lead. Lead wasn't just in gasoline—it was in lead ('tin') cans, lead pipes, lead arsenate pesticide, lead-glazed ceramics, lead paint. Americans ingested about 20 tons per year of lead.

This all changed, in part because of a chemist named Clair Patterson. A native of Iowa, tall and intensely earnest, "Pat" Patterson worked on the Manhattan project at Oak Ridge National Laboratories. There he learned mass spectroscopy techniques used to isolate uranium isotopes, and argued against use of the atomic bomb on a Japanese city. He returned to University of Chicago, for his Ph.D. in chemistry, where he became interested in using isotopes, particularly the decay of uranium to lead, to measure the age of the earth. For 7 years he had problems measuring lead in old meteorites, mainly because his laboratory was incredibly contaminated with lead. In the process of decontaminating his lab, he discovered that everyday materials contained much more lead than previously thought. In 1963, he published a landmark paper showing that even snowfall in pristine national parklands was heavily contaminated with lead from human activity. Three days after publication of Patterson's article, Ethyl executives showed up at his office. After a brief discussion, the men left. Shortly thereafter, Patterson's research grants with the U.S. Public Health Service were not renewed.

The pressure to abandon his work on lead contamination was severe. But a chance meeting with a toxicologist, Harriet Hardy, convinced Patterson that children were dying from lead poisoning. Patterson traveled to the ends of the earth, demonstrating that snow and ice deposited during the twentieth century at the north and south poles were heavily contaminated with industrial lead. In 1965, Patterson published an article lambasting the lead additive industry, claiming that lead levels in humans had risen dramatically over the last 20 years, that all people, not just factory workers and paint-eating toddlers, were exposed to dangerously high levels of lead. Contrary to accepted wisdom, he argued that sublethal lead exposure levels were damaging. Patterson was attacked viciously for this article. The chief of toxicology of the U.S. Public Health Service disparaged the work by saying "Is Patterson trying to be a second Rachel Carson?"

The battle lines were drawn. A conference on environmental lead, held in late 1965, ended with Ethyl Corporation releasing a statement emphasizing the "economic hardships that would be caused by the disappearance of lead antiknocks." Government hearings brought to the attention of the U.S. public that virtually all of the lead toxicity research had been sponsored by industries with economic ties to the use of lead, and that they had held rights to refuse publication of any research results. The public mood shifted toward a call for government regulation. In 1970, Congress passed the Clean Air Act, and established the Environmental Protection Agency. Companies such as General Motors installed catalytic converters to control emissions. Lead was a catalyst poison, so leaded gasoline could not be used in these new cars. New technologies were developed to "crack" petroleum into high-octane fuel, rich in branched and aromatic hydrocarbons. As the public replaced old cars with new, use of leaded gasoline plummeted. So too did the levels of lead in the blood, and in the polar ice caps.

Controversies over gasoline additives continue. Gasoline made by cracking is high in benzene, a known carcinogen. Oxygenated fuel additives, like methyl t-butyl ether (MTBE), were mixed into gasoline to reduce harmful emissions, but because of water pollution problems MTBE has been withdrawn. Ethanol, made from corn, is added to gasoline in many Midwestern states, but heavy government subsidies paid to ethanol producers have been attacked. The search for safe and efficient transportation fuels continues.

Quick Quiz Answers

6.1 Energy/time or $[M \cdot L^2/t^3]$. Energy or $[M \cdot L^2/t^2]$.

6.2 330 ft lb$_f$. Jumping makes no difference in net potential energy change.

6.3 Pressure is force/area, or $[M \cdot L/t^2 \cdot L^2]$ or $[M/t^2 \cdot L]$. Volume is $[L^3]$. Pressure times volume is therefore $[M \cdot L^2/t^2]$, which is the dimension of energy.

6.4 151.83°C and 2748.1 kJ/kg. Subcooled.

6.5 Not quite; both U and H decrease a bit with increasing pressure. Pretty close, but not perfect.

6.6 About 150 J/g, or about 6600 J/gmol.

6.7 2014.5 kJ/kg; different because change in pressure of vapor was ignored in calculation. Zero.

6.8 Energy per unit time, or $[M \cdot L^2/t^3]$. Zero (if thermos is well-insulated). Negative.

6.9 $d\left(E_{k,\text{sys}} + E_{p,\text{sys}} + U_{\text{sys}}\right)/dt$;

$$\sum_{\text{all } j \text{ in}} \dot{m}_j \left(\hat{E}_{k,j} + \hat{E}_{p,j} + \hat{H}_j\right) - \sum_{\text{all } j \text{ out}} \dot{m}_j \left(\hat{E}_{k,j} + \hat{E}_{p,j} + \hat{H}_j\right); \sum_j \dot{Q}_j + \sum_j \dot{W}_{s,j}.$$

Energy/time or $[M \cdot L^2/t^3]$

6.10 $\left(E_{k,\text{sys}} + E_{p,\text{sys}} + U_{\text{sys}}\right) - \left(E_{k,\text{sys}} + E_{p,\text{sys}} + U_{\text{sys}}\right)$;
$\left[E_{k,\text{in}} - E_{k,\text{out}}\right] + \left[E_{p,\text{in}} - E_{p,\text{out}}\right] + \left[H_{\text{in}} - H_{\text{out}}\right]; Q + W_s$. Energy or $[M \cdot L^2/t^2]$.

6.11 Lower, some of the enthalpy released in the reaction goes to warm up the N_2 in the air.

References and Recommended Readings

1. Schematics and descriptions of process equipment such as heat exchangers, compressors, and pumps are available in *Perry's Chemical Engineers' Handbook*. The *Visual Encyclopedia of Chemical Engineering Equipment*, developed by Dr. Susan Montgomery, is a compact disc available for purchase at a nominal fee from CACHE Corporation at www.che.utexas.edu/cache/products.html.

2. Tables of thermodynamic data (U, H) and physical property data are available from numerous sources, including *Perry's Chemical Engineers' Handbook*, *CRC Handbook of Chemistry and Physics*, and *Lange's Handbook of Chemistry*. Another useful resource is the *DIPPR (Design Institute for Physical Property Data) Chemical Database*. The National Institute of Standards and Technology (NIST) provides free online access to the NIST Chemistry WebBook at webbook.nist.gov, which contains physical property data (e.g, enthalpy of formation or heat capacity) for thousands of compounds, and detailed thermodynamic data for selected fluids, including steam.

3. For more information on methods to estimate explosive potential, see M. R. Murphy, S. K. Singh, and E. S. Shanley, *Chem. Eng. Prog* 98, 54–62 (2003).

4. See *Prometheans in the Lab,* by S.B. McGrayne, for more on Thomas Midgley and Clair Patterson.

Chapter 6 Problems

Warm-Ups

P6.1 Eating an apple provides about 100 kcal. Calculate the energy value of an apple in kJ and in Btu. Combustion of 1 gal of gasoline releases about 125,000 kJ. About how many apples does it take to equal 1 gal of gasoline?

P6.2 List three everyday items that use electricity. Calculate the energy usage rate in W, hp, kJ/s, kcal/min, and Btu/h.

P6.3 If gasoline sells for $2.00/gallon and combustion of 1 gal of gasoline releases about 125,000 kJ, what is its price per Btu? How does this compare to the price of natural gas or coal, given in Sec. 6.1?

P6.4 305 gal of a 75 wt% ethanol/25 wt% water solution (specific gravity = 0.877) are mixed with 72 lbmol of a 24 mol% ethanol /76 mol% water solution (specific gravity = 0.945). The molecular weight of ethanol is 46.07 and that of water is 18.016. Suppose that the mixture were taken to Phobos, the inner moon of Mars, where the acceleration of gravity is 3.78 ft/s^2. What would be the mass in lb and the weight in lb$_f$ of the mixture?

P6.5 In the early 1400s construction began on the magnificant Duomo in Florence, Italy. One of many challenges was the delivery of heavy loads of building materials to the top of the dome. Fillippo Brunelleschi invented a hoist powered by oxen. It has been estimated that one ox could lift a half-ton load 200 ft high in 13 min. Use this to determine a conversion factor for oxpower to horsepower.

P6.6 A person biking for 30 min to work can cover about 5 miles and will burn about 100 kcal, equivalent to eating an apple. Compare the energy expenditure, in equivalent apples, of a person driving the same distance in a pickup truck, if the truck gets 20 miles per gallon and combustion of one gallon of gas releases 125,000 kJ.

P6.7 Energy flow from the sun at temperate latitudes is about 96.4 mW/cm^2. New photovoltaic devices for harnessing solar power are under development (*Science* 295:2245, 2002). The devices consist of cadmium selenide rods and the polymer poly-3(hexylthiophene). These new devices have a power conversion of 1.7%. What surface area (ft^2) of the devices is required to light a light bulb or run a hair dryer?

P6.8 A 100-lb downhill skier races down a 30° slope and quickly reaches a constant speed of 60 mph. It takes 32 seconds for her to drop 1200 ft in elevation. She generates the same heat from friction as if she had three 100-W light bulbs under each ski. Compare ΔE_k, ΔE_p and energy dissipated by friction for this skier.

P6.9 Meteorites enter the atmosphere at about 11 km/s but decelerate within a few seconds to a terminal velocity of 100 m/s because of collision with air molecules. What is the change in the kinetic energy of a 3-kg meteorite? What happens to this energy?

P6.10 How many kg of water can be heated from 50°C to 75°C by condensing 1 kg of steam at 100°C? Use Table 6.1.

P6.11 Calculate the change in enthalpy in taking a saturated liquid ammonia-water solution containing 0.65 kg NH_3 per kg solution from 213 K and 0.1 bar to 253 K and 1 bar. Use Fig. 6.6.

P6.12 As a cup of coffee cools from 70°C to room temperature, about how much heat is lost? Calculate in cal, J, and Btu. Make any reasonable approximations.

P6.13 The refrigerant Freon R-22 (CCl_2F_2) has a very low heat capacity compared to water (C_p about 0.6 J/g °C, versus about 4 J/g°C for water). Suppose you are considering Freon-22 versus water as a cooling fluid. If 100 J heat must be removed by using 10 g fluid, compare the temperature increase in Freon-22 to that in water.

P6.14 Look up polynomial expressions for C_p of CO_2 and H_2O vapor in App. B. Calculate C_p at several temperatures from 100°C to 500°C and plot C_p versus T. Compare these accurate expressions to the rule-of-thumb numbers given in the text.

P6.15 Calculate the enthalpy change associated with dissolving 1 gmol of citric acid in a large beaker of water. Will the beaker feel warm or cool? (Refer to App. B for enthalpy of solution data.)

P6.16 Ethanol (C_2H_5OH) and acetic acid (CH_3COOH) react in the liquid phase to make ethyl acetate ($C_4H_8O_2$) and water. What is the standard enthalpy of reaction (kJ/gmol)?

P6.17 Use $\Delta\hat{H}_f^\circ$ to calculate $\Delta\hat{H}_r^\circ$ for the reaction:

$$C_2H_6\,(g) + 3.5\,O_2\,(g) \rightarrow 2\,CO_2\,(g) + 3\,H_2O\,(l).$$

Compare to $\Delta\hat{H}_c^\circ$ for C_2H_6 (App. B).

P6.18 A hard candy contains 21 g sucrose ($C_{12}H_{22}O_{11}$). If you eat it, how many calories will you consume? $\Delta\hat{H}_f^\circ = -2221$ kJ/gmol for sucrose (s).

P6.19 A general college chemistry text suggests a simple experiment you can do at home. In a nutshell: Water is added to an empty soda can to a level of 1 cm. The can is then heated on a hot plate until the water boils and steam is emitted for about 1 minute. With steam coming out the top of the can, the can is grabbed and turned upside down and immediately plunged into a bowl containing ice-cold water.

You don't need to do the experiment to know what will happen, because you know all about conservation of energy. Write an energy balance equation, using the can and its contents as the system, at two points: (a) as the can is being heated and (b) as the can is plunging into the ice-cold water. Is there a heat term in (a) or (b)? If yes, is it positive or negative? Is there a work term in (a) or (b)? If yes, is it positive or negative?

P.6.20 Simplify the energy balance equation for each of the following cases. State whether Q or W is positive, negative, or zero in each case. Briefly explain your answer. (*Hint*: be clear on your system choice!)

(a) Steam enters a rotary turbine and turns a shaft connected to a generator. The inlet and outlet steam ports are at the same height. Some energy is lost to the surroundings as heat.

(b) A liquid stream is heated from 25°C to 80°C. The inlet and outlet pipes have the same diameter, and there is no change in elevation between these points.

(c) Water passes through the sluice gate of a dam and falls on a turbine rotor, which turns a shaft connected to a generator. The fluid velocity on both sides of the dam is negligible, and the water undergoes insignificant pressure and temperature changes between the inlet and outlet.

(d) Crude oil is pumped through a cross-country pipeline. The pipe inlet is 200 m higher than the outlet, the pipe diameter is constant, and the pump is located near the midpoint of the pipeline. Heat generated by friction in the line is lost through the wall.

(e) N_2 and H_2 are fed to a reactor that contains a catalyst that converts N_2 and H_2 to ammonia. The ammonia exits the reactor.

(f) Water in a lake falls 200 m through a 6-in-diameter pipe and exits into a river.

(g) Yeast digest sugar in bread dough.

(h) Acetic acid flows through the tubes of a furnace, where it reacts to ketene and water.

(i) Cold milk is added to a hot cup of coffee.

(j) A student pours water into concentrated sulfuric acid in a beaker.

(k) A student sleeps in a classroom.

Drills and Skills

P6.21 Is enthalpy really a state function? (a) Use Table 6.1 to calculate the change in enthalpy of H_2O as it changes from liquid, 1 bar, 75°C to vapor, 10 bar, 200°C. (b) Use Table 6.1 to find the change in enthalpy of H_2O for (i) increasing temperature from 75°C to 99.6°C, as a liquid at 1 bar, (ii) changing from liquid to vapor phase at 99.6°C and 1 bar, (iii) increasing temperature of vapor at 1 bar from 99.6°C to 200°C and (iv) increasing pressure of vapor from 1 bar to 10 bar at 200°C. Sum up the individual enthalpy changes and compare to the number you got in part (a).

P6.22 Refer to Table 6.1. Plot the data as $\hat{H}$ versus T at $P = 1$, 10, and 100 bar, from 50°C to 350°C. From your plot, estimate (a) C_p for steam and (b) C_p for liquid water, at 1, 10, and 100 bar. Find $\Delta\hat{H}_v$ at each pressure. How does $\Delta\hat{H}_v$ depend on pressure? What happens to $\Delta\hat{H}_v$ as the pressure approaches critical pressure P_c?

P6.23 1 kg water at 1 bar and 50°C is pressurized and heated to 10 bar and 100°C. What is the change in enthalpy associated with this process? Calculate first using model equations, then compare to steam tables.

P6.24 20 kg of water vapor at 1 bar is heated from 100 to 320°C. What is the enthalpy change calculated (a) from the steam table, (b) using the approximate C_p given in App. B, and (c) using the polynomial expression for C_p given in App. B?

P6.25 Freshwater can be reclaimed from seawater by several different methods. Let's compare two methods: one in which seawater is frozen to make ice and a salty brine, and the other in which water is evaporated off. Calculate the energy required to freeze 1 kg of water and to evaporate 1 kg of water at 1 atm, starting at 20°C in both cases. Estimate energy cost of each process ($/metric ton), using the data in App. B and Sect. 6.1.

P6.26 Calculate the energy change, in J, for 1 gmol of H_2O undergoing the following changes:
(a) accelerating from 0 to 100 km/h
(b) rising 100 meters
(c) increasing in pressure at 100°C from 0.1 bar to 1 bar
(d) increasing in temperature from 0 to 100°C at 5 bar
(e) dissociating to H_2 and O_2 at 25°C
(f) undergoing a nuclear reaction where all mass converts to energy ($E = mc^2$)

P6.27 Water flows through a pipe at 10 lb/h. The water temperature is 70°F and the pressure is 1 atm. Suppose I want to heat the water to make saturated steam at 1 atm. Can I do it with a hair dryer?

P6.28 $\Delta \hat{H}_v$ for acetone is 29.1 kJ/gmol, at T_b(1 atm) = 56.5°C. Use the Antoine equation to find the temperature where acetone boils at 1.5 atm. Calculate $\Delta \hat{H}_v$ at that temperature, using polynomial C_p data given in App. B. Generalize your strategy to obtain an equation to calculate $\Delta \hat{H}_v$ for acetone at any temperature in the range 20 to 100°C. Use the results of your calculations to plot $\Delta \hat{H}_v$ versus T. Also plot $\Delta \hat{H}_v$ versus P. Compare your two plots and comment on the change in $\Delta \hat{H}_v$ with T and with P.

P6.29 Streams of 40 gmol/s CO and 60 gmol/s CO_2 at 200 kPa and 50°C are mixed, then heated to 200°C. What heat input (kJ/s) is required?

P6.30 You've got a solid that you know is sodium carbonate (Na_2CO_3), but you're not sure if it is in a hydrated or nonhydrated form. You dissolve 1.00 g of the powder in 5 L of water in an insulated container. Initially the water and the powder are both at 20.00°C. You very carefully measure the temperature after dissolution of the powder. What temperature would you measure if the solid were (a) Na_2CO_3, (b) $Na_2CO_3 \cdot H_2O$, (c) $Na_2CO_3 \cdot 7\ H_2O$, or (d) $Na_2CO_3 \cdot 10\ H_2O$? Could you use this test to identify the hydration state of your solid?

P6.31 Calculate the fuel value (kJ/kg burned) and CO_2 emissions (g/kJ) for ethanol (C_2H_5OH) and hydrazine (rocket fuel, N_2H_4). Compare to the values for more conventional fuels, reported in Sec. 6.1.2.

P6.32 In a bomb calorimeter, a sample is placed into an inner combustion chamber and combusted to CO_2 and H_2O. The combustion chamber sits inside another chamber which is filled with an exact amount of water. The temperature of the outer chamber is carefully monitored. In one experiment, 2.7 g of a proprietary organic chemical with the molecular formula $C_8H_{10}O_3$ is placed in a bomb calorimeter and combusted. In the outer chamber, which contains 650 g water, the temperature increases from 23.6°C to 36.2°C. Use these data to calculate an enthalpy of formation (kJ/g) for the unknown chemical.

P6.33 Cold packs are used by athletes and trainers to reduce swelling and inflammation after injuries. A cold pack is activated by initiating dissolution of ammonium nitrate (NH_4NO_3) into water. In one product, 25 g ammonium nitrate is mixed with 250 g water, initially at room temperature (about 22°C). About how cold will the pack get? The enthalpy of solution at infinite dilution is 6.47 kcal/gmol ammonium nitrate.

P6.34 Pocket-sized hot packs are popular as handwarmers with skiers, other winter-sports enthusiasts, and people who need to work outside in the cold. With one type of disposable hotpack, Fe filings are placed in a sealed bag. When the seal is broken by the user, air leaks in. Oxygen reacts with the iron, producing iron oxides, mainly Fe_2O_3. Marketing studies suggest that consumers want a slow heat release of about 100 kJ/h and would like the hotpacks to last about 6 h. How much Fe would you put in the hot pack? How much would the mass of the hotpack change over the course of its use (total g and %)?

P6.35 Heat is applied to a tank filled with water, generating saturated steam at 1 bar from cold water. The tank contains 25 kg of water at 20°C at the start, and the heater supplies 15 kW. Saturated steam leaves through a vent for further superheating in another unit. How long does it take for the first steam to be produced, and how long will it take to vaporize all the water?

P6.36 We need to calculate a standard-state $\Delta\widehat{H}_r^\circ$ for the oxidation of *p*-xylene (C_8H_{10}) to terephthalic acid (TPA, $C_8O_4H_6$), which is used in making plastic 2-L soda bottles. However, heats of formation of the compounds from the respective elements are not available. Reference tables do have information on heats for combustion to CO_2 (g) and H_2O (l) at 25°C, and show that the energy released is 1089 kcal/mol for *p*-xylene (l) and 770 kcal/mol for TPA (s).

$$C_8H_{10} \text{ (l)} + 3\,O_2 \text{ (g)} \rightarrow C_8O_4H_6 \text{ (s)} + 2\,H_2O \text{ (l)}$$

Determine a numerical value for $\Delta\widehat{H}_r^\circ$. Given that $\Delta\widehat{H}_f^\circ$ for $CO_2 = -94$ kcal/mol and $\Delta\widehat{H}_f^\circ$ for liquid $H_2O = -68$ kcal/mol, calculate $\Delta\widehat{H}_f^\circ$ for *p*-xylene and TPA.

P6.37 A pure sodium crystal (100 g) sits in a steel cup. One drop (1 mL) water is added. Hydrogen gas is produced. What else happens? Explain, as quantitatively as possible, why it is dangerous to add water to pure metals like sodium.

P6.38 Yeast cells use glucose ($C_6H_{12}O_6$) as their predominant energy source. Aerobic metabolism of glucose leads to complete oxidation of glucose to carbon dioxide and water. (*Aerobic* means that the metabolism requires oxygen from air.)

$$C_6H_{12}O_6 + 6\,O_2 \rightarrow 6\,CO_2 + 6\,H_2O$$

What is $\Delta \widehat{G}_r^\circ$ and $\Delta \widehat{H}_r^\circ$ of glucose oxidation?

This reaction is coupled to the synthesis of ATP (adenosine triphosphate) from inorganic phosphate and ADP (adenosine diphosphate). $\Delta \widehat{G}_r^\circ$ of this reaction is $+7.3$ kcal/gmol.

$$ADP + H_3PO_4 \rightarrow ATP$$

If 6 moles of ADP are converted to ATP for every mole of oxygen consumed by aerobic glucose, what is the efficiency of conversion of the chemical energy of glucose into chemical energy of the phosophamide bond in ATP? If 5 g of glucose is dissolved in a fully insulated beaker containing 10 mL water and yeast, estimate the water temperature after all the glucose is consumed. (Assume that the beaker itself has no heat capacity. Is this a good assumption?)

For problems P6.39 through P6.51, explicitly use the 12 Easy Steps.

P6.39 6000 kg/h steam at 10 bar and 400° C is expanded adiabatically to 0.5 bar in a turbine that generates power. The steam leaving the turbine is cooled by removing heat at the rate of 1.25×10^7 kJ/h to produce a saturated liquid at 0.5 bar. How much work is produced in the turbine (in kW)? What is the quality of the steam leaving the turbine? (Steam quality is the fraction of the steam that is vapor. Steam can have some entrained liquid droplets.)

P6.40 100 lbmol/h air is preheated from ambient temperature (70°F) to 400°F in a heat exchanger by condensing saturated steam at 500 psia to saturated liquid. What is the steam flow rate (lbmol/h) through the heat exchanger? C_p of air is 7 Btu/lbmol °F.

P6.41 100 kg saturated steam at 10 bar, 20 kg ice, and 80 kg water at 20°C are mixed together in an insulated pot. The pressure is kept constant at 10 bar. What is the temperature and phase of the mixture in the pot at equilibrium?

P6.42 1000 kg/min of liquid water at 50°C and 1 bar is pumped isothermally to 100 bar and then fed at constant pressure to a boiler, where saturated steam is produced. The saturated steam then passes through an expansion valve, across which the pressure drops to 60 bar. Neglect friction and changes in kinetic energy, and assume that the expansion valve operates adiabatically. How much work must be supplied by the pump? How much heat must be supplied by the boiler? What is the final temperature, pressure, and phase? (Use steam tables.)

P6.43 Hot oil is pumped from a vessel in a crude oil distillation plant through an 8-in-diameter pipe to a storage tank on a hill 300 feet above the plant. The oil must be cooled from 200°F to 80°F prior to storage. The flow rate is 1400 gal/min. Assume that the density of the oil is 50 lb/ft³ and the heat capacity is 0.63 Btu/lb °F. Calculate the amount of cooling required in the cooler (Btu/h) and the work done by the pump (hp). You may neglect frictional losses. Assume that all equipment is well insulated.

P6.44 Saturated steam at 29.0 psig is to be used to heat a stream of ethane at constant pressure. The ethane enters the heat exchanger at 60°F and 20 psia at a rate of 27,890 ft³/min and is heated to 200°F. The steam condenses and leaves the exchanger as a liquid at 80°F. How much heat (Btu/min) must be provided to heat the ethane from 60°F to 200°F? At what rate (ft³/min) must steam be supplied to the heat exchanger?

P6.45 We need to cool a hot vapor stream for further processing, and wish to recover some of the heat as steam that will be usable elsewhere. The process stream is *n*-octane, flowing at 150 gmol/min, 1 atm, and 350°C. It is to be cooled to produce liquid octane at 100°C. You are to make saturated steam at 5 bar from a supply of water available at 60°C and 5 bar. What is the temperature of the steam being produced? How much steam can be produced (kg/min)? (Try solving with just the rules of thumb for C_p and $\Delta \hat{H}_v$. Then, look up physical properties and compare the exact solution to the back-of-the-envelope calculations.)

P6.46 A hot benzene stream (1000 gmol/h) at 500°C and 1 bar pressure is rapidly cooled by mixing with cold liquid benzene at 25°C and 1 bar pressure. The final temperature of the combined benzene streams must be 200°C. How much liquid benzene must be added?

P6.47 Acetylene is used in welding torches, but it is much more expensive than methane. Compare the adiabatic flame temperature of acetylene to that of methane and see why acetylene is chosen instead of methane. (Assume stoichometric amount of air is mixed with the fuel.)

P6.48 A waste gas stream contains 40 mol% ethane (C_2H_6), 19 mol% N_2, 1 mol% H_2S, 10 mol% O_2, and 30 mol% CO. The gas is fed to a furnace along with 20% excess air and combusted. The waste stream entering the furnace is at 25°C. The air enters the furnace at 25°C. The flue gas exits the furnace stack at 150°C. Complete combustion can be assumed. The heat generated in the furnace is used to generate saturated steam at 10 bar, using boiler feedwater at 50°C and 10 bar. Calculate the amount of steam generated in the furnace per gmol waste gas fed (kg steam per gmol gas).

P6.49 Methane (CH_4) is combusted with 25% excess air in an adiabatic burner operating at 1 atm pressure. The methane and air enter the burner at 25°C. Assume that combustion proceeds to completion. The combustion gases are then cooled to 80°C and discharged to the atmosphere. What is the temperature of the gases leaving the burner? What fraction of the water condenses in the cooler? What is the minimum temperature in the cooler to avoid any condensation?

P6.50 Miner's lamps contain calcium carbide and water, which react to produce acetylene and calcium hydroxide. The acetylene is then burned as a torch. The reaction is

$$CaC_2 \text{ (s)} + 2 \text{ } H_2O \text{ (l)} \rightarrow C_2H_2 \text{ (g)} + Ca \text{ (OH)}_2 \text{ (aq)}.$$

If the acetylene should be burned at a rate equivalent to a 100 W light bulb, what should be the reaction rate (gmol/s) of the above reaction?

P6.51 1000 kgmol/hr acetic acid at 3 atm pressure and 50°C is heated in a heat exchanger to 170°C by condensing saturated steam at 10 bar. What is the steam flow rate (kg/h)? A colleague suggests that saturated steam at 5 bar would be a better choice for the heat source than saturated steam at 10 bar, because the required steam flow rate would be less. What do you think of this idea?

Scrimmage

P6.52 Use government sources to find data on energy consumption in the United States and internationally over the past 50 years. Imagine that you are the technical consultant to a U.S. senator. Write a position paper, based on your research, on an energy-related issue for the senator's use.

P6.53 A liquid mixture (100 gmol/s) of 40 mol% *n*-hexane/60 mol% *n*-heptane at 10 atm pressure and 25°C is sent through an expansion valve, where the pressure drops to 1 atm, and into a flash tank, which is heated. Vapor and liquid are separated in the tank and leave at 1 atm and 85°C. Calculate the compositions and flow rates of the vapor and liquid streams leaving the tank. Calculate the amount of heat (kJ/s) added to the tank.

P6.54 Printers have to clean up solvent-laden exhaust fumes from their press-rooms. Let's consider condensing the solvent to recover it for recycle. The gas stream is modeled as 20 mol% isopropanol in air, flowing at 0.5 m³/s at 1 atm and 50°C.
  (a) Sketch a flow sheet for a condenser that will recover 99% of the iso-propanol as liquid condensate. Determine the condenser temperature required, and solve and label all stream flow rates, compositions, and temperatures. Calculate the rate of cooling (kJ/s) in the condenser.
  (b) Let's imagine an ideally effective heat recovery exchanger that uses the product streams to provide some of the cooling. Calculate the net cooling needed for a modified system when the exiting air and condensate streams leave the process at 25°C. Calculate the cooling required in the condenser.

P6.55 Lactic acid can be hydrogenated to propylene glycol in the vapor phase over a copper catalyst at 200°C. In this process a liquid aqueous solution of lactic acid is combined with gaseous H_2, both at 30°C and 1 atm (760 mmHg) pressure. The combined stream is then heated to 200°C in a heater before being fed to the reactor. Assume the feed stream to the heater contains 46.51 mol% lactic acid, 41.86 mol% water, and 11.63 mol% hydrogen, and the molar flowrate of the feed stream is 215 gmol/min. Determine whether the stream leaving the heater is saturated vapor, superheated vapor, saturated liquid, superheated liquid, or vapor-liquid mixture. Determine the rate of energy supplied by the heater (kW).

P6.56 A local brewery operation has offered to donate, free of charge, spent grains from its operations to a heating plant at a college campus. As the heating plant engineer, your job is to see whether it is technically feasible to use this material. The spent grains are a mix of starches, sugars, alcohols, water, and other materials. Chemical analysis of the grains gives the following elemental composition: 14.4 wt% C, 6.2 wt% H, 78.8 wt% O, 0.6 wt% S. Compare the heating value (kJ/kg) of these grains to natural gas. Would you accept the donation? To complete your calculations, estimate $\Delta \hat{H}_c^\circ$ of the grains by assuming that the elements in the grains are in their native state; that is, that those grains are simply a mix of C, H_2, O_2, and S. Will this assumption give an upper or lower bound on $\Delta \hat{H}_c^\circ$?

P6.57 Combustion of 1 gal of gasoline to CO_2 and H_2O produces about 125,000 kJ. If a small internal combustion engine were attached to a 20-lb toy rocket, and 0.005 gal (about 19 mL) of gasoline were burned to propel the rocket upward, how far would the rocket go before it stopped ascending? Neglect air resistance, and neglect the change in weight as the gasoline burns.

P6.58 At a plant that manufactures styrene, the reactor outlet (100 gmol/s, 1000 mmHg, and 500°C) contains 50 mol% H_2, 30 mol% ethylbenzene (C_8H_{10}), and 20 mol% styrene (C_8H_8). The reactor outlet stream is cooled and partially condensed in a heat exchanger, then separated into vapor and liquid streams in a flash drum, which operates as a single equilibrium stage. The vapor stream is sent to a naphtha desulfurization plant elsewhere in the facility. The vapor must contain no more than 0.1 mol% styrene. Calculate the flow rates and compositions of the vapor and liquid streams leaving the flash drum. At what temperature must the flash drum operate? About how much heat (kJ/s) must be removed in the heat exchanger?

P6.59 You own the Finer Furniture Factory, supplier of beautiful painted-wood tables, chairs, bookcases, and the like. Your painters are complaining about headaches from solvent vapors. So, you install a fan to blow clean air through the painting room. Now the neighbors are complaining about the "chemical" smell. A local contractor suggests that you purchase a condenser from him, to remove the solvent vapors. Here's some data: The exhaust air leaving the paint booth is typically 2.5 kgmol/min and 60°C, and contains about 10 mol% solvents, mostly heptane. You can use cold water at 15°C to condense the vapors. How much heat is removed in the condenser? What's the maximum possible percent reduction in the heptane content of the exhaust air? Will this be enough to keep the neighbors happy? (*Hint*: smells reach your nose through the air, as vapors.)

P6.60 Your new apartment is crawling with bugs. You know that ethyl formate ($HCOOC_2H_5$) is a great bug killer, and you can buy a jug of it at the local gardening center. You decide to design a do-it-yourself fumigator. You need the fumigator to produce an air-formate vapor mix containing 12 mol % ethyl formate, and to deliver 25 gmol/min vapor to spray around the apartment. Could you use a household item, like a light bulb or a hair blowdryer, to provide the needed energy? Some data on ethyl formate: C_p (l) = 0.508 cal/g °C. $\Delta \hat{H}_v$ = 97.18 cal/g at T_b = 54.3°C. At 76°C, P^{sat} = 2 atm and at 110.5°C, P^{sat} = 5 atm.

P6.61 It's a hot day, and your windows won't open. (Air conditioning? Not at the rent you're paying!). You decide to cool your room by pulling all the shades and placing a bucket full of ice in the center of the room. Assume that your room starts at 95°F, and it is 10 ft by 12 ft, with 9-ft ceilings. What's the maximum temperature drop you could achieve? Make any reasonable assumptions.

P6.62 In the Mojave desert sit 1926 mirrors surrounding a 100-m tower. The tower has numerous pipes coated with energy-absorbing paint. The adjustable mirrors focus the sunlight on the pipes, moving as the sun crosses the sky. Molten sodium nitrate and potassium nitrate circulate through the pipes and are heated to 565°C. The hot molten salts are used to produce steam and thence electricity. The system generates 10 MW electricity, enough for a city of 10,000 homes. What is the average circulating rate of the molten salts? Make any reasonable approximations, but explain your reasoning. Why do you think molten salts circulate in the pipes rather than steam?

P6.63 Liquefied natural gas (LNG) containing 85 mol% methane and 15% ethane is stored at 273 K and 260 atm pressure. A process requires 10,000 gmol/h of vaporized natural gas at 50 atm pressure and 500 K. To feed this process, the LNG passes through an expansion valve, where the pressure drops to 50 atm, and then through a heat exchanger, where the temperature increases to 500 K. Saturated steam at 30 bar pressure is condensed in the heat exchanger to heat the natural gas. Assume that the expansion valve operates adiabatically, that the change in kinetic energy is negligible, that the heat exchanger is well insulated, and that the vapor obeys the ideal gas law. What is the condensation rate of the steam (kg/h)?

Saturated steam at 3 bar pressure has a higher $\Delta \hat{H}_v$ than saturated steam at 30 bar, so less steam would be required. Also, the lower presssure steam is less expensive. Why isn't steam at 3 bar used in this process?

P6.64 A dilute aqueous solution of benzoic acid (1.6 g benzoic acid per kg solution) needs to be treated before the water can be disposed of in a nearby lake. Benzoic acid (BA) is a solid below 122 °C, so the polluted stream is treated by simply evaporating all the water off, using saturated steam at 5 bar pressure. The water vapor is then condensed and cooled to 60°C by mixing it with cold (15°C) water, and the stream is dumped. The benzoic acid is taken off as a solid and is sold.

Evaporating water takes a lot of energy, so an alternative process has been proposed (see sketch). The dilute benzoic acid solution is contacted with benzene in an extraction unit. Two streams are produced. The extract contains benzene (B) and benzoic acid, which is sent to an evaporator. The benzene is evaporated off, using saturated steam at 5 bar, then condensed in a heat exchanger using cooling water and recycled to the extraction unit. The benzoic acid is taken off as a solid for sale. The raffinate stream contains the benzoic acid that was not recovered, at 0.02 g benzoic acid per kg solution. Also, benzene is slightly soluble in water, so the raffinate stream contains some benzene.

Assume the feed solution flow rate rate is 3785 kg/h, and the benzene recirculation rate in the proposed process is 3327 kg/h benzene. Both processes operate at 1 atm. Estimate the steam required (kg/h) to vaporize water in the existing process and to vaporize benzene in the proposed process (neglect sensible heat contributions). Which process requires less energy input? Compare the environmental impact of the two processes, listing one advantage and one disadvantage for each process.

Can you think of another way to reduce the steam and cooling water requirements of the existing process without using benzene?

Existing process

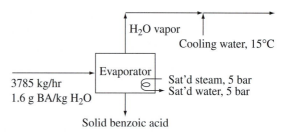

Proposed process

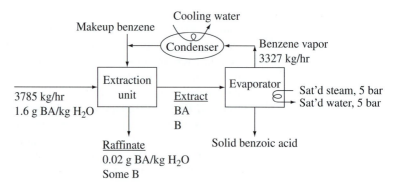

P6.65 Coke (basically pure carbon) is burned with air in a boiler to produce saturated steam at 60 bar pressure. The reaction is

$$C(s) + O_2(g) \rightarrow CO_2(g)$$

The water fed to the boiler to produce the steam is first preheated by the combustion gases from the boiler (see sketch). The water enters the preheater at 50°C and 60 bar. 1000 gmol/s of carbon enters the boiler at 25°C and is burned completely to CO_2, and air (21 mol% O_2/79 mol% N_2) enters the boiler at 25°C and is fed at a rate of 5952 gmol/s. The combustion gases leave the preheater at 100°C and the water fed to the boiler is saturated water at 60 bar. Assume that the boiler and the preheater are well insulated. Calculate the production rate of saturated steam in kg/s and the temperature of the combustion gases leaving the boiler.

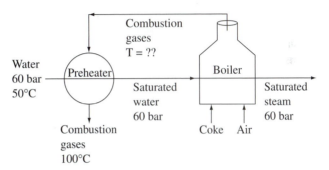

P6.66 Fermentation broth can be heat sterilized by several different methods. In the direct injection method, steam is directly injected into the broth to raise the temperature of the broth from 37 to 140°C. How much 15 psig steam must be injected into 15,000 L of broth? In the batch mode, the broth is heated quickly (10 min) from 37 to 121°C and held at that temperature for 20 min. Heating is supplied by coils placed in the broth; 15 psig steam is condensed in these coils. Since the vat holding the broth is not insulated, there is some heat loss from the vat. The heat loss rate is proportional to the broth temperature: 210 kJ/min °C × (T_{broth} − 25°C). Plot the steam flow rate versus time for the heating and sterilization periods. Calculate the total steam required.

P6.67 A truck filled with expensive drugs sits in the loading dock from 6 AM to 6 PM because of a temporary strike. Normally the truck's contents are held at 50°F, but with the cooling system shut down, the truck begins to warm up. You need to determine whether the drugs are still OK to use. They cannot be exposed to temperatures above 140°F for longer than 3 h. The temperature in a truck sitting in the sun varies because of the regular pattern of the sun. The net heat into the truck is modeled to vary as

$$\dot{Q} = 2.4 \times 10^5 \sin\left(\frac{\pi t}{24}\right)\cos\left(\frac{\pi t}{24}\right)$$

with $\dot{Q}$ in kJ/h and t (time) in h, with $t = 0$ at 6 AM.

The truck contains 8000 kg of drugs. The heat capacity of the drugs is 2 kJ/kg °C. Can the drugs be saved or must they be thrown away?

P6.68 Methanol is synthesized from CO and H_2 over a copper-based catalyst according to

$$CO + 2\ H_2 \rightarrow CH_3OH$$

A side reaction also occurs:

$$CO + 3\ H_2 \rightarrow CH_4 + H_2O$$

The reactants (75 gmol/s H_2, 25 gmol/s CO) enter the plant at 25°C and are heated to 275°C prior to being fed to the reactor. The reactor operates at 50 atm and is adiabatic. 80% of the CO fed to the reactor is converted to products. The production rate of CH_3OH out of the reactor is 18.7 gmol/s. The reactor products are cooled by heat exchange with the reactants entering the reactor. The stream is cooled further to 25°C with cooling water, then sent to a condenser drum, where liquid and vapor streams are separated and processed further. (Refer to sketch.) What is the temperature at the reactor outlet? What is the temperature of the product stream leaving the heat exchanger E-1?

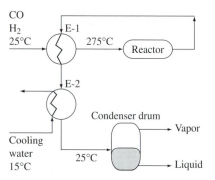

P6.69 You are on a process development team charged with generating a preliminary process flow diagram for the dehydrogenation of ethanol (C_2H_5OH) to make acetaldehyde (CH_3CHO):

$$C_2H_5OH \rightarrow CH_3CHO + H_2$$

The reaction occurs over a catalyst at a reactor pressure of 10 atm and a reactor temperature of 300°C or higher and goes essentially to completion. The ethanol feed and the acetaldehyde product are stored at 25°C and 10 atm. Hydrogen is used elsewhere in the facility.

(a) What is $\Delta \hat{H}_r$ (kJ/gmol) at 300°C?

(b) One of your team members suggests taking the hot product stream leaving the reactor and using that to heat up the cold ethanol feed, as shown below. What do you think of this idea? Explain your reasoning.

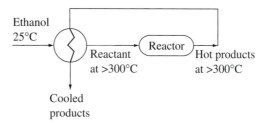

P6.70 The commercial blasting agent ANFO is basically ammonium nitrate (NH_4NO_3) mixed with fuel oil; purchase of ANFO is strictly controlled even though both ammonium nitrate and fuel oil individually are readily available. (Farmers purchase ammonium nitrate as fertilizer.) Pure ammonium nitrate is stable at room temperature. Combine ammonium nitrate with fuel oil and detonate, and you'll have explosive release of N_2, H_2O, and CO_2. To evaluate the explosive reaction of ANFO, we'll model fuel oil as eicosane, $C_{20}H_{42}$.
  (a) Calculate $\Delta \hat{H}^\circ_f$ (kJ/gmol) of eicosane.
  (b) Calculate $\Delta \hat{H}^\circ_r$ for the anaerobic (no air) decomposition of a mixture of 4880 g ammonium nitrate and 282 g eicosane at 25°C to N_2, H_2O, and CO_2.
  (c) About 2000 kg of ammonium nitrate was detonated in the Oklahoma City bombing. If the detonation happened in 1 second, what was the estimated energy release in horsepower (hp)? Data:

  Eicosane ($C_{20}H_{42}$) $\Delta \hat{H}^\circ_c$ (liquid) $= -12{,}390$ kJ/gmol
  Ammonium nitrate (NH_4NO_3) $\Delta \hat{H}^\circ_f$ (solid) $= -365$ kJ/gmol

P6.71 Your company is coming out with a new polymer. The synthesis reaction is slightly exothermic, with $\Delta \hat{H}^\circ_r = -2$ kJ/g. In the laboratory, the reaction was carried out in a small cylindrical reactor of diameter 2.5 cm and length 100 cm. You are ready to scale up the reactor for a 1000-fold greater production. You plan to keep the reactor length the same and just increase the reactor diameter. The rate of heat loss from the reactor wall is estimated at 3.8 W/cm² which was enough to keep the small reactor isothermal. Should you worry about providing additional cooling for the production-scale reactor?

P6.72 Nitroglycerin [$C_3H_5(NO_3)_3$] is produced by adding concentrated nitric acid to glycerin. It used to be carried out as a semibatch operation in large stirred pots, each containing about 1 ton of material. The pot was initially filled with glycerin and the acid was then added slowly over a 2-h period.
  (a) Calculate $\Delta \hat{H}^\circ_r$ for nitroglycerin production from nitric acid and glycerin.
  (b) If heat is not removed, an uncontrollable reaction is followed by explosive decomposition of the nitroglycerin. Nitroglycerin decomposes to CO_2, H_2O, N_2, and O_2. What is $\Delta \hat{H}^\circ_r$ of nitroglycerin decomposition?

(The story is told that operators used to sit on one-legged stools to make sure they didn't fall asleep while watching the pot. Now, a safer continuous-flow process has been developed that reduces the size of the reactor and the reaction time.)

P6.73 Toluene is to be partially oxidized to benzoic acid over a selective catalyst in a gas-phase flow reactor. The reactor feed is a mixture of toluene and oxygen in steam, flowing at 1 atm and 125°C. The toluene:oxygen feed ratio is controlled to 1:0.6 (mol toluene/mol oxygen). The oxygen reacts completely. The reactor is adiabatic.

(a) Explain why oxygen is chosen as the limiting reactant. Use what you know about chemistry to guess whether or not the reaction is exothermic or endothermic. Then, calculate $\Delta \hat{H}^{\circ}_r$. Did you guess right?

(b) The catalyst is deactivated at temperatures above 350°C, and the steam diluent flow rate is adjusted to keep the reactor temperature from exceeding that limit. What steam rate (gmol steam/gmol toluene fed) do you recommend?

P6.74 A company (http://www/milleniumcell/com/solutions/white.html) claims to have a great idea for a new battery design. The overall reaction is that of sodium borohydride with oxygen:

$$NaBH_4 \text{ (ag)} + 2O_2 \text{ (g)} \rightarrow NaBO_2 \text{ (aq)} + 2H_2O \text{ (l)}$$

where (aq) indicates the compound is dissolved in water.
The reaction occurs in two steps, at the anode:

$$BH_4^- + 8OH^- \rightarrow BO_2^- + 6H_2O + 8e^-$$

and at the cathode:

$$8e^- + 4H_2O + 2O_2 \rightarrow 8OH^-$$

If 33% of the chemical energy released in this reaction is converted to electricity, how much electrical energy can be obtained (kJ/g $NaBH_4$)? What do you think of the proposed sodium borohydride battery compared to conventional batteries?

P6.75 In a highly secret facility deep in the woods, a reactor is operating. The reaction is exothermic and liquid phase; the reactant flow rate is 52 g/h, and the reactor is well mixed. $\Delta \hat{H}^{\circ}_r$ is -4.2 kJ/g, and C_p of the liquid is 4 J/g °C. Heat is removed by cooling coils at 21 kJ/h. At steady state, the reactants enter the reactor at 298 K and leave at 298 K and all is well.

(a) What is the reaction rate (g/h)?

(b) The rate of reaction (g/h) increases with temperature as

$$\dot{R} = \dot{R}_{298} \exp\left[460\left(\frac{1}{298} - \frac{1}{T}\right)\right]$$

where $\dot{R}_{298}$ is the rate of reaction (g/h) at 298 K. Generate a plot showing the rate of reaction versus temperature, from 298 K to 600 K.

(c) You are a highly trained secret agent whose job it is to shut down this illegal operation. You find the site, and see that it has been only recently abandoned (the coffee in the mug is still warm). Your first step is to shut down the cooling coils. If the temperature in the reactor rises above 400 K, the place explodes. Do you have much time left on this planet?

P6.76 Acetaldehyde (CH_3CHO) is produced by dehydrogenation of ethanol (C_2H_5OH):

$$C_2H_5OH \rightarrow CH_3CHO + H_2$$

An undesired side reaction occurs, which produces ethyl acetate:

$$2C_2H_5OH \rightarrow CH_3COOC_2H_5 + 2H_2$$

In a pilot plant reactor, 90% of the ethanol fed to the reactor is converted to products, and there is a 65% yield of acetaldehyde (based on the total amount of ethanol fed to the reactor). The ethanol enters the reactor as a vapor at 300°C and 5 bar pressure, and at a flow rate of 100 gmol/h. How much heat must be added/removed (state which) to maintain the reactor outlet temperature at 300°C?

P6.77 High-purity silicon for semiconductor manufacture is made from cheap ingredients—sand (SiO_2) and coke. Here are the main reactions:

$$SiO_2(s) + C(s) \rightarrow Si(s) + CO(g)$$

$$Si(s) + Cl_2(g) \rightarrow SiCl_4(g)$$

$$SiCl_4(g) + Mg(s) \rightarrow MgCl_2(s) + Si(s)$$

The reactions as written are not balanced. Figure out the appropriate stoichiometric ratios. Then, calculate the quantity (grams) of sand and other reactants required to make one gram of silicon, the quantity of byproducts, and the energy needed or removed (state which) for producing 1 g of silicon.

P6.78 Isobutane (iC_4H_{10}) is dehydrogenated to isobutene (iC_4H_8) and hydrogen (H_2) over a platinum catalyst in an adiabatic reactor at 1 atm. At a reactor outlet temperature of 500°C, the equilibrium constant K_a for this reaction is 0.1216 atm. The inlet to the reactor contains a 2-to-1 molar ratio of hydrogen to isobutane. Calculate the fractional conversion of isobutane across the reactor if the reactor outlet composition is equilibrated. Calculate the reactor inlet temperature for this equilibrated system if $\Delta \hat{H}°_r$ = 139.5 kJ/gmol at 500°C.

P6.79 Eddie Entrepreneur runs a luxury hotel on a tropical island. He's got a problem: He needs cool dry air to pamper his wealthy guests. They like their rooms to be at a comfortable 68°F with a 45°F dew point. The tropical air, however, is hot and humid: typically 95°F with a dew point of 90°F. But Eddie doesn't want to install an air conditioner, because repair parts for the compressors are hard to come by in this remote location.

Ima Innovator proposes a crazy idea: Use liquid water to cool air and decrease humidity. This is a sketch of the equipment she plans to sell to Eddie:

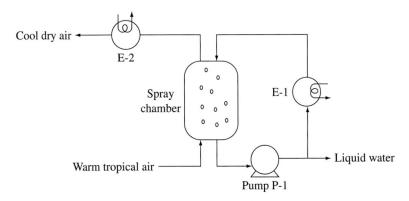

The system works like this (according to Ima): Warm tropical air flows into the bottom of a spray chamber. Liquid water is sprayed from a nozzle at the top of the spray chamber, and water is collected at the bottom. Most of the water is recirculated through a heat exchanger, but some is removed. Air leaving the top of the chamber is passed through another heat exchanger and then sent to cool the guestrooms. The chamber operates at 760 mmHg (1 atm). Ima says the spray chamber operates as a single equilibrium stage (but Eddie, with just a Harvard M.B.A., has no idea what this means).

Eddie knows that pumps require a lot less maintenance than compressors, so he's intrigued by the idea, but he's not really sure this system could work. He's asked you to evaluate Ima's system.

(a) Use as a basis 1000 gmol/h of cool dry air (includes air plus any water vapor) produced. How much liquid water (gmol/h) is removed from the system?

(b) Estimate the value of $\dot{Q}$ (J/h) for the two heat exchangers E-1 and E-2. State whether $\dot{Q}$ is positive or negative for each exchanger. You can assume that the work input from the pump P-1 is negligibly small.

(c) Ima proposes using water (60°F) from an underground well as the heat exchange fluid in E-1. Will this work? Why or why not?

P6.80 Sulfur dioxide can be manufactured by direct oxidation of sulfur:

$$S + O_2 \rightarrow SO_2$$

The reaction is fast and irreversible and generates a lot of heat. All the sulfur is consumed, and the reactor is operated adiabatically. Because of materials limitations, the reactor temperature must be no higher than 450°C.

First, calculate the temperature of the reactor if air is fed at stoichiometric quantities. Then, consider two solutions to the heat management problem: (a) feed air and sulfur at stoichiometric proportion to the burner, then separate out the nitrogen and recycle it; (b) separate the nitrogen from the oxygen in the air, feed pure oxygen and sulfur to the burner, cool some of the product sulfur dioxide, and recycle it to the reactor. Which process scheme is better? Consider in particular the difficulty of any separations and the tolerance of the process to sudden changes in feed flow rate or quality. Can you come up with any better ideas?

P6.81 Hydrogen is produced from propane by a method called steam reforming:

$$C_3H_8 \text{ (g)} + 3H_2O \text{ (g)} \rightarrow 3CO \text{ (g)} + 7H_2 \text{(g)}$$

The water-gas shift reaction also takes place in the reactor, leading to further hydrogen production:

$$CO \text{ (g)} + H_2O \text{ (g)} \rightarrow CO_2 \text{ (g)} + H_2 \text{ (g)}$$

The reaction is carried out over a nickel catalyst in a shell-and-tube reactor (see sketch). The feed to the reactor contains propane at 10 gmol/s and saturated steam at 60 gmol/s and 125°C. The products emerge from the reactor at 800°C. The excess steam in the feed assures essentially complete combustion of the propane. Heat is added to the reactor by passing a hot gas over the outside of the tubes that contain the catalyst. The heating gas is fed at 360 gmol/s, enters the unit at 1400°C, and leaves the unit at 900°C. The heat capacity of the heating gas is 0.040 kJ/gmol °C. The product gases from the reactor are sent to a heat exchanger, where the gases are cooled to 25°C and the water is condensed.

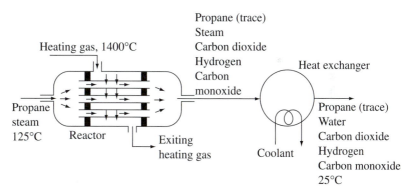

(a) Calculate the molar composition of the product gas and the amount of heat that must be removed in the heat exchanger. Is the heat removal rate in the heat exchanger greater than or less than the heat input from the heating gas? Why?

(b) Over time, dust in the heating gas deposits on the tubes of the reactor. This process is called *fouling* and reduces the rate of heat transfer. The result is that the temperature of the heating gas leaving the reactor increases with time. Because of the characteristics of the heat exchanger, there is a constant temperature difference of 100°C between the heating gas leaving the reactor and the product gases leaving the reactor, no matter what the temperature of the heating gas leaving the reactor is. Assume that over time the heating gas exit temperature rises from 900°C to 1000°C because of fouling. Show how the molar composition of the product gas and the rate of heat removal in the heat exchanger change with time. Can you explain your results?

P6.82 Superheated steam at 17 bars and 250°C is to be produced at a rate of 12,100 kg/h. Methane will be burned completely (to CO_2 and H_2O) with dry air to generate heat for the production of superheated steam. In order to ensure complete combustion of the methane, the air (21 mol% O_2, 79 mol% N_2) will be added such that the oxygen flow rate is 20% above the stoichiometrically required amount. Assume that the air, methane, and water all enter the boiler at 25°C. The minimum temperature for the hot combustion gases anywhere in the process is 150°C, to prevent condensation of water in the gas.

Come up with an energy-efficient design. Calculate the amount of methane burned. On your design, show the composition and flow rates of all streams and all relevant intermediate temperatures.

P6.83 In the refrigeration cycle shown below, write energy balance equations around each piece of equipment. Where appropriate, indicate whether heat and work terms are positive or negative. Sketch a plot with enthalpy on the y axis and pressure on the x axis. Use the plot to show how the pressure and enthalpy changes as the refrigerant is moved around the cycle. Connect the points with lines, and identify each line with a piece of equipment.

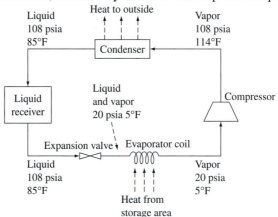

P6.84 In an open-loop air chiller, air at room temperature and pressure is compressed to 200°C and 8 atm, the hot high-pressure air is cooled in a heat exchanger to 50°C, and then the air is expanded to 1 atm, which produces work and cools the air further to −50°C. Suppose the air feed rate is 2 kg/min. Calculate the work needed to operate the compressor and the work provided in the expander, in watts. Find the heat transferred in the heat exchanger. Determine the overall enthalpy change of the air as it goes through the chiller. Compare this to the net work input of the process. Is this an efficient design? Compare the open-loop air chiller to a refrigeration cycle.

P6.85 A large pump at a sewage treatment facility is to be powered by a steam engine (see sketch). For the steam engine, boiler feedwater (10 kg/s) at 1 bar pressure and 50°C is pumped to a pressure of 10 bar, then into a boiler. The heat input into the fluid from the boiler is 30,000 kJ/s, which is generated by methane combustion. The boiler produces steam at 10 bar, which then enters the turbine. The exhaust leaving the turbine is saturated steam at 1 bar pressure.

(a) Calculate the work required to operate the feedwater pump and the work output at the turbine (in kJ/s). What is the net work output of this steam engine (difference between work generated and work consumed)? What is the efficiency η of the steam engine for conversion of thermal energy to mechanical energy, if η = (net work produced)/(net thermal energy input)?

(b) Methane is used as fuel in the boiler burners. Air (79 mol% N_2, 21 mol% O_2) is fed to the burner at 20% excess O_2. Assume complete combustion to CO_2. If the flue gas leaving the boiler stack must be hot enough to prevent water condensation in the stack, how much methane must be fed to the boiler? Assume the methane and air enter the boiler at 25°C. What is the maximum heat available for transfer to the steam by combustion of the methane, if the flue gases exits the boiler at 25°C and all the water is condensed to liquid? What is the efficiency η_T of the boiler for conversion of chemical energy to thermal energy, if η_T = (actual heat transferred/maximum heat available for transfer)?

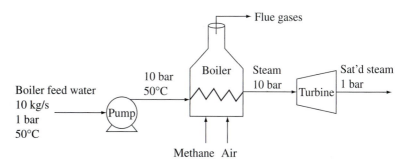

P6.86 Heat is to be transferred between two liquid streams, using a spare heat exchanger. Stream A is flowing in at 20 kg/min and 130°C, while stream B enters the exchanger at 15 kg/min and 15°C. The heat capacities are C_{PA} = 2.4 kJ/kg °C and C_{PB} = 4.5 kJ/kg °C, respectively. The heat exchanger has an exchange area of 4.0 m² and a heat transfer coefficient of 400 kJ/m²h °C. Calculate the outlet temperatures and the overall heat transfer rate for this exchanger, operated in countercurrent flow.

P6.87 A hot oil stream is to be cooled from 200 to 100°F by cooling water available at 60°F. The overall heat transfer coefficient for the heat exchanger is U = 100 Btu/h °F ft². Data on the streams are as follows:

	Hot oil	Cooling water
Flow rate, lb/h	10,000	6000
Heat capacity C_p, Btu/lb °F	0.8	1.0
Inlet temperature, °F	200	60
Outlet temperature, °F	100	?

(a) How much heat must be exchanged, what is the outlet temperature of the cooling water, and how large a heat exchanger is required?

(b) The water exit temperature is quite high, and you are to give consideration to the benefits of increasing the cooling water flow rate. Repeat your heat exchanger design calculations for a range of water flow rates between 6000 and 25,000 lb/h. Present your results in graphical form, and comment.

P6.88 The following three streams need to be heated or cooled as indicated. All three streams are in the vapor phase at all temperatures. A preliminary heat exchanger flow sheet has been designed and is shown below.

Stream	T_{in}, °F	T_{out}, °F	Flow, lb/h	C_p, Btu/lb °F
A	140	320	20,000	0.7
B	350	310	30,000	0.7
C	320	120	10,000	0.7

Steam is available at 250, 350, and 550°F, with the cost increasing with increasing temperature. Cooling water is available at 80°F. Point out one thing seriously wrong with this network. Suggest at least two ways to improve this heat exchanger network. Sketch out your revised flow sheet, and indicate temperatures of entering and exiting streams. Explain your reasoning.

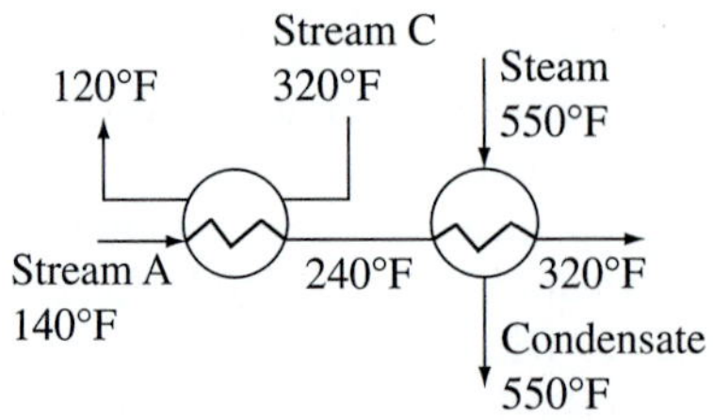

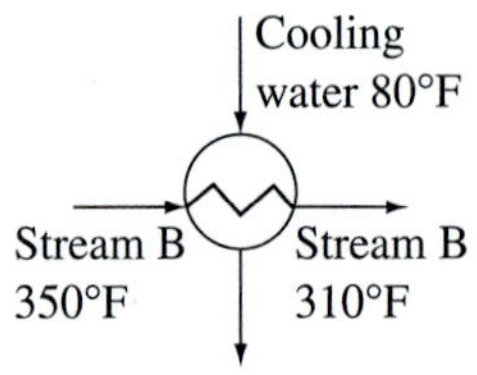

P6.89 Suppose you had the following four streams to be heated/cooled:

Stream	Flow rate, kmol/h	Inlet T, °C	Outlet T, °C	Cp (J/gmol °C)
A	2,000	240	50	80
B	10,000	25	200	29
C	8,000	150	25	123
D	20,000	40	100	72

Saturated steam at 40 bar pressure and cooling water at 20°C are available at the plant site.

Design an efficient heat exchanger network for this system. Assume that the heat transfer coefficient U_h is 300 kJ/m² hr °C for all of your heat exchangers, and calculate the total surface area and the flow rates of steam or cooling water required.

P6.90 Conversion of methanol (CH₃OH) to formaldehyde (HCHO) is an endothermic reaction. In an existing process, methanol vapor at 100°C is further preheated to 250°C by 60 bar saturated steam, then fed to the reactor. Sufficient steam heating is added through coils in the catalyst bed to maintain an outlet temperature of 150°C. The steam coils have been a constant maintenance headache; the coils often leak and the plant needs to be shut down for repairs about six times a year. The plant manager would like to eliminate the need for steam supply to the reactor, to avoid the unplanned shutdowns.

You propose that, instead of supplying steam to the reactor, some of the hydrogen that is produced in the reactor be burned (see sketch):

$$H_2 + \tfrac{1}{2} O_2 \rightarrow H_2O$$

To do this, air (79% N_2, 21% O_2) at 100°C will be mixed with the methanol vapor at 100°C. The reactor feed and outlet temperatures remain the same. Assume that all the added oxygen is consumed.

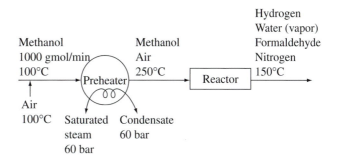

(a) Calculate the amount of hydrogen (gmol/min) that needs to be burned in order to completely eliminate the need for addition of heat to the reactor.

(b) Identify one safety concern associated with the proposed change in operation, and suggest a method to reduce this hazard. What is the environmental impact of the proposed change? Will there be a net change in CO_2 emissions?

(c) Comment on any other necessary changes in the process. Would you recommend that this process change be implemented? Why or why not?

Game Day

P6.91 Styrene (C_8H_8) is produced by catalytic dehydrogenation of ethylbenzene (C_8H_{10}):

$$C_8H_{10} \text{ (g)} \rightarrow C_8H_8 \text{ (g)} + H_2 \text{ (g)}$$

The reaction takes place in an adiabatic reactor at inlet conditions of 650°C and 1 atm. At these conditions, 40% of the ethylbenzene fed to the reactor is converted to styrene. Steam is mixed with the ethylbenzene fed to the reactor at a 10:1 steam:ethylbenzene molar ratio. (This is done to prevent side reactions and to keep the catalyst clean—the steam does not actually react with anything.)

Generate a process flow diagram that is *efficient in its use of raw materials and energy.* Show all major process units (reactors, heat exchangers, distillation columns, furnaces, etc.).

Complete material and energy balances for your process. Specify the mass or molar flow rates, temperatures, and/or heat inputs/outputs wherever possible (including steam or cooling water flow rates). Clearly indicate any assumptions you make.

In your process, assume that fresh ethylbenzene is supplied at 25°C and that styrene product is sent to storage tanks at 25°C. Hydrogen is used elsewhere in the facility—it can be piped out at any convenient temperature. Ethylbenzene and styrene can be separated into essentially pure streams by distillation. Water and hydrocarbon liquids are immiscible. Utilities available on site include cooling water at 15°C and saturated steam at 1 bar, 10 bar, and 30 bar.

P6.92 Mr. Big, a senior executive at ABC Industrial Gases, has decided that new outlets for the company's gas products are needed for continuing corporate growth. The market for methanol is expanding, and you, as the process engineer for ABC, have been assigned the job of coming up with a process for production of 21,500 metric tons per year of methanol from carbon monoxide and hydrogen. After an initial study of the situation, you've come up with the following:

1. Available onsite are carbon monoxide and hydrogen. The carbon monoxide is contaminated with 6 mol% nitrogen. The hydrogen is pure.

2. Methanol can be produced from carbon monoxide and hydrogen by the following reversible reaction:

$$CO + 2H_2 \rightleftarrows CH_3OH$$

The reaction occurs in the vapor phase over a proprietary catalyst and reaches equilibrium rapidly if the temperature is over 100°C.

3. If the temperature in the reactor gets much below 100°C, the reaction rate slows down to the point where equilibrium is no longer reached.

4. Cooling water at 20°C is available on site in virtually unlimited quantites. Inexpensive steam for heating is available on site, usually at 198°C but sometimes the temperature drops a little bit.

5. You can't have any liquid present in the reactor.

6. Smaller reactors are cheaper than bigger reactors. Increasing the reactor size by 100% increases the cost by about 50%. The volume of the reactor V varies as

$$V = \frac{\dot{n}RT}{P}$$

where $\dot{n}$ is proportional to the molar flow rate through the reactor, T is the absolute temperature, P is the pressure, and R is a constant. Reactor cost increases slowly with increasing pressure, except if the pressure goes below 1 atm the cost goes way up.

7. The plant should operate 24 h/day, 350 days/yr.

Your job is to put together a report for Mr. Big on a process to produce methanol. He doesn't care that you've already made plans for the weekend, or that the research director, Ms. Organic, is expecting results from some ongoing laboratory experiments by Tuesday.

Your solution should be in the form of a report as follows:

(a) Brief written summary of the process (1 or 2 pages). This should describe the goal of the process and how the process works. This problem is not completely defined. You will have to make choices along the way. Justify your choices, and explain what additional information you would need to further optimize your choice (e.g., cost to construct and install a reactor). This report is written to Mr. Big, who is very busy and very impatient. His favorite question is "so what's the bottom line?" Mr. Big got a B.S. degree in chemical engineering 25 years ago.

(b) Process flow diagram with all streams labeled.

(c) Appendix summarizing calculations and other supporting data.

P6.93 You are on a process development team charged with generating a preliminary process flow diagram for the production of acetone from isopropanol by the following reaction:

$$C_3H_7OH \rightarrow C_3H_6O + H_2$$

The isopropanol feed and the acetone product are stored at ambient conditions (about 25°C and 1 atm). Hydrogen is used elsewhere in the facility.

The reaction temperature must be at or above 300°C for the reaction to occur at a reasonable rate; at that temperature it is possible to get complete conversion of isopropanol to products. Assume that the reactor pressure is 5 atm. Utilities available in the plant include cooling water and saturated steam. Assume that the following utilities are available at the process plant:

   450 psig saturated steam at $18/1000 kg
   150 psig saturated steam at $6/1000 kg
   15 psig saturated steam at $1/1000 kg
   60°F. cooling water at $0.10/1000 kg
   Ammonia liquid refrigerant at −10°F and 23 psia at $12/1000 kg

Come up with a reasonable process flow sheet that includes a reactor, separators, and heat exchangers as needed. Use a basis of 100 gmol/s isopropanol feed into the plant. Specify flow rates and temperatures of each process stream as well as flows of any utilities (steam, cooling water, etc.) required.

Mathematical Methods

A.1 Using Matrix Methods to Solve Simultaneous Linear Equations

A.1.1 Matrix Equations and Gaussian Elimination

As we see throughout this text, situations often arise where we have to solve a system of linear equations. Systems of linear equations occur, for example, in balancing chemical reactions (Chap. 1) and in developing steady-state models of process flow sheets (Chap. 3). Linear algebra is the branch of mathematics concerned with efficient solution of systems of linear equations. Here we will briefly discuss some methods for solving such systems of equations, without describing the theoretical underpinnings. For an in-depth introduction to the topic, read Chap. 1 of *Linear Algebra and Its Applications* (G. N. Strang, 1980, Academic Press, New York).

Linear equations are those where the variables (the unknowns) are all raised to the first power. For example,

$$2u + 3v + 4w = 10$$

is a linear equation in variables u, v and w, whereas

$$2u^2 + 3\sqrt{v} + 4w = 10$$

is linear in w but not in u or v. We are interested in the case where we have many linear equations written in many variables. As a simple example, consider a set of three equations written in three variables (taken from G. N. Strang, *Linear Algebra and its Applications*):

$$u + v + w = 6$$
$$u + 2v + 2w = 11 \tag{A1.1}$$
$$2u + 3v - 4w = 3$$

There are really two questions of importance (to us): (1) Is there a solution? (2) What is the solution? There is a simple algorithm we can use to answer these questions, called *Gaussian elimination*, if we employ matrix notation.

We write this system of equations in matrix notation by placing the coefficients of the variables in a 3×3 matrix, the variables u, v and w in a column vector, and the numbers on the right-hand side in another column vector:

$$\begin{bmatrix} 1 & 1 & 1 \\ 1 & 2 & 2 \\ 2 & 3 & -4 \end{bmatrix} \begin{bmatrix} u \\ v \\ w \end{bmatrix} = \begin{bmatrix} 6 \\ 11 \\ 3 \end{bmatrix} \tag{A1.2}$$

The coefficient matrix is called $\mathbf{A}$. Each entry in $\mathbf{A}$ is identified by row and column number as a_{ij}, where i is the row and j is the column. The column of variables is called the vector $\mathbf{x}$. Each entry in $\mathbf{x}$ is identified as x_j. We'll name the column vector on the right-hand side the vector $\mathbf{b}$. Each entry in $\mathbf{b}$ is identified as b_i. Equation (A1.2) is an example of the general matrix equation

$$\mathbf{Ax} = \mathbf{b} \tag{A1.3}$$

To multiple a matrix and a vector, we multiple a *row* in the matrix $\mathbf{A}$ by the *column* in the vector $\mathbf{x}$. The result equals the value in the same *row* of the vector $\mathbf{b}$. In other words,

$$\sum_j a_{ij} x_j = b_i \tag{A1.4}$$

In our example, if we apply Eq. (A1.4) three times, for $i = 1$, 2, and 3, we recover the three linear equations (Eq. A1.1).

Our job is to find the solution $\mathbf{x}$. Is there a unique (one and only one) solution to Eq. (A1.2)? There are a couple of tests we can apply. First, the number of rows in $\mathbf{A}$ must equal the number of columns. (This is equivalent to saying that the number of unknowns equals the number of equations.) In other words, $\mathbf{A}$ must be a square matrix. Second, the "reduced-row echelon form" of $\mathbf{A}$ must have all nonzero numbers on the diagonal (e.g, all values of a_{ij} where $i = j$, must be nonzero), and there must be all zeros below the diagonal (e.g, all values of a_{ij} where $i > j$, must be zero). Third, the vector $\mathbf{b}$ must not be the null vector—in other words, at least one value of b_i must not equal zero. (There are some additional considerations, but these three tests are good enough for our purposes.) If these tests are not met, then the problem is either underspecified or overspecified. (Row echelon reduction is essentially equivalent to degree of freedom analysis!)

To find the reduced-row echelon form, we apply Gaussian elimination to matrix $\mathbf{A}$ until we get 1 on the diagonal and 0 everywhere below the diagonal. We'll describe this method in just a paragraph or two.

To find the solution to Eq. (A1.3), we use one of two methods. In one method we find the inverse of the matrix A, $\mathbf{A}^{-1}$, such that

$$\mathbf{x} = \mathbf{A}^{-1}\mathbf{b}$$

$\mathbf{A}^{-1}$ is *not* simply a matrix where each entry is a_{ij}^{-1}! This is unfortunate but true. It is relatively easy to find $\mathbf{A}^{-1}$ with a calculator, but it is not so easy to do by hand.

The other method for finding the solution is Gaussian elimination. To apply this method, we

1. Multiply the first row of numbers by a common factor (called a *pivot*) and subtract the result from the second row. Choose the pivot such that the $a_{21} = 0$ when you are done. Write down the new matrix: *the first row remains the same, only the second row changes.*
2. Multiply b_1 by the same pivot and subtract from b_2.
3. Repeat steps 1 and 2 for all remaining rows, choosing a different pivot, multiplying the first row by that pivot and subtracting the result until every number in the first column is zero except a_{11}.
4. Multiply the second row of numbers by a pivot and subtract from the third row, choosing the pivot such that $a_{32} = 0$.
5. Multiply b_2 by the same pivot and subtract from b_3.
6. Repeat steps 4 and 5 for all remaining rows, choosing a different pivot each time, until every number in the second column below a_{22} is zero.

It is also okay to exchange columns; sometimes that produces a solution more quickly. The only difference between finding the row-reduced echelon form of **A** and finding the solution to **Ax** = **b** by Gaussian elimination is that in finding reduced-row the echelon form case we apply these steps only to **A** and in finding the solution we apply to both **A** and **b**.

We'll illustrate by using Eq. (A1.2).

1. We choose 1 as the pivot. Multiplying each entry in the first row by 1 and then subtracting from the second row gives

$$\begin{bmatrix} 1 & 1 & 1 \\ 0 & 1 & 1 \\ 2 & 3 & -4 \end{bmatrix}$$

2. We multiply b_1 by 1 and subtract from b_2. The equation becomes

$$\begin{bmatrix} 1 & 1 & 1 \\ 0 & 1 & 1 \\ 2 & 3 & -4 \end{bmatrix} \begin{bmatrix} u \\ v \\ w \end{bmatrix} = \begin{bmatrix} 6 \\ 5 \\ 3 \end{bmatrix}$$

3. We repeat, choosing 2 as the pivot. Multiplying each entry in the first row by 2 and then subtracting the result from the third row gives

$$\begin{bmatrix} 1 & 1 & 1 \\ 0 & 1 & 1 \\ 0 & 1 & -6 \end{bmatrix} \begin{bmatrix} u \\ v \\ w \end{bmatrix} = \begin{bmatrix} 3 \\ 1 \\ -9 \end{bmatrix}$$

4. Every entry in column 1 below a_{11} equals zero. We move on to row 2. We choose 1 as the pivot, multiply the second row by 1 and subtract the result from the third row:

$$\begin{bmatrix} 1 & 1 & 1 \\ 0 & 1 & 1 \\ 0 & 0 & -5 \end{bmatrix} \begin{bmatrix} u \\ v \\ w \end{bmatrix} = \begin{bmatrix} 3 \\ 1 \\ -10 \end{bmatrix}$$

We now have a coefficient matrix where all the entries below the diagonal are zero. (If we went on to divide row 3 by –5, we would produce the reduced-row echelon form.) Finding the solution is easy: we start at the bottom:

$$-5w = -10$$
$$w = 2$$

Now we move up one row, substituting in the value of each variable:

$$v + w = 1$$
$$v + 2 = 1$$
$$v = -1$$

Finally we get to the top row:

$$u + v + w = 3$$
$$u - 1 + 2 = 3$$
$$u = 2$$

What if we have a system of equations such as

$$\begin{bmatrix} 0 & 1 & 1 \\ 1 & 1 & 1 \\ 2 & 3 & -4 \end{bmatrix} \begin{bmatrix} u \\ v \\ w \end{bmatrix} = \begin{bmatrix} 5 \\ 6 \\ 3 \end{bmatrix}$$

At first glance, this looks like it presents a problem, because of that zero in the top row. In these cases, we simply switch the positions of rows. *If we switch a row in the matrix* **A** *we must switch the corresponding row in the vector* **b**. If there are a lot of zeros in matrix **A**, the problem becomes much simpler to solve by judicious juggling of rows to put as many zeros below the diagonal as possible, before we proceed to find pivots and solve.

Example A.1   **Gaussian Elimination in Balanced Chemical Equations**

Use Gaussian elimination to find the stoichiometric coefficients needed to balance the reaction between cyclohexanol and nitric acid to make adipic acid, nitrogen oxide, and water.

$$C_6H_{12}O + HNO_3 \rightarrow C_6H_{10}O_4 + NO + H_2O$$

Solution

There are four elements and five compounds. We'll select cyclohexanol as our base compound (stoichiometric coefficient of 1) and write four element balance equations for C, H, O, and N, in the four unknown stoichiometric coefficients. In matrix form, these equations become:

$$
\begin{array}{c} C \\ H \\ O \\ N \end{array}
\begin{bmatrix} 0 & 6 & 0 & 0 \\ 1 & 10 & 0 & 2 \\ 3 & 4 & 1 & 1 \\ 1 & 0 & 1 & 0 \end{bmatrix}
\begin{bmatrix} v_2 \\ v_3 \\ v_4 \\ v_5 \end{bmatrix}
=
\begin{bmatrix} 6 \\ 12 \\ 1 \\ 0 \end{bmatrix}
$$

Note that the CHON formula for the base compound, cyclohexanol, forms the matrix **b**, and the formulas for the second, third, fourth, and fifth compounds form the columns of the matrix A! So it is simple to write down the matrix equation by inspection!

Now we apply Gaussian elimination. We exchange the first and second rows:

$$
\begin{bmatrix} 1 & 10 & 0 & 2 \\ 0 & 6 & 0 & 0 \\ 3 & 4 & 1 & 1 \\ 1 & 0 & 1 & 0 \end{bmatrix}
\begin{bmatrix} v_2 \\ v_3 \\ v_4 \\ v_5 \end{bmatrix}
=
\begin{bmatrix} 12 \\ 6 \\ 1 \\ 0 \end{bmatrix}
$$

and then proceed. We don't *have* to do the next step, but it does simplify the calculations—we exchange other rows to put the rows with the most zeros and the smallest numbers towards the top of the matrix. For this case, we'll exchange rows 1 and 4 (in both **A** and **b**):

$$
\begin{bmatrix} 1 & 0 & 1 & 0 \\ 0 & 6 & 0 & 0 \\ 3 & 4 & 1 & 1 \\ 1 & 10 & 0 & 2 \end{bmatrix}
\begin{bmatrix} v_2 \\ v_3 \\ v_4 \\ v_5 \end{bmatrix}
=
\begin{bmatrix} 0 \\ 6 \\ 1 \\ 12 \end{bmatrix}
$$

The second row already has a zero in the first column, so we move immediately to the third and fourth rows, to zero out the first column entries. The pivot is 3 (for the third row) and 1 (for the fourth row):

$$
\begin{bmatrix} 1 & 0 & 1 & 0 \\ 0 & 6 & 0 & 0 \\ 0 & 4 & -2 & 1 \\ 0 & 10 & -1 & 2 \end{bmatrix}
\begin{bmatrix} v_2 \\ v_3 \\ v_4 \\ v_5 \end{bmatrix}
=
\begin{bmatrix} 0 \\ 6 \\ 1 \\ 12 \end{bmatrix}
$$

Moving on to zero out the entries in the second column below the diagonal, the pivot is $(4/6)$ for the third row and $(10/6)$ for the fourth row:

$$
\begin{bmatrix} 1 & 0 & 1 & 0 \\ 0 & 6 & 0 & 0 \\ 0 & 0 & -2 & 1 \\ 0 & 0 & -1 & 2 \end{bmatrix}
\begin{bmatrix} v_2 \\ v_3 \\ v_4 \\ v_5 \end{bmatrix}
=
\begin{bmatrix} 0 \\ 6 \\ -3 \\ 2 \end{bmatrix}
$$

To clear out the third column below the diagonal, we use the pivot (1/2) on the third row, and subtract from the fourth row:

$$
\begin{bmatrix} 1 & 0 & 1 & 0 \\ 0 & 6 & 0 & 0 \\ 0 & 0 & -2 & 1 \\ 0 & 0 & 0 & 3/2 \end{bmatrix}
\begin{bmatrix} v_2 \\ v_3 \\ v_4 \\ v_5 \end{bmatrix}
=
\begin{bmatrix} 0 \\ 6 \\ -3 \\ 7/2 \end{bmatrix}
$$

Now we work backward to find:

$$
\begin{bmatrix} v_2 \\ v_3 \\ v_4 \\ v_5 \end{bmatrix}
=
\begin{bmatrix} -8/3 \\ 1 \\ 8/3 \\ 7/3 \end{bmatrix}
$$

or

$$
C_6H_{12}O + \tfrac{8}{3} HNO_3 \rightarrow C_6H_{10}O_4 + \tfrac{8}{3} NO + \tfrac{7}{3} H_2O
$$

A.1.2 Using a TI-83 Plus Calculator to Solve Systems of Linear Equations

Gaussian elimination works well for manual solution of a few simultaneous linear equations, but gets tedious (and prone to error) as the number of equations increases. Equation-solving software programs take the tedium out of solving matrix equations. But you don't need a computer; you can solve matrix equations with inexpensive scientific calculators. If you have a good strategy for setting up the equations, solution becomes very simple and easy!

Here is how to use your TI-83 Plus to perform two operations: row echelon reduction (to check for the existence of a solution), and matrix inversion (to find the solution). We will illustrate with a chemical equation balancing example, where the unbalanced reaction is

$$
NH_3 + CH_2O + HCN \rightarrow C_4H_5N_3 + H_2O
$$

There are 4 elements and 5 compounds, therefore there are 4 element balance equations (C, H, O, N) and 5 unknown stoichiometric coefficients. The four element balance equations are:

$$
0v_1 + 1v_2 + 1v_3 + 4v_4 + 0v_5 = 0
$$
$$
3v_1 + 2v_2 + 1v_3 + 5v_4 + 2v_5 = 0
$$
$$
0v_1 + 1v_2 + 0v_3 + 0v_4 + 1v_5 = 0
$$
$$
1v_1 + 0v_2 + 1v_3 + 3v_4 + 0v_5 = 0
$$

Choosing NH_3 as our basis compound, so $\nu_1 = -1$, we write this system of linear equations as the matrix equation:

$$\begin{bmatrix} 1 & 1 & 4 & 0 \\ 2 & 1 & 5 & 2 \\ 1 & 0 & 0 & 1 \\ 0 & 1 & 3 & 0 \end{bmatrix} \begin{bmatrix} \nu_2 \\ \nu_3 \\ \nu_4 \\ \nu_5 \end{bmatrix} = \begin{bmatrix} 0 \\ 3 \\ 0 \\ 1 \end{bmatrix}$$

To determine if there is a solution, we find the reduced-row echelon form of **A**. We set up a matrix in the calculator that contains all the entries of **A**, then we use the matrix math function "rref" to find the reduced-row echelon form. To implement this solution strategy on a TI-83 Plus, we

Press $\boxed{2^{\text{nd}}}$ $\boxed{\text{matrix}}$, then use arrows $\gg$ to highlight "edit"

Press $\boxed{1}$, $\boxed{4}$, $\boxed{\text{enter}}$, $\boxed{4}$, $\boxed{\text{enter}}$. This sets up matrix **A** as a 4×4 matrix.

Reading across the rows, enter each number in the matrix by pressing each $\boxed{\text{number}}$, then $\boxed{\text{enter}}$. This fills the elements of the matrix with the coefficients.

Press $\boxed{2^{\text{nd}}}$ $\boxed{\text{quit}}$, then press $\boxed{\text{clear}}$ if necessary. This says that you are done with setting up the matrix.

Press $\boxed{2^{\text{nd}}}$ $\boxed{\text{matrix}}$, then use arrow $>$ to highlight "math," then scroll down with arrows unitl you get to "rref." Press $\boxed{\text{enter}}$. This says that you want to perform the reduced row echelon function.

Press $\boxed{2^{\text{nd}}}$ $\boxed{\text{matrix}}$, $\boxed{1}$, $\boxed{\text{enter}}$. This applies row reduction to matrix **A**. You should see:

$$\begin{bmatrix} 1 & 0 & 0 & 0 \\ 0 & 1 & 0 & 0 \\ 0 & 0 & 1 & 0 \\ 0 & 0 & 0 & 1 \end{bmatrix}$$

There is a 1 in every diagonal spot and 0 below the diagonal. (In this particular case we got the identity matrix: 0 everywhere except 1 on the diagonal. This won't always happen.) This assures us that there is likely one and only one solution to our system of equations. To find the solution, we simply take the inverse of **A**, and multiply the inverse by **b**. You have already entered **A** into the calculator. Now you need to enter **b**. To do this,

Press $\boxed{2^{\text{nd}}}$ $\boxed{\text{matrix}}$, then use arrows $\gg$ to highlight "edit"

Scroll down to highlight 2 then press $\boxed{\text{enter}}$, $\boxed{4}$, $\boxed{\text{enter}}$, $\boxed{1}$, $\boxed{\text{enter}}$. This sets up matrix **b** as a 4×1 column vector. (You want to leave matrix **A** where it is, which is why you highlight 2 instead of 1.)

Reading down, enter each number in **b** by pressing each number, then enter. This fills the elements of the matrix with the coefficients.

Press 2nd quit, then press clear if necessary. This says that you are done with setting up the matrix.

Press 2nd matrix 1, x^{-1}, enter. This calculates the inverse of matrix **A**.

Press × 2nd matrix 2 enter. This multiplies $\mathbf{A}^{-1}$ by **b**.

You should get back the vector containing the stoichiometric coefficients:

$$\mathbf{x} = \begin{bmatrix} -2 \\ -2 \\ 1 \\ 2 \end{bmatrix}$$

A.2 Finding Roots of Nonlinear Equations

The problem of finding roots of nonlinear equations comes up frequently, and the matrix methods described in App. A.1 are no longer helpful. Let's consider a couple of examples:

Illustration 1:

In Example 4.17 we needed to find the extent of reaction $\dot{\xi}$ at equilibrium for the ammonia synthesis reaction, given a numerical value for K_a

$$K_a = 6.6 \times 10^5 \,\text{atm}^{-2} = \frac{(2\,\dot{\xi})^2(4000 - 2\,\dot{\xi})^2}{(1000 - \dot{\xi})(3000 - 3\,\dot{\xi})^3} \frac{1}{(1\,\text{atm})^2}$$

Illustration 2:

In Example 5.12, we needed to find the dew point temperature T of a 40% hexane/60% heptane mixture at 1520 mmHg, with saturation pressure given by Antoine's equation:

$$x_6 + x_7 = 1 = \frac{y_6 P}{P_6^{\text{sat}}} + \frac{y_7 P}{P_7^{\text{sat}}}$$

$$= \left[\frac{0.4}{10^{6.87601\,(1171.17/T + 224.41)}} + \frac{0.6}{10^{6.89677\,(1264.90/T + 216.54)}} \right] 1520$$

It is possible to find $\dot{\xi}$ (Illustration 1) or T (Illustration 2) by trial and error: we guess a value for $\dot{\xi}$ or T, evaluate the equation at the guessed value, and see if the equality holds. If not, then we make a new guess, reevaluate the equation at the new value, and continue until we are comfortably close to finding a value that satisfies the equation. For example, in our first illustration we could guess that $\dot{\xi} = 0.1$. Evaluating the equation we'd find that the right-hand side equals 2.37×10^{-8}, a far cry from the value we were looking for. Trial and error works better if we first examine the equation for clues as to reasonable values for the root. Again, in our first illustration we see that $\dot{\xi}$ must be less than 1000, because if $\dot{\xi} = 1000$ we'd end up with zero in the demoninator, and if $\dot{\xi} > 1000$ we'd get a negative number. We also know that $\dot{\xi} > 0$, and, given that the equilibrium constant is quite large, we suspect that the equilibrium conversion is fairly high and therefore $\dot{\xi}$ is rather large. These clues might guide us to guess something like $\dot{\xi} = 900$. At this guessed value, we'd find that the right-hand side equals 5808, closer to but still smaller than the actual K_a. Further trials could be carried out on a graphing calculator or a spreadsheet.

But there's no need to do trial and error by hand. Spreadsheets such as Microsoft Excel and equation-solving software such as EES quickly and easily carry out these iterative calculations. Here we will illustrate two tools that can be implemented in Excel: *Goal Seek* and *Solver*. (The details may be slightly different depending on the version of the program you have. Consult the help files for your program if you have trouble.)

Goal Seek is the simplest way to find the root of an equation in Excel. To find the extent of reaction in our example, we set up cells in the spreadsheet and type in the equation for K_a in cell **B2** as shown. (*Italics* indicate text or numbers that we type in. We will later enter values for extent in cell **B1**.)

	A	**B**
1	*extent*	
2	*Ka*	*=((2*B1*2*B1)*(4000-2*B1)*(4000-2*B1))/((1000-B1)*((3000-3*B1)^3))*

Next, we enter a reasonable first guess for the extent of reaction into cell **B1**, e.g., *900*. The equation in **B2** is evaluated at this value; the spreadsheet returns the value of 5808.

	A	**B**
1	*extent*	*900*
2	*Ka*	5808

Now we go to the *Tools* pull-down menu and select *Goal Seek*. This will open a window in which we:

Set cell:	*B2* {the cell that carries the equation for Ka}
To value:	*660000* {the desired numerical value for Ka}
By changing cell:	*B1* {the cell that carries the value for the root}

Finally we click OK and almost instantaneously we find that *Goal Seek* found a solution, and it is sitting right in cell **B1**! The solution is 969.23.

(See what happens if you enter a poor initial guess, like –900, or 1400. A good initial guess often makes the difference between success and failure!) It's also a good practice to check the solution on your calculator, to make sure you typed the equation into the cell correctly.

You can use *Goal Seek* with more than one equation, as long as there is only one variable. Let's examine this with our second illustration. Instead of working through the math to come up with one equation, we simply create separate cells to calculate the saturation pressures of hexane (Psat6) and heptane (Psat7) as a function of temperature T using the Antoine equations, and we create cells to input known values of pressure P and vapor mole fractions (y_6 and y_7). Then we create a cell that calculates $x_6 + x_7$, and we set up *Goal Seek* to adjust T until $x_6 + x_7 = 1$. This way, we could reuse the spreadsheet to find dew point temperatures at different pressures, if we wanted to. The setup of the spreadsheet is:

	A	**B**
1	T	
2	P	*1520*
3	*y6*	*0.4*
4	*y7*	*=1-B3*
5	*Psat6*	*=10^(6.87601-(1171.17/(B1+224.41)))*
6	*Psat7*	*=10^(6.89677-(1264.90/(B1+216.54)))* *-(1268.115/(B1+216.9)))*
7	*x6*	*=B3*B2/B5*
8	*x7*	*=B4*B2/B6*
9	*sum*	*=B7+B8*

(Our strategy is to independently calculate $x_6 = y_6 P / P_6^{sat}$ and $x_7 = y_7 P / P_7^{sat}$, then check to see if they sum to 1.0)

We make an initial guess for T (100°C is a reasonable place to start), we input the initial guess into **B1**, then in Goal Seek we

Set cell:	*B9*
To value:	*1.0*
By changing cell:	*B1*

The solution comes back quickly: $T = 114$.

Solver in Excel is another root-finding tool that is more powerful (and a little more complicated) than *Goal Seek*. You find *Solver* on the *Tools* pull-down menu. (If you don't see *Solver* there, you need to install it.) Any problem that *Goal Seek* can solve can be solved in *Solver*. There are a couple of advantages of *Solver* over *Goal Seek*: (1) you can enter constraints on the allowable solution, (2) you can find the value of a parameter that maximizes or minimizes a function, and (3) you can find a solution involving more than one equation and one variable.

To use *Solver* to find the dew point, we set up the spreadsheet in exactly the same manner as we did previously with *Goal Seek*. From the *Tools* pull-down menu we select *Solver*. In the window that opens we

Set target cell:	*B9*
Equal to (click on button that says "Value of"):	*1.0*
By changing cells:	*B1*

Then click on *Solve*. The solution appears in cell **B1** in our spreadsheet. Click on *Close* to go back to the spreadsheet.

Now let's look at the more advanced kinds of problems that Solver can handle.

Entering constraints: Quadratic and cubic equations have two and three roots, respectively. We might want only the positive root, or only the largest root. To avoid getting a root we don't want, we can solve subject to constraints. To illustrate, we return to Example 3.9, where we want to solve the quadratic equation:

$$60 = 2t_f - 0.015t_f^2$$

We know from the evaporation rate information in the problem that t_f must be less than 66.67 minutes. To use *Solver*, we set up a simple spreadsheet:

	A	**B**
1	*Time*	
2	*Equation*	=2*B1-0.015*B1*B1

Now we open *Solver* and

Set target cell:	*B2*
Equal to (click on "Value of"):	*60*
By changing cells:	*B1*

Still with the *Solver* window open, we move to the box that says "Subject to the Constraints" and click on "Add." This opens an *Add Constraint* window. We enter *B1* under *Cell Reference,* adjust the next panel to $< =$, then enter *66.7* under *Constraint*. This says that the value in **B1** must be less than or equal to 66.7. Then we click OK, which returns us to the *Solver* window. Finally, we click "Solve." (Try solving this problem without any constraints and using an initial guess of 66.7 minutes, and then repeat with an initial guess of 66.6 minutes. You'll get two different answers! Always check your solution for reasonableness.)

Maximizing/minimizing the value of a function: Situations arise frequently when we want to find an optimum temperature, pressure, flow rate, etc., perhaps based on maximizing profit or minimizing cost. We can easily do this in *Solver*. To illustrate, in Prob. P5.77 we analyze a process where we used benzene as a solvent to recover benzoic acid. Our goal is to find the optimum benzene flow rate that maximizes our profit, considering the value of the benzoic acid and the cost of the benzene. The profit can be shown to be:

$$\$ = 1.35 m_{BAe} - 0.03(m_{Be} + 686.3)$$

where the first term is the value of the benzoic acid recovered and the second is the cost of the benzene used. Recovery and solvent use are subject to a constraint due to phase equilibrium, which turns out to be

$$K_D = 4 = \frac{m_{BAe}/m_{Be}}{(196 - m_{BAe})/10490.3}$$

Our situation is that we want to find the maximum $, subject to the phase equilibrium constraint. In our spreadsheet we enter:

	A	**B**
1	*mBAe*	
2	*mBe*	
3	*$*	=1.35*B1-0.03*(B2+686.3)
4	*Kd*	=(B1/B2)/((196-B1)/10490.3)

A perusal of the equilibrium constraint will lead us to the conclusion that m_{BAe} must be less than 196, and it must be greater than zero; we'll make an initial guess of $m_{BAe} = 100$. We also note that m_{Be} must be greater than m_{BAe}; we'll make an initial guess of $m_{Be} = 500$. Our strategy is to allow both m_{BAe} and m_B to vary, subject to the constraint that $K_D = 4$.

Now, we open *Solver* and we

Set target cell: B3
Equal to (click on "Max" button):
By changing cells: B1, B2

We then click "Add" next to the *Subject to the Constraints* box, bringing up the *Add Constraints* window. We enter the constraint that B4 = 4.0. Then we close the *Add Constraints* window to return to the *Solver* window, and click "Solve." Solver quickly returns the answer that the maximum profit is $34.12, and that $m_{BAe} = 89.1$ and $m_{Be} = 2186$ at maximum profit.

Finding roots to simultaneous nonlinear equations: Imagine a situation where an endothermic reaction occurs in a reactor with a known amount of heat input, and we wish to know the extent of reaction at equilibrium along with the outlet temperature. Of course, the chemical reaction equilibrium constant depends on temperature, but through the energy balance, the temperature depends on the extent of reaction! In this case, we would have two equations, one describing the chemical reaction equilibrium constant, and one from the energy balance, in two unknowns: the extent of reaction $\dot\xi$, and the temperature T. Imagine for a specific case we derive an equation or the equilibrium constant K_a

$$K_a = \frac{(2\,\dot\xi)^2}{(1.75 - \dot\xi)(5.75 + \dot\xi)} = \exp\left(20.399 - \frac{20{,}495}{T}\right)$$

and we find for the energy balance

$$\dot\xi(170.4) + 0.041(10.3 + \dot\xi)(T - 298) = 550$$

where the first term is the enthalpy of reaction, the second is an $nC_p\Delta T$ term, and the right-hand side is the heat input to the reactor. It's not easy to find the solution to these two equations by hand! We set up a spreadsheet with our two variables and our two equations. It's a good idea to break the equations down, because this makes troubleshooting easier. Here is our spreadsheet:

	A	B
1	*extent1*	
2	*T*	
3	*Ka*	*=exp(20.399-(20495/B2))*
4	*Ka*	*=(2*B1*2*B1)/((1.75-B1)*(5.75+B1))*
5	*Energy bal*	*=170.4*B1+0.041*(10.3+B1)*(B2-298)*

Let's first try some initial guesses and evaluate the equations at those initial guessed values. A quick observation of the equations leads us to surmise that $\dot\xi < 1.75$. So let's guess $\dot\xi = 1$. If this is a chemical process, usually the temperature is fairly high, so let's start with a guess $T \approx 800$ K (about 527°C.)

We enter these values into our spreadsheet, and get back initial evaluation of the equations at these guessed values (our entries are in italics, the calculated values are not italicized):

	A	B
1	*extent1*	*1*
2	*T*	*800*
3	*Ka*	0.0054
4	*Ka*	0.790
5	*Energy bal*	402.98

It's useful to examine this interim evaluation. K_a calculated on the basis of T (cell B3) is much lower than K_a calculated based on $\dot{\xi}$ (cell B4). This indicates that the extent of reaction 1 is likely too low an estimate. Also, the energy balance, which should sum to 550, is a bit too low. This suggests that we might need a higher temperature. Still, our initial guesses are in the right ballpark.

With the correct solution, K_a calculated on the basis of T (cell B3) will equal K_a calculated on the basis of $\dot{\xi}$ (cell B4). We indicate this by setting up another cell, where the ratio of these two calculated values is determined:

	A	B
1	extent1	1
2	T	800
3	Ka	0.0054
4	Ka	0.790
5	Energy bal	402.98
6	ratio	=B3/B4

Our strategy will be to find the solution where cell **B5** (the energy balance) equals 550, subject to the constraint that cell **B6** equals 1.0. We open Solver and

Set target cell: B5
Equal to (click on "Value of" button):   550
By changing cells: B1, B2

We "Add" one constraint as described previously:

$$B6 = 1$$

Finally, we click "Solve". Solver lets us know that it found a solution, and after we click OK we see on our spreadsheet:

	A	B
1	extent1	1.209
2	T	1026.9
3	Ka	1.554
4	Ka	1.554
5	Energy bal	550
6	ratio	1.0000

It is always a good idea to check the solution, first by making sure the numbers make physical sense, and second by inserting the values back into the original equations and evaluating on your calculator.

Solving simultaneous nonlinear equations can be tricky; it may be hard to find a feasible solution. If you have difficulty, first check that the equations are set up correctly and typed into the spreadsheet correctly. Here are some other ideas that *may* help:

Try simplifying the problem and see if you can first solve the simpler problem.

Look at limiting cases. Consider for example what happens at complete conversion, or no conversion.

Examine how each function varies with the different variables. See which functions are very sensitive, or very insensitive, to small changes in a given variable.

Add more constraints, such as that the variables must be nonnegative.

Rewrite the equations. For example, try evaluating $\ln K_a$ rather than K_a, as it varies more smoothly with T. Try to avoid situations where the spreadsheet may try to divide by zero, or where the variables become extremely large or extremely small.

A.3   Fitting Equations to Data: Linear Interpolation and Nonlinear Regression

In chemical process engineering, we handle data and model equations frequently. We'll discuss two methods here: linear interpolation of tabulated data and finding parameter values when we are fitting data to model equations.

A.3.1   Linear Interpolation

In Example 6.5, we found the enthalpy of steam at a temperature that was not on the steam table by linear interpolation. Here we will explicitly show how linear interpolation works.

Suppose you need to know $\hat{H}$ for steam at 1 bar and 127°C. This state is not listed on Table 6.1 or in App. B, but we do know $\hat{H}$ for steam at 1 bar and several other temperatures. With linear interpolation, we look at two temperatures (T_1 and T_2), where $\hat{H}$ is known, that flank the desired temperature. In our example, these two temperatures are $T_1 = 100°C$ ($\hat{H}_1 = 2675.8$ kJ/kg), and $T_2 = 150°C$ ($\hat{H}_2 = 2776.6$ kJ/kg). We then assume that $\hat{H}$ is a linear function of temperature T:

$$\hat{H} = a + bT$$

where a and b are constants. Now, we substitute in the enthalpy values at the two known temperatures:

$$2675.8 = a + b\,(100°C)$$

$$2776.6 = a + b\,(150°C)$$

and solving we find that $a = 2474.2$, $b = 2.02$, and

$$\hat{H} = 2474.2 + 2.02T$$

where $\hat{H}$ is in kJ/kg and T is in °C. Therefore, at 127°C

$$\hat{H} = 2474.2 + 2.02\,(127) = 2730.7 \text{ kJ/kg}$$

Another way to interpolate linearly is to use the idea of proportionality:

$$\frac{\hat{H} - \hat{H}_1}{\hat{H}_2 - \hat{H}_1} = \frac{T - T_1}{T_2 - T_1}$$

or

$$\hat{H} = \frac{T - T_1}{T_2 - T_1} \times (\hat{H}_2 - \hat{H}_1) + \hat{H}_1$$

or, for our example,

$$\hat{H} = \frac{127 - 100}{150 - 100}(2776.6 - 2675.8) + 2776.6 = 2730.7 \text{ kJ/kg.}$$

Is it justifiable to assume that $\hat{H}$ is a linear function of temperature T? To answer this question, it's a good idea to plot the data. We did this for the steam table data in Fig. 6.7. A visual inspection confirms that indeed it is fair to draw a straight line to represent the temperature dependence of $\hat{H}$.

A.3.2   Fitting Data to Model Equations

There are many examples of model equations throughout this text. We will focus particularly on equations that correlate data and allow prediction of physical properties. The Antoine equation, used to calculate saturation pressures at a given temperature, is an example of such an equation. Another example is the use of polynomial expressions for the effect of temperature on heat capacity. The constants in these equations come from regression fit of data to the model equation. In regression, we aim to find the "best" fit of the equation to the data. The best fit is defined as those model parameters that minimize the sum of the square of the residuals. A residual is simply the difference between the data point and the predicted value at a specific condition (e.g., the difference between the true saturation pressure at 300 K and the saturation pressure calculated by the Antoine equation). We calculate the residuals, square them, sum the squares, and try to find the parameter values in the equation that give us the minimum sum.

"Linear" regression simply means that the model equation is written as a linear equation; "nonlinear" regression means that the model equation is a nonlinear equation.

Many software programs can handle linear and nonlinear regression, including Microsoft Excel. Here we'll introduce a different program, called KaleidaGraph, that is relatively inexpensive, available in PC and Macintosh platforms, easy to use, and more powerful than Excel at fitting data to model equations and at generating good-looking plots.

We'll illustrate with an Antoine equation example. *Perry's Chemical Engineers' Handbook* has tables of vapor pressure versus temperature data for many compounds. For example, *Perry's* reports the following data for *d*-limonene (a compound with a lovely citrus fragrance):

T, °C	14.0	40.4	53.8	68.2	84.3	94.6	108.3	128.5	151.4	175.0
P, mmHg	1	5	10	20	40	60	100	200	400	760

We'd like to find an Antoine equation that characterizes the saturation pressure of *d*-limonene. Recall that the Antoine equation is

$$\log_{10}P^{\text{sat}} = A - \frac{B}{T + C}$$

where A, B, and C are constants specific for the compound of interest.

To find values of A, B, and C for *d*-limonene, using KaleidaGraph, we first open a "data" file and input the temperature and pressure values in the table. We can name the first column *Temperature* by going to the *Data* pull-down menu, clicking on *Column Format,* then selecting column A in the box and typing in Temperature. We can similarly select B and type in Pressure. (You can also adjust the format of your numbers, if you like.) After you click on OK the data table will look like this:

	0 Temperature	1 Pressure
0	14.000	1.000
1	40.400	5.000
2	53.800	10.000
3	68.200	20.000
4	84.300	40.000
5	94.600	60.000
6	108.30	100.00
7	128.50	200.00
8	151.40	400.00
9	175.00	760.00

In order to fit the data to the Antoine equation as written, we take the log of the saturation pressure. This is done by going to the *Windows* pull-down menu and clicking on "Formula Entry." This opens a window into which we can type an equation. We type $c2 = log(c1)$, then click on "Run." What this command does is fill the second column ($c2$) with the log (base 10) of the column ($c1$). (If you later change a number in column $c0$, you will have to rerun the calculation; KaleidaGraph does not automatically update, unlike Excel.) Under Data/Column Format we name the third column "log P." Now our data table looks like this:

	0 Temperature	1 Pressure	2 log P
0	14.000	1.000	0.000
1	40.400	5.000	0.699
2	53.800	10.000	1.000
3	68.200	20.000	1.301
4	84.300	40.000	1.602
5	94.600	60.000	1.778
6	108.30	100.00	2.000
7	128.50	200.00	2.301
8	151.40	400.00	2.602
9	175.00	760.00	2.881

Now we create a plot of our data. Under the pull-down menu *Gallery* we select Linear/Scatter. This opens up a window in which we can select the X and Y values for the plot. We select Temperature for X and Log P for Y. Now we click on "New Plot." A plot is immediately generated, and by going to the *Plot/Styles* or *Plot/Axis Option* windows, we can change symbols, tick marks, axes labels, etc. But we are interested in fitting an equation to the data. With the plot showing, go to the *Curve Fit* menu. There are options for various kinds of fits (linear, polynomial, exponential, logarithmic), or we can simply "connect the dots." But we wish to fit a very specific form of an equation to our data. In KaleidaGraph, you can fit the data to an equation of any form. To do this, we Select Curve Fit/General/Edit General. This opens a *General Curve Fits* window. We click on "Add" to add a new equation, and then select New Fit in the left-hand side window to name the equation (e.g., "Antoine"). Click OK, which will return us to our plot. Now we select again *Curve Fit/General* and should see Antoine as one of the choices, which we select. We click on "Define." This opens a window called *General Curve Fit Definition*, in which we enter the model equation. The format is a bit awkward. There is no equals sign, we use M1, M2, M3, etc. for the model parameters, and M0 instead of X. In our illustration, then the function (Y) to be fit is log P, the independent variable X (or M0) is Temperature, and the Antoine constants *A*, *B*, and *C*, are M1, M2, M3. We type in *M1-(M2/(M3+M0))* in the window (which in KaleidaGraph language is equivalent to A-B/(C+T)). It is a good idea to insert initial guesses for the parameter values. A perusal of the Antoine constant list in App. B would tell us that the constant *A* (or M1) is usually around 6 or 8, the constant *B* (or M2) is around 1000 and the constant C is roughly 200. We use semicolons to separate the model equation from the initial guesses. To input both the model equation and the initial guesses, we type in the *General Curve Fit Definition* window

$$M1\text{-}(M2/(M3+M0));M1=6;M2=1000;M3=200.$$

Clicking OK returns us to the *Curve Fit Selections* window. Now we select the data to be fitted by clicking in the appropriate box (marked log P in our example). A click of OK produces almost instantaneously a graph with both data and equation plotted, and a table of the model parameters.

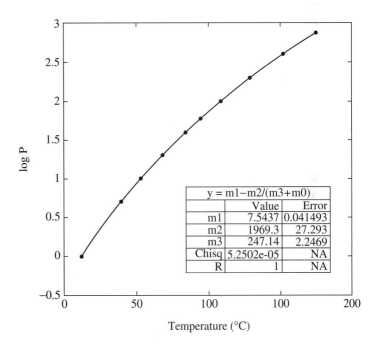

Our fit looks very good (a low Chi Sq and a R close to 1 are indicators of the goodness of fit). The Antoine equation for *d*-limonene is

$$\log_{10}P^{\text{sat}} = 7.544 - \frac{1970}{T + 247}$$

where we rounded off parameter values to be consistent with the indicated error.

B

Physical Properties

B.22 Enthalpy of Solution of Inorganic Solids Dissolved in Water,
   ΔH_{soln}, at Indicated Dilution and 18°C, per gmol Solute   671–672
B.23 Enthalpy of Mixing of Liquids or Gases with Water at 25°C   672

More extensive tabulation of physical property data is available in reference books such as:

> *Perry's Chemical Engineers' Handbook,* edited by R. H. Perry and D. W. Green, McGraw-Hill, Inc., New York, NY.
>
> *CRC Handbook of Chemistry and Physics,* CRC Press, Boca Raton, FL.
>
> *Lange's Handbook of Chemistry,* J. A. Dean, McGraw-Hill, Inc., New York, NY.
>
> *Chemical Properties Handbook,* C. L. Yaws, (1999) McGraw-Hill, Inc., New York, NY.
>
> *Physical and Thermodynamic Properties of Pure Chemicals: Evaluated Process Design Data,* T. E. Daubert et al., (1999), Taylor & Francis, Philadelphia, PA.
>
> *NIST Chemistry Webbook,* edited by P. J. Linstrom and W. G. Mallard, (2005) National Institute of Standards and Technology, Gaithersburg, MD. (http://webbook.nist.gov.)

The Knovel scientific and engineering online database (www.knovel.com) provides searchable access to many reference books but requires a subscription.

The data in this appendix were compiled from these and other sources. For critical applications, you should consult one or more of the original sources.

B.1   Atomic Mass and Number of the Elements

Table B.1	Atomic Mass and Number of the First 100 Elements						
Element	**Symbol**	**Atomic number**	**Atomic mass**	**Element**	**Symbol**	**Atomic number**	**Atomic mass**
Hydrogen	H	1	1.00794	Oxygen	O	8	15.9994
Helium	He	2	4.00260	Fluorine	F	9	18.99840
Lithium	Li	3	6.941	Neon	Ne	10	20.1797
Beryllium	Be	4	9.012182	Sodium	Na	11	22.989768
Boron	B	5	10.811	Magnesium	Mg	12	24.3050
Carbon	C	6	12.011	Aluminum	Al	13	26.981539
Nitrogen	N	7	14.00674	Silicon	Si	14	28.0855

Element	Symbol	Atomic number	Atomic mass	Element	Symbol	Atomic number	Atomic mass
Phosphorous	P	15	30.97362	Cadmium	Cd	48	112.411
Sulfur	S	16	32.066	Indium	In	49	114.82
Chlorine	Cl	17	35.4527	Tin	Sn	50	118.71
Argon	Ar	18	39.948	Antimony	Sb	51	121.75
Potassium	K	19	39.0983	Tellurium	Te	52	127.60
Calcium	Ca	20	40.078	Iodine	I	53	126.90447
Scandium	Sc	21	44.95591	Xenon	Xe	54	131.29
Titanium	Ti	22	47.88	Cesium	Cs	55	132.90543
Vanadium	V	23	50.9415	Barium	Ba	56	137.327
Chromium	Cr	24	51.9961	Lanthanum	La	57	138.9055
Manganese	Mn	25	54.93085	Cerium	Ce	58	140.115
Iron	Fe	26	55.847	Praseodymium	Pr	59	140.90765
Cobalt	Co	27	58.9332	Neodymium	Nd	60	144.24
Nickel	Ni	28	58.69	Promethium	Pm	61	(145)
Copper	Cu	29	63.546	Samarium	Sm	62	150.36
Zinc	Zn	30	65.39	Europium	Eu	63	151.965
Gallium	Ga	31	69.723	Gadolinium	Gd	64	157.25
Germanium	Ge	32	72.61	Terbium	Tb	65	158.92534
Arsenic	As	33	74.92159	Dysprosium	Dy	66	162.50
Selenium	Se	34	78.96	Holmium	Ho	67	164.93032
Bromine	Br	35	79.904	Erbium	Er	68	167.26
Krypton	Kr	36	83.80	Thulium	Tm	69	168.93421
Rubidium	Rb	37	85.4678	Ytterbium	Yb	70	173.04
Strontium	Sr	38	87.62	Lutetium	Lu	71	174.967
Yttrium	Y	39	88.90585	Hafnium	Hf	72	178.49
Zirconium	Zr	40	91.224	Tantalum	Ta	73	180.9479
Niobium	Nb	41	92.90638	Wolfram	W	74	183.85
Molybdenum	Mo	42	95.94	Rhenium	Re	75	186.207
Technetium	Tc	43	(98)	Osmium	Os	76	190.2
Ruthenium	Ru	44	101.07	Iridium	Ir	77	192.22
Rhodium	Rh	45	102.9055	Platinum	Pt	78	195.09
Palladium	Pd	46	106.42	Gold	Au	79	196.96654
Silver	Ag	47	107.8682	Mercury	Hg	80	200.59

(*continued*)

Table B.1   (continued)

Element	Symbol	Atomic number	Atomic mass	Element	Symbol	Atomic number	Atomic mass
Thallium	Tl	81	204.3833	Protactinium	Pa	91	231.03588
Lead	Pb	82	207.2	Uranium	U	92	238.0289
Bismuth	Bi	83	208.98037	Neptunium	Np	93	237.0482
Polonium	Po	84	(209)	Plutonium	Pu	94	(244)
Astatine	At	85	(210)	Americium	Am	95	(243)
Radon	Rn	86	(222)	Curium	Cm	96	(247)
Francium	Fr	87	(223)	Berkelium	Bk	97	(247)
Radium	Ra	88	226.025	Californium	Cf	98	(251)
Actinium	Ac	89	227.028	Einsteinium	Es	99	(252)
Thorium	Th	90	232.0381	Fermium	Fm	100	(257)

Mass numbers in parentheses are those for the most stable or best known isotope.
Source: *CRC Handbook of Chemistry and Physics,* 70th edition; *Perry's Chemical Engineers' Handbook,* 6th ed.

B.2   Nonideal Gas Model Equation and Critical Properties

One way to write the ideal gas law is

$$\frac{P\widehat{V}}{RT} = 1$$

where P = pressure, T = temperature, R = ideal gas constant, and $\widehat{V}$ = specific molar volume (volume per mole). The ideal gas law is a very useful model equation for calculating specific volumes (or, equivalently, densities) of gases at low to moderate pressures. For accurate calculations at higher pressures, either experimental data or more complicated model equations are required. Many such equations have been proposed; *Perry's Chemical Engineers' Handbook* or any chemical engineering thermodynamics textbook is a good source of information. Although differing in detail and complexity, these equations share the common feature of calculating a value for the compressibility factor Z, where

$$Z = \frac{P\widehat{V}}{RT}$$

For an ideal gas, $Z = 1$. Most of the time, for real gases $Z < 1$. (You will typically see values of roughly $0.7 < Z < 1$.) One of the most widely used model equations for predicting specific volumes of real gases is the Redlich-Kwong equation:

$$Z^3 - Z^2 + (A - B^2 - B)Z - AB = 0$$

where

$$A = \frac{aP}{R^2 T^{2.5}} \qquad a = \frac{\Omega_a R^2 T_c^{2.5}}{P_c} \qquad \Omega_a = \frac{1}{9(\sqrt[3]{2} - 1)}$$

$$B = \frac{bP}{RT} \qquad b = \frac{\Omega_b RT_c}{P_c} \qquad \Omega_b = \frac{\sqrt[3]{2} - 1}{3}$$

Knowing just the critical temperature T_c and critical pressure P_c for the compound of interest is sufficient to calculate Z (and hence specific volume or density) for that gas at a given T and P. Since the Redlich-Kwong equation is a cubic equation, there are three roots. The largest real root is the correct value of Z for a gas. Critical temperatures and pressures for selected compounds are in Table B.2. Convert T_c to an absolute temperature scale before using in the Redlich-Kwong equation.

Table B.2   Critical Temperature T_c and Critical Pressure P_c of Selected Compounds

Compound	Formula	T_c, °C	P_c, atm	Compound	Formula	T_c, °C	P_c, atm
Acetaldehyde	C_2H_4O	188.0		Carbon disulfide	CS_2	273.0	76.0
Acetic acid	$C_2H_4O_2$	321.6	57.2	Carbon monoxide	CO	−139	35.0
Acetic anhydride	$C_4H_6O_3$	296.0	46.0	Chlorine	Cl_2	144.0	76.1
Acetone	C_3H_6O	235.0	47.0	Diethylamine	$(C_2H_5)_2NH$	223.5	36.2
Acetonitrile	C_2H_3N	274.7	47.7	Dimethylamine	$(CH_3)_2NH$	164.6	51.7
Acetylene	C_2H_2	36.0	62.0	Ethane	C_2H_6	32.1	48.8
Air		−140.7	37.2	Ethyl acetate	$CH_3COOC_2H_5$	250.1	37.8
Ammonia	NH_3	132.4	111.5	Ethanol	C_2H_5OH	243.1	63.1
Argon	Ar	−122	48.0	Ethylene	C_2H_4	9.7	50.5
Benzene	C_6H_6	288.5	47.7	Ethylene oxide	C_2H_4O	192.0	
Bromine	Br_2	311	102	Fluorine	F	−155	25.0
Butadiene, 1,3	C_4H_6	152	42.7	Helium	He	−267.9	2.26
n-butane	C_4H_{10}	153	36.0	Heptane	C_7H_{16}	266.8	26.8
Carbon dioxide	CO_2	31.1	73.0	Hydrazine	N_2H_4	380.0	145.0

(*continued*)

Table B.2   (continued)

Compound	Formula	T_c, °C	P_c, atm	Compound	Formula	T_c, °C	P_c, atm
Hydrogen	H_2	−239.9	12.8	n-Pentane	C_5H_{12}	197.2	33.0
Hydrogen chloride	HCl	51.4	81.6	Phenol	C_6H_5OH	419.0	60.5
Hydrogen cyanide	HCN	183.5	53.2	Phosgene	$COCl_2$	182.0	56.0
Hydrogen sulfide	H_2S	100.4	88.9	n-Propane	C_3H_8	96.8	42.0
Isobutane	C_4H_{10}	134.0	37.0	Propionic acid	C_2H_5COOH	339.5	53.0
Isopentane	C_5H_{12}	187.8	32.8	n-Propanol	C_3H_7OH	263.7	49.95
Mercury	Hg	>1550	>200	Propylene	C_3H_6	92.3	45.0
Methyl acetate	CH_3COOCH_3	233.7	46.3	Pyridine	C_6H_5N	344.0	60.0
Methanol	CH_3OH	240.0	78.7	Radon	Rn	104.0	62.0
Methyl ethyl ether	$CH_3OC_2H_5$	164.7	43.4	Sodium	Na	2546	343
Neon	Ne	−228.7	25.9	Silicon tetrafluoride	SiF_4	−1.5	50.0
Nitric oxide	NO	−94.0	65.0	Sulfur dioxide	SO_2	157.2	77.7
Nitrogen	N_2	−147.1	33.5	Sulfur trioxide	SO_3	218.3	83.6
Nitrogen tetroxide	N_2O_4	158.0	100	Toluene	$C_6H_5CH_3$	320.6	41.6
Nitrous oxide	N_2O	36.5	71.7	Triethylamine	$(C_2H_5)_3N$	262.0	30.0
n-Octane	C_8H_{18}	296.0	24.6	Trimethylamine	$(CH_3)_3N$	161.0	41.0
Oxygen	O_2	−118.8	49.7	Water	H_2O	374.15	218.4

To convert to T (K), add 273.15.
To convert to P (bar), divide by 1.01325.
Source: *Perry's Chemical Engineers' Handbook*, 6th ed.

B.3   Gibbs Energy, Enthalpy of Formation, and Enthalpy of Combustion

The standard Gibbs energy of formation is useful for calculating the Gibbs energy change with reaction at 298 K, as in Eq. (4.14):

$$\Delta \widehat{G}^\circ_r = \sum_i v_i \Delta \widehat{G}^\circ_{i,f}$$

To a good approximation, we can calculate the Gibbs energy change at any temperature T by using the van't Hoff expression, Eq. (4.15):

$$\ln K_{a,T} = -\frac{\Delta \hat{G}_T}{RT} = -\frac{1}{R}\left[\frac{\Delta \hat{G}_r^\circ - \Delta \hat{H}_r^\circ}{298} + \frac{\Delta \hat{H}_r^\circ}{T}\right]$$

where

$$\Delta \hat{H}_r^\circ = \sum v_i \Delta \hat{H}_{i,f}^\circ$$

or

$$\Delta \hat{H}_r^\circ = -\sum v_i \Delta \hat{H}_{i,c}^\circ$$

Be sure to use an absolute temperature scale (K) in the van't Hoff equation.

Table B.3   Standard Gibbs Energy of Formation $\Delta \hat{G}_f^\circ$, Enthalpy of Formation $\Delta \hat{H}_f^\circ$, and Enthalpy of Combustion $\Delta \hat{H}_c^\circ$ at 298 K

Compound	Formula	$\Delta \hat{G}_f^\circ$ kJ/gmol	$\Delta \hat{H}_f^\circ$ kJ/gmol	$\Delta \hat{H}_c^\circ$ kJ/gmol
Acetaldehyde (g)	C_2H_4O	−133.1	−166.2	−1104.5
Acetic acid (g)	$C_2H_4O_2$	−374.6	−432.8	−814.6
(l)		−392.5	−486.18	
Acetic anhydride (g)	$C_4H_6O_3$	−473.4	−572.5	−1675
Acetone (g)	C_3H_6O	−151.3	−215.7	−1659
(l)		−155.5	−248.2	
Acetonitrile (g)	C_2H_3N	91.868	74.04	−1190.4
Acetylene (g)	C_2H_2	210.68	228.2	−1257
Adipic acid (l)	$C_6H_{10}O_4$	−985.4	−741.3	
Ammonia (g)	NH_3	−16.6	−46.15	−316.8
Ammonium nitrate (s)	$N_2H_5NO_3$	−184	−365.56	
(aq)		−190.7	−340	
Argon (g)	Ar	0	0	0
Benzene (g)	C_6H_6	129.6	82.88	−3136
Butadiene, 1,3 (g)	C_4H_6	149.7	109.24	−2409
n-Butane (g)	C_4H_{10}	−15.707	−124.73	−2657.3
Calcium carbonate (s)	$CaCO_3$	−1133.0	−1211.3	

(*continued*)

Table B.3   (continued)

Compound	Formula	$\Delta \hat{G}^\circ_f$ kJ/gmol	$\Delta \hat{H}^\circ_f$ kJ/gmol	$\Delta \hat{H}^\circ_c$ kJ/gmol
Calcium chloride (s)	CaCl$_2$	−752.28	−797.47	
Carbon dioxide (g)	CO$_2$	−394.37	−393.5	0
Carbon disulfide (g)	CS$_2$	66.8	116.9	−1076.9
Carbon monoxide (g)	CO	−137.27	−110.53	−283
Carbonyl sulfide (g)	COS	−165.5	−141.5	
Chlorine (g)	Cl$_2$	0	0	0
Chlorobenzene (l)	C$_6$H$_5$Cl	89.2	11.5	
Chloroform (g)	CHCl$_3$	−103.61	−70.1	
Cyclohexane (g)	C$_6$H$_{12}$	31.8	−123.1	
(l)		26.7	−156.2	
Diethylamine (g)	(C$_2$H$_5$)$_2$NH	73.08	−71.42	−2800.3
Diethyl ether (g)	(C$_2$H$_5$)$_2$O		−252.7	
(l)		−116.1	−272.8	
Dimethylamine (g)	(CH$_3$)$_2$NH	68.0	−18.6	
Dimethyl carbonate	C$_3$H$_6$O$_3$	−452.4	−570.1	
Dimethyl ether (g)	(CH$_3$)$_2$O	−109.0	−184.1	−28.84
Ethane (g)	C$_2$H$_6$	−31.92	−83.82	−1428.6
Ethanol (g)	C$_2$H$_5$OH	−167.85	−234.95	−1235
(l)		−174.72	−277.61	
Ethyl acetate (g)	CH$_3$COOC$_2$H$_5$	−328.0	−444.5	−2061
(l)		−318.4	−463.3	
Ethylamine (g)	C$_2$H$_5$NH$_2$	36.16	−47.15	−1587.4
Ethylbenzene (g)	C$_8$H$_{10}$	130.73	29.92	−4345
(l)		119.7	−12.5	
Ethylene (g)	C$_2$H$_4$	68.44	52.51	−1323
Ethylene glycol (g)	C$_2$HO$_2$	−302.6	−387.5	−1059
(l)		−319.8	−451.5	
Ethylene oxide (g)	C$_2$H$_4$O	−13.23	−52.63	−1218
Formaldehyde (g)	CH$_2$O	−102.6	−108.6	−526.8
Formic acid (g)	CH$_2$O$_2$	−351.0	−378.6	−211.5

Compound	Formula	$\Delta \hat{G}_f^{\circ}$ kJ/gmol	$\Delta \hat{H}_f^{\circ}$ kJ/gmol	$\Delta \hat{H}_c^{\circ}$ kJ/gmol
Gallium nitride (s)	GaN		−109.6	
Glycerol (glycerin) (g)	$C_3H_8O_3$		−577.9	
(l)		−475.5	−665.9	
n-Heptane (g)	C_7H_{16}	8.165	−187.8	−4464.7
(l)		1.757	−224.4	
Hexamethylenediamine (g)	$C_6H_{16}N_2$	120.96	−127.9	
n-Hexane (g)	C_6H_{14}	−0.066	−166.94	−3855.1
(l)		−3.81	−198.8	
Hydrazine (g)	N_2H_4	159.17	95.353	−5342
(l)			50.46	
Hydrogen (g)	H_2	0	0	0
Hydrogen chloride (g)	HCl	−95.30	−92.31	−28.6
Hydrogen peroxide (g)	H_2O_2	−105.48	−136.11	
(l)		−118.11	−188.95	
Hydrogen cyanide (g)	HCN	124.7	135.14	−623.3
Hydrogen sulfide (g)	H_2S	−32.84	−19.96	
Iron oxide (ferrous) (s)	FeO	−248.45	−270.37	
(ferric, hematite) (s)	Fe_2O_3	−749.35	−830.5	
(magnetite) (s)	Fe_3O_4	−1013.8	−1116.7	
Isobutane (g)	C_4H_{10}	−20.76	−131.418	−2649
Isobutene (g)	C_4H_8	70.27	−0.54	−2540.8
Isopentane (g)	C_5H_{12}	−14.05	−153.7	−3239.5
Magnesium chloride (s)	$MgCl_2$	−601.5	−641.1	
Methane (g)	CH_4	−50.49	−74.52	−802.6
Methyl acetate (g)	CH_3COOCH_3		−410.0	
Methanol (g)	CH_3OH	−162.32	−200.94	
(l)		−166.12	−238.655	−638.46
Methyl ethyl ether	$CH_3OC_2H_5$	−117.1	−216.4	−1931.4
Naphthalene (g)	C_8H_{10}	224.08	150.58	−498.09
Nitric acid (g)	HNO_3	−73.51	−133.85	
(l)		−79.91	−173.22	

(*continued*)

Table B.3 (continued)

Compound	Formula	$\Delta \widehat{G}^\circ_f$ kJ/gmol	$\Delta \widehat{H}^\circ_f$ kJ/gmol	$\Delta \widehat{H}^\circ_c$ kJ/gmol
Nitric oxide (g)	NO	86.57	90.25	−90.2
Nitroglycerin	$C_3H_5(NO_3)_3$		−279.1	
Nitrogen (g)	N_2	0	0	0
Nitrogen dioxide (g)	NO_2	51.3	33.3	
Nitrogen tetroxide (g)	N_2O_4	97.95	9.33	
Nitrous oxide (g)	N_2O	104.16	82.05	−82
n-Octane (g)	C_8H_{18}	16.0	−208.75	−5074.2
(l)		7.4	−249.95	
Oxygen (g)	O_2	0	0	0
n-Pentane (g)	C_5H_{12}	−8.81	−146.76	−3244.9
(l)		−9.25	−173.05	
Phenol (g)	C_6H_5OH	−32.637	−96.399	−2921
(l)		−46.11	−158.16	
Phosgene (g)	$COCl_2$	−206.8	−220.1	
n-Propane (g)	C_3H_8	−24.39	−104.68	−2043.1
Propionic acid (g)	C_2H_5COOH	−366.7	−453.5	−1395
(l)		−383.5	−509.2	
n-Propanol (g)	C_3H_7OH	−159.9	−255.2	−1843.8
(l)		−166.69	−300.70	
Propylene (g)	C_3H_6	62.15	19.71	−1925.7
Silicon tetrachloride (l)	$SiCl_4$	−560.24	−627	
Silicon dioxide (c,quartz)	SiO_2	−796.6	−850.8	
Sodium borohydride (aq)	$NaBH_4$	−147.61	−199.6	
Sodium carbonate (c)	Na_2CO_3	−1044.12	−1127.42	
Sodium chloride (c)	NaCl	−384.485	−411.375	
Sodium cyanide (c)	NaCN		−94.0	
Sodium hydroxide (s)	NaOH	−379.4	−425.9	
(aq)		−419.2	−469.15	
Sodium metaborate (aq)	$NaBO_2$	−940.81	−1012.49	
Styrene	C_8H_8	213.9	147.4	−4219

Compound	Formula	$\Delta \widehat{G}_f^\circ$ kJ/gmol	$\Delta \widehat{H}_f^\circ$ kJ/gmol	$\Delta \widehat{H}_c^\circ$ kJ/gmol
Sulfur dioxide (g)	SO_2	−299.9	−296.81	0
Sulfur trioxide (g)	SO_3	−370.66	−394.93	
Sulfuric acid (l)	H_2SO_4		−810.4	
(aq)			−887.1	
Toluene (g)	$C_6H_5CH_3$	122.0	50.17	−3734
(l)		114.148	11.996	
Triethylamine (g)	$(C_2H_5)_3N$	114.1	−95.8	−4040.5
Trimethylamine	$(CH_3)_3N$	98.99	−243.1	−2244.9
Trinitrotoluene (g)	$C_7H_5(NO_2)_3$		24.1	
(s)			−65.6	
Urea (g)	$(NH_2)_2CO$	−152.7	−235.5	
(l)		−194.3	−324.5	
(s)		−196.8	−333.6	
Vinyl chloride (g)	C_2H_3Cl	41.95	28.45	−1178
Water (g)	H_2O	−228.59	−241.83	0
(l)		−237.19	−285.84	−44.0
o-Xylene (g)	C_8H_{10}	122.2	19.08	−4333.0
(l)		110.33	−24.44	
m-Xylene (g)	C_8H_{10}	118.76	17.32	−4331.8
(l)		107.654	−25.418	
p-Xylene (g)	C_8H_{10}	121.4	18.03	−4333.0
(l)		110.08	−24.246	

$\Delta \widehat{H}_c^\circ$ is the enthalpy change associated with combustion of the compound in the gas phase, with CO_2 (g), H_2O (g), Cl_2 (g), N_2 (g), and SO_2 (g) as products. With H_2O (l) as product, $\Delta \widehat{H}_c^\circ$ decreases (becomes more negative) by $44.0n$ kJ/gmol, where n is the number of moles of H_2O. $-\Delta \widehat{H}_c^\circ$ is sometimes called the lower heating value with water vapor and the higher heating value with liquid water as the product.

Source: Compiled from data in *Perry's Chemical Engineers' Handbook,* 6th and 7th eds., *Lange's Handbook of Chemistry,* 14th ed., and NIST Chemistry Webbook.

B.4   Antoine Equation Constants

The Antoine equation

$$\log_{10} P^{\text{sat}} (\text{mmHg}) = A - \frac{B}{T\,(^{\circ}\text{C}) + C}$$

is a useful equation for modeling saturation pressures of liquids and solids. The constants should not be used outside the indicated temperature range.

Table B.4   Antoine Equation Constants for Selected Compounds

Compound	Formula	Range, °C	A	B	C
Acetaldehyde	CH_3CHO	−45 to +70	8.0055	1600	291.8
Acetic acid	CH_3COOH		7.38782	1533.313	222.309
Acetic anhydride	$C_4H_6O_3$		7.14948	1444.718	199.817
Acetone	CH_3COCH_3		7.02447	1161.0	224
Acetonitrile	CH_3CN		7.11988	1314.4	230
Acrylonitrile	C_3H_3N	−20 to +140	7.03855	1232.53	222.47
Ammonia	NH_3	−83 to +60	7.36050	926.132	240.17
Benzene	C_6H_6	+8 to +103	6.90565	1211.033	220.790
Benzoic acid	C_6H_5COOH	96 to 250	7.3533	1771.4	145.67
Bromine	Br_2		6.87780	1119.68	221.38
n-Butanol	C_4H_9OH	+15 to +131	7.47680	1362.39	178.77
Butadiene, 1,3	C_4H_6	−58 to +15	6.84999	930.546	238.854
Carbon disulfide	CS_2	3 to 80	6.94279	1169.11	241.59
Chlorine	Cl_2		6.93790	861.34	246.33
Chloroform	$CHCl_3$	−35 to 61	6.4934	929.44	196.03
Diethanolamine	$(C_2H_5O)_2NH$	194 to 241	8.1388	2327.9	174.4
Diethylamine	$(C_2H_5)_2NH$	31 to 61	5.8016	583.30	144.1
Dimethylamine	$(CH_3)_2NH$	−72 to +6.9	7.08212	960.242	221.67
Ethanol	C_2H_5OH	−2 to +100	8.04494	1554.3	222.65
Ethanolamine	C_2H_7ON	65 to 171	7.4568	1577.67	173.37
Ethyl acetate	$CH_3COOC_2H_5$	15 to 76	7.10179	1244.95	217.88
Ethylamine	$C_2H_5NH_2$	−20 to +90	7.05413	987.31	220.0
Ethylbenzene	C_8H_{10}	26 to 164	6.95719	1424.255	213.21
Ethylene glycol	$C_2H_6O_2$	50 to 200	8.0908	2088.9	203.5

Compound	Formula	Range, °C	A	B	C
Ethylene oxide	C_2H_4O	−49 to +12	7.12843	1054.54	237.76
Formic acid	CH_2O_2	37 to 101	7.5818	1699.2	260.7
Glycerol	$C_3H_8O_3$	183 to 260	6.165	1036	28
n-Heptane	C_7H_{16}	−2 to +124	6.89677	1264.90	216.54
n-Hexane	C_6H_{14}	−25 to 92	6.87601	1171.17	224.41
Hydrogen cyanide	HCN	−16 to 46	7.5282	1329.5	260.4
Hydrogen peroxide	H_2O_2		7.96917	1886.76	220.6
Isopentane	C_5H_{12}		6.78967	1020.012	233.097
Isopropanol	C_3H_7OH	0 to 100	8.11778	1580.92	219.61
Lactic acid	$C_3H_6O_3$		8.06	1823.7	134
Methanol	CH_3OH	−14 to 65	7.89750	1474.08	229.13
		65 to 110	7.97328	1515.14	232.85
Methyl acetate	CH_3COOCH_3	1 to 56	7.0652	1157.63	219.73
Methyl ethyl ketone	$CH_3COC_2H_5$		6.97	1210	216
Naphthalene (s)	$C_{10}H_8$	86 to 250	7.01065	1733.71	201.86
(l)		125 to 218	6.8181	1585.86	184.82
Nitrogen	N_2		6.49457	255.68	266.55
n-Octane	C_8H_{18}	19 to 152	6.91868	1351.99	209.15
Oxygen	O_2		6.69144	319.013	266.697
n-Pentane	C_5H_{12}	−50 to 58	6.85221	1064.63	233.01
Phosgene	$COCl_2$	−68 to 68	6.84297	941.25	230
Phenol	C_6H_5OH	107 to 182	7.133	1516.79	174.95
n-Propanol	C_3H_7OH	2 to 120	7.84767	1499.21	204.64
Propionic acid	C_2H_5COOH	56 to 139	6.403	950.2	130.3
Silicon tetrachloride	$SiCl_4$	0 to 53	6.85726	1138.92	228.88
Styrene	C_8H_8	32 to 82	7.14016	1574.51	224.09
Tetramethyl lead	$C_4H_{12}Pb$	0 to 60	6.9377	1335.3	219.1
Toluene	C_7H_8	6 to 137	6.95464	1344.8	219.48
Water	H_2O	0 to 60	8.10765	1750.286	235.0
		60 to 150	7.96681	1668.21	228.0
n-Xylene	C_8H_{10}	32 to 172	6.99891	1474.679	213.69
m-Xylene	C_8H_{10}	28 to 166	7.00908	1462.266	215.11
p-Xylene	C_8H_{10}	27 to 166	6.99052	1453.43	215.31

Source: *Lange's Handbook of Chemistry,* 14th ed and NIST Chemistry Webbook.

B.5   Phase Equilibrium Data

Table B.5	Henry's Law Constant (atm), $H_i = \frac{y_i P}{x_i} = \frac{p_i}{x_i}$, for Gas Dissolved in Water					
	0°C	**10°C**	**20°C**	**30°C**	**40°C**	**50°C**
He	129,000	126,000	125,000	124,000	121,000	115,000
H_2	57,900	63,600	68,300	72,900	75,100	76,500
N_2	52,900	66,800	80,400	92,400	104,000	113,000
CO	35,200	44,200	53,600	62,000	69,600	76,100
O_2	25,500	32,700	40,100	47,500	53,500	58,800
CH_4	22,400	29,700	37,600	44,900	52,000	57,700
C_2H_6	12,600	18,900	26,300	34,200	42,300	50,000
C_2H_4	5,520	7,680	10,200	12,700		
CO_2	728	1,040	1,420	1,860	2,330	2,830
H_2S	268	367	483	609	745	884

Adapted from Hines and Maddox, *Mass Transfer Fundamentals and Applications,* 1985.

Table B.6	Partial Pressures of SO_2 in Equilibrium with Dissolved SO_2 in Water									
	Partial pressure of SO_2, p_{SO_2}, mmHg									
Grams SO_2 per 100 grams water	**10°C**	**20°C**	**30°C**	**40°C**	**50°C**	**60°C**	**70°C**	**80°C**	**90°C**	**100°C**
0.5	21	29	42	60	83	111	144	182	225	274
1.0	42	59	85	120	164	217	281	356	445	548
1.5	64	90	129	181	247	328	426	543	684	850
2.0	86	123	176	245	333	444	581	756	940	
2.5	108	157	224	311	421	562	739	956		
3.0	130	191	273	378	511	682	897			
3.5	153	227	324	447	603	804				
4.0	176	264	376	518	698					
4.5	199	300	428	588	793					
5.0	223	338	482	661						
5.5	247	375	536	733						

| Grams SO₂ per 100 grams water | Partial pressure of SO₂, p_{SO_2}, mmHg | | | | | | | | | |
	10°C	20°C	30°C	40°C	50°C	60°C	70°C	80°C	90°C	100°C
6.0	271	411	588	804						
6.5	295	448	642							
7.0	320	486	698							
8.0	370	562	806							
9.0	421	638								
10.0	473	714								
11.0	526	789								
12.0	580									
13.0	635									
14.0	689									
15.0	743									
16.0	799									

Source: *Perry's Chemical Engineers' Handbook,* 6th ed.

Table B.7	Partial Pressures of NH₃ in Equilibrium with Dissolved NH₃ in Water

| Grams NH₃ per 100 grams solution | Partial Pressure of NH₃, p_{NH_3}, mmHg | | | | | | | | |
	0°C	10°C	21°C	32°C	43°C	54°C	65.5°C	77°C	88°C
4.74	13.4	24.3	42.9	70	111	170	247	349	477
9.5	26.9	46	78.6	130	207	315	461	655	
14.3	46.5	78	134	220	344	520	760		
19.1	78	131	221	356	550				
23.9	138	215	355	563					
28.8	221	343	556						
33.7	338	530							
38.6	462	788							
43.6	731								

Source: Adapted from data in *Perry's Chemical Engineers' Handbook,* 6th ed.

Table B.8   Solubility of Salts in Water

Compound	Formula	0°C	10°C	20°C	30°C	40°C	50°C	60°C	70°C	80°C	90°C	100°C
Calcium bicarbonate	$Ca(HCO_3)_2$	16.15		16.6		17.05		17.50		17.95		18.4
Magnesium chloride	$MgCl_2 \cdot 6H_2O$	52.8	53.5	54.5		57.5		61.0		66.0		73.0
Potassium nitrate	KNO_3	13.3	20.9	31.6	45.8	63.9	85.5	110.0	138	169	202	246
Potassium sulfate	K_2SO_4	7.35	9.22	11.11	12.97	14.76	16.50	18.17	19.75	21.4	22.8	24.1
Sodium chloride	$NaCl$	35.7	35.8	36.0	36.3	36.6	37.0	37.3	37.8	38.4	39.0	39.8
Sodium sulfate	$Na_2SO_4 \cdot 10H_2O$	5.0	9.0	19.4	40.8							
	$Na_2SO_4 \cdot 7H_2O$	19.5	30	44								
	Na_2SO_4					48.8	46.7	45.3		43.7		42.5

Data are listed as grams of anhydrous substance per 100 g water, in a saturated liquid solution. The formula shows the solid phase (hydrated or anhydrous) that is in equilibrium with the saturated solution.
Source: *Perry's Chemical Engineers' Handbook,* 6th ed.

Table B.9   Benzene-Naphthalene Solid-Liquid Equilibrium

Mole fraction naphthalene in liquid phase, x_n	Temperature, °C	Solid phase
0.0	5.5	Benzene
0.023	4	Benzene
0.039	3	Benzene
0.083	0	Benzene
0.135	−3	Benzene
0.148	0	Naphthalene
0.17	5	Naphthalene
0.20	10	Naphthalene
0.26	20	Naphthalene
0.34	30	Naphthalene
0.43	40	Naphthalene
0.54	50	Naphthalene
0.67	60	Naphthalene
0.82	70	Naphthalene
1.0	80.2	Naphthalene

Saturated liquid solution of benzene and naphthalene in equilibrium with a single-component solid phase. (Calculated by assuming ideal solution behavior and using melting points and enthalpies of melting of pure components.)

Table B.10   *m*-Xylene–*p*-Xylene Solid-Liquid Equilibrium

Mole fraction *p*-xylene in liquid phase, x_p	Temperature, °C	Solid phase
0.0	−47.2	*m*-xylene
0.074	−50	*m*-xylene
0.125	−52	*m*-xylene
0.165	−45	*p*-xylene
0.20	−40	*p*-xylene
0.285	−30	*p*-xylene
0.396	−20	*p*-xylene
0.536	−10	*p*-xylene
0.71	0	*p*-xylene
0.923	10	*p*-xylene
1.0	13.2	*p*-xylene

Saturated liquid solution of *m*-xylene and *p*-xylene in equilibrium with a single-component solid phase. (Calculated by assuming ideal solution behavior and using melting points and enthalpies of melting of pure components.)

Table B.11   Ethanol-Water Vapor-Liquid Equilibrium at 1 atm

Temperature, °C	Mole fraction ethanol in liquid phase, x_e	Mole fraction ethanol in vapor phase, y_e
100.0	0.000	0.000
95.5	0.019	0.17
89.0	0.0721	0.3891
86.7	0.0966	0.4375
85.3	0.1238	0.4704
84.1	0.1661	0.5089
82.7	0.2337	0.5445
82.3	0.2608	0.5580
81.5	0.3273	0.5826
80.7	0.3965	0.6122
79.8	0.5079	0.6564
79.7	0.5198	0.6599
79.3	0.5732	0.6841
78.74	0.6763	0.7385
78.41	0.7472	0.7815
78.15	0.8943	0.8943

Source: *Perry's Chemical Engineers' Handbook*, 6th ed.

Table B.12   Methanol-Benzene Vapor-Liquid Equilibrium at 1 atm

Temperature, °C	Mole fraction methanol in liquid phase, x_m	Mole fraction methanol in vapor phase, y_m
70.67	0.026	0.267
66.44	0.050	0.371
62.87	0.088	0.457
60.20	0.164	0.526
58.64	0.333	0.559
58.02	0.549	0.595
58.10	0.699	0.633
58.47	0.782	0.665
59.90	0.898	0.760
62.71	0.973	0.907

Source: *Perry's Chemical Engineers' Handbook*, 6th ed.

Table B.13   Water-Acetic Acid-Methyl Isobutyl Ketone Liquid-Liquid Equilibrium, at 25°C

Weight % in raffinate			Weight % in extract		
Water	Acetic acid	MIBK	Water	Acetic acid	MIBK
98.45	0	1.55	2.12	0	97.88
95.46	2.85	1.7	2.80	1.87	95.33
85.8	11.7	2.5	5.4	8.9	85.7
75.7	20.5	3.8	9.2	17.3	73.5
67.8	26.2	6.0	14.5	24.6	60.9
55.0	32.8	12.2	22.0	30.8	47.2
42.9	34.6	22.5	31.0	33.6	35.4

Each row shows the compositions of the raffinate and extract phases at equilibrium.
Source: *Perry's Chemical Engineers' Handbook*, 6th ed.

Table B.14   Ethylbenzene-Styrene-Ethylene Glycol Liquid-Liquid Equilibrium, at 25°C

Weight % in raffinate			Weight % in extract		
Ethylbenzene	**Styrene**	**Ethylene glycol**	**Ethylbenzene**	**Styrene**	**Ethylene glycol**
90.56	8.63	0.81	9.85	1.64	88.51
80.40	18.67	0.93	9.31	3.49	87.20
70.49	28.51	1.00	8.72	5.48	85.80
60.93	37.98	1.09	8.07	7.45	84.48
53.55	45.25	1.20	7.35	9.25	83.40
52.96	45.84	1.20	7.31	9.49	83.20
43.29	55.32	1.39	6.30	12.00	81.70
41.51	57.09	1.40	6.06	12.54	81.40

Each row shows the compositions of the raffinate and extract phases at equilibrium.
Source: *Perry's Chemical Engineers' Handbook*, 6th ed.

Table B.15   Distribution Coefficient, $K_D = x_{A, \text{phase II}}/x_{A, \text{phase I}}$, for Solute a Distributing between Two Immiscible Liquids

Solute A	Solvent phase I	Solvent phase II	K_D
Acetic acid	Water	Methyl acetate	1.273
Acetic acid	Water	Furfural	0.787 (26.7°C)
Acetic acid	Water	Heptadecanol	0.312
Acetic Acid	Water	Benzene	0.0328
Acetic Acid	Water	1-Butanol	1.613 (26.7°C)
Oleic acid	Cottonseed oil	Propane	0.150 (85°C)
Chlorine	Water	Carbon tetrachloride	5.0
Bromine	Water	Carbon tetrachloride	27
Iodine	Water	Carbon tetrachloride	55
Ammonia	Water	Carbon tetrachloride	0.0042
Diethylamine	Water	Chloroform	2.2

(continued)

Table B.15	(continued)		
Solute A	**Solvent phase I**	**Solvent phase II**	K_D
Diethylamine	Water	Benzene	1.8
Diethylamine	Water	Toluene	0.63
Diethylamine	Water	Xylene	0.20
Ethanol	Water	Benzene	0.1191
Ethanol	Water	Heptadecanol	0.270
Ethanol	Water	*n*-Butanol	3.00 (20°C)
Methyl ethyl ketone	Water	Gasoline	1.686
Methyl ethyl ketone	Water	2-Methyl furan	84.0
Penicillin F	Water (pH 6.0)	Amyl acetate	0.06
Penicillin F	Water (pH 4.0)	Amyl acetate	32

Data at 25°C unless otherwise noted. Reliable only at dilute solute concentrations.
Compiled from data in *Perry's Chemical Engineers' Handbook*, 6th ed., *Biochemical and Biotechnology Handbook*, 1991, 2nd ed., and *Process Synthesis*, D. F. Rudd, G. J. Powers and J. J. Siiroia, 1973.

B.6   Steam Tables

$\hat{H}$ and $\hat{U}$ are given in units of kJ/kg, with the reference condition as the triple point of liquid water (273.15 K, 0.00611 bar). $\hat{V}$ is given in units of m³/kg.
Source: E. W. Lemmon, M. O. McLinden and D. G. Friend, "Thermophysical Properties of Fluid Systems" in *NIST Chemistry WebBook, NIST Standard Reference Database Number 69*, Eds. P. J. Linstrom and W. G. Mallard, June 2005, National Institute of Standards and Technology, Gaithersburg MD, 20899 (http://webbook.nist.gov).

(See table on next page.)

Table B.16 Specific Enthalpy $\hat{H}$, Internal Energy $\hat{U}$, and Volume $\hat{V}$ of H_2O at Several Temperatures and Pressures

P, bar (T^{sat}, °C)		Sat'd liquid	Sat'd vapor	Temperature (°C) 50	100	150	200	250	300	350
0.006116	$\hat{H}$	0.00	2500.9	2594.5	2688.6	2783.7	2880.0	2977.8	3077.0	3177.7
(0.01)	$\hat{U}$	0.00	2374.9	2445.4	2516.4	2588.4	2661.7	2736.3	2812.5	2890.1
	$\hat{V}$	0.00100	206.55	244.45	282.30	320.14	357.98	395.81	433.64	470.69
0.1	$\hat{H}$	191.81	2583.9	2592.0	2687.5	2783.1	2879.6	2977.5	3076.8	3177.6
(45.806)	$\hat{U}$	191.80	2437.2	2443.3	2515.5	2587.9	2661.4	2736.1	2812.3	2890.0
	$\hat{V}$	0.00101	14.670	14.867	17.197	19.514	21.826	24.137	26.446	28.755
1.0	$\hat{H}$	417.50	2674.9	209.46	2675.8	2776.6	2875.5	2974.5	3074.6	3175.8
(99.606)	$\hat{U}$	417.40	2505.6	209.36	2506.2	2583.0	2658.2	2733.9	2810.7	2888.7
	$\hat{V}$	0.00104	1.6939	0.00101	1.6959	1.9367	2.1725	2.4062	2.6389	2.8710
5.0	$\hat{H}$	640.09	2748.1	209.80	419.51	632.24	2855.9	2961.1	3064.6	3168.1
(151.83)	$\hat{U}$	639.54	2560.7	209.30	418.99	631.69	2643.3	2723.8	2803.3	2883.0
	$\hat{V}$	0.00109	0.37481	0.00101	0.00104	0.00109	0.4250	0.4744	0.5226	0.57016
10.0	$\hat{H}$	762.52	2777.1	210.19	419.84	632.5	2828.3	2943.1	3051.6	3158.2
(179.88)	$\hat{U}$	761.39	2582.7	209.18	418.80	631.41	2622.2	2710.4	2793.6	2875.7
	$\hat{V}$	0.00113	0.1944	0.00101	0.00104	0.00109	0.2060	0.2328	0.2580	0.2825
20.0	$\hat{H}$	908.5	2798.3	211.06	420.59	633.12	852.45	2903.2	3024.2	3137.7
(212.38)	$\hat{U}$	906.14	2599.1	209.03	418.51	630.94	850.14	2680.2	2773.2	2860.5
	$\hat{V}$	0.00118	0.0996	0.00101	0.00104	0.00109	0.00116	0.1115	0.1255	0.1386
40.0	$\hat{H}$	1087.5	2800.8	212.78	422.10	634.36	853.27	1085.8	2961.7	3093.3
(250.35)	$\hat{U}$	1082.5	2601.7	208.74	417.93	630.01	848.65	1080.8	2726.2	2827.4
	$\hat{V}$	0.00125	0.04978	0.00101	0.00104	0.00109	0.00115	0.00125	0.0589	0.0665
60.0	$\hat{H}$	1213.9	2784.6	214.50	423.60	635.61	854.09	1085.7	2885.5	3043.9
(275.58)	$\hat{U}$	1206.0	2589.9	208.44	417.36	629.08	847.18	1078.2	2668.4	2790.4
	$\hat{V}$	0.00132	0.03245	0.00101	0.00104	0.00109	0.00115	0.00125	0.0362	0.0423
100.0	$\hat{H}$	1408.1	2725.5	217.94	426.62	638.11	855.8	1085.8	1343.3	2924.0
(311.00)	$\hat{U}$	1393.5	2545.2	207.86	416.23	627.27	844.31	1073.4	1329.4	2699.6
	$\hat{V}$	0.00145	0.0180	0.00101	0.00104	0.00108	0.00115	0.00124	0.00140	0.0224
150.0	$\hat{H}$	1610.2	2610.7	222.23	430.39	641.27	857.99	1086.1	1338.3	2693.1
(342.16)	$\hat{U}$	1585.3	2455.6	207.15	414.85	625.05	840.84	1067.6	1317.6	2520.9
	$\hat{V}$	0.00166	0.01034	0.00101	0.00104	0.00108	0.00114	0.00123	0.00138	0.0115
200	$\hat{H}$	1827.2	2412.3	226.51	434.17	644.45	860.27	1086.7	1334.4	1646.0
(365.75)	$\hat{U}$	1786.4	2295.0	206.44	413.50	622.89	837.49	1062.2	1307.1	1612.7
	$\hat{V}$	0.00204	0.00586	0.00100	0.00103	0.00108	0.00114	0.00123	0.00136	0.00166
220.64	$\hat{H}$	2084.3	2084.3	228.28	435.73	645.77	861.23	1087.0	1333.0	1635.6
(373.95)	$\hat{U}$	2015.7	2015.7	206.16	412.95	622.01	836.14	1060.0	1303.1	1599.6
	$\hat{V}$	0.00311	0.00311	0.00100	0.00103	0.00108	0.00114	0.00122	0.00135	0.00163

(continued)

Table B.16   (continued)

P, bar (T^{sat}, °C)		Sat'd liquid	Sat'd vapor	Temperature (°C)						
				400	**500**	**600**	**700**	**800**	**900**	**1000**
0.006116	$\hat{H}$	0	2500.9	3280.1	3489.8	3706.3	3930	4160.7	4398.4	4642.8
(0.01)	$\hat{U}$	0.00	2374.9	2969.4	3133	3303.4	3480.8	3665.4	3856.9	4055.3
	$\hat{V}$	0.00100	206.55	507.96	583.42	658.88	734.35	809.81	885.27	960.73
0.1	$\hat{H}$	191.81	2583.9	3279.9	3489.7	3706.3	3929.9	4160.6	4398.3	4642.8
(45.806)	$\hat{U}$	191.80	2437.2	2969.3	3132.9	3303.3	3480.8	3665.3	3856.9	4055.2
	$\hat{V}$	0.00101	14.670	31.063	35.680	40.296	44.911	49.527	54.142	58.758
1.0	$\hat{H}$	417.50	2674.9	3278.6	3488.7	3705.6	3929.4	4160.2	4398.0	4642.6
(99.606)	$\hat{U}$	417.40	2505.6	2968.3	3132.2	3302.8	3480.4	3665.0	3856.6	4055.0
	$\hat{V}$	0.00104	1.6939	3.1027	3.5655	4.0279	4.4900	4.9519	5.4137	5.8754
5.0	$\hat{H}$	640.09	2748.1	3272.3	3484.5	3702.5	3927.0	4158.4	4396.6	4641.4
(151.83)	$\hat{U}$	639.54	2560.7	2963.7	3129.0	3300.4	3478.5	3663.6	3855.4	4054.0
	$\hat{V}$	0.00109	0.37481	0.6173	0.7109	0.8041	0.897.0	0.9897	1.0823	1.1748
10.0	$\hat{H}$	762.52	2777.1	3264.5	3479.1	3698.6	3924.1	4156.1	4394.8	4639.9
(179.88)	$\hat{U}$	761.39	2582.7	2957.9	3125.0	3297.5	3476.2	3661.7	3853.9	4052.7
	$\hat{V}$	0.00113	0.1944	0.3066	0.3541	0.4011	0.4478	0.4944	0.5408	0.5872
20.0	$\hat{H}$	908.5	2798.3	3248.3	3468.2	3690.7	3918.2	4151.5	4391.1	4637.0
(212.38)	$\hat{U}$	906.14	2599.1	2945.9	3116.9	3291.5	3471.6	3658.0	3850.9	4050.2
	$\hat{V}$	0.00118	0.0996	0.1512	0.1757	0.1996	0.2233	0.2467	0.2701	0.2934
40.0	$\hat{H}$	1087.5	2800.8	3214.5	3446.0	3674.9	3906.3	4142.3	4383.9	4631.2
(250.35)	$\hat{U}$	1082.5	2601.7	2920.7	3100.3	3279.4	3462.4	3650.6	3844.8	4045.1
	$\hat{V}$	0.00125	0.04978	0.0734	0.0864	0.0989	0.1110	0.1229	0.1348	0.1465
60.0	$\hat{H}$	1213.9	2784.6	3178.2	3423.1	3658.7	3894.3	4133.1	4376.6	4325.4
(275.58)	$\hat{U}$	1206.0	2589.9	2893.7	3083.1	3267.2	3453.0	3643.2	3838.8	4040.1
	$\hat{V}$	0.00132	0.03245	0.0474	0.0567	0.0653	0.0735	0.0816	0.0896	0.0976
100.0	$\hat{H}$	1408.1	2725.5	3097.4	3375.1	3625.8	3870.0	4114.5	4362.0	4613.8
(311.00)	$\hat{U}$	1393.5	2545.2	2833.1	3047.0	3242.0	3434.0	3628.2	3826.5	4029.9
	$\hat{V}$	0.00145	0.0180	0.0264	0.0328	0.0384	0.0436	0.0486	0.0535	0.0584
150.0	$\hat{H}$	1610.2	2610.7	2975.7	3310.8	3583.1	3839.1	4091.1	4343.7	4599.2
(342.16)	$\hat{U}$	1585.3	2455.6	2740.6	2998.4	3209.3	3409.8	3609.2	3811.2	4017.1
	$\hat{V}$	0.00166	0.01034	0.0157	0.0208	0.0249	0.0286	0.0321	0.0355	0.0388
200.0	$\hat{H}$	1827.2	2412.3	2816.9	3241.2	3539.0	3807.8	4067.5	4325.4	4584.7
(365.75)	$\hat{U}$	1786.4	2295.0	2617.9	2945.3	3175.3	3385.1	3590.1	3795.7	4004.3
	$\hat{V}$	0.00204	0.00586	0.00995	0.0148	0.0182	0.0211	0.0239	0.0265	0.0290
220.64	$\hat{H}$	2084.3	2084.3	2732.9	3210.8	3520.4	3794.7	4057.7	4317.8	4578.8
(373.95)	$\hat{U}$	2015.7	2015.7	2551.9	2922.0	3160.9	3374.7	3582.1	3789.3	3999.0
	$\hat{V}$	0.00311	0.00311	0.0082	0.0131	0.0163	0.019.0	0.0216	0.0239	0.0263

B.7   Heat Capacities

Table B.17   Heat Capacity C_p of Selected Liquids and Vapors

Compound	Formula	C_p (approx.)	A	B	C	D
Acetaldehyde (g)	C_2H_4O	54.7				
(l)		89.05				
Acetic acid (g)	$C_2H_4O_2$	66.5	4.840	0.2549	−1.753e-4	4.949e-8
(l)		124.4				
Acetone (g)	C_3H_6O	74.5	6.301	0.2606	−1.253e-4	2.038e-8
(l)			72.2	0.186		
Acetonitrile (g)	C_2H_3N	52.2	20.48	0.1196	−4.492e-5	3.203e-9
Acetylene (g)	C_2H_2	44.2	26.82	0.07578	−5.007e-5	1.412e-8
Ammonia (g)	NH_3	35.6	27.31	0.02383	1.707e-5	−1.185e-8
Argon (g)	Ar	20.8	20.8			
Benzene (g)	C_6H_6	81.7	−33.92	0.4739	−3.017e-4	7.13e-8
(l)		134.3	−6.2106	0.5650	−3.141e-4	
Bromine (g)	Br_2	36.3	33.86	0.01125	−1.192e-5	4.534e-9
Butadiene, 1,3 (g)	C_4H_6	79.5	−1.687	0.3419	−2.340e-4	6.335e-8
n-Butane (g)	C_4H_{10}	98.9	9.487	0.3313	−1.108e-4	−2.822e-9
Carbon dioxide (g)	CO_2	37.0	19.80	0.07344	−5.602e-5	1.7115e-8
Carbon disulfide (g)	CS_2	34.2	27.44	0.08127	−7.666e-5	2.673e-8
Carbon monoxide (g)	CO	29.1	30.87	−0.01285	2.789e-5	−1.272e-8
Carbon tetrachloride (g)	CCl_4	84.0	40.72	0.2049	−2.270e-4	8.843e-8
Chlorine (g)	Cl_2	34.0	26.93	0.03348	−3.869e-5	1.547e-8
Chloroform (g)	$CHCl_3$	65.8	24.00	.1893	−1.841e-4	6.657e-8
(l)		114.8	159.75	−0.3566	6.902e-4	
Chlorobenzene (l)	C_6H_5Cl	150.8	93.77	0.2732	−2.652e-4	
Cyclohexane (l)	C_6H_{12}	155.9	−75.225	1.1754	−1.344e-3	
Diethylamine (g)	$(C_2H_5)_2NH$	119.5				
(l)		172.5				
Diethyl ether (g)	$(C_2H_5)_2O$	112.5	21.42	.3359	−1.035e-4	−9.357e-9
Dimethylamine (g)	$(CH_3)_2NH$	115.7				
(l)		136.8				

(continued)

Table B.17 (continued)

Compound	Formula	C_p (approx.)	A	B	C	D
Dimethyl ether (g)	$(CH_3)_2O$	65.6				
(l)		102.3				
Ethane (g)	C_2H_6	52.5	5.409	0.1781	$-6.94e\text{-}5$	8.71e-9
(l)		68				
Ethanol (g)	C_2H_5OH	65.5	9.014	0.2141	$-8.39e\text{-}5$	1.373e-9
(l)		112.0				
Ethyl acetate (g)	$CH_3COOC_2H_5$	113.6				
(l)		169.9				
Ethylbenzene (g)	C_8H_{10}	128.4	-43.10	0.7072	$-4.811e\text{-}4$	1.301e-7
(l)		185.6				
Ethylene (g)	C_2H_4	43.7	3.806	0.1566	$-8.348e\text{-}5$	1.755e-8
Ethylene glycol (g)	C_2HO_2	78.0				
Ethylene oxide (g)	C_2H_4O	48.2	-7.519	0.2222	$-1.256e\text{-}4$	2.592e-8
Formaldehyde (g)	CH_2O	35.4				
Glycerol (glycerin) (l)	$C_3H_8O_3$	150.2				
n-Heptane (g)	C_7H_{16}	165.9	-5.146	0.6762	$-3.651e\text{-}4$	7.658e-8
(l)		212				
n-Hexane (g)	C_6H_{14}	143.1	-4.413	0.528	$-3.119e\text{-}4$	6.498e-8
(l)		189.1				
Hydrazine (g)	N_2H_4	45.5	9.768	0.1895	$-1.657e\text{-}4$	6.025e-8
(l)		98.9				
Hydrogen (g)	H_2	29.1	27.14	0.0093	$-1.381e\text{-}5$	7.645e-9
Hydrogen chloride (g)	HCl	29.5	30.67	-0.0072	1.246e-5	$-3.898e\text{-}9$
Hydrogen cyanide (g)	HCN	36.0	21.86	0.06062	$-4.961e\text{-}5$	1.815e-8
Hydrogen sulfide (g)	H_2S	34.2	31.94	0.001436	2.432e-5	$-1.176e\text{-}8$
Isobutane (g)	C_4H_{10}	97.2	-1.390	0.3847	$-1.846e\text{-}4$	2.895e-8
Isobutene (g)	C_4H_8	89.9	16.05	0.2804	$-1.091e\text{-}4$	9.098e-9
Isopentane (g)	C_5H_{12}	118.7	-9.525	0.5066	$-2.729e\text{-}4$	5.723e-8
Isopropanol (g)	C_3H_7OH	80	32.43	0.1885	6.406e-5	$-9.261e\text{-}8$
(l)		155				
Lactic acid (g)	$C_3H_6O_3$	145				
(l)		262				

Compound	Formula	C_p (approx.)	A	B	C	D
Methane (g)	CH_4	35.7	19.25	0.05213	1.197e-5	−1.132e-8
Methyl acetate (l)	CH_3COOCH_3	155.6				
Methanol (g)	CH_3OH	43.9	21.15	0.07092	2.587e-5	−2.852e-8
(l)		81.2				
Nitric oxide (g)	NO	29.8	29.35	−9.378e-4	9.747e-6	−4.187e-9
Nitrogen (g)	N_2	29.1	31.15	−1.357e-2	2.680e-5	−1.168e-8
Nitrogen dioxide (g)	NO_2	36.97				
Nitrogen tetroxide (g)	N_2O_4	77.26				
(l)		142.51				
Nitrous oxide (g)	N_2O	38.5	21.62	7.281e-2	−5.778e-5	1.830e-8
n-Octane (g)	C_8H_{18}	188.7	−6.096	0.7712	−4.195e-4	8.855e-8
(l)		255				
Oxygen	O_2	29.3	28.11	−3.68e-6	1.746e-5	−1.065e-8
n-Pentane (g)	C_5H_{12}	120.1	−3.626	0.4873	−2.58e-4	5.305e-8
(l)		168.6				
Phenol (g)	C_6H_5OH	103.6				
Phosgene (g)	$COCl_2$	57.7				
Potassium nitrate (l)	KNO_3	123.4				
n-Propane (g)	C_3H_8	73.6	−4.224	0.3063	−1.586e-4	3.215e-8
n-Propanol (g)	C_3H_7OH	87.3	2.470	0.3325	−1.855e-4	4.296e-8
(l)		150.9	346.30	−1.749	3.552e-3	
Propylene (g)	C_3H_6	63.9	3.710	0.2345	−1.160e-4	2.205e-8
Silicon tetrachloride (l)	$SiCl_4$	135.6				
Sodium nitrate (l)	$NaNO_3$	155.6				
Styrene (g)	C_8H_8	122.1	−28.25	0.6159	−4.023e-4	9.935e-8
(l)		182.6				
Sulfur (g)	S_8	156.1				
(l)	S	32				
Sulfur dioxide (g)	SO_2	39.9	23.85	0.06699	−4.961e-5	1.328e-8
Sulfur trioxide (g)	SO_3	50.8	19.21	0.1374	−1.176e-4	3.700e-8
Toluene (g)	$C_6H_5CH_3$	103.8	−24.35	0.5125	−2.765e-4	4.991e-8
(l)		157.2	125.8	0.0565	1.3593-4	
Triethylamine (g)	$(C_2H_5)_3N$	160.9				

(*continued*)

Table B.17 (continued)

Compound	Formula	C_p (approx.)	A	B	C	D
Trimethylamine (g)	$(CH_3)_3N$	91.8				
Water (g)	H_2O	33.6	32.24	0.001924	1.055e-5	−3.596e-9
(l)		75.4	72.43	0.0104		
o-Xylene (g)	C_8H_{10}	133.3	−15.85	0.5962	−3.443e-4	7.528e-8
m-Xylene (g)	C_8H_{10}	127.6	−29.27	0.6297	−3.747e-4	8.478e-8
p-Xylene (g)	C_8H_{10}	126.9	−25.09	0.6042	−3.374e-4	6.820e-8

For approximate calculations, use the number in the column labeled "C_p (approx.)," which is the heat capacity at 25°C. For more accurate calculations, use the polynomial expression $C_p = A + BT + CT^2 + DT^3$, where C_p is in J/gmol K (or J/gmol °C) and T is in K. To convert to cal/gmol K or to Btu/lbmol °F, multiply by 0.239.

Source: Compiled from data in *Introductory Chemical Engineering Thermodynamics*, J. R. Elliott and C. T. Lira, Prentice-Hall, 1999; *Perry's Chemical Engineers' Handbook*, 6th ed.; and *Lange's Handbook of Chemistry*, 14th ed.

Table B.18 Heat Capacity C_p of Selected Solids

Compound	Formula	C_p, J/gmol K (with T in K)
Benzoic acid	C_6H_5COOH	147
Calcium carbonate	$CaCO_3$	$82.3 + 0.497T − 1.287e6/T^2$
Carbon (graphite)	C	$11.2 + 0.0109T − 4.89e5/T^2$
Glucose	$C_6H_{12}O_6$	226 (25°C)
Gold	Au	$23.47 + 0.006T$
Iron oxide	FeO	$52.8 + 0.006T − 3.188e5/T^2$
	Fe_2O_3	$103.4 + 0.69T − 1.77e6/T^2$
	Fe_3O_4	$172.3 + 0.0787T − 4.1e-6/T^2$
Magnesium chloride	$MgCl_2$	$72.4 + 0.0158T$
Naphthalene	$C_{10}H_8$	$150.5 + 0.6T$
Phenol	C_6H_5OH	220.6 (at 20°C)
Silicon	Si	$24.0 + 0.0025T − 4.225e5/T^2$
Silicon dioxide (quartz)	SiO_2	$45.5 + 0.036T − 1.01e6/T^2$
Sodium chloride	NaCl	$45.15 + 0.0176T$
Sucrose	$C_{12}H_{22}O_{11}$	428 (at 20°C)
Titanium dioxide	TiO_2	$49.4 + 0.0315T − 1.75e5/T^2$
Urea	CH_4N_2O	80.3 (at 20°C)

Source: Compiled from data in *Perry's Chemical Engineers' Handbook*, 6th ed. and NIST Chemistry Webbook.

Table B.19	Heat Capacity C_p of Miscellaneous Materials
Material	C_p, **J/g K**
Cellulose	1.34
Clay	0.94
Coal	1.09 to 1.55
Concrete	0.65
Diamond	0.61
Fireclay brick	1.25 (1500°C)
Glass (pyrex)	0.8
Limestone	0.91
Rubber	1.74
Sand	0.8
Silk	1.38
Steel	0.50
Wood	1.9 to 2.7
Wool	1.36

Source: *Perry's Chemical Engineers' Handbook*, 6th ed.

B.8   Temperature and Enthalpy of Phase Change

Table B.20 Enthalpy of Melting $\Delta\hat{H}_m$ at the Normal Melting Temperature T_m and Enthalpy of Vaporization $\Delta\hat{H}_v$ at the Normal Boiling Temperature T_b at 1.0 atm

Compound	Formula	T_m (°C)	$\Delta\hat{H}_m$ kJ/gmol	T_b (°C)	$\Delta\hat{H}_v$ kJ/gmol
Acetaldehyde	C_2H_4O	−123.5	3.2	21	25.8
Acetic acid	$C_2H_4O_2$	16.6	11.535	118.3	23.7
Acetic anhydride	$C_4H_6O_3$	−73.1	10.5	137	41.2
Acetone	C_3H_6O	−94.8	5.691	56.5	29.1
Acetonitrile	C_2H_3N	−45	8.2	81.6	34.2
Acetylene	C_2H_2	−80.8	3.8	−84	17.0
Acrylonitrile	C_3H_3N	−83.5	6.23	77.3	32.6
Adipic acid	$C_6H_{10}O_4$	153.2	34.58		

(*continued*)

Table B.20 (continued)

Compound	Formula	T_m (°C)	$\Delta\hat{H}_m$ kJ/gmol	T_b (°C)	$\Delta\hat{H}_v$ kJ/gmol
Ammonia	NH_3	−77.7	5.66	−33.4	23.35
Argon	Ar	−189.3	1.2	−185.8	6.65
Benzene	C_6H_6	5.5	9.951	80.1	30.7
Benzoic acid	C_6H_5COOH	122	18.0	249	90.6
Bromine	Br_2	−7.2	10.79	58.0	31.045
Bromoform	$CHBr_3$	−6	11.1	149	39.66
Butadiene, 1,3	C_4H_6	−108.9	7.984	−4.4	22.5
n-Butane	C_4H_{10}	−138.3	4.664	−0.5	22.4
Calcium carbonate	$CaCO_3$	1282	(53)		
Calcium chloride	$CaCl_2$	782	25.5		
Carbon (graphite)	C	3600	46		
Carbon dioxide	CO_2	−57.5	7.95	−78.4	25.23
Carbon disulfide	CS_2	−112	4.395	46.5	
Carbon monoxide	CO	−205.0	0.837	−191.5	6.04
Carbon tetrachloride	CCl_4	−24.0	2.69	77	30.46
Chlorine	Cl_2	−100.98	6.41	−34.6	20.41
Chlorobenzene	C_6H_5Cl	−45	9.55	131.8	35.19
Chloroform	$CHCl_3$	−63.6	8.80	61	29.24
Diethylamine	$(C_2H_5)_2NH$	−93	5.94	58	29.1
Diethyl ether	$(C_2H_5)_2O$	−116.3	7.272	34.6	27.39
Dimethylamine	$(CH_3)_2NH$	−92.2	5.943	7.8	26.4
Dimethyl carbonate	$C_3H_6O_3$	−5.6		90	33.2
Dimethyl ether	$(CH_3)_2O$	−141.6	4.94	−24	21.51
Ethane	C_2H_6	−183.3	2.859	−88.9	14.7
Ethyl acetate	$CH_3COOC_2H_5$	−83.6	10.481	77.1	31.9
Ethylbenzene	C_8H_{10}	−94	9.2	136.2	35.6
Ethanol	C_2H_5OH	−114.5	5.021	78.3	38.6
Ethylamine	$C_2H_5NH_2$	−81		15	27.5
Ethylene	C_2H_4	−169.4	3.4	−103.7	14.45
Ethylene glycol	$C_2H_6O_2$	−11.5	11.234	197	49.6
Ethylene oxide	C_2H_4O	−112.44	5.2	13	25.5
Formaldehyde	CH_2O	−92		−19	23.3

Compound	Formula	T_m (°C)	$\Delta \hat{H}_m$ kJ/gmol	T_b (°C)	$\Delta \hat{H}_v$ kJ/gmol
Formic acid	CH_2O_2	8.3	12.72	101	22.7
Gallium	Ga	29.8	5.59	2071	
Glycerol	$C_3H_8O_3$	18.2	8.475	287	91.7
n-Heptane	C_7H_{16}	−90.6	14.162	98.4	31.8
n-Hexane	C_6H_{14}	−95.3	13.078	68.74	28.9
Hydrazine	N_2H_4	2	12.7	113.3	45.3
Hydrogen	H_2	−259.2	0.117	−252.7	0.904
Hydrogen chloride	HCl	−114.2	1.99	−85.0	16.15
Hydrogen peroxide	H_2O_2	−2	10.54	158	42.97
Hydrogen cyanide	HCN	−13.4	8.412	25.7	25.217
Hydrogen sulfide	H_2S	−85.5	2.376	−60.3	18.67
Iron	Fe	1530	14.895	2735	354
Iron oxide	FeO	1380			
Isobutane	C_4H_{10}	−159.42	4.611	−11.7	
Isopentane	C_5H_{12}	−159.9	5.147	27.86	
Isopropanol	C_3H_8O	−89.5	5.373	82.3	39.9
Lactic acid	$C_3H_6O_3$	16.8		217	63.4
Methane	CH_4	−182.5	0.937	−161.4	8.535
Methyl acetate	CH_3COOCH_3	−98		56.3	30.3
Methanol	CH_3OH	−97.9	3.177	64.7	37.8
Methylamine	CH_2NH_3	−93.5	6.133		25.6
Methyl ethyl ether	$CH_3OC_2H_5$			10.8	26.7
Naphthalene	$C_{10}H_8$	80.2	19.123	218	43.3
Nitric acid	HNO_3	−47	2.51	83	
Nitric oxide	NO	−163.6	2.3	−151.7	13.83
Nitrogen	N_2	−210.0	0.720	−195.8	5.59
Nitrogen dioxide	NO_2			21.93	
Nitrogen tetroxide	N_2O_4	−13	23.2	30	29.5
Nitroglycerin	$C_3H_5N_3O_9$	13		256	92
Nitrous oxide	N_2O	−90.8	6.5	−88.5	16.53
n-Octane	C_8H_{18}	−56.8	20.652	125.66	34.4
Oxygen	O_2	−218.9	0.444	−182.9	6.816

(*continued*)

Table B.20 (continued)

Compound	Formula	T_m (°C)	$\Delta \hat{H}_m$ kJ/gmol	T_b (°C)	$\Delta \hat{H}_v$ kJ/gmol
n-Pentane	C_5H_{12}	−129.7	8.419	36.08	25.8
Phenol	C_6H_5OH	40.9	11.289	181.8	45.9
Phosgene	$COCl_2$	−127.9	5.74	7.6	24.4
Propane	C_3H_8	−181.7	3.526	−42.1	19.0
Propionic acid	C_2H_5COOH	−21	10.66	139.3	55
n-Propanol	C_3H_7OH	−126.1	5.195	97.2	41.4
Propylene	C_3H_6	−185.3	3.004	−47.7	18.4
Silicon	Si	1427	39.6	2290	
Silicon tetrachloride	$SiCl_4$	−67.6	7.7	56.8	28.7
Silicon dioxide (quartz)	SiO_2	1470	14.226	2230	
Sodium carbonate	Na_2CO_3	854	29		
Sodium chloride	NaCl	747	25.69	1392	158.78
Sodium cyanide	NaCN	562		1500	155.98
Sodium hydroxide	NaOH	322	8.4	1378	
Sulfur	S	114	1.727	444.6	9.20
Sulfur dioxide	SO_2	−75.5	7.401	−5	24.94
Sulfur trioxide	SO_3	17			
Sulfuric acid	H_2SO_4	10.5	9.87		
Styrene	C_8H_8	−30.6	11.0	145.1	37.05
Toluene	$C_6H_5CH_3$	−95	6.851	110.6	33.2
Triethylamine	$(C_2H_5)_3N$	−114		89.6	31.0
Trimethylamine	$(CH_3)_3N$	−117.1	6.5	2.9	22.9
Trinitrotoluene	$C_7H_5N_3O_6$	79	23.4	explodes	
Urea	CH_4N_2O	133	14.5	decomposes	87.9 (sublim.)
Water	H_2O	0.0	6.008	100.0	40.65
o-Xylene	C_8H_{10}	−25.2	13.611	144.4	36.2
m-Xylene	C_8H_{10}	−47.2	11.554	139.1	36.7
p-Xylene	C_8H_{10}	13.2	16.805	138.4	35.7

Source: Compiled from data in *Perry's Chemical Engineers' Handbook*, 6th ed., *CRC Handbook of Chemistry and Physics*, 70th ed., *Lange's Handbook of Chemistry*, 14th ed.

B.9   Enthalpies of Solution and of Mixing

Table B.21   Enthalpy of Solution of Organic Solids Dissolved in Water, $\Delta \hat{H}_{soln}$, at Infinite Dilution and 25°C

Compound	Formula	$\Delta \hat{H}_{soln}$ kJ/gmol solute
Acetic acid	$C_2H_4O_2$	−9.418
Citric acid	$C_6H_8O_7$	−22.598
Lactose	$C_{11}H_{22}O_{11} \cdot H_2O$	−15.50
Maleic acid	$C_4H_4O_4$	−18.58
Menthol	$C_{10}H_{20}O$	0
Phenol	C_6H_5OH	−10.9
Phthalic acid	$C_8H_6O_4$	−20.38
Picric acid	$C_6H_3N_3O_7$	−29.7
Potassium citrate		+11.8
Sodium citrate(tri)		+22.05
Sucrose	$C_{12}H_{22}O_{11}$	−5.518
Urea	CH_4N_2O	−15.1
Vanillin		−21.8

+ denotes heat evolved (exothermic), − denotes heat absorbed (endothermic).
Source: Compiled from data in *Perry's Chemical Engineers' Handbook.*

Table B.22   Enthalpy of Solution of Inorganic Solids Dissolved in Water, $\Delta \hat{H}_{soln}$, at Indicated Dilution and 18°C

Compound	Formula	Dilution, gmol water per g substance	$\Delta \hat{H}_{soln}$ kJ/gmol solute
Aluminum chloride	$AlCl_3$	600	−325.9
Ammonium chloride	NH_4Cl	∞	+15.98
Ammonium sulfate	$(NH_4)_2SO_4$	∞	+11.5
Calcium chloride	$CaCl_2$	∞	−20.5
Calcium chloride	$CaCl_2.H_2O$	∞	−51.46

(*continued*)

Table B.22 (continued)

Compound	Formula	Dilution, gmol water per g substance	$\Delta \hat{H}_{soln}$ kJ/gmol solute
Ferric chloride	$FeCl_2$	1000	-132.6
Phosphoric acid	H_3PO_4	400	-11.67
Sodium bicarbonate	$NaHCO_3$	1800	$+17.15$
Sodium carbonate	Na_2CO_3	∞	-23.30
Sodium carbonate	$Na_2CO_3.H_2O$	∞	-9.16
Sodium carbonate	$Na_2CO_3.7H_2O$	∞	$+45.22$
Sodium carbonate	$Na_2CO_3.10H_2O$	∞	$+67.86$
Sodium hydroxide	$NaOH$	∞	-42.59

$-$ denotes heat evolved (exothermic), $+$ denotes heat absorbed (endothermic).
Note: $\Delta \hat{H}_{soln}$ is very sensitive to waters of hydration and to dilution factor.
Source: Compiled from data in *Perry's Chemical Engineers' Handbook.*

Table B.23 Enthalpy of Mixing of Liquids or Gases with Water at 25°C

Compound	Formula	$\Delta \hat{H}_{mix}$ kJ/gmol solute
Acetic acid (l)	CH_3COOH	-1.506
Ammonia (g)	HN_3	-30.5
Formic acid (l)	$HCOOH$	-0.85
Hydrogen chloride (g)	HCl	-74.84
Nitric acid (l)	HNO_3	-33.27

$-$ denotes heat evolved.
Source: *Perry's Chemical Engineers' Handbook*, 6th ed.

Answers to Select Problems

Chapter 1

P1.1 $SiO_2 + 3C \rightarrow 2CO + SiC$

P1.3 $(NH_4)_2PtCl_6 \rightarrow Pt + \frac{2}{3} NH_4Cl + \frac{2}{3} N_2 + \frac{16}{3} HCl$

P1.5 1.32 lb, 272,400 g

P1.7 3.1 g CO_2 and 1.46 g H_2O per g hexane, 1.47 g CO_2 and 0.6 g H_2O per g glucose

P1.9 $204/ton Cl_2, $240/ton NH_3

P1.11 460,000 tonmol/yr, 4.1×10^{13} grams/yr, 15 lb/person/yr, $3.4 billion/yr

P1.13 Some possibilities: greater number of reactions in pathway, higher purity requirements, less pressure to trim costs

P1.15 $HNO_3 + \frac{10}{3} CH_3OH \rightarrow C_3H_7NO_2 + \frac{1}{3} CO_2 + \frac{11}{3} H_2O$, 748 mg methanol.

P1.17 13 wt%, 25 wt%, 11 wt%, 6.7 wt%

P1.19 50 gmol CH_4, 50 gmol Cl_2, 100 gmol phenol, 50 gmol CH_3COCH_3, 100 gmol NaOH

P1.21 0.063

P1.23 $0.22/kg acetic acid

P1.25 0.737 kg butene, 0.395 kg formaldehyde, 0.105 kg oxygen, 0.237 kg water byproduct

P1.27 $CH_{1.66}O_{0.5}N_{0.166}$; about one-quarter glucose consumed in yeast production

P1.29 (a) lose $0.317/kg without a market for galactose; (b) make $0.283/kg

P1.31 $0.92/lb

P1.33 Pathway 2, raw material costs are marginally higher but safety is much improved.

P1.35 Pathway 2, avoids use of highly toxic HCN

P1.37 Atom economies: Strecker, 1.00, Bucherer, 0.43

P1.39 EO: +$945/1000 kg, PO: +$988/1000 kg, both economically attractive

P1.41 Best combination gives $2.64/mol PVC profit, atom economy of 0.87

Chapter 2

P2.1 (a) mass fractions = 0.33, mol fractions = 0.953, 0.024, 0.023

(b) mol fractions = 0.33, mass fractions = 0.012, 0.476, 0.512

P2.3 28.8 g/gmol; 76.6 wt% N_2

P2.5 48.8 g

P2.7 About 100 billion. Compare to the world's human population!

P2.9 4.4×10^{-5} g/cm³, about 5.9 million tons, 4.4×10^{-3} kg, 3.5 kg, 67.6 kg

P2.11 83% decrease

P2.13 3.59 lbmol/min, 345 ft³/min, 1290 ft³/min at STP

P2.15 About 0.5 g

P2.17 Raw materials: gasoline, oxygen (air). Products: mainly CO_2 and H_2O.
Continuous flow (when the engine is running) but transient.

P2.19 (a) Input = accumulation

(b) − output = accumulation

(c) Input – consumption = output

(d) Output = generation

(e) Input = output

P2.21 Depends on your faucet. About 6 gal/min, 380 g/s, 170 lbmol/h, or 13,000 tons/yr. Small commodity plant.

P2.23 0.156 tons ore or 6,250,000 tons seawater per ounce gold.

Worldwide, about 80 million tons gold.

P2.25 14,400 kg/min, 1.94 million L/min, 0.111 mass fraction methane

P2.27 (a) 8.21 tons/h benzene; (b) 0.014 mol% hydrogen, 0.062 mol% methane, 99.6 mol% benzene, (c) 75% toluene converted, 80% gas-phase stream split, 5% gas-phase stream split

P2.31 Milk: input = output. Live bacteria: input – consumption = output (assume short enough time so no reproduction therefore no generation). Dead bacteria: input + generation = output (probably some dead bacteria already in the milk).

P2.33 42 students

P2.35 (a) 24.5 lb/h; (b) 4.1 lb/h; (c) 24 lb

P2.37 10,540 lb air/day, 95 mol% N_2.

P2.39 8.4×10^4 ft³/h.

P2.41 303 g/h, 33.1 wt% $ZnCl_2$, 63.5 wt% $SiCl_4$, 3.4 wt% Si

P2.43 0.056 lb

P2.45 About 3.5 g CO_2, 4.6 g glucose.

P2.47   14.3 lb NH_4NO_3, 16.6 lb $Ca(H_2PO_4)_2$, 7.9 lb KCl, 61% filler.

P2.49   0.334 kg H_2/day, 100.3 kg liquid/day, 69.8 wt% water, 24.2 wt% sorbitol, 6.0 wt% glucose

P2.53 N_2 is pretty close, CO_2 condenses, liquid water is definitely *not* an ideal gas.

P2.55   64 tons S/day, 40 tons water/day, 428,000 SCF air/day.

P2.57   0.50 oz NaCN and 0.046 oz H_2O consumed. 8.8 cm^3

P2.59   Glucose, 180 h, 60 g nitrates/L, 4.3 g phosphates/L

P2.61   1859 lb ZnO, 755 L, 187 kg/day

P2.63   0.63 lb NaN_3

P2.65   If 10% NaCl: 667 kg A, 100 kg B, 66.7 kg C, 167 kg D. 9%, 12.5%.

P2.67   3.8 tons millscale, 41.1 tons oyster, 29.1 tons clay, 79.3 tons limestone about $98/ton

P2.69   25 stream variables, 2 reaction variables, 1 flow, 4 stream compositions (one from splitter), 2 system performances, 20 material balance equations

P2.71   1297 lb sugar crystals/h, 253 lb molasses/h, 2500 lb water/h added

---

Chapter 3

P3.1   wt fraction = 0.894, mol fraction = 0.0099

P3.3   6 kg, 90.4 kg

P3.5 $\dot{\xi}$ = 22.5 gmol/min, 155 gmol/min

P3.7   No.

P3.9   0.2, 18 lb/h

P3.11   0.75

P3.13   161 kgmol/h, 5.5 mol% ethyl acetate, 1.6 mol% acetic acid, 5.7 mol% ethanol, 87.2 mol% water

P3.15   97 kgmol/h, 0.45 mol% ethanol, 3.35 mol% acetic acid, 4.95 mol% ethyl acetate, 91.25 mol% water

P3.17   72.3 kg/h

P3.19   1.17 seconds

P3.21   About 6 hours, probably not much

P3.23   About 99% gone

P3.25   967.4 L, 0.43 mol% SO_2, 3.0 gmol O_2/kg grain

P3.27   Glucose at 180 h, 4.28 g phosphate/L, 8.57 g nitrate/L

P3.29   26,580 molecules

P3.31 $\dot{m}$ = 2433 $(0.9)^{d-1}$ juice/day

P3.33   100 lb total, 0.255 lb Na_2SO_4 per lb solution

P3.35 $6NaClO_3 + 6H_2SO_4 + CH_3OH \rightarrow 6ClO_2 + 6NaHSO_4 + CO_2 + 5H_2O$

P3.37 Only three are independent.

P3.39 Outlet is 30,000 kgmol/h, 50% ethylene, 3.3% O_2, 6.7% ethylene oxide, 20% CO_2, 20% H_2O

P3.41 5.1 kg

P3.43 15.35 h for 90% release

P3.45 Example: if $N = 5$ and $f_c = 0.9$, 349 kg drug/day are made.

P3.47 $f_s = 4/5$

P3.49 (a) at 0 ppm salt, 4.9 wt% salt in brine; (b) at 1 wt% salt in feed, 3364 tons/day and 1.4 wt% salt in brine

---

Chapter 4

P4.1 If there are contaminants in the feed that cannot easily be separated, yet recycle is needed

P4.3 Propane is limiting, 20% excess oxygen

P4.5 0.25

P4.7 $K_a \approx 1$, $\Delta G_r \approx 0$

P4.9 1008 gmol/h, 192 gmol/h, 0.84

P4.11 44 min

P4.13 2.6% ethylene oxide, 69.8% water, 25.3% ethylene glycol, 2.3% DEG; yield = 0.78

P4.15 Conversion = 0.42, yield = 0.36, total flow = 192.5 gmol/min

P4.17 430% excess is minimum

P4.19 4.6% O_2, 70.9% N_2, 3% CO, 7.7% CO_2, 13.5% H_2O, 0.22% SO_2 (all mol%)

P4.21 1180 kgmol raw materials, 1077 kgmol waste products

P4.23 18.9 h, 44,900 mg produced

P4.25 814 kg/day fatty acid, 312 kg/day water removed

P4.27 0.66 moles formaldehyde per mole methanol fed

P4.29 Conversion = 0.52, selectivity = 0.96

P4.31 38% efficient

P4.33 25 MeOH, 5 CO, 15 H_2, 5 N_2

P4.35 about 190 days

P4.37 (a) 0.90, (b) 890, (c) 0.039, (d) 2.6

P4.39 50 lbmol/h fresh gas feed, 12.5 lbmol/h pure gas containing 4 mol% CO_2.

P4.41 0.856, 7.2 gmol/h lactic acid, 64.4 gmol/h H_2, 42.8 gmol/h PD, 42.8 gmol/h water, 100 gmol/h inert

P4.43   490°C

P4.45   158.4 gmol/min, 26.3% octane, 5.3% isobutene, 68.4 % isobutane

P4.47 $k_f = 0.09$ min^{-1}

P4.49   Atom economy: 0.82 for conventional, 0.54 for Connie's process.

P4.51   512 gmol through reactor, 50 gmol formaldehyde, 30 gmol formic acid

P4.53   (b), (a), (c)

P4.55   28 mL/h, 10.9 mL/h

Chapter 5

P5.1   Crystallization: freezing point; adsorption: binding to solid; extraction: solubility in two liquids; distillation: volatility; filtration: size; absorption: solubility of gas

P5.3   100°C

P5.5 P^{sat} increases with T [or plot log P^{sat} versus $(1/T)$ for straight line], $P^{sat} = 0.25$ atm at 20°C , $P^{sat} = 1$ atm at 56°C, estimate $P^{sat} = 3.6$ atm at 100°C

P5.7 $x_E = 0.09$, $x_W = 0.91$; $y_E = 0.41$, $y_W = 0.59$

P5.9 $x_{SO_2} = 0.011$, $x_{SO_2} = 0.057$

P5.11   Two components are present, not just one; $y_W = 0.026$

P5.13   Top—extracted into less-dense organic phase

P5.15   Extraction, condensation, sedimentation, leaching, adsorption, vaporization, absorption, distillation, filtration, crystallization

P5.17   Convert CHCl$_3$ plus bromine to CHBr$_3$ by reaction, avoids difficult separation of bromine from chloroform

P5.19   (a) 2969 lb/h; (b) 81.7 wt%; (c) 0.73

P5.21   123,805 L/h, 12.4 vol%

P5.23   95.7 wt% pure, 59% recovery, separation factor = 127. Recycle doesn't change anything!

P5.25 P^{sat} decreases a lot with increasing molecular weight. Alcohols have much lower P^{sat} at similar C number, because of hydrogen-bonding interactions. Water has an anomalously low saturation pressure for its molecular weight.

P5.27   (a) 173.6°C, (b) 164.0°C, (c) 169°C, 168.5°C

P5.29   (a) 179.6°C, (b) 50.5 mol% benzene in vapor, 32% vaporized

P5.31   (a) 5.94 mol% isopropanol in vapor, 90.5% recovered; (b) 2.3°C

P5.33   (a) 92°C; (b) $x_E = 0.05$; (c) 82°C; (d) $y_E = 0.57$

P5.35   1530 mmHg, eight stages

P5.37   (a) 2.97 million kmol/h; (b) 1.58×10^5 kmol/h

P5.39   1160 kg extract, 11% acetic acid, 84% MIBK, 5% water, 840 kg raffinate, 14% acetic acid, 3% MIBK, 83% water

P5.41   about 95%

P5.43   10 g beads

P5.47   Ideas: remove salt with ion exchange adsorption; concentrate lactic acid, lactose, protein by RO filtration; ultrafilter to separate and concentrate proteins

P5.49   One idea: distillation first to separate propylene, HCl, Cl_2 from heavier components; use water to extract HCl from Cl_2 and propylene; use second distillation to separate allyl chloride from other heavies.

P5.51   Scheme B is better choice. 476 gmol/min air plus 100 gmol/min S fed to reactor; 376 gmol/min N_2 and 100 gmol/min SO_2 leave process.

P5.53   (c) Fractional recovery of acetic acid is 0.903.

P5.55   Sludge is 54.5% limestone. Liquid is 0.217% NaCl. Separation factor is 92,585.

P5.57   (a) 84.6% pure; (b) 97.4% pure.

P5.59   Hints: cool, crystallize and filter out naphthalene as solid, send benzene-rich solution to distillation column; distillate is 99% pure benzene, recycle bottoms to crystallizer. Don't use solvent extraction.

P5.61   (a) 1975 lb/hr crystals, 7828 lb/hr evaporated, 4177 lb/hr recycled; (b) bad idea.

P5.63   About 17.8°C, $y_{EB} = 0.0048$

P5.65   All gas

P5.67   (a) 74.7°C, 53% recovery of isopentane, 52% recovery of n-pentane; (b) 40 stages; (c) 70°C

P5.71   Absorber: $T = 40$°C, $P = 10,000$ mmHg; stripper: $T = 100$°C, $P = 760$ mmHg, MEA flow rate $= 1043$ kmol/h

P5.73   (a) 8.1×10^4 gmol/s

P5.75   (a) 0.385; (b) 88.4% recovery with Zooey's idea

P5.77   About 2900 benzene, $34.10 profit

P5.79   (a) 10,555 L; (b) two steps

Chapter 6

P6.1   About 300 apples

P6.3 $16.88/million Btu

P6.5   0.47 hp

P6.7   66 ft^2 for 100-W lightbulb, 984 ft^2 for 1500-W hairdryer

P6.9   −181,500 kJ

P6.11   +230 kJ/kg solution

P6.13   16.7°C for Freon, 2.39°C for water

P6.15   +22.6 kJ, will feel cool

P6.17   −1559.9 kJ/gmol

P6.19   (a) $Q > 0$, $W = 0$; (b) $Q < 0$, $W < 0$

P6.21   (a) 2513 kJ/kg

P6.23   210.4 kJ

P6.25 $7.38/metric ton, $12.70/metric ton

P6.27   Need about 11,000 Btu/h, 1500-W hairdryer supplies only about 5100 Btu/h

P6.29   506 kJ/s

P6.31   Ethanol: 26,840 kJ/kg, 1.9 kg CO_2/kg. Hydrazine: 141,700 kJ/kg, 0 kg CO_2/kg

P6.33   About 15°C.

P6.35   555 s, 3760 s

P6.37   Rapid heating to about 63°C

P6.39 $\dot{W} = -1400$ kW, 99% vapor

P6.41   179.9°C, 31% vapor, 69% liquid

P6.43   −707,000 Btu/min, 85 hp

P6.45   5.3 kg/min steam

P6.47   About 3400°C versus about 2500°C

P6.49   1740°C, none condenses, about 55°C

P6.51   7410 kg/h, 5 bar steam isn't hot enough

P6.53   21.7 gmol/s vapor at 57 mol% hexane, 78.3 gmol/s liquid at 35.3 mol% hexane, 2180 kJ/s

P6.55   Superheated vapor, 260 kW.

P6.57   About 7000 meters (over 4 miles!)

P6.59   8.35 kJ/min, about 67%

P6.61   about 22°C with 4 kg ice

P6.63   about 98 kg/h, steam must be >500 K or 227°C and 3 bar steam isn't that hot.

P6.65   147 kg/s, about 900°C

P6.67   maximum T reached is about 130°F, so drugs are OK

P6.69   (a) +73.8 kJ/gmol (b) bad idea! Reaction is highly endothermic.

P6.71   Yes. Energy generated by reaction increases by 1000-fold but heat loss from reactor (reactor surface area) increases only by 10-fold; (b) bad idea! Endothermic reaction.

P6.73 (a) $\Delta \hat{H}_r^{\circ} = -588.2$ kJ/gmol. (b) about 26 gmol steam per gmol toluene fed.

P6.75 (a) 5 g/h

P6.77 per g Si: reactants are 2.14 g SiO_2, 0.86 g C, 1.7 g Mg, 5.1 g Cl_2, byproducts are 2 g CO, 6.8 g $MgCl_2$. $\Delta \hat{H}_r^{\circ} = -23.3$ kJ/g

P6.79 (a) 40 gmol water. (b) $Q = -2.6 \times 10^6$ J in E-1, $Q = +3.7 \times 10^5$ J in E-2. (c) air must be cooled to 45°F; 60°F water is not cool enough.

P6.81 (a) 76 mol% H_2, 8 mol% each CO, CO_2, H_2O. 2870 kJ/s removed in heat exchanger. (b) as Q from heating gas decreases, but reactor outlet T increases, get greater extent of reaction for the exothermic reaction (R2).

P6.83 Condenser: $\dot{Q} = \dot{m}(\hat{H}_{out} - \hat{H}_{in})$ $\dot{Q}$ is negative. Receiver: $\hat{H}_{out} - \hat{H}_{in}$. Expansion valve: $\hat{H}_{out} - \hat{H}_{in}$. Evaporater: $\dot{Q} = \dot{m}(\hat{H}_{out} - \hat{H}_{in})$ $\dot{Q}$ is positive. Compressor: $\dot{W} = \dot{m}(\hat{H}_{out} - \hat{H}_{in})$ $\dot{W}$ is positive.

P6.85 (a) 7.3 kJ/s at pump, -5350 kJ/s at turbine, $\eta = 0.18$. (b) 890.6 kJ/gmol methane, $\eta_T = 0.89$.

P6.87 86.7°C, 46°C, 2075 kJ/min

Glossary

A

Absorption   An equilibrium-based separation process in which a gas mixture is separated into its components by addition of a liquid. The liquid is chosen such that the components of the gas have differing solubilities in the liquid.

Adsorption   An equilibrium-based separation process in which a fluid is contacted with a solid material, and a component in the fluid preferentially sticks to the solid material.

Atom economy   A simple measure of the efficiency of a reaction pathway in converting reactants to products. The fractional atom economy is the mass of desired product generated divided by the total mass of reactants consumed.

B

Balanced chemical reaction equation   An equation that shows the molecular formula and relative number of moles of all reactants and products of a chemical reaction.

Basis   A quantity or flow rate that indicates the size of a process.

Batch   A processing mode in which input streams enter all at once, physical and chemical changes occur over time within the process, and output streams are removed all at once at some later time.

Block flow diagram   A process flow sheet that shows the major process units as blocks and shows how process streams move between the process units.

C

Catalyst   A material that speeds up the rate of reaction without changing the reaction equilibrium or the stoichiometric coefficients, and without being consumed by reaction.

Chemical kinetics   The study of the rates of chemical reactions.

Chemical process synthesis   The art and science of choosing appropriate raw materials and chemical reaction pathways, and developing efficient, economical, reliable, and safe chemical processes.

Chemical reaction equilibrium   In a reacting system at constant temperature and pressure, the condition at which the concentrations of reactants and products do not change with time, characterized by a chemical reaction equilibrium constant K_a.

Component   An element, compound, or composite material. The material balance equation is written for a defined component.

Continuous flow   A processing mode in which input streams flow continuously into the process and output streams flow continuously out of the process.

Crystallization   An equilibrium-based separation process in which a liquid feed is adjusted to cause formation of a second solid phase; the solid phase differs in composition from the solution.

D

Degree of freedom analysis   A systematic and rapid method for determining if a problem has an equal number of equations and unknowns.

Distillation   An equilibrium-based separation that exploits differences in relative volatility to separate components. The process is carried out in a multistage column, where repeated evaporation and condensation occurs.

E

Energy balance equation   A statement of the law of conservation of energy, summed up as input − output = accumulation. Sometimes referred to as the First Law of Thermodynamics.

Enthalpy   A convenient measure of the energy of a system that equals the internal energy plus the PV energy.

Enthalpy of mixing   The change in enthalpy of a system due to mixing of two or more fluids, at constant temperature, pressure, and phase.

Enthalpy of phase change The change in enthalpy of a system due to a change in phase, at constant temperature, pressure, and composition.

Enthalpy of reaction The change in enthalpy of a system due to chemical reaction, at constant temperature, pressure, and phase.

Enthalpy of solution The change in enthalpy of a system due to dissolution of a solid in a liquid, at constant temperature and pressure.

Equilibrium stage The fundamental concept for understanding equilibrium-based separations. In an equilibrium stage, a multicomponent feed mixture is adjusted, through the use of a separating agent, to form two (or more) phases, and the phases reach equilibrium and are separated from each other.

Equilibrium-based separation A separation process in which the feed is multicomponent but single phase. Within the separation unit a second phase is generated and the compositions of the two phases are different from each other.

Excess reactant A reactant fed at greater than its stoichiometric ratio relative to other reactants.

Extent of reaction The ratio of the molar reaction rate of a compound to its stoichiometric coefficient.

Extraction An equilibrium-based separation process in which a partially or totally immiscible liquid is added to a liquid feed and the desired solute partitions between the feed and added solvent.

F

Filtration A mechanical separation in which a multiphase mixture passes across a porous barrier; large particles are retained while small particles and fluids pass through the barrier.

Fractional conversion A system performance specification for a reactor: the moles of reactant consumed by reaction divided by the moles of reactant fed to the reactor.

Fractional recovery A system performance specification for a separator: the moles of a component in an output stream from the separator divided by the moles of that component in the input stream to the separator.

Fractional split A system performance specification for a splitter: The ratio of the quantity or flow rate in one output stream from the splitter to the quantity or flow rate of the input stream to the splitter.

G

Generation-consumption analysis A systematic method for synthesizing reaction pathways involving multiple chemical reactions. From a generation-consumption analysis, we calculate the moles of raw materials consumed in generating a mole of product, and determine the moles of all byproducts generated per mole of desired product.

H

Heat Flow of energy across system boundaries due to a difference in temperature between the system and surroundings in the absence of material flow.

Heat capacity The constant-pressure heat capacity is a proportionality constant defined as the change in enthalpy with a change in temperature, at constant pressure, phase, and composition. The constant-volume heat capacity is a proportionality constant defined as the change in enthalpy with a change in temperature, at constant volume, phase, and composition.

I

Input-output diagram The simplest form of a process flow sheet, with a single block representing all physical and chemical operations in the process, and showing the raw materials fed to the process and the products and byproducts leaving the process.

Internal energy Energy that a system possesses because of its molecular activity. Includes energy stored in molecules in the form of covalent chemical bonds, noncovalent intermolecular forces, and thermal motion.

K

Key components The two components in a multicomponent feed that a separator is designed to separate from each other.

Kinetic energy Energy that a system possesses because of its velocity.

L

Linear equation An equation in which all variables are raised to the first power.

Linear model A system of linear equations describing a process flow sheet.

Limiting reactant A reactant fed at less than its stoichiometric ratio relative to other reactants.

M

Material balance equation A statement of the law of conservation of mass, summed up as input − output + generation − consumption = accumulation.

Mechanical separation A separation process in which the feed contains two or more phases, and differences in size or density are exploited to separate the two phases from each other.

Mixer A process unit with two or more process streams as input and a single mixed process stream as output.

P

Phase A homogeneous, physically distinct, and mechanically separable portion of matter.

Phase equilibrium In a system containing more than one phase at constant temperature and pressure, the condition at which the quantities and compositions of each phase do not change with time.

Potential energy Energy that a system possesses because of its position in a force field such as gravity.

Process flow diagram (PFD) A process flow sheet that shows all major pieces of process equipment representationally, and indicates how process streams connect the pieces together. A PFD usually includes (1) equipment for moving material or energy around; (2) major utilities, and

(3) quantity or flow, composition, temperature, and pressure of all process streams.

Process stream Material flow into or out of process units.

Process topology A map of a process, showing how process units are connected together and the direction of flow of material between process units.

Process unit A block or other icon on a process flow sheet that serves a specific process function. Materials enter the process unit, undergo physical and/or chemical changes, and leave the unit.

Process variables Variables such as quantity, flow rates, composition, pressure, and temperature that define the state of a process stream or process unit.

Purge A process stream split off from a recycle stream that leaves the process. Typically used to remove inert components from a reactor system in which recycle is used.

Purity A stream composition specification for a separator, equal to the moles or mass of a component in a stream divided by the total moles or mass of that stream.

R

Rate-based separation A separation process that separates components in a single-phase stream based on differences in the rate of transport of components through a medium, usually under an applied force field.

Reactor A process unit that provides conditions allowing a chemical reaction to take place. In the simplest case, there is a single input

stream containing reactants and a single output stream containing reaction products.

Recycle Return of material from the output of a process unit back to that unit's input. Commonly used when a reactor's conversion is low.

Reference state A state of a system (temperature, pressure, composition, phase, velocity, and position) at which the energy of that system is arbitrarily set equal to zero.

S

Scale factor A multiplier applied to all quantities or flow rates in a process to adjust the results of generation-consumption analysis or other process flow calculations to a new desired basis.

Selectivity A system performance specification for a reactor: the moles of reactant converted to desired product divided by the moles of reactant consumed by all reactions.

Separating agent An agent, either energy or material, added to an equilibrium stage in order to generate a second phase.

Separator A process unit in which one or more input streams undergo physical changes by which the material is separated into two or more output streams with differing compositions.

Splitter A process unit with a single input stream that splits the stream into two or more output streams without changing the composition of the streams.

Stoichiometric coefficient A number in a balanced chemical reaction equation that indicates the relative number of moles of reactant consumed or product generated. The stoichiometric coefficient for a reactant is negative and that for a product is positive.

Steady state An operating condition under which process variables do not change with time.

Stream composition specification Specified values for the composition (mass or mole percent, mass or mole fraction, or concentration) of a process stream.

System A specified volume with well-defined boundaries.

System performance specification Specified values for the performance of a system; the extent to which physical and/or chemical changes occur within a process unit.

T

Transient An operating condition under which process variables change with time.

W

Work Flow of mechanical energy across system boundaries due to driving forces other than temperature. Shaft work is work done on or by the system that involves rotating or reciprocating equipment.

Y

Yield A system performance specification for a reactor: the moles of reactant converted to desired product divided by the moles of reactant fed to the reactor.

Index